MOTOR

AUTOMOTIVE MECHANICS

MOTOR

AUTOMOTIVE

MECHANICS

ANTHONY E. SCHWALLER

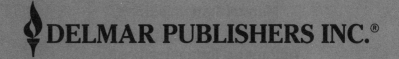

DELMAR PUBLISHERS INC.®

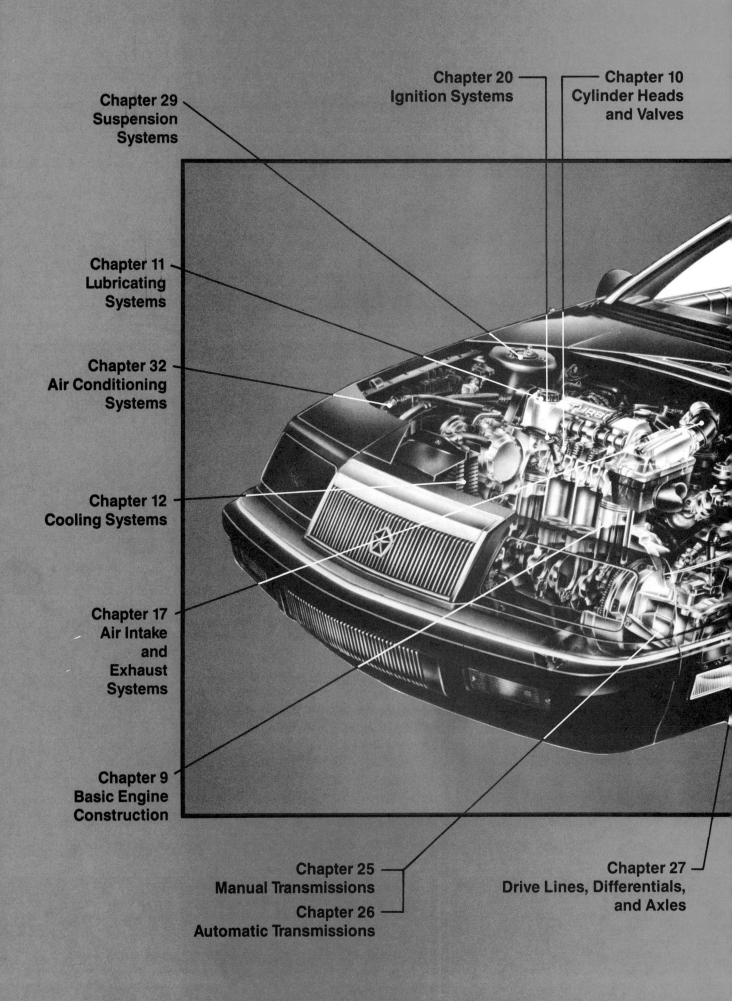

Chapter 20
Ignition Systems

Chapter 10
**Cylinder Heads
and Valves**

Chapter 29
**Suspension
Systems**

Chapter 11
**Lubricating
Systems**

Chapter 32
**Air Conditioning
Systems**

Chapter 12
Cooling Systems

Chapter 17
**Air Intake
and
Exhaust
Systems**

Chapter 9
**Basic Engine
Construction**

Chapter 25
Manual Transmissions

Chapter 26
Automatic Transmissions

Chapter 27
**Drive Lines, Differentials,
and Axles**

Chapter 33
Heating and
Ventilation Systems

Chapter 30
Steering Systems

Chapter 34
Auxiliary and
Electrical Systems

Chapter 31
Tires and Wheels

Chapter 28
Braking Systems

OFFICIAL PACE CAR
71ST INDIANAPOLIS 500 MAY 24, 1987

Courtesy of Chrysler News Photo

NOTICE TO THE READER

Publisher does not warrant or guarantee any of the products described herein or perform any independent analysis in connection with any of the product information contained herein. Publisher does not assume, and expressly disclaims, any obligation to obtain and include information other than that provided to it by the manufacturer.

The reader is expressly warned to consider and adopt all safety precautions that might be indicated by the activities described herein and to avoid all potential hazards. By following the instructions contained herein, the reader willingly assumes all risks in connection with such instructions.

The publisher makes no representation or warranties of any kind, including but not limited to, the warranties of fitness for particular purpose or merchantability, nor are any such representations implied with respect to the material set forth herein, and the publisher takes no responsibility with respect to such material. The publisher shall not be liable for any special, consequential or exemplary damages resulting, in whole or in part, from the readers' use of, or reliance upon, this material.

DEDICATION

This book is dedicated to several people. Special thanks should go to Mr. William Gasper who helped the author learn the basics of the automobile and who helped to develop the interest and motivation to stay within the field over the past years. The author is also grateful to his parents, Omer and Garnett Schwaller, for the drive and motivation that was instilled in him as a child. Without these ingredients, the writing of this book would not have been possible. The author would also like to thank his sons, Matthew and Joshua, and his wife, Renee, for their continued support and understanding during the writing of this textbook.

Delmar Staff
Editor-in-Chief: Mark W. Huth
Developmental Editor: Marjorie A. Bruce
Managing Editor: Barbara A. Christie
Production Editor: Eleanor Isenhart
Design Coordinator: Susan C. Mathews
Production Coordinator: Linda Helfrich

For information, address Delmar Publishers Inc.,
2 Computer Drive West, Box 15-015,
Albany, New York 12212

COPYRIGHT © 1988
BY DELMAR PUBLISHERS INC.

Printed in the United States of America
Published simultaneously in Canada
by Nelson Canada
A Division of International Thomson Limited

10 9 8 7 6 5 4 3 2

Library of Congress Cataloging-in-Publication Data

Schwaller, Anthony E.
    Motor automotive mechanics/Anthony E. Schwaller.
        p. cm.
    Includes index.
    ISBN 0-8273-2542-8.   ISBN 0-8273-2543-6 (Instructor's guide).
    ISBN 0-8273-2545-2 (Transparencies).   ISBN 0-8273-2539-8 (Motor-Delmar guide).
    ISBN 0-8273-2544-4 (Workbook).
    1. Automobiles.   2. Automobiles — Maintenance and repair.
I. Title.
TL146.S38   1988
629.28'7 — dc19   87-32956
                                                           CIP

WE ENCOURAGE PROFESSIONALISM

ASE CERTIFIED

THROUGH TECHNICIAN CERTIFICATION

CONTENTS

SECTION 3   MECHANICAL/FLUID ENGINE SYSTEMS

SECTION 4   ELECTRICAL ENGINE SYSTEMS

SECTION 5   EMISSION CONTROL SYSTEMS

PREFACE

INTRODUCTION

The importance of the automobile in our society has been growing steadily over the last several decades. As a prime mover of people, the automobile contributes daily to the maintenance of our economic and social systems. This basic form of transportation gives people the freedom to travel further from their homes to pursue careers, indulge in leisure time activities, and act as consumers of a wider variety of goods and services.

Increased usage of the automobile over the years has resulted in a continual evolution in design to achieve faster, more streamlined, cleaner, more economical, more reliable, and safer vehicles. The pattern of change continues and is accelerating. Each year innovations resulting from advancing technology appear on new automobile models. The modern automobile is vastly different from the auto of 10–20 years ago. Several significant changes have occurred in this period. Beginning in 1968, automobiles have been designed to produce fewer emissions containing a decreased number of pollutants. Monitoring systems provide a means of checking that emission control is working. In 1973, the energy crisis directed the automotive industry to the search for more fuel-efficient cars. Changes in engine design and vehicle configuration helped to improve fuel economy. Safety considerations have resulted in further changes in vehicle design in recent years.

The increasing sophistication of electronic control systems has led to wider applications in automobiles, resulting in more efficient and safer operation. The development of on-board computers to monitor and control performance is also resulting in improved reliability and safety. Each new year will bring further changes in automobile and accessory design. Future developments in communications, electronics, materials, manufacturing processes, and energy will have significant effects on the automobile.

SCOPE OF THE TEXT

The automobile is a composite of many complex systems. These systems require routine maintenance and service, and at times may require more extensive service. The service technician faces the challenge of understanding each of the systems found in the automobile and the interrelationships of these systems. This understanding is based on a knowledge of basic physical principles. Another challenge the technician faces is the need to stay current with changes as each new model year appears.

Motor Automotive Mechanics is a basic text that is designed to help students achieve the necessary understanding of automotive principles. Upon this foundation the student will build skills through actual lab and shop work. All of the following essential information is thoroughly explained for each automotive system: science principles, theory of operation, safety consideration, diagnosis, troubleshooting, and service. Where appropriate for a system, computer controls are discussed.

The content is divided into eight major sections and 34 chapters. Each chapter includes a number of learning aids to help students in their study of automotive mechanics.

Introduction — provides a statement of the intent of the chapter.

Objectives — state the expected learning that will take place as a result of studying the chapter.

Content — provided in a logical sequence with many heads to divide content into manageable segments of information.

Figures — numerous line drawings and photographs to illustrate concepts and show current equipment, components, and systems.

Tables — summarize important points, measurements, statistics, troubleshooting.

Diagnosis and Service — procedures to familiarize students with a logical approach to troubleshooting a problem and effectively performing service to correct the problem.

Terms to Know — highlights important vocabulary to be learned; each term is highlighted and defined in the chapter. Definitions are also provided in the glossary at the back of the text.

Chapter Summary — highlights the important concepts covered in the chapter; also serves as a ready reference.

Review Questions — a tool to reinforce and test students' comprehension of content. Three types of questions are provided: standard multiple choice, ASE-type multiple choice, and essay.

Reading Level — appropriate for students in vocational automotive programs.

Other important features of the text include:

Vignettes — articles covering interesting topics drawn from the automotive industry, supporting industries, and the use of automobiles in recreation and sports; each vignette contains one or more photographs.

Color Inserts — full color photos illustrating interesting aspects of automotive technology, including high-speed racing, drag racing, computer design of vehicles, and testing of vehicles.

Car Clinics — describe a common problem with an automobile, followed by a tip for diagnosing the problem or the solution to the problem.

Safety Tips — for each diagnostic and service procedure, specific safety precautions are listed.

Glossary of Terms — provides definitions of all terms introduced in the chapters.

Appendix listing MOTOR Service Manuals — available for student use in conjunction with the text. These manuals provide diagnosis and service information for specific components and systems of current automobile models.

Appendix of Abbreviations — a listing of abbreviations commonly used in automotive technology, including abbreviations relating to newer technologies, such as computer control.

TEXT SUPPLEMENTS

The text is supported by several well-designed supplements which will assist the instructor in presenting the content. In addition, the student will find that the supplements provide many additional opportunities for learning and for hands-on application of the theory to actual shop situations. The supplements consist of a *Student Workbook*, an *Instructor's Guide*, a set of *Transparency Masters*, and a *MOTOR Reference Guide* cross-referencing MOTOR repair manuals to the text content.

- The *Student Workbook* provides numerous questions to help reinforce student learning. The questions consist of multiple choice, matching and identification questions, as well as additional ASE-type multiple choice questions. The workbook also provides task worksheets covering typical service procedures. Each worksheet includes a statement of the objective of the task, appropriate references (to the related text chapter and shop manuals), equipment and materials needed, safety precautions, step-by-step procedures, tables to record data where appropriate, and questions relating to the service procedures, as required.

- The *Instructor's Guide* suggests methods of teaching automotive mechanics technology. It provides a listing of addresses of auto manufacturers and other companies from which product literature and technical information can be obtained, and a listing of sources of audio-visual materials. The guide also contains the correct answers for the review questions in the text and answers for the activities in the Student Workbook. Answers for the task worksheet questions are also included in the Instructor's Guide.

- More than 100 *Transparency Masters* are packaged separately for the instructor's convenience in presenting the content. Each transparency is taken from the text and is referenced to the appropriate text content.

- The *Motor Reference Guide* instructs students on how to use the MOTOR manuals most efficiently with this textbook. The cross-reference guide includes a chart listing selected major topics from the text and 11 MOTOR manuals with check marks indicating each manual that contains information for the specific text topic; an explanation of how to find information in the MOTOR manuals, and a table of contents and a brief description for each of the 11 manuals included on the comparison grid. This guide provides valuable instructions for students and will help save time in using the MOTOR library of repair manuals.

The combination of *Motor Automotive Mechanics*, supplements and MOTOR manuals forms a comprehensive, unequaled teaching/learning package. The text supplies the basic automotive theory and general diagnosis and service procedures. The MOTOR manuals provide the practical "how-to" information to complete any servicing task.

ABOUT THE AUTHOR

Anthony Schwaller has been involved with automotive technology for many years, beginning as an automotive technician. He also worked as a technical trainer for General Motors in Detroit, Michigan. After leaving Detroit, he taught automotive mechanics at Eastern Illinois University, Charleston, IL, and St. Cloud University, St. Cloud, MN, where he is currently serving both as a professor and administrator. The author received his B.S. and M.S. from the University of Wisconsin — Stout and his Ph.D. from Indiana State University. He has authored two other textbooks, over 35 articles, and has presented more than 40 papers and addresses at various conferences in the field of technology.

ACKNOWLEDGMENTS

Three individuals on the staff of MOTOR Publications provided technical assistance and support in the development of the text and the *MOTOR Reference Guide*. The author and the staff at Delmar Publishers wish to thank Philip Shalala for his vision in recognizing the benefits of a Delmar-MOTOR association. Special appreciation is extended to Claude Milot, Marketing Director and Louis C. Forier, Editorial Director, for their cooperation and enthusiasm which contributed greatly to the successful completion of the project.

There are many organizations and people that contributed to the development of this text. Among the companies that provided valuable technical information, photographs, and illustrations are:

Sun Electric Corporation
Dana Corporation
EIS Brake Parts, Division Standard Motor Products, Inc.
Champion Spark Plug Company
Sellstrom Manufacturing Co.
CR Industries
OTC Division of Sealed Power Corp.
American Isuzu Motors Inc.
Tire Industry Safety Council
Hunter Engineering Company
Stanadyne, Diesel Systems Division
First Brands Corporation
   (Formerly Union Carbide Corp.)
Mazda Motor Corporation
Austin Rover Group Limited
General Motors Corporation
   Buick Motor Division
   Chevrolet-Pontiac-Canada Group
   Harrison Radiator Division
   Delco Moraine
   Delco Remy
   United Delco
   Rochester Products
   General Motors Product Service Training
   Oldsmobile Division
   General Motors Proving Grounds

Davis Publications, Inc.
Chrysler Motors
Lucas Electrical, Parts and Service
Volkswagen United State, Inc.
Tune-up Manufacturers Institute
Federal-Mogul Corporation
Ingersoll-Rand Power Tool Division
Snap-on Tools Corporation
School Products Co., Inc.
Breton Publishers
Society of Automotive Engineers, Inc.
Robertshaw Controls Company
Hastings Manufacturing Compnay
Peugeot Motors of America, Inc.
Sachs Automotive Products Company
Echlin Incorporated
Firestone Tire and Rubber Company
United Technologies (Formerly American Bosch)
ASE (Automotive Service Excellence)
Clayton Industries
Goodyear Tire and Rubber Company
DCA Educational Products, Inc.
Ford Motor Company
Allied Aftermarket Division
Volvo Cars of North America
Nissan Motor Corporation in U.S.A.

The author would also like to thank several people who contributed valuable information for the textbook: Anthony Gilberti, Bob Nelsen, and Mike Meyer. Thanks also to James Rennich for providing photos. Special thanks should go to Steve Morgan, who helped with the student workbook, to Douglas Gossett, who wrote the task worksheets, and to Pattie Patit, for help in preparing the manuscript.

A number of instructors reviewed the text and provided suggestion for improvements. Their assistance is appreciated.

John Borovsky, Pioneer High School, Ann Arbor, MI 48103

John H. Kerstetter, Sun Area Vocational-Technical School, New Berlin, PA 17855

Anthony Greco, Automotive High School, Brooklyn, NY 11222

William W. Ruiter, North Eugene High School, Eugene, OR 97402

Brian F. McCleish, Bloom Trail High School, Chicago Heights, IL 60411

Sampson J. Smith, Jr., Ventura High School, Ventura, CA 93003

Douglas F. Gossett, Jefferson State Vocational Technical School & Manpower Skill Center, Louisville, KY 40203

David Johnson, Anoka Area Vocational Technical Institute, Anoka, MN 55303

Charles Barnett, Northside Independent School District, San Antonio, TX 78238

Henry Brooks, Texas State Technical Institute, Waco, TX 76705

Edward Hunkins, Linden McKinley High School, Columbus, OH 43214

Richard Proulx, Mid-Florida Technical Institute, Orlando, FL 32809

Donald P. Kessler, Cy-Fair High School, Houston, TX 77040

Allen J. Reed, George Brown, Michael Blackman, Joseph Costa, A. J. Taylor, Charles Peterson, R. H. Myhand, Arizona Automotive Institute, Glendale, AZ 85301

David R. Hollands, Centennial College, Scarborough, Ontario, Canada

Blaine Thorsley, Northern Alberta Institute of Technology, Edmonton, Alberta, Canada

CHAPTER 1

Introduction to the Automobile

INTRODUCTION

The automobile has become a very important part of our technological society. People rely on vehicles for travel more each day. The automobile has become one of the most important technological innovations within our society. This chapter is designed to introduce the automobile to you and to identify its importance within our society.

OBJECTIVES

After reading this chapter, you will be able to:

- Define the basic design of the automobile, including the body, frame, engine, drive lines, running gear, and suspension systems.
- Identify the importance and influence of the automobile within our society.
- Categorize the many careers within the automotive industry.
- Examine the automotive industry and its supporting service and sales organizations.

CHAPTER HIGHLIGHTS

1.1 VEHICLE DESIGN
  A. Body and Frame
  B. Engine
  C. Drive Lines and Running Gear
  D. Suspension System

1.2 AUTOMOBILES AND SOCIETY
  A. Automobile Use
  B. The Automotive Industry
  C. Regional Offices and Distributorships
  D. Dealerships
  E. Fleet Service and Maintenance
  F. Independent Service
  G. Supporting Specialty Shops
  H. Parts Distribution

1.3 CAREERS IN THE AUTOMOTIVE INDUSTRY
  A. General Mechanic/Technician
  B. Foreman
  C. Service Manager
  D. Parts Manager
  E. Marketing and Sales
  F. Company Representatives — Sales and Service
  G. Supporting Careers

SUMMARY

CAR CLINIC

PROBLEM: TROUBLESHOOTING QUESTIONS TO ASK

Quite often a problem develops in a vehicle. Many times the technician will replace parts until the cause of the problem is found. This costs money and time. What would be a good procedure to start with before trying to fix a problem in the car?

SOLUTION:

The first step in trying to find the problem in a car is to ask questions. This will help narrow down the problem to one of the major systems in the vehicle or engine. Some of the more important questions to ask the owner of the car include:

1. When was the trouble first noticed?
2. Did the problem develop quickly or over a period of time?
3. Was the complaint or problem recorded at an earlier time?
4. Was the problem noticed at all speeds and loads or only at certain speeds and loads?
5. What type of vehicle and engine is the problem associated with?
6. Did any unusual noise develop from the problem?
7. Has the cooling temperature increased?
8. Has there been any work done on the car recently? Did the problem start before or after the work was done?
9. Is the oil consumption normal?
10. How does the engine respond to acceleration or deceleration?
11. Is there excessive blue (oil) or black (rich) smoke from the exhaust?
12. Does the engine start easily?
13. Does the engine surge or hunt at idle, high idle, or full load?
14. How many miles are on the engine or vehicle?
15. What are the normal driving conditions of the vehicle?

The answers to these questions will help guide the technician and troubleshooter to a more logical solution to many of the problems that are found in the automobile.

1.1 VEHICLE DESIGN

In order to study the automobile, it is important to review the basic parts of the vehicle. See *Figure 1–1*. The vehicle can be subdivided into several major categories: the body and frame, the engine or power source, the drive lines and running gear, and the suspension systems.

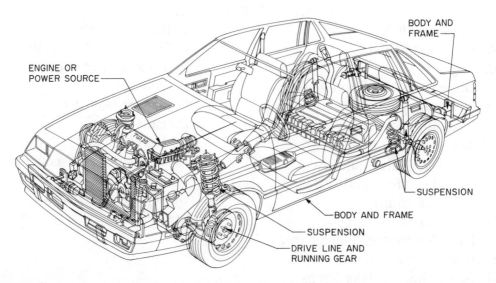

FIGURE 1–1. The basic parts of the automobile include the body, frame, running gear, transmission, suspension, engine, and power train. *(Courtesy of Chrysler Corporation, Sales & Marketing PR)*

Body and Frame

The *body and frame* section of the automobile is the basic foundation of the vehicle. All other components and systems are attached to the body and frame. The frame supports the car body, engine, power train and wheels, and the drive lines and running gear. *Figure 1–2* illustrates the body and frame of a typical vehicle.

There are two types of body and frame configurations. The separate body and frame construction has been used for the longest time. This is illustrated in Figure 1–2 (A). The second type of construction is called the unitized body. This type of vehicle is designed with the frame and body in one unit. See Figure 1–2 (B).

Engine

The *engine* in the automobile is used to power the vehicle. The engine is also called the power source or motor. The word *motor* is defined as that which imparts motion. So a motor can be any device that produces power. However, the power source in the automobile is usually referred to as the engine.

Most automobiles use the gasoline engine as a power source. However, newer power sources are being tested and introduced into the vehicle every year. For example, the diesel engine is also being used as a power source today. In addition, some automotive engineers predict the use of gas turbines and stirling engines for future power sources in the automobile.

Most automobiles use the reciprocating piston or Otto cycle engine. See *Figure 1–3*. However, certain car manufacturers are now offering rotary design engines as an optional power source. *Figure 1–4* illustrates the rotary engine now used on certain vehicles. Although it looks similar to the reciprocating piston engine, internally there are many differences.

The engine is typically located in the front of the vehicle. But certain car manufacturers are also designing vehicles with rear engines. In addition, certain manufacturers have developed engines that are placed in the middle of the body and frame.

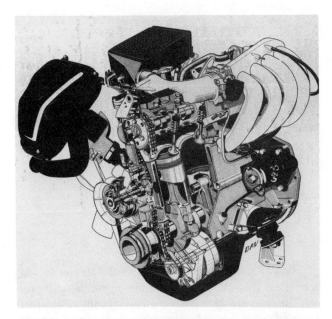

FIGURE 1–3. The automobile engine is called a reciprocating piston engine. (*Courtesy of Peugeot Motors of America, Inc.*)

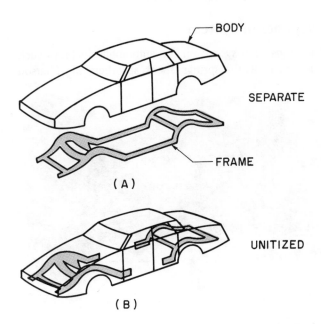

FIGURE 1–2. Two types of body and frame configurations are used in the modern automobile. They include (A) separate body and frame, and (B) unitized body.

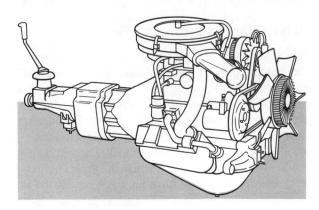

FIGURE 1–4. Some car manufacturers now offer the rotary engine as an alternative power source.

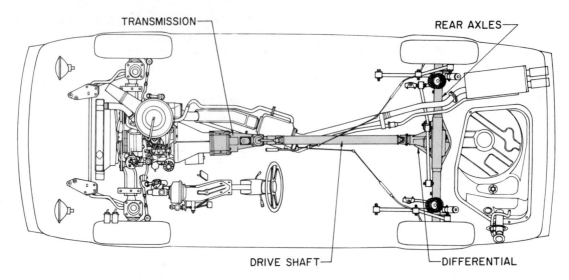

FIGURE 1–5. Many vehicles are using the rear wheel drive line system. (Courtesy of Mazda Motor Corporation)

All power sources or engines are designed to have several supporting technical systems. These include:

1. The fuel system (may include computer systems)
2. The ignition system
3. The starting and charging system
4. The lubricating system
5. The cooling system

The *fuel system* is designed to mix the air and fuel in the engine. This will produce an efficient combustion process. The *ignition system* is designed to ignite the air and fuel that has been mixed. The *starting and charging system* is designed to start the engine and to keep the battery charged during operation. The *lubricating system* is designed to keep all engine parts lubricated so that friction is reduced internally. Finally, the *cooling system* is designed to keep the engine at the most efficient operating temperature. Each of these systems will be studied in depth in this textbook.

Drive Lines and Running Gear

The *drive lines* are defined as those components that transmit the power from the engine to the wheels. This action will propel the vehicle in a forward or reverse direction. As shown in **Figure 1–5**, the drive lines include components such as the transmission, drive shafts, differential, and rear axles. Each of these components is discussed in an individual chapter within this textbook.

There are two methods in which the drive lines can be designed: the rear wheel drive lines system and the front wheel drive lines system. The rear wheel drive lines system is shown in Figure 1–5. In this system, the engine is in the front of the body and frame. The transmission of power is

then sent to the rear of the vehicle for propulsion.

The front wheel drive lines system has been used on certain newer vehicles in the past few years. In this system the engine is in the front of the vehicle. The drive lines and running gear are also in the front. Both systems have advantages and disadvantages, and are equally reliable in their operation. See **Figure 1–6**.

The *running gear* consists of components on the automobile that are used to control the vehicle. The running gear is defined as the braking systems, the wheels and tires, and the steering systems. There is a chapter devoted to each of these systems within this textbook.

Suspension System

The *suspension system* on the automobile includes such components as the springs, shock absorbers, MacPherson

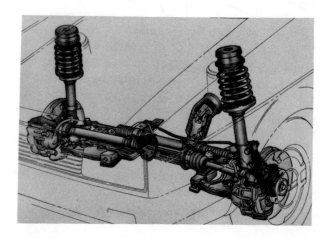

FIGURE 1–6. Certain car manufacturers place the drive line in the front of the vehicle. (Courtesy of Volkswagen of America, Inc.)

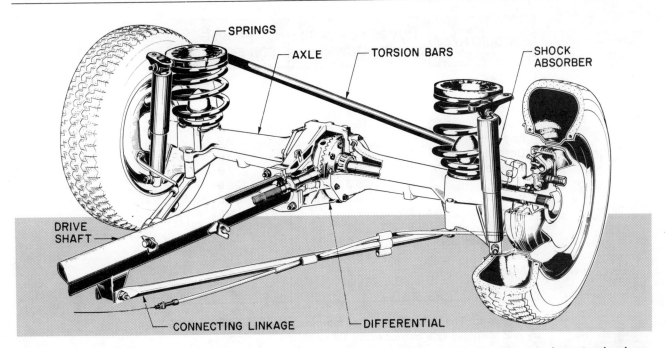

FIGURE 1-7. All vehicles have suspension components. These include the shock absorbers, springs, torsion bars, axles, and connecting linkages used to support the total vehicle. *(Courtesy of Peugeot Motors of America, Inc.)*

struts, torsion bars, axles, and connecting linkages. These components are designed to support the body and frame, the engine, and the drive lines on the road. Without these systems, the comfort and ease of driving of the vehicle would be reduced. *Figure 1-7* illustrates some of the components that are used on the suspension system.

The springs and torsion bars are used to support the axles of the vehicle. The two types of springs commonly used are the leaf spring and the coil spring. Figure 1-7 illustrates the more contemporary form of coil spring.

Torsion bars are made of long spring steel rods. One end of the rod is connected to the frame while the other end is connected to the movable parts of the axles. As the axles move up and down, the rod is twisted and acts as a spring.

Shock absorbers are used to slow down the upward and downward movement of the vehicle. This action occurs when the car goes over a rough road. Shock absorbers will be studied in detail in a later chapter.

The axles and connecting linkages are those components that connect the springs, torsion bars, and shock absorbers to the vehicle frame and to the wheels.

1.2 AUTOMOBILES AND SOCIETY

Automobile Use

People today do not have to be told how important the automobile is to their lives. The automobile is used to take people to work, deliver food and other commodities to stores, and to move people, services, and products throughout society. Our society is taking part in a revolution that has been going on for over 80 years. This is called the automotive revolution. The mass production of automobiles has affected American social history in the 20th century more than any other invention. Today, the automobile is so significant that it is consuming more than 52% of the total energy used in the transportation sector of society. This is shown on the graph illustrated in *Figure 1-8*.

ENERGY USE IN THE THE TRANSPORTATION SECTOR

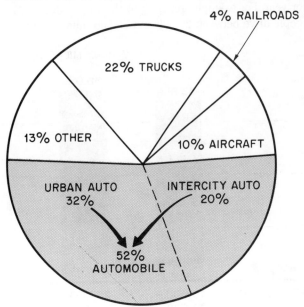

FIGURE 1-8. Over 52% of all energy within the transportation sector is consumed by the automobile.

COUNTRY	NUMBER OF CARS (MILLIONS)	PEOPLE FOR EACH CAR
United States	121.7	1.9
Japan	23.7	4.9
Germany	23.2	2.6
France	19.2	2.8
Italy	17.7	3.2
United Kingdom	15.4	3.6
Canada	10.3	2.3
USSR	8.3	32.0
Brazil	8.2	15.0
Spain	7.6	5.0
Australia	5.9	2.5
Netherlands	4.3	3.3
Mexico	3.4	21.0
Argentina	3.2	8.4
Belgium	3.2	3.1

FIGURE 1–9. The United States has more cars than any other country and averages one car for every 1.9 people.

The United States remains by far the most mobile society in the world. Statistics that illustrate this mobility are shown in **Figure 1–9**. There is one car for every 1.9 people. The United States is also the leader in car population. Cars in the United States accounted for more than 38% of the 321 million cars on the road worldwide in the early 1980s.

The effects of the automobile have been staggering. The automobile has given people the freedom of mobility. With this mobility people are able to go where they want, when they want, and in most any kind of vehicle they want. **Figure 1–10** illustrates how many miles Americans drive per year. For example, referring to this figure, 36% of the people drive between 20,000 and 30,000 miles per year.

Today, cities and suburbs are being designed on the assumption that the use of the automobile will continue to expand. Millions of dollars are spent to build our society around the automobile. For example, highways have been built to support the millions of vehicles that travel on them. Families can move to the suburbs because they have transportation to work in the city. Theaters are designed to keep people in the cars during the movies. Restaurants are designed to have drive-through facilities. In fact, our entire society has been designed around the automobile.

Each working day more than 100,000 automobiles roll off assembly lines around the world. Now the world's largest manufacturing industry, automotive manufacturing has

HOW MANY MILES DO AMERICANS DRIVE ANNUALLY ?

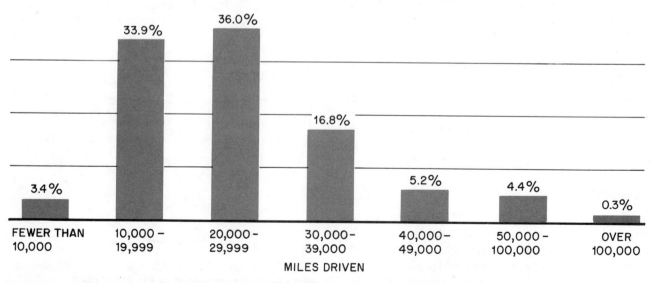

FIGURE 1–10. This chart illustrates the percentage of Americans driving in the different mileage per year ranges.

strongly influenced the economic and social evolution of modern industrial societies. To illustrate this point, the following industries have grown because of the development of the automobile:

1. Petroleum refining.
2. Road construction and maintenance.
3. Motor vehicle manufacturing.
4. Parts manufacturing and distribution.
5. Automobile sales and servicing.
6. Providing passenger transportation.
7. Insurance companies.
8. Support companies such as plastics, steel, electronics, rubber, glass, and fabric manufacturers, and many others.

The United States leads other countries in how much of the gross national product is spent on automobile transportation. More than 10% of the gross national product in the United States has been spent on automobile transportation. It is estimated that 30 million people around the world depend upon the automotive industry for their jobs. Close to half of these are in the United States. In fact, 22% of the work force is working in the automotive sector of society. A recent study by Hertz Corporation indicated that Americans spend 15% of their personal income on automobile transportation.

The Automotive Industry

There are several major automobile manufacturers currently producing cars in the United States. They are:

1. General Motors Corporation
2. Chrysler Corporation
3. Ford Motor Company
4. American Motors Corporation

In addition to these manufacturers, many foreign corporations are now manufacturing and selling cars in the United States. Some of the larger foreign companies are shown in *Figure 1-11*.

Within the automotive industry, there are many suborganizations and departments that are needed. First, the automobile must be designed. To do this, there are many electrical and mechanical engineers that primarily design the vehicle. From the design process, the vehicle is developed into a prototype or a working model. At this stage the automobile designs are tested to determine their reliability and feasibility. The automobile is then ready for manufacturing. All parts must be made first at many supporting plants and companies throughout the U.S. After the parts are manufactured, the vehicle is assembled. The total manufacturing process is very complex and involves thousands of people.

Regional Offices and Distributorships

The next level within the automotive organization is called the *regional office*, sometimes referred to as the *distributorship*. These offices are branches of the main automotive manufacturer. Regional offices are concerned with the selling and service procedures of the company. People who work in the regional offices are employed by the main car manufacturer. They are actually considered the link between the car manufacturer and the dealerships where cars are sold and serviced. Regional offices are geographically located throughout the United States.

Dealerships

After the vehicle has been manufactured, it is sent to the automotive dealer. There are about 25,000 dealers in the United States. These vary from 2-3 people to 20-40 people working for the *dealership*. The dealership is called a *franchised dealer*. This means that the dealership has a contract with the main car manufacturer to sell and service its

VEHICLE	EXPORTING COUNTRY	NUMBER SOLD
Toyota	Japan	441,307
Datsun	Japan	389,105
Honda	Japan	308,161
Mazda	Japan	135,153
Subaru	Japan	125,229
Volvo	Sweden	63,525
Volkswagen	W. Germany	54,710
Mercedes-Benz	W. Germany	50,057
BMW	W. Germany	40,332
Audi	W. Germany	38,095

FIGURE 1-11. 28% of all cars purchased in the United States come from a foreign supplier. The chart shows some of the major suppliers of foreign automobiles.

FIGURE 1–12. Car dealerships throughout the United States sell and service automobiles.

vehicles. The dealerships are privately owned by the dealer. Those who own dealerships are not employed by the car manufacturing company, but they have a contract to sell and service their products.

The dealership is the main link between the car manufacturing company, through the regional offices, to the customer. All sales and service provided by the dealership is controlled by the policies of the car manufacturing company. Warranty problems are also taken care of through the dealership. See *Figure 1–12*.

Fleet Service and Maintenance

In certain cases, large *fleet service* companies can offer transportation for goods, services, and products throughout our society. These fleet offices have so many vehicles, they usually have their own service and maintenance organization. For example, companies such as car rental companies have fleet service available for all of their vehicles. Usually, the service and maintenance is done on a schedule. Having scheduled service periods for the vehicles is referred to as *preventative maintenance*.

Independent Service

There are an estimated 180,000 *independent service* repair and maintenance garages within the United States. These vary from small shops with 2–3 employees to larger shops

with 20–30 employees. Typically, a car owner will have the vehicle serviced at the dealership until the warranty has expired. Then, depending upon the person's preference, future service may be given to the independent service garages. Cars that have had two to three owners often use the services of the independent garages. Gasoline stations, see *Figure 1–13*, fall within this category of service shops.

Supporting Specialty Shops

Because the automobile has become so technologically complex in the past few years, *specialty shops* have been developed. Rather than learning the total vehicle systems and components, the mechanic may specialize in one area. For example, various carburetor, transmission, and muffler shops have been developed over the past few years. Others may include generator, starter, tire, brake, or body shops. The advantage of the specialty shop is that the mechanic has a great deal of technical knowledge about a specific area within the automobile. Prices may be lower as well.

Parts Distribution

In order for all of the service, repair, and maintenance shops to continue operating, parts must be readily available. This means that parts must be distributed throughout the United States for all cars that are driven. *Parts distribution* means parts are sold from parts warehouses to independent parts

FIGURE 1-13. Gasoline stations are considered independent service shops.

dealers. Some parts dealers operate on a national level while some are strictly local. Examples of parts dealers are NAPA, *Figure 1-14*, Automotive Parts and Accessories, Champion Auto Stores, and Crown Auto. Obviously, there are many more.

All of these parts stores are set up as retailers and are usually independently owned. Most parts stores today are also set up on a computer inventory system. Because of the many parts that must be stocked, computer inventory makes the parts systems much easier to operate and run to produce a profit.

Other stores also have automotive parts available. For example, many department stores and hardware stores carry a line of high volume parts such as oil, air and fuel filters, batteries, and so on. See *Figure 1-15*.

FIGURE 1-14. Parts dealers supply the service shops with quality parts.

FIGURE 1-15. Some department stores also carry a line of commonly used automotive parts.

1.3 CAREERS IN THE AUTOMOTIVE INDUSTRY

When studying the automobile industry, it is important to be aware of the many careers and jobs that are currently available in it. The automobile, because of its great impact on society, has touched every part of our lives. Many of the careers in the automobile industry are related to servicing the vehicles. For example, auto mechanics, service managers, and technicians all deal with the automobile in a service function. However, there are many other careers that are also supported by the automobile industry. These include marketing and sales positions, parts distribution positions, distributorships, and company sales and service representatives.

These careers are typically available in a variety of working areas. These may include:

a. Dealerships for selling and servicing.
b. Independent garages for servicing.
c. Service stations for general service.
d. Tire and battery dealers.
e. Specialty shops which handle wheel alignment, transmissions, body repair, and tune-up work.
f. Service shops owned by large stores such as Sears, K Mart, etc.
g. Fleet repair shops such as truck, bus, and automobile fleets.
h. Parts stores and parts distribution centers.
i. Recycling and salvage yard operations.

General Mechanic/Technician

Probably one of the most important careers in the automobile industry is that of the *general mechanic* or *technician*. This person is expected to diagnose, service, and competently repair any problem on the automobile. Usually, these skills come from a sound foundation in auto mechanics, experience, and continuous upgrading and training on new technologies. In a study done by *Motor Age*, the question was asked, "Which profession do you think takes the most training and technical updating?" ***Figure 1–16*** illustrates the results.

This person must also be able to solve problems associated with the automobile. Customers usually have a complaint. The general auto mechanic must use the customer's information, his past experience, and his educational training to determine exactly what the problem is. This person must then repair the problem correctly.

Foreman

In all automotive shops, there is usually a person called the *foreman*. The foreman is considered a supervisory position.

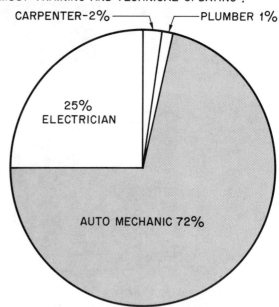

WHICH PROFESSION DO YOU THINK TAKES THE MOST TRAINING AND TECHNICAL UPDATING ?

CARPENTER–2%   PLUMBER 1%

25% ELECTRICIAN

AUTO MECHANIC 72%

FIGURE 1–16. Seventy-two percent of the people surveyed by *Motor Age* indicated that the auto mechanic requires the most training and technical updating.

In this case, the foreman is responsible for organizing the work schedules of the general mechanics as well as making sure the service function runs smoothly. This person must also be able to work with people, including the customers and the mechanics. In addition, the foreman must have a good deal of technical experience with the automobile. This is because the foreman is expected at times to work on complex technical problems with the mechanics.

Service Manager

The *service manager* is responsible for the entire service operation of the dealership. The responsibilities of the service manager include making sure the customer gets proper service, working with the foreman and mechanics to train and update them, and carrying out the factory policies, warranties, and so on. The service manager must have good human relations skills, and have a sound knowledge of the automobile and its technical components.

Parts Manager

If a service organization is to operate smoothly, parts for the vehicles must be available. The *parts manager* is responsible for making sure the customers' parts are immediately available. This person is then responsible for ordering, stocking, inventorying, and selling high-quality replacement parts and accessories. This person must also have available a network of other distributors and suppliers who can also get the needed parts quickly.

Marketing and Sales

In the automotive industry there are many positions available in the area of marketing and sales. People who sell automobiles usually have to have a broad background of automotive expertise. The salesperson must be able to understand the basic components of the vehicle as well as how these components interrelate with each other. This person must also be interested in dealing directly with people in the selling and marketing of the automobile.

Company Representatives — Sales and Service

Besides careers in the sales and service areas of dealers and distributors, there are a great many careers working directly for the automotive manufacturers. For example, all major automotive manufacturers have *sales* and *service representatives*. These people are many times the link between the automotive manufacturer and the sales and service dealers throughout the country.

The service representatives' responsibilities include working as a link between the dealerships and the specific company, training mechanics on new technologies, working with warranty problems, and generally acting as a technical expert in complex service problems. Because of these responsibilities, the service representative needs to have a very detailed exposure to the automobile, including all of the components used.

The sales representatives are the selling link between the dealer or distributor and the specific manufacturer. The sales representatives work with sales and marketing problems that need company attention. Because of these responsibilities, the sales representatives must have a broad knowledge of the automotive industry as well as an understanding of the total automobile and all of the systems and components.

Supporting Careers

Because the automobile industry is so large, there are many supporting career areas available to those who are interested in automobiles. Such careers include:

1. Claims adjusters — working with vehicles that have been in accidents.
2. Vocational instructors — teaching automotive technology.
3. Auto body repair — repairing vehicles that have been in accidents.

4. Frame and alignment repair — repairing the front end steering systems and straightening frames after accidents.
5. Specialty shops — repairing specific components such as tires, carburetors, fuel injection, muffler and exhaust systems, transmissions, etc.

As you can see, the automobile industry is very large and is capable of supporting thousands of careers. This is true for those who are interested and have a good understanding of the automobile.

SUMMARY

The overall purpose of this chapter was to introduce the automobile, define basic components, explain how the automobile is used in society, and identify various careers within the automotive industry.

The automobile is subdivided into several major components. These include the body and frame, the engine, also called the power source, the drive lines and running gear, and the suspension systems. These components make up the basic foundation of the automobile. All supporting systems such as the brakes, shocks, fuel systems, lubrication systems, and so on are subsystems. This textbook is concerned with studying the subsystems of the automobile.

The automobile is one of the most influential technological developments of our time. The average number of people per cars in the United States is said to be 1.9 people per vehicle. This means that the United States uses the automobile more than any other country. Because of its influence, the automobile has generated many offices and service establishments. These include regional and distributorship offices, fleet service and maintenance shops, independent service garages, supporting specialty shops, and parts distribution shops.

Because the automobile has had such an influence on our society, many careers are available for those who want to work with the automobile. Some of the major careers include the general mechanic, the foreman, the service manager, the parts manager, the marketing and sales representative, the service company representative, and many additional supporting careers. Each of these careers requires some expertise in the automotive field. Some require extensive technical background, while others require both technical expertise and good human relations skills.

The automobile has affected all aspects of our lives. Because of this effect, it becomes extremely important to understand the technical components of the contemporary automobile. This textbook will help to meet this objective.

TERMS TO KNOW

Can you explain each of the following terms? Review the chapter until you can use each term correctly.

Body and frame

Engine

Fuel system

Ignition system

Starting and charging
system

Lubricating system

Cooling system

Drive lines

Running gear

Suspension systems

Regional offices and
distributorships

Dealership

Franchised dealer

Fleet service

Independent service

Specialty shops

Parts distribution

General mechanic

Foreman

Service manager

Parts manager

Service representative

Supporting careers

REVIEW QUESTIONS

Multiple Choice

1. Which of the following is considered the mechanical foundation of the automobile?
   a. Drive lines
   b. Suspension systems
   c. Body and frame
   d. All of the above
   e. None of the above

2. Which type of body and frame is all in one unit?
   a. Separate
   b. Unitized
   c. Stub frame
   d. Nose frame
   e. Stabilized

3. Which of the following types of power sources are currently used?
   a. Diesel power sources
   b. Rotary power sources
   c. Piston engine (gasoline) power sources
   d. All of the above
   e. None of the above

4. What subsystem on the engine is used to mix the air and fuel correctly?
   a. Lubrication
   b. Cooling
   c. Fuel
   d. Drive line
   e. Starting

5. The components on the vehicle used to transmit power are referred to as:
   a. Running gear
   b. Drive lines
   c. Suspension systems
   d. Shock absorbers
   e. Transfer gear

6. Springs, shock absorbers, and torsion bars are part of which system?
   a. Drive lines
   b. Running gear
   c. Suspension systems
   d. Transfer gear
   e. Brake systems

7. What percentage of the energy in the transportation sector is used by the automobile?
   a. 15%
   b. 52%
   c. 95%
   d. 45%
   e. 22%

8. Approximately what percent of all cars purchased are bought from foreign suppliers?
   a. 1–4%
   b. 25–30%
   c. 82–93%
   d. 50–52%
   e. 71–76%

9. Which of the following offices are often considered a franchised office?
   a. Dealerships
   b. Independent service garages
   c. Fleet operations
   d. Company parts distribution
   e. All of the above

10. Which of the following careers would require the most technical training and background about the total automobile?
    a. Parts manager
    b. Sales representative
    c. General mechanic
    d. Automotive driver
    e. All of the above

11. Which of the following systems on the engine is used to provide the spark to burn the air and fuel mixture?
    a. Lubrication system
    b. Cooling system
    c. Ignition system
    d. Starting system
    e. Suspension system

12. Which of the following systems are used to slow down the upward and downward movement of the vehicle?
    a. Springs
    b. Axles
    c. Shock absorbers
    d. Torsion springs
    e. None of the above

13. What is the average number of people per car within the United States?
    a. 1.9
    b. 3.8
    c. 9.2
    d. 10.1
    e. 11.2

14. What percentage of the work force is working in the automobile sector of society?
    a. 22–25%
    b. 35–45%
    c. 50–60%
    d. 65–70%
    e. None of the above

15. An automotive shop that works only on carburetors would be classified as a (an):
    a. Independent service shop
    b. Supporting specialty shop
    c. Dealership
    d. Company specialty shop
    e. None of the above

The following questions are similar in format to ASE (Automotive Service Excellence) test questions.

16. Technician A says that the springs, shock absorbers, and torsion bars are part of the suspension system. Technician B says the springs, shock absorbers, and torsion bars are part of the running gear. Who is right?
    a. A only
    b. B only
    c. Both A and B
    d. Neither A nor B

17. Technician A says the parts used to transmit power to propel the vehicle forward are called the running gear. Technician B says the parts used to transmit power to propel the vehicle forward are called the drive line. Who is right?
    a. A only
    b. B only
    c. Both A and B
    d. Neither A nor B

18. Technician A says that only a small percentage of people, about 2%, work in the automotive sector of society. Technician B says that about 75% of people work in the automotive sector of society. Who is right?
    a. A only
    b. B only
    c. Both A and B
    d. Neither A nor B

Essay

19. Identify at least five careers in the automotive field.

20. Describe the person that is in charge of the entire service operation of a dealership.

21. What are some of the supporting careers in the automotive industry?

22. Describe the difference between the suspension system and the drive line of an automobile.

CHAPTER 2

Safety in the Automotive Shop

INTRODUCTION

Much of the service and maintenance work done on the automobile is completed in the automotive shop. Because of the complexity of the automobile, many tools, instruments, and machines are used for service. In addition, there are many people in the service area. Complex tools, machines, and instruments, coupled with many people, make the automotive shop a likely place for accidents to happen. In addition, the automobile has very explosive fuel that can be dangerous. Safety in the automobile shop has become a very important aspect in the total study of automotive technology.

Safety has become such an important part of our society and industry that the Federal Government established the Occupational Safety and Health Act of 1970. This Act is known as *OSHA*. This Act makes safety and health on the job a matter of law for four million American businesses. This Act also applies to automotive service shops.

The Act provides several things. It establishes standards and regulations for safety. It improves unsafe and unhealthful working conditions. It also assists in establishing plans for safe working conditions.

OBJECTIVES

After reading this chapter, you will be able to:
- Define and illustrate common safety equipment used in the automobile service area.
- List various safety rules used in any automobile service area.
- Develop certain safety attitudes concerning safety in the automobile service area.
- List the possible danger areas for common accidents in the automobile service area.

CHAPTER HIGHLIGHTS

2.1 SAFETY EQUIPMENT
   A. Safety Glasses
   B. Fire Extinguishers
   C. Airtight Containers
   D. Gasoline Containers

   E. Gloves
   F. Types of Clothing and Shop Coats
   G. Shoes
   H. First Aid Boxes
   I. Ear Protectors

2.1 SAFETY EQUIPMENT

To have safe working habits in the automotive shop, it is important to know the safety equipment. There are several important types of safety equipment used. Some examples of safety equipment are safety glasses, fire extinguishers, airtight containers, gasoline containers, gloves, first aid boxes, and ear protectors.

Safety Glasses

One of the most important aspects of safety is to have all shop personnel wear *safety glasses*. Many service mechanics and technicians have been permanently blinded because they thought safety glasses were not important.

There are many types of safety glasses used today. An important rule to remember is that all safety glasses should have safety glass, and some sort of side protection. *Figure 2-1* illustrates several types of approved safety glasses.

When purchasing safety glasses, always remember to buy a pair that feels comfortable. If the glasses are not comfortable, people have a tendency to either remove them or wear them on the top of the head. Both situations leave the eyes totally unprotected.

Fire Extinguishers

Another important piece of safety equipment is the fire extinguisher. All personnel should look around the shop and determine where each fire extinguisher is located. This may be very important, especially if you are in a hurry to put out a fire. Once the location of each fire extinguisher is known, learn how to operate them.

There are several types of fire extinguishers. *Figure 2-2* illustrates the types of extinguishers and classifications of fires. Fires are classified into three types.

Type A fires: Fires from ordinary combustible materials. These include materials such as wood, paper,

FIGURE 2-1. Many types of safety glasses are approved for use in industry. Select a pair that has safety glass and side protectors, and is very comfortable so they can remain on without discomfort. Certain models of safety glasses have vents to reduce fogging. *(Courtesy of Sellstrom Manufacturing)*

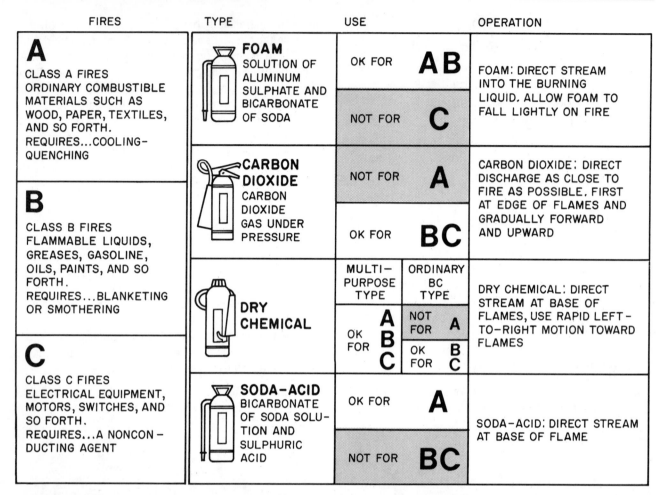

FIRES	TYPE	USE		OPERATION
A CLASS A FIRES ORDINARY COMBUSTIBLE MATERIALS SUCH AS WOOD, PAPER, TEXTILES, AND SO FORTH. REQUIRES...COOLING-QUENCHING	**FOAM** SOLUTION OF ALUMINUM SULPHATE AND BICARBONATE OF SODA	OK FOR **AB**		FOAM: DIRECT STREAM INTO THE BURNING LIQUID. ALLOW FOAM TO FALL LIGHTLY ON FIRE
		NOT FOR **C**		
B CLASS B FIRES FLAMMABLE LIQUIDS, GREASES, GASOLINE, OILS, PAINTS, AND SO FORTH. REQUIRES...BLANKETING OR SMOTHERING	**CARBON DIOXIDE** CARBON DIOXIDE GAS UNDER PRESSURE	NOT FOR **A**		CARBON DIOXIDE: DIRECT DISCHARGE AS CLOSE TO FIRE AS POSSIBLE. FIRST AT EDGE OF FLAMES AND GRADUALLY FORWARD AND UPWARD
		OK FOR **BC**		
	DRY CHEMICAL	MULTI-PURPOSE TYPE OK FOR **ABC**	ORDINARY BC TYPE NOT FOR **A** OK FOR **BC**	DRY CHEMICAL: DIRECT STREAM AT BASE OF FLAMES, USE RAPID LEFT-TO-RIGHT MOTION TOWARD FLAMES
C CLASS C FIRES ELECTRICAL EQUIPMENT, MOTORS, SWITCHES, AND SO FORTH. REQUIRES...A NONCON-DUCTING AGENT	**SODA-ACID** BICARBONATE OF SODA SOLU-TION AND SULPHURIC ACID	OK FOR **A**		SODA-ACID: DIRECT STREAM AT BASE OF FLAME
		NOT FOR **BC**		

FIGURE 2-2. There are many types of fire extinguishers. The automotive shop should be equipped to handle all fires classified as A, B, and C types.

textiles, and clothing. This type of fire usually requires cooling and quenching.

Type B fires: Fires from flammable liquids, greases, gasoline, oils, paints, and other liquids. This type of fire requires smothering and blanketing.

Type C fires: Fires started from electrical equipment malfunctions, motors, switches, and wires. This type of fire requires a nonconducting agent to put it out.

Referring to Figure 2-2, four types of fire extinguishers are shown. The foam type is used for A and B type fires but not for C type fires. Carbon dioxide fire extinguishers are mostly used for B and C type fires but not for A type fires. Figure 2-2 illustrates the types of fires that can be controlled with dry chemicals as well as with soda-acid fire extinguishers.

Airtight Containers

In the automotive shop there are many oily rags. These rags can cause a fire from spontaneous combustion. Spontaneous combustion is a chemical reaction. It is produced from a slow generation of heat from the oxidation of the oil in the rag. This occurs until the ignition temperature is reached. The fuel then begins to burn, causing a fire. Because of this danger, *airtight containers* are used to hold oily rags. *Figure 2-3* shows an airtight waste container for holding waste rags.

Gasoline Containers

Gasoline is a very explosive fuel. Because of this danger, gasoline should always be kept in an approved container. The container should always be painted red and should have approved openings for pouring and venting. *Figure 2-4* shows the gasoline container approved by OSHA.

FIGURE 2-3. Airtight containers should be used to hold oily rags.

FIGURE 2-4. All gasoline should be kept in an approved gasoline container. The container has the correct pouring and venting system.

Another type of gasoline container is used to store many types of fuel and oil. This container is designed to contain any type of explosion. Any fuel, oil, or other flammable liquid should be stored in the proper small container. The small containers should then be stored in *explosion-proof cabinets*. See *Figure 2-5*.

Gloves

Many activities done in the automotive shop may require the use of protective gloves. For example, always use protec-

tive gloves when grinding metal, working with caustic cleaning solutions, and when welding or working with hot metals.

There are many types of protective gloves to use. *Figure 2-6* illustrates rubber-coated gloves for heat protection.

FIGURE 2-5. All fuel, oil, or other flammable liquid should be stored in a nonflammable, explosion-proof cabinet.

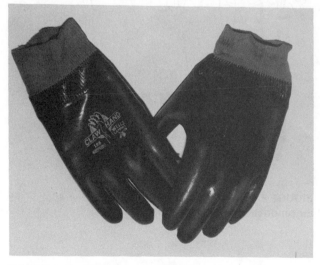

FIGURE 2-6. Always wear protective gloves when using solvents, working with hot metal, and when grinding metal.

Types of Clothing and Shop Coats

When working in the automotive service area, the type of clothing that is worn is important. Always wear clothes that you know can get dirty. A person working in the automotive shop that tries not to get dirty will probably not complete the service correctly.

There are many types of shop coats available for working on automobiles. See *Figure 2–7*. Shop coats can protect clothing and make the person more comfortable in the automotive shop. The only concern with shop clothes is that the clothing should not fit loosely. Any loose fitting clothing may get caught on machines or parts of the automobile. This may cause a person to fall.

This is also true if the person is wearing a tie. If free to move around, ties can get caught in any machine or part that is rotating. When wearing a tie under a shop coat, always make sure the tie is pinned to the shirt or held tightly in place under the coat.

Shoes

When working in the automobile shop, it is important not to wear sandals or summer-type shoes. When servicing the automobile, heavy parts are often lifted from one spot to another. Heavy objects such as cylinder heads, manifolds,

brake drums, and so forth may be dropped accidentally. The person working in the automobile shop should always wear either steel-tipped shoes or shoes that are strong enough to withstand heavy objects without damage or injury to the feet.

First Aid Boxes

All automotive service areas should have several *first aid boxes*. These boxes, as shown in *Figure 2–8*, should be located for easy access. Each person who works in the shop should know where each first aid box is located.

Ear Protectors

Excessive noise in the automotive shop should be eliminated or reduced whenever possible. In certain cases, noise protection will have to be used. For example, if engines are run on dynamometers, or put under heavy loads, noise may be above recommended levels.

Noise is measured in units called *decibels*. Various levels of noise are shown in *Figure 2–9*. Damage to the ears could result from constant high levels of noise. It is generally accepted that approximately 90–100 decibels could damage the ear.

Because of possible damage, it is recommended that some sort of ear protection be used in the automotive shop. Ear protection can be obtained by using commercially available earplugs, or ear muffs as shown in *Figure 2–10*.

FIGURE 2–7. Always wear shop coats when working in the automotive service area.

FIGURE 2–8. Every automotive service area should have several first aid boxes. The box should include bandages and other necessary first aid equipment.

REPRESENTATIVE SOUND LEVELS

SOUND LEVEL, dB	OPERATION OR EQUIPMENT	
150	Jet engine test cell	
145	_____ Threshold of Pain	
130	Pneumatic press (close range) Pneumatic rock drill Riveting steel tank	
125	Pneumatic chipper Pneumatic riveter	
120	_____ Threshold of Discomfort Turbine generator	**DANGER ZONE**
112	Punch press Sandblasting	
110	Drills, shovels, operating trucks Drop hammer	
105	Circular saw Wire braiders, stranding machine Pin routers Riveting machines	
100	Can-manufacturing plant Portable grinders Ram turret lathes Automatic screw machine	
90	Welding equipment Weaving mill Milling machine Pneumatic diesel compressor Engine lathes Portable sanders	**RISK ZONE**
85	California freeway traffic (overpass)	

HEARING DAMAGE IF CONTINUED EXPOSURE ABOVE THIS LEVEL

80	Tabulating machines, electric typewriters	
75	Stenographic room	
70	Electronics assembly plant	
65	Department store	
60	Conversation	
35	Quiet home forced air heating	**SAFE ZONE**
10	Whisper	

FIGURE 2–9. Noise is measured in decibels (dB). Different decibel levels are shown. Damage to the ear can result from prolonged decibel levels of 90–100.

FIGURE 2–10. Earmuffs provide protection from high decibel levels when working on the automobile.

CAR CLINIC

PROBLEM: BREAK-IN PROCEDURE

A customer has been told several methods used to break in a new engine. The driver is keeping the engine at a constant 55 mph as much as possible. Also, it was suggested that the oil be changed after the first 500 miles. What is the best procedure for breaking in an engine?

SOLUTION:

The best method is to follow the manufacturer's recommendation. This usually includes driving moderately between 50 and 60 miles per hour for about 2,000 miles under normal loads. Use the manufacturer's break-in oil. The oil is designed to aid the engine parts in wearing in to the mating surfaces. Also, the engine needs about 50–60 cycles of from cold to hot to help wear in the parts. Change the break-in oil according to the owner's manual for new cars.

2.2 SAFETY RULES IN THE SHOP

There are many safety rules that make working in the automobile shop much easier and safer. These rules are concerned with shop layout, lifting and carrying, good housekeeping, smoking, and being aware of carbon monoxide.

Shop Layout

There are many kinds of automotive shops. Each shop will have a different shop layout. The shop layout shows where the equipment is located, where the cars are worked on, and where special repairs take place. It is important for the service mechanic or student to become familiar with the total shop layout so each person can work with high efficiency. Typical shop layouts are shown in *Figure 2–11*.

In Figure 2–11, the auto bays are used for most service work on the automobile. Parts washers are nearby so that travel is minimum to wash the needed parts. Tools are located on boards and in special areas shown. Many mechanics have their own sets of tools that can be rolled directly up to the vehicle. The service manager's office is located in a central location to provide easy access.

Most automotive shops are designed to have certain types of repair in only one place. For example, shops may have an alignment area, body and fender repair area, painting area, tune-up area, and/or general repair and maintenance area. Others may include valve grinding, block boring, etc. This area is called the machining area. Large shops have all of these, while smaller shops may only have certain areas. It is important for the service person to become completely familiar with the total shop layout.

Lifting and Carrying

When working around the automobile, it may be necessary to lift heavy objects. Many back injuries have occurred because the service mechanic did not lift properly. The following procedure should always be used when lifting any heavy object.

1. Consider the size, weight, and shape of the object.
2. Set feet solidly, with one foot slightly ahead of the other for stability.
3. Get as close to the object as possible.
4. Keep the back as straight as possible and bend legs.
5. Grip the object firmly.
6. Straighten the legs to lift the object, bringing back to a vertical position.
7. Never carry a load you cannot see over or around.
8. Setting down the object requires just the reverse procedure.

In addition, safety should always be observed when lifting heavy objects with a chain hoist and car hoist. Make sure the chain hoist is securely connected to the engine before lifting. Bolts should be of sufficient strength to hold the engine and its weight. In addition, when lifting the vehicle on a car hoist, follow the manufacturer's recommended procedure. This usually includes making sure the car is positioned correctly on the hoist as well as making sure the hoist contacts the correct part of the vehicle.

Good Housekeeping

A clean and orderly shop makes employees respect the equipment and working area. A customer also has more confidence in the work being done if the shop is clean.

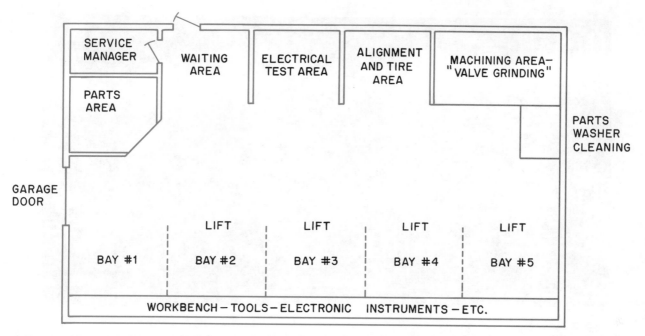

FIGURE 2–11. It is important to be familiar with the automotive shop layout. When the layout is known, improved safety will result.

Following is a list of housekeeping items that should always be observed:

1. Is proper light provided?
2. Are walls and windows clean?
3. Are stairs clean and well lighted?
4. Are floor surfaces clean of loose material?
5. Are floors free of oil, grease, etc?
6. Are containers provided for refuse?
7. Are aisles free of obstructions?
8. Are there safe and free passages to fire extinguishers?
9. Are tools arranged in proper places? See **Figure 2–12**.
10. Are oil rags in the proper container?
11. Are tools free of grease and oil?
12. Are tools in good working condition?
13. Are proper guards provided on all machinery?
14. Are benches and seats clean and in good condition?
15. Are parts and materials in proper location?
16. Has the creeper (rolling platform for crawling under car) been put away?

If all shop employees are aware of these items, the automotive service area will become a safer place to work.

Smoking in the Shop

In the automotive shop, fuel is often nearby to run the engines. It is very important to know when and where to smoke and not to smoke. Always be aware of the "smoking" as well as the "no smoking" areas. This becomes the responsibility of the automotive mechanic as well as the management. Management will help to enforce these smoking rules.

Carbon Monoxide

It is important to be familiar with *carbon monoxide* when working with the automobile. Carbon monoxide is usually given off from the exhaust of running engines. Most new cars have exhaust emission controls. However, many older cars still give off carbon monoxide. The gas is odorless and colorless. You may not even be aware that it is there. If taken in through normal breathing, carbon monoxide can cause death. The presence of carbon monoxide in a person can be noticed by the following symptoms:

1. Headaches
2. Nausea
3. Ringing in the ears
4. Tiredness
5. Fluttering heart

If any of these symptoms appear, it is very important to get fresh air immediately.

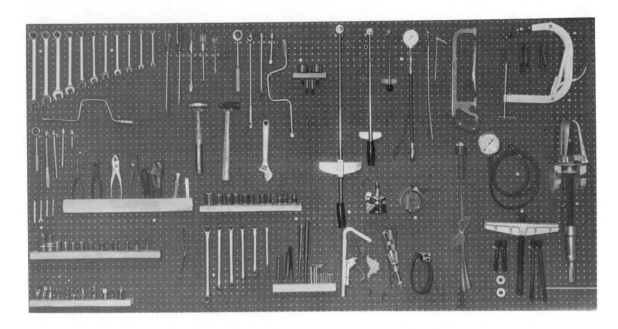

FiGURE 2–12. Tools should be well organized and always replaced to their original position.

To eliminate the possibility of carbon monoxide, always have good ventilation and make sure the engine is properly exhausted in the shop. This is done by using the proper exhaust systems as shown in *Figure 2–13*.

Other Rules to Follow

The following list of rules should always be observed when working in an automotive service area.

1. Make sure all electrical appliances such as drills, electric motors, grinders, and so forth have a three-wire cord for correct grounding. Also, make sure the cords are not frayed or damaged.

2. Make sure all hand tools are in good condition. Using a damaged hand tool or the incorrect tool for the job may result in a severe hand injury.

3. Whenever a car is lifted by using jack stands, always make sure the jack is centered. When the vehicle is raised, always use safety jack stands under the car. Never go under a car without safety jack stands. A set of safety jack stands is shown in *Figure 2–14*.

FIGURE 2–13. Always use the proper exhaust piping when running a vehicle in the shop. Also, remember to turn on the fan to remove exhaust from the vehicle. *Reminder:* Carbon monoxide is a colorless, odorless gas which can cause death.

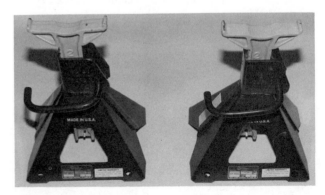

FIGURE 2–14. Always support the raised vehicle with safety jack stands. Never work under a vehicle without the proper safety jack stands.

4. Never wear jewelry such as rings, bracelets, or watches when working on a car. These items can easily catch on moving parts or cause an electrical short and cause serious injury.

5. Never use compressed air to remove dirt from your clothing or you may get dirt in your eyes. Also, never spin bearings with compressed air. If the bearing is damaged, one of the steel balls may come loose and cause serious injury. Damage to the bearing may also result.

6. Always be careful where welding sparks are falling. Sparks can cause a fire or explosion if dropped on flammable materials.

7. When using any machine such as hydraulic presses, hoists, drill presses, or special equipment, make absolutely certain that all operational procedures are studied first. Get checked out on the machine from the instructor first.

CAR CLINIC

PROBLEM: HOW TO START A COLD ENGINE

It seems that the best way to start and run a cold engine is to start it up and then let it idle until reaching operating temperature. But it always takes a long time and a lot of gasoline to get the car to operating temperature. Is this the correct way to start a cold engine?

SOLUTION:

The best way to start a cold engine is to start the engine and wait about a minute at idle. This will give the oil time to get to all parts of the engine. Then slowly start out, going through all the gears. An engine under a slight to moderate load will reach operating temperature much quicker than an engine at idle.

SUMMARY

The purpose of this chapter was to introduce safety in the automotive service area. Safety has become a very important part of working on the automobile. With the complexity of the vehicle today, many new tools and machines are used. This makes safety even more important.

Safety can be improved in the shop by using the right safety equipment. Safety glasses should always be worn, no matter where you are in the shop. Know excactly where the fire extinguishers are located and how to use them. Also, know what type of fire extinguisher should be used on Type A, B, and C fires.

The automotive shop always has oily rags that must be stored or disposed of. Always use an airtight container to store these rags. This will eliminate the possibility of spontaneous combustion. In addition, gasoline containers should always be stored in an explosion-proof cabinet.

Other safety equipment that should be used in the shop includes proper gloves and correct types of clothing, including shop coats and shoes. In addition, know where each first aid box is located. Then, if an accident does happen, the first aid equipment will be easily found.

Ear protectors are also important as safety equipment. When loud noises are produced in the shop, usually above 100 decibels, always use the recommended ear plugs or ear muffs. This will protect the ears from progressive damage from loud noises.

There are many other rules that should be followed when working in the automotive shop. Always know the shop layout. Know where all the equipment is located. Know what each part of the shop is used for.

When lifting and carrying heavy objects, always lift by using the legs rather than the back. Serious back injury may result from lifting incorrectly.

Any automotive shop will be a safer place to work if good housekeeping is maintained. For example, make sure there is proper light, and walls and windows are clean. Also, make sure floor surfaces are free of grease and loose materials. Keep tools clean, and place all tools and parts in their proper location.

Other rules include: 1) smoke only in designated areas, 2) be aware of carbon monoxide and always exhaust cars properly, 3) make sure all electrical equipment is in proper working order, 4) always use jack stands when working under the vehicle, and 5) never wear jewelry — it may get caught on moving parts and cause injury.

There are many rules and regulations that must be understood and adhered to in the automotive shop. If these rules and regulations are followed, the process of working on your automobile will be a satisfying and rewarding experience.

TERMS TO KNOW

Can you explain each of the following terms? Review the chapter until you can use each term correctly.

OSHA (Occupational Safety and Health Act of 1970)	First aid boxes
	Decibel
Safety glasses	Back injury
Type A fires	Housekeeping
Type B fires	Smoking rules
Type C fires	Carbon monoxide
Airtight containers	Jack stand
Explosion-proof cabinets	

REVIEW QUESTIONS

Multiple Choice

1. In what year did the federal government establish the Occupational Safety and Health Act?
   a. 1925
   b. 1945
   c. 1970
   d. 1980
   e. 1985

2. All safety glasses should have:
   a. Side protectors
   b. Safety glass
   c. Steel rims
   d. A and b of the above
   e. All of the above

3. Which type of fire is the burning of wood, paper, and clothing?
   a. Type A fires
   b. Type B fires
   c. Type C fires
   d. Type D fires
   e. Type F fires

4. Airtight containers are used to hold:
   a. Wrenches that have grease on them
   b. Oily rags
   c. Shop parts
   d. Gaskets
   e. Solvent

5. All gasoline containers should be:
   a. Approved by OSHA
   b. Hand made
   c. Painted black
   d. Painted green
   e. Never used

6. Which of the following shoes are not acceptable to wear in the automotive shop?
   a. Steel-tipped shoes
   b. Strong leather shoes
   c. Summer sandals
   d. Approved work shoes
   e. All of the above

7. The unit of measurement for sound is called the:
   a. Millimeter
   b. Gram
   c. Decibel
   d. Horsepower
   e. Inch

8. What level of sound may be damaging to your ears?
   a. 90–100
   b. 30–40
   c. 10–20
   d. 0–5
   e. 5–9

9. When lifting heavy objects, always lift with the:
   a. Legs
   b. Back
   c. Neck
   d. Arms only
   e. Stomach

10. Carbon monoxide is produced from:
    a. Transmission oil
    b. Gas engine exhaust
    c. Radiator coolant
    d. Diesel fuel
    e. None of the above

11. When working in the automotive shop, always:
    a. Wear shop coats
    b. Run the vehicle without proper exhaust
    c. Wear good clothing
    d. Wear white shirts
    e. Keep tools greasy

12. Which of the following is/are important when working in the automotive shop?
    a. Know the shop layout
    b. Wear a shop coat
    c. Wear steel-tipped shoes
    d. All of the above
    e. None of the above

13. When working in the automotive shop, it is important to:
    a. Put oil rags in the proper container
    b. Leave tools greasy for later clean-up
    c. Let the shop foreman put tools away
    d. Let the service manager fix the vehicle
    e. None of the above

14. Which of the following is considered dangerous exhaust from automobiles?
    a. Water
    b. Carbon monoxide
    c. Carbon dioxide
    d. Nitrogen
    e. Oxygen

15. When working in the automotive shop, never
    a. Wear rings or other jewelry
    b. Use compressed air to remove dirt from clothing
    c. Use tools that are in bad condition
    d. All of the above
    e. None of the above

The following questions are similar in format to ASE (Automotive Service Excellence) test questions.

16. Technician A says it is OK to inhale carbon monoxide and other fumes from the automobile exhaust. Technician B says it is OK to inhale carbon monoxide fumes but not carbon dioxide fumes from the automobile exhaust. Who is right?
    a. A only
    b. B only
    c. Both A and B
    d. Neither A nor B

17. Technician A says a decibel level of 92 may be damaging to one's ears. Technician B says a decibel level of 98 may be damaging to one's ears. Who is right?
    a. A only
    b. B only
    c. Both A and B
    d. Neither A nor B

18. Technician A says that one should use foam-type extinguishers for Type A fires. Technician B says that one should use foam-type extinguishers for Type C fires. Who is right?
    a. A only
    b. B only
    c. Both A and B
    d. Neither A nor B

Essay

19. Discuss some of the safety features of a well designed shop layout.

20. List the procedure for lifting heavy objects.

21. Identify at least five safety (housekeeping) items that should be observed in the automotive shop.

22. What are the symptoms of the presence of carbon monoxide in a person?

23. State three safety rules to be followed when working in the automotive shop.

CHAPTER 3

Fasteners, Gaskets, Sealants, and Bearings

INTRODUCTION

The automobile is a complex combination of many parts. These parts are held together by *fasteners* such as screws, bolts, and rivets. When parts are placed together in the presence of oil, gaskets, sealants, and seals are used to control the oil from leaking. Also, many vehicle parts move next to stationary parts. Bearings must be used to reduce friction between these parts. The overall purpose of this chapter is to examine the many types of fasteners, gaskets, sealants, seals, and bearings used on the automobile.

OBJECTIVES

After reading this chapter, you will be able to:

- Identify the different types of bolts, washers, splines, keyways, snap rings, and screws used on the automobile.
- Determine how to select torque specifications for different size bolts.
- Examine the purposes and styles of different types of gaskets.
- Define the purposes of and the types of seals used on the automobile.
- Define the parts of a bearing.
- Identify the types of bearings used on the automobile, including ball, roller, and thrust bearings.
- Identify various service and diagnostic procedures used when working with fasteners, gaskets, seals, and bearings.

CHAPTER HIGHLIGHTS

3.1 FASTENERS
    A. Threaded Fasteners
    B. Types of Threaded Fasteners
    C. Bolt Identification
    D. Unified and American National Thread Sizes
    E. Bolt Hardness and Strength
    F. Metric Threads

    G. Determining Torque on Bolts
    H. Nuts
    I. Washers
    J. Snap Rings
    K. Splines
    L. Keyways
    M. Helicoils

3.1 FASTENERS

Fasteners are defined as those objects that secure or hold together parts of the automobile. Examples include bolts, nuts, washers, snap rings, splines, keyways, rivets, and setscrews. Actually, there are many types of fasteners. In this chapter certain fasteners will be defined. Their applications will also be illustrated.

Threaded Fasteners

One of the most popular type of fastener is called the *threaded fastener*. Threaded fasteners include bolts, nuts, screws, and similar items that allow the mechanic to install or remove parts easily. Examples of threaded fasteners are shown in *Figure 3–1*.

Types of Threaded Fasteners

Figure 3–2 shows a few of the more common types of fasteners used on the automobile. These include bolts, studs, setscrews, cap screws, machine screws, and self-tapping screws.

Bolts have a head on one end and threads on the other. Their length is measured from the bottom surface of the head to the end of the threads. Most automotive bolts have a hexagon head.

Studs are rods with threads on both ends. They are used where bolts are not suitable. For example, studs are used on parts that must be removed frequently for service such as the exhaust manifold. One end of the stud is screwed into a threaded or tapped hole in the exhaust manifold. The other end of the stud passes through the flange on the exhaust pipe. A nut is then used on the projecting end of the stud to hold the parts together.

FIGURE 3–1. Many types of threaded fasteners are used on the automobile today. Bolts, nuts, and screws are all considered threaded fasteners.

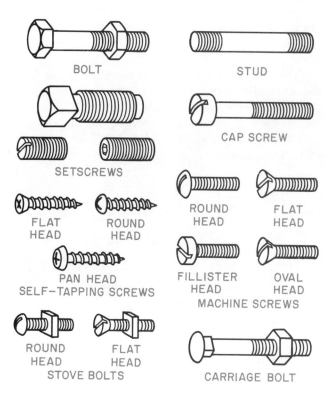

FIGURE 3–2. Many types of threaded fasteners are used on the automobile.

Setscrews are used to prevent rotary motion between two parts such as a pulley and shaft. Setscrews are either headless or have a square head. As shown in Figure 3–2, they are available with a variety of points.

Cap screws pass through a clearance hole in one member of a part. They then screw into a threaded or tapped hole in another part. The point of the cap screw has a 45 degree chamfer.

Machine screws are similar to cap screws, but they have a flat point. The threads on a machine screw run the entire length of the stem or shank. Several types of heads are used, including the round, flat, fillister torx, and oval heads.

Self-tapping screws are used to fasten sheet-metal parts or to join light metal, wood, or plastic together. These screws form their own threads in the material as they are turned. They are available with different head shapes and points.

Bolt Identification

To identify the type of threads on a bolt, bolt terminology must be defined. The bolt has several parts, as shown in *Figure 3–3*. The head is used to *torque* or tighten the bolt. A socket fits over the head which enables the bolts to be tightened. Common USC (U.S. Customary) and metric sizes for bolt heads include those shown in *Figure 3–4*. The sizes are given in fractions of an inch and in millimeters (metric). Some of the USC and metric sockets are very close in size. It is important not to use metric sizes for USC bolts, or USC sizes for metric bolts. The bolt heads may be damaged.

The second part of the bolt is called the *shank*. The shank of the bolt is measured by its diameter. The definition of the shank is illustrated in Figure 3–3. It is the part of the bolt between the threads and the head. Common shank sizes include 1/4, 5/16, 3/8, 7/16, 1/2, 9/16, and 5/8 inches in diameter.

Another way of identifying the size of a bolt is by the number of *threads per inch* as shown in *Figure 3–5*. This can be determined by using a ruler and counting the number of threads per inch for each bolt.

COMMON ENGLISH (U.S. CUSTOMARY) HEAD SIZES	COMMON METRIC HEAD SIZES
WRENCH SIZE	WRENCH SIZE
3/8″	9 mm
7/16″	10 mm
1/2″	11 mm
9/16″	12 mm
5/8″	13 mm
11/16″	14 mm
3/4″	15 mm
13/16″	16 mm
7/8″	17 mm
15/16″	18 mm
1″	19 mm
1 1/16″	20 mm
1 1/8″	21 mm
1 3/16″	22 mm
1 1/4″	23 mm
1 5/16″	24 mm
1 3/8″	26 mm
1 7/16″	27 mm
1 1/2″	29 mm
	30 mm
	32 mm

FIGURE 3–4. There are many standard bolt head sizes. Both USC and metric sizes are shown.

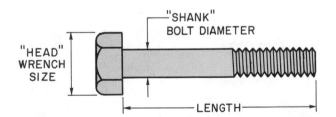

FIGURE 3–3. Bolts are identified by defining the head size, the shank size, the number of threads per inch, and the length. Tensile strength markings are also indicated by the number of lines on the head of the bolt.

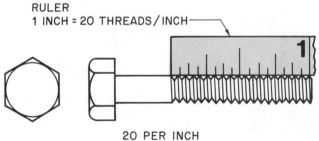

FIGURE 3–5. Threads on bolts can be measured by using a ruler and counting the number of threads per inch.

The number of threads per inch can also be measured by a *screw-pitch gauge*, **Figure 3–6**. The tool consists of numerous blades with thread-shaped teeth on one edge. The blades are inserted over the thread until one that fits the thread exactly is found. The number stamped on the blade that matches the thread indicates the number of threads per inch of the bolt.

The size of a bolt can also be expressed in terms of its length. A ruler is used to measure the length from the end of the bolt to the bottom surface of the head. Bolts are commonly manufactured in lengths with 1/2 inch increments or sizes.

Figure 3–7 shows how the identifying features of a bolt are combined together to specify a bolt size. For example, a bolt identified as 3/8 × 16 × 1 1/2 means that the bolt has a shank diameter of 3/8 inch, there are 16 threads per inch, and the length of the bolt is 1 1/2 inches.

Bolts and nuts can have different numbers of threads per inch and still have the same shank size. For example, a 3/8 inch bolt can have either 16 or 24 threads per inch.

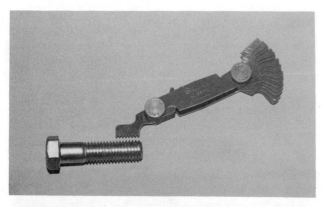

FIGURE 3–6. A screw-pitch gauge is used to determine the exact number of threads per inch.

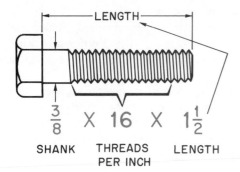

FIGURE 3–7. When a bolt is identified, the shank size, the number of threads per inch, and the length are all used to determine its size.

The greater the number of threads per inch, the finer the thread. The finer the thread, the greater its holding ability.

Unified and American National Thread Sizes

Threads can also be identified by referring to them as either coarse threads or fine threads. Coarse threads identified as *NC* or *UNC* (National Coarse or Unified National Coarse) are used for general-purpose work. They are very adaptable for cast iron and soft metals where rapid assembly or disassembly is required. Fine threads identified as *NF* or *UNF* (National Fine or Unified National Fine) are used where greater resistance to vibration is required. NF threads are also used where greater strength or holding force is necessary.

Bolt Hardness and Strength

Bolts are made from different materials having various degrees of hardness to the metal. Softer metal or harder metal can be used to manufacture the bolts. Under certain conditions, the standard hardness used in a particular situation is not sufficient. Therefore, bolts are made with different hardnesses and strengths for use in different situations. Bolts are marked with lines on the top of the head to identify *bolt hardness* as shown in **Figure 3–8**. The number of lines on the head of the bolt is related to the tensile strength. As the number of lines increases so does the *tensile strength*. Tensile strength is the amount of pressure per square inch the bolt can withstand just before breaking when being pulled apart. The harder or stronger the bolt, the greater the tensile strength.

Metric Threads

Metric threads are being used increasingly on all cars manufactured in the U.S. This means that metric bolts will also be used more often. Certain vehicles today have transmissions and other specific components made in metric sizes. Because metric threads and fasteners are being used more often, it is important that they not be mixed with bolts that have standard threads.

For metric bolts, the shank and length are measured in millimeters. In addition, to determine the type of thread used, the distance between threads is measured in millimeters. Thus, the bolts are considerably different as compared to inch or USC sizes.

The hardness or strength of metric bolts is indicated by using a *property class* number stamped on the head of the bolt as shown in **Figure 3–9**.

SAE GRADE MARKINGS					
DEFINITION	No lines — unmarked indeterminate quality SAE Grades 0-1-2	3 Lines — common commercial quality Automotive & AN Bolts SAE Grade 5	4 Lines — medium commercial quality Automotive & AN Bolts SAE Grade 6	5 Lines — rarely used SAE Grade 7	6 Lines — best commercial quality N.A.S. & Aircraft Screws SAE Grade 8
MATERIAL	Low Carbon Steel	Med. Carbon Steel Tempered	Med. Carbon Steel Quenched & Tempered	Med. Carbon Alloy Steel	Med. Carbon Alloy Steel Quenched & Tempered
TENSILE STRENGTH	65,000 p.s.i.	120,000 p.s.i.	140,000 p.s.i.	140,000 p.s.i.	150,000 p.s.i.

FIGURE 3–8. Bolts are identified by hardness. The more lines shown on the head of the bolt, the stronger the bolt.

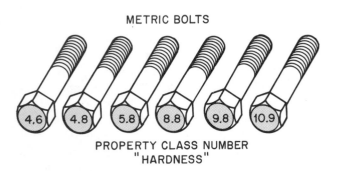

METRIC BOLTS

PROPERTY CLASS NUMBER "HARDNESS"

FIGURE 3–9. Metric bolts are rated in hardness by using the property class number.

Determining Torque on Bolts

All bolts and nuts used on the automobile should be torqued or tightened to specifications. Many repair manuals list all common torque specifications. A torque wrench is used to measure the tightening force. These specifications are known as *standard bolt and nut torque specifications* shown in *Figure 3–10*. The correct torque in both inch pounds and/or foot pounds should always be followed when tightening bolts and nuts. If the bolt is tightened to a value less than that given by the specification, the bolt may vibrate loose. The bolt can also be tightened too much. This may cause the threads to strip or the bolt to break. In either case, damage to other automobile parts can result. Always torque all bolts and nuts to the manufacturer's specifications. If there are no specifications listed, use the torque specifications shown in Figure 3–10.

Nuts

Many types and styles of nuts are used on the automobile. *Figure 3–11* illustrates some common types of nuts. The most common type is called the hex style.

Hex nuts are classified as regular or heavy. They are used on all high-quality work. These nuts are easy to tighten with wrenches in close or tight places.

Slotted hexagon nuts, also called castellated nuts, are used where there is danger of the nuts coming off. For example, vibration may cause the nuts to loosen. A cotter pin is used to lock the nut in place.

Jam hexagon nuts are thinner than regular nuts. They are used where height is restricted or as a means of locking the working nut in place.

Square nuts are regular or heavy unfinished nuts. They are used with square-head bolts in rough assembly work such as assembly of body parts.

Lock nuts have a self-contained locking feature to prevent back-off rotation. They are designed with undersized threads, plastic or fiber inserts. This design acts as a gripping force.

Free-running seating lock nuts are applied over hexagon nuts. These nuts have a concave surface which flattens when it contacts the top of the hexagon nut. This causes the threads to deflect. The nut binds on the bolt and prevents it from coming loose.

Spring nuts are made of thin spring metal. They are designed with formed prongs as shown in Figure 3–11. Spring nuts are used in sheet-metal construction where high torque is not required.

STANDARD BOLT AND NUT TORQUE SPECIFICATIONS

SIZE NUT OR BOLT	TORQUE (lb-ft)	SIZE NUT OR BOLT	TORQUE (lb-ft)	SIZE NUT OR BOLT	TORQUE (lb-ft)
1/4 – 20	7–9	7/16 – 20	57–61	3/4 – 10	240–250
1/4 – 28	8–10	1/2 – 13	71–75	3/4 – 16	290–300
5/16 – 18	13–17	1/2 – 20	83–93	7/8 – 9	410–420
5/16 – 24	15–19	9/16 – 12	90–100	7/8 – 14	475–485
3/8 – 16	30–35	9/16 – 18	107–117	1 – 8	580–590
3/8 – 24	35–39	5/8 – 11	137–147	1 – 14	685–695
7/16 – 14	46–50	5/8 – 18	168–178		

FIGURE 3–10. All bolts and nuts have a standard bolt and nut torque specification. All bolts and nuts should be properly torqued unless stated differently in the maintenance manual.

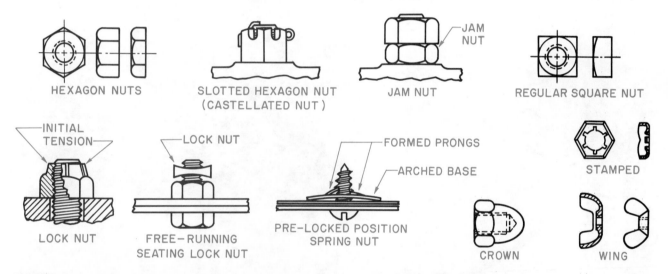

FIGURE 3–11. Many styles of nuts are used on the automobile. Each style has a specific purpose and application.

Crown nuts are used where the ends of the external threaded part must be concealed or hidden, or where the nut must be attractive.

Wing nuts have two arms (or projections) to aid hand tightening and loosening. Typically, these nuts are tightened and loosened frequently. High torque is not a consideration.

Washers

Most washers are placed on bolts to lock the bolt and keep it from coming loose. Several types of washers are used for locking bolts in place as shown in ***Figure 3–12***. Washers

FIGURE 3–12. Washers are typically used to lock the bolts to the structure to keep them from coming loose and to prevent damage to softer metal parts.

are also used for other reasons. For example, copper washers are used to aid in sealing the bolt to a structure. These are called *compression washers*. This may help reduce oil leakage when the threads of the bolt are in or near oil. Certain cars use copper washers on oil pans to eliminate leakage. Flat washers are also used to help spread out the load of tightening the nut. This helps to prevent the nut from digging into the material when it is tightened.

Snap Rings

Snap rings are used to prevent gears and pulleys from sliding off the shaft. There are two types of snap ring: the external and internal snap ring. *Figure 3–13* illustrates snap ring pliers being used to install a snap ring.

Splines

Splines are defined as external or internal teeth cut in a shaft. When a shaft must be inserted into a gear, pulley, or other part, and the part must be able to move on the shaft, a spline will be used. The output shaft on the transmission has an external spline. The end of the drive shaft connected to the transmission has an internal spline. This allows the drive shaft to move along the axis when the rear wheels go over a bump in the road. See *Figure 3–14*.

Keyways

Keyways are used to lock parts together by fitting a key between a slot on the shaft and a pulley. *Figure 3–15* shows a keyway application. Keyways are used to lock the front drive pulley on the engine to the engine crankshaft. The only difference between a keyway and a spline is that on a spline, the hub or pulley can move parallel or axially along the shaft.

FIGURE 3–13. Snap rings can be external or internal types. They are used to keep gears and pulleys on the shaft.

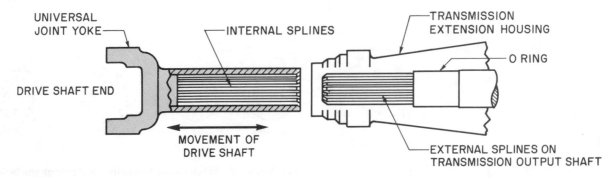

FIGURE 3–14. Splines, both external and internal, are used to lock rotating shafts together while allowing slide (axial) movement between shafts.

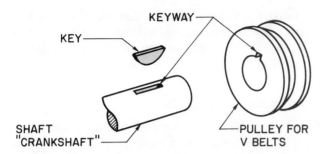

FIGURE 3–15. Keyways are used to hold parts to shafts so that the part and shaft rotate as one unit.

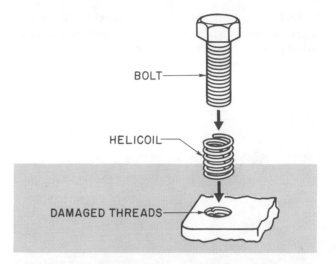

Helicoils

A common problem is threads stripping inside an engine block, cylinder head, or other structure. This problem is usually caused by too high a torque or by threading the bolt into the hole incorrectly. Rather than replacing the block or cylinder head, the threads can be replaced by the use of threaded inserts or *helicoils*. *Figure 3–16* illustrates a helicoil.

The procedure for replacing damaged threads is as follows: (1) the damaged threads are drilled out to a specific size depending upon the helicoil size; (2) the hole is tapped, using the outside diameter of the helicoil; (3) the helicoil is threaded into the new, larger threads. The inside of the helicoil provides the new threads for the bolt. There are many sizes of helicoils. The tap, drill sizes, and necessary tools are all included in the helicoil kit.

FIGURE 3–16. Helicoils are used to replace damaged internal threads in engine blocks, cylinder heads, and other structures.

CAR CLINIC ▬▬▬

PROBLEM: SYNTHETIC OIL

Is it OK to mix synthetic oil with regular oil in a vehicle? A customer has been using synthetic oil for about 15,000 miles in the vehicle. Now it is noticed that the engine is using a small amount of oil. Is it acceptable to mix regular oil with synthetic oil to keep the oil to the correct level?

SOLUTION:

The manufacturers of synthetic oil recommend that synthetic oil NOT be mixed with regular oil. More importantly, if the engine is using oil, it may have to be rebuilt. New rings may have to be put on the pistons to eliminate oil consumption. Also, check for valve guide seals leaking into the intake or exhaust. A compression leakdown test will help determine which cause is producing the oil burning.

3.2 GASKETS, SEALS, AND SEALANTS

Purpose of Gaskets

Gaskets are used on the automobile to prevent leakage of gases, liquids or greases between two parts bolted together.

Examples of gaskets are shown in *Figure 3–17*. Gaskets are placed between two mating machined surfaces. Bolts or fasteners are then tightened according to standard specifications. Gaskets are used between the cylinder head and the block of the engine or the water pump housing and the engine.

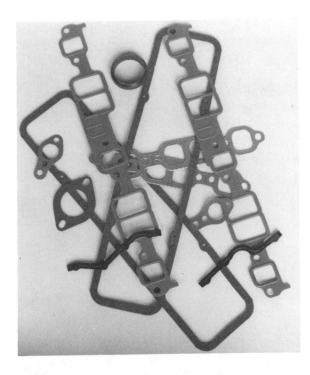

FIGURE 3–17. Many gaskets are used on the automobile engine to seal parts that are bolted together. Gaskets keep oil, grease, and gases from escaping and dirt from entering the engine.

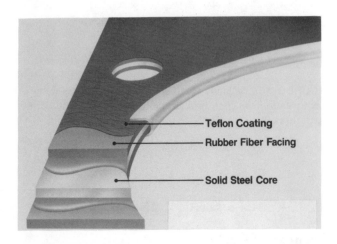

FIGURE 3–18. Compression gaskets are used to seal high pressures produced from the cylinder during the compression and power strokes. (Courtesy of Fel-Pro Incorporated)

Gaskets are designed for a particular job. For this reason, gaskets are made from different materials. Some of the more common materials used are cork, synthetic rubber, steel, copper, and asbestos.

Compression Gaskets

Compression gaskets are designed to be squeezed during tightening, *Figure 3–18*. Tightening a compression gasket seals the parts to contain high pressure gases. Compression gaskets are used most often for head gaskets. As the cylinder head is torqued to specifications, the head gasket is squeezed. This forms a good seal for high compression during combustion. Some spark plugs also use compression gaskets as washers to seal in the high-compression gases from compression and combustion.

Seals

There are many uses for *seals* in the automobile. Seals are devices placed on rotating shafts to prevent oil, gases, and other fluids from escaping. For example, seals are used on the front and back axles, crankshafts, water pump shafts, and many other locations throughout the vehicle.

Seals are designed to withstand high pressures and to seal fluids in the engine. Seals are made of felt, synthetics, rubber, fiber, or leather. The parts of the common seal are shown in *Figure 3–19*.

The outer case is usually pressed into the housing which contains the stationary part. The inner case holds the parts of the seal. Seals are sometimes designed with one or two lips. The lips are pressed against the rotating part. This pressure causes the sealing action. Springs may be added to the lip tension so that higher pressures can be contained.

Seals should always be installed according to the manufacturer's specifications. In general, the sealing lip

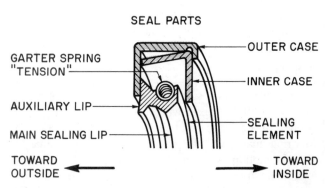

FIGURE 3–19. Seals must be placed in such a direction as to have the main sealing lip pointed toward the liquid, gas, or pressure being contained.

should always be placed toward the fluid or pressure being contained.

Sealants

There are several sealants that also can be used in the automotive industry. *Sealants* are similar to a thick liquid that hardens after being placed on the metal. Some sealants are called form-a-gaskets. Certain sealants are used to seal between metal surfaces. There are many brands of seals available, made from many materials. Silicon is one of the most popular sealant materials. Only use sealants where the manufacturer recommends them. If not recommended, use the correct gasket for the parts to be sealed.

Static and Dynamic Seals

Seals are classified as either static or dynamic. O rings are examples of static seals. An *O ring* is placed between two stationary parts where a fluid passes between the parts. The O rings shown in *Figure 3–20* keep the fluid from leaking out of the two stationary objects.

Dynamic seals provide sealing between a stationary part and a rotating part. A dynamic seal is shown in *Figure 3–21*. In this case, the fluid is contained by the seals. Note that the lip rides directly on the rotating shaft.

Labyrinth Seals

Labyrinth seals are used on high-speed shafts or where the lips of the seal can cause excessive wear on the shaft. A labyrinth seal is shown in *Figure 3–22*. As liquid is moved toward the sealing area, it contacts the first labyrinth. Centrifugal force causes the fluid to spin outward into the upper grove. If the pressure is great enough to push the fluid further, the next labyrinth seal will also produce a sealing effect. Because of centrifugal forces, the fluid has difficulty going toward the center of the shaft. This action keeps the fluid from leaking out of the area. The labyrinth seals will not work if a fluid touches the seal directly when the shaft is not rotating. There is no contact of the seal to the shaft; therefore, oil will leak out. Certain foreign car manufacturers use labyrinth seals on the crankshaft to seal oil in the crankcase area. In this case, the seal will not wear on the metal, eliminating possible damage to this area of the crankshaft.

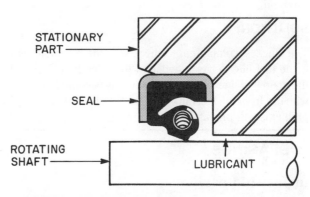

FIGURE 3–21. Dynamic seals are used to seal fluids where one part is stationary and one part is rotating. *(Courtesy of CR Industries)*

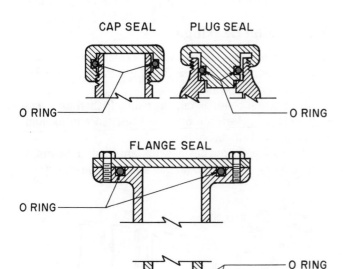

FIGURE 3–20. O rings are considered static seals. Static seals are placed between two stationary parts to eliminate leakage. *(Courtesy of Federal-Mogul Corporation)*

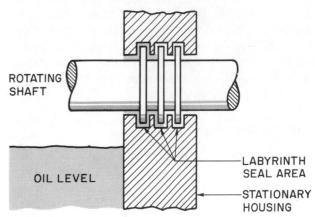

FIGURE 3–22. Labyrinth seals use centrifugal force to spin the fluid outward. Because of this action, the fluid cannot pass through to the outside.

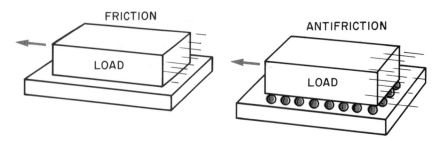

FIGURE 3–23. There are two types of bearings. Friction bearings have oil between the moving and stationary parts. Antifriction bearings have balls or rollers between the moving and stationary parts.

CAR CLINIC

PROBLEM: BAD FUEL FILTER

An engine is becoming consistently more erratic and the engine power is reducing. At times the engine shuts down. It seems that the engine isn't getting enough fuel. What would be a quick component to check with these symptoms?

SOLUTION:

The quickest component to check with these symptoms is the fuel filter. Over a period of time, the fuel filter in the main fuel line to the carburetor may become plugged with dirt from the fuel tank. Replace the fuel filter and the problems will probably be eliminated.

3.3 BEARINGS

Purpose of Bearings

Bearings are used in the automobile to reduce friction between moving parts and stationary parts. A secondary purpose is to remove the heat produced by unavoidable friction.

There are two types of bearings: *friction bearings* and *antifriction bearings*. Friction bearings have no rotating or moving parts, whereas antifriction bearings use small rollers or steel balls to reduce friction. This is shown in *Figure 3–23*. With friction bearings, rotating or moving parts slide on the stationary part. Antifriction bearings contain balls or rollers to support the rotating part.

Friction Bearings

Friction bearings will produce more heat and frictional losses in the engine than other types of bearings. However, this type of bearing usually requires less maintenance and can be replaced easily. When friction bearings are used, the load of the moving part is supported by a layer of oil between the load and the stationary part. The oil molecules act like small ball bearings.

There are two types of friction bearings, *Figure 3–24*.

One type is a two-piece bearing which is used on the crankshaft of most gasoline engines. The second type is a one-piece bearing. It is commonly known as a *bushing*.

Bushings are held in place by pressing them into the block or stationary part. In an automobile, bushings are used on the camshaft, some generator and alternator shafts, the starter armature shaft, and the distributor shaft.

Friction Bearing Design

Friction bearings have a strong back or shell, generally made of steel, *Figure 3–25*. A thin intermediate layer of copper-lead is added to the shell. On top of this is an antifriction material called Babbitt. Babbitt is composed mostly of tin or lead, occasionally silver, or it may be an aluminum alloy. This material may be only 1/1000 of an inch thick.

Antifriction Bearings

Antifriction bearings are used where a greater load is placed on the rotating shaft. *Figure 3–26* illustrates the parts of the common ball bearing.

Antifriction bearings rely on rolling friction for operation. This type of friction offers less resistance to rotation, resulting in less frictional loss.

There are many designs for antifriction bearings. *Figure 3–27* illustrates both ball and roller bearing types.

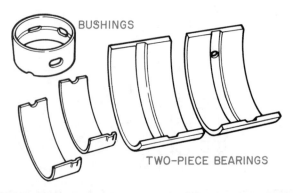

FIGURE 3–24. Bushings are designed as one piece, while bearings are designed with two pieces. *(Courtesy of Federal-Mogul Corporation)*

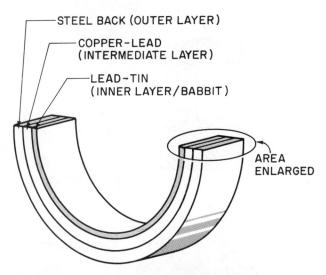

STEEL BACK (OUTER LAYER)
COPPER-LEAD (INTERMEDIATE LAYER)
LEAD-TIN (INNER LAYER/BABBIT)
AREA ENLARGED

FIGURE 3–25. Friction bearings are made of many materials. The more common materials are shown.

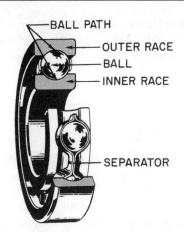

BALL PATH
OUTER RACE
BALL
INNER RACE
SEPARATOR

FIGURE 3–26. Parts of a standard ball bearing *(Courtesy of CR Industries)*

1. BALL BEARINGS:
ECONOMICAL, WIDELY USED

SINGLE ROW RADIAL
FOR RADIAL LOADS.

SINGLE ROW ANGULAR CONTACT
FOR RADIAL AND AXIAL LOADS.

AXIAL THRUST
FOR AXIAL LOADS.

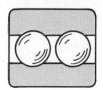

DOUBLE ROW
FOR HEAVIER RADIAL LOADS.

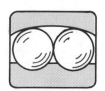

SELF-ALIGNING
FOR RADIAL AND AXIAL LOADS,
LARGE AMOUNTS OF ANGULAR
MISALIGNMENT.

2. ROLLER BEARINGS:
FOR SHOCK, HEAVY LOAD

CYLINDRICAL
FOR RELATIVELY HIGH SPEEDS.

NEEDLE
FOR LOW SPEEDS,
INTERMITTENT LOADS.

TAPERED
FOR HEAVY AXIAL LOADING.

SPHERICAL
FOR THRUST LOADS AND LARGE
AMOUNTS OF ANGULAR
MISALIGNMENT.

SPHERICAL THRUST
TO MAINTAIN ALIGNMENT UNDER
HIGH THRUST LOADS AND
HIGH SPEEDS.

FIGURE 3–27. Both roller and ball bearings are considered antifriction. Many styles and types are used in the automobile.

In reference to Figure 3–27, a radial load is a load perpendicular to the axis of the shaft. An *axial load* is a load parallel to the axis of the shaft. An axial load may also be called a *thrust load*. *Thrust bearings* are designed to reduce friction on the rotating shaft when thrust forces are produced parallel to the axis of the shaft. See ***Figure 3–28***.

Sealed Bearings

At times bearings are used in applications where the bearings must seal as well as reduce friction. Sealed bearings not only reduce friction but also keep dust, dirt, or other debris away from the internal parts of the bearing. ***Figure 3–29*** shows a type of sealed bearing in which the seal is housed within the bearing. A plate on the outside of the bearing keeps out dirt. The seal has four spots that seal. The seal also uses a labyrinth to keep oil in the bearing and dirt out of the bearing.

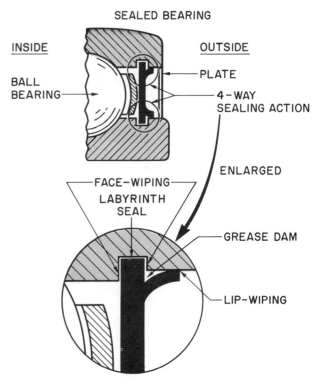

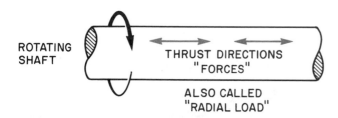

FIGURE 3–28. Thrust is produced when there is a pressure or load that is parallel to the axis of the rotating shaft.

FIGURE 3–29. Seals are also used in bearings to keep dirt out and lubrication in. In this case, a labyrinth seal design is used.

DIAGNOSIS and SERVICE

SAFETY TIPS
1. *When tightening bolts and nuts, make sure the wrench is always placed solidly on them. If the wrench slips off the bolt when tightening, the knuckles and fingers may be injured.*
2. *Always wear safety glasses when tightening or loosening bolts and nuts, especially when using air tools.*
3. *When using pullers, remember that you are applying very high forces and pressures on the bolts, nuts, and equipment. Always be sure the pullers are securely fastened before beginning.*

4. *Never use an air gun with high pressure to dry off or spin roller or ball bearings. The rollers or balls may fly out under the high speed and cause injury.*
5. *Make sure any grease that is dropped on the floor is immediately cleaned up so that you won't slip and fall, possibly injuring yourself.*

1. Always use the Standard Bolt and Nut Torque Specifications when tightening all bolts and nuts.
2. When replacing bolts that are stripped or broken, always use a bolt having the same hardness.
3. Never mix metric and USC threads.

4. Never use a USC wrench with a metric bolt or nut.

5. Always determine the torque of the bolt by the number of threads per inch and the shank size of the bolt. Never use the head size of the bolt to determine the torque specification.

6. Never use an NC bolt in place of an NF bolt of the same shank size.

7. Always use a lock washer on bolts that may vibrate loose.

8. When removing parts that use a keyway, always store the key in a safe place.

9. Use a helicoil to replace damaged threads.

10. Always scrape the machined surfaces clean of all old gasket material.

11. Never use a compression gasket a second time.

12. Always place a seal on a rotating shaft so the lip of the seal is toward the fluid being contained.

13. When installing bearings or bushings, always keep parts well lubricated during assembly.

14. When leaks occur near covers such as valve and timing chain covers, always check the covers for bent sides. If the sides are bent slightly, an oil leak may occur that may not be sealed by the gasket. Bend or straighten the cover to eliminate the leak.

15. Oil leaks can also occur from seals that are working on rotating shafts. For example, seals are used on rotating shafts in transmissions, rear drive shafts, and drive shafts. When a leak is observed, make sure the shaft is not damaged where the seal rides on the metal. At times, this area may wear so that even a new seal may not stop an oil leak. Replace the shaft if necessary.

16. Sometimes when gaskets are being installed, they slip out of place. To help keep the gasket in place during installation, use a thick grease or tie the gasket in place with very thin wire that can be easily removed. Another technique for keeping a gasket in place is shown in *Figure 3–30*. Use small wires or cut down paper clips to hold the bolts in place. The gasket is held in place with the bolts. This is an example of an oil pan gasket.

17. Roller and ball bearings should be repacked with grease at regular intervals. *Repacking of bearings* usually includes the following procedure:
    a. Clean the bearing thoroughly with solvent.
    b. Dry the bearing completely.
    c. Force grease into the bearing by placing a sufficient amount of grease in one hand. With the bearing in the other hand, force the grease into the bearing. Make sure the grease has com-

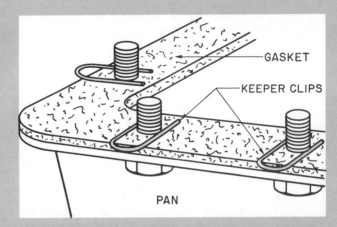

FIGURE 3–30. Gaskets can be held in place by using small paper clips cut to hold the bolts of the cover in place. The oil pan, gasket, and bolts are shown here.

pletely surrounded the balls or rollers and comes out the other side.
    d. Repeat the previous step on the opposite side of the bearing.

18. When installing seals, make sure that the seal is not cocked or installed at an angle. A cocked seal may leak oil readily. When a seal has been cocked or replaced incorrectly, it must be removed and replaced with a new seal.

19. Seals generally harden with age. Under normal conditions, however, seals should not harden and should remain flexible. If a seal hardens, it may be due to excessive temperature near the seal. Always investigate if the excessive temperature caused the seal to harden.

20. Brinelling, small dents in the bearing race, can be caused by improper bearing installation, such as impacts on the outer race. Brinelling will normally cause more severe bearing damage.

21. Contamination of bearings usually shows up as small scratches, pitting, and scoring along the raceway. To prevent contamination, always keep all parts of the bearing free of dirt, dust, and other particles.

22. Misalignment is another cause of bearing failure. When the balls or rollers of a bearing are running from one side of the race to the other, one race may be misaligned. This causes uneven load distribution which causes excess friction and heat to build up. In this case, the bearing was probably installed incorrectly. To prevent misalignment, always follow the manufacturer's recommended procedure for installation.

SUMMARY

This chapter shows the many types of fasteners, gaskets, seals, and bearings used on the automobile. Threaded fasteners such as bolts, nuts, and a variety of screws are used in many locations on the automobile. Bolts are identified by the shank size, hardness, number of threads per inch, and whether the threads are coarse or fine. This information is used as the basis for selecting the proper torque values as listed on the *Standard Bolt and Nut Torque Specification* chart.

In addition to these fasteners, many types of locking washers, snap rings, splines, and keyways are also used on the automobile. Snap rings, splines, and keyways are used to hold parts to rotating shafts.

When threads are damaged by incorrect tightening or incorrect threading of the bolt into the hole, helicoils can be used. Helicoils provide new threads for the bolts.

Gaskets and seals are used on the engine, differentials, and axles. The purpose of gaskets and seals is to seal gases, oil, greases, and other fluids in the vehicle. This is especially true when oil is near or in contact with a rotating shaft. Compression gaskets are used when high-pressure gases such as those produced by combustion must be contained. Both static and dynamic seals are used. Labyrinth seals are used when the lips of the seal lips could cause wear on the shaft.

Whenever a rotating shaft is to be supported in a stationary part, a bearing is used. There are many types of bearings. Friction and antifriction bearings are the most common types. Friction bearings have oil between the rotating shaft and the stationary part.

Two types of friction bearings are commonly used. One-piece bearings, also called bushings, are used for camshafts, alternator and generator shafts, and similar applications. Split or two-piece bearings are used primarily on crankshafts.

Antifriction bearings use rollers or balls between the stationary part and the rotating shaft. There are many types of antifriction bearings. Many are designed to absorb thrust or axial loads.

Some bearings are also sealed. Sealed bearings are used when a seal and a bearing are needed together. Sealed bearings are used on rear axles, on wheel bearings, and in other applications where oil or grease are in direct contact with the bearing.

There are several service tips and diagnostic procedures used with fasteners, gaskets, and bearings. Important areas include using correct bolts and nuts, making sure bearings are maintained and serviced properly, and installing gaskets, bearings, and fasteners correctly.

TERMS TO KNOW

Can you explain each of the following terms? Review the chapter until you can use each term correctly.

Fasteners	Snap ring
Threaded fasteners	Spline
Torque	Keyway
Shank	Helicoil
Threads per inch	Gasket
National Fine (NF)	Seal
National Coarse (NC)	O ring
Screw-pitch gauge	Labyrinth seal
Bolt hardness	Bearing
Tensile strength	Friction bearing
Metric threads	Bushing
Property class	Antifriction bearing
Standard bolt and nut torque specification	Axial load
	Thrust
Compression washer	Thrust bearing

REVIEW QUESTIONS

Multiple Choice

1. Which of the following are used to keep a pulley on a shaft?
   a. Washers
   b. Snap rings
   c. Gaskets
   d. Sealants
   e. Bearings

2. Which of the following is used to seal liquid flowing through two stationary parts?
   a. Labyrinth seal
   b. Dynamic seal
   c. O rings
   d. All of the above
   e. None of the above

3. In order to replace threads, a _____ is commonly used.
   a. Helicoil
   b. Stud
   c. Snap ring
   d. Sealant
   e. Bearing

4. Another word for axial load is _____.
   a. Force
   b. Thrust
   c. Bushing
   d. End forces
   e. Expansion force

5. A gasket used to seal high pressures and gases is called a:
   a. Compression gasket
   b. Thrust gasket
   c. Labyrinth gasket
   d. Sealant
   e. Bearing

6. Torque for bolts should always be determined by the:
   a. Socket size of the bolt
   b. Length of the bolt
   c. Shank size of the bolt
   d. All of the above
   e. None of the above

7. The greater the number of lines on the bolt head, the:
   a. Softer the bolt
   b. More threads per inch on the bolt
   c. Harder the bolt
   d. Easier it is to turn
   e. Harder it is to turn

8. Which bolt has the greatest ability to hold two objects together?
   a. An NC bolt
   b. An NF bolt
   c. A UNC bolt
   d. All of the above
   e. None of the above

9. The lip of a dynamic seal should always be pointed toward the:
   a. Outside or dirty area
   b. Fluid to be contained
   c. Stationary part
   d. Air
   e. None of the above

10. Which of the following allows two shafts to be locked together yet able to slide axially?
    a. Keyways
    b. Splines
    c. Helicoils
    d. Bearings
    e. None of the above

11. A compression gasket should be used:
    a. Only once
    b. After the gasket has been compressed
    c. To seal oil and low pressures
    d. To seal air only
    e. None of the above

12. Which of the following are used to prevent motion between two rotating parts?
    a. Setscrews
    b. Snap rings
    c. Machine screws
    d. Bearings
    e. Sealants

13. Which of the following are used to join light metal or plastic?
    a. Self-tapping screws
    b. Splines
    c. Setscrews
    d. Bearing seals
    e. Compression gaskets

14. Helicoils are used to:
    a. Replace external threads
    b. Replace internal threads
    c. Lock two shafts together
    d. Seal surfaces together
    e. Tighten bolts correctly

15. A bolt identified as $3/8 \times 16 \times 1\ 1/2$ means:
    a. 3/8 inch shank size
    b. 16 millimeters long
    c. 1 1/2 inch socket size
    d. All of the above
    e. None of the above

16. Which type of bearing uses a set of rollers or balls between the moving and stationary parts?
    a. Antifriction
    b. Friction
    c. Bushing
    d. Compression bearings
    e. None of the above

17. Which type of bearing produces more heat and frictional losses in the engine?
    a. Antifriction
    b. Friction
    c. Radial
    d. Roller
    e. Ball bearing

18. Which type of bearing is used on camshafts, crankshafts, and alternator shafts?
    a. Antifriction bearings
    b. Bushings
    c. Thrust roller bearings
    d. Sealed ball bearings
    e. None of the above

The following questions are similar in format to ASE (Automotive Service Excellence) test questions.

19. Technician A says that O rings are used to form a compression gasket for high pressures. Technician B says that O rings are used to improve the accuracy of torquing a bolt. Who is right?
    a. A only
    b. B only
    c. Both A and B
    d. Neither A nor B

20. A bolt is identified as 3/8 × 16 × 1 1/2. Technician A says the bolt has 16 threads per inch. Technician B says the bolt is 3/8 inch in shank size. Who is right?
    a. A only
    b. B only
    c. Both A and B
    d. Neither A nor B

21. Technician A says the tensile strength of a bolt is identified by the type of chrome on the bolt. Technician B says the tensile strength of a bolt is identified by the number of lines on the top of the bolt. Who is right?
    a. A only
    b. B only
    c. Both A and B
    d. Neither A nor B

Essay

22. Describe how the torque is determined on a bolt.

23. Describe the difference between a spline and a snap ring.

24. What is the purpose of using a helicoil?

25. What is the difference between friction and antifriction bearings?

26. In which direction should the lip of a seal be positioned? Why?

CHAPTER 4

Measuring Instruments and Common Hand Tools

INTRODUCTION

In the automotive shop, it is necessary to use many tools. In some shops, each mechanic has his/her own set of tools. In some cases, the tool set may include several hundred tools used to work on the automobile.

Tools are used to make service easier for the mechanic. They are used for pulling parts off of shafts, tightening bolts and nuts to correct specifications, measuring various distances and clearances, and lifting heavy objects. These tools maximize the mechanic's effort by multiplying forces. Today's automotive mechanic should be familiar with the use of many tools. Proper tool selection will improve both the quality and quantity of the service done on the automobile.

The purpose of this chapter is to introduce the common tools used on the automobile. Because of the number of tools now available, only identification of these tools will be presented.

OBJECTIVES

After reading this chapter, you will be able to:
- Compare the USC and metric systems of measurement.
- Identify common hand tools used in the shop, including hammers, pliers, screwdrivers, wrenches, taps and dies, chisels and punches, screw extractors, and sockets.
- Analyze the common measuring tools, including micrometers, vernier calipers, feeler gauges, torque wrenches, dial indicators, and pressure and vacuum gauges.
- Analyze power and pressing tools such as pulleys, bushing/bearing/seal installers, hydraulic tools, and impact wrenches.

CHAPTER HIGHLIGHTS

4.1 MEASURING SYSTEMS
- A. USC Measurements
- B. Metric System
- C. Measuring Length in Meters
- D. Metric Prefixes
- E. Measuring Volume
- F. Measuring Mass
- G. Other Metric Units
- H. Conversion Between USC and Metric Systems

4.1 MEASURING SYSTEMS

The United States is now committed to using two measuring systems. These are called the United States Customary system, known as *USC*, and the *metric system*. In the past, all parts, tools, and measuring instruments made in the United States were designed using the USC measurements. However, in the past several years, the United States has been increasingly using the metric system. The manufacturers are slowly converting their measurements to the metric system. This is because many foreign countries use the metric system. Many American automobile manufacturers now design certain components in metric sizes. Because of these changes, it is important to study both measurement systems.

USC Measurements

The measuring system used most often in the United States is called the U.S. Customary system (USC). This system measures length, volume, and mass as shown in *Figure 4–1*.

When working with tools in the automobile industry, many length measurements are used. For example, wrench and socket sizes typically increase in 1/64, 1/32, 1/16, and 1/8 inch sizes, *Figure 4–2*. In addition, many small measurements and clearances are listed in 1/1,000th of an inch. This can be shown in decimal form as 0.001 of an inch.

Metric System

When you are familiar with it, the metric system is much easier to use than the USC measurements. This is because all of the units such as length, volume, and mass use a *base of 10*. This means that all units are multiples of 10. This is not true with USC measurements. The foot has a base of 12 (12 inches in a foot), the yard has a base of 3 (3 feet in a yard), the pound has a base of 16, (16 ounces in a pound), and so on.

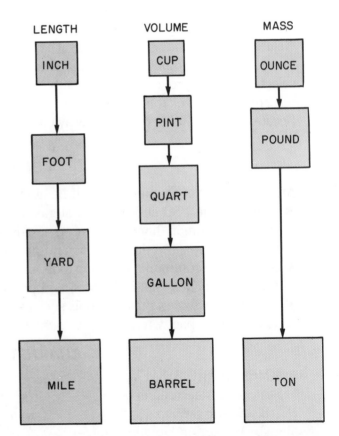

FIGURE 4–1. The USC measurements are shown. Length, volume, and mass all have specific units for measurement.

COMMON WRENCH SIZES
1/4"
5/16"
3/8"
7/16"
1/2"
9/16"
5/8"
11/16"
3/4"
13/16"
7/8"
15/16"
1"

FIGURE 4–2. Many length measurements are taken on the automobile. The USC system uses these common sizes.

Measuring Length in Meters

The base unit in the metric system is the *meter*. The distance of a meter is defined as the length equal to 1,650,763.73 wavelengths of krypton in a vacuum. This distance is always constant and can easily be duplicated under laboratory conditions. In relation to the USC measurement, 1 meter is equal to 39.37 inches.

Metric Prefixes

Instead of using such phrases as 1/16 of an inch, the metric system uses a set of *prefixes*. These prefixes are shown in *Figure 4–3*. For example, the prefix *kilo* means 1,000. If a distance is measured as 1,000 meters, it can also be called a kilometer. In terms of small distances, the prefix *milli* means one thousandth of a unit. If the unit is the meter, one millimeter is equal to 1/1000th of a meter. When looking at a meter stick as shown in *Figure 4–4*, each small increment is a measure of one millimeter. One thousand millimeters make up one complete meter. A decimeter is 1/10 of a meter, while a centimeter is 1/100 of a meter.

Measuring Volume

In the metric system, volume is measured by the *liter*. One tenth of a meter is called a decimeter. This is about the width of a person's fist. If a cube is made with each edge one decimeter long, the cube will have a volume of one cubic decimeter. A cubic decimeter is also equal to a liter. See *Figure 4–5*. Note also that the length of one decimeter is equal to 10 centimeters. If one centimeter is cubed, the unit is called a *cubic centimeter*. Both liters and cubic centimeters are used to measure the volume of an engine.

Measuring Mass

The unit used to measure mass in the metric system is called the *gram*. Kilogram is also used because the gram is so small. If one cubic centimeter is filled with water, the mass

SOME COMMON PREFIXES			
NAME	SYMBOL	MEANING	MULTIPLIER
mega	M	one million	1 000 000
kilo	k	one thousand	1 000
hecto	h	one hundred	100
deca	da	ten	10
deci	d	one tenth of a	0.1
centi	c	one hundredth of a	0.01
milli	m	one thousandth of a	0.001
micro	μ	one millionth of a	0.000 001

FIGURE 4–3. The metric system uses these prefixes to aid in defining measurements.

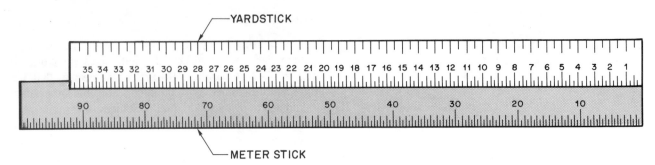

FIGURE 4–4. A meter stick is composed of 1,000 small increments. Each increment is called one millimeter.

ONE CUBIC DECIMETER =

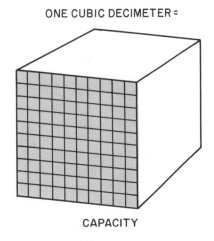

CAPACITY

FIGURE 4–5. A cubic decimeter, also called a liter, is a volume measurement in the metric system. Engines are sized by indicating the number of ccs (cubic centimeters) or liters.

is one gram. If a liter, 1,000 cubic centimeters, is filled with water, the mass is 1,000 grams or 1 kilogram.

Other Metric Units

Other units are used in the metric system. *Figure 4–6* illustrates a complete listing of common metric units. The quantity, the unit, and the common symbols are shown. Becoming familiar with these units will aid the automotive mechanic in his/her work.

Conversion Between USC and Metric Systems

At times it may be necessary to convert from USC units to metric system units or from metric system units to USC units. If this is necessary, the chart shown in *Figure 4–7* gives some of the more common *conversion factors*. Also, there are many slide rules and calculators that are available for conversion.

4.2 COMMON HAND TOOLS

There are many types of tools used by the automotive mechanic today. Hand tools are designed to make the mechanic's work much easier. Hand tools receive energy from the mechanic's hand and transform this energy into productive work. Hand tools multiply forces to accomplish work. A person cannot loosen a bolt by hand. However, the bolt can be loosened easily by using the correct wrench. Tools accomplish this work in two ways: 1) by multiplying forces, as in prying a heavy object with a bar, and 2) by concentrating applied force into a small area, as in using a wrench.

QUANTITY	UNIT	SYMBOL
Length	millimeter (one thousandth of a meter) meter kilometer (one thousand meters)	mm m km
Area	square meter hectare (ten thousand square meters)	m² ha
Volume	cubic centimeter cubic meter milliliter (one thousandth of a liter) liter (one thousandth of a cubic meter)	cm³ m³ ml l
Mass	gram (one thousandth of a kilogram) kilogram ton (one thousand kilograms)	g kg t
Time	second minute, hour, day, etc.	s . . .
Speed	meter per second kilometer per hour	m/s km/h
Power	watt kilowatt (one thousand watts)	W kW
Energy	joule kilowatt hour	J kW•h
Electric potential difference	volt	V
Electric current	ampere	A
Electric resistance	ohm	Ω
Frequency	hertz	Hz
Temperature	degree Celsius	°C

FIGURE 4–6. Many units are used in the metric system. Note that each has a symbol to represent the unit.

Wrenches

There are many types of wrenches used today. Wrenches are used to turn threaded fasteners with primarily hexagonal heads. Three types of wrenches are commonly used. They are the open end, box, and combination wrenches. *Figure 4–8* illustrates the differences between these ends of the wrench. The box end wrench is closed, the open end wrench is open, and the combination wrench has both box and open end styles. *Figure 4–9* illustrates many of the different sizes of combination, open end, and box wrenches available today. Note that wrench sets such as these can be purchased in both English (USC) and metric sizes.

There are other styles of wrenches used on the automobile as well. The adjustable wrench is used when certain size wrenches are not available. This wrench is able to fit bolts and nuts of different sizes by adjusting the size of the jaws. Adjustable wrenches come in different lengths and sizes as well. See *Figure 4–10*.

Hammers

The automotive shop is not complete unless there are hammers available for various service jobs. There are several types of hammers used in the shop. Using a hammer correctly, matched to the job in style, size, and weight, helps get the work done faster and safer. The common types of hammers include the ball peen, plastic tip, rubber tip, and bronze tip hammers. See *Figure 4–11*.

UNIT	CONVERSION FACTOR
Length	1 inch = **25.4** mm 1 foot = **30.48** cm 1 yard = **0.9144** m 1 mile = **1.609 344** km
Area	1 square inch = **6.4516** cm² 1 square foot = **9.290 304** dm² 1 square yard = **0.836 127 4** m² 1 acre = **0.404 685 6** ha 1 square mile = **2.589 988** km²
Volume	1 cubic inch = **16.387 064** cm³ 1 cubic foot = **28.316 85** dm³ (or liters) 1 cubic yard = **0.764 555** m³ 1 fluid ounce = **28.413 062** cm³ 1 gallon = **4.546 090** dm³ (or liters)
Mass	1 ounce (avoirdupois) = **28.349 523** g 1 pound (avoirdupois) = **0.453 592 37** kg 1 ton (short, 2,000 lb) = **907.184 74** kg
Temperature	**(5/9)** × (number of degrees Fahrenheit – **32**) = number of degrees Celsius
Speed	1 mile per hour = **0.447 04** m/s = **1.609 344** km/h
Force	1 pound-force = **4.448 222** N 1 kilogram-force = **9.806 65** N
Pressure	1 pound-force per square inch (psi) = **6.894 757** kPa 1 inch of mercury (0°C) = **3.386 39** kPa 1 mm of mercury (0°C) = **133.322** Pa 1 standard atmosphere (atm) = **101.325** kPa
Energy, Work	1 British thermal unit (Btu) = **1055.06** J 1 foot pound force = **1.355 818** J 1 calorie (international) = **4.1868** J 1 kilowatt hour (kW·h) = **3.6** MJ
Power	1 horsepower (550 ft.lbf/s) = **745.6999** W 1 horsepower (electric) = **746** W

FIGURE 4–7. Some of the more common conversion factors are listed. Calculators can be used if conversion is necessary.

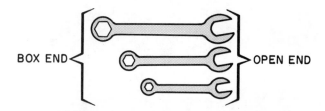

FIGURE 4–8. The three most common styles of wrenches are the box, open end, and combination wrenches.

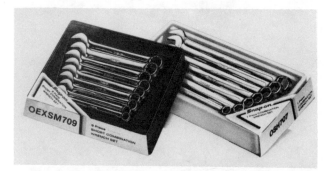

FIGURE 4–9. Wrenches come in many sizes. Tool manufacturers will sell different sizes to make up different sets of tools. (Courtesy of Snap-On Tools Corporation)

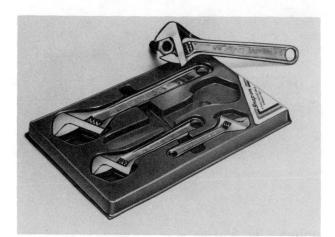

FIGURE 4–10. Adjustable wrenches are used for a variety of service jobs. Always make sure the wrench is adjusted as tight as possible or damage to the bolt head may occur. (Courtesy of Snap-On Tools Corporation)

Socket Sets and Drives

Socket sets, drives, and extensions are used to tighten and loosen bolts and nuts. They are usually faster than using open end or box wrenches. Socket sets can also loosen bolts and nuts that standard wrenches cannot get at. Socket sets include various sizes of sockets, drives, extensions, and ratchets. They are identified by the drive size. The *drive* on a socket set is the square area that connects the ratchet to the socket. The most common drive sizes are 1/4, 3/8, and 1/2 inch.

Socket Points

The sockets can be purchased with different *socket points*. Either 6-point, 8-point, or 12-point styles are available. See *Figure 4–12*. If a bolt is positioned so that only a small amount of rotation is available with the ratchet, then a 12-point socket should be used. Twelve-point sockets can be repositioned every 30 degrees. Eight-point sockets can be repositioned every 45 degrees and 6-point sockets can be repositioned every 60 degrees. In addition to these uses, 4-point and 8-point sockets are typically used for square nuts. Six-point and twelve-point sockets are used for hex nuts. *Figure 4–13* illustrates a standard socket set showing the sockets, drives and extensions, ratchets, and other accessories.

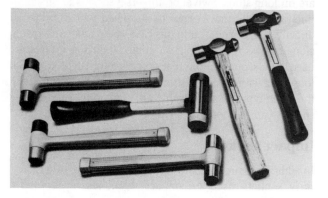

FIGURE 4–11. Many types of hammers are used in the automotive shop. The softer hammers with rubber tips should be used on material that may be damaged easily. (Courtesy of Snap-On Tools Corporation)

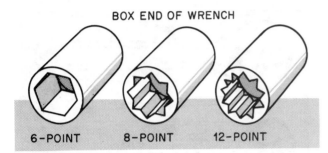

FIGURE 4–12. Sockets can be identified by stating the number of points in the socket. Twelve-point sockets are used for applications where the degree of rotation is restricted.

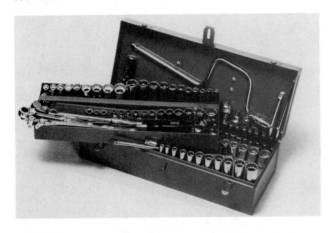

FIGURE 4–13. A standard socket set includes various sockets, ratchets, drives, and extensions, and several accessories. (Courtesy of Snap-On Tools Corporation)

FIGURE 4–14. Three types of sockets are commonly used in the automotive shop: deep length, swivel or universal, and standard sockets. (Courtesy of Snap-On Tools Corporation)

Socket Types

Sockets are also designed as *deep length*, swivel or universal, and standard. Deep length sockets are used for nuts that are on long studs. Swivel or universal sockets are used for nuts and bolts that cannot be loosened with the ratchet directly above the bolts. In this case, the ratchet and extensions can be rotated from an angle. Standard sockets are used in most applications that simply require tightening and loosening without any restrictions. *Figure 4–14* illustrates the three types of sockets that are commonly used. These sockets can be identified by their chrome covering.

Impact Sockets

In addition to the standard socket sets, there are also heavy-duty sockets used on impact wrenches or wrenches powered by electricity or air. These sockets can be identified by their black color. These sockets should be used where heavy-duty work is required, such as removing nuts from a wheel and rim. *Figure 4–15* shows a set of impact sockets with 6-point style.

Screwdrivers

Screwdrivers come in a variety of sizes and shapes. Screwdrivers are identified by the type of tip they have. The two most common types of screwdriver tips are the slotted tip and the Phillips tip. These are shown in *Figure 4–16*. Always make sure the tip of the screwdriver fits the screw head correctly.

Other tips are also designed on screwdrivers. There are several special screwdrivers that are also used on the automobile. These include the Torx (star) tip, magnetic tip, offset tip, Reed and Prince tip (similar to the Phillips tip), and flexible shaft screwdriver.

Pliers

Pliers are used to grip objects of various sizes and shapes, and for cutting. Because of the many uses of pliers, there are many styles. *Figure 4–17* illustrates the pliers commonly used in the automotive shop. Note that pliers should not be used in place of a wrench. Damage to the pliers and the head of the bolt or nut may occur.

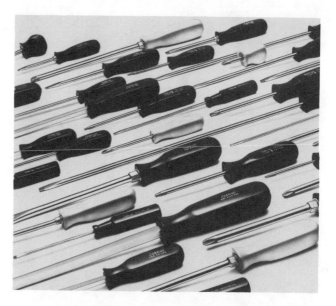

FIGURE 4–15. Impact sockets are used where high torque is required. These sockets are made with the 6-point style. They are stronger and capable of more torque. Standard sockets may easily break under these conditions. *(Courtesy of Snap-On Tools Corporation)*

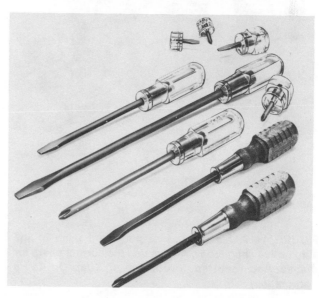

FIGURE 4–16. Of the many types of screwdriver tips, the slotted and the Phillips tip are the most common. *(Courtesy of Snap-On Tools Corporation)*

The different types of pliers give the mechanic a variety of options. Some pliers are used for gripping, some for electrical work, some for cutting, and some for special purposes.

The following types of pliers are commonly used in the automotive shop: *Slip-joint* pliers are used for common gripping applications. Slip-joint pliers have a joint to allow for two different sizes. *Needle nose* pliers have long, slim jaws used for holding small objects such as pins and electrical components. *Adjustable joint* pliers have an adjustable jaw that interlocks in several positions. These pliers have the ability to grip large or small objects depending upon where the jaw is locked. *Cutter* pliers enable the mechanic to cut wire and other small objects. *Electrical service* pliers have a wire cutter, stripper, and clamping device built into one plier. *Retaining ring* pliers are designed to install and remove either internal or external retaining rings. *Vise-grip* pliers are used to grip and hold an object in place after removing one's hand. The vice-grip pliers use a locking mechanism to hold the object independently. *Wire stripping* pliers are used to strip the insulation from electrical wires so that proper electrical connections can be made.

Taps and Dies

Threads can either be internal or external. Internal threads are cut with a tap. External threads are formed or cut with a die. *Figure 4–18* shows a complete tap and die set for use in the automotive shop.

A *tap* is a hardened piece of steel with threads on the outside. There are three types of taps. The *taper tap* has a long taper. It is used to cut threads completely through open holes. It is also used to start threads in blind and partly closed holes.

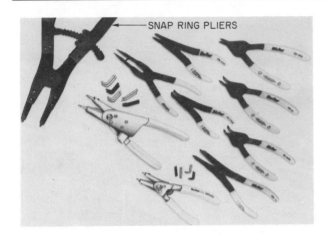

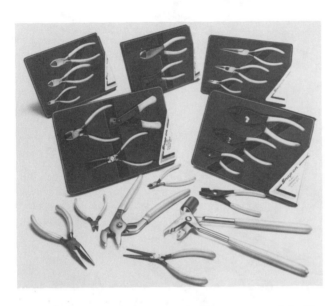

FIGURE 4-17. Many types of pliers are used in the automotive shop today. Always use the correct pliers for the service required. *(Courtesy of Snap-On Tools Corporation)*

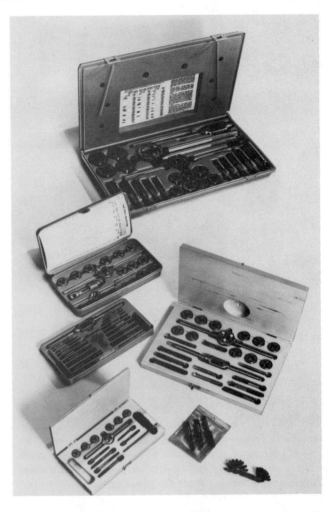

FIGURE 4-18. Taps are used to cut internal threads such as in a block. Dies are used to form external threads such as on a bolt. *(Courtesy of Snap-On Tools Corporation)*

The *plug tap* has a shorter taper and is used after the taper tap to provide fuller threads in blind holes.

The *bottoming tap* is used after the taper and plug taps. It is used to cut a full thread to the very bottom of the hole.

A *threading die* is a round, hardened steel block with a hole containing threads. The threads are slightly tapered on one side to make it easier to start the cutting.

Chisels and Punches

Chisels and punches are used in the automotive shop for a variety of jobs, from punching pins through an object to removing a bearing. A good set of punches and chisels should always be available for the mechanic to use. *Figure 4-19* shows a complete set of chisels and punches.

Screw Extractor

There are many occasions where bolts are broken off. A broken bolt that is not made of hardened steel may be removed from a hole with a screw extractor. See *Figure 4-20*. First a hole must be drilled into the broken bolt. Then the correct size screw extractor is placed into the hole. The screw extractor is then turned counterclockwise with a tap wrench. The screw extractor acts much like a corkscrew to remove the broken bolt.

There are many other hand tools that are used in the automotive shop. Some of these tools include hex head (Allen wrenches), pipe wrenches, ratchet box wrenches, prybars, and crowfoot wrenches. All of these hand tools make the automotive mechanic's job much easier and safer.

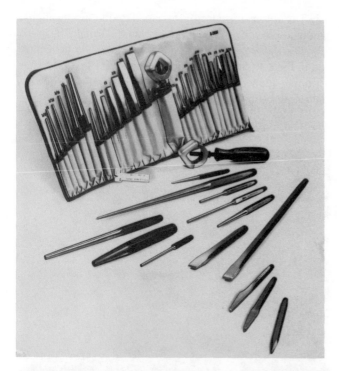

FIGURE 4–19. A quality set of chisels and punches should be available for use in all automotive shops. *(Courtesy of Snap-On Tools Corporation)*

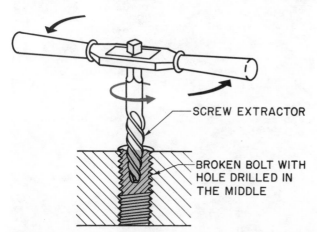

SCREW EXTRACTOR

BROKEN BOLT WITH HOLE DRILLED IN THE MIDDLE

FIGURE 4–20. A screw extractor can be used to remove broken bolts from inside an object.

CAR CLINIC

PROBLEM: STICKY CARBURETOR NEEDLE VALVE

A vehicle has been stored over winter. When taken out of storage in the spring, the engine doesn't start. There seems to be a lack of fuel available at the carburetor. A check of the fuel pump shows that the pump is working. What could be the problem?

SOLUTION:

Quite often, the cause of this problem is a sticky needle valve in the carburetor bowl area. Because the engine and carburetor have not been used regularly, the needle valve may stick in the closed position. Then there is a lack of fuel in the carburetor float area. If there is no fuel in the bowl area, the engine will not be able to start. Check by removing the float bowl to check for gasoline. If the bowl is dry, then the problem is the carburetor float valve.

4.3 MEASURING TOOLS

When servicing an automobile, there are certain dimensions, specifications, and clearances that must be measured. These may include torque specifications and various clearances. Measuring tools aid the automotive mechanic in checking these specifications and clearances.

Torque Wrenches

Torque wrenches are used to tighten bolts and nuts to their correct torque specification. Torque specifications were given in Chapter 3, as part of threaded fasteners. Torque wrenches are used to control the amount of tension on a bolt by measuring the amount of twist (torque) developed while tightening the bolt. Torque wrenches are designed to match sockets using 1/4, 3/8, and 1/2 inch drives. Other drives are also available but are not as common. They include 3/4, 1, and 1 1/2 inch drives for heavy-duty service. Torque wrenches typically have a dial or scale that indicates the amount of torque in inch pounds (in. lbs.) or in foot pounds (ft. lbs.). *Figure 4–21* shows torque wrenches used in the automotive shop.

Types of Torque Wrenches

A common type of torque wrench is called the *adjustable click* type. A specified torque value is adjusted on the torque wrench. When that torque level is reached, the wrench clicks so the operator knows the correct value of torque has been reached.

The *dial* torque wrench has a dial on top so that as the torque is being applied to the bolt an exact reading can be obtained immediately. *Figure 4–22* shows the types of dials used on this type of torque wrench.

T handle torque wrenches are used with standard ratchet wrenches. A dial is also used to indicate the exact amount of torque being applied. A socket is placed on the bottom of the torque wrench while the ratchet is placed on top.

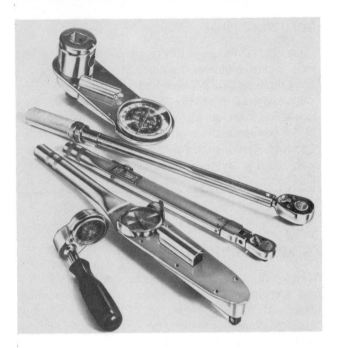

FIGURE 4–21. Torque wrenches are used to tighten bolts and nuts. Several types are available, including dial, adjustable click, and torque drives. *(Courtesy of Snap-On Tools Corporation)*

The *torque driver* is used for small torque specifications, where a screwdriver or nut driver would normally be used.

The *deflecting beam* torque wrench uses a deflecting rod in the center of the wrench to measure the exact value of torque applied to the bolt or nut. ***Figure 4–23*** shows a deflecting beam torque wrench.

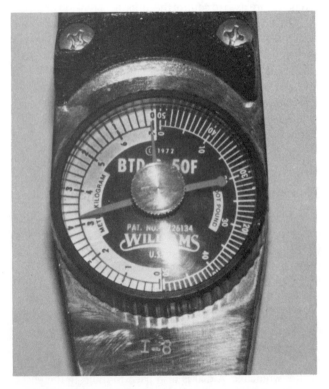

FIGURE 4–22. Several types of dials are used on torque wrenches. Some read in inch pounds, some in foot pounds, and some read both.

FIGURE 4–23. Another kind of torque wrench is called the deflecting beam torque wrench. As the bolt is tightened, the pointer reads the correct torque on the scale.

Micrometers

Micrometers are made to measure very small, accurate clearances. A micrometer is capable of measuring length in 0.001 of an inch. There are several types of micrometers. *Outside micrometers* measure outside dimensions such as shaft diameters, bearing thickness, shim thickness, etc. *Inside micrometers* are designed to measure internal dimensions such as engine cylinders and small holes. Inside diameters can also be measured with telescoping gauges. These gauges are used to measure the inside diameter of a bore. They are then measured with an outside micrometer for exact bore diameters.

Micrometers come in many sizes, *Figure 4– 24*. Common sizes include 0″–1″, 1″–2″, 2″–3″, 3″–4″, and 4″–5″. In some cases, the micrometer set may include several adapters and extensions.

Parts of a Micrometer

Figure 4–25 shows the basic parts of the micrometer. In this case, an outside micrometer is used.

All outside micrometers are the same except the frame. When larger dimensions are used, the frame is made larger to fit larger dimensions.

Reading a Micrometer

There are three steps to reading a micrometer. First, using the ratchet, slowly and gently turn the thimble until there is no clearance between the object being measured and the spindle and frame. At this point, note the highest figure on the barrel that is uncovered by the thimble. This number is the first figure to the right of the decimal point. For example, this number would be 0.200 on the micrometer shown in *Figure 4–26*.

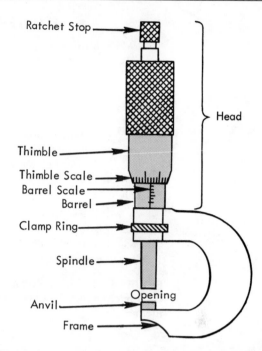

FIGURE 4–25. The outside micrometer has several parts. The object to be measured is placed between the anvil and the spindle. The measurement is taken on the sleeve and thimble.

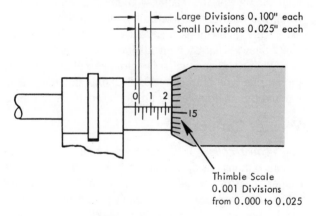

READING A STANDARD MICROMETER
Barrel Scale

Large Divisions 0.100″ each
Small Divisions 0.025″ each

Thimble Scale
0.001 Divisions
from 0.000 to 0.025

3 steps to read, add together: Example Above:

1. Large barrel divisions ------- X 0.100 = 0.200″
2. Small barrel divisions ------- X 0.025 = 0.025″
3. Thimble divisions --------- X 0.001 = 0.016″
   Reading ---------------- 0.241″

FIGURE 4–26. A micrometer is read by simply totaling the number of whole divisions on the barrel scale and adding the thousandths from the thimble scale.

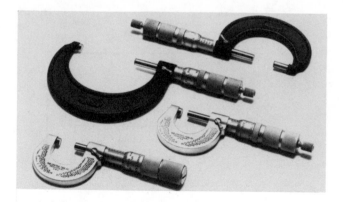

FIGURE 4–24. Micrometers are designed in many sizes to match many types of measurements. *(Courtesy of Snap-On Tools Corporation)*

READING METRIC MICROMETERS

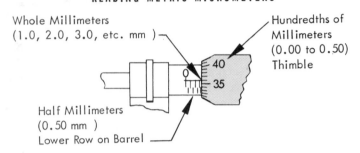

Whole Millimeters
(1.0, 2.0, 3.0, etc. mm)

Hundredths of
Millimeters
(0.00 to 0.50)
Thimble

Half Millimeters
(0.50 mm)
Lower Row on Barrel

Example:

Whole mm lines visible on barrel	3 =	3.00 mm
Additional half mm line (lower) visible on barrel	1 =	.50 mm
Lines on thimble which have passed long line on barrel	36 =	.36 mm
Reading of measurement Total	=	3.86 mm

FIGURE 4–27. The metric micrometer is read the same way as the standard micrometer. The graduations now read in millimeters rather than inches.

Second, note the whole number of graduations between the 0.200 mark and the thimble. In the case shown in Figure 4–26, there is one complete division after the 0.200 mark. Each of these graduations represents 0.025 of an inch. Therefore, add 0.025 to the 0.200 found in step one. This equals 0.225 of an inch.

Third, read the thimble opposite the index on the barrel. The graduations on the thimble represent 0.001 inch. In this case, add 0.016 of an inch to the reading in the first and second steps. The total reading shown in Figure 4–26 is 0.241 of an inch. Note that if the micrometer were a 2–3 inch micrometer, the reading would be 2.241 inches. When using the metric micrometer, the procedure is the same except the graduations represent different values. *Figure 4–27* shows what each graduation represents.

Other styles and types of micrometers read the same as the outside type of micrometer. The only difference is the method the micrometer fits on the object to be measured.

Vernier Caliper

The *vernier caliper* is a very useful tool for measuring various dimensions on the automobile engine. The vernier caliper measures length to 0.001 of an inch. A vernier caliper can measure inside or outside and in some cases depth measurements. See *Figure 4–28*. Vernier calipers can also be used to measure the inside diameter along with the telescoping gauges.

Reading a Vernier Caliper

There are several steps to reading a vernier caliper. The first step, as shown in *Figure 4–29*, is to read the number of whole inches on the main scale, left of the zero on the vernier. In this case, the reading is 1.000.

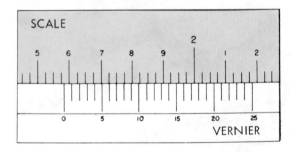

FIGURE 4–28. A vernier caliper is a very useful tool for the automotive mechanic. Clearances, both inside and outside, can be measured to 0.001 of an inch.

VERNIER SCALE

FIGURE 4–29. There are several steps to reading the vernier caliper. Vernier scales are much the same as a micrometer reading except the scale is spread out on the vernier.

Second, read the highest numbered graduation on the main scale that also lies to the right of the index zero, on the *vernier scale*. Each graduation is 0.100 of an inch. Add this reading to the reading in step one. The total now is 1.500 inches.

Third, read the highest number of whole minor divisions to the right of the index. Each of these graduations are 0.025 of an inch. Add this number (three minor divisions or 0.075 of an inch) to the reading in step two. The total so far is 1.575 inches.

The last step is to find the vernier graduation that most perfectly coincides with any graduation on the main scale. Each of these divisions represents 0.001 inch. This value (0.007) is then added to the value in step three. The total reading is 1.582. Once a vernier caliper is mastered, readings can be taken with ease and accuracy.

Feeler Gauges

Feeler gauges are used to help check or adjust clearances to a specific measurement. They are made of thin metal blades or wires, each of which is designed to be a different thickness in thousandths of an inch. The set of feeler gauges shown in *Figure 4-30* ranges in thickness from 0.0015-0.025 of an inch. When an adjustment is needed, such as the clearances between valves and rocker arms, the correct gauge is placed between the two objects. The clearance is adjusted so that a small amount of drag can be felt when pulling the feeler gauge out of the area of adjustment.

Figure 4-31 shows another common type of feeler gauge used on the automobile engine. Wire gap gauges are used mostly for setting spark plug clearances. Certain feeler gauges are made of nonmetallic metal such as brass. These feeler gauges are used to measure clearances on magnetic pickup coils on electronic ignition systems.

Dial Indicators

Figure 4-32 shows a *dial indicator* set. A dial indicator is a measurement tool used to determine clearances between two objects. For example, gear backlash (the clearances between the teeth on two gears in mesh) may have to be measured. Dial indicators are also used to check ball joints,

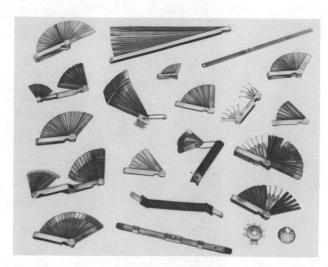

FIGURE 4-31. A wire feeler gauge is used to set the spark plug gap. (Courtesy of Snap-On Tools Corporation)

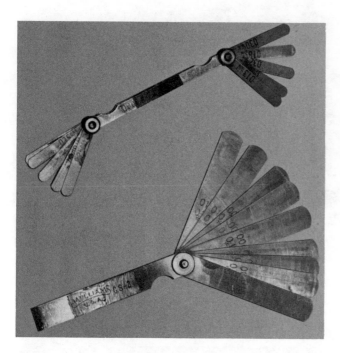

FIGURE 4-30. Feeler gauges are used to check or adjust clearances to a specific dimension.

FIGURE 4-32. Dial indicators are used to measure small clearances such as crankshaft end play, flywheel runout, and so on. (Courtesy of Snap-On Tools Corporation)

A MECHANIC'S TOOL SET

A skilled mechanic needs a wide variety of tools. A high-quality set of tools will help the mechanic accurately and completely diagnose problems and service the automobile. Many tool companies provide a complete line of tools for use on the automobile. (Courtesy of Snap-On Tools Corporation)

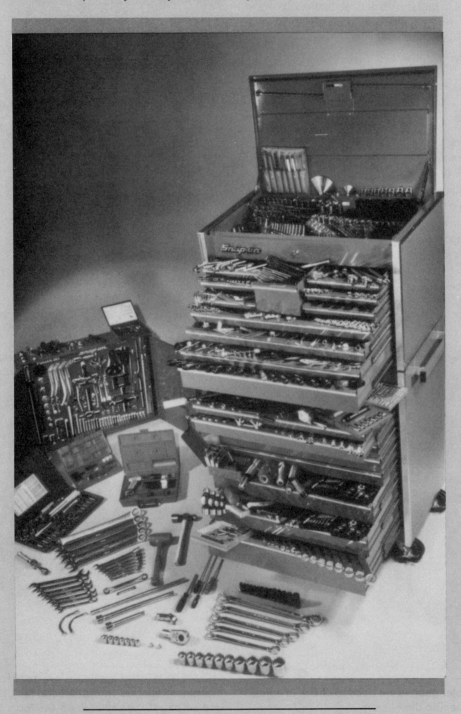

SPECIAL TOOLS

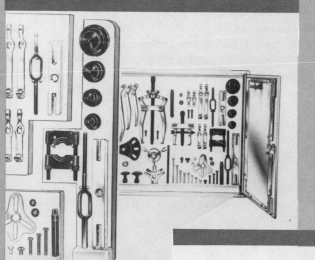

As the automobile becomes more and more sophisticated, many special tools are needed to service it. These photos show an assortment of pullers, special mirrors, and instruments that aid the mechanic in servicing the automobile. (Courtesy of Owatonna Tool Company)

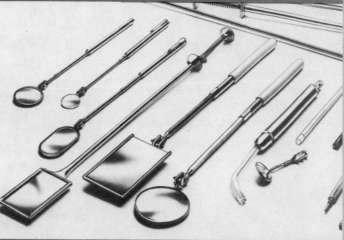

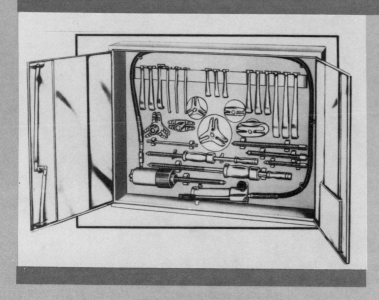

tie rods, cam and valve guide wear, crankshaft end play, and disc brake runout. Dial indicators measure clearances in 0.001 of an inch.

When using the dial indicator for checking end play of a shaft, position the dial indicator on the end of the shaft. This is done by using the adapters in the set. When positioned correctly, set the indicator dial at zero. Proceed to move the shaft back and forth axially. See *Figure 4–33*. While moving this shaft, read the clearance in 0.001 of an inch.

Vacuum/Pressure Gauges

Vacuum and *pressure gauges* are used to help troubleshoot, or when necessary, check various pressures and vacuum on the engine. *Figure 4–34* illustrates a complete set of pressure and vacuum gauges. The more common gauges include the engine oil pressure gauge, transmission oil pressure gauge, engine vacuum gauge, fuel pump pressure gauge, fuel injection pressure gauge, and the vacuum pump and gauge. Service and diagnosis using vacuum and pressure gauges are shown in many of the following chapters.

4.4 POWER AND PRESSING TOOLS

The automotive mechanic uses power and pressing tools in the shop. These make the heavy jobs that must be done on the automobile much easier, safer, and quicker. These tools include pullers, presses, impact wrenches, and bushing, bearing, and seal installers.

Pullers

Pullers are used to accomplish three types of action. These actions are pulling an object off of a shaft, pulling an object out of a hole, and pulling a shaft out of an object. These are illustrated in *Figure 4–35*.

The first example represents pulling a gear, wheel, or bearing, off of a shaft. The second example represents removing bearing cups, retainers, or seals from holes. The third action represents gripping a shaft and bracing against the housing to remove the shaft.

The pullers shown in *Figure 4–36* illustrate how pullers are designed for these uses. The jaw-type puller is used to pull a gear off a shaft. The internal puller is used to remove bearing cups and seals. The push-puller is used to pull a shaft out of a stationary housing.

Because there are so many types, styles, and sizes of seals, pulleys, and shafts, pullers must be designed to fit many applications. *Figure 4–37* illustrates a set of gear and

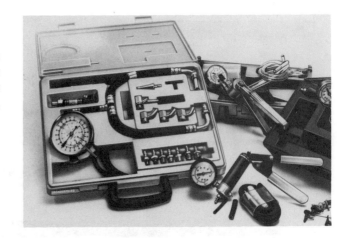

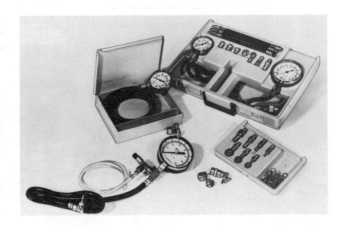

FIGURE 4–34. Pressure and vacuum gauges are used to test the automobile during service. *(Courtesy of Snap-On Tools Corporation)*

FIGURE 4–33. A dial indicator can read runout. The movement is shown on the dial indicator in 0.001 inch increments. *(Courtesy of Motor Magazine)*

bearing pullers that are commonly used on the automobile. There are many sizes and shapes of pullers necessary to the variety of jobs needed.

Bushing, Bearing, and Seal Installers

Installing bushings, bearings, and seals can be a very difficult job. During installation, these components must be aligned correctly. Even pressure must be applied as the component is installed. *Bushing installers* are used to perform this job. **Figure 4–38** shows a bushing, bearing, and seal driver set. These sets include discs and handles to provide a pilot. There are several spacers and drivers to help apply an even force on the part being installed. **Figure 4–39** shows a three-step process for installation. Discs range in size from 1/2 inch through 4 1/2 inches in diameter.

Hydraulic Presses

There are many times when working on automotive parts that pressing is required. For example, presses are used to work on rear axle bearings and piston pins, press out studs, straighten parts, and press in bearings. These jobs are typically done with a hydraulic press. These presses are capable of producing 50 tons or more of pressure on a part. **Figure 4–40** is an example of a hydraulic press, showing some of the press accessories.

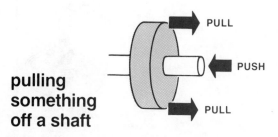

pulling something off a shaft

Removing a gear, bearing, wheel, pulley, etc., to replace it or get at another part.

pulling something out of a hole

Internal bearing cups, retainers or oil seals are usually press-fitted and are difficult to remove.

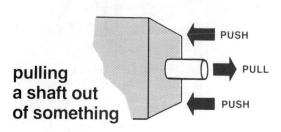

pulling a shaft out of something

A transmission shaft or pinion shaft is often hard to remove from a bore or housing.

FIGURE 4–35. Pullers are used for many problems in the automotive shop. Pullers are designed for all three actions shown. *(Courtesy of Owatonna Tool Company)*

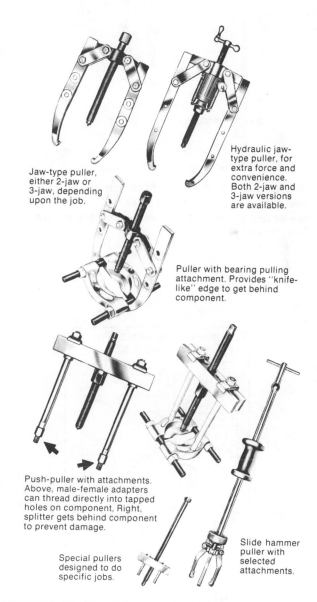

Jaw-type puller, either 2-jaw or 3-jaw, depending upon the job.

Hydraulic jaw-type puller, for extra force and convenience. Both 2-jaw and 3-jaw versions are available.

Puller with bearing pulling attachment. Provides "knife-like" edge to get behind component.

Push-puller with attachments. Above, male-female adapters can thread directly into tapped holes on component, Right, splitter gets behind component to prevent damage.

Special pullers designed to do specific jobs.

Slide hammer puller with selected attachments.

FIGURE 4–36. The different types of pullers are used to accomplish the necessary forces to pull gears, remove seals, and pull shafts from housings. *(Courtesy of Owatonna Tool Company)*

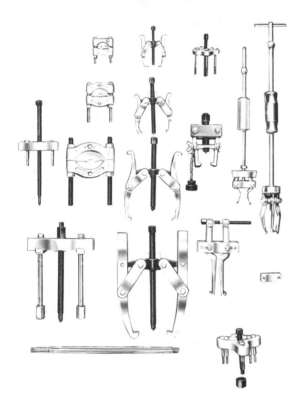

FIGURE 4–37. Many types of pullers are used on the automobile. This is because there is a variety of pulling jobs necessary on the automobile. *(Courtesy of Owatonna Tool Company)*

Impact Wrenches

An *impact wrench* is a powered wrench. It can be operated by using either air (called pneumatic) or electric power. The wrench works using the principle of impact rotation. Impact rotation is a pounding or impact force to aid in loosening or

1. Select the proper-size components

2. Assemble your driver tool

3. Perform the job easily

FIGURE 4–39. Three-step process in using the driver set: (1) select proper size, (2) assemble driver tool, and (3) perform the installation. *(Courtesy of Owatonna Tool Company)*

BUSHING, BEARING AND SEAL DRIVER SET

FIGURE 4–38. A bushing, bearing, and seal driver set is used to correctly install these components. Correct alignment and even force distribution are provided by these tools. *(Courtesy of Owatoona Tool Company, Sealed Power Division)*

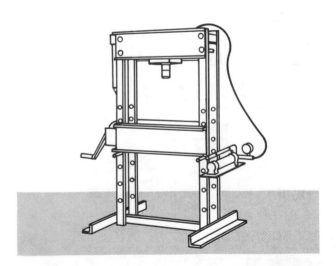

FIGURE 4–40. Hydraulic presses are used for a variety of pressing jobs. These may include work on piston pins, axle bearings, and other heavy pressing applications. *(Courtesy of Owatoona Tool Company, Sealed Power Division)*

FIGURE 4–41. Impact wrenches are used to produce impact forces during rotation. They are easy to use and can speed up the process of loosening and tightening bolts and nuts. *(Courtesy of Ingersoll-Rand)*

tightening nuts or bolts. Impact wrenches speed up the process of tightening or loosening bolts and nuts. *Figure 4–41* shows an air-type impact wrench. The drive can have 3/8, 1/2, 3/4, or 1 inch ends. An internal valve regulates the power output. Also, both forward and reverse directions are selected easily.

SUMMARY

This chapter introduced common hand tools and measuring tools that are used on the automobile. In order to study these tools, a sound knowledge is needed in measuring systems.

There are two types of measuring systems. They are the USC, or English system, and the metric system. USC measurements use feet, inches, miles, and so on to represent length measurements. Volume is measured by quarts, pints, etc. Mass is measured by pounds, ounces, etc. This system uses several bases such as 12, 3, and 64.

The metric system is based on one unit called the meter. All other units are derived from the meter using a base of 10. That is, one meter has 10 decimeters, 100 centimeters, and 1,000 millimeters. Both the USC measuring units and metric system measurements will be used where appropriate in the textbook.

Once the basic measurement system is known, the tools used in the automotive shop can be explained. Two types of tools were defined in this chapter. These are common hand tools and measuring tools.

Common hand tools include wrenches, hammers, socket sets and drives, screwdrivers, pliers, taps and dies, chisels and punches, and screw extractors.

Wrenches come in three common styles. They are the open end, box end, and combination wrench. Adjustable wrenches are also used in the automotive shop.

There are several types of hammers used in the automotive shop. The most common is called the ball peen hammer. Other styles that have rubber, plastic, and bronze tips are used. These should be used where damage may occur on the material being hammered.

There are several types of sockets, drives, and extensions. Swivel, deep length, and impact are all used in the automotive shop. The sockets are designed to have a certain number of points in the socket. Common point numbers include 6, 8, and 12. When a bolt is positioned so that only a small amount of ratchet rotation is possible, use the 12-point socket. It can be repositioned every 30 degrees.

Screwdrivers are an important tool for the auto mechanic. Screwdrivers are identified by the type of tip they have. Common tip styles include the slotted and Phillips tip. Other tips include Torx, magnetic, offset, and Reed and Prince.

Pliers are designed in a variety of styles and types. Pliers are used for cutting and gripping various objects. Common types include slip-joint, needle nose, cutter, electrical service, and retaining ring pliers.

Measuring tools are used to measure various lengths, torque bolts, adjust clearances, and check pressures and vacuums. These tools are torque wrenches, micrometers, vernier calipers, feeler gauges, dial indicators, and vacuum/pressure gauges.

Torque wrenches are used to tighten bolts and nuts to their correct torque specification. They usually read torque in inch pounds or in foot pounds. One common type of torque wrench uses a dial to indicate the exact torque being applied to a bolt.

Micrometers are used to measure clearances up to 0.001 of an inch. Micrometers come in any size. Common sizes are 1–2, 2–3, and 3–4 inch micrometers. Measurements are taken by reading a series of numbers on the barrel of the micrometer.

Vernier calipers measure clearances up to 0.001 of an inch. Vernier calipers are read by using a vernier scale placed on the instrument.

Feeler gauges are used to help check and adjust small clearances in a range from 0.0015–0.080 of an inch. Clearances such as spark plugs, points, and valves can all be adjusted by using feeler gauges.

Dial indicators are used to measure clearances between two objects. When a feeler gauge cannot be inserted between two objects, a dial indicator may be used. An example might be the clearance between two gears or crankshaft end play. Dial indicators can also read in 0.001 of an inch.

Vacuum and pressure gauges are used when troubleshooting a vehicle or when certain vacuum and pressure readings must be checked. Common gauges include engine oil pressure, transmission oil pressure, engine vacuum, fuel pump pressure, and fuel injection pressure.

Certain tools are identified as power and pressing tools. These are pullers, bushing, bearing, and seal installers, hydraulic presses, and impact wrenches.

Pullers are used to remove gears, bearings, and seals in a variety of sizes and styles. Bushing, bearing, and seal installers are used to provide correct alignment and force when installing these components. Hydraulic presses are used to press bearings on shafts, remove studs, and straighten objects. Impact wrenches are used to produce an impact force during rotation when loosening or tightening nuts and bolts.

The tools mentioned in this chapter are a small portion of the many tools used by the auto mechanic/technician. Many variations of these tools are also available. A solid foundation in the types, styles, and uses of tools will make the service mechanic more productive and safe in his/her work.

TERMS TO KNOW

Can you explain each of the following terms? Review the chapter until you can use each term correctly.

USC measurements	Torque wrench
Metric system	Micrometer
Base 10	Vernier caliper
Prefix	Vernier scale
Liter	Feeler gauge
Cubic centimeter	Dial indicator
Conversion factor	Vacuum gauge
Socket point	Pressure gauge
Impact socket	Puller
Deep length (well) socket	Bushing installer
Pliers	Impact wrench

REVIEW QUESTIONS

Multiple Choice

1. Which of the following systems of measurement use the base of 10?
   a. The USC
   b. The metric system
   c. The English system
   d. The SSC system
   e. The OSC system

2. What is the common unit of measure in the metric system?
   a. The foot
   b. The meter
   c. The cubic centimeter
   d. The liter
   e. The inch

3. What type of wrench has the end closed off rather than open?
   a. Open end
   b. Adjustable
   c. Box end
   d. Closed end
   e. None of the above

4. When there is only a small amount of room for ratchet rotation which type of socket should be used?
   a. 6 point
   b. 8 point
   c. 12 point
   d. 24 point
   e. 36 point

5. Which type of socket should be used for high torque when using a power wrench?
   a. Impact socket
   b. Standard socket
   c. Deep length socket
   d. Chrome socket
   e. None of the above

6. A _____ is used to cut external threads on a shaft.
   a. Tap
   b. Die
   c. Chisel
   d. Wrench
   e. Socket

7. Which measurement tool is used to tighten bolts to correct specifications?
   a. Dial indicator
   b. Adjustable wrench
   c. Torque wrench
   d. Pressure gauge
   e. Vacuum gauge

8. Which measurement tool can accurately measure 0.001 of an inch?
   a. Vernier caliper
   b. Vacuum gauge
   c. Pressure gauge
   d. Micrometer
   e. A and D

9. Which of the following tools should be used to adjust or measure small clearances of 0.001 of an inch?
   a. Feeler gauge
   b. Torque wrench
   c. Micrometer
   d. Pressure gauge
   e. Vacuum gauge

10. Which measuring tool has a scale in which two graduations or lines are lined up with each other to obtain the reading?
   a. Micrometer
   b. Vernier caliper
   c. Feeler gauge
   d. Torque wrench
   e. None of the above

11. Which of the following tools would be used to install a bearing on a shaft with a press fit?
   a. Bearing puller
   b. Hydraulic press
   c. Impact wrench
   d. Torque wrench
   e. None of the above

12. Which of the following tools would be used to remove a gear from a shaft?
   a. Puller
   b. Hydraulic press
   c. Impact wrench
   d. Torque wrench
   e. All of the above

13. Which of the following are controls that can be adjusted on the impact wrench?
   a. Power or force
   b. Forward or reverse
   c. Angle of operation
   d. A and B
   e. All of the above

14. When using a bearing or bushing installer, it is important to:
   a. Apply the correct distribution of forces
   b. Make sure the bushing is not aligned
   c. Never use a driver or spacer
   d. Apply oil to the surface
   e. None of the above

15. Which type of tap is used to start threads?
   a. Plug tap
   b. Taper tap
   c. Bottoming tap
   d. Top tap
   e. All of the above

16. When a bolt is broken off and is still partly in the hole, which tool should be used?
   a. Tap and die
   b. Impact wrench
   c. Screw extractor
   d. Torque adapter
   e. Vacuum adapter

17. Which of the following wrenches uses a long pointer that points to a scale as the bolt is tightened?
   a. Deflection beam torque wrench
   b. Dial torque wrench
   c. Dial indicator
   d. Vernier caliper
   e. Micrometer

18. Which prefix represents the unit in 1,000?
   a. Milli
   b. Centi
   c. Deci
   d. Kilo
   e. None of the above

19. Which of the following prefixes represents the unit in 1/1,000?
   a. Milli
   b. Centi
   c. Deci
   d. Kilo
   e. Mega

20. When one foot is subdivided into 12 inches, a base of _____ is used.
   a. 12
   b. 10
   c. 1/12
   d. 120
   e. 15

The following questions are similar in format to ASE (Automotive Service Excellence) test questions.

21. Technician A says that a feeler gauge can be used to measure small clearances of 0.001 of an inch. Technician B says that a micrometer can be used to measure clearances with an accuracy of 0.001 of an inch. Who is right?
   a. A only
   b. B only
   c. Both A and B
   d. Neither A nor B

22. Technician A says a tap is used to cut internal threads. Technician B says a die is used to cut internal threads. Who is right?
    a. A only
    b. B only
    c. Both A and B
    d. Neither A nor B

23. Technician A says the metric system uses a base of 12. Technician B says the USC system uses a base of 10. Who is right?
    a. A only
    b. B only
    c. Both A and B
    d. Neither A nor B

Essay

24. Describe the relationship between a liter and a cubic centimeter.

25. What is the difference between 6-, 8-, and 12-point sockets?

26. Describe the different types of taps used to cut threads.

27. What is the difference between a vernier caliper and a micrometer?

28. Which type of measuring tool would be used to measure small clearances such as valve clearances?

CHAPTER 5

Manuals and Specifications

INTRODUCTION

When an automobile is brought to the service shop for repair, it is the mechanic's responsibility to repair the vehicle. The automotive mechanic must have many specifications and service procedures available. The technical data is different for each automobile. It is impossible for any person to remember all of this technical information. Because of the number of vehicles, many specifications and service procedures are needed. This includes new and old vehicles, many manufacturers, both foreign and American, and many different problems.

The service manuals available to the automotive mechanic become one of the most important tools used. It is important to know what is available, how to locate information, what this information means, and how to use specifications correctly. This chapter is designed to address these areas.

OBJECTIVES

After reading this chapter, you will be able to:
■ List the different types of information in service manuals.
■ List different types of specifications.
■ Recognize where technical data can be obtained.
■ State the common publishers of service data.
■ Define the use of service bulletins.

CHAPTER HIGHLIGHTS

5.1 SERVICE MANUALS
    A. Diagnosing Information
    B. Procedure Information
    C. Specifications
    D. Types of Specifications

5.2 SOURCES OF INFORMATION
    A. Independent Publishers
    B. Manufacturer's Service Manuals
    C. Service Bulletins
SUMMARY

5.1 SERVICE MANUALS

Service manuals for automobiles include technical data, procedures, and service descriptions. This information is needed to troubleshoot and diagnose, service, and repair components on the automobile. This includes anything from a description of how to remove a part, to finding the problem, to determining the exact measurement for a clearance.

Typically, maintenance or service manuals do not include much information on theory or operational characteristics. They are designed and written to help the service mechanic service the vehicle. The best procedure and method of service is shown. In addition, special factory tools and their part numbers are shown in some service manuals.

Diagnosing Information

Because the automobile is so complex, service manuals include information to aid the mechanic in diagnosing and *troubleshooting* the vehicle. Diagnosing or troubleshooting is defined as identifying a problem, selecting a cause, and correcting the problem. *Diagnosis* is one of the most important skills that the mechanic must have.

Figure 5–1 illustrates an example of a fuel system diagnostic sheet. Three things are given. The mechanic must first identify the condition, then select a possible cause. When the cause has been identified, the correction is then provided. There are usually several causes for a particular condition. This gives the mechanic the option of using past experience when diagnosing a problem. This diagnostic guide is very easy to read. This guide also is especially helpful for identifying major problem areas on a vehicle.

A second method used to help diagnose a problem in a vehicle is shown in *Figure 5–2*. Here a flowchart is shown. The mechanic follows through each level, taking various readings and measurements. Based upon the results, the mechanic follows the flowchart until an incorrect reading is obtained. Finally, a suggested repair helps solve

FUEL SYSTEM DIAGNOSIS

The following diagnostic procedures are for fuel system problems and their effects on vehicle performance. Other systems of the vehicle can also cause similar problems and should be checked when listed on the chart. The problem areas described are:

1. Engine cranks normally. Will not start.
2. Engine starts and stalls.
3. Engine starts hard.
4. Engine idles abnormally and/or stalls.
5. Inconsistent engine idle speeds.
6. Engine diesels (after-run) upon shut off.
7. Engine hesitates on acceleration.
8. Engine has less than normal power at low speeds.
9. Engine has less than normal power on heavy acceleration or at high speed.
10. Engine surges.
11. Poor gas mileage.

CONDITION	POSSIBLE CAUSE	CORRECTION
Engine Cranks Normally — Will Not Start	Improper starting procedure used.	Check with the customer to determine if proper starting procedure is used, as outlined in the Owner's Manual.
	Choke valve not operating properly.	Check the choke valve and/or linkage as necessary. Replace parts if defective. If caused by foreign material and gum, clean with suitable non-oil base solvent.
	No fuel in carburetor.	• Perform fuel pump flow test. • Inspect fuel inlet filter. If plugged, replace. • If fuel filter is okay, remove air horn and check for a bind in the float mechanism.

FIGURE 5–1. Service manuals provide information about troubleshooting and diagnosing technical problems. The procedure to be followed is: Identify the problem (condition), select a possible cause, and then correct the problem (remedy).

FLOW DIAGRAM

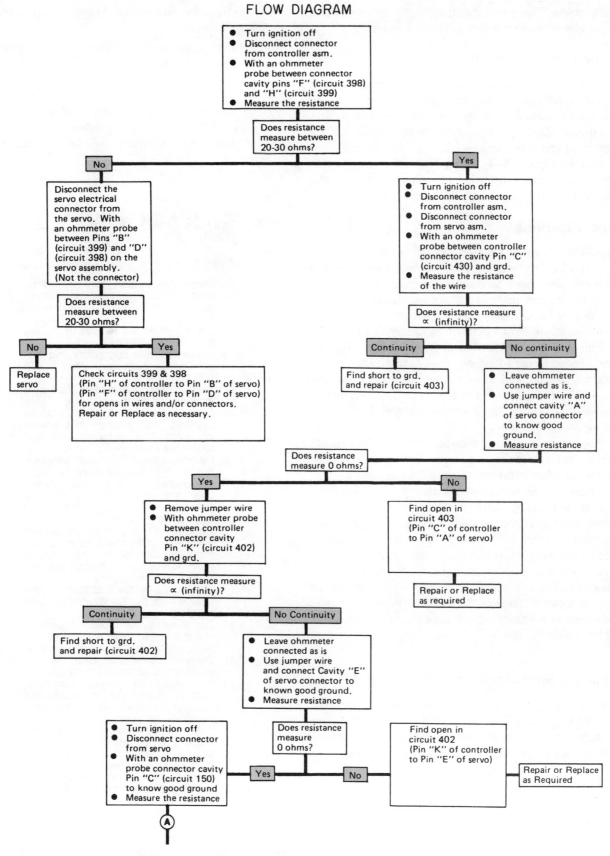

FIGURE 5–2. This flow diagram presents a detailed system for troubleshooting complex technical systems. Measurements are taken. Based on the result of each step, the service technician makes a decision as to what to do next. *(Courtesy of Motor Publications, Auto Repair Manual, 1981–1987)*

the problem. This format has more detailed explanations, especially for more complex technical systems.

Procedure Information

The correct procedure should always be followed when servicing an automobile. Suggested procedures are shown in the service manual to aid the mechanic. These procedures are used to speed up the mechanic's work. This makes the job easier and safer. An example of a set of procedures is shown in *Figure 5–3*. This procedure shows how to remove and replace the cylinder head on a particular vehicle. As indicated earlier, very little theory is given.

Specifications

Specifications are included as part of the service manual. Specifications are technical data, numbers, clearances, and measurements used to diagnose and adjust automobile components. Specifications can be referred to as specs. They are usually considered precise measurements under standard conditions. Most specifications can be measured with the measuring tools described in Chapter 4. Specifications are supplied by the automotive manufacturer. Examples of selected specifications include valve clearance, spark plug gaps, tire pressure, number of quarts of oil, ignition timing, and size of engine. There are many other specifications as well.

Specifications are not necessarily the law. They should be used as guides to show the mechanic how the automobile was set up when it was manufactured. A mechanic working on a new car should follow the factory specifications as closely as possible. However, when an automobile gets older, some specifications may not be exactly right for best operation. In this case, the manufacturer's specifications must be considered with the mechanic's experience. This may result in a departure from the factory's specifications. However, it is important to stay as close to the original specifications as possible. In addition, many mechanics subscribe to magazines that feature technical updates on older vehicles.

Types of Specifications

There are many types of specifications. Following is a list of common specifications used on the automobile. The manufacturers and maintenance manual publishers may or may not list the specifications using these categories.

1. *General Engine Specifications*: These specifications identify the size and style of the engine. They include cubic inch displacement, engine codes, carburetor type, bore and stroke, horsepower, torque, compression ratio, and normal oil pressure. See *Figure 5–4*. Each publisher may include different data in this section. For example, engine numbers,

firing order, and number of cylinders may also be included.

2. *Tune-up Specifications*: These specifications help identify adjustments necessary for a tune-up on the vehicle. These include spark plug gap, firing order, degrees of ignition timing, carburetor settings, and fuel pump pressure. *Figure 5–5* shows a set of tune-up specifications. Publishers may include different data for tune-up. Certain tune-up information

CYLINDER HEAD
REPLACE

4-151

1. Raise and support front of vehicle, then drain cooling system and disconnect exhaust pipe from exhaust manifold.
2. Lower vehicle and remove oil dipstick tube and air cleaner.
3. Disconnect wire connectors and vacuum hoses from carburetor or TBI unit.
4. Remove EGR valve base plate from intake manifold, if applicable.
5. Disconnect heater hose from intake manifold, then remove AIR system discharge tube attaching bolt from intake manifold.
6. Remove ignition coil lower attaching bolt, then disconnect wiring from coil.
7. Disconnect all wiring from cylinder head and intake manifold, then remove engine upper support attaching bolt from engine strut.
8. Remove A/C compressor and position aside with refrigerant lines attached.
9. Remove alternator drive belt, then remove AIR pump bracket bolt from engine block, if equipped.
10. Disconnect throttle and throttle valve cables from throttle lever and intake manifold.
11. Disconnect upper radiator hose from cylinder head, then disconnect AIR hose from tube assembly, if equipped.
12. Remove rocker arm cover, then remove rocker arms and push rods.
13. Remove cylinder head attaching bolts, then lift cylinder head and intake and exhaust manifolds as an assembly from cylinder block.
14. Reverse procedure to install. Coat heads and threads of cylinder bolts with a suitable sealing compound, then install bolts finger tight. Tighten cylinder head bolts in sequence shown in **Fig. 5**. Tighten intake manifold bolts in sequence shown in **Fig. 6**, if necessary.

FIGURE 5–3. The correct procedures are written in service manuals. This information will aid the mechanic in performing service in a timely and safe manner. *(Courtesy of Motor Publications, Auto Repair Manual, 1981–1987)*

General Engine Specifications

Year	Engine CID①/Liter	Engine VIN Code②	Carburetor	Bore and Stroke	Compression Ratio	Net H.P. @ RPM③	Maximum Torque Ft. Lbs. @ RPM	Normal Oil Pressure Pounds
1981	V6-252, 4.1L⑥	4	E4ME, 4 Bbl④	3.96 × 3.40	8.0	125 @ 3800	210 @ 2000	37
	V8-350, 5.7L⑤	N	Fuel Injection	4.06 × 3.38	22.5	105 @ 3200	200 @ 1600	30–45
	V8-368, 6.0L	9	Fuel Injection	3.80 × 4.06	8.2	140 @ 3800	265 @ 1400	30–45
1982	V6-252, 4.1L⑥	4	E4ME, 4 Bbl.④	3.96 × 3.40	8.0	125 @ 4000	205 @ 2000	37
	V8-250, 4.1L	8	Fuel Injection	3.465 × 3.307	8.5	135 @ 4600	200 @ 1600	30
	V8-350, 5.7L⑤	N	Fuel Injection	4.06 × 3.38	22.5	105 @ 3200	200 @ 1600	30–45
1983	V8-250, 4.1L	8	Fuel Injection	3.465 × 3.307	8.5	135 @ 4200	200 @ 2200	30
	V8-350, 5.7L⑤	N	Fuel Injection	4.06 × 3.38	22.5	105 @ 3200	200 @ 1600	30–45
1984–85	V8-250, 4.1L	8	Fuel Injection	3.465 × 3.307	8.5	135 @ 4400	200 @ 2200	30
	V8-350, 5.7L⑤	N	Fuel Injection	4.06 × 3.38	22.7	105 @ 3200	200 @ 1600	30–45
1986	V8-307, 5.0L	Y	E4MC, 4 Bbl	3.8 × 3.4	8.0	140 @ 3200	255 @ 2000	30–45

①—CID—Cubic inch displacement.
②—On 1980 vehicles, the fifth digit in the VIN denotes the engine code. On 1981–86 vehicles, the eighth digit in the VIN denotes the engine code.
③—Net rating—as installed in the vehicle.
④—Rochester.
⑤—Oldsmobile built diesel engine.
⑥—Buick built engine.

FIGURE 5–4. One type of specification is called "General Engine Specifications." It gives information about the engine to help the mechanic identify the exact type of engine in a vehicle. *(Courtesy of Motor Publications, Auto Repair Manual, 1981–1987)*

is also printed on a decal, affixed in the engine compartment. See *Figure 5–6*.

3. *Capacity Specifications*: These specifications include measurements needed to identify the capacity of different fluids on the vehicle. These include cooling capacity, number of quarts of oil, fuel tank size, transmission capacity, and rear axle capacity. *Figure 5–7* shows a standard list of capacity specifications.

4. *Overhaul and Maintenance Specifications*: These include specifications used to aid the mechanic in servicing the vehicle. They include distributor advance at different speeds, valve seat angles, valve stem clearances, piston measurements, ring end gaps, bearing clearances, shaft end play, and many more. *Figure 5–8* shows common specifications for cylinder block, pistons, pins, and rings. These specifications help the mechanic determine how much wear has occurred. The mechanic is then able to decide whether or not to replace the component in question. Usually maximum or minimum clearances are given for this purpose. In some cases, specifications are also given for a new vehicle or a used vehicle.

5. *Operational Specifications*: These specifications tell how the vehicle is to operate, what type of oil to use, and so on. Some of them are found in the owner's manual. For example, *Figure 5–9* shows the break-in speed limit taken from an owner's manual. Some specifications are found in magazines and other technical literature. For example, *Figure*

5–10 shows a performance comparison with several vehicles. Other specifications include tire inflation, type of gasoline to use, tire size, and general information for the operator of the vehicle.

6. *Torque Specifications*: It was mentioned in an earlier chapter that it is important to torque each bolt or nut when replacing or installing a component on the automobile. Torque specifications are used for this purpose. *Figure 5–11* shows an example of several specific torque specifications on a particular manufacturer's engine. These torque specifications should be used in place of any standard bolt and nut torque specifications.

CAR CLINIC

PROBLEM: FAN RUNS AFTER ENGINE TURNS OFF

A customer complains that every time the engine is shut off, a small fan continues to run under the hood. Is there a problem with the wiring or is the fan designed to shut off after the engine cools?

SOLUTION:

Many manufacturers have added fans to smaller and intermediate size cars to keep the fuel from boiling, possibly causing vapor lock. Normally, these fans are designed to run up to 10 minutes after a hot vehicle has been shut off.

TUNE UP SPECIFICATIONS

Year & Engine/VIN Code	Spark Plug Gap	Ignition Timing BTDC[1]★				Curb Idle Speed[2]		Fast Idle Speed		Fuel Pump Pressure
		Firing Order Fig. ▲	Man. Trans.	Auto. Trans.	Mark Fig.	Man. Trans.	Auto. Trans.	Man. Trans.	Auto. Trans.	
1981										
4-140/A	.034	D	6°	12°	E	850[9]	750D[22]	2000[3]	2300[3]	5-7[20]
6-200/B[10][12]	.050	F	—	10°	C	—	600D	—	2000[3]	5-7[20]
6-200/B[11][12]	.050	F	—	12°	C	—	600D	—	2000[3]	5-7[20]
V8-255/D Exc. Calif.[16]	.050	G	—	10°	B	—	500D	—	1600[3]	6-8
V8-255/D Exc. Calif.[23]	.050	G	—	10°	B	—	500D	—	1700[3]	6-8
V8-255/D Calif.[16]	.050	G	—	8°	B	—	500/650D	—	1500[3]	6-8
V8-255/D Calif.[23]	.050	G	—	12°	B	—	500/650D	—	1500[3]	6-8
V8-302/F Exc. Calif.	.050	G	—	10°	B	—	500D	—	1600[3]	6-8
V8-302/F Calif.	.050	G	—	8°	B	—	500/675D	—	1600[3]	6-8
1982										
4-140/A Exc. Calif.	.034	D	6°	12°	E	850	800D	1800[3]	2000[3]	5-7[20]
4-140/A Calif.	.034	D	4°	12°	E	850	800D	1600[3]	1800[3]	5-7[20]
6-200/B Exc. High Alt.	.050	F	—	10°	C	—	450/600D	—	2000[3]	5-7[20]
6-200/B High Alt.	.050	F	—	12°	C	—	450/600D	—	2000[3]	5-7[20]
V6-232/3 Exc. Calif. & High Alt.	.044	I	—	[24]	A	—	550/650D	—	2200[4]	6-8
V6-232/3 Calif.	.044	I	—	12°	A	—	550/650D	—	1200[3]	6-8
V6-232/3 High Alt.	.044	I	—	12°	A	—	550/650D	—	2100[3]	6-8
V8-255/D Exc. High Alt.	.050	G	—	8°	B	—	500/650D	—	1500[3]	6-8
V8-255/D High Alt.	.050	G	—	14°	B	—	500/650D	—	1700[3]	6-8
1983										
4-140/A[13]	.044	D	9°	9°	E	850	800	1800[3]	2000[3]	5-7[20]
4-140/W[14]	.034	D	—	—	E	—	—	—	—	—
6-200/X Exc. High Alt.[21]	.050	F	—	10°	C	—	450/600	—	2000[3]	6-8
6-200/X High Alt.[21]	.050	F	—	12°	C	—	450/600	—	2000[3]	6-8
6-200/X Exc. Calif.[25]	.050	F	—	10°	C	—	450/550	—	2200[3]	6-8
6-200/X Calif.[25]	.050	F	—	10°	C	—	450/550	—	2100[3]	6-8
V6-232/3 Exc. Calif. & High Alt.	.044	I	—	12°	A	—	550/650	—	2200[4]	6-8[20]
V6-232/3 Calif.	.044	I	—	10°	A	—	550/700	—	2200[4]	6-8[20]
V6-232/3 High Alt.	.044	I	—	12°	A	—	550/700	—	2100[4]	6-8[20]
V8-302/F	.050	G	—	—	B	—	550	—	—	—
1984										
4-140/A[13]	.044	D	10°	10°	E	850	750D	2000	2200	5-7[20]
4-140/W[14]	.034	D	10°	10°	E	825-975[5]	825-975[5]	[5]	[5]	—
V6-232/3	.044	I	—	10°	A	—	—	—	—	—
V8-302/F	.050	G	—	—	B	—	550D	—	[6][12]	—
V8-302 H.O./M	.044	H	—	—	B	—	550D	—	—	—
1985										
4-140/A[13]	.044	D	—	10°	E	—	710-790D	2200[3]	2200[3]	5-7[20]
4-140/W[14]	.034	D	10°	10°	E	825-975	925-1075N	[5]	[5]	—
V6-232/C[8][12]	.044	I	—	10°	A	—	500-600D	—	[5]	—
V6-232/C[12][15]	.054	I	—	10°	A	—	500-600D	—	[5]	—
V6-232/C[12][17]	.054	I	—	7°	A	—	500-600D	—	[5]	—
V8-302/F[12][18]	.050	G	—	7°	B	—	550D	—	2400[4]	—
V8-302/F[12][19]	.050	G	—	10°	B	—	550D	—	2400[4]	—
1986										
4-140/A[13]	.044	D	—	—	E	800	750D	—	—	5-7[20]
4-140/W[14]	.034	D	—	—	E	[5]	—	[5]	[5]	—
V6-232/3	.054	I	—	—	A	—	550-625D	—	[5]	—
V8-302/F	.050	G	—	—	B	—	—	—	[5]	—

FIGURE 5-5. Tune-up specifications are used to help the service technician tune-up the vehicle. Only those specification for tune-up are shown. (*Courtesy of Motor Publications, Auto Repair Manual, 1981-1987*)

FIGURE 5-6. Some tune-up specifications are located on a decal, which is affixed to the engine compartment.

Cooling System & Capacity Data

Year	Model or Engine/V.I.N.	Cooling Capacity, Qts.		Radiator Cap Relief Pressure, Lbs.	Thermo. Opening Temp.	Fuel Tank Gals.	Engine Oil Refill Qts. ①	Transmission Oil			Rear Axle Oil Pints
		Less A/C	With A/C					3 Speed Pints	4 Speed Pints	Auto. Trans. Qts. ②	
1981	Eldorado V6-252/4	—	13.1	15	195	21.1	4	—	—	⑦	3⅛⑥
	Eldorado V8-368 D.E.F.I./9	—	22.4	15	178	20.3	4	—	—	⑦	3⅛⑥
	Eldorado V8-350 Diesel/N	—	18.4	15	195	22.8	7⑧	—	—	⑦	3⅛⑥
	Seville V6-252/4	—	13.1	15	195	21.1	4	—	—	⑦	3⅛⑥
	Seville V8-368 D.E.F.I./9	—	22.4	15	178	20.3	4	—	—	⑦	3⅛⑥
	Seville V8-350 Diesel/N	—	18.4	15	195	22.8	7⑧	—	—	⑦	3⅛⑥
	Others V6-252/4	—	18.2	15	195	25	4	—	—	⑩	4¼
	Others V8-368/9	—	21.4	15	195	25	4	—	—	⑩	4¼
	Others V8-350 Diesel/N	—	23.7	15	195	27	7⑧	—	—	⑩	4¼
1982	Eldorado V6-252/4	—	13.1	15	195	21.1	4	—	—	⑤	3⅛⑥
	Seville V6-252/4	—	13.1	15	195	21.1	4	—	—	⑤	3⅛⑥
	Others V6-252/4	—	18.2	15	195	25	4	—	—	⑨	4¼
	Eldorado V8-250/8	—	11.8	15	195	20.3	5③	—	—	⑤	3⅛⑥
	Seville V8-250/8	—	11.8	15	195	20.3	5③	—	—	⑤	3⅛⑥
	Others V8-250/8	—	10.8	15	195	25	4③	—	—	⑨	4¼
	Eldorado V8-350 Diesel/N	—	18.4	15	195	22.8	7⑧	—	—	⑤	3⅛⑥
	Seville V8-350 Diesel/N	—	18.4	15	195	22.8	7⑧	—	—	⑤	3⅛⑥
	Others V8-350 Diesel/N	—	23.7	15	195	27	7⑧	—	—	⑨	4¼
1983	Eldorado V8-250/8	—	10.9	15	195	20.3	5③	—	—	④	3⅛⑥
	Seville V8-250/8	—	10.9	15	195	20.3	5③	—	—	④	3⅛⑥
	Others V8-250/8	—	11.0	15	195	24.5	4③	—	—	⑪	4¼
	Eldorado Diesel/N	—	18.4	15	195	22.8	7⑦	—	—	④	3⅛⑥
	Seville Diesel/N	—	18.4	15	195	22.8	7⑧	—	—	④	3⅛⑥
	Others Diesel/N	—	23.7	15	195	26	7⑧	—	—	⑪	4¼
1984-85	Eldorado V8-250/8	—	—	15	195	20.3	5③	—	—	—	3⅛⑥
	Seville V8-250/8	—	—	15	195	20.3	5③	—	—	—	3⅛⑥
	Others V8-250/8	—	—	15	195	24.5	4③	—	—	—	4¼
	Eldorado Diesel/N	—	—	15	195	22.8	7⑧	—	—	—	3⅛⑥
	Seville Diesel/N	—	—	15	195	22.8	7⑧	—	—	—	3⅛⑥
	Others Diesel/N	—	—	15	195	26	7⑧	—	—	—	4¼
1986	Brougham	—	—	15	195	24.5	4③	—	—	—	4¼

①—Add one quart with filter change.
②—Approximate. Make final check with dipstick.
③—Includes filter.
④—Oil pan 5 qts. Total capacity 13 qts.
⑤—Oil pan 5 qts. Total capacity 11¾ qts.
⑥—Front drive axle.

⑦—Oil pan 5 qts. Total capacity 12 qts.
⑧—Includes oil filter. Recommended diesel engine oil—use oil designated SF/CD, SF/CC or SE/CC on all except 1982-85 models or SF/CD, SF/CC on 1982-85 models.
⑨—Oil pan 3½ qts. Total capacity 11 qts.

⑩—THM 200C—oil pan 3½ qts., total capacity 11 qts.; THM 350 & 350C—oil pan 3.15 qts., total capacity 10 qts.; THM 400—oil pan 3 qts., total capacity 10 qts.
⑪—Oil pan, 5 qts. Total capacity, 11 qts.

FIGURE 5-7. Capacity specifications tell the mechanic how much fluid is required in such components as the fuel tank, cooling system, engine, transmissions, etc. *(Courtesy of Motor Publications, Auto Repair Manual, 1981-1987)*

Cylinder Block, Pistons, Pins & Rings

Engine CID/Liter	Year	Cylinder Bore Diameter (Std.)	Cylinder Bore Taper Max.	Cylinder Bore Out of Round Max.	Piston Diameter (Std.)	Piston Clearance	Piston Pin Diameter	Piston Pin To Pin Clearance	Piston End Ring Gap①		Piston Ring Side Clearance	
									Comp.	Oil	Comp.	Oil
4-97/1.6L	1981-85	3.15	.01	.005	②	③	.8119-.8124	.0003-.0005	.012	.016	④	—
4-116/1.9L	1985-86	3.23	.01	.005	⑤	.0016-.0024	.8119-.8124	.0003-.0005	.010	.016	.0015-.0032	—
4-121/2.0L⑥	1984-86	3.3859-3.3867	.0059	.0059	3.3842-3.3852	.0013-.0020	.9843-1.023	—	.0179	.0079	⑦	—
4-140/2.3L⑧	1981-83	3.7795-3.7831	.01	.005	⑨	⑩	.9118-.9124	.0002-.0004	.010	.015	.002-.004	—
4-140/2.3L⑧	1984-86	3.7795-3.7825	.01	.005	⑪	.0030-.0038	.9118-.9124	.0003-.0005	.010	.015	.002-.004	—
4-140/2.3L⑫	1984-86	3.679-3.683	.009	.004	⑬	.0013-.0021	.9119-.9124	.0002-.0005	.008	.015	.002-.004	—
4-153/2.5L⑫	1986	3.679-3.683	.009	.004	⑭	.0013-.0021	.9119-.9124	.0002-.0005	.008	.015	.002-.004	—
6-149/2.4L⑥	1985	3.1496-3.150	.0006	.0006	⑮	⑯	—	—	.008	.010	⑰	.0012-.0024
6-200/3.3L	1981-83	3.6800-3.6848	.010	.005	⑱	.0013-.0021	.9119-.9124	.0003-.0005	.008	.015	.002-.004	—
V6-182/3.0L	1986	3.504	.002	.002	⑲	.0012-.0023	.9119-.9124	.0002-.0005	.010	.010	.0016-.0037	—
V6-232/3.8L	1982-86	3.811	.002	.002	⑳	.0014-.0032	.9119-.9124	.0002-.0005	.010	.010	.0016-.0037	—
V8-255/4.2L	1981-82	3.6800-3.6945	.010	.005	㉑	.0014-.0024	.9119-.9124	.0003-.0005	.010	.015	.002-.004	—
V8-302/5.0L	1981-86	4.004-4.0052	.010	.005	㉒	.0018-.0026	.9119-.9124	.0002-.0004	.010	.015	.002-.004	—
V8-351/5.8L	1981-86	4.0000-4.0048	.010	.005	㉓	.0018-.0026	.9119-.9124	.0003-.0005	.010	.015	.002-.004	—

N/A—Not applicable.
①—Minimum.
②—Coded red, 3.1463-3.157 inches; coded blue, 3.1468-3.1474 inches.
③—1981-82, .0012-.002 inch; 1983-85, .0018-.0026 inch.
④—Top, .002-.0032 inch; second, .0016-.0032 inch.
⑤—Coded red, 3.224-3.225 inches; coded blue, 3.226-3.227 inches.
⑥—Diesel engine.
⑦—Standard: top, .0020-.0035 inch; second, .0016-.0031 inch. Service limit, .0079 inch.
⑧—Overhead Cam (OHC) engine.
⑨—Coded red, non-turbo, 3.7780-3.7786 inches; turbo, 3.7760-3.7766 inches. Coded blue, non-turbo, 3.7792- 3.7798 inches; turbo, 3.7772-3.7778 inches.
⑩—Non-turbo, .0014-.0022 inch; turbo, .0034-.0042 inch.
⑪—Coded red, 3.7764-3.7770 inches; coded blue, 3.7776-3.7782 inches.
⑫—High Swirl Combustion (HSC) engine.
⑬—Coded red, 3.6784-3.6790 inches; coded blue, 3.6796-3.6802 inches.
⑭—Coded red, 3.6783-3.6789 inches; coded blue, 3.6795-3.6801 inches; coded yellow, 3.6807-3.6811 inches.
⑮—Alcan, 3.1442 inches; KS, 3.1441 inches; Mahle, 3.1447 inches.
⑯—Standard: Alcan, .001-.0021 inch; KS, .0013-.003 inch; Mahle, .0018-.0029 inch. Service limit: All, .006 inch.
⑰—Top, .0024-.0055 inch; second, .0020- .0031 inch.
⑱—Coded red, 3.6784-3.6790 inches; coded blue, 3.6796-3.6802 inches.
⑲—Coded red, 3.5024-3.5031 inches; coded blue, 3.5035-3.5041 inches; coded yellow, 3.5045-3.5051 inches.
⑳—Coded red, 3.8095-3.8101 inches; coded blue, 3.8107-3.8113 inches; coded yellow, 3.8119-3.8125 inches.
㉑—Coded red, 3.6784-3.6790 inches; coded blue, 3.6798-3.6804 inches; coded yellow, 3.6812-3.6818 inches.
㉒—Coded red, 3.9984-3.9990 inches; coded blue, 3.9996-4.0002 inches; coded yellow, 4.0020-4.0026 inches.
㉓—Coded red, 3.9978-3.9984 inches; coded blue, 3.9990-3.9996 inches; coded yellow, 4.0014-4.0020 inches.

FIGURE 5–8. Overhaul and maintenance specifications tell the mechanic the exact measurements of the internal parts of the engine. (Courtesy of Motor Publications, Auto Repair Manual, 1981–1987)

BREAK-IN SPEED LIMIT MPH (KM/H)

		1ST	2ND	3RD	4TH	5TH
MANUAL TRANSAXLE	4-speed	0 to 22 (0 to 35)	12 to 37 (20 to 60)	20 to 55 (30 to 90)	25 to 75 (40 to 120)	
	5-speed	0 to 22 (0 to 35)	10 to 37 (15 to 60)	15 to 53 (25 to 85)	22 to 68 (35 to 110)	28 to 80 (45 to 130)
AUTOMATIC TRANSAXLE		"1" Low		"2" Second		"D" Drive
		0 to 30 (0 to 50)		0 to 53 (0 to 85)		0 to 75 (0 to 120)

FIGURE 5–9. Operational specifications show how the vehicle operates. In this case, the specifications are showing the break-in speed limit to operate the vehicle. *(Courtesy of Nissan Motor Corporation in USA)*

	MPG (CITY DRIVING)	ACCELERATION 0–60 MPH (SEC.)	BRAKES 60–0 MPH (HOT) (FT.)	HANDLING (MPH)	MANEUVERABILITY (MPH)	NOISE @ 60 MPH (dBA)
Chevrolet Camaro	16	10.5	192	61.9	27.7	71
Toyota in-line 6	21	11	172	66.5	28.5	71
Chevrolet Corvette V8	14	8.4	184	59.5	27.6+	74
Datsun turbo in-line 6	21	9.8	148	65+	27.6+	73
Mazda RX7 rotary	21	11.8	209	65+	27.6+	75
Porsche 924 turbo in-line 4	20	9.5	150	65+	27.6	74

FIGURE 5–10. Certain operational specifications show how vehicles compare to others on acceleration, miles per gallon, and so on.

Engine Tightening Specifications★

★ Torque specifications are for clean and lightly lubricated threads only. Dry or dirty threads produce increased friction which prevents accurate measurement of tightness.

Year	Engine Model/V.I.N.	Spark Plugs Ft. Lbs.	Cylinder Head Bolts Ft. Lbs.	Intake Manifold Ft. Lbs.	Exhaust Manifold Ft. Lbs.	Rocker Arm Shaft Bracket Ft. Lbs.	Rocker Arm Cover Ft. Lbs.	Connecting Rod Cap Bolts Ft. Lbs.	Main Bearing Cap Bolts Ft. Lbs.	Flywheel to Crankshaft Ft. Lbs.	Vibration Damper or Pulley Ft. Lbs.
1983–84	4-150/U	27	85②	23	23	19③	28①	33	80	⑤	80④
1981–82	4-151/B	11	92	37	37	20③	7	30	65	68	160
1983	4-151/B	11	92	26	37	20③	7	30	65	68	162
1981	6-258/C	11	85	23	23	19③	28①	33	65	105	80④
1982–86	6-258/C	11	85②	23	23	19③	28①	33	80	105	80④

①—Inch pounds.
②—Coat underside of cylinder head bolt heads & threads with a suitable sealing compound.
③—Rocker arm cap screw.
④—Lubricate bolt threads lightly before assembly.
⑤—Torque bolts to 50 ft. lbs., then tighten bolts an additional 60 degrees.

FIGURE 5–11. Torque specifications are used to help the mechanic determine the exact torque for certain bolts and nuts. *(Courtesy of Motor Publications, Auto Repair Manual, 1981-1987)*

SPACE FRAME

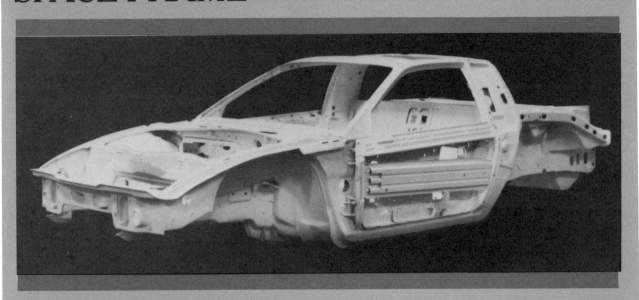

The space frame (top) is the structural framework of the Pontiac Fiero. It can be compared to the roll cage of a race car, surrounding the passengers within the steel framework. During production, the space frame, chassis, and powertrain are assembled into a drivable chassis (bottom), making the car basically complete. Electrical and mechanical components are then tested before exterior panels are attached. (Courtesy of Pontiac Motor Division, General Motors Corporation)

5.2 SOURCES OF INFORMATION

There are many sources from which specifications, *service procedures*, and troubleshooting information can be obtained. Both *independent publishers* and automotive manufacturers write service manuals. Parts manufacturers also distribute booklets with specifications and procedures. This is done so their parts will be installed correctly.

Independent Publishers

One of the more popular service and repair manuals is a series published by The Hearst Corporation. These are referred to as *Motor* manuals. See *Figure 5–12*. In these Motor manuals, all specifications, service procedures, and troubleshooting guides are printed. They are easy to read and illustrate the service procedures clearly. These manuals usually include information about cars for the past seven years.

Figure 5–13 shows a typical contents for one of the Motor auto repair manuals. There are several sections. A troubleshooting section gives various tips on how to diagnose the major systems of the vehicle. Then there is an extensive section (vehicle information) on each of the major cars being produced. In each car section, there is also general service information that is common to a particular vehicle. Common service topics include alternator service, disc brakes, power brakes, transmissions, steering columns, and universal joints. This section is very helpful when searching for specific service information, tests to make in each area, and general disassembly and reassembly procedures. In addition there are also several other sections on specifications and special service tools.

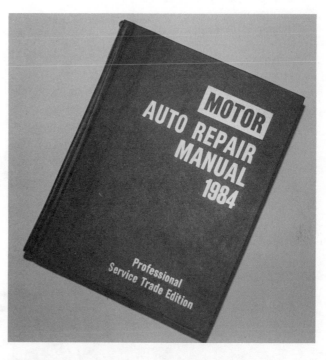

FIGURE 5–12. The Motor Manuals are among the most commonly used manuals in the automotive shop.

TABLE OF CONTENTS

This edition covers mechanical specifications and service procedures on 1981–87 models available at time of publication. Data reported in this manual is subject to change. To report current additions or revisions between revisions of the manual, a "Manual Update" page is published every month for your guidance in MOTOR Magazine.

"The data reported herein has been compiled from authoritative sources. While every effort is made by the editors to attain accuracy, manufacturing changes as well as typographical errors and omissions may occur. The publisher then cannot be responsible for such omissions, errors or changes."

FIGURE 5–13. The Motor Auto Repair Manuals have complete sections on vehicle information, specifications, special tools, and troubleshooting. (Courtesy of Motor Publications, Auto Repair Manual, 1981–1987)

Each of the vehicle sections also includes extensive information about specifications. Typical specifications in each section include:

a. Engine Identification Code
b. General Engine Specifications
c. Tune-up Specifications
d. Valve Specifications
e. Distributor Specifications
f. Alternator Specifications
g. Pistons, Pins, Rings, Crankshaft, and Bearing Specifications
h. Engine Tightening Specifications

i. Wheel Alignment Specifications
j. Starting Motor Specifications
k. Cooling System and Capacity Data
l. Drive Axle Specifications

Motor also publishes a series of manuals that can be used on special parts of the vehicle. *Figure 5–14* illustrates the variety of service manuals published by Motor. They include emission controls, automatic transmissions, air conditioners, heaters, and so on. These manuals are extremely helpful for service and diagnosis of special systems. For example, *Figure 5–15* shows the contents for the Motor *Emission Control Manual*. This manual includes all service

FIGURE 5–14. Motor auto repair manuals provide a complete listing of repairs, specifications, and troubleshooting for all American and foreign vehicles. *(Courtesy of Motor Publications)*

TABLE OF CONTENTS

This edition covers specifications and service procedures on 1985–86 models available at time of publication. Data reported in this manual is subject to change. To report current additions or revisions between revisions of the manual, a "Manual Update" page is published every month for your guidance in MOTOR Magazine.

FIGURE 5–15. The Motor *Emission Control Manual* has extensive information about emission controls for domestic cars, domestic trucks, and imports. *(Courtesy of Motor Publications, Emission Control Manual, 1985–1986)*

information, specifications, and diagnosis procedures for the many emission controls that are and have been used on vehicles to date. The information is subdivided by Domestic Cars, Domestic Trucks, and Imports. Motor also has a complete line of foreign car manuals as well.

Motor also publishes a monthly magazine entitled *Motor*. This magazine includes information about new automotive products, various special service and diagnosis information, special tools, and various interesting articles about performance and testing of vehicles. See *Figure 5–16*.

Chilton Auto Repair Manuals, *Figure 5–17*, are also a very common source of technical information for automo-

bile service shops. Auto manufacturers now produce automobiles in major lines or "families." Internally, different models from a single manufacturer are quite similar. They share common systems and components. When dealing with large numbers of automobiles and components, it is easier sometimes to locate information about a technical system. Chilton Manuals, for example, have a section on fuel systems and carburetor systems. This information would be the same for one family of vehicles. Chilton manuals carry service information over a seven-year period.

Mitchell Manuals approach the service information from a repair, rather than a particular vehicle manufacturer. There are volumes for each area of repair. The three volumes include Tune-up, Mechanical, and Transmission. Other manuals are available, including wiring diagrams, alternators, starters, air conditioning, and emission controls. Within each section or volume, vehicles are listed with corresponding specifications and necessary information. A set of Mitchell Manuals is shown in *Figure 5–18*.

Another publisher that prints service manuals is *Haynes Publishing Group*. These manuals, as shown in *Figure 5–19*, can be purchased at parts stores. Manuals are written to cover specifications, procedures, and troubleshooting for a specific family of vehicles. Manuals are also written for older imported vehicles. These manuals may include more than just service procedures and specifi-

FIGURE 5–16. Motor publishes a monthly magazine entitled *Motor* that has new product information, service and diagnosis procedures, new tools, and vehicle tests. *(Courtesy of Motor Publications)*

FIGURE 5–18. Mitchell manuals are another source of comprehensive technical information for the automobile. The volumes of the manual include Tune-up, Mechanical, and Diagnosis. *(Courtesy of Mitchell)*

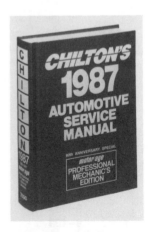

FIGURE 5–17. Chilton auto repair manuals offer a variety of technical data for automobiles. Each vehicle section is subdivided into technical systems such as brakes, engine, suspension, and so on. *(Courtesy of Chilton Book Company)*

cations. They may also include principles of operation and theory.

Manufacturer's Service Manuals

The automotive manufacturers also provide service manuals for their dealerships. *Manufacturer's service manuals* include principles of operation and some theory. These service manuals are written for one family or type of

FIGURE 5-19. Haynes Publishing Group provides specialized manuals on certain types of older vehicles. They can be purchased in most automotive parts stores.

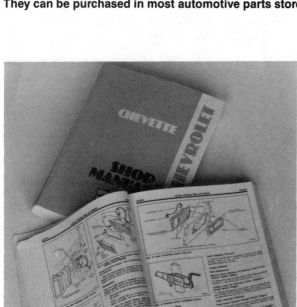

FIGURE 5-20. Service manuals can also be obtained directly from the manufacturer. These manuals usually cover a specific family of vehicles or a particular vehicle.

vehicle. *Figure 5-20* shows a set of a service manuals from the manufacturer. Manuals are also available for a particular year and model of vehicle. Common manuals are:

1. Chassis Service Manual
2. Body Manual
3 Electrical Troubleshooting Manual
4. Owner's Manual
5. Wiring Diagram Manual
6. Do-it-yourself Service Guide

These manuals can be obtained by writing to the manufacturer of the particular vehicle. Also, the owner's manual, which is usually kept in the vehicle, may have an address to write to for specific manuals.

Service Bulletins

There are numerous technical changes on specific vehicles each year. Therefore, the service manuals from the manufacturer must be updated. Updates are published using *service bulletins*. Service bulletins, shown in *Figure 5-21*, have changes in specifications and/or repair procedures during the model year. These changes do not appear in the service manual until the next year. The car manufacturer provides these bulletins to the dealers and repair facilities on a regular basis.

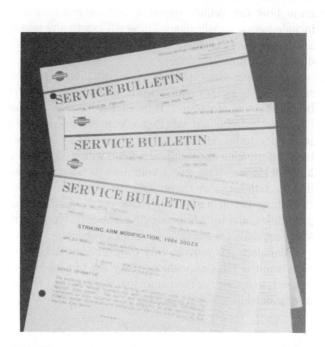

FIGURE 5-21. Service bulletins provide updated information and changes not included in the service manuals.

CAR CLINIC

PROBLEM: ENGINE VIBRATION

A customer has noticed that the engine seems to have excessive vibration at idle. At higher speeds there is also solid vibration felt throughout the vehicle. What could be the problem?

SOLUTION:

The most likely cause of engine vibration as described in this problem is a bad or broken motor mount. This assumes, of course, that the engine itself is operating correctly and that the engine is not missing due to ignition or fuel problems. Make a visual observation of all motor mounts and replace where necessary.

SUMMARY

When working on the automobile, manuals and specifications become an important tool. This chapter introduced manuals and specifications that are used by the service technician.

Service manuals include diagnosis information, troubleshooting, and procedure information. There are also many vehicle specifications in service manuals. These include general engine specifications, tune-up specifications, and capacity and overhaul specifications. The car owner may also need operational specifications to determine how the vehicle operates. Service manuals also include torque specifications. These tell exactly the torque and horsepower the engine can produce under certain conditions and rpm.

There are many sources of information to obtain specifications. Motor's manuals provide many service and troubleshooting procedures. They are easy to read and are illustrated well. Chilton Auto Repair Manuals also provide a source for technical information. These service manuals break information into technical systems within all vehicles. Mitchell Manuals have several volumes which include tune-up, mechanical, and transmission information. Haynes Publishing Group has manuals that cover specifications, procedures, and troubleshooting for a specific type of car. Many manuals are available for foreign vehicles.

Manufacturers provide their own service manuals for each family of vehicles produced. They also provide service bulletins to update the mechanics that work on the vehicles in a dealership.

TERMS TO KNOW

Can you explain each of the following terms? Review the chapter until you can use each term correctly.

Service manual
Diagnosis
Troubleshooting
Service procedures
Specifications
General engine specifications
Tune-up specifications
Capacity specifications

Overhaul and maintenance specifications
Operational specifications
Torque specifications
Independent publishers
Manufacturer's service manuals
Service bulletins

REVIEW QUESTIONS

Multiple Choice

1. Which of the following information is not included in a service manual?
   a. Specifications
   b. Sales and promotion information
   c. Disassembly procedures
   d. Detailed theoretical information
   e. Vehicle colors and styling

2. Which of the following information can be shown in a service manual using a flow diagram form?
   a. Specifications
   b. Troubleshooting and diagnostic information
   c. Overhaul information
   d. Special tools
   e. Types of vehicles

3. Which of the following information is usually shown as measurements, clearances, and numbers?
   a. Procedure information
   b. Troubleshooting information
   c. Specification information
   d. Overhaul procedures
   e. Special tools

4. Specifications that include spark plug gap, ignition timing, and carburetor adjustments are called:
   a. Tune-up specifications
   b. Capacity specifications
   c. Overhaul and maintenance specifications
   d. Procedure specifications
   e. None of the above

5. Specifications that show bearing clearances, shaft end play, ring gaps, etc. are called:
   a. Tune-up specifications
   b. General engine specifications
   c. Overhaul and maintenance specifications
   d. Capacity specifications
   e. Troubleshooting specifications

6. Torque specifications are identified as:
   a. Torque on nuts and bolts
   b. Engine torque at certain speeds
   c. Torque applied to the crankshaft during operation
   d. Torque applied to the generator
   e. Torque applied on turns

7. Motor manuals are published by:
   a. The manufacturer of the automobile
   b. An independent publisher
   c. Dealerships
   d. Mechanics
   e. Universities

8. Which is not a common title of a service manual provided by a publisher?
   a. Emission Control Manual
   b. Chassis Service Manual
   c. Wiring Diagram Manual
   d. Water Pump Manual
   e. Fasteners Manual

9. Approximately how many years do the Chilton and Motor manuals cover?
   a. 3 years
   b. 7 years
   c. 15 years
   d. 18 years
   e. None of the above

10. Which of the following types of technical information is sent to the service dealers for updates on the service manuals?
    a. Service bulletins
    b. Update bulletins
    c. New data bulletins
    d. Technique bulletins
    e. Sales bulletins

The following questions are similar in format to ASE (Automotive Service Excellence) test questions.

11. When repairing an engine, Technician A says overhaul procedures can be found in the service manual. Technician B says overhaul procedures can be found in the sales literature. Who is right?
    a. A only
    b. B only
    c. Both A and B
    d. Neither A nor B

12. Technician A says that plug gaps, clearances, and torque on bolts cannot be found in a service manual. Technician B says that plug gaps, clearances, and torque on bolts can be found in a service manual. Who is right?
    a. A only
    b. B only
    c. Both A and B
    d. Neither A nor B

13. Technician A says that diagnosis information can be found in the service manuals. Technician B says that diagnosis information can only be found in literature obtained from the manufacturer. Who is right?
    a. A only
    b. B only
    c. Both A and B
    d. Neither A nor B

Essay

14. What are examples of capacity specifications in the service manuals?

15. What is the purpose of service bulletins?

16. What are the major service manual names?

17. What is the purpose of having diagnosis sheets?

18. What type of specifications might be found in the "general engine specifications"?

CHAPTER 6

Engine Principles

INTRODUCTION

The component in an automobile that powers the vehicle is called the *engine*. The engine is a device that is constructed to do one major thing. It is designed to convert energy in fuel into power so the vehicle can be moved. This chapter is designed to introduce you to engine principles. The principles of engine design are presented.

OBJECTIVES

After reading this chapter, you will be able to:
- Define how energy is converted in the automobile.
- Classify engines according to their design.
- Analyze the major components needed to make an engine run.
- Specify the requirements for combustion.
- Compare the strokes of a four-stroke cycle engine.
- Compare gasoline to diesel engine principles.
- Analyze the operation of the Wankel engine.

CHAPTER HIGHLIGHTS

6.1 ENERGY CONVERSION
   A. Energy Defined
   B. Power Defined
   C. Forms of Energy
   D. Energy Conversion
   E. Chemical to Thermal Conversion
   F. Thermal to Mechanical Conversion
   G. Mechanical to Electrical Conversion
   H. Electrical to Mechanical Conversion
   I. Electrical to Radiant Conversion

6.2 ENGINE CLASSIFICATIONS
   A. Internal Combustion Engines
   B. External Combustion Engines
   C. Intermittent Combustion Engines
   D. Continuous Combustion Engines
   E. Reciprocating Engines
   F. Rotary Engines
   G. Other Classification Methods
   H. Classification of Engines Used in Automobiles

6.3 BASIC ENGINE TERMINOLOGY
   A. Cylinder Block
   B. Cylinders
   C. Pistons
   D. Connecting Rod and Crankshaft
   E. Cylinder Head
   F. Combustion Chamber
   G. Valves
   H. Camshaft
   I. Flywheel
   J. Carburetor

6.1 ENERGY CONVERSION

Energy Defined

The engine in an automobile is designed to accomplish one thing. This is to convert *energy* from one form to another. Energy is defined as the ability to do work. This means there is energy within the fuel that is put into the engine. The engine takes the energy from the fuel and converts it into a form of power. The power is used to propel the vehicle.

Power Defined

Power is defined as a measure of the work being done. Power is the final output of the engine after it has converted the energy in the fuel into work. A more common term used today is horsepower. Horsepower is a measure of the work being done by the engine.

Forms of Energy

Energy can take on one of six forms. Referring to *Figure 6–1*, they include chemical, electrical, radiant, mechanical, nuclear, and thermal forms of energy. The automobile uses all of the preceding energy forms except nuclear energy.

1. *Chemical energy* is defined as energy contained in molecules of different atoms. Examples of chemical energy are gasoline, diesel fuel, coal, wood, chemicals inside a battery, and food.
2. *Electrical energy* is defined as the ability to move electrons within a wire. Electrical energy uses voltage, wattage, resistance, and so on for operation. Many of the components on a car, including the radio, horn, lights, and starter, utilize electrical energy.
3. *Mechanical energy* is defined as the ability to physically move objects. Examples include water falling over a dam, the ability to move a vehicle forward, and gravity. The starter motor on a car takes electrical energy and converts it into mechanical energy to start the engine.
4. *Thermal energy* is defined as heat. This form of energy is released when fuel burns. The radiator removes excess thermal energy from the engine. The combustion of fuel produces thermal energy.
5. *Radiant energy* is defined as light energy. It is measured by frequencies. Examples of radiant energy include the energy coming to the earth from the sun, the energy from a light bulb, and the energy from anything that glows.

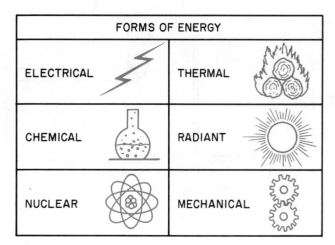

FIGURE 6–1. There are several forms of energy, including chemical, electrical, radiant, mechanical, nuclear, and thermal. All of these forms of energy are used on the automobile except the nuclear form. (Courtesy of DCA Educational Products)

6. *Nuclear energy* is defined as the energy within atoms when they are split apart or combined. Uranium, the fuel used in a nuclear power plant, has nuclear energy internally. Nuclear energy is not used when studying the automobile.

Energy Conversion

Energy conversion is defined as changing one form of energy to another. Energy usually does not come in the right form. Therefore, it must be converted to a form we can use. For example, the vehicle uses mechanical energy to go forward, electrical energy for the radio, and radiant energy from the light bulbs. Gasoline or diesel fuel is the main source of energy on the vehicle. It is in the form of chemical energy. The engine is designed to convert the chemical energy into the correct forms of energy needed.

Chemical to Thermal Conversion

When any fuel is burned, it changes the energy from chemical (fuel) to thermal (heat). See *Figure 6–2*. This process happens when the fuel burns in an engine. However, thermal energy is not really needed. Mechanical energy is what is needed from the engine.

Thermal to Mechanical Conversion

Once the thermal energy is produced by burning the fuel, the thermal energy causes rapid expansion of the gases within the engine. This rapid expansion is called mechanical energy. The combustion process on any engine converts chemical to thermal, and thermal to mechanical energy. This mechanical energy is then used to help propel the vehicle.

Mechanical to Electrical Conversion

The alternator or generator, *Figure 6–3*, is designed to convert some of the mechanical energy into electrical energy. The electrical energy is used to operate the radio, start the car, provide ignition, and operate other electrical appliances on the vehicle. The storage battery is used to store any excess electrical energy, especially if needed when the car is not running.

Electrical to Mechanical Conversion

The starter motor, *Figure 6–4*, is designed to convert electrical energy into mechanical energy to start the vehicle. This device is called a motor. All motors, including windshield wiper motors, heater fans, and starters, convert electrical energy to mechanical energy.

Electrical to Radiant Conversion

Electrical to radiant conversion occurs when light bulbs are used. The energy coming out of a light bulb is radiant energy. Electrical energy is used to operate a light bulb. A light bulb, then, converts electrical energy to radiant energy.

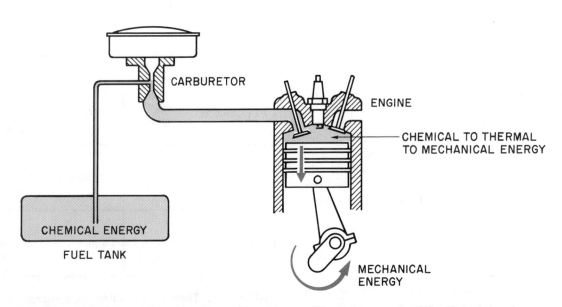

FIGURE 6–2. The internal combustion engine converts chemical energy in the fuel into thermal energy in the combustion area. The thermal energy is then converted into mechanical energy by the piston and crankshaft.

FIGURE 6–3. The alternator on a car engine is used to convert mechanical energy into electrical energy.

FIGURE 6–4. The starter motor on a car engine is used to convert electrical energy into the mechanical energy required to start the engine.

CAR CLINIC

PROBLEM: HIGH EXHAUST BACK PRESSURE

A customer owns a car with a diesel engine in it. Recently a mechanic suggested that the rough idle noticed by the driver could be caused by high exhaust back pressure. What are the effects of high exhaust back pressure on diesel engines?

SOLUTION:

High exhaust back pressure is developed by some type of restriction in the exhaust system. A rusted muffler, plugged tail pipe, or bent exhaust system may have caused the high back pressure. The symptoms on the engine of having high back pressure include:

1. higher engine temperature
2. a decrease in power
3. exhaust smoke becoming denser
4. rough idle
5. excessive carbon buildup on the valves, injectors, and pistons
6. contaminated oil

If these symptoms appear, the exhaust system must be inspected and repaired or replaced where necessary.

6.2 ENGINE CLASSIFICATIONS

Engines can be classified several ways. These include 1) by the location of the combustion, 2) by the type of combustion, and 3) by the type of internal motion.

Internal Combustion Engines

In an *internal combustion* engine (ICE), the combustion occurs within the engine. The combustion process occurs directly on the parts that must be moved to produce mechanical energy. The fuel is burned within the engine. See *Figure 6–5*. A gasoline engine is an internal combus-

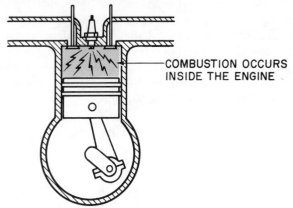

FIGURE 6–5. In an internal combustion engine, the combustion of gases occurs inside the engine, touching the moving parts.

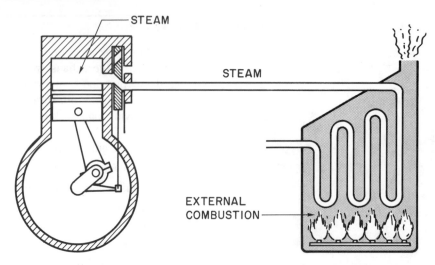

FIGURE 6-6. An external combustion engine has the combustion area removed from the pistons.

tion engine. Small lawn mower engines, snowmobile engines, and motorcycle engines are also internal combustion engines.

External Combustion Engines

In an *external combustion* engine, the combustion is removed from the parts that must be moved. See *Figure 6-6*. For example, the boiler in a steam engine is external. It is not touching the piston. Actually, the thermal energy in an external combustion engine heats another fluid. In this case, it is water. Water, converted to steam, pushes against the piston.

There has been some research to determine if an external combustion engine could work in an automobile application. So far, this type of engine has not proven successful in the automobile market.

Intermittent Combustion Engines

Intermittent combustion means that the combustion within the engine starts and stops. A standard gasoline engine has an intermittent combustion design. The combustion starts and stops many times during operation. Diesel engines are called intermittent combustion engines as well. Diesel engines have been used by several automobile manufacturers in the past years in the automobile.

Continuous Combustion Engines

A *continuous combustion* engine has combustion that continues all of the time. The combustion does not stop. It keeps burning continuously. A blow torch is an example of continuous combustion. See *Figure 6-7*. Engines that use continuous combustion include turbine engines, rocket engines, stirling engines, and jet engines. Research has shown that turbines could be used in the automobile, but they are very costly for this purpose.

Reciprocating Engines

In a *reciprocating* engine, the motion produced from the energy within the fuel moves parts up and down. The motion reciprocates, or moves back and forth. Gasoline and diesel engines are reciprocating engines. In this case, the power from the air and fuel burning starts the piston moving. The piston starts, then stops, then starts, then stops, etc. In this engine, the reciprocating motion must then be changed to rotary motion. A crankshaft is designed to change this motion. Refer to *Figure 6-8*.

Rotary Engines

In a *rotary engine*, the parts that are moving rotate continually. For example, a turbine and a Wankel engine are considered rotary engines. The mechanical movement of the parts takes the shape of a circle. Referring to Figure 6-8, the crankshaft is also an example of rotary motion.

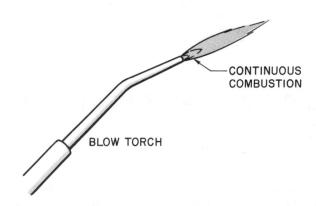

FIGURE 6-7. An example of continuous combustion is a blow torch. The combustion is continuous, not intermittent. Turbine engines have a continuous combustion process.

ROBOTS IN MANUFACTURING

Car manufacturing has become very automated. Robots are used to weld many of the sheet metal panels to the vehicle. The use of robots makes the automobile production process much more efficient and less costly. (Courtesy of Pontiac Motor Division, General Motors Corporation)

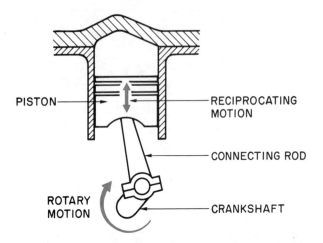

FIGURE 6–8. The reciprocating motion means the parts are moving up and down. Gasoline and diesel engines are considered reciprocating engines.

Other Classification Methods

Engines can also be classified by the following methods:

1. *By cycles* — There are two- and four-cycle engines.
2. *By cooling systems* — There are liquid-cooled and air-cooled engines.
3. *By fuel systems* — There are gasoline-fueled and diesel-fueled engines.
4. *By ignition systems* — There are spark-ignition and compression-ignition engines.

All of these methods of classifying engines can be combined with those previously mentioned.

Classification of Engines Used in Automobiles

The automobile uses engines with several of the classifications just listed. The gasoline and diesel engines used in cars are considered internal combustion, intermittent combustion, reciprocating engine designs. If the rotary (Wankel) engine is used, it is considered an internal combustion, intermittent combustion, rotary design. Many alternative designs have been tested for use in the automobile. The gasoline, diesel, and rotary (Wankel) engines are the ones used today. The gasoline engine is still the most popular form used. The gasoline engine is considered a four-cycle, spark-ignition engine. It can be either liquid-cooled or air-cooled.

6.3 BASIC ENGINE TERMINOLOGY

To understand the principles of automobile engines, certain parts must be defined. These parts are considered the major components of the engine. They include the cylinder block, cylinders, pistons, connecting rods and crankshaft, cylinder head, combustion chamber, valves, flywheel, and carburetor.

Cylinder Block

The *cylinder block* is defined as the foundation of the engine. See *Figure 6–9*. The cylinder block is most often made of cast iron or aluminum. All other components of the engine are attached to the cylinder block. The cylinder block has several internal passageways to let cooling fluid circulate around the block. It also has several large holes machined into the block where the combustion occurs.

Cylinders

The *cylinders* are defined as internal holes in the cylinder block. See Figure 6–9. These holes are used for combustion. The holes tell the number of cylinders used on an engine. For example, on small gasoline engines, such as lawn mowers, there is one cylinder. Automobiles usually use four, six, or eight cylinders. Some engines used on heavy equipment have as many as 24 cylinders.

Pistons

Pistons are defined as the round object that slides up and down in a cylinder. Refer to *Figure 6–10*. There is one piston for each cylinder. Pistons are made of light material such as high-quality aluminum that can withstand high

FIGURE 6–9. The cylinder block is the foundation of the block engine. All other parts are attached to the cylinder block. *(Courtesy of Peugeot Motors of America, Inc.)*

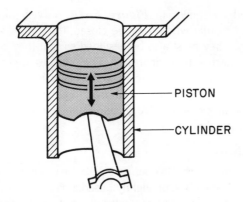

FIGURE 6-10. The piston slides up and down inside the cylinder.

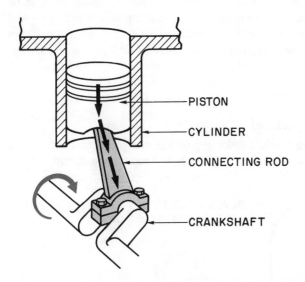

FIGURE 6-11. The connecting rod is attached to the bottom of the piston. The crankshaft changes reciprocating motion to rotary motion.

temperatures. If fuel and air ignite to cause expansion above the piston, this expansion will force the piston downward. The motion converts the energy in the fuel into mechanical energy (piston moving downward). The piston must also have seals or rings on it. These seals stop any combustion from passing by the piston.

Connecting Rod and Crankshaft

Attached to the bottom of the piston is the *connecting rod*. This is shown in *Figure 6-11*. Its main purpose is to attach the piston to a device known as the *crankshaft*. The crankshaft is used to change the reciprocating motion of the piston and connecting rod to rotary motion. Rotary motion is used as the output power of the engine. The piston, connecting rod, and crankshaft parts are shown in *Figure 6-12*.

FIGURE 6-12. A piston, connecting rod, and crankshaft for a four-cylinder engine are shown. All parts are connected together to change reciprocating motion to rotary motion. *(Courtesy of Peugeot Motors of America, Inc.)*

Cylinder Head

The *cylinder head* is the part that fits over the top of the cylinder block. See *Figure 6-13*. It usually houses the ports/valves that allow fuel and air to enter into the cylinder. The spark plug is also attached to the cylinder head. The cylinder head is made of cast iron or aluminum. When it is bolted to the cylinder block, it seals the cylinders so that air and fuel can be controlled in and out of the cylinder.

Combustion Chamber

The *combustion chamber* is where the combustion takes place inside the cylinder. When the cylinder head has been attached, the area inside of the cylinder head and block is called the combustion chamber. On some engines, the combustion chamber is located inside the head. Other engines have the combustion chamber located inside the top of the piston. This is especially true on diesel engines.

Valves

Valves are placed inside the cylinder head to allow air and fuel to enter and leave the combustion area. The valves are

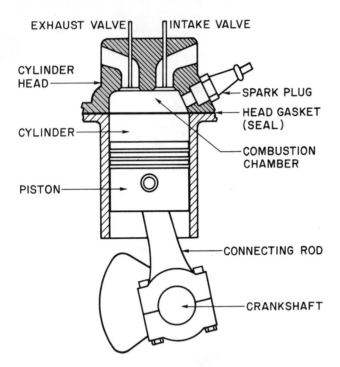

EXHAUST VALVE   INTAKE VALVE

CYLINDER HEAD

SPARK PLUG

HEAD GASKET (SEAL)

CYLINDER

COMBUSTION CHAMBER

PISTON

CONNECTING ROD

CRANKSHAFT

FIGURE 6-13. A cylinder head seals the top of the engine. It houses the intake and exhaust ports, valves, and the spark plug. The combustion chamber is located in the head area. The valves for intake and exhaust are shown.

designed as shown in *Figure 6-14*. Valves must be designed so that when they are closed, the port is sealed perfectly. They must also be designed so that they can be opened exactly at the right time. These valves are opened by using a camshaft and are closed by using springs. There is an intake valve to allow fuel and air to enter the cylinder. There is an exhaust valve to allow the burned gases to escape the cylinder.

Camshaft

The *camshaft* is used to open and close the valves at the correct time. Cam lobes, or slightly raised areas, are machined on the camshaft to open the valves so that air and fuel can enter the cylinder. The valves are then closed by springs on each valve. The camshaft is driven by the crankshaft. Refer to *Figure 6-15*. This means that the camshaft must be timed to the crankshaft so that the valves will open and close in correct time with the position of the piston. There is one lobe placed on the camshaft for each valve that must be opened and closed. The camshaft can also be placed or mounted directly on top of the cylinder head. This design is called an overhead camshaft (OHC).

Flywheel

The *flywheel* is located on the end of the crankshaft. It is designed to act as a weight to keep the crankshaft rotating

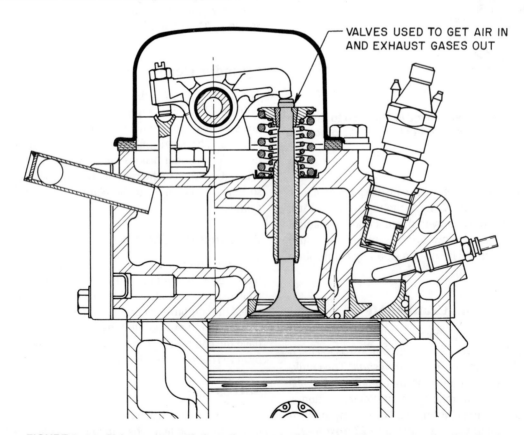

VALVES USED TO GET AIR IN AND EXHAUST GASES OUT

FIGURE 6-14. Valves open and close the ports in the combustion chamber to allow intake and exhaust gases to enter and leave the engine. *(Courtesy of Peugeot Motors of America, Inc.)*

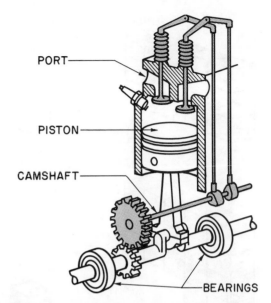

FIGURE 6–15. The camshaft opens and closes the valves at the correct time. The camshaft is driven by the two gears on the cam and crankshaft.

once power has been applied to the piston. The flywheel is usually heavy. It smooths out any intermittent motion from the power pulses. See *Figure 6–16*.

Carburetor

The *carburetor* is placed on the engine to mix the air and fuel in the correct proportion. This is called the fuel induction system. On many cars today, fuel induction can also be done by fuel injection. Air and fuel must be mixed correctly for the engine to operate efficiently. The carburetor's or fuel injector's job is to mix the air and fuel during cold weather, warm weather, high altitudes, high humidity, and low-speed and high-speed conditions and acceleration.

6.4 COMBUSTION REQUIREMENTS

Air, Fuel, and Ignition

The internal combustion engine has certain requirements for efficient operation. Any engine requires three things for its operation. There must be sufficient air for combustion, correct amounts of fuel mixed with the air, and some type of ignition to start combustion. When these three ingredients are present, *Figure 6–17*, an explosion will take place. This explosion will change chemical energy in the fuel to thermal energy. The thermal energy will then cause rapid expansion of gases. This expansion pushes the piston downward. The downward force on the piston makes the crankshaft rotate. This rotary power can be used for pushing the vehicle forward. If any one of these three ingredients is missing, the engine will not run.

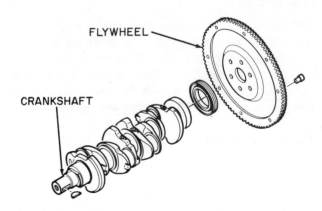

FIGURE 6–16. The flywheel is a heavy weight on the end of the crankshaft. It smooths out any power pulses and helps return the piston to the top of the cylinder. *(Courtesy of Federal-Mogul Corporation)*

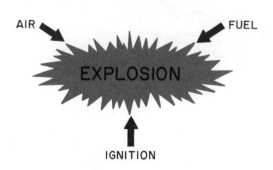

FIGURE 6–17. Three things are needed for combustion: air, fuel, and ignition. If one ingredient is missing, combustion will not happen.

Timing

Timing is defined as the process of identifying when the air, fuel, and ignition occur. This is done in relationship to the position of the piston and crankshaft. For the engine to operate efficiently, the air and fuel mixture must enter the cylinder at the correct time. This means that the intake valve must be opened and closed at the correct time. The exhaust valve must also be opened and closed at the correct time.

The ignition must also be timed. Ignition of the air and fuel must occur at a precise time. The timing of the ignition changes with speed and load. When the intake and exhaust valves are correctly timed and the ignition occurs at the correct time, maximum power will be obtained in converting chemical energy into mechanical energy.

Air-fuel Ratio

Air-fuel ratio is defined as the ratio of air to fuel mixed by the carburetor. The air and fuel must be thoroughly mixed. Each molecule of fuel must have enough air surrounding it to be completely burned. If the two are not mixed in the correct ratio, engine efficiency will drop and exhaust emission levels will increase.

The standard air-fuel ratio should be near 15 parts of air to 1 part of fuel. This measurement is calculated by weight. Actually, the most efficient ratio is stated as 14.7–1. For every pound of fuel used, 15 pounds of air would be needed, *Figure 6–18*. In terms of size, this is equal to burning 1 gallon of fuel to 9,000 gallons of air.

Rich and Lean Mixtures

A low ratio of around 12–1 suggests a *rich mixture* of fuel. A mixture of 17–1 suggests a *lean mixture*. See *Figure 6–19*. Generally, rich mixtures are less efficient during combustion. The rich mixture is used during cold weather and starting conditions. The lean mixture burns hotter than

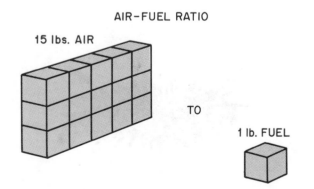

AIR-FUEL RATIO

15 lbs. AIR

TO

1 lb. FUEL

FIGURE 6–18. The air-fuel ratio should be as close as possible to 15 parts of air to 1 part of fuel. This is measured by the weight of the air and fuel.

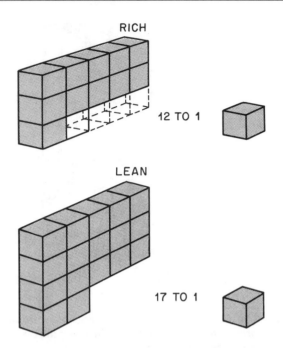

RICH

12 TO 1

LEAN

17 TO 1

FIGURE 6–19. A rich mixture means less air and too much fuel (12 to 1). A lean mixture means too much air and not enough fuel (17 to 1).

a rich mixture. Normally, the fuel acts as a coolant in the combustion process. With less fuel to cool, the combustion process gets hotter. This can cause severe damage to the pistons and valves if not corrected.

Much has been done in the past few years to control the air-fuel ratio to exact requirements. New carburetors are able to keep the mixture under better control with the use of computers and special types of fuel injection. By controlling air-fuel mixtures accurately, fuel mileage can be increased well into the 40–50 miles per gallon range for smaller sized engines.

TDC and BDC

TDC stands for top dead center. *BDC* stands for bottom dead center. TDC indicates the position of the piston when it is located at the top of its motion. When the piston is at the bottom of its travel, it is at bottom dead center (BDC). These two terms are used to help identify the position of the piston during some of the timing processes. See *Figure 6–20*.

Bore and Stroke

The *bore* and *stroke* of an engine help identify its size. The bore of the engine is defined as the diameter of the cylinder. See *Figure 6–21*. The stroke of the engine is a measurement of the distance the piston travels from the top to the bottom of its movement. It is the distance from TDC to BDC.

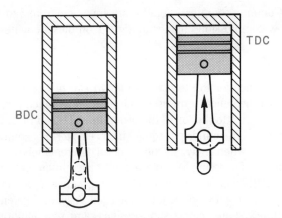

FIGURE 6-20. TDC means top dead center and is the highest point in the piston travel. BDC means bottom dead center and is the lowest point in the piston travel.

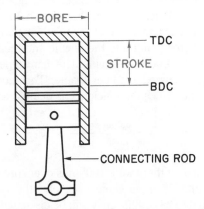

FIGURE 6-21. The bore and stroke of the engine help determine the size of the engine. Bore is the diameter of the cylinder and stroke is the distance from TDC to BDC.

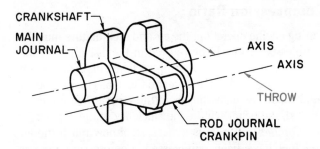

FIGURE 6-22. The distance from the crankshaft center to the crankpin center is called the throw. When this distance is multiplied by 2, the result is the stroke.

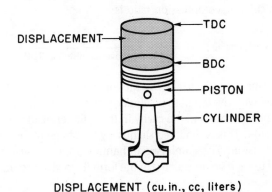

DISPLACEMENT (cu. in., cc, liters)

FIGURE 6-23. Cubic inch displacement is the volume of the cylinder from BDC to TDC. It is also stated in cubic centimeters and in liters.

The stroke is determined by the design of the crankshaft. The distance from the center of the crankshaft to the center of the crankpin is called the *throw*. See ***Figure 6-22***. If multiplied by 2, this dimension will be the same distance as the stroke. If the stroke is changed on the engine, the crankshaft will have a different length throw.

Engine Displacement

Engine *displacement* is defined as the volume of air in all of the cylinders of an engine. Each cylinder has a certain displacement. It can be determined by using the following formula:

$$\text{Displacement} = 0.785 \times \text{bore}^2 \times \text{Stroke}$$

This formula will tell the exact displacement of one piston from top dead center to bottom dead center. See ***Figure 6-23***. If there is more than one cylinder, the total displacement would be multiplied by the number of cylinders.

From this information, and using this formula, what is the displacement of an engine that has six cylinders, a bore of 3.5 inches, and a stroke of 3.70 inches?

Solution: Displacement $= 0.785 \times 3.5^2 \times 3.7 \times 6$
Displacement $= 213.4$ cubic inches

This formula calculates displacement in cubic inches. Today, however, many engines are sized by cubic centimeters (cc or cm²) and by liters. For example, today's engines are identified as 2.5 liters, 850 cc, etc. The conversion from cubic inches to cubic centimeters is:

1 cubic inch = 16.387 cc
and
1,000 cc = 1 liter

The same formula is used to calculate the displacement in metric units. In this case, the bore and stroke are measured in centimeters.

Compression Ratio

During engine operation, the air and fuel mixture must be compressed. This will be covered in the discussion of the four-cycle principle following this section. This compression helps squeeze and mix the air and fuel molecules for better combustion. Actually, the more the air and fuel are compressed, the better will be the efficiency of the engine.

Compression ratio is a measure of how much the air and fuel have been compressed. Compression ratio is defined as the ratio of the volume in the cylinder above the piston when the piston is at BDC, to the volume in the cylinder above the piston when the piston is at TDC. The compression ratio is shown in *Figure 6–24*. The formula for calculating compression ratio is:

$$\text{Compression Ratio} = \frac{\text{Volume above the piston at BDC}}{\text{Volume above the piston at TDC}}$$

In many engines, at TDC, the top of the piston is even or level with the top of the cylinder block. The combustion chamber volume is in the cavity in the cylinder head above the piston. This is modified slightly by the shape of the top of the piston. The combustion chamber volume must be added to each volume stated in the formula to give accurate results.

With this information, calculate the compression ratio if piston displacement is 45 cubic inches, and combustion chamber volume is 5.5 cubic inches.

$$\text{Compression ratio} = \frac{45 + 5.5}{5.5}$$

$$\text{Compression ratio} = 9.1 \text{ to } 1$$

Common compression ratios are anywhere from 8 to 1 on low compression engines to 25 to 1 on diesel engines.

BMEP

BMEP stands for the term *brake mean effective pressure*. This is a theoretical term used to indicate how much pressure is applied to the top of the piston from TDC to BDC. It is measured in pounds per square inch. This term becomes very useful when analyzing the results of different fuels used in engines. For example, if diesel fuel is used in an engine, more BMEP will be produced. This will produce more output power than if gasoline fuel were used. Also, as different injection systems, combustion designs, and new ignition systems are added, the BMEP of the engine is affected.

Engine Efficiency

The term *efficiency* can be used to indicate the quality of different machines. Efficiency can also pertain to engines. Engine efficiency is a measure of the relationship between the amount of energy put into the engine and the amount of energy available out of the engine. The many types of efficiency will be discussed in a later chapter. For understanding basic engine principles, efficiency is defined as:

$$\text{Efficiency} = \frac{\text{output energy}}{\text{input energy}} \times 100$$

For example, if there were 100 units of energy put into the engine, and 28 units came back out, the efficiency would be equal to 28%. Efficiency will be discussed later in this chapter.

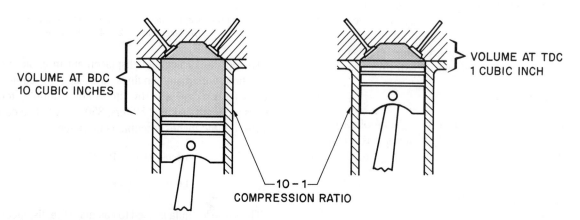

FIGURE 6–24. Compression ratio is the ratio between the volume above the piston at BDC to the volume above the piston at TDC.

ROBOTS USED FOR SEALING

Robots are used for a variety of jobs in the process of manufacturing the automobile. They are used for jobs that require high accuracy where the procedure must be repeated many times. These robots are programmed to apply sealant between the body parts of a vehicle. (Courtesy of GMF Robotics Corporation)

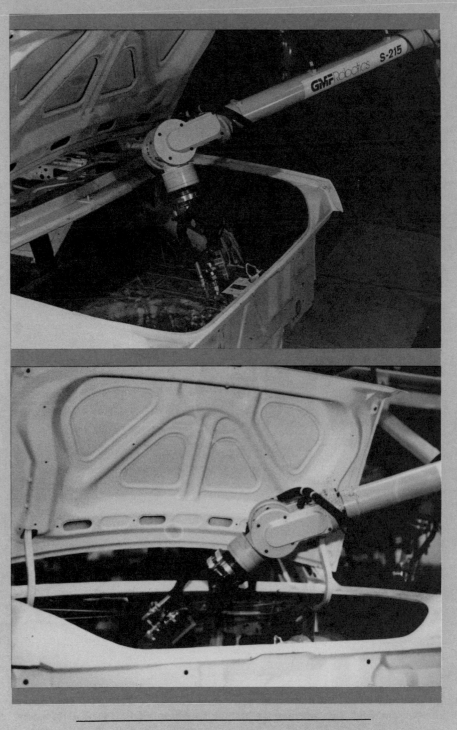

6.5 FOUR-CYCLE ENGINE DESIGN

Automotive vehicles use four-stroke engines. A four-stroke engine can also be called a *four-cycle engine*. The term *stroke* or *cycle* can be used. A four-stroke engine has a very distinct operation. These four strokes are titled intake, compression, power, and exhaust. In this section, the four-stroke cycle gasoline engine will be explained.

Intake Stroke

Refer to *Figure 6–25*. To start, the location of the piston is near TDC. Note that the intake valve is open. As the piston is cranked downward, (called the intake stroke), air and fuel are drawn into the cylinder. This occurs because as the piston moves down, a vacuum is created. When any object is removed from an area, a vacuum is created. This vacuum (lower than atmospheric pressure) draws fresh air and fuel into the cylinder. It can also be said that the atmospheric pressure pushes the air and fuel into the cylinder.

The air is first drawn through the carburetor. Here the air is mixed with the fuel at the correct air-fuel ratio (14.7–1). When the piston gets to BDC, the intake valve starts to close. With the valve closed, the air and fuel mixture is trapped in the cylinder area.

Compression Stroke

The piston now travels from BDC to TDC with air and fuel in the cylinder. This action is called the compression stroke. See *Figure 6–26*. The compression stroke takes the air-fuel mixture and compresses it according to the compression ratio of the engine. This compression causes the air and fuel to be mixed very effectively. Actually, the higher the compression ratio, the greater the mixing of air and fuel. This leads to improved engine efficiency.

It is very important that there be no leaks for the compression gases to escape. Leaks may occur in the valves, the gasket between the head and cylinder block, and past the rings on the piston. Note that at the end of the compression stroke, the crankshaft has revolved 360 degrees or one revolution.

During the compression stroke, the air and fuel is actually heated from the action of compression. It is like using an air pump to pump up a tire. As the air at the bottom of the pump is compressed, the air gets hotter. If the compression ratio is too high, temperatures within the combustion chamber may ignite the fuel. This process is referred to as preignition and can cause pinging. This means that the explosion in the combustion chamber started before the piston got to TDC.

It would be very helpful if compression ratios were increased. However, as long as air and fuel are being com-

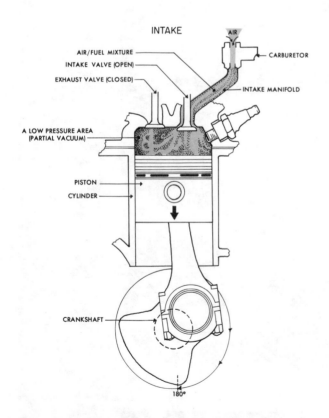

FIGURE 6–25. During the intake stroke, the piston moves down, bringing in fresh air and fuel. The intake valve is open until BDC. *(Courtesy of Breton Publishers)*

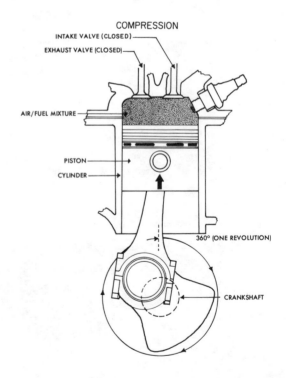

FIGURE 6–26. During the compression stroke, the piston moves from BDC to TDC. This action compresses the air and fuel in the cylinder. *(Courtesy of Breton Publishers)*

pressed, the compression ratios must be low so the air and fuel don't preignite. Higher compression ratio engines will be discussed in the diesel section of this chapter.

Power Stroke

During the power stroke, *Figure 6–27*, both the intake and exhaust valves remain closed. When the piston is coming up on the compression stroke, spark will occur very near TDC. At this point, air, fuel, and ignition are present. This causes the air and fuel to explode. When this happens, the gases expanding during the explosion push down on the top of the piston. This pressure pushes the piston downward through the power stroke. This is also when BMEP is created.

Again, it is very important for the entire combustion chamber to be sealed without any leaks. Leaks mean that some of the energy in the fuel may escape. This reduces the amount of power pushing down on the piston.

Exhaust Stroke

The last stroke in the four-cycle design is called the exhaust stroke. The exhaust stroke, *Figure 6–28*, starts when the piston starts moving upward again. The crankshaft will continue to rotate because of the flywheel weight. At the begin-

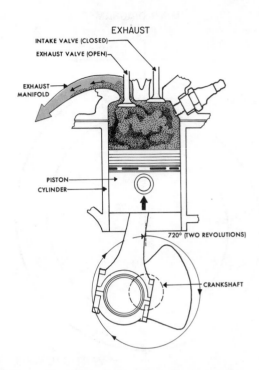

FIGURE 6–28. During the exhaust stroke, the exhaust valve opens. The upward motion of the piston pushes the exhaust gases out of the engine. *(Courtesy of Breton Publishers)*

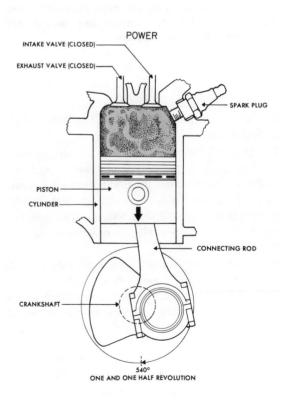

FIGURE 6–27. The power stroke is shown. Ignition occurs slightly before TDC. The combustion pushes the piston downward through the power stroke. *(Courtesy of Breton Publishers)*

ning of the exhaust stroke, the exhaust valve opens. As the piston travels upward, it pushes the burned or spent gases out of the cylinder into the atmosphere.

Near the top of the exhaust stroke, the exhaust valve starts to close. At this point, the intake valve is already starting to open for the next intake stroke. It is important to note that the crankshaft has revolved two revolutions at this point. Only one power stroke has occurred. If the engine is running at 4,000 rpm (revolutions per minute), then there are 2,000 power pulses for each cylinder per minute.

Timing Diagrams

A *timing diagram* is a method used to identify time in which all of the four stroke events occur. A timing diagram is shown in *Figure 6–29*. The diagram is set on a vertical and horizontal axis. There are 360 degrees around the axis. On the circle, events of the four-cycle engine can be graphed. One way to look at the diagram is to think of these events in terms of the position of the crankshaft and 360 degrees rotation. For example, at the top of the diagram, the piston would be located exactly at TDC. Any event that happens before TDC is referred to as BTDC (before top dead center). Any event that happens after top dead center is called ATDC (after top dead center). The mark at the bottom of the graph would illustrate the position of the piston at BDC. Two circles

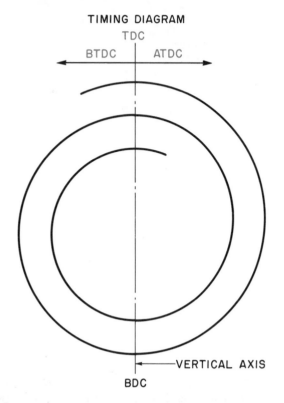

FIGURE 6–29. A timing diagram helps to identify when all of the four-stroke cycle events occur. The 360 degrees rotation represents the revolutions of the crankshaft.

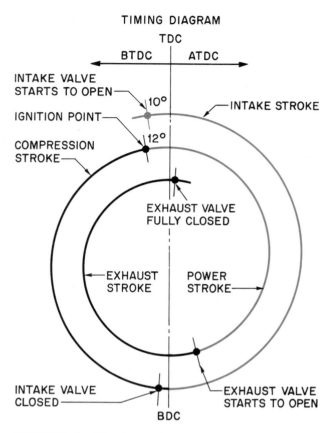

FIGURE 6–30. The timing diagram shows the events of the four-stroke engine. Intake, compression, power, and exhaust are plotted on the diagram.

are shown to represent two complete revolutions of the crankshaft. During the four strokes of operation, the crankshaft revolves two complete revolutions or 720 degrees of rotation.

Four-stroke Timing Diagram

Referring to *Figure 6–30*, follow through the four-stroke design on the timing diagram. Note that these events and degrees may vary with each engine and manufacturer. The cycle starts with the intake valve opening slightly before TDC. It should be fully open at TDC. It takes this many degrees of crankshaft rotation to open the intake valve completely.

As the piston travels downward on the intake stroke, the intake valve starts to close shortly before BDC. It is fully closed slightly after BDC. At this point, the intake stroke is completed.

The compression stroke starts when the intake valve is fully closed. The piston travels upward, compressing the air and fuel mixture. As the piston is traveling upward, the air-fuel mixture is being mixed by the compression of gases. Also, the temperature is rising inside of the combustion chamber. About 12 degrees before TDC, ignition from a spark plug occurs. The point of ignition is several degrees

before TDC. It takes about 12 degrees for the explosion or expansion to actually build up to a maximum. At TDC the expansion is at a maximum point. Now the piston is ready to be pushed downward.

If the timing of the ignition were sooner, or more degrees before TDC, then the explosion would occur too soon. This would then reduce the BMEP during the power stroke. If the timing of the ignition were too late, or after TDC, then the BMEP would also be less. It is important that maximum power from the explosion of gases occurs just when the piston is at TDC.

The power stroke starts when the piston starts downward. In this case, the power stroke is shown on the inside circle of the timing diagram. As the explosion occurs, the gases expand very rapidly. This expansion causes the piston to be forced down. This action produces the power for the engine.

Near BDC, at the end of the power stroke, the exhaust valve starts to open. By the time the piston gets to BDC, the exhaust valve is fully open. As the crankshaft continues to turn, the piston travels upward. This action forces the burned gases out of the exhaust valve into the atmosphere. The exhaust valve is fully closed a few degrees after TDC.

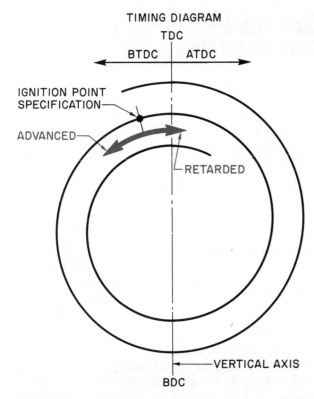

FIGURE 6–31. If the ignition timing moves farther before TDC, the engine is advanced. If the ignition timing moves toward TDC or after, the ignition is retarded.

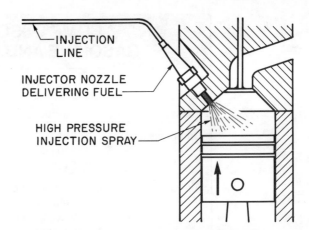

FIGURE 6–32. In a diesel engine, air and fuel are mixed near TDC. Fuel is injected into the combustion chamber under high pressure. The heat of compression then ignites the air and fuel mixture.

The time during which both the intake valve and the exhaust valve are open (near TDC) is called valve overlap.

Advance and Retarded Timing

The only part of the timing on the engine that is adjustable is the timing of the ignition. If the ignition time is moved or adjusted more BTDC, the condition is called advance timing. If the ignition time is moved or adjusted toward or after TDC (ATDC), the condition is called retarded timing. See *Figure 6–31*.

6.6 DIESEL ENGINE PRINCIPLES

The *diesel engine* is much the same as the gasoline engine in many of its principles. It is considered a four-stroke engine. The diesel engine is considered an internal combustion engine. It is also considered a compression ignition rather than a spark ignition engine.

Diesel Compression Ratio

One major difference between a diesel engine and a gasoline engine is that the diesel engine has a very high compression ratio. Compression ratios from 20 to 1 up to 25

to 1 are very common. This high compression ratio means that any fuel that is in the cylinder during compression will be ignited. Therefore, only air is brought into the cylinder during the intake stroke. There is no carburetor needed to mix the air and fuel. Fuel is injected in a diesel engine. With high compression ratios, temperatures inside the combustion chamber may be as high as 1,000 degrees F. This would be enough to ignite most fuels. This is why the diesel engine is called a compression ignition engine.

Fuel Injection

At or slightly before TDC, a fuel injector injects fuel into the combustion chamber on diesel engines. This is called *fuel injection*. A fuel injector is a device that pressurizes fuel near 20,000 psi. This fuel is injected into the combustion chamber. See *Figure 6–32*. At this point, all three ingredients are there to produce combustion. The power and exhaust strokes are the same as in the gasoline engine.

Comparison of Diesel to Gasoline Engine

Figure 6–33 shows common comparisons between the diesel and gasoline four-stroke cycle engines.

1. The intake on the gasoline engine is an air-fuel mixture. The diesel engine only has air during the intake stroke.

2. The compression pressures on the gasoline engine are lower. This is because the compression ratios are also lower.

3. The air and fuel mixing point on the gasoline engine is at the carburetor. The mixing point on the diesel engine is near top dead center or slightly BTDC.

COMPARISON BETWEEN GASOLINE AND DIESEL ENGINES		
	GASOLINE	**DIESEL**
Intake	Air-Fuel	Air
Compression	8–10 to 1 130 psi 545°F	16–20 to 1 400–600 psi 1,000°F
Air-Fuel Mixing Point	Carburetor or Before Intake Valve With Fuel Injection	Near Top Dead Center By Injection
Combustion	Spark Ignition	Compression Ignition
Power	464 psi	1,200 psi
Exhaust	1,300°–1,800°F CO = 3%	700°–900°F CO = 0.5%
Efficiency	22–28%	32–38%

FIGURE 6–33. There are several differences between a gasoline engine and a diesel engine. Differences are in the intake, exhaust, power, compression, efficiency, combustion, and air-fuel mixing point.

4. Combustion is caused by a spark plug on the gasoline engine. The diesel engine uses compression ignition.

5. The power stroke on the gasoline engine produces around 460 psi. On the diesel engine, the power stroke produces near 1,200 psi. This is because there is more energy in diesel fuel than in gasoline.

6. The exhaust temperature of the gasoline engine is much higher than that of the diesel engine. This is because some of the fuel is still burning when it is being exhausted.

7. The efficiency of the diesel engine is about 10% higher than that of the gasoline engine. This is mostly because the compression ratios are higher in the diesel engine and there is more energy in a gallon of diesel fuel.

6.7 ROTARY DESIGN (WANKEL)

In the late 1960s, several new engine designs were introduced into the automotive market. One such engine is called the *rotary engine*. The rotary design has been in existence for some time. Lately there has been a renewed interest in developing the rotary engine for automotive applications. In this converter, a rotor, instead of a piston, is used to convert chemical energy into mechanical energy. The engine is an intermittent combustion, spark ignition, rotary design (not reciprocating).

Rotary Cycle Operation

Refer to *Figure 6–34*. This is called position 1. The upper port is called the intake port. The lower port is called the exhaust port. There are no valves. The position of the center rotor opens and closes the ports much like a valve would.

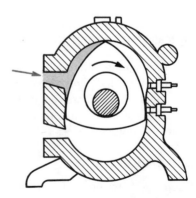

FIGURE 6–34. Intake on the rotary engine is produced because the area in position 1 becomes larger.

The rotor moves inside of an elongated circle. Because of the shape of the housing, certain areas are enlarged or compressed during rotation. As the rotor is turned, an internal gear causes the center shaft to rotate. This is the output power.

When the leading edge of the rotor face sweeps past the inlet port, the intake cycle begins. Gasoline and air (14.7–1 air-fuel ratio) is drawn into the enlarging area. This continues until the trailing edge passes the intake port.

As the rotor continues to rotate, the enlarged area is now being compressed. This is called position 2. See *Figure 6–35*. The compression ratio is very close to that of a standard gasoline engine. This is because both air and fuel are being mixed.

When the rotor travels to position 3, *Figure 6–36*, the air and fuel are completely compressed. At this point, ignition occurs from two spark plugs. There are two spark plugs for better ignition. All three ingredients are now available for correct combustion. The air and fuel explode. The explosion causes expansion of gases. This expansion pushes the rotor face downward, causing the rotor to receive a power pulse.

Figure 6–37 illustrates position 4. As the rotor continues to travel or rotate, the leading edge uncovers the exhaust port. The rotor's movement within the housing causes the exhaust gases to be forced out of the engine.

So far, only one side of the rotor has been analyzed. Note that while intake is occurring in position 1, compression is occurring on another face of the rotor, and exhaust is occurring on the third side of the rotor. This means that there are three power pulses for each rotation of the rotor.

The rest of the rotary engine uses many of the same components as the standard gasoline engine. The carburetor design is the same. The starter, alternator, and external components are the same. A complete rotary engine is shown in *Figure 6–38*.

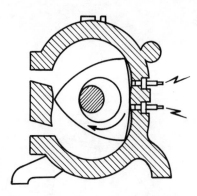

FIGURE 6–36. During the power stroke on the rotary engine, the expansion of gases causes pressure against the rotor face. This causes the rotor to continue turning.

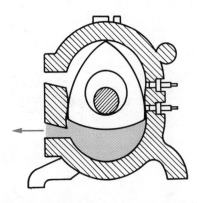

FIGURE 6–37. As the rotor continues to rotate, the leading edge uncovers an exhaust port. Exhaust gases are then forced out of the chamber.

SUMMARY

The purpose of this chapter was to introduce various engine principles. Many terms were defined and engine principles were introduced.

The engine in the automobile converts energy in the fuel to power for moving the vehicle. In this process, energy takes on several forms, including chemical, electrical, mechanical, thermal, and radiant.

Engines are classified in several ways. Combustion occurs inside the internal combustion engine. In the external combustion engine, combustion is removed from the center of the engine. Intermittent combustion starts and stops. Continuous combustion means the combustion continues all of the time. Reciprocating engines have motion that moves parts up and down. Rotary engines have continuous rotation of parts.

There are several terms that are helpful in understanding engine principles. The cylinder block is the foundation of the engine. Inside the cylinder block, there are cylinders

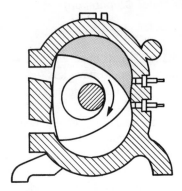

FIGURE 6–35. As the rotor continues to turn, the air and fuel are compressed to a standard compression ratio.

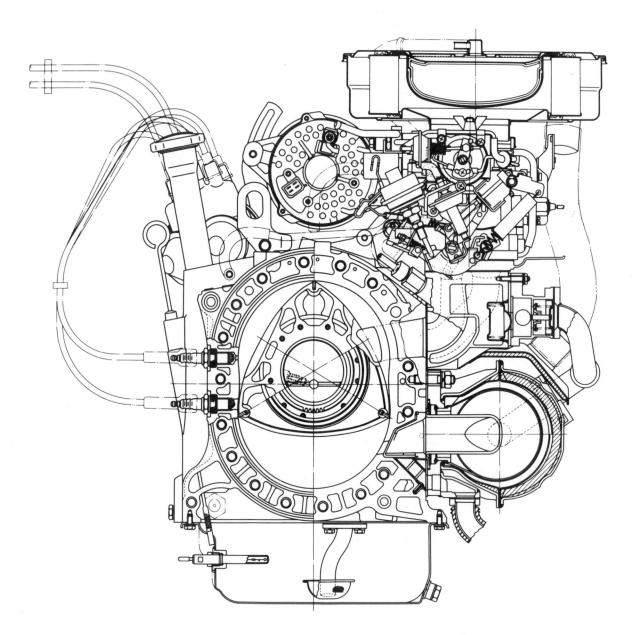

FIGURE 6-38. The rotary engine uses many of the same components as the gasoline reciprocating engine. This is a cross-cut section showing many internal parts. (Courtesy of Mazda Motor Corporation)

for pistons to fit into. The connecting rod connects the piston to the crankshaft. The cylinder head fits over the top of the engine. The combustion chamber is where combustion of the fuel takes place. The valves open and close ports to let the air and fuel in and out. The camshaft is used to help open and close the valve. The flywheel is a weight that keeps the crankshaft rotating. The carburetor mixes the air and fuel within the engine.

For an engine to work properly, there are certain combustion requirements. There must be the right air and fuel mixture. Normally 15 parts of air will mix with one part of fuel. A lean mixture would have less fuel. A rich mixture would have more fuel. This air-fuel mixture must be ignited at the correct time.

The bore and stroke of the engine help determine the size and power of the engine. Bore is the diameter of the cylinder, while stroke is the distance from TDC to BDC. These are used to determine engine displacement. Compression ratio is a measure of how much the air and fuel mixture has been compressed in the cylinder. BMEP stands for Brake Mean Effective Pressure. It is a theoretical term that indicates the amount of pressure applied to the top of a piston. Engine efficiency is a measure of the output power related to the input power. Usually gasoline engines are about 28% efficient.

The automobile engine is called a four-cycle engine. This means that there are intake, compression, power, and exhaust strokes while the engine is operating. The intake stroke brings in fresh air and fuel. The compression stroke compresses the air-fuel mixture. The mixture is ignited and the power stroke occurs. As the piston comes up again, exhaust gases are pushed out of the engine. This is the exhaust stroke. Timing diagrams are used to help show the events of the four-cycle engine. Both advance and retarded timing can be seen on the timing diagram.

The diesel engine is also being used in some automobiles today. The compression ratio is much higher, near 25–1. Fuel is injected near TDC. No carburetor or spark plug is used. Combustion starts by the heat of compression rather than by using the spark plug.

The rotary engine uses a triangular rotor to cause the intake, compression, power, and exhaust strokes. Air and fuel are mixed in a carburetor and the compression ratios are about the same as in a gasoline engine.

TERMS TO KNOW

Can you explain each of the following terms? Review the chapter until you can use each term correctly.

Energy	Carburetor
Power	Timing
Internal combustion	Air-fuel ratio
External combustion	Rich mixture
Intermittent combustion	Lean mixture
Continuous combustion	TDC and BDC
Reciprocating	Bore
Rotary	Stroke
Cylinder block	Displacement
Cylinder	Compression ratio
Piston	BMEP (brake mean effective pressure)
Crankshaft	
Connecting rod	Efficiency
Cylinder head	Four-cycle engine
Combustion chamber	Timing diagram
Valve	Diesel engine
Camshaft	Fuel injection
Flywheel	Rotary engine

REVIEW QUESTIONS

Multiple Choice

1. The ability to do work is defined as _____.
   a. Power
   b. Energy
   c. Pressure
   d. Force
   e. Work done

2. A measure of the work being done is defined as _____.
   a. Energy
   b. Power
   c. Thermal Energy
   d. Force
   e. Work started

3. The energy used to start the vehicle is called:
   a. Radiant energy
   b. Mechanical energy
   c. Thermal energy
   d. Electrical energy
   e. Forced energy

4. Which form of energy is defined as the ability to move electrons in a wire?
   a. Thermal
   b. Electrical
   c. Mechanical
   d. Radiant
   e. None of the above

5. Which type of energy does a gasoline engine convert the chemical energy to?
   a. Thermal
   b. Mechanical
   c. Nuclear
   d. A and B
   e. All of the above

6. The gasoline engine is considered which type of engine?
   a. External combustion
   b. Continuous combustion
   c. Internal combustion
   d. Rotary
   e. All of the above

7. Other methods are used to classify engines. Which of the following methods is used to classify engines?
   a. By the cycles
   b. By the ignition system
   c. By the fuel system
   d. All of the above
   e. None of the above

8. The object that is the foundation of the engine and the point to which all other parts are attached is called the:
   a. Piston
   b. Crankshaft
   c. Carburetor
   d. Block
   e. Cylinders

9. The part in the engine that moves up and down in a cylinder is called the:
   a. Piston
   b. Crankshaft
   c. Carburetor
   d. Flywheel
   e. Camshaft

10. Which part of the gasoline engine houses the valves?
    a. Carburetor
    b. Camshaft
    c. Cylinders
    d. Cylinder head
    e. Piston

11. Which object is used to keep the engine turning when the power is not being applied?
    a. Flywheel
    b. Crankshaft
    c. Camshaft
    d. Piston
    e. Connecting rod

12. Which of the following is used to mix the air and fuel?
    a. Starter
    b. Carburetor
    c. Alternator
    d. Cooling fan
    e. Cylinder head

13. What is the best air-fuel ratio for a gasoline engine?
    a. 14.7–1
    b. 18.3–1
    c. 13.2–1
    d. 12.1–3
    e. 20.0–1

14. Which of the following is considered a "lean" mixture?
    a. 13–1
    b. 10–1
    c. 17–1
    d. 11–1
    e. 12–1

15. The diameter of the cylinder is called the _____.
    a. Stroke
    b. Throw
    c. Bore
    d. Torque
    e. Force

16. Engine displacement can be measured in:
    a. Liters
    b. Cubic inches
    c. Cubic centimeters
    d. All of the above
    e. None of the above

17. When the volume above the piston at BDC is divided by the volume above the piston at TDC, the result is called:
    a. Air-fuel ratio
    b. Compression ratio
    c. Fuel injection
    d. Engine displacement
    e. Rotary displacement

18. The pressure on top of the piston during the power stroke is called:
    a. BMEP
    b. Air pressure
    c. Combustion
    d. Force
    e. All of the above

19. Which of the following is not called one of the strokes on the four-stroke cycle engine?
    a. Intake
    b. Spark
    c. Power
    d. Exhaust
    e. Compression

20. Which of the following events on the four-stroke engine is/are adjustable?
    a. Ignition time
    b. Exhaust valve opening time
    c. Intake valve opening time
    d. Top dead center
    e. Bottom dead center

21. What is the compression ratio on some diesel engines?
    a. 25–1
    b. 8–1
    c. 10–1
    d. 6–1
    e. 5–1

22. The diesel engine uses _____ to mix the air and fuel correctly.
    a. Fuel injection
    b. Carburetors
    c. Spark plugs
    d. Camshafts
    e. Electric motors

23. The diesel engine ignites the fuel from:
    a. The heat of compression
    b. A spark plug
    c. The fuel injection
    d. The igniter
    e. Exhaust from the first cylinder

24. How many power pulses are there per rotor revolution on the rotary engine?
    a. 1
    b. 2
    c. 3
    d. 4
    e. 5

25. What object opens and closes the ports on the rotary (Wankel) engine?
    a. The valves
    b. The position of the rotor
    c. The carburetor
    d. The spark plug
    e. The camshaft

The following questions are similar in format to ASE (Automotive Service Excellence) test questions.

26. Technician A says that an automobile engine converts chemical energy to thermal energy. Technician B says that an automobile engine converts thermal energy to mechanical energy. Who is right?
    a. A only
    b. B only
    c. Both A and B
    d. Neither A nor B

27. Technician A says the compression ratio on a diesel engine is higher than on a gasoline engine. Technician B says the compression ratio on a diesel engine is lower than on a gasoline engine. Who is right?
    a. A only
    b. B only
    c. Both A and B
    d. Neither A nor B

28. Technician A says the order of strokes on a four-cycle engine is intake, power, compression, and exhaust. Technician B says the order of strokes on a four-cycle engine is intake, compression, power, and exhaust. Who is right?
    a. A only
    b. B only
    c. Both A and B
    d. Neither A nor B

29. Technician A says the gas engine used in a car is called an external combustion engine. Technician B says the gas engine used in a car is called an internal combustion engine. Who is right?
    a. A only
    b. B only
    c. Both A and B
    d. Neither A nor B

Essay

30. Describe how the intake, compression, power, and exhaust strokes occur in a Wankel engine.

31. What is the purpose of using a cylinder head on an engine?

32. Define the term *timing* on an engine.

33. Why is it so important to get the exact air-fuel ratio on an engine?

34. How is engine displacement calculated?

35. Define compression ratio and describe how it can be increased and decreased.

CHAPTER 7

Engine Measurements

INTRODUCTION

Over the past years, automotive manufacturers have built many sizes and types of engines. These engines differ in the amount of power they can produce. Horsepower, torque, fuel consumption, and efficiency have changed and improved. The purpose of this chapter is to define the many terms used to measure engine performance and operation.

OBJECTIVES

After reading this chapter, you will be able to:
- Define the term horsepower.
- Compare the different types of horsepower.
- Relate torque to horsepower.
- Examine the use of dynamometers.
- Analyze performance charts.
- Compare different types of engine efficiency.

CHAPTER HIGHLIGHTS

7.1 TYPES OF HORSEPOWER
   A. Work
   B. Torque
   C. Horsepower Defined
   D. Brake Horsepower
   E. Indicated Horsepower
   F. Frictional Horsepower
   G. Reducing Frictional Horsepower
   H. Road Horsepower

7.2 DYNAMOMETERS
   A. Definition of Dynamometer
   B. Performance Charts

7.3 ENGINE EFFICIENCY
   A. Mechanical Efficiency
   B. Volumetric Efficiency
   C. Thermal Efficiency

SUMMARY

7.1 TYPES OF HORSEPOWER

Many types of horsepower are used for automobile engines. When comparing engines, brake horsepower is used. When discussing efficiency, frictional and indicated horsepower is used. When analyzing gasoline mileage, road horsepower is used. These and other horsepower definitions should be analyzed. To do this, the term work must first be discussed.

Work

Work is defined as the result of applying a force. This force is created by a source of energy. When the force moves a certain mass a certain distance, work is produced. Work is defined as shown in *Figure 7–1*.

Force is measured in pounds. Distance is measured in feet. When the two units are put together, ft. lbs. are measured. Work, then, is measured in ft. lbs. For example,

as shown in *Figure 7–2*, if a vehicle were moved 50 feet with a force of 20 pounds, then 1,000 ft. lbs. of work would be produced.

Torque

Torque is one way to measure work. Torque is defined as twisting force. See *Figure 7–3*. This force is produced in an engine because of the combustion of fuel. Combustion pushes the piston down. This causes the crankshaft to rotate, producing torque. This force causes the wheels to rotate.

Torque is actually available at the rear of the engine. Torque is expressed in foot pounds (energy needed to move a certain number of pounds one foot). An engine is said to have 500 ft. lbs. of torque at a certain speed. Speed on a gasoline or diesel engine is measured in *rpm*. This term means revolutions per minute. Torque can be measured

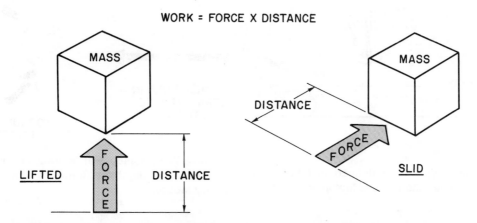

FIGURE 7–1. Work is defined as the result of moving a certain mass a certain distance. It is measured in ft. lb. The movement can be lifting or sliding motion.

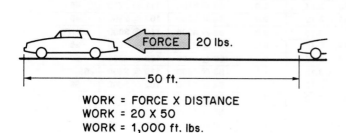

FIGURE 7–2. When a vehicle is pushed by 20 pounds of force a distance of 50 feet, 1,000 foot pounds of work are created.

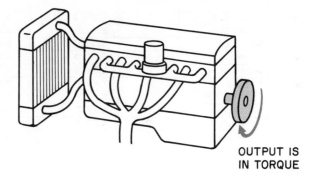

FIGURE 7–3. Torque is defined as twisting force. The work that an engine produces is measured as torque at the back of the engine. *(Courtesy of DCA Educational Products)*

directly from a rotating shaft by using a dynamometer. Dynamometers will be discussed later in this chapter.

Horsepower Defined

Horsepower (hp) is also a measure of the work being done. It is a unit of work or a measure of work done within a certain time. Horsepower is defined as the *rate* at which work is being done. When anything is measured by a rate, time is considered. Therefore, horsepower is defined as how long it takes to do work.

One horsepower is defined as the amount of work needed to lift 550 pounds one foot in one second. See *Figure 7–4*. If this work is measured per minute (rather than per second), one horsepower is defined as the amount of work needed to lift 33,000 pounds one foot in one minute.

These two definitions are the standard way of defining horsepower. It is important to note that the direction of motion when horsepower is applied is in a straight line. However, torque is always related to rotation. See *Figure 7–5*.

Figure 7–6 shows the horsepower and torque specifications from a service manual. Note that both the torque and horsepower are stated at a specific rpm.

Brake Horsepower

Brake horsepower (bhp) is defined as the actual horsepower measured at the rear of the engine under normal conditions. It is called brake horsepower because a brake is used to slow down the shaft inside of a dynamometer. Brake horsepower is often used to compare engines and their characteristics. Automotive manufacturers use brake horsepower to show differences between engines. For example, a 235 cubic inch engine will produce less bhp than a 350 cubic inch engine.

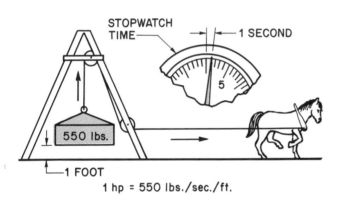

FIGURE 7–4. One horsepower is defined as the amount of work required to raise 550 lbs. one foot in one second.

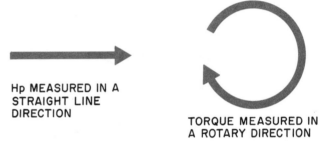

FIGURE 7–5. Horsepower is always a measure of work applied in a straight line. Torque is a measure of force in rotary motion.

General Engine Specifications

Year	Engine		Carburetor	Bore and Stroke	Compression Ratio	Net H.P. @ R.P.M.③	Maximum Torque Ft. Lbs. @ R.P.M.	Normal Oil Pressure Pounds
	CID①/Liter	V.I.N. Code②						
1982	4-112/1.8L⑤	G	E2SE, 2 Bbl.④	3.50 × 2.90	9.0	85 @ 5100	100 @ 2800	45
	4-110/1.8L⑥	O	E.F.I.⑦	3.34 × 3.12	9.0	84 @ 5200	102 @ 2800	65
	4-121/2.0L	B	E2SE, 2 Bbl.④	3.50 × 3.14	9.0	90 @ 5100	111 @ 2700	45
1983	4-110/1.8L⑥	O	E.F.I.⑦	3.34 × 3.12	9.0	84 @ 5200	102 @ 2800	45
	4-121/2.0L	P	E.F.I.⑦	3.50 × 3.15	9.3	86 @ 4900	110 @ 3000	63–77
1984	4-110/1.8L⑥	O	E.F.I.⑦	3.34 × 3.12	9.0	84 @ 5200	102 @ 2800	65
	4-110/1.8L⑥⑧	J	E.F.I.⑦	3.34 × 3.12	8.0	150 @ 5600	150 @ 2800	65
	4-121/2.0L	P	E.F.I.⑦	3.50 × 3.15	9.3	86 @ 4900	110 @ 2400	63–77
1985	4-110/1.8L⑥	O	E.F.I.⑦	3.34 × 3.12	9.0	84 @ 5200	102 @ 2800	65
	4-110/1.8L⑥⑧	J	E.F.I.⑦	3.34 × 3.12	8.0	150 @ 5600	150 @ 2800	65
	4-121/2.0L	P	E.F.I.⑦	3.50 × 3.15	9.3	86 @ 4900	110 @ 2400	63–77
	V6-173/2.8L	W	E.F.I.⑦	3.50 × 2.99	8.9	130 @ 4800	160 @ 3600	50–55
1986	4-110/1.8L⑥	O	E.F.I.⑦	3.34 × 3.13	8.8	84 @ 5200	98 @ 2800	65
	4-110/1.8L⑥⑧	J	E.F.I.⑦	3.34 × 3.13	8.0	150 @ 5600	150 @ 2800	65
	4-121/2.0L	P	E.F.I.⑦	3.50 × 3.15	9.0	88 @ 4800	110 @ 2400	63–77
	V6-173/2.8L	W	E.F.I.⑦	3.50 × 2.99	8.5	120 @ 4800	155 @ 3600	50–55

FIGURE 7–6. Horsepower and torque are usually stated for each engine type in a service manual. *(Courtesy of Motor Publications, Auto Repair Manual, 1981–1987)*

Other factors that may change bhp include type of carburetor, quality of combustion, compression ratio, type of fuel, and air-fuel ratio.

Indicated Horsepower

Indicated horsepower (ihp) is defined as theoretical horsepower. Indicated horsepower has been calculated by the automotive manufacturers. Ihp represents the maximum horsepower available from the engine under ideal or perfect conditions. Ihp is calculated based on the engine size, displacement, operational speed, and the pressure developed theoretically in the cylinder. Indicated horsepower will always be more than bhp.

Frictional Horsepower

Frictional horsepower (fhp) is defined as the horsepower used to overcome internal friction. Any time two objects touch each other while moving, friction is produced. Friction must be overcome with more energy. This happens within an engine. Sources of frictional horsepower include bearings, pistons sliding inside the cylinder, the compression stroke, the generator, fan, water pump, belts, air conditioner, and so on. Refer to *Figure 7–7*.

Other sources of frictional horsepower losses outside the engine include the wind, tire rolling resistance, road conditions, and so on. All of these have a tendency to slow down the engine. They make up the frictional horsepower.

It is advantageous to reduce frictional horsepower as much as possible. The more frictional horsepower that the engine must overcome, the more that will be needed to operate the vehicle. This means poorer fuel mileage on the vehicle.

Reducing Frictional Horsepower

Frictional horsepower has been analyzed very carefully in the past few years. Research efforts have found that poor gasoline mileage occurred because of large amounts of frictional horsepower. Considerable changes have been made in the automobile to reduce frictional losses. Some of these changes include:

1. Reducing the rolling resistance on tires. Tires are designed by using computers to reduce the rolling resistance. Radial tires also reduce rolling resistance.
2. Reducing the air drag on a vehicle. Making the vehicle have less wind resistance.

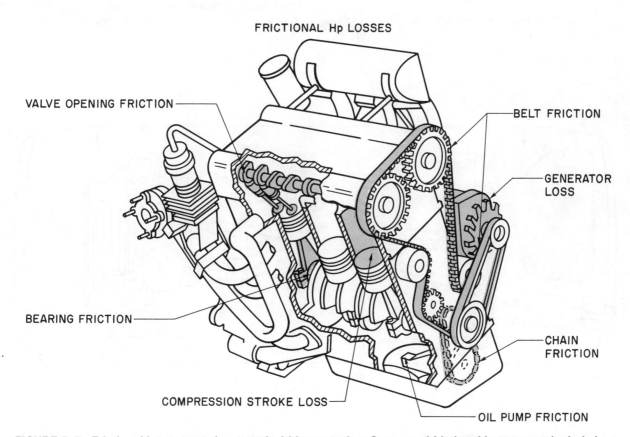

FIGURE 7–7. Frictional horsepower is created within an engine. Sources of frictional horsepower include bearing friction, the compression stroke, belts, generator, valve opening, and so on. *(Courtesy of Peugeot Motors of America, Inc.)*

3. Running the cooling fan on an electric motor, rather than from the engine. The fan now operates only when needed. Also, some fans turn off and on by using a clutch system.

4. Making the vehicle lighter. On the average, there is one mile per gallon (mpg) lost in fuel for every 400 pounds on the vehicle.

5. Changing the undercarriage of the vehicle to reduce air drag on the bottom.

These and other designs have enabled the automotive manufacturers to improve gasoline mileage from 12–15 miles per gallon to over 50 miles per gallon on some vehicles.

Road Horsepower

Road horsepower is defined as the horsepower available at the drive wheels of the vehicle. Road horsepower will always be less than bhp. The difference between road horsepower and brake horsepower is the result of frictional horsepower. Frictional horsepower losses are also produced from the friction in the transmission, drive shaft, and rear differential assemblies. See **Figure 7–8**. Road horsepower can be shown as:

Road Hp = bhp − fhp through the drive train

CAR CLINIC

PROBLEM: THROTTLE STICKS

The carburetor on an engine seems to continually gum up and the throttle gets sticky, especially when the engine is cold. The carburetor has been cleaned several times. Over the years, the owner has used many cans of carburetor cleaner in the tank and often uses gasohol. What could cause the sticky throttle?

SOLUTION:

The gum-up problem could be caused by too much alcohol in the fuel. Gasoline would not normally gum up that badly. Also, check for worn piston rings and valve guides, which could also cause gummy deposits in the carburetor. The combustion gases produced by bad rings or valve guides are routed through the carburetor by the PCV system.

7.2 DYNAMOMETERS

Definition of Dynamometer

At times it may be necessary to *load* the engine while operating. Loading the engine is the same as pulling a trailer up a steep hill. A *dynamometer* is a device attached to the

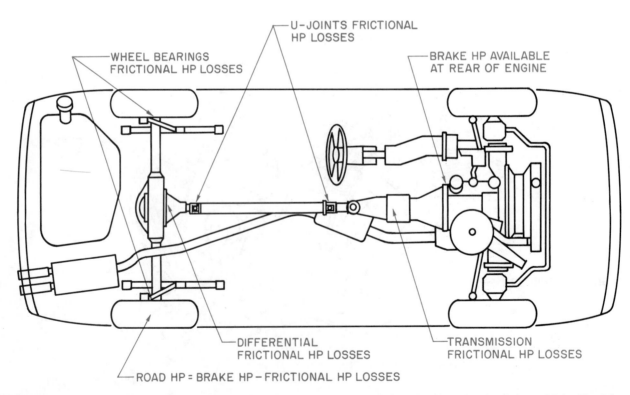

FIGURE 7–8. Road horsepower is defined as the horsepower available at the drive wheels of the vehicle. The frictional horsepower from the drive train is subtracted from the brake horsepower of the engine. The result is road horsepower. (Courtesy of Mazda Motors of America, Inc.)

DYNO TESTING

Engine and vehicle performance can be tested while the car is in the shop. A dynamometer is used to load down the car. The front drive wheels are placed on two rollers that are built into the floor. A turbine is attached under the floor, to the rollers. The rollers can be slowed down by increasing the amount of liquid the turbine must flow through. This acts as a load on the drive wheels. Performance testing can give the mechanic much more information when troubleshooting an engine.

ENGINE PERFORMANCE CURVE

Engine performance curves are used to illustrate horsepower and torque at different rpm. This performance chart shows the difference between a non-turbocharged and a turbocharged engine. (Courtesy of Chrysler Corporation)

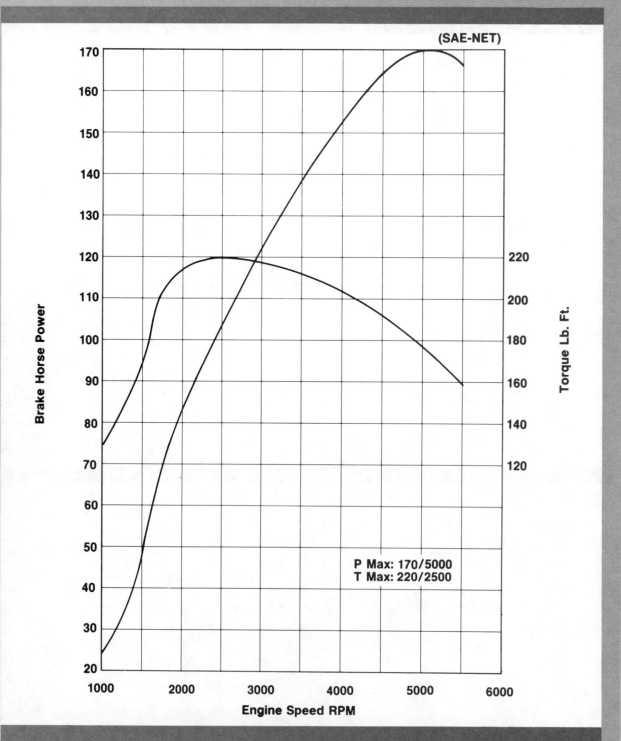

back of the engine to absorb the power being created by the engine. When the engine is at idle, it is impossible to determine how much horsepower or torque can be produced. If an engine is run on a dynamometer, it can be loaded down to simulate actual driving conditions. *Figure 7–9* shows a dynamometer attached to the rear of an engine on a stand. The *engine dynamometer* measures brake horsepower and torque at the output of the engine.

Another type of dynamometer can be used to measure road horsepower. It is called a *chassis dynamometer*. A chassis dynamometer measures the horsepower and torque available at the drive wheels of the vehicle. This dynamometer measures road horsepower. *Figure 7–10* shows the layout of a chassis dynamometer. In this case, the tires roll on two rollers. These are called the idle roll and the drive roll. The power absorption unit absorbs the energy. This unit acts as the load on the vehicle. Both speed (rpm) and torque are measured on the scales shown.

Performance Charts

Gasoline and diesel engines have certain operating characteristics. This means that they have different torque, horsepower, and fuel consumption at different rpm. By using a dynamometer, a *performance chart* (also called a characteristic curve) can be developed. *Figure 7–11* shows a standard performance chart. The bottom axis shows the rpm of the engine. The left axis shows the bhp on the engine. The right axis shows the torque being produced on the engine. Also, note that on the bottom right, a fuel consumption scale is included.

When the engine is loaded with a dynamometer, a certain maximum torque and horsepower can be produced at a specific rpm. For example, referring to Figure 7–11, this particular engine is capable of producing about 130 ft. lbs. of torque at 1,500 rpm. Also, this engine can produce about

65 bhp at 3,500 rpm. This chart shows the characteristics of the engine throughout its rpm range. A fuel consumption curve is also shown. This indicates the amount of fuel used in pounds per brake horsepower per hour. This unit is sometimes referred to as BSFC or brake specific fuel consumption.

It should be noted that a dynamometer can only measure the torque being produced at the rear of the engine or at the drive wheels. The dynamometer does not measure horsepower. The following formula is used to obtain horsepower readings so the curve can be made:

$$\text{Horsepower} = \frac{\text{Torque} \times \text{rpm}}{5,252}$$

The number 5,252 is called a constant and is related to the definition of one horsepower (33,000 foot pounds per minute).

CAR CLINIC

PROBLEM: FOULED PLUG

A customer has an engine that consistently fouls out on number 5 cylinder. The vehicle has 76,000 miles on it. The number 5 plug is very wet, oily, and carbon filled. Even a new plug will foul out in a short time. Why would only one plug foul out and not the others?

SOLUTION:

The most common cause of having just one plug foul out is a bad spark plug or spark plug wire. Since the plug has been changed, it would be wise to check the spark plug wires. This can be done on an electronic engine diagnosis scope or analyzer. When was the last time the spark plug wires were changed? They do eventually go bad. Replace the spark plug wires with a new set to eliminate the problem.

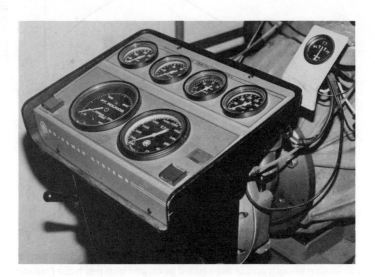

FIGURE 7–9. An engine dynamometer, used to load the engine, can be attached to the rear of the engine.

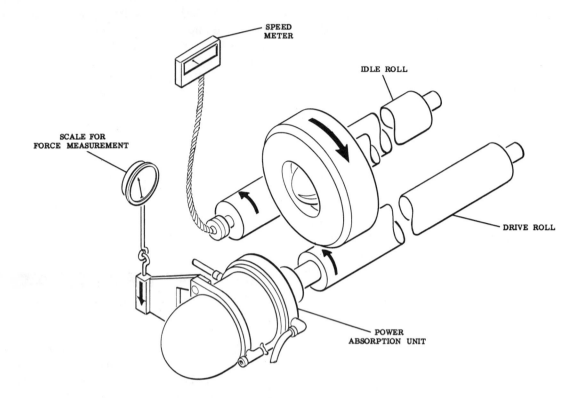

FIGURE 7–10. A chassis dynamometer is used to measure the road horsepower. The tires roll directly on the idle and drive roll. The absorption unit loads the system. *(Courtesy of Clayton Manufacturing Company)*

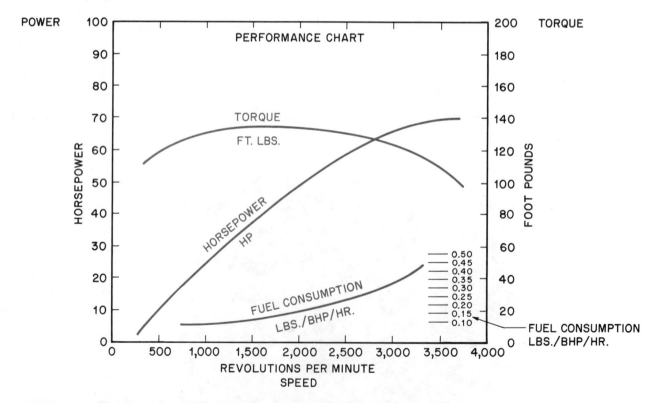

FIGURE 7–11. Performance charts show the amount of horsepower and torque that an engine can produce for a range of rpm. Fuel consumption is also shown and is measured in lb. per bhp per hour (pounds per brake horsepower per hour). *(Courtesy of DCA Educational Products)*

7.3 ENGINE EFFICIENCY

The term *efficiency* means many things in the automotive field. Efficiency generally refers to how well a particular job can be done. It is usually expressed as a ratio of input to output. There are, however, other types of efficiency, including mechanical efficiency, volumetric efficiency, and thermal efficiency.

Efficiencies are expressed as percentages. They are always less than 100%. The difference between the efficiency and 100% is the percentage lost during the process.

Mechanical Efficiency

One way to show efficiency is by measuring the mechanical systems of the machine. This measures how efficient the mechanical systems are in a machine. The machine we are concerned about is the gasoline engine. *Mechanical efficiency* is a relationship between the theoretical (ihp) amount of work required to do a certain job and the actual (bhp) amount of work required to do the job. For example, if a certain car requires 185 actual horsepower and the theoretical horsepower required to do the same amount of work is 205, the mechanical efficiency can be calculated. The formula to calculate mechanical efficiency is:

$$\text{Mechanical Efficiency} = \frac{\text{Actual horsepower}}{\text{Theoretical horsepower}} \times 100$$

$$\text{Mechanical Efficiency} = \frac{185}{205}$$

$$\text{Mechanical Efficiency} = 90\%$$

The losses on any mechanical system are due primarily to friction. If frictional horsepower can be reduced on an engine, the mechanical efficiency will increase. If frictional horsepower increases on an engine, the mechanical efficiency will decrease.

Volumetric Efficiency

Another way to measure the efficiency of an engine is related to how easily air flows in and out of the engine. As the piston starts down on the intake stroke, air (and fuel) flow into the engine. As the engine increases in revolutions per minute, the intake valves are not open for as long a time. This means that the amount of air per time period may be less. *Volumetric efficiency* measures this condition. The formula for measuring volumetric efficiency is:

$$\text{Volumetric Efficiency} = \frac{\text{Actual air used}}{\text{Maximum air possible}} \times 100$$

For example, at a certain engine speed, 40 cubic inches of air-fuel mixture enter the cylinders. However, to completely fill the cylinder, 55 cubic inches should enter. Using these two numbers:

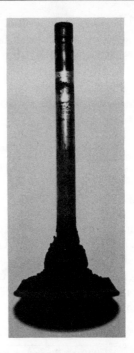

FIGURE 7–12. Volumetric efficiency of an engine can be reduced by restrictions to the air flow in and out of the engine. The valve shows a buildup of carbon deposits. This condition would reduce the volumetric efficiency of the engine.

$$\text{Volumetric Efficiency} = \frac{40}{55} \times 100$$

$$\text{Volumetric Efficiency} = 72\%$$

One way to improve the volumetric efficiency is to improve the scavenging of the cylinder. This means to improve the ease at which air and fuel can enter the engine. Over a period of time the valves may have a buildup of carbon deposits. This is shown in *Figure 7–12*. It is obvious that this condition will reduce the volumetric efficiency of the engine.

Other factors that will affect volumetric efficiency are:

1. Exhaust restriction.
2. Air cleaner restrictions.
3. Carbon deposits on cylinders and valves.
4. Shape and design of valves.
5. Amount of restriction in the intake and exhaust ports by curves. Ports can be polished to reduce friction.

Thermal Efficiency

A more specific form of efficiency is called *thermal efficiency*. Thermal efficiency tells how effectively an engine converts the heat energy in its fuel into actual power at the output shaft. It takes into account all of the losses on the engine, including thermal losses, mechanical losses, and volumetric losses. For this reason, thermal efficiency is sometimes called "overall efficiency." It is the most common form of efficiency used to compare engines.

Thermal efficiency is found by using the following formula:

$$\text{Thermal Efficiency} = \frac{\text{Actual output}}{\text{Heat input}}$$

When using this formula, always make sure the units of input and output are the same. The heat input is expressed in Btu's. A Btu is an amount of heat needed to raise one pound of water one degree F. A gallon of gasoline has approximately 110,000 Btu's.

To get the actual output in the same unit, note that one horsepower is equal to 42.5 Btu/min. Therefore, the formula can be shown as:

$$\text{Thermal Efficiency} = \frac{\text{bhp} \times 42.4 \text{ Btu/min.}}{110,000 \text{ Btu/gal.} \times \text{gal. per min. (gpm)}}$$

Relating this efficiency to a gasoline engine, approximately 25% of the input energy is available at the output. Referring to **Figure 7–13**, the remaining part of the input

EFFICIENCIES OF DIFFERENT ENGINES	
Gasoline Engine	25–28%
Diesel Engine	35–38%
Aircraft Gas Turbine	33–35%
Liquid Fuel Rocket	46–47%
Rotary Engine	20–22%
Steam Locomotive	10–12%

FIGURE 7–14. All machines have an overall efficiency. Gasoline engines are approximately 25% efficient. Diesel engines are about 36% efficient. Other engine efficiencies are shown for comparison.

energy is lost through various ways. The cooling system absorbs a certain percentage of the input energy. The exhaust system carries away a certain amount of energy. Nine percent of the input energy is lost through radiation. When all of these losses are added together, the output energy drops to about 25%.

Different machines have different efficiencies. **Figure 7–14** shows some of the more common machines and their thermal efficiencies. Diesel engines are about 10% more efficient than gasoline engines.

Many things affect efficiency. Some of these factors include the amount of energy in the fuel, the quality of combustion, the amount of frictional loss, and the mechanical quality of the machine. It is important to improve the efficiency of the automobile so that fuel mileage can be increased.

SUMMARY

The purpose of this chapter was to introduce different engine measurements. Horsepower, torque, dynamometers, and engine efficiency are all part of different engine measurements.

Work is defined as the result of applying a force. It is part of the definition of horsepower. Torque is also a way to measure work. Torque is defined as twisting force. When torque and rpm are used, horsepower can be calculated. Horsepower is a measure of the work being done. It is the rate of work being done.

Several types of horsepower are used to describe engine performance. These include brake horsepower, indicated horsepower, frictional horsepower, and road horsepower. Frictional horsepower can be reduced in several ways.

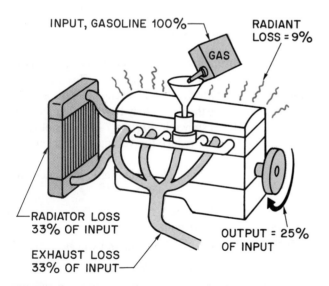

FIGURE 7–13. A gasoline engine loses much of its energy to other systems. 33% of the input energy is lost through the radiator. 33% is lost to the exhaust. 9% is lost from radiation. Approximately 25% is left for power to the rear wheels. (Courtesy of DCA Educational Products)

These include reducing rolling resistance and air drag, making the vehicle lighter, and operating small loads such as the cooling fan from electrical sources.

Horsepower and torque are measured on a dynamometer. This is a device that loads the engine so horsepower can be produced. There are two types of dynamometer: the engine dynamometer and the chassis dynamometer.

When an engine is loaded on the dynamometer, a performance chart is made from the data. The dynamometer can only measure torque and rpm. Horsepower has to be calculated for different rpm.

Engine efficiency is also a way to indicate engine measurements. There are several types of efficiency. Mechanical efficiency measures the mechanical systems. It is the relationship between actual horsepower and theoretical horsepower. Volumetric efficiency measures how easily the air flows in and out of the engine. It is the relationship between the actual air used and the maximum air possible. Thermal efficiency measures how well the engine converts heat energy into actual power at the output shaft. It is the relationship between actual power output and heat input.

TERMS TO KNOW

Can you explain each of the following terms? Review the chapter until you can use each term correctly.

Work	Load
Torque	Dynamometer
Horsepower	Chassis dynamometer
Revolutions per minute (rpm)	Engine dynamometer
	Performance chart
Brake horsepower	Efficiency
Indicated horsepower	Mechanical efficiency
Frictional horsepower	Volumetric efficiency
Road horsepower	Thermal efficiency

REVIEW QUESTIONS

Multiple Choice

1. When force is multiplied by distance the result is called _____ .
   a. Work
   b. Rpm
   c. Torque
   d. Distance
   e. Pressure

2. Which of the following work units are measured in a straight line?
   a. Torque
   b. Horsepower
   c. Rpm
   d. All of the above
   e. None of the above

3. Which of the following is true about horsepower and torque?
   a. Torque is a measure of rotation; horsepower is measured in a straight line.
   b. Horsepower is measured in time.
   c. Torque is measured in foot pounds.
   d. All of the above.
   e. None of the above.

4. When 550 pounds are lifted in one _____ a distance of one foot, one horsepower is created.
   a. Second
   b. Minute
   c. Hour
   d. Day
   e. Month

5. Horsepower measured at the rear of the engine is called:
   a. Road horsepower
   b. Brake horsepower
   c. Indicated horsepower
   d. Rear engine horsepower
   e. Frictional horsepower

6. Horsepower measured at the drive wheels of a vehicle is called:
   a. Road horsepower
   b. Brake horsepower
   c. Indicated horsepower
   d. Frictional horsepower
   e. Theoretical horsepower

7. Theoretical horsepower is referred to as:
   a. Road horsepower
   b. Frictional horsepower
   c. Indicated horsepower
   d. Brake horsepower
   e. Torque

8. Horsepower lost because of friction in the engine is called _____ horsepower.
   a. Frictional
   b. Chassis
   c. Indicated
   d. Heat horsepower
   e. All of the above

9. The greater the frictional horsepower, the
   a. Better the gasoline mileage
   b. Better the efficiency
   c. Less the gasoline mileage
   d. More the power available
   e. Better the performance of the vehicle

10. Which chart is not shown on a performance curve?
   a. Brake horsepower
   b. Torque
   c. Frictional horsepower
   d. Fuel consumption
   e. Indicated horsepower

11. How is a load applied to a vehicle to measure road horsepower?
   a. By a chassis dynamometer
   b. By an engine dynamometer
   c. By an rpm gauge
   d. Road horsepower cannot be measured
   e. By a pressure gauge

12. Fuel consumption is measured by what unit?
   a. Lbs./ihp/hour
   b. Lbs./bhp/hour
   c. Lbs./fhp/hour
   d. Gallons per day
   e. Quarts per hour

13. Which efficiency measures the air flow in an engine?
   a. Volumetric
   b. Thermal
   c. Mechanical
   d. Air flow efficiency
   e. Electrical efficiency

14. Which efficiency measures the theoretical horsepower versus the actual horsepower?
   a. Thermal
   b. Mechanical
   c. Volumetric
   d. Indicated
   e. Electrical

15. Which of the following could have a negative effect on volumetric efficiency?
   a. An exhaust restriction
   b. A dirty air cleaner
   c. Heavy carbon deposits on the valves
   d. All of the above
   e. None of the above

The following questions are similar in format to ASE (Automotive Service Excellence) test questions.

16. Technician A says as frictional horsepower increases fuel consumption decreases. Technician B says as frictional horsepower increases engine efficiency decreases. Who is right?
   a. A only
   b. B only
   c. Both A and B
   d. Neither A nor B

17. Technician A says that if a vehicle has a dirty air cleaner, the volumetric efficiency will decrease. Technician B says that if a vehicle has a dirty air cleaner, the mechanical efficiency will decrease. Who is right?
   a. A only
   b. B only
   c. Both A and B
   d. Neither A nor B

18. Technician A says that torque and horsepower are different terms and have different meanings. Technician B says the two terms are the same. Who is right?
   a. A only
   b. B only
   c. Both A and B
   d. Neither A nor B

19. The actual horsepower is 80 and the theoretical horsepower is 100. Technician A says the mechanical efficiency of the engine is 80 percent. Technician B says the mechanical efficiency of the engine is 100 percent. Who is right?
   a. A only
   b. B only
   c. Both A and B
   d. Neither A nor B

Essay

20. Identify several ways in which frictional horsepower can be reduced.

21. What is the formula for calculating horsepower when both torque and rpm are known?

22. What is the difference between mechanical and volumetric efficiency?

23. Define thermal efficiency.

24. Define frictional horsepower.

CHAPTER 8

Types of Engine Design

INTRODUCTION

Gasoline and diesel engines that are used on automobiles can be designed in different styles, types, and configurations. *Configuration* means the figure, shape, or form of the engine. This means that some engines may have an overhead cam, while others may have valves in the block. The principles of the engine design remain the same, but the location and configuration change. Some of the shapes the engine can take include in-line or V, slant or opposed, number of cylinders, types of head design, location of valves, and shape of the block. The purpose of this chapter is to investigate different types and styles of engines used in the automobile. Alternative engine types will also be studied.

OBJECTIVES

After reading this chapter, you will be able to:
- Identify the difference between in-line, V, slant, and opposed piston and cylinder arrangement.
- Define the differences between the 4, 6, 8, and 12 cylinder engines.
- Compare the differences between the L, I, F, and T head design.
- Analyze the difference between overhead, in-block, and dual camshaft design engines.
- Compare two-valve and four-valve engines.
- Examine the operation of the stratified charged engine.
- Identify the operation of the stirling cycle engine.
- Analyze the design and operation of the gas turbine engine.

CHAPTER HIGHLIGHTS

8.1 CYLINDERS AND ARRANGEMENT
    A. In-line Engines
    B. V-configuration Engines
    C. Slant Cylinder Engines
    D. Opposed Cylinder Engines
    E. Radial Cylinder Engines
    F. Number of Cylinders
    G. Five-cylinder and Three-cylinder Engines

8.2 VALVE AND HEAD ARRANGEMENT
    A. I-head Design
    B. L-head Design
    C. F-head and T-head Designs
    D. In-block Camshaft
    E. Overhead Camshaft
    F. Dual Camshafts
    G. Four-valve Engines
    H. Stratified Charged Engines

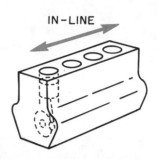

8.1 CYLINDERS AND ARRANGEMENT

Automobiles use many styles and types of engines. One style used is identified by its shape and number of cylinders. Depending upon the vehicle, either an in-line, V, slant, or opposed cylinder arrangement can be used. Common engine designs are the 4-, 6-, 8-, or 12-cylinder engines.

In-line Engines

Engines can be designed as an *in-line* style. This means the cylinders are all placed in a single row, *Figure 8–1*. There is one crankshaft and one cylinder head for all of the cylinders. The block is a single cast piece with all cylinders located in an upright position.

In-line engine designs have certain advantages and disadvantages. They are easy to manufacture, which brings the cost down somewhat. They are very easy to work on and to perform maintenance on. In-line engines have adequate room under the hood to work on other vehicle parts. However, because the cylinders are positioned vertically, the front of the vehicle must be higher. This affects the *aerodynamic* design of the car. Aerodynamic design refers to the ease at which the car can move through the air. The front of the vehicle cannot be made lower as with other engines. This means that the aerodynamic design of the car cannot be improved easily.

V-configuration Engines

The *V-configuration* cylinder design has two rows of cylinders. See *Figure 8–2*. These cylinder rows are approximately 90 degrees from each other. This is the angle in most V configurations. However, other angles ranging from 60–90 degrees are used.

This design utilizes one crankshaft that operates the cylinders on both sides. Two connecting rods are attached to each journal on the crankshaft. However, there must be two cylinder heads for this type of engine.

One advantage of using a V configuration is that the engine is not as vertically high as with the in-line configuration. The front of a vehicle can now be made lower. This design improves the outside aerodynamics of the vehicle.

If eight cylinders are needed for power, a V configuration makes the engine much shorter and more compact.

FIGURE 8–1. Engines can be designed using an in-line configuration. This means the cylinders are all in a line, and vertical.

Manufacturers used to make in-line 8-cylinder engines. This made the engine rather long. The vehicle was hard to design around this long engine. The long crankshaft also caused increase torsional vibrations in the engine.

Slant Cylinder Engines

Another way of arranging the cylinders is in the *slant* configuration. This is shown in *Figure 8–3*. It is much like an in-line engine except the entire block has been placed at a slant. The slant engine was designed to reduce the distance from the top to the bottom of the engine. Vehicles using the slant engine can be designed more aerodynamically.

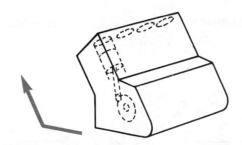

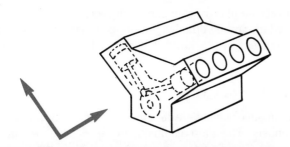

FIGURE 8-3. A slant cylinder configuration reduces the distance from the top of the engine to the bottom of the engine.

FIGURE 8-2. Engines can be designed in a V configuration. This means two rows of cylinders are in the shape of a V.

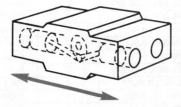

FIGURE 8-4. Opposed cylinder engines are used when there is very little vertical room for the engine. They are used mostly as rear-mounted engines on smaller vehicles.

Opposed Cylinder Engines

Several manufacturers have designed engines called *opposed cylinder* engines. An example is shown in **Figure 8-4**. Opposed cylinder engines are used in applications where there is very little vertical room for the engine. For this reason, opposed cylinder designs are commonly used on vehicles that have the engine in the rear. The angle between the two cylinders is typically 180 degrees. One crankshaft is used with two cylinder heads. There are two connecting rods attached to each journal on the crankshaft. Several car manufacturers, both foreign and American, have used this type of engine, mostly in smaller vehicles.

Radial Cylinder Engines

In a *radial cylinder* engine, **Figure 8-5**, all of the cylinders are set in a circle. All cylinders point toward the center of the circle. The connecting rods of all pistons work on a single journal of the crankshaft. The journal rotates around the center of the circle. The name *radial* is given because each cylinder is set on a radius of the circle. There is usually

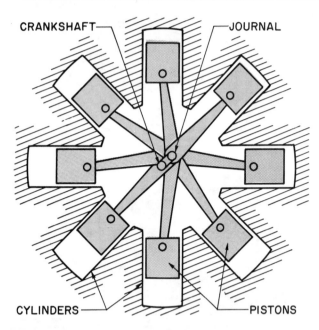

FIGURE 8–5. Gasoline engines can also be designed using a radial configuration. However, these engines are mostly used in the aircraft industry, and not in automobiles.

an odd number of cylinders. The radial engine occupies very little space. Although this design is used quite often in aircraft engines, it has not been used as an automobile engine.

Number of Cylinders

Automotive engines are designed using a variety of cylinder numbers. The most common are 4-, 6-, and 8-cylinder engines. The differences are in the horsepower and torque needed for the vehicle. For example, average horsepower and torque figures are shown for several engines in **_Figure 8–6_**. The differences are caused by the number of cylinders used on the engine.

An advantage to using fewer cylinders is reduced _fuel consumption_. Over the past few years, automotive manufacturers have made many changes in the engine to reduce fuel consumption. One change is designing engines that have fewer cylinders. With the current concern for a clean environment and improved gasoline mileage, it is difficult to justify designing a vehicle with eight cylinders. The horsepower and torque developed by an 8-cylinder engine is usually well above that which is needed for most driving

General Engine Specifications

Year	Engine CID①/Liter	V.I.N. Code②	Carburetor	Bore and Stroke	Compression Ratio	Net Brake H.P. @ RPM③	Maximum Torque Ft. Lbs. @ RPM	Normal Oil Pressure Pounds
1982	4-151, 2.5L⑭	2	Fuel Injection	4.00 × 3.00	8.2	90 @ 4000	132 @ 2800	37.5
	V6-173, 2.8L	1	E2SE, 2 Bbl.④	3.50 × 2.99	8.5	102 @ 4800	142 @ 2400	50–65
	V6-229, 3.8L⑪	K	E2ME, 2 Bbl.④	3.736 × 3.480	8.6	110 @ 4200	170 @ 2000	50–65
	V6-231, 3.8L⑤	A	E2ME, 2 Bbl.④	3.80 × 3.40	8.0	110 @ 3800	190 @ 1600	50–65
	V8-262, 4.3L⑨⑩	V	Fuel Injection	4.057 × 3.385	22.5	85 @ 3600	165 @ 1600	30–45
	V8-267, 4.4L⑪	J	E2ME, 2 Bbl.④	3.50 × 3.48	8.3	115 @ 4000	200 @ 2400	45
	V8-305, 5.0L	H	E4ME, 4 Bbl.④	3.736 × 3.480	8.6	145 @ 4000	240 @ 2400	50–65
	V8-305, 5.0L	7	Fuel Injection	3.736 × 3.480	9.5	165 @ 4200	240 @ 2400	50–65
	V8-350, 5.7L	6	Fuel Injection	4.00 × 3.48	9.0	200 @ 4200	285 @ 2800	45
	V8-350, 5.7L⑨⑩	N	Fuel Injection	4.057 × 3.385	22.5	105 @ 3200	200 @ 1600	30–45
1983	4-151, 2.5L⑭	2	Fuel Injection	4.00 × 3.00	8.2	92 @ 4000	134 @ 2800	37.5
	V6-173, 2.8L	1	E2SE, 2 Bbl.④	3.50 × 2.99	8.5	107 @ 4800	145 @ 2100	50–65
	V6-229, 3.8L	9	E2ME, 2 Bbl.④	3.736 × 3.480	8.6	110 @ 4200	170 @ 2000	50–65
	V6-231, 3.8L⑤	A	E2ME, 2 Bbl.④	3.80 × 3.40	8.0	110 @ 3800	190 @ 1600	45
	V6-262, 4.3L⑨⑩	V	Fuel Injection	4.057 × 3.385	21.6	85 @ 3600	165 @ 1600	30–45
	V8-305, 5.0L	H	E4ME, 4 Bbl.④	3.736 × 3.480	8.6	150 @ 3800	240 @ 2400	50–65
	V8-305, 5.0L	S	Fuel Injection	3.736 × 3.480	9.5	175 @ 4200	250 @ 2800	45
	V8-350, 5.7L⑨⑩	N	Fuel Injection	4.057 × 3.385	21.6	105 @ 3200	200 @ 1600	30–45
1984	4-151, 2.5L⑭	2	Fuel Injection	4.00 × 3.00	9.0	90 @ 4000	132 @ 2800	37.5
	V6-173, 2.8L	1	E2SE, 2 Bbl.④	3.50 × 2.99	8.5	112 @ 5100	148 @ 2400	50–65
	V6-229, 3.8L	9	E2ME, 2 Bbl.④	3.736 × 3.480	8.6	110 @ 4200	170 @ 2000	50–65
	V6-231, 3.8L⑤	A	E2ME, 2 Bbl.④	3.80 × 3.40	8.0	110 @ 3800	190 @ 1600	45
	V6-262, 4.3L⑨⑩	V	Fuel Injection	4.057 × 3.385	21.6	85 @ 3600	165 @ 1600	30–45
	V8-305, 5.0L	H	E4ME, 4 Bbl.④	3.736 × 3.480	8.6	150 @ 3800	240 @ 2400	50–65
	V8-305 H.O. 5.0L	G	E4ME, 4 Bbl.④	3.736 × 3.480	9.5	190 @ 4800	240 @ 3200	50–65
	V8-350, 5.7L	8	Fuel Injection	4.00 × 3.48	9.0	205 @ 4300	290 @ 2800	50–65
	V8-350, 5.7L⑨⑩	N	Fuel Injection	4.057 × 3.385	21.6	105 @ 4200	200 @ 1600	30–45

continued

FIGURE 8–6. Automotive engines are primarily designed using 4, 6, and 8 cylinders. The difference in engines is the horsepower and torque available from each engine. _(Courtesy of Motor Publications, Auto Repair Manual, 1981–1987)_

GENERAL ENGINE SPECIFICATIONS—Continued

Year	Engine CID①/Liter	Engine V.I.N. Code②	Carburetor	Bore and Stroke	Compression Ratio	Net Brake H.P. @ RPM③	Maximum Torque Ft. Lbs. @ RPM	Normal Oil Pressure Pounds
1985	4-151/2.5L⑭	2	Fuel Injection	4.00 × 3.00	9.0	88 @ 4400	132 @ 2800	37.5
	V6-173/2.8	S	Fuel Injection	3.50 × 2.99	8.9	135 @ 5100	165 @ 3600	50–65
	V6-262/4.3L	Z	Fuel Injection	4.00 × 3.48	9.3	130 @ 3600	210 @ 2000	50–65
	V8-305/5.0L	F	Fuel Injection	3.74 × 3.48	9.5	215 @ 4400	275 @ 3200	—
	V8-305/5.0L⑮	G	E4ME, 4 Bbl.④	3.74 × 3.48	9.5	190 @ 4800	240 @ 3200	50–65
	V8-305/5.0L⑯	G	E4ME, 4 Bbl.④	3.74 × 3.48	9.5	180 @ 4800	235 @ 3200	50–65
	V8-305/5.0L⑮	H	E4ME, 4 Bbl.④	3.74 × 3.48	8.6	165 @ 4400	250 @ 2000	50–65
	V8-305/5.0L⑦	H	E4ME, 4 Bbl.④	3.74 × 3.48	8.6	150 @ 4000	240 @ 2000	50–65
	V8-305/5.0L⑧	H	E4ME, 4 Bbl.④	3.74 × 3.48	8.6	165 @ 4200	245 @ 2400	50–65
	V8-350/5.7L⑫	8	Fuel Injection	4.00 × 3.48	9.0	230 @ 4000	330 @ 3200	50–65
	V8-350/5.7L	N	Fuel Injection	4.00 × 3.39	22.1	105 @ 3200	200 @ 1600	30–45
1986	4-151/2.5L⑭	2	Fuel Injection	4.00 × 3.48	9.0	88 @ 4800	132 @ 2800	37.5
	V6-173/2.8L	S	Fuel Injection	3.50 × 2.99	8.9	135 @ 5100	160 @ 3900	50–65
	V6-262/4.3L	Z	Fuel Injection	4.00 × 3.48	9.3	140 @ 4000	225 @ 2000	50–65
	V8-305/5.0L	F	Fuel Injection	3.74 × 3.48	9.5	190 @ 4000	285 @ 2800	50–65
	V8-305/5.0L⑮	G	E4ME, 4 Bbl.④	3.74 × 3.48	9.5	190 @ 4800	240 @ 3200	50–65
	V8-305/5.0L⑯	G	E4ME, 4 Bbl.④	3.74 × 3.48	9.5	180 @ 4400	225 @ 3200	50–65
	V8-305/5.0L⑮	H	E4ME, 4 Bbl.④	3.74 × 3.48	9.5	⑰	⑱	50–65
	V8-305/5.0L⑧	H	E4ME, 4 Bbl.④	3.74 × 3.48	9.5	165 @ 4200	245 @ 2400	50–65
	V8-305/5.0L⑦	H	E4ME, 4 Bbl.④	3.74 × 3.48	9.5	150 @ 4000	240 @ 2000	50–65
	V8-350/5.7L⑫	8	Fuel Injection	4.00 × 3.48	9.5	230 @ 4000	330 @ 3200	50–65

①—CID—Cubic inch displacement.
②—V.I.N.—On 1981–85 vehicles, the 8th digit in the V.I.N. denotes engine code.
③—Ratings are net—As installed in the vehicle.
④—Rochester.
⑤—For service procedures on this engine, see Buick chapter.
⑥—Turbocharged engine.
⑦—Monte Carlo.
⑧—Caprice/Impala.
⑨—For service procedures on this engine, see Oldsmobile chapter.
⑩—Diesel.
⑪—Exc. Calif.
⑫—Corvette.
⑬—Camaro Z28.
⑭—For service procedures on this engine, refer to Pontiac chapter.
⑮—Camaro.
⑯—Monte Carlo SS.
⑰—Exc. IROC-Z & Z28, 155 @ 4200; IROC-Z & Z28, 165 @ 4400.
⑱—Exc. IROC-Z & Z28, 245 @ 2000; IROC-Z & Z28, 250 @ 2000.

FIGURE 8–6. (CONTINUED)

applications. Applications that require heavy hauling, such as trailers and boats, may require the additional horsepower and torque of an 8-cylinder engine.

Combining the number and type of cylinders, today's manufacturers design the following common types of engines:

a. In-line (4- and 6-cylinder engines, most common, as well as 3- and 5-cylinder engines)
b. V configuration (6- and 8-cylinder engines)
c. Slant (4- and 6-cylinder engines)
d. Opposed (4- and 6-cylinder engines)

In past years, other common engine configurations included:

a. In-line (6- and 8-cylinder engines)
b. V configuration (8- and 12-cylinder engines)

Five-cylinder and Three-cylinder Engines

Some five-cylinder and three-cylinder engines have been built for automobiles by both American and foreign manufacturers. Although these engines are not commonly found in the automotive field, they are used.

CAR CLINIC

PROBLEM: OIL COMES OUT OF DIPSTICK

A customer has a car that leaks oil out of the dipstick tube. The dipstick is still in the tube, but it is usually slightly lifted out of its tube. The engine has about 75,000 miles on it. What could be the problem?

SOLUTION:

A common cause of oil vapor coming out of the dipstick is high crankcase pressure. Depending upon the year and make of the vehicle, the PCV valve could be stuck closed or there could be excessive blowby. With bad rings, compression pressures escape past the bad rings and increase the crankcase pressure. The pressure then needs to escape. If the crankcase venting is plugged, the pressure and crankcase vapors will go out through the dipstick tube. Check the PCV valve for correct operation and make sure the crankcase venting is not plugged.

MUD RACING

This truck is involved in mud racing. In this sport, the truck is modified with a high-horsepower and high-torque engine, and special wheels needed to get through a mud pit. The object of the competition is to start at the edge and go as far as possible through the mud pit. Most vehicles cannot get completely through the pit. The truck is then pulled back out of the pit with a large steel cable attached to a tractor. (Courtesy of Jim Rennich, Bloomington, MN)

8.2 VALVE AND HEAD ARRANGEMENT

Engines used in the automobile can be designed with different valve and head design. L-, I-, F-, and T- (LIFT) head designs have been used. Also, overhead, dual, and in-block camshafts have been used. In addition, stratified charged engines are now becoming popular.

I-head Design

The *I-head* valve design is the most common arrangement of valves. Referring to *Figure 8–7*, I-head means the valves are directly above the piston (overhead valves). The valves are located in the cylinder head. The design allows easy breathing of the engine. Air and fuel can move easily into and out of the cylinder with little restriction. This process improves the volumetric efficiency of the engine. The I-head is also easy to maintain. For example, if a valve is damaged, it can be replaced easily. Adjusting the valves is easier too.

The valves can be on both sides of the piston, on top, or on one side only. *Figure 8–8* shows an example of a head with an I design. This type of valve arrangement is also called overhead valves, sometimes referred to as OHV. A rather complex mechanical system must be used to open the valves from the camshaft. It includes the *lifters*, *pushrods*, *rocker arms*, and valves.

L-head Design

Another type of valve arrangement is called the *L-head*. See *Figure 8–9*. The valves are located in the block. The inlet and outlet ports are shorter. The head does not have any mechanical valves located within its structure. This head is referred to as a "flat head." Older vehicles, especially from Ford Motor Company, utilized the flat-head design with the valves built within the block. These engines were commonly called the "flat-head V-8." They were common in the 1930s, 40s, and 50s. *Figure 8–10* shows the head on this type of engine.

FIGURE 8–8. On an I-head design, the valves are located in the head.

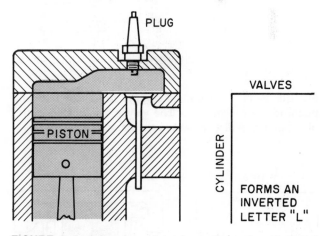

FIGURE 8–9. L-head engines have the valves located within the block. These engines are commonly called the "flat-head engine."

FIGURE 8–10. The L-head engine uses a flat head.

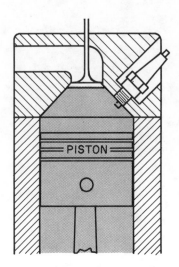

FIGURE 8–7. An I-head engine has the valves located above the piston in the head. The shape of the engine is an I.

In the L-head design, fewer mechanical parts are needed to operate the valves from the camshaft. Here the rocker arms and pushrods have been eliminated. See *Figure 8–11*. The valves and lifters are operated directly on top of the camshaft. One disadvantage is that if any damage occurs around the valve port, the block may have to be replaced. It is also more difficult to adjust the valves when they are located in the block.

F-head and T-head Designs

There have been other types of valve arrangements as well. These include the *F-head* and *T-head* designs. See *Figure 8–12*. Actually, both of these designs combine the I-head and L-head designs. The T-head has the valves located within the block. The difference between this design and the L-head design is that two camshafts are needed. Because of this extra expense, T-head designs are not commonly used in the automotive industry today.

The F-head design is a combination of the I-head and the L-head designs. There are valves located in the head as well as in the block. It has some of the advantages of the L-head and I-head designs. However, the increased cost of parts is a disadvantage.

In-block Camshaft

Many engines today have the camshaft located directly within the block. This type of camshaft can be used on L-, T-, F-, and I-head engines. *Figure 8–13* shows where the camshaft is located in this design. An advantage is that the

FIGURE 8–13. Many engines have the camshaft located directly within the block.

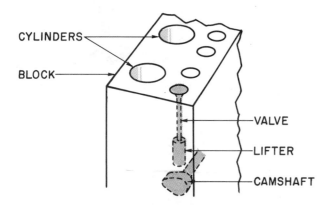

FIGURE 8–11. The L-head engine does not use any pushrods or rocker arms. The valves are located in the block and ride directly above the camshaft.

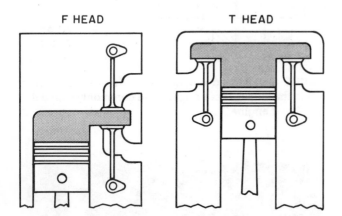

FIGURE 8–12. Other engine configurations are available, such as the F head and T head. However, these engines are not as common in the automotive market.

FIGURE 8–14. The camshaft is driven directly from the crankshaft when located within the block. Either gears or a chain drive can be used.

camshaft can be driven directly off of the crankshaft. A standard gear set or a gear and chain arrangement can be used. See *Figure 8–14*.

A disadvantage of in-block camshafts is the linkage needed to open and close the valves. This includes the lifters, pushrods, and rocker arms. All of these parts can become worn, causing more chance of failure on the parts. Another disadvantage of this system is that the maximum rpm range of the engine is limited by the valve mechanism. The valves have a tendency to float (not close completely) at higher speeds.

This system must also have some way of accounting for the clearance between all of these parts. There is an adjustment called *valve clearance*. As the engine and parts heat up, the parts expand. If there were no valve clearances, the heated parts would expand and keep the valves open during the compression and power strokes. If this happened, the engine would not run correctly. The valve clearance can be accounted for by using hydraulic lifters or by adjusting the valve clearance. Valve clearance will be discussed in a later chapter.

Overhead Camshaft

Many manufacturers are now designing engines that have an *overhead camshaft*. The camshaft is placed directly above the valves in the head. This design is used on I-head valve designs. See *Figure 8–15*. One advantage of this design is that the cam operates directly on the valves. There are fewer parts that could wear, causing failure. Also, having fewer valve train parts means the engine has a higher rpm range. In addition, there is less valve clearance needed for expansion as the engine heats up. There is also a more positive opening and closing of the valves. The valve movement, therefore, responds more quickly to the cam shape.

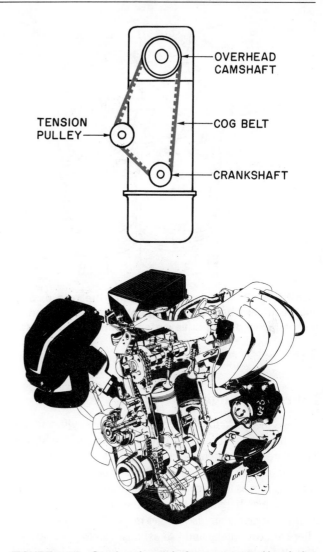

FIGURE 8–16. Overhead camshafts are operated by chains or belts driven from the crankshaft. The chain or cogs on the belt keep the cam and crankshaft in time with each other. *(Courtesy of Peugeot Motors of America, Inc.)*

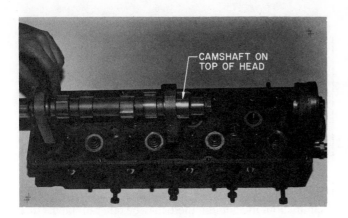

FIGURE 8–15. Engine manufacturers also design overhead camshafts (OHC). The camshaft is mounted on top of the cylinder head.

This may not be true for in-block camshafts because of the extra linkage. The camshaft is now driven by either a chain or a belt as shown in *Figure 8–16*.

Dual Camshafts

Some engines use more than one camshaft. This type of engine is called a *dual camshaft* engine. Dual camshaft engines are designed so that one camshaft operates the intake valves, while the second camshaft operates the exhaust valves. Several foreign manufacturers design engines using dual camshafts. See *Figure 8–17*. This design is used often in sports cars and on racing engines. Designs such as this are usually more expensive and are not as common on passenger cars.

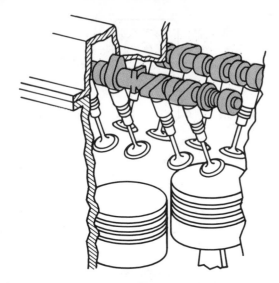

FIGURE 8–17. Certain engines utilize the twin or dual overhead camshaft style. Here, both cams are driven by a chain drive or belt drive system from the crankshaft.

FIGURE 8–18. Four-valve heads are used where there is a need for additional air and fuel to enter the engine.

Four-valve Engines

Most engines used today have two valves per cylinder. One is for the intake stroke and one is for the exhaust stroke. There are times when more air and fuel must be brought into the engine. This is done with a *supercharger*. Supercharging means to force more air into the cylinders for more power. Supercharging will be discussed in a later chapter. *Four-valve heads* are used to get more air into the engine. There are two intake and two exhaust valves per cylinder. With the additional valves, more intake air and fuel and exhaust gases can be moved through the engine. Although the cost is somewhat higher, four-valve engines operate with higher volumetric efficiency. *Figure 8–18* shows a four-valve cylinder head.

Stratified Charged Engine

Certain engines used in the automobile are called stratified charged engines. When the mixture within the combustion chamber is thoroughly mixed, the charge is called "homogeneous." When the mixture is not evenly mixed, the charge is said to be *stratified* or in layers. A *stratified charged engine* has a second, small area for combustion in the cylinder head. See *Figure 8–19*. There is a spark plug in the chamber. A rich mixture of air and fuel enters the stratified chamber. A lean mixture of air and fuel enters the major combustion chamber. The overall air-fuel mixture is leaner. This is done primarily to reduce pollution from the engine. When the spark plug ignites the rich air-fuel mixture, this burning mixture is used to ignite the lean mixture.

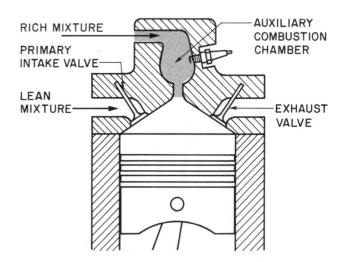

FIGURE 8–19. A stratified charged engine has an auxiliary combustion chamber that ignites the air/fuel mixture. This allows a leaner mixture to be burned in the regular combustion chamber.

CAR CLINIC

PROBLEM: REMOVING ENGINE SLUDGE

An engine has a heavy amount of sludge inside. What is the best way to remove sludge from an engine?

SOLUTION:

The best way to remove sludge from an engine is to disassemble the engine, clean all the parts, replace worn or damaged parts, and rebuild the engine. Chemicals on the market that remove sludge may also cause oil contamination. In addition, there may be other problems in the engine. The removing of the sludge will not fix these problems.

SWASH PLATE

(Side View)

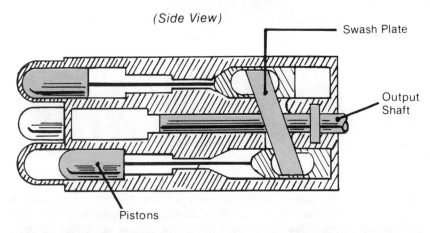

FIGURE 8–20. A stirling engine uses four pistons, a swash plate, and an external combustion chamber. (Courtesy of Davis Publications, Inc.)

8.3 ALTERNATIVE ENGINE TYPES

Although gasoline and diesel engines are the most common, engineers and scientists are working on new types of engines. Two engines being studied for possible use in the automobile are the stirling engine and the gas turbine engine.

Stirling Engine

The *stirling engine* operates very smoothly with complete combustion and low emission characteristics. Both General Motors and Ford Motor Company have studied stirling engines. Several designs have been tested. The most popular stirling engine uses a swash plate design.

There are four cylinders in the swash plate design as shown in *Figure 8–20*. An external combustion chamber is used. The combustion is considered continuous. The heat from combustion causes the four pistons to be forced downward. Each piston is attached to a *swash plate*. Mechanically, the swash plate is an angular disk attached to the output shaft of the engine. As the pistons are forced downward, the connecting rods move sequentially, pushing the swash plate in a rotary motion. The power pulses must occur in the correct order. Referring to *Figure 8–21*, as number 1 piston fires, it pushes the swash plate clockwise. Then number 4 piston must fire, followed by number 3, number 2, and so on. The engine runs smoothly because of the swash plate. The swash plate is doing the same thing as a crankshaft on a conventional internal combustion engine. It changes reciprocating motion to rotary motion.

Stirling Gas Cycle

The stirling gas cycle is shown in *Figure 8–22*. Thermal energy from any resource, such as coal, oil, or diesel fuel,

is applied to the heater. Although they appear to be separate in the diagram, all heaters are connected together. The heat causes the gases to expand above the number 1 piston. The area below the number 1 piston is connected to the cooler near number 4 piston. This causes the gases below number 1 cylinder to contract. The difference in pressure across number 1 cylinder forces the piston downward.

As number 1 piston moves down, the gases below number 1 piston are forced through the cooler, generator, and heater to number 4 piston. As the gases pass through the cooler, they contract. The gases are then heated by the generator and heater. The gases expand and build up pressure above number 4 piston. At the same time, the gases below number 4 piston are near the number 3 cooler. This contracts the gases. Now there is a pressure differential across number 3 piston. This process continues to number

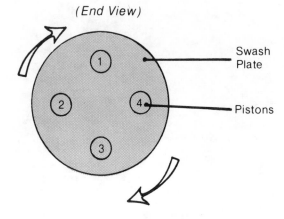

FIGURE 8–21. The swash plate is used to convert the downward motion of the piston to rotary motion needed for pushing a vehicle forward. (Courtesy of Davis Publications, Inc.)

STIRLING CYCLE

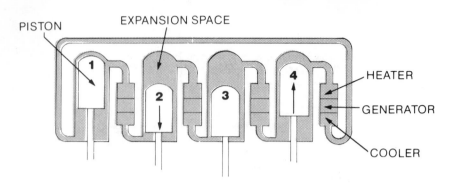

PISTON

EXPANSION SPACE

HEATER

GENERATOR

COOLER

FIGURE 8–22. The stirling engine has four pistons. The firing order is 1, 4, 3, 2. For example, as number 1 piston is forced down, the pressure on the underside of the piston helps push down number 4 piston. *(Courtesy of Davis Publications, Inc.)*

2 piston, then number 1 piston again. The firing order, then, is number 4, 3, 2, and 1.

The stirling engine gets four power pulses per revolution against the swash plate. There is also a suction on the bottom of the piston. This design has great potential for higher efficiency. A stirling engine is shown in *Figure 8–23*.

Gas Turbine Engine

Another type of engine being studied for use in the automotive market is called the *gas turbine engine*. This engine uses continuous combustion. The combustion is classified as internal. The motion produced by the turbine is considered rotary motion. The gas turbine engine has been tested by several manufacturers in the past 15 years. Because of its high cost, however, it still has not been used extensively in the automotive market.

Gas Turbine Cycle

The gas turbine burns diesel fuel. *Figure 8–24* is an example of a gas turbine. There is a centrifugal air *compressor*. Rotating at 35,000 rpm, this compressor forces pure air (not air and fuel) into the engine. The heat of compression increases the temperature to about 500 degrees F.

Air passes through the compressor and through a *regenerator*. The regenerator is designed to pick up excess heat from the exhaust. As the exhaust gases flow through the regenerator, the heat is conducted into the metal of the regenerator. The regenerator turns only 18 revolutions per minute so heat can easily be absorbed. As the intake air passes through the regenerator, it picks up this excess heat. The process brings the air temperature up to about 1,200 degrees F.

The air is then sent into the *burner*. This is the combustion chamber. Air and fuel are added to an already burning flame. Coming out of the burner, the gas is near 2,100 degrees F. This hot gas is sent through the first turbine called the compressor turbine. The temperature of the gases at the first turbine is about 2,200 degrees F. The compressor turbine turns near 35,000 rpm. Its major purpose is to turn the air compressor at that speed.

The remaining energy in the gases from the first turbine enter the second turbine. This is called the power turbine. It is connected to the back wheels of the vehicle. The temperature of the gases going into the power turbine is near 1,400 degrees F. This energy turns the power turbine to produce power for the wheels of the vehicle. The gases then

FIGURE 8–23. This burner-powered version of the stirling engine runs smoothly on any liquid fuel and produces few pollutants. *(Courtesy of Ford Motor Company)*

GAS TURBINE FLOW DIAGRAM

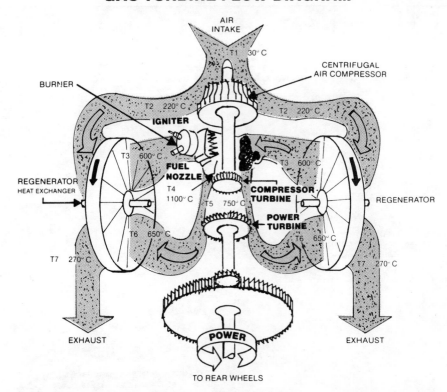

FIGURE 8–24. The gas turbine engine uses a compressor, two turbines, two regenerators, and a continuous combustion process. Efficiency is near 48%. *(Courtesy of Davis Publications, Inc.)*

pass through the regenerator and out into the atmosphere. The layout of turbine parts is shown in ***Figure 8–25***.

The turbine engine has potential for heavy-duty applications in the automobile. It averages about 45% efficiency. However, the cost of this engine is still well above a comparable gasoline or diesel engine. The advantages include smooth running, multiple fuels, higher efficiency, and no cooling system. The disadvantages include high cost, no dealerships for repair, no parts distribution systems available, and too much power for the average vehicle. Automotive manufacturers, however, are downsizing these engines to be competitive with the gasoline engine.

GAS TURBINE FLOW CHART

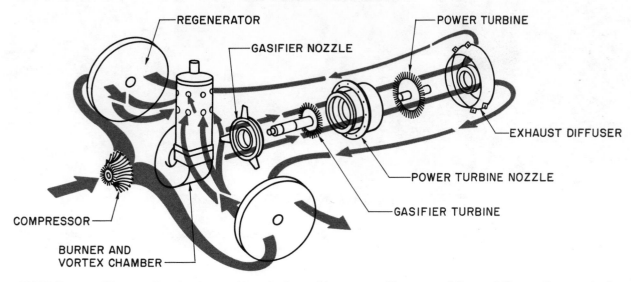

FIGURE 8–25. The gas flow is shown through the turbine parts. *(Courtesy of General Motors Corporation)*

DRAG RACING

The automobile has been involved in many forms and types of sports activities. Here several vehicles are used for drag racing. Engines are modified to increase horsepower and torque, different gear ratios are used, and tires that obtain maximum traction are installed. (Courtesy of Mike Meyers)

SUMMARY

The purpose of this chapter was to introduce various types of engines. There are many types of engines. For example, some have the cylinders in-line or in a row, and some have the cylinders in a V configuration. Common in-line engines have either 4, 6, or 8 cylinders. Common V-type engines are V-6 and V-8 styles. In addition, the cylinders can be arranged in a slant configuration. These are built so they have a low profile. This makes the vehicle lower and better aerodynamically.

Engine manufacturers also build opposed cylinder engines. This engine has the cylinders arranged horizontally. The angle between the two sides is near 180 degrees.

Another method used to identify engine types is by the valve and head arrangement. I-head engines have the valves located in the cylinder head above the cylinders and pistons. L-head designs have the valves located to the side of the pistons, built in the block. There are no valves in the head. F- and T-head designs combine the best designs of the I-head and L-head, but are not as common today.

Camshafts can be located either in the block or overhead and above the valves. Both methods are used in engines today. Some engines have dual camshafts. One camshaft operates the intake valves, and one camshaft operates the exhaust valves.

When more air is needed in the engine, four-valve heads can be used. There are two exhaust valves and two intake valves. Engines that are supercharged may use this design.

The stratified charged engine is becoming very popular. A small combustion chamber has a very rich air-fuel mixture that is ignited by the spark plug. This combustion is used to ignite a lean air-fuel mixture in the main combustion chamber. The result is better fuel mileage.

Other types of engines, including the stirling engine and the gas turbine engine, are being tested for use in the automobile. The stirling engine uses a swash plate which replaces the crankshaft. Four pistons are used to push the swash plate in a circular motion.

The gas turbine engine uses continuous combustion. Air is forced into the engine by a large compressor. The air is heated by a regenerator and sent to the combustion chamber. Fuel is added and the hot exhaust gases pass through turbines that rotate. This motion is used to turn the wheels of the vehicle.

TERMS TO KNOW

Can you explain each of the following terms? Review the chapter until you can use each term correctly.

Configuration

Valve clearance

In-line

V configuration

Slant

Opposed cylinder

Aerodynamic

Fuel consumption

I-head

Lifters

Rocker arms

Pushrods

L-head

Four-valve heads

Overhead camshaft

Supercharger

Stratified

Stratified charged engine

Stirling engine

Swash plate

Turbine

Gas turbine engine

Regenerator

Compressor

Burner

REVIEW QUESTIONS

Multiple Choice

1. The _____ engine design has the cylinders in a vertical row.
   a. Slant
   b. In-line
   c. V-configuration
   d. Rotary
   e. Gas turbine

2. What is the approximate angle between the V on a V-type engine?
   a. 60–90 degrees
   b. 150–160 degrees
   c. 180–190 degrees
   d. 195–205 degrees
   e. 350–360 degrees

3. The slant engine is used because:
   a. The engine has a lower profile
   b. The engine has more power
   c. The engine is easier to manufacture
   d. The engine is lighter
   e. All of the above

4. Which type of engine would be best suited for placement in the rear of the vehicle?
   a. Slant
   b. Opposed
   c. V configuration
   d. In-line
   e. Slant, V configuration

5. One of the biggest differences between 4-, 6-, and 8-cylinder engines is:
   a. Each has different horsepower and torque
   b. Only 6-cylinder engines can be used on small vehicles
   c. Only 4 cylinders can be used in automobiles
   d. Eight cylinders are lighter and more efficient
   e. Four cylinders have more power

6. Which of the following is the most common valve arrangement today?
   a. L-head
   b. T-head
   c. I-head
   d. P-head
   e. V-head

7. Which type of valve arrangement uses a rocker arm, pushrods, and lifters?
   a. L-head
   b. F-head
   c. I-head
   d. V-head
   e. T-head

8. The reason for having a four-valve engine is:
   a. So more fuel and air can enter the engine
   b. To get better fuel mileage
   c. To reduce the power available on an engine
   d. To heat up the engine more
   e. To cool down the engine more

9. What type of engine uses a small additional combustion chamber with a rich air-fuel mixture?
   a. Stirling engine
   b. Stratified charged engine
   c. Gas turbine
   d. All of the above
   e. None of the above

10. Which engine uses a swash plate?
    a. Gas turbine engine
    b. Stratified charged engine
    c. Stirling engine
    d. All of the above
    e. None of the above

11. How many pistons are used on the stirling engine?
    a. 4 pistons
    b. 6 pistons
    c. 8 pistons
    d. 10 pistons
    e. 12 pistons

12. The stirling engine is considered a/an:
    a. Continuous combustion engine
    b. External combustion engine
    c. Multifuel engine
    d. All of the above
    e. None of the above

13. What part on the gas turbine takes the heat of exhaust and puts it into the intake?
    a. Compressor
    b. Turbine
    c. Regenerator
    d. Fuel injector
    e. Exhaust manifold

14. The gas turbine engine is considered a/an:
    a. Continuous combustion engine
    b. Internal combustion engine
    c. Rotary engine
    d. All of the above
    e. None of the above

15. Which device on the gas turbine is used to extract the power from the gases?
    a. Turbine
    b. Compressor
    c. Burner
    d. Fuel injector
    e. Exhaust manifold

The following questions are similar in format to ASE (Automotive Service Excellence) test questions.

16. Technician A says the type of engine most commonly used in vehicles is the in-line type. Technician B says the type of engine most commonly used in vehicles is the opposed type. Who is right?
    a. A only
    b. B only
    c. Both A and B
    d. Neither A nor B

17. Technician A says that a stratified charged engine means the engine also has a turbocharger. Technician B says that a stratified charged engine means the engine has a precombustion chamber for burning more efficiently. Who is right?
    a. A only
    b. B only
    c. Both A and B
    d. Neither A nor B

18. Technician A says that the reason for having a valve clearance is to help lubricate the parts. Technician B says the reason for a valve clearance is to improve the efficiency of the engine. Who is right?
    a. A only
    b. B only
    c. Both A and B
    d. Neither A nor B

19. Technician A says that a stratified charged engine is the same as a four-valve engine. Technician B says that a stratified charged engine means the spark plugs are firing sooner before top dead center. Who is right?
    a. A only
    b. B only
    c. Both A and B
    d. Neither A nor B

Essay

20. Describe the design and operation of a stirling engine.

21. What is the purpose of a swash plate?

22. What is the advantage of using an overhead camshaft type engine?

23. Describe the operation of a gas turbine engine.

24. Compare the in-line and V-configuration types of engines.

25. Compare the differences between the I-head and the L-head design of valve arrangements.

CHAPTER 9

Basic Engine Construction

INTRODUCTION

Basic engine construction is a prerequisite when studying the automotive power plant. This section will introduce the basic parts of the engine. The cylinder block, crankshaft, pistons, rods, and camshafts will be discussed.

OBJECTIVES

After reading this chapter, you will be able to:
- Identify the major parts of the cylinder block.
- Recognize the purpose of core plugs.
- Identify cylinder block differences such as aluminum versus cast iron, types of sleeves, and water jackets.
- List the parts of the crankshaft assembly.
- Identify the purpose of bearings and caps, oil passageways, vibration dampers, flywheels, and thrust surfaces on the crankshaft.
- Identify crankshaft seals and their purpose.
- List the parts of the piston and rod assembly.
- State the purpose and operation of the rings, piston, pins, and bearings.
- Identify the effect of pressures and temperatures on the design of the piston.
- Identify various service and diagnosis tips and procedures on the block, pistons, crankshaft, and bearings.

CHAPTER HIGHLIGHTS

9.1 CYLINDER BLOCK
   A. Block Design
   B. Block Manufacturing
   C. Core Plugs
   D. Cylinder Sleeves
   E. Aluminum vs. Cast Iron Blocks

9.2 CRANKSHAFT ASSEMBLY
   A. Crankshaft Design
   B. Crankshaft Alignment

C. Crankshaft Vibration
D. Vibration Dampers
E. Bearings and Bearing Caps
F. Bearing Design
G. Thrust Bearings
H. Oil Passageways
I. Bearing Clearance
J. Crankshaft Seals

9.1 CYLINDER BLOCK

Block Design

The engine block is the main supporting structure to which all other engine parts are attached. A cylinder block is a large cast-iron or aluminum casting. It has two main sections: the cylinder section and the crankcase section. *Figure 9–1* shows a cylinder block for a 4-cylinder engine. The cylinder section is designed for the pistons to move up and down during operation. The surfaces are machined to allow the pistons to move with minimum wear and friction.

The *crankcase* section is used to house the crankshaft, oil pump, oil pan, and the oil during operation. Cooling passageways are built within the block. These passageways surround the cylinder. They allow coolant to circulate throughout the cylinder area to keep the engine cool. There is also a drilled passageway within some blocks for the camshaft. Many oil holes are drilled internally so that engine parts can be adequately lubricated. Other holes are also drilled to allow other parts to be attached to the cylinder block.

Block Manufacturing

The first step in building a block is to design a pattern. Sand is then formed around the pattern. When the pattern is removed, sand cores are placed within the cavity. These sand cores will eventually be the cooling passageways and cylinders. Molten metal is poured into the cavity made by the sand. After the metal has cooled, the sand is removed and the cores are broken so they can be removed. This design is called a *cast* block. The metal is usually a gray cast iron with several special metals added to it. This increases the strength and wear characteristics of the block. The extra metals also help to reduce shrinkage and warpage from the heat produced by combustion.

Once the block is cast, and after it has been cooled and cured, surfaces are machined so other parts can be attached to the block. These surfaces include the cylinders, top of the block (deck), camshaft bore, crankshaft bore, and oil pan surfaces. The front and rear of the block and engine mounts are also machined so that parts can be attached and sealed correctly. See *Figure 9–2*.

Certain smaller engines can also be *die cast*. This means that the liquid metal is forced into a metal rather than sand mold. This kind of casting gives smoother surfaces, and more precise shapes can be made. Less machining is needed on this type of block.

Core Plugs

All cylinder blocks use *core plugs*. These are also called freeze or expansion plugs. During the manufacturing process, sand cores are used. These cores are partly broken and dissolved when the hot metal is poured into the mold. However, holes have to be placed in the block to get the sand out of the internal passageways. These are called core holes. The holes are machined and core plugs are placed into these holes. See *Figure 9–3*.

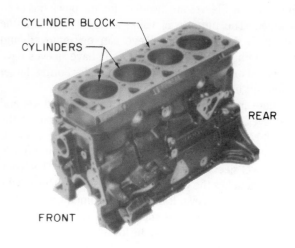

FIGURE 9–1. The cylinder block is the main structure of the engine. All other engine parts are attached to the cylinder block. *(Courtesy of Peugeot Motors of America, Inc.)*

FIGURE 9–2. The cylinder block has many machined surfaces to which other parts are bolted. (Courtesy of Peugeot Motors of America, Inc.)

FIGURE 9–3. Freeze or expansion plugs are used to protect the block if the coolant freezes. Rather than cracking the block from expansion, the freeze plugs are popped out.

Core plugs are made of soft metal. They can also protect the block from cracking. If the coolant in the block freezes near the core plugs, the coolant will expand. This may cause the block to expand and crack. Rather than having the block crack, the core plugs may pop out and possibly save the block.

Cylinder Sleeves

Some manufacturers use *cylinder sleeves*. Rather than casting the cylinder bores directly into the block, they insert a machined sleeve. *Figure 9–4* shows a sleeve for a cylinder block. Sleeves are inserted after the block has been machined. The purpose of using a sleeve is that if the cylinder is damaged, the sleeve can be removed and replaced rather easily. Blocks that don't have sleeves have to be bored out to remove any damage. After boring, larger pistons will be needed.

There are two types of sleeves. They are called wet and dry sleeves. The dry sleeve is pressed into a hole in the block. It can be machined quite thin because the sleeve is supported from the top to the bottom by the cast iron block.

The wet sleeve is also pressed into the block. The cooling water touches the center part of the sleeve. This is why it is called a wet sleeve. It must be machined thicker than the dry sleeve because it is supported only on the top and bottom. Seals must be used on the top and bottom of the wet sleeve. Seals are used to keep the cooling water from leaking out of the cooling system. Wet sleeves are used on some larger diesel engines.

Aluminum vs. Cast Iron Blocks

Blocks can be made from either cast iron or aluminum. In the past, most blocks were made of cast iron. This improved

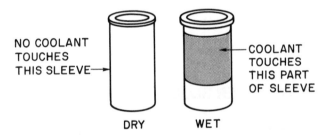

FIGURE 9–4. Cylinder sleeves are used on some engines. The sleeve is inserted into the block after being manufactured. The sleeve acts as the cylinder bore. There are both wet and dry sleeves.

strength and controlled warpage from heat. With the increased concern for improved gasoline mileage, however, car manufacturers are trying to make the vehicle lighter. One way is to reduce the weight of the block. Aluminum is used for this purpose. Aluminum is a very light metal. Certain materials are added to the metal before it is poured into the mold. These metals are used to make the aluminum stronger and less likely to warp when heat is applied from combustion. Aluminum blocks must also have a sleeve or steel liner placed in the block. Steel liners are placed in the mold before the metal is poured. After the metal is poured, the steel liner cannot be removed.

Silicon is also added to the aluminum. Through a special process, the silicon is concentrated on the cylinder walls. This process eliminates the need for a steel liner. This is called "silicon-impregnated cylinder walls." One problem with this design is that it requires the use of very high-quality engine oils. Because of owner neglect, this engine does not usually survive its intended service life.

NEW ENGINE DESIGNS

Many new engine designs are being tested to improve efficiency. This V-6 engine uses a 90 degree V, rather than the usual 60 degree V. This makes the engine more compact and lowers the center of gravity, which is important for aerodynamics and handling. The crankpins are offset 30 degrees to retain an even firing sequence. (Courtesy of Peugeot Motors of America)

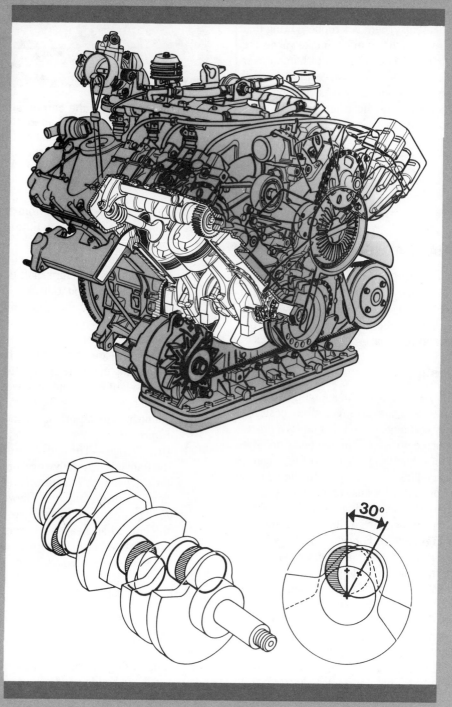

142 MOTOR AUTOMOTIVE MECHANICS

CAR CLINIC

PROBLEM: CAR SMOKES IMMEDIATELY AFTER START UP

A Honda engine has heavy smoking immediately after start up. The smoke lasts about 10 minutes. The rings and valves have been checked and seem OK. What could be the problem?

SOLUTION:

This is a common problem with engines that have precombustion chambers. There is a rubber O ring that seals the chamber where the auxiliary intake valves are located. The O rings have been known to harden over time. The hardened O rings do not seal very well and oil can seep into the combustion chambers. The problem goes away when the engine heats up and the parts expand enough to make a seal. The *solution is to replace the rubber O rings.*

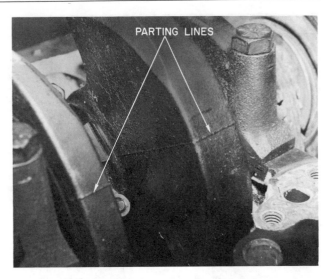

FIGURE 9–5. Crankshafts can be either forged or cast. Forged crankshafts can be identified by a ground-off separation line and cast crankshafts have a parting line.

9.2 CRANKSHAFT ASSEMBLY

The crankshaft is designed to change the reciprocating motion of the piston to rotary motion. It is bolted to the bottom of the cylinder block. The crankshaft assembly includes the crankshaft, bearings, flywheel, harmonic balancer, timing gear, and front and rear seals.

Crankshaft Design

The crankshaft is manufactured by either forging or casting. *Forged* steel crankshafts are stronger than cast iron crankshafts, but they cost more money. Forging is a process where metal is heated to a certain temperature, then stamped or forged into a particular shape. Casting means to heat the metal to its melting point and pour the liquid metal into a form made from sand. Because of the improvements in casting, more crankshafts are being cast today. Cast and forged crankshafts can be identified by the flashing and parting lines. Refer to *Figure 9–5*. Forged crankshafts have a ground-off separation line. Cast crankshafts have a small parting line where the mold came together.

After the crankshaft is cast or forged, it must be *heat treated.* This means the outer surfaces must be made harder so that the crankshaft will not wear on the bearing surfaces. Heat treating is done by heating the outer part of the crankshaft to 1,600–1,800 degrees F. Then the metal is cooled rapidly in oil, water, or brine (salt water). The rapid cooling causes the outer part of the crankshaft (0.060 inch) to be hardened.

The parts of the crankshaft include:

a. Throw — The distance from the centerline of the main journal to the centerline of the connecting rod journal.
b. The main journal — The position on the crankshaft that connects the crankshaft to the block.
c. The connecting rod journal — The position on the crankshaft where the connecting rod is attached.
d. The *counterweights* — Weights are cast or forged into the crankshaft for balance. For each throw, there is a counterweight to balance the motion. Depending upon the engine, the counterweight can be a weight or another connecting rod journal.
e. Thrust surfaces — Surfaces machined on the crankshaft to absorb axial motion or thrust. See *Figure 9–6*.
f. Drive flange — The end of the crankshaft that drives the transmission. A flywheel or flexplate is bolted to the drive flange. The transmission is attached to the flexplate or driven by the flywheel.
g. *Fillets* — Small rounded areas that help strengthen the crankshaft near inside corners. Stress tends to concentrate at sharp corners and drilled passageways. Fillets help reduce this stress. See *Figure 9–7*.

Crankshaft Alignment

When the crankshaft fits into the block, it must be exactly in line with the holes in the block. This is called *alignment.* When the block is bored for the crankshaft main bearing bore, a line boring machine is used. To do this, the main bearing caps are bolted and torqued to the block. The line borer then machines each main bearing area in line with

FIGURE 9–6. Thrust surfaces are machined into the crankshaft to absorb axial motion or thrust motion on the crankshaft.

FIGURE 9–7. Fillets are used to strengthen the crankshaft. Fillets are rounded corners where stresses are increased.

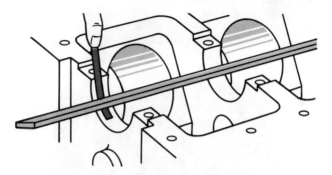

FIGURE 9–8. The crankshaft can be checked for alignment by using a straight bar and a set of feeler gauges. *(Courtesy of Federal-Mogul Corporation)*

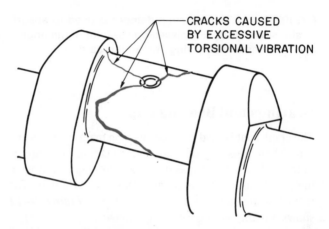

FIGURE 9–9. Torsional vibration within the crankshaft can cause severe cracking.

the others. Over a period of time, the block may warp. This could cause the alignment to be incorrect. This condition could cause excessive wear on certain parts of the crankshaft. Alignment can be checked by using a feeler gauge and a straight-edge bar as shown in *Figure 9–8*.

Crankshaft Vibration

During normal operation, the crankshaft is twisted and turned, producing constant vibration within the crankshaft. For example, when one cylinder is on a compression stroke, that part of the crankshaft tries to slow down. At the same time, other cylinders may have full pressure from the power stroke. This causes the crankshaft to partially twist and snap back during each revolution. This effect is called *torsional vibration*. Additional torsional vibration can also be caused from using the wrong flywheel, converter drive plate, or torque converter in the transmission. *Figure 9–9* illus-

trates the results of having too much torsional vibration. Crankshafts typically crack near the connecting rod journal of number 1 cylinder.

Vibration Dampers

Vibration dampers are used to compensate for the torsional vibration. See *Figure 9–10*. The vibration damper is also called a "harmonic balancer." It is constructed by using an inertia ring and a rubber ring. The inertia ring is used to help dampen the internal vibrations. The two are bonded together and attached to the front of the crankshaft. As the crankshaft twists back and forth, the inertia ring has a dragging or slowing down effect. As torsional vibrations occur, the rubber and inertia rings absorb the vibration.

Note that the weight of the inertia ring is sized to a particular engine. If the wrong vibration damper is used, the crankshaft may be damaged. Incorrect vibration dampers can be identified by observing the timing marks. The timing marks on the vibration damper may not line up correctly with the timing tag on the front of the block.

FIGURE 9–10. The vibration damper is used to absorb torsional vibration. It is placed on the front of the engine and includes a rubber ring and an inertia ring.

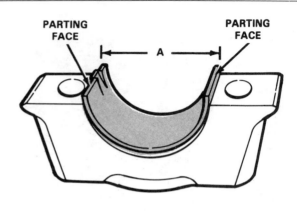

A = FREE SPREAD DIAMETER

FIGURE 9–12. Spread (dimension A) helps keep the bearing in place in the block and bearing cap. *(Courtesy of Federal-Mogul Corporation)*

Bearings and Bearing Caps

The crankshaft is held in place by using main bearings and caps. There is usually one more main bearing cap than the number of cylinders. Depending upon the engine, however, there may be less main bearing caps than cylinders. The main bearing caps are bolted to the block. *Figure 9–11* shows a main bearing cap being removed.

When in place, the main bearings hold the crankshaft securely in place to allow for rotation. On some high-performance gas engines and diesel engines, there may be four bolts, rather than two, holding the bearing cap to the block. The extra support is needed because these engines produce more torque and higher loads on the crankshaft.

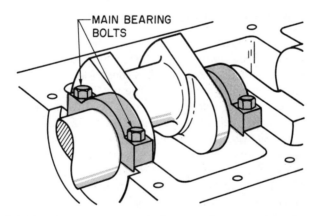

FIGURE 9–11. Main bearing caps are used to hold the crankshaft in place in the cylinder block. They can be removed by removing the bearing bolts. *(Courtesy of Federal-Mogul Corporation)*

Bearing Design

Insert bearings are placed between the main bearing caps and the crankshaft. As discussed in an earlier chapter, these bearings are designed as two-piece friction-type bearings. Main bearings have what is called spread. They are slightly larger than the housing into which they fit. See *Figure 9–12*. The spread allows them to snap into place. Half the bearing is placed into the block. The other half is placed in the bearing cap.

The main bearings have several parts. All bearings have a steel backing. This provides the strength and support for the bearing. There are also several soft metals used on the bearing surface. Soft metals such as copper-lead, *Babbitt*, aluminum, and tin allow a certain amount of dirt to be embedded into the soft metal. They also help the bearing to form to the shape of the crankshaft journals. *Figure 9–13* shows a common bearing with several metals in its design.

Bearings must be designed to accomplish several things. These include:

a. Load-carrying capacity — The ability to withstand pressure loads from combustion.
b. Fatigue resistance — The ability to withstand constant bending.
c. Embedability — The ability to permit foreign particles to embed or be absorbed into the metal.
d. Conformability — The ability to be shaped to the small variations in shaft alignment. See *Figure 9–14*.
e. Corrosion resistance — The ability to resist the by-products of combustion that are carried to the bearing by the oil.
f. Wear rate — The ability to be strong enough to eliminate excessive wear.

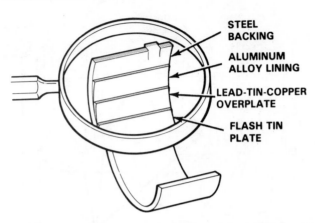

FIGURE 9-13. Main bearings have a steel backing and several soft metals such as copper, lead, and aluminum. *(Courtesy of Federal-Mogul Corporation)*

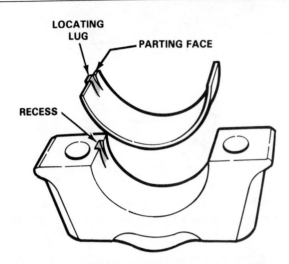

FIGURE 9-15. Bearings have locating lugs. These are used to hold the bearing in place and eliminate the bearing spinning in its bore. *(Courtesy of Federal-Mogul Corporation)*

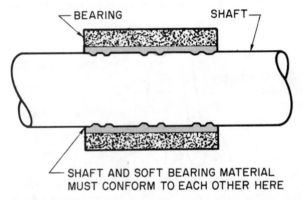

FIGURE 9-14. Bearings must be able to conform to the shape of the crankshaft. This is called conformability. *(Courtesy of Federal-Mogul Corporation)*

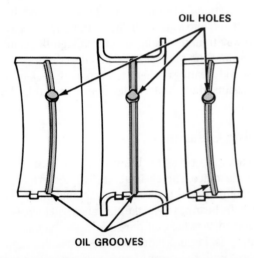

FIGURE 9-16. Oil holes and grooves are used to help lubricate the bearings. *(Courtesy of Federal-Mogul Corporation)*

The bearing insert also has a locating lug. See *Figure 9-15*. This lug holds the bearings in place within the block and cap so that the bearing cannot spin. Oil grooves and holes are also machined into the bearing insert. Oil passes from the center of the crankshaft through the hole in the bearing, and circles around the crankshaft in the oil groove. This design provides complete lubrication of the crankshaft journals. See *Figure 9-16*.

There must also be a clearance between the crankshaft journals and the insert bearings. This clearance is called the *main bearing clearance.*

Thrust Bearings

The crankshaft moves back and forth axially (movement parallel to the axis of the crankshaft) during operation. This movement can be created from angled gears on the front of the crankshaft. This axial motion, called thrust, could cause

the crankshaft to wear heavily on the block. Thrust bearings are used to compensate for this motion. One of the main bearings is designed for thrust absorption. The center bearing in Figure 9-16 is a thrust bearing. The thrust bearing has a thrust face where a machined surface on the crankshaft rubs against it. On most engines, the thrust bearing is the center main bearing. On some engines, however, the thrust bearing is the flange on the rear main bearing. Other engines use separate thrust washers instead of the flanged type. See *Figure 9-17*.

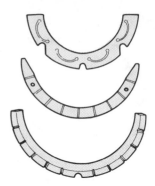

FIGURE 9–17. Separate thrust bearings are used on some older engines. *(Courtesy of Federal-Mogul Corporation)*

Oil Passageways

For the crankshaft to receive proper lubrication, oil must pass through the crankshaft to the bearings. *Figure 9–18* illustrates the internal passageways drilled for oil. Oil pressure from the lubrication system is fed through the block to each main bearing. The oil goes through the bearing insert and into a groove in the bearing. There is a drilled passageway from the main bearing through the crankshaft to the connecting rod bearing. Oil is then fed into the connecting rod bearings where it eventually leaks out and sprays against the cylinder walls. The oil then drips back to the oil pan.

Bearing Clearance

The clearance between the bearing and the journal is called bearing clearance. This clearance is designed so that just the right amount of oil is allowed to flow through it. See *Figure 9–19*. If the bearings wear, the clearance will increase. This condition may reduce oil pressure. If the connecting rod clearance is larger, more oil may pass through this clearance. With the correct bearing clearance, the amount of oil throw-off from the rotating shaft is minimal. When the clearance is doubled, oil throw-off is five times greater. As the clearance increases, oil throw-off is increased even more. Under these conditions, piston rings are unable to scrape this excessive oil from the cylinder walls. Oil will then enter the combustion chamber and be burned.

Crankshaft Seals

A seal is used at the front and rear of the crankshaft to keep the oil in the engine. See *Figure 9–20*. The rear crankshaft seals are placed in the rear main bearing cap and the block. There are several designs. Both lip-type synthetic rubber seals and graphite-impregnated wick or rope-type seals are used on the rear of the crankshaft. Some are a two-piece design, others are a one-piece insert design. One-piece lip-type seals that ride on the vibration damper are used on the front of the engine.

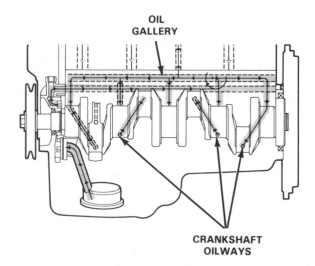

FIGURE 9–18. Oil passageways are drilled into the crankshaft to allow oil to flow from the main bearing to the connecting rod bearing journal. *(Courtesy of Federal-Mogul Corporation)*

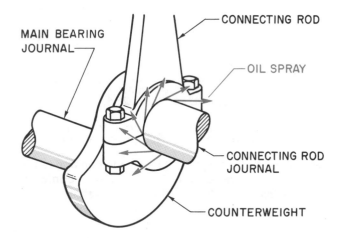

FIGURE 9–19. The bearing clearance determines how much oil will flow past and out of the bearing area. Too much clearance will reduce oil pressure and increase the amount of oil on the cylinder walls.

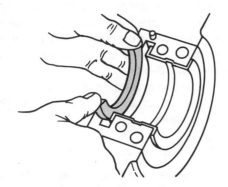

FIGURE 9–20. Seals are installed in the front and rear of the crankshaft to keep oil inside the engine. *(Courtesy of Federal-Mogul Corporation)*

CAR CLINIC ▰▰▰▰▰▰

PROBLEM: CAMSHAFT DAMAGE

A car has about 30,000 miles on it. During this time the camshaft has been replaced two times. Why would a camshaft go bad in so few miles?

SOLUTION:

Several conditions cause excessive camshaft wear. The friction and bearing loads on a camshaft are very great. Under these conditions motor oil may only be good for about 5,000 miles. The oil breaks down and cannot lubricate under high stresses. Frequent oil changes will improve this situation. Several manufacturers have also reduced the valve spring tension so as not to wear the camshaft as much. Remember also that high rpm on the engine can also increase the stresses on the camshaft.

9.3 PISTON AND ROD ASSEMBLY

The piston and rod assembly is designed to transmit the power from combustion to the crankshaft. There are several parts on this assembly. The piston and rod assembly is shown in *Figure 9–21*. It consists of the following parts:

   a. compression rings
   b. oil control rings
   c. piston
   d. piston pin (wrist pin) and lock ring (if used)
   e. connecting rod and bushing
   f. bolts
   g. bearing shells
   h. cap
   i. nuts

When this assembly is placed into the cylinder block, the downward motion from combustion is transmitted to the crankshaft.

Piston Parts

The piston is defined as a hollow aluminum cylinder. It is closed on the top and open on the bottom. It fits closely within the engine cylinder or sleeve, and is able to be driven alternately up and down in the cylinder. The piston serves as a carrier for the piston rings.

The parts of the piston are shown in *Figure 9–22*. Each part is defined.

   a. Land — That part of the piston above the top ring or between ring grooves. The lands confine and support the piston rings in their grooves.

   b. Heat dam — A narrow groove cut in the top land of some pistons to reduce heat flow to the top ring groove. This groove fills with carbon during engine operation and reduces heat flow to the top ring. Heat dams are also designed as cast slots in the piston.
   c. Piston head — The top piston surface against which the combustion gases exert pressure. The piston head may be flat, concave, convex, or of irregular shape.
   d. Piston pins (wrist pins or gudgeon pins) — Connections between the upper end of the connecting rod and the piston. They can be 1) anchored to the piston and floating in the connecting rod,

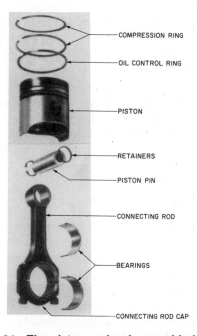

FIGURE 9–21. The piston and rod assembly is used to transmit the power from combustion to the crankshaft. *(Courtesy of Peugeot Motors of America, Inc.)*

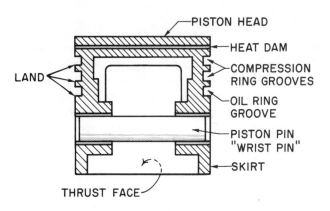

FIGURE 9–22. The piston has many parts to it. Some of the more important parts are shown.

2) anchored to the connecting rod and floating in the piston, or 3) full floating in both connecting rod and piston. Number 3 requires a lock ring to hold the pin in place.

e. Skirt — That part of the piston located between the first ring groove above the wrist pin hole and the bottom of the piston. The skirt forms a bearing area in contact with the cylinder wall and is 90 degrees opposite the piston pin.

f. Thrust face — That portion of the piston skirt which carries the thrust load of the piston against the cylinder wall.

g. Compression ring groove — A groove cut into the piston around its circumference to hold the compression rings.

h. Oil ring groove — A groove cut into the piston around its circumference. Oil ring grooves are usually wider than compression ring grooves. They generally have holes or slots through the bottom of the groove for oil drainage back to the crankcase area.

i. Piston pin bushing — A bushing fitted between the piston pin and the piston. It acts as a bearing material and is used mostly on cast iron pistons. This bushing can also be located in the small end of the connecting rod assembly. It is usually made of bronze.

Piston Requirements

The piston assembly must be designed to operate under severe conditions. The temperatures produced on top of the piston are very high. This causes stress and expansion problems. The piston is moved up and down many times per minute, which produces high pressures and stress. To handle these conditions, most pistons are made of aluminum. Aluminum makes the piston lighter. However, some larger engines, especially certain diesels, may use a cast iron piston. In this case, the rpm would be lower. The lighter pistons can operate much more effectively in today's gasoline engines, which run in excess of 5,000 rpm.

Piston Expansion

When combustion occurs on the top of the piston, some of the heat is transmitted down through the piston body. This causes the piston to expand. If the expansion were too great, the piston might wear the cylinder to a point of damage. To compensate for expansion, older pistons have a split skirt. See *Figure 9–23*. When the piston skirt expands, the slot closes rather than increasing in size.

The T slot, which is also used on older engines, is another method of controlling expansion. In this case the T slot tends to hold back the transfer of heat from the head to the skirt. It also allows for expansion within the slot.

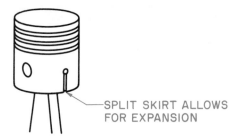

SPLIT SKIRT ALLOWS FOR EXPANSION

FIGURE 9–23. Piston with a split skirt to allow for expansion when heated.

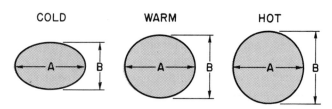

COLD   WARM   HOT

FIGURE 9–24. Cam ground pistons also help for expansion. As the piston heats up, it becomes round. This makes the piston fit more accurately in the cylinder bore.

Some pistons use steel rings, which are cast directly into the piston. These steel rings will not expand as much as the aluminum. The steel rings have a tendency to control or minimize expansion.

Cam ground pistons are also used to control the expansion in a gasoline engine. Refer to *Figure 9–24*. The pistons are ground in the shape of a cam or egg. As the piston heats up during operation, it becomes round. The piston is designed so that maximum expansion takes place on dimension B. Dimension A remains about the same.

Head Shapes

The shape of the piston head varies according to the engine. Head shapes are used to create turbulence and change compression ratios. Generally, small low-cost engines use the "flat-top." This head comes so close to the valves on some engines that there must be a recessed area in the piston for the valves. Another type of head is called the "raised dome" or pop-up head. This type is used to increase the compression ratio. The "dished" head can also be used to alter the compression ratio. *Figure 9–25* illustrates different types of piston head design. Other types of piston heads are used, but only for special applications.

Piston Skirt

In the past several years, it has become important to make the engine as small as possible, yet still powerful. One way to do this is to keep the height of the piston and connecting

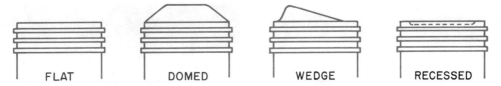

FIGURE 9–25. Pistons are designed with different head shapes. These are the most common types.

rod to a minimum. This is done by shortening the connecting rod. A *slipper skirt* is used. Part of the piston skirt is removed so the counterweights won't hit the piston. This design means there can be a smaller distance between the center of the crankshaft and the top of the piston. The output power of the engine is not affected. See *Figure 9–26*.

Skirt Finish

The surface of the skirt is somewhat rough. Small grooves are machined on the skirt so that lubricating oil will be carried in the grooves. See *Figure 9–27*. This helps lubricate the piston skirt as it moves up and down in the cylinder. If the engine overheats, however, the oil will thin out and excessive piston wear may occur.

Piston Rings

There are two types of sealing problems on the piston. Compression pressures must not be allowed to escape past the

rings. This is defined as *blow-by*. See *Figure 9–28*. If blow-by occurs, there is a loss of power. If there is excessive blow-by, too much oil might be forced off of the cylinder walls, causing excessive *scuffing* and wear on the cylinder walls and piston rings.

FIGURE 9–27. Small grooves are sometimes placed on the skirt of the piston. These grooves help to keep oil on the walls, improving lubrication.

FIGURE 9–26. A "slipper" skirt is used to make the connecting rod shorter. This makes the engine smaller and lighter.

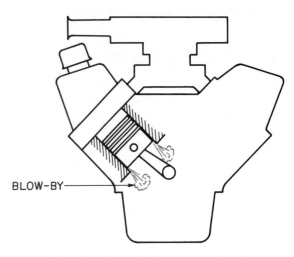

FIGURE 9–28. Blow-by is defined as combustion gases that escape past the rings and into the crankcase area.

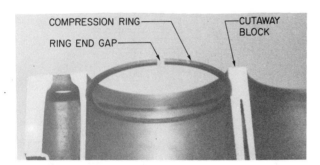

FIGURE 9-29. Piston ring end gap is the distance between the ends of the compression ring when placed in the cylinder.

The rings must also keep the oil below the combustion chamber. If not, this may result in excessive oil consumption. The moving piston is sealed with compression and oil control rings. The rings are slightly larger than the piston. When installed, they push out against the cylinder walls. Since they contact the cylinder wall, they seal against pressure losses and oil loss. Most engines use two compression rings and one oil control ring. In certain diesel engines, however, more compression rings may be used to seal the higher pressures.

Piston Ring End Gap

When the compression ring is placed in the cylinder, there is a gap at the ends of the ring. See **Figure 9–29**. This gap is referred to as piston ring end gap. The piston ring end gap gets smaller as the ring increases in temperature. If it were too small, the ring would bind in the cylinder and break, causing excessive scoring of the cylinder walls. If the piston ring end gap were too large, excessive blow-by would result. Piston ring end gap should be checked when rebuilding engines. **Figure 9–30** illustrates the measuring of piston ring end gap.

Compression Ring Material

Compression rings are made of cast iron. This material is very brittle and can break easily if bent. However, the brittle material wears very well. Certain heavy-duty engines and some diesel engines use ductile iron as piston ring material. This material is stronger and resists breaking, but the cost of these rings is higher. Some high-quality piston rings have a fused outside layer of chromium or molybdenum. Chromium or molybdenum reduces wear on the rings and cylinder walls. See **Figure 9–31**.

Ring Design

There are several types of compression rings. **Figure 9–32** shows some of the more common rings. The plain or rectangular ring fits flat against the cylinder walls. The taper-

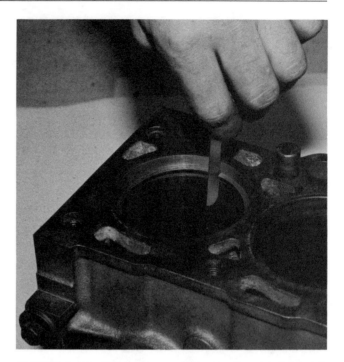

FIGURE 9-30. Piston ring end gap can be checked with a feeler gauge when placed in the cylinder.

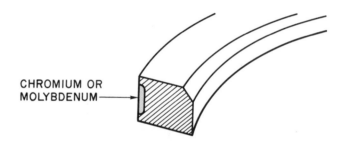

CHROMIUM OR MOLYBDENUM

FIGURE 9-31. A chromium or molybdenum layer is placed on some compression rings to improve the wearing characteristics.

faced ring improves scraping ability on the down stroke. Other rings such as the corner-grooved and reverse-beveled rings are designed as *torsional rings*. They have either a chamfer or counterbore machined into the rings.

Torsional rings are also shown in Figure 9–32. Any chamfer or counterbore causes internal stresses in the ring. These stresses cause the ring to twist slightly as shown in **Figure 9–33**. This only happens when the ring is compressed inside the cylinder. Twist is used to form *line contact* sealing on the cylinder wall and the piston ring groove. Line contact improves the sealing and scraping characteristics of the rings. This is also called static tension.

When the ring is in this position, there is no downward pressure on the ring. This happens during the intake, compression, and exhaust strokes of the engine. High pressure is only applied to the ring on the power stroke. On the intake

COMPRESSION RINGS
POPULAR RING TYPES

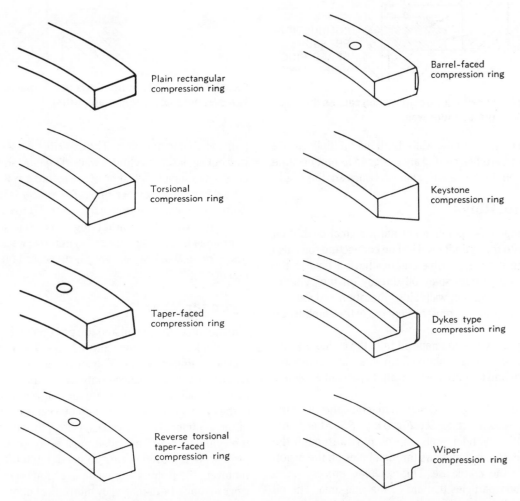

Plain rectangular
compression ring

Torsional
compression ring

Taper-faced
compression ring

Reverse torsional
taper-faced
compression ring

Barrel-faced
compression ring

Keystone
compression ring

Dykes type
compression ring

Wiper
compression ring

FIGURE 9–32. Compression rings can be designed in several ways. Note the different cross-sectional shapes used. *(Courtesy of Hastings Manufacturing Company)*

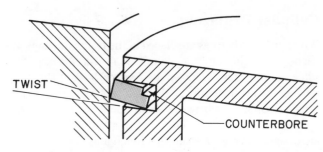

TWIST

COUNTERBORE

FIGURE 9–33. When a chamfer or counterbore is put on a compression ring, it causes the ring to twist in the groove. This helps to seal the ring to the cylinder bore during operation.

stroke, the twist forces the bottom corner of the ring to act as a scraper against the cylinder walls. This aids in removing any excess oil on the cylinder walls. On the compression stroke, the ring still retains the twist. This allows the ring to glide over any oil still on the cylinder walls rather than carrying it to the combustion chamber.

As the piston rises, compression pressures help flatten the ring for better sealing. On the power stroke, hot gases from the combustion chamber enter the ring groove. The ring is forced out and flat against the cylinder wall. See *Figure 9–34*. Now there is a good seal during the power stroke. This is also called "dynamic sealing." During the exhaust stroke, static conditions are present again. The ring has a twist again. This twist causes the ring to glide again

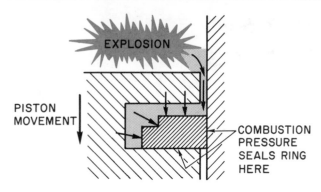

FIGURE 9–34. Pressure from combustion causes the rings to seal against the cylinder wall.

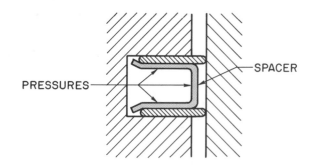

FIGURE 9–36. Oil control rings use a spacer to keep the two scrapers apart from each other.

over any oil on the cylinder walls. Both the multiple-groove and radius or barrel-faced ring are designed to produce line contact as well. Better sealing characteristics are the result.

Oil Control Rings

When the engine is operating normally, a great deal of oil is thrown onto the cylinder walls. The connecting rods also splash oil on the walls. Some engines have a hole in the connecting rod to help spray oil directly on the cylinder walls. Oil on the cylinder walls aids lubrication and reduces wear. This oil, however, must be kept out of the combustion chamber.

Oil rings are made to scrape oil from the cylinder walls. They are also used to stop any oil from entering the combustion chamber and to lubricate the walls to prevent excessive wear.

All oil control rings are designed to scrape the oil off the walls on the down stroke. See *Figure 9–35*. After being scraped off the cylinder walls, the oil passes through the center of the ring. It then flows through holes on the piston and back to the crankcase. This scraping process helps remove carbon particles that are in the ring area. The oil flow also helps cool and seal the piston.

Oil control rings are made of two, three, or four parts. These usually include an expander, a top rail, a spacer, and a bottom rail. On some rings, several of these parts may be

built together in one piece. The expander is used to push the ring out against the cylinder walls. The top and bottom rails are used to scrape the oil off of the cylinder walls. These are sometimes called scraper rings. The spacer is used to keep the two scrapers apart from each other. *Figure 9–36* shows a common type of oil control ring. On certain scraper rings, a chrome-plated section is used to improve wear characteristics. *Figure 9–37* shows several rings used on automobile engines.

Cylinder Wear

The motion of the piston and the position of the rings cause the cylinder to wear unevenly. *Figure 9–38* shows how a typical cylinder wears. The cylinder develops a taper. *Cylinder taper* is produced only where the rings touch the cylinder walls. The greatest amount of wear is near the top of the cylinder. The least amount of wear is near the bottom of the cylinder. This produces a ridge in the upper part of the cylinder bore. This ridge must be removed before the pistons are removed during an overhaul. If the ridge is not removed, the pistons will be damaged upon removal. New rings may also be damaged by hitting the bottom of the ridge.

Manufacturer's specifications as listed in maintenance manuals only allow a certain amount of taper on the cylinder walls. Too much taper will affect the piston ring end gap as shown in *Figure 9–39*. Too much end gap, as shown on the top of the cylinder walls, will produce excessive blow-by.

Cylinder Deglazing

When new piston rings are placed in the piston, the outside shape of the ring and the cylinder wall will not have the same shape. This is shown in *Figure 9–40*. The ring touches the cylinder only on the high spots of the ring. This causes poor sealing between the ring and cylinder walls.

Because of this condition, the ring and cylinder walls are designed to have a somewhat rough finish. Then, as the ring and cylinder walls begin to wear, the high spots on the rough surfaces will wear first. The two objects will then take on the same shape. This is known as "breaking-in, seating-in, or wearing-in" an engine. Deglazing is the opera-

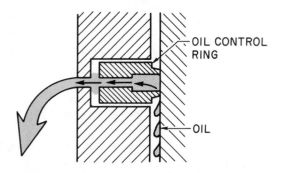

FIGURE 9–35. The oil being scraped off the cylinder wall passes through the ring, through holes in the piston, and back to the crankcase.

TYPES OF OIL CONTROL RINGS

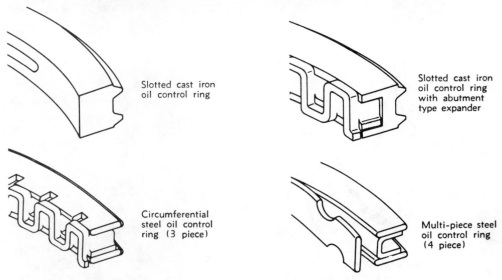

Slotted cast iron oil control ring

Slotted cast iron oil control ring with abutment type expander

Circumferential steel oil control ring (3 piece)

Multi-piece steel oil control ring (4 piece)

FIGURE 9–37. Many styles of oil rings are used on the piston. Each has a special purpose and design. *(Courtesy of Hastings Manufacturing Company)*

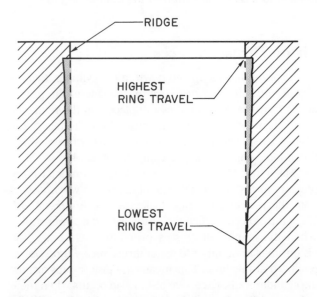

RIDGE

HIGHEST RING TRAVEL

LOWEST RING TRAVEL

FIGURE 9–38. Over a period of time, the rings on the piston wear the cylinder bore to a tapered shape.

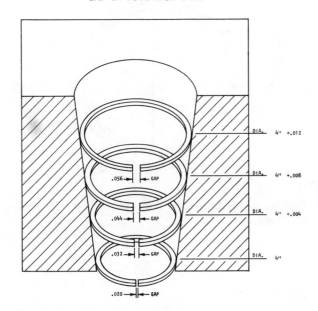

EFFECT OF .012 CYLINDER TAPER ON RING GAP IN FOUR INCH BORE

DIA. 4" +.012

DIA. 4" +.008

.056 ← → GAP

DIA. 4" +.004

.044 ← → GAP

DIA. 4"

.032 ← → GAP

.020 ← → GAP

FIGURE 9–39. A tapered cylinder bore can change the ring end gap. This may cause excessive blow-by. *(Courtesy of Hastings Manufacturing Company)*

tion used to make the cylinder rough. See *Figure 9–41*. Deglazing also tends to help retain some oil on the cylinder walls. The extra oil helps to lubricate the new piston rings and aids in the break-in process.

Honing is another process used to improve the cylinder walls. Honing is somewhat different from deglazing. Deglazing is used to roughen up the cylinder walls to remove the glaze, while honing is used to make the cylinder more round. Honing removes more material than deglazing, although the same size piston can be used. The tool used

to hone has stones positioned in a very rigid tool. Usually there are four stones on a hone and only three stones on a deglazer. Note that cylinders can also be bored using a boring tool. In this case, boring removes more material so that larger pistons and rings must be used.

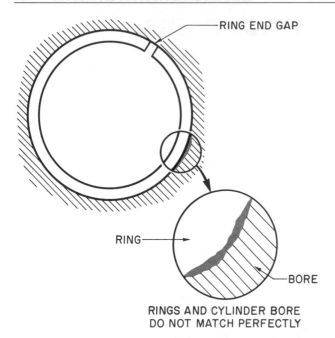

FIGURE 9-40. When new rings are placed in an engine, they are not seated. The shape of the ring is not exactly the same shape as the cylinder bore.

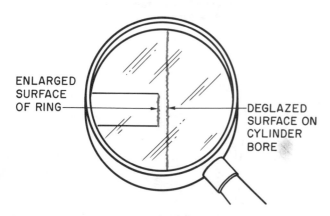

FIGURE 9-41. Deglazing a cylinder will aid in seating the rings.

Piston Pins

Piston pins, also known as wrist pins, are used to attach the piston to the connecting rod. *Figure 9-42* shows a common piston pin made of high-quality steel in the shape of a tube. Piston pins are both strong and lightweight. They are *case-hardened* to provide long-wearing operation. Case-hardened means the outer surface of the piston pin is hardened. The inside is still considered soft metal. Hardening a metal makes it very brittle and could cause the piston pin to break easier. The softer metal on the inside prevents the piston pin from cracking.

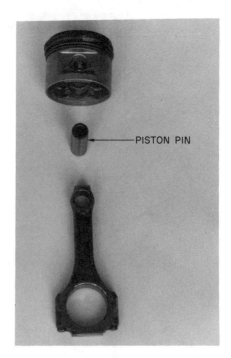

FIGURE 9-42. The piston pin is also called wrist pin and is used to attach the piston to the connecting rod.

Piston Thrust Surfaces

As the piston moves up and down in the cylinder, there is a *major and minor thrust* force on the side of the piston. Minor thrust force is the pressure placed on the right side of the piston when viewing the piston from the rear of the engine. This is when the engine is turning clockwise. Minor thrust force happens on the compression stroke.

Major thrust force is the reverse. Major thrust force is the pressure placed on the left side of the piston when viewed from the rear and turning clockwise. See *Figure 9-43*. Major thrust occurs when the piston is on the power stroke. When the crankshaft crosses over top dead center, the piston shifts from minor to major thrust force. *Piston slap* is produced at this time. This means the piston slaps against the cylinder walls. Excessive piston slap occurs when there is too much clearance between the piston and the cylinder walls. This can cause noise and wear on the piston and cylinder walls.

Piston Pin Offset

To eliminate piston slap, the piston pin is located slightly off center. The piston pin is located closer to the minor thrust surface. Because the mechanics of movement have been changed, piston slap is reduced.

Connecting Rod

The connecting rod connects the piston to the crankshaft. One end is attached to the piston pin. The other end is

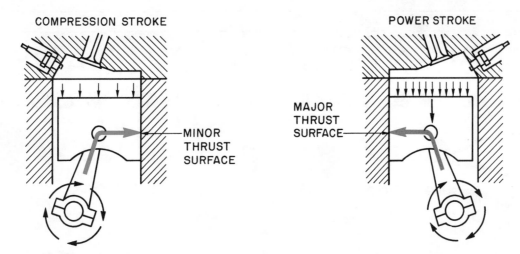

FIGURE 9–43. Major and minor thrust surfaces are built into the piston to absorb piston thrusts.

attached to the crankshaft rod journal. *Figure 9–44* shows the parts of the connecting rod. These include the connecting rod, the rod cap, two bearing inserts, the connecting rod bolts, and the nuts. Surfaces called *bosses* are forged into the connecting rod to balance it. They are machined until a perfect balance is obtained.

The connecting rod and cap are line bored. The cap is attached to the connecting rod when the inside bore is machined. It is important to keep the connecting rod cap matched to the connecting rod during service. If they are ever mismatched, the bore may be incorrect. This will cause the connecting rod and bearing cap to be misaligned, causing damage or excessive wear on the bearings.

The caps and connecting rods are marked with numbers to keep the caps matched to the connecting rods. See *Figure 9–45*. Always match these numbers when rebuilding an engine. On certain engines, the numbers have been omitted, while on others the same number may appear on the connecting rod. A good rule is to always keep the cap matched to the correct connecting rod. Some mechanics also use a punch to mark each cap and cylinder rod accordingly. This assures the mechanic that the cap and connecting rod will be assembled correctly.

Some connecting rods have an oil squirt hole in their lower section. This hole directs oil to the cylinder walls for improved lubrication. Other designs have a squirt hole in the cap mating surface. Some designs have no squirt hole.

Connecting Rod Bearings

The connecting rod bearings are designed the same as the main bearings. Each bearing is made of two pieces. Small locating lugs are used to locate the bearing in the proper position and to prevent the bearing from spinning in the bore. These lugs fit into a slot machined into the connecting rod cap and connecting rod. As in main bearings, the connecting rods are made of a steel back, copper and lead inside, and a thin coating of pure tin.

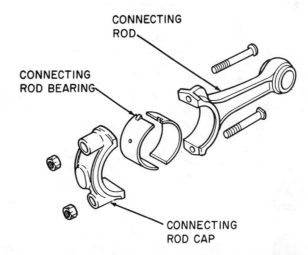

FIGURE 9–44. The connecting rod assembly includes the connecting rod, the connecting rod bearing cap, two bearing inserts, and the bolts and nuts. (Courtesy of Federal-Mogul Corporation)

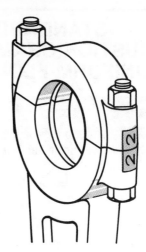

FIGURE 9–45. Connecting rod caps and connecting rods should always be kept together. Numbers are stamped on them to keep them in order. (Courtesy of Federal-Mogul Corporation)

FAST BURN COMBUSTION CHAMBER

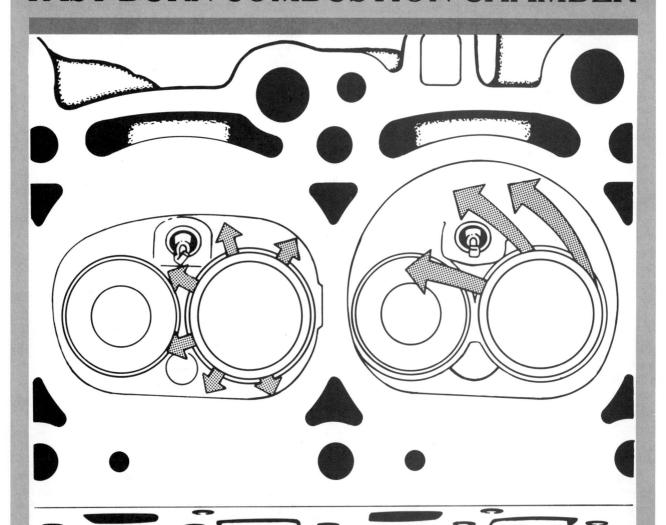

**STANDARD
COMBUSTION CHAMBER
• UNIFORM FLOW**

**FAST BURN
COMBUSTION CHAMBER
• TANGENTIAL FLOW**

A fast burn combustion chamber is used on certain four-cylinder engines to improve engine efficiency. Faster combustion is achieved by directing air flow tangentially through the intake valves to create turbulence. Fast burn combustion chambers also decrease potential engine "knock." With less potential for knock, compression ratios can be increased without increasing fuel octane requirements. (Courtesy of Chrysler Corporation)

DIAGNOSIS and SERVICE

1. A *compression test* is made to determine the condition of the engine parts that affect cylinder pressure. This test should always be performed before a tune up or whenever there is a complaint about engine performance, or excessive oil or fuel consumption. A compression test is a performance comparison between the cylinders, which is compared to manufacturer specifications. The specifications are listed in many auto maintenance manuals. All conditions during the test should be the same. This includes the cranking speed, position of throttle and choke, and temperature of engine during the entire test. There are two types of compression test: the dry and wet compression tests. The preparation for each test includes:
   a. The engine must be at normal operation temperature.

   b. Check battery for being fully charged.
   c. Check starter for operation at the same speed.
   d. Remove all spark plugs. **Note:** *When removing the spark plug wires, pull only on the boot.* Pulling on the wires may damage the carbon-impregnated core and cause the engine to misfire.
   e. Connect a jumper wire between the high-tension terminal of the coil and ground. This will prevent high-tension sparking while cranking the engine.
   f. Connect a remote starter switch so the engine can be cranked from the engine compartment.

2. The *dry compression test* is performed as follows:
   a. Hold the carburetor linkage and choke wide open.
   b. Screw the compression tester into the spark plug hole, only finger tight. Do not use a wrench. See ***Figure 9–46***.
   c. Crank the engine until the compression gauge reaches the highest reading. This should be reached with three or four revolutions of the engine. Write down the highest reading.
   d. Repeat this procedure with the same number of revolutions on all cylinders. Be sure to release the compression gauge between each test.
   e. Compare the readings of all cylinders.
   f. A 20% variation between the cylinder readings is usually satisfactory. If the readings vary more than 20%, the cause should be determined. Refer to the manufacturer's specifications for the exact amount of variation on each engine.

3. A *wet compression test* is taken to determine if low readings are due to compression leakage past the valves or piston rings. Squirt a small amount of oil in each cylinder through the spark plug hole. Now perform the compression test as outlined in Number 2. The oil should temporarily form a seal between the piston and cylinder wall. If the read-

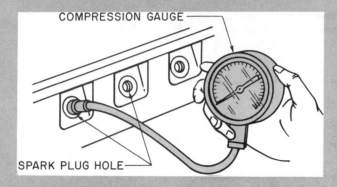

FIGURE 9–47. A deglazer is used to roughen up the cylinder walls to aid in break-in when rebuilding an engine. A ridge reamer, which is also shown, is used to remove the small ridge on the top of a cylinder. *(Courtesy of Snap-On Tools Company)*

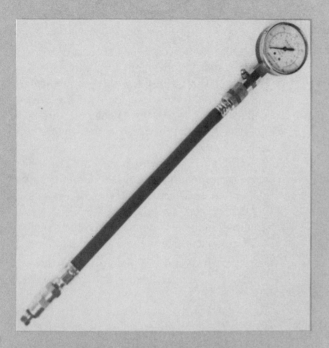

FIGURE 9–46. Both a wet and dry compression test can be made using the compression tester.

ings increase 10 pounds or more than during a dry test, compression is probably leaking by the rings. If there is no increase in the readings, a compression loss is probably caused by compression leaking past the valves. Note that when taking a compression test on new rings, it may not be possible to obtain the full compression until the rings are seated.

4. Always deglaze the cylinder bore when using new piston rings. A *deglazing tool* is shown in **Figure 9–47**.
   a. With the piston removed from the cylinder, place the deglazer in the cylinder bore.
   b. Place a small amount of cutting oil in the cylinder.
   c. Using a drill, rotate the deglazer slowly, making sure the deglazer does not hit the crankshaft. This may break the deglazer stones.

d. While rotating the deglazer, move it up and down to obtain a cross-hatched pattern in the cylinder bore.
   e. Remove the deglazer and clean the bore.
   f. Check the bore for any spots on the cylinder bore that were not evenly deglazed.
   g. If there are spots that have not been deglazed, try deglazing again. If the spots are too large, however, the cylinder may have to be honed. Honing takes more material off. Again, make sure the cylinder bore has been completely cleaned after deglazing, using clean rags.

5. Always check the top of the *block for flatness*. Warpage can be created from excessive heat.
   a. Use a long, steel, straight-edged ruler.
   b. Place the straight edge on top of the block.
   c. Using a feeler gauge, check for clearance under the straight edge. If the space is greater than the specifications listed in the repair manual, the block may have to be machined.

6. Always check the block for small cracks between the cylinder wall and the valve ports. Excessive heat can cause the block to crack at this point.

7. When installing core plugs, make sure the plug is inserted evenly. Use a small amount of sealant between the plug and the block.

8. Never use a vibration damper from another engine. It may not be designed for the engine and could cause the crankshaft to break.

9. When installing main bearing caps, make sure the caps are placed in the correct order. The caps should be inserted exactly in the same position as when disassembled.

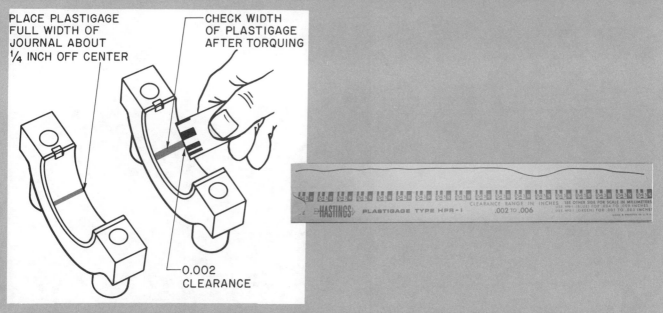

FIGURE 9–48. Plastigage is used to check the clearance of the main and connecting rod bearings. As the cap is torqued to specifications, the small plastic strip flattens out. The width of the strip determines the clearance.

10. Always check *bearing clearance* with plastigage. See *Figure 9–48*.
    a. Place a small amount of plastigage on the bearing journal.
    b. Place the bearing and bearing cap on the journal and torque to maintenance manual specifications.
    c. Remove the bearing cap and bearing.
    d. Measure the width that the plastigage has been flattened. This can be done using a vernier caliper or the gauge furnished on the plastigage container.
    e. The width of the flattened plastigage is a measure of the bearing clearance.

11. Remember to check the oil passageways for foreign matter blocking the hole. See *Figure 9–49*.

12. Whenever rebuilding an engine, replace both the front and rear crankshaft seals.

13. When pistons have been removed, replace them in the same cylinder and in the correct position.

14. Use a *ring groove cleaner*, *Figure 9–50*, to clean the carbon built up in the ring grooves before replacing the piston rings.
    a. Select the correct cutting tool for the right size ring groove.
    b. Place the ring groove cleaner in place with the cutting tool in the ring groove.
    c. Rotate the ring groove cleaner until all carbon is removed from the groove.
    d. Continue cleaning until all of the carbon has been cleaned from the ring grooves.

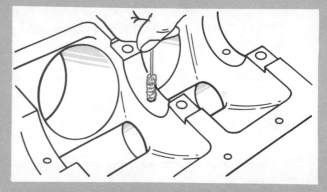

FIGURE 9–49. All oil passages on the crankshaft should be cleaned before installing. *(Courtesy of Federal-Mogul Corporation)*

    e. Clean the piston completely before installing new rings.

15. When pistons are removed, check for cracking between the piston pin and the top of the piston.

16. Always check *piston ring end gap* on new piston compression rings. End gap measurements are typically about 0.004 inch for each inch diameter of the piston.
    a. Place the compression ring in the cylinder bore without the piston.
    b. Make sure the compression ring is placed evenly in the bore.
    c. Using a feeler gauge, measure the gap produced by the ring in the cylinder.

17. When installing a ring on the piston, make sure the ring is placed in the correct groove and installed with the correct side up.

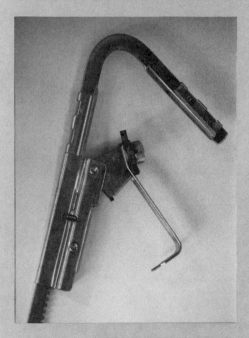

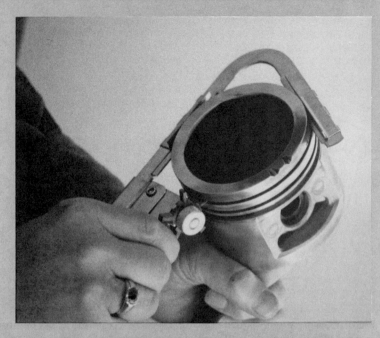

FIGURE 9–50. The ring groove cleaner is used to remove carbon from the ring grooves.

18. When pistons have been removed from the cylinder bore, check the amount of *cylinder taper*.
    a. Taper can be checked by using a bore gauge.
    b. Taper can also be checked using an inside diameter micrometer.
    c. Take two readings at several points down the cylinders (take one reading, then rotate the gauge 90 degrees and take a second reading).
    d. Record all readings.
19. Check the piston pin for excessive wear in the piston.

20. Check the connecting rod bearing clearance with plastigage. Use the same procedure as with main bearings.
21. Whenever any part is installed on the block, always torque the fasteners to the correct torque specification.
22. When installing pistons, use a *piston ring compressor*.
    a. *Figure 9–51* shows a ring compressor.
    b. Make sure there is sufficient lubrication on the new rings and cylinder walls.

FIGURE 9–51. The ring compressor is used to compress the rings so the piston can be inserted into the cylinder bore.

c. Place the bearings on the connecting rod.

d. Place the connecting rod gently in a vise.

e. Make sure the end gaps of the rings are not aligned together. The end gaps are typically staggered around the piston (120 degrees apart).

f. Place the piston ring compressor around the top of the piston and rings.

g. Tighten the ring compressor to compress the rings.

h. Place the assembly in the cylinder bore making sure the piston and connecting rod are located correctly.

i. Push the piston down into the bore with a soft rubber hammer until the ring compressor is no longer needed. *Note:* Do not force the piston into the cylinder. There should be little resistance as the piston goes into the cylinder. If there is resistance, the ring compressor may not be aligned correctly. Remove the assembly and start over.

23. When installing the crankshaft, use a feeler gauge to check for *crankshaft end play.* See **Figure 9–52**.

a. Place the crankshaft in the block.

b. Move the crankshaft fully to one end of its end play.

c. Move the crankshaft to the other end of its movement. Notice the distance moved.

d. Select the correct size feeler gauge to measure the amount of movement axially on the crankshaft.

e. Check the clearance between the thrust bearing (usually the center bearing) and the thrust surfaces on the crankshaft.

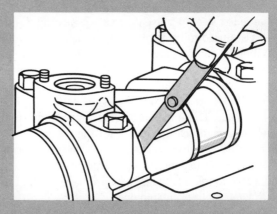

FIGURE 9–52. A feeler gauge can be used to check the crankshaft end play. (Courtesy of Federal-Mogul Corporation)

24. Care must always be taken to install main and connecting rod bearings correctly. All oil holes must be correctly lined up with oil passages in the block assembly.

25. Before removing pistons from the cylinder bore, remove the ridge that has been produced with a *ridge remover.*

a. A ridge reamer such as the one shown at the top of Figure 9–47 can be used to remove the ridge.

b. Bolt the ridge reamer on the block.

c. Adjust the ridge reamer to take a very small cut.

d. After getting the feel of rotating the ridge reamer, adjust it to take a larger cut.

e. Be careful that the ridge reamer doesn't produce chatter. This may cause vertical grooves to be cut in the cylinder.

f. Continue removing the ridge until there is no ridge left on the cylinder wall.

SUMMARY

The purpose of this chapter was to investigate the design of the basic engine. The basic engine includes the cylinder block, crankshaft, bearings, seals, piston and rod assembly, and the camshaft.

The cylinder block is the basic structure of the engine. All other parts are attached to the cylinder block. It is made with large cylinder holes for the pistons. The lower section contains the crankshaft. There are many drilled passageways for oil to lubricate other parts of the engine.

Cylinder blocks are made by pouring liquid metal into a mold. Special metals are added to the metal to make it stronger and able to withstand more wear. After the block is cast, it is machined so other parts can be attached to it. Core or expansion plugs are installed into the cylinder block. These are used so that if the coolant freezes and expands, the plugs will pop out. This will prevent the block from cracking.

On certain engines, the cylinders are machined so the piston can fit correctly. Other engines use a machined sleeve that is pressed into the block. Both wet and dry sleeves are used. Wet sleeves have coolant touching them. Dry sleeves do not have coolant touching them. Certain aluminum blocks also have sleeves. A steel sleeve is placed into the mold before the aluminum is poured. This sleeve, called a liner, cannot be removed.

The crankshaft assembly is used to convert the downward motion of the piston to rotary motion. It includes bearings, the flywheel, vibration damper, timing gears and sprockets, and seals. The crankshaft can either be forged or cast. In addition, the outside surface of the bearing journals are hardened steel to prevent excessive wearing. Counterweights are used to balance the crankshaft as it revolves.

Vibration dampers are used to reduce internal vibration in the crankshaft. They are also called harmonic balances. These devices help reduce torsional vibration.

The bearings used on the crankshaft are made of several metals. These metals provide support and strength as well as a good bearing surface for the crankshaft. Bearings must be designed to carry heavy loads, be corrosion resistant, be able to conform to the shape of the crankshaft, and be able to resist fatigue. They also have small locating lugs to keep them from spinning.

Thrust bearings are another part of the crankshaft assembly. They help absorb thrust forces from the crankshaft. The crankshaft has holes drilled from each main bearing journal to the connecting rod journal. These passages feed oil to the connecting rod journal.

It is very important to keep the bearing clearances within the manufacturer's specifications. Too much clearance will reduce oil pressure and increase oil thrown on the cylinder walls.

There are seals on both ends of the crankshaft. These seals are used to keep the oil contained within the engine.

The piston and rod assembly includes compression and oil control rings, the piston, piston pin, connecting rod and bushing, bolts and nuts, and bearing inserts. The piston is made of aluminum. It must be designed to withstand many forces, including pressures on top of the piston, high temperatures, and various thrust forces. Pistons have heat dams to control the flow of heat through the piston. They are designed to allow for expansion as the piston heats up.

Piston rings are used on the piston to help seal the pressures of compression and power. Oil must also be scraped from the cylinder walls. To do this, two compression rings are used, along with one set of oil control rings. There are many types of compression rings. The most common are rings designed to provide a twist when placed in the cylinder bore. This helps to create a seal during engine operation. Rings should also be checked for piston ring end gap. Too little or too large ring end gap may cause damage and poor performance.

During normal operation, the cylinder bore wears in the shape of a taper. The rings wear the cylinder bore into this shape. The taper should be checked to determine if it is within specifications. When new rings are placed on the piston, deglazing is performed to help the rings seat into the cylinder bore.

As the piston moves up and down in the cylinder, thrust forces push the piston from side to side. Piston pins are offset to help reduce the side thrust. Because of this offset, it is important that the piston be replaced in the correct position.

There are many service tips that should be followed when working on an engine. Always follow the manufac-

turer's recommended procedures when servicing the engine.

TERMS TO KNOW

Can you explain each of the following terms? Review the chapter until you can use each term correctly.

Crankcase	Cam ground pistons
Cast	Slipper skirt
Forged	Blow-by
Die cast	Scuffing
Core plugs	Torsional ring
Cylinder sleeve	Line contact
Heat treated	Cylinder taper
Counterweights	Case-harden
Fillet	Major thrust
Torsional vibration	Minor thrust
Babbitt	Piston slap
Main bearing clearance	Boss
Heat dam	

REVIEW QUESTIONS

Multiple Choice

1. The part of an engine that is the basic structure to which other parts are bolted is called the:
   a. Crankcase
   b. Crankshaft
   c. Cylinder block
   d. Camshafts
   e. Cylinder head

2. When a cylinder block is cast, the internal passageways have a _____ core.
   a. Metal
   b. Sand
   c. Cast
   d. Plastic
   e. Paper

3. _____ are used to seal up holes (used to remove sand cores during casting) in the block and can protect the block at times from cracking if the coolant freezes.
   a. Core plugs
   b. Metal seals
   c. Rubber seals
   d. Sand plugs
   e. All of the above

4. Which of the following is a type of sleeve or liner used on a cylinder block?
   a. Wet sleeve
   b. Dry sleeve
   c. Cast in liner
   d. All of the above
   e. None of the above

5. Cylinder blocks can be made of:
   a. Cast iron
   b. Brass
   c. Aluminum
   d. Plastic
   e. A and C

6. The device used to change reciprocating motion to rotary motion on the engine is called the:
   a. Crankshaft
   b. Piston
   c. Piston rings
   d. Connecting rod
   e. Cylinder head

7. Thrust surfaces are machined on the crankshaft to:
   a. Absorb rotary thrust
   b. Absorb axial thrust
   c. Help balance the crankshaft
   d. Improve friction
   e. Reduce oil flow

8. Which device is used to reduce torsional vibration?
   a. Crankshaft
   b. Piston rings
   c. Vibration damper
   d. All of the above
   e. None of the above

9. Main bearings are designed for:
   a. Fatigue resistance
   b. Embedability
   c. Conformability
   d. All of the above
   e. None of the above

10. The small lug on the bearing insert is used to:
    a. Keep the bearing from spinning in the bore
    b. Increase lubrication to the bearings
    c. Keep the bearing in balance
    d. Help in manufacturing
    e. None of the above

11. _____ are used to absorb the axial thrust produced on the crankshaft.
    a. Thrust weights
    b. Thrust bearings
    c. Expansion rings
    d. Core plugs
    e. Radial bearings

12. As the amount of main bearing clearance increases, which of the following will happen?
    a. Oil pressure will increase
    b. Oil pressure will decrease
    c. Blow-by will decrease
    d. Blow-by will increase
    e. Compression will increase

13. A/an _____ is used to reduce the heat flow to the top piston ring.
    a. Oil control ring
    b. Heat dam
    c. Piston pin
    d. Compression ring
    e. Cylinder heat sink

14. The part that connects the piston to the connecting rod is called a(an) _____.
    a. Compression ring
    b. Heat dam
    c. Wrist pin
    d. Expansion pin
    e. Connecting pin

15. Pistons are designed to have a major and minor _____.
    a. Pressure area
    b. Heat sink
    c. Wear surface
    d. Thrust surface
    e. Speed surface

16. Which of the following are used as a means to control the expansion of the piston (both new and older pistons)?
    a. Cam-shaped pistons
    b. Split or slotted skirt
    c. Oil control ring gap
    d. T slot in piston
    e. A, B, and D of the above

17. As the end gap of the compression rings increases,
    a. Blow-by increases
    b. Power from combustion increases
    c. Oil consumption decreases
    d. Compression ratio increases
    e. All of the above

18. Piston rings that have a chamfer or counterbore cut into them are called:
    a. Oil control rings
    b. Torsional rings
    c. High-pressure rings
    d. Low-pressure rings
    e. Lubrication rings

19. The oil scraped off of the cylinder walls by the oil control rings is:
    a. Sent back to the crankcase through small tubes
    b. Sent back to the crankcase through small holes in the ring groove
    c. Held in the ring groove for further lubrication
    d. Not used again
    e. Sent to the cylinder head

20. The greatest amount of cylinder wear is found:
    a. Near the top of the cylinder bore
    b. In the middle of the cylinder bore
    c. On the bottom of the cylinder bore
    d. On the extreme bottom of the cylinder bore
    e. None of the above

21. To get new piston rings to fit the cylinder bore correctly, it is necessary to:
    a. Bore all cylinders
    b. Deglaze each cylinder
    c. Polish each cylinder
    d. Replace the cylinder each time
    e. Insert a new piston each time

22. Which of the following tools is used to check the clearance between the main bearing and the main bearing journals?
    a. Plastigage
    b. Micrometer
    c. Ruler
    d. Dial indicator
    e. Vernier caliper

The following questions are similar in format to ASE (Automotive Service Excellence) test questions.

23. A piston is removed from the vehicle. After careful inspection, carbon is observed under the rings and in the ring grooves. Technician A says the carbon in the ring grooves should be removed and new rings should be installed. Technician B says the piston can and should be replaced in the cylinder bore without removing the carbon buildup. Who is right?
    a. A only
    b. B only

c. Both A and B
d. Neither A nor B

24. When taking a compression test, the psi reading on all four cylinders is within 30% of each other. Technician A says all cylinders are OK. Technician B says all are bad. Who is right?
    a. A only
    b. B only
    c. Both A and B
    d. Neither A nor B

25. After removing the cylinder heads, a small ridge is noticed on the top of the cylinder. Technician A says to leave the ridge there and remove it with a cylinder deglazer, after the pistons have been removed. Technician B says to remove the ridge with a ridge reamer before removing the pistons. Who is right?
    a. A only
    b. B only
    c. Both A and B
    d. Neither A nor B

26. Technician A says it is not OK to exchange vibration dampers from different engines. Technician B says it is OK to exchange vibration dampers from different engines. Who is right?
    a. A only
    b. B only
    c. Both A and B
    d. Neither A nor B

Essay

27. Explain how a forged crankshaft can be identified.

28. List the parts of the crankshaft.

29. What is torsional vibration?

30. What are some of the characteristics in the design of bearings?

31. What is the purpose of a thrust face on a piston?

32. What would be the result if the piston ring end gap were too small during installation?

CHAPTER 10

Cylinder Heads and Valves

INTRODUCTION

All automotive engines use a cylinder head and a set of valves to operate the engine correctly. The cylinder head acts as a cover or top to the engine. The valves allow the air and fuel to enter and exhaust the engine at the correct time. The components make up the cylinder head and valve mechanism. This chapter is designed to acquaint you with the parts used on the cylinder head and valves.

OBJECTIVES

After reading this chapter, you will be able to:
- Identify the purpose of the cylinder head.
- List all parts on the cylinder head.
- Identify the designs used for combustion chambers.
- List the parts of the valve assembly.
- Analyze the purposes of the valve guides, seals, seats, springs, keepers, retainers, and rotators.
- Define the purpose and parts of the camshaft, including the thrust plate and bushings.
- Analyze the design of cam lobes.
- Determine how to time the camshaft to the crankshaft.
- Identify the parts in the valve operating mechanism, including lifters, pushrods, and rocker arms.
- Analyze several diagnosis and service procedures used on the cylinder head and valve assemblies.

CHAPTER HIGHLIGHTS

10.1 CYLINDER HEAD DESIGN
    A. Cylinder Head Manufacturing
    B. Intake and Exhaust Ports
    C. Coolant Passages
    D. Combustion Chamber
    E. Wedge Combustion Chamber
    F. Hemispherical Combustion Chamber
    G. Diesel Combustion Chamber

10.2 VALVE ASSEMBLY
    A. Valve Parts
    B. Valve Material
    C. Valve Guides
    D. Valve Seals
    E. Valve Seats
    F. Valve Springs
    G. Valve Spring Vibration
    H. Valve Keepers and Retainers
    I. Valve Rotators

10.1 CYLINDER HEAD DESIGN

The cylinder head has several purposes. It acts as a cap or seal for the top of the engine. It holds the valves, and it has ports to allow air, fuel, and exhaust to move through the engine. The cylinder head on many engines also contains the combustion chamber for each cylinder or piston. *Figure 10–1* shows a cylinder head for a V-8 engine.

Cylinder Head Manufacturing

Cylinder heads can be made from cast iron or aluminum. Aluminum is used to make the engine lighter, but it transfers heat more rapidly and expands more than cast iron with the addition of heat. This may cause warpage. Both aluminum and cast iron objects are made by poring hot liquid metal into a sand mold.

The cylinder head must have coolant passages. This means that sand cores also have to be used in the casting process. In addition, passages must be cast for intake and exhaust ports. Figure 10–1 shows some of the ports that are cast into the cylinder head.

After the cylinder head has been cast, it must be machined. Areas must be machined so that intake and exhaust manifolds can be attached, valves can be seated, spark plugs and injectors can be installed, and a good seal can be provided to the block.

Intake and Exhaust Ports

Intake and exhaust ports must be cast into the cylinder head. These ports are made so the air and fuel can pass through the cylinder head into the combustion chamber. It would be ideal if one port could be used for each valve. Because of space, however, ports are sometimes combined. These ports are called *siamese ports*. See *Figure 10–2*. Siamese ports can be used because each cylinder uses the port at a different time.

Crossflow ports are used on some engines. Crossflow heads have the intake and exhaust ports on the opposite sides.

Coolant Passages

Large openings that allow coolant to pass through the head are cast into the cylinder head. Coolant must circulate throughout the cylinder head so excess heat can be removed. The coolant flows from passages in the cylinder block through the head gasket and into the cylinder head. Depending upon the engine configuration, the coolant then passes back to other parts of the cooling system.

Combustion Chamber

The shape of the combustion chamber affects the operating efficiency of the engine. Two types of combustion chamber

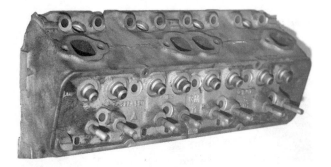

FIGURE 10–1. The cylinder head is used to seal off the top of the engine and to hold valves, allow air and fuel to flow through, and to hold the injector or spark plug.

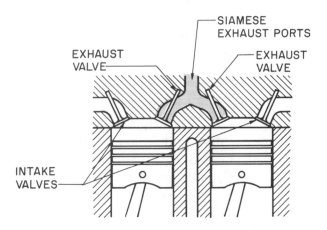

FIGURE 10–2. Siamese ports are used on the cylinder head. This means that two cylinders will feed the same exhaust port.

designs are commonly used. They are the *wedge-shaped combustion chamber* and the *hemispherical combustion chamber.*

Several terms are used to describe combustion chambers. *Turbulence* is a very rapid movement of gases. When gases move, they make contact with the combustion chamber walls and pistons. Turbulence causes better combustion because the air and fuel are mixed better. *Quenching* is the cooling of gases by pressing them into a thin area. The area in which gases are thinned is called the quench area.

Wedge Combustion Chamber

The wedge-shaped combustion chamber was used on cars until about 1968. Refer to **Figure 10–3**. As the piston comes up on the compression stroke, the air and fuel mixture is squashed in the quench area. The quench area causes the air and fuel to be mixed thoroughly before combustion. This helps to improve the combustion efficiency of the engine. Spark plugs are positioned to get the greatest advantage for combustion. When the spark occurs, smooth and rapid burning moves from the spark plug outward. The wedge-shaped combustion chamber is also called a turbulence-type combustion chamber. On newer model cars, the quench area has been reduced, which helps reduce exhaust emissions.

Hemispherical Combustion Chamber

The hemispherical combustion chamber gets its name from the chamber shape. Hemi is defined as half, and spherical means circle. The combustion chamber is shaped like a half circle. This type of chamber is also called the "hemi-head."

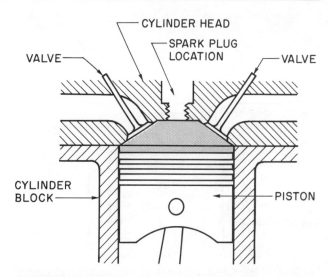

FIGURE 10–4. The hemispherical combustion chamber is shaped like a half circle. The valves are placed on both sides of the spark plug.

The valves are located as shown in **Figure 10–4**. One distinct advantage is that larger valves can be used. This improves the volumetric efficiency of the engine.

The hemispherical combustion chamber is considered a nonturbulence-type combustion chamber. Little or no turbulence is produced in this chamber. The air and fuel mixture is compressed evenly on the compression stroke. When flat top pistons are used, little turbulence can be created. The spark plug is located directly in the center of the valves. Combustion radiates evenly from the spark plug, completely burning the air-fuel mixture.

One of the more important advantages of the hemispherical combustion chamber is that air and fuel can enter the chamber very easily. The wedge combustion chamber restricts the flow of air and fuel to a certain extent. This is called *shrouding*. **Figure 10–5** shows the valve very close

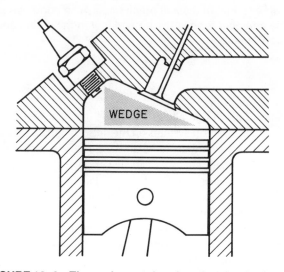

FIGURE 10–3. The wedge combustion chamber is shaped like a wedge to improve the turbulence within the chamber.

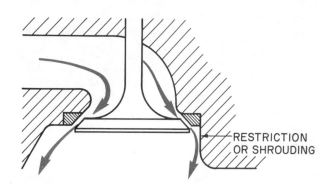

FIGURE 10–5. Shrouding is defined as a restriction in the flow of intake gases caused by the shape of the combustion chamber.

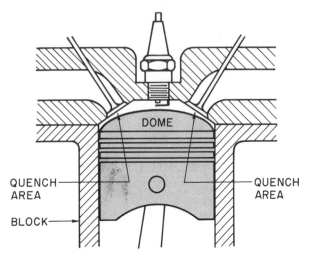

FIGURE 10–6. Certain pistons are dome shaped on the top. This design improves the efficiency of the hemispherical combustion chamber by producing a quench area.

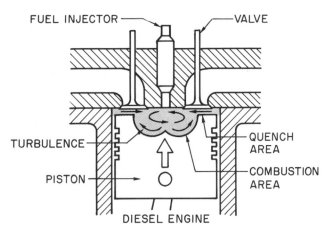

FIGURE 10–7. On certain diesel applications, the open combustion chamber is located directly within the top of the piston. This design has both a quench and turbulence area.

to the side of the combustion chamber, which causes the air and fuel to be restricted. Volumetric efficiency is reduced. Hemispherical combustion chambers do not have this restriction. Hemispherical combustion chambers are used on many high-performance applications. This is especially true when large quantities of air and fuel are needed in the cylinder.

Some high-performance engines use a domed piston. This type of piston has a quench area to improve turbulence. See *Figure 10–6*. Several variations of this design are used by different engine manufacturers.

Diesel Combustion Chamber

Diesel combustion chambers are different from gasoline combustion chambers. Diesel fuel burns differently so the combustion chamber must be different. Three types of combustion chambers are used in diesel engines: the open combustion chamber, the precombustion chamber, and the turbulence combustion chamber.

The open combustion chamber has the combustion chamber located directly inside the piston. *Figure 10–7* shows the open combustion chamber with diesel fuel being injected directly into the center of the chamber. The shape of the chamber and the quench area produces turbulence.

The *precombustion chamber* shown in *Figure 10–8* is used on both the gas and diesel engines. A smaller, second chamber is connected to the main combustion chamber. On the power stroke, fuel is injected into the small chamber. Combustion is started and then spreads to the main

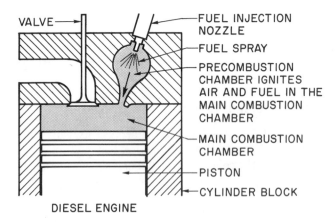

FIGURE 10–8. Precombustion chambers are used to ignite air and fuel in a small prechamber. The combustion in this chamber ignites the air and fuel in the main combustion chamber.

chamber. This design allows lower fuel injection pressures and simpler injection systems on the diesel engines. On gas engines, the precombustion chamber has a very rich mixture, but the main chamber can be very lean. The overall effect is a leaner engine, producing better fuel economy.

The turbulence combustion chamber is shown in *Figure 10–9*. The chamber is designed to create an increase in air velocity or turbulence in the combustion chamber. The fuel is injected into the turbulent air and burns more completely.

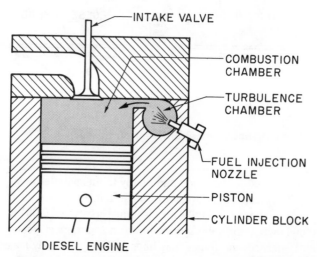

FIGURE 10-9. The turbulence combustion chamber is used on certain diesel applications to increase turbulence of air and fuel.

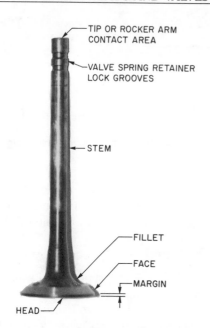

FIGURE 10-10. The parts of a typical valve.

CAR CLINIC

PROBLEM: ENGINE MISSES ON ONE CYLINDER

An engine has developed a steady miss on one cylinder. The engine has 45,000 miles on it. The engine developed the miss suddenly. The spark plugs and wires have been checked and are all OK. What might be a good thing to check?

SOLUTION:

The key is that the engine has developed a steady miss and that it happened suddenly. A common cause of this problem is a broken valve spring. To check which cylinder is missing, run the engine and carefully remove and replace each spark plug wire with the appropriate spark plug wire pliers. Note the cylinder where there is no change in the rpm. This is the cylinder that probably has a broken valve spring. Remove the valve cover and check the condition of the valve spring. Use tools designed to replace the valve spring without removing the cylinder head.

10.2 VALVE ASSEMBLY

The valves are located within the cylinder head on most engines designed today. There are two valves for each cylinder. One is for the intake and one for the exhaust. Certain high-performance engines, however, use four valves per cylinder.

The valve and associated parts include the valves, valve seats, valve guides, springs, retainers, and seals. The intake valve is usually larger in diameter than the exhaust valve because the intake valve and port handle a slow-moving air-fuel mixture. On the exhaust stroke, the gases move more easily from the pressure of the piston forcing them out. This type of valve is called a *poppet-type valve*. It is usually smaller in diameter than the intake valve.

Valve Parts

The valve has several parts. *Figure 10–10* shows the parts of the valve. The head of the valve is the part that is inside the combustion chamber. It must withstand extremely high temperatures, in the range of 1,300 to 1,500 degrees F.

The valves can be designed with different head shapes. *Figure 10–11* shows some of the shapes used on engines today. The more metal on the head of the valve, the more rigid the valve. Less metal means the valve will be able to conform to the seat more effectively. The valve is said to be elastic. Rigid valves last longer, but they don't seal as well. Elastic valves seat better, but they may not last as long.

The *valve face* is the area that touches and seals the valve to the cylinder head. This is the area that must be machined if the valve is damaged. The valve stem is used to support the valve in the cylinder head. The valve spring retainer lock grooves are used to keep the spring attached to the valve during operation. The valve margin is the distance between the face of the valve and the head. The margin is reduced whenever the valve is machined or ground down. If this margin is too small (see manufacturer's

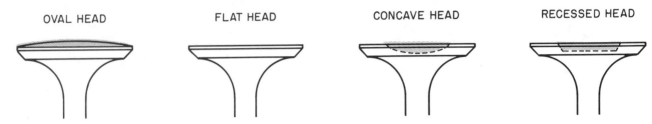

OVAL HEAD   FLAT HEAD   CONCAVE HEAD   RECESSED HEAD

FIGURE 10-11. Valves are designed with different head shapes to meet different purposes.

specification), the valve may burn easily. The fillet is the curved area between the stem and the inner edge of the face. The fillet provides extra strength for the valve.

Valve Material

The valve is made of very strong metal with nickel, chromium, and small amounts of manganese and other materials. The metal must be able to transfer heat very rapidly. If heat is not transferred rapidly, the valve will burn and become damaged. Certain valves also use a metallic sodium inside the stem. The sodium becomes a liquid at operating temperature. The liquid sodium then helps to transfer the heat from the stem to the valve guide more rapidly.

Valve Guides

The *valve guide* is defined as the hole that supports the valve in the cylinder head. It acts as a bushing for the valve stem to slide in. The valve guide is part of the cylinder head. The valve guide helps to support and center the valve so that correct *seating* can be obtained. It also helps to dissipate the heat produced within the combustion chamber through the valve. *Figure 10-12* shows a valve guide and how heat

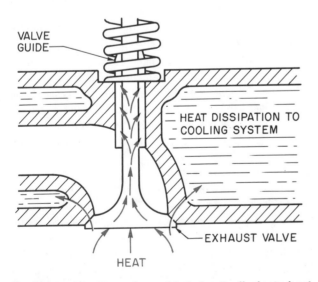

VALVE GUIDE

HEAT DISSIPATION TO COOLING SYSTEM

EXHAUST VALVE

HEAT

FIGURE 10-12. The valve guide helps to dissipate heat from the valve.

is transferred to the guide and finally to the cylinder head.

The clearance between the valve stem and the guide is very important. Generally, the clearance is between 0.001 to 0.004 inch. If the valve guides are worn and there is greater clearance, several things may happen. The valve may leak air, causing the air-fuel ratio to be altered. Oil may leak past the valve guide, causing high oil consumption. And the valve may not seat evenly, causing the valve seat to wear rapidly.

There are two types of valve guides: integral and insert. *Integral guides* are those that are machined directly into the cylinder head. The guide is part of the cast cylinder head. *Insert guides* are small cast cylinders that are pressed into the cylinder head. When valve guides are worn excessively, insert guides can be removed and replaced with new insert guides. This cannot be done with integral guides without expensive machining.

Valve Seals

Because there is a clearance necessary between the valve stem and the valve guide, oil control methods must be used. Oil deflectors are placed on the valve stem or spring. These deflectors shed oil from the valve stem and prevent oil from collecting on the top of the guide.

Positive guide seals are another means of controlling oil flow. Positive guide seals are small seals that fit snugly around the valve stem. The seal is held to the stem with small springs or clamps. See *Figure 10-13*. These seals restrict most oil that would normally pass through the valve guide.

Some car manufacturers use *passive seals*, including O rings and umbrella-type valve stem seals. These seals are used mostly on new engines. They do not work as well on older engines that have more wear on the valve stem.

Valve Seats

Valve seats are defined as circular-shaped surfaces that are machined into the cylinder block or head. The valve face seals or seats against the valve seat, *Figure 10-14*. These seats provide a surface for the intake and exhaust valves to seal for gas leakage. The seats also help to dissipate the heat built up in the valve.

There are two types of valve seats: integral and insert. As with valve guides, the integral type is cast directly as part

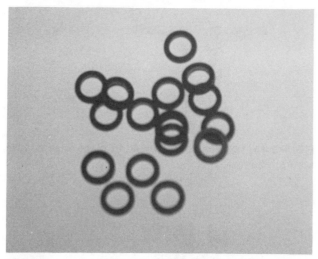

FIGURE 10–13. Valve seals help to control oil flow into the valve guide.

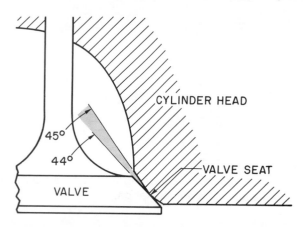

FIGURE 10–15. An interference angle is cut between the valve face angle and the seat angle. This helps the valves to seat quicker to the cylinder head.

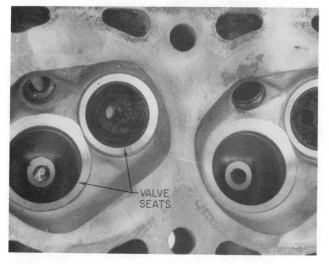

FIGURE 10–14. Valve seats help the valves seal to the cylinder block or head. They are machined into the block or head.

of the cylinder head. The insert type uses a metal ring as the seat. It is pressed into the cylinder block and ground to the correct angle. The insert type of valve seat is used most often on engines with aluminum cylinder heads. Insert valve seats can be made from cast iron, hardened cast iron, hi-chrome steel, and *stellite* (very hard steel).

Valve seats are ground to a specific angle for correct operation. They are either 30 or 45 degrees. An *interference angle* is becoming very common when grinding valves. An interference angle is obtained by grinding the valve face about one degree less than the valve seat. This is shown in *Figure 10–15*. In this example, the valve is ground to 44 degrees while the seat is ground to 45 degrees. The interference angle tends to cut through any deposits that have been formed on the seat. It also produces a more positive seal. As the engine is run and the valve seats wear, the interference angle is gradually eliminated. The result is good *line contact*, which helps transmit excess heat away from the valve.

Valve Springs

The valve springs are designed to keep the valves closed when the camshaft is not lifting the valve. The valve springs are held to the valve stem with various types of *keepers*. The spring must be designed to close the valve correctly. If the valve spring is weak, there may be *valve float*. Valve float means the valve stays open slightly longer than it is designed to. This usually happens when the valve springs are weak and the engine is operating at high speeds. *Valve bounce* can also occur if the spring is weak and operated at high speeds. Valve bounce occurs when the valve slams against the seat, causing it to bounce slightly.

Valve springs are made from several types of wire materials. These include carbon wire, chrome vanadium, and chrome silicon. The stress or load temperature and aging qualities determine exactly what type of material is used.

Valve Spring Vibration

During normal operation, valve springs may develop vibration. This vibration is known as *harmonics*. At times, these harmonics cause the spring to function incorrectly. Several

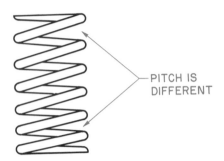

FIGURE 10–16. Variable pitch springs reduce vibration within the spring during operation.

FIGURE 10–17. A damper spring placed inside the valve spring reduces harmonics during spring vibration.

designs are used to reduce harmonics. These include using stronger springs, variable *pitch* springs, dual springs, varying the outside diameter of the spring, and placing small vibration dampers inside the springs. ***Figure 10–16*** shows the variable pitch spring. The end with the closer spacing should always be installed toward the cylinder head. ***Figure 10–17*** shows a valve spring with a damper spring inserted inside.

Valve Keepers and Retainers

Keepers and retainers are used to keep the valve spring secured to the valve stem. The retainer acts as a washer and seat for the top of the spring. See ***Figure 10–18***. The retainer sits on top of the spring. The valve then passes through the retainer.

Valve keepers are used to hold the retainers to the valve stem. See ***Figure 10–19***. Valve keepers are designed as tapered keys or locks. As the spring pressure pushes up on the retainer, the keepers are pinched or wedged into the retainer. This action causes the spring to be firmly attached to the valve during all operation. Several types of keepers are used, but the most common is the split type. The split type is easy to remove yet maintains a positive lock.

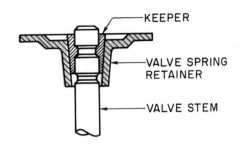

FIGURE 10–18. The valve is held in place with a valve spring retainer.

FIGURE 10–19. Valve keepers are used to keep the valve spring and retainer in place.

Valve Rotators

Valve rotators are used on certain engines. If valves are rotated a small amount each time they are opened, valve life will be extended. This is especially true when using leaded fuels. (The effect of using rotators with unleaded fuels is still being studied.) Rotating the valves:

1. minimizes deposits of carbon on the stem of the valve.
2. keeps the valve face and seat cleaner.
3. prevents valve burning caused by localized hot spots.
4. prevents valve edge distortion.
5. helps to maintain uniform valve head temperatures.
6. helps to maintain even valve stem tip wear.
7. helps to improve lubrication on the valve stem.

There are several types of valve rotators. Most operate on the principle that as the valve is compressed, small balls inside the rotator roll up an *inclined surface*. This action causes the valve to rotate slightly as it is being compressed.

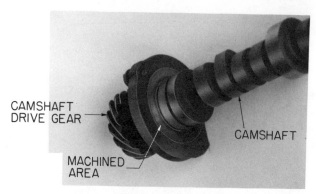

FIGURE 10–21. The thrust plate on a camshaft helps to absorb thrust forces.

CAR CLINIC

PROBLEM: ENGINE MISSES WHEN COLD

A vehicle misses very severely when it is cold. The car has about 80,000 miles on it. After the engine warms to operating temperature, the miss disappears. The problem started at about 45,000 miles.

SOLUTION:

Misses such as this are usually caused by a cracked head. When the engine warms up, the crack is sealed and the miss disappears. Give the engine a cylinder leakdown test when it is cold to see if the compression is leaking. Crank the engine at night and look for stray sparks jumping from one spark plug wire to another. Some other possible causes are too rich or lean a fuel mixture during cold running, or a compression ring or valve sticking when cold.

10.3 CAMSHAFT ASSEMBLY

The camshaft is used to open and close the intake and exhaust valves throughout the four strokes of the engine. To do this, the camshaft is driven by the crankshaft. The camshaft assembly includes the camshaft, camshaft timing gear, camshaft bearings, and timing chain, belt, or gears, if used.

Camshaft Design

The camshaft is a long shaft that fits into the block or head on overhead cam engines. Two cams or *lobes* are machined

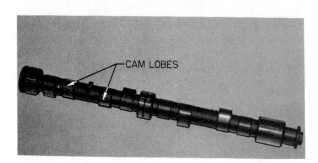

FIGURE 10–20. Cam lobes are machined on the camshaft to lift the valves open at the right time.

on the camshaft for each cylinder. As the camshaft turns, the lobes open and close the valves. See *Figure 10–20*. Several bearing surfaces are also machined on the camshaft. These surfaces are used to support the camshaft at several places throughout its length. The typical camshaft is made of cast or forged steel. The cam and bearing surfaces are hardened to provide protection from excessive wear. In addition, a gear is placed on the camshaft. This gear is used to drive the distributor and the oil pump.

Camshaft Thrust

Camshafts need some means to control the shaft end thrust. One method is to use a *thrust plate* between the camshaft gear and a flange machined into the camshaft. See *Figure 10–21*. The thrust plate is bolted to the block to contain any thrust movement of the camshaft. On certain overhead camshaft designs, the thrust plate is bolted directly to the head.

Camshaft Lobe Design

There are many designs used on camshaft lobes. The contour of the camshaft lobe determines how the valves open and close. Camshaft lobes play an important part in volumetric efficiency. The speed and amount of opening, and speed of closing are controlled by the shape of the cam lobe.

The cam lobe has several parts. See *Figure 10–22*. The amount of lift is measured by the height of the cam lobe. The shape of the ramp determines how rapidly or slowly the valves open or close. For example, on certain camshafts the ramp is designed to open the valve gradually. The valve should close as rapidly as possible, but it should not bounce on the ramp. When operating on the heel, the valve is not opened. The nose is designed to keep the valve open for a certain length of time.

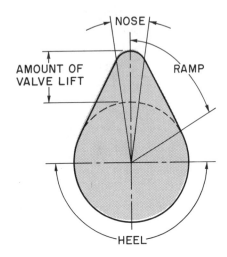

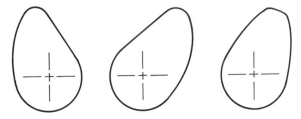

FIGURE 10-22. The shape of the camshaft lobe is designed to raise and lower the valve at a specific time.

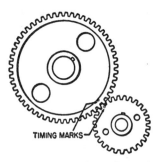

FIGURE 10-23. The camshaft and crankshaft must be timed correctly. Marks must be lined up to be installed correctly. *(Courtesy of Motor Publications, Auto Repair Manual, 1981–1987)*

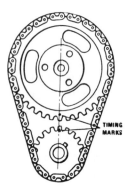

FIGURE 10-24. When a timing chain is used, the two marks on the gears must be aligned. This will time the camshaft to the crankshaft. *(Courtesy of Motor Publications, Auto Repair Manual, 1981–1987)*

Camshaft Timing to Crankshaft

For the valves to open and close in correct relation to the position of the crankshaft, the camshaft must be timed to the crankshaft. This means that the two shafts must be assembled so that the lobes open the valves at a precise time in relation to the position of the piston and crankshaft. Several methods are used to do this.

There is a set of timing gears on the crankshaft and camshaft. These gears are located on the shaft by using a *keyway*. The keyway locates the gear on the shaft in the correct position. The camshaft and crankshaft are assembled so that two dots line up. See *Figure 10-23*. If they are assembled this way, the camshaft and crankshaft will be in time with each other. Because of the four-stroke design, the camshaft always rotates half as fast as the crankshaft.

Some engines use a timing chain to connect the camshaft and crankshaft. In this case, two marks are again lined up during assembly. See *Figure 10-24*. When a chain drive is used on some engines, a spring-loaded damper pad is used to keep the chain tight.

Overhead camshafts are also timed by lining up marks shown on the two shafts. It is important to always review the procedures listed in the maintenance manual when timing the camshaft and crankshaft. A belt tension device is used to keep the belt tight.

Bushings

The camshaft is supported in the cylinder block by several bushings. These bushings are friction-type bearings and are called camshaft bearings by some mechanics. They are designed as one piece and are typically pressed into the camshaft bore in the block, *Figure 10-25*.

FIGURE 10-25. A camshaft bearing is a full round design, pressed into the block. *(Courtesy of Federal-Mogul Corporation)*

COMPUTER GRAPHICS

Computer graphics play a major role today in designing the interiors of cars. The engineer shown working at the computer screen is using computer graphics to help determine the optimum seating positions for the driver and passengers. The information generated via computer-aided design is used to help develop the seating room that simulates an actual car interior. This design permits individuals of varying sizes to evaluate the location of the steering wheel and various control knobs and levers. (Courtesy of Ford Motor Company)

VALVE TRAIN

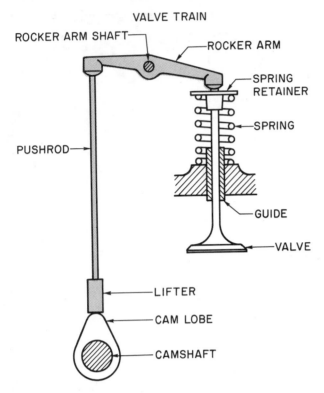

FIGURE 10-26. The lifters, pushrods, and rocker arms make up the valve operation mechanism.

FIGURE 10-27. Adjustments can be made on the rocker arms to control the valve train clearance.

10.4 VALVE OPERATING MECHANISM

The valve operating mechanism consists of the lifters which ride on the camshaft, the pushrods, and the rocker arms. These components make up the mechanism used to transfer the camshaft lift to the valve assembly, *Figure 10-26*.

As the engine heats up during operation, the parts of the valve mechanism expand. Because of this expansion, there must be a clearance in the valve mechanism. This clearance is called *valve train clearance*. All engine valve mechanisms must have this clearance.

Lifters

The valve lifters, also called *tappets*, are designed to follow the shape of the camshaft and lobes. The valve lifters ride directly on the camshaft. Lifters are used to change the rotary motion of the camshaft to reciprocating motion to open and close the valve.

Several types of lifters are used. They include solid lifters, cam followers, and hydraulic lifters.

Solid Lifters

Solid lifters (used on older engines) transfer motion as a solid piece from the camshaft to the pushrod. Solid lifters are designed as lightweight cylinders. Some are designed as a hollow tube, while others are designed as a solid piece. Solid lifters are used primarily on older engines or on high-performance engines. A disadvantage of solid lifters is that they are noisy and have a distinct tapping sound.

All solid lifters must have some means of adjusting the valve clearance. Adjustments are made on the rocker arm section of the valve mechanism. See *Figure 10-27*.

Cam Followers

Cam followers are sometimes used instead of lifters. Cam followers are also called "roller lifters" by some mechanics. Cam followers resemble solid lifters. The only difference is that followers have rollers that contact the camshaft, rather than metal sliding on the camshaft. Cam followers reduce friction and distribute the load more evenly. They are used primarily on high-compression engines such as diesel and racing engines. The roller is on one end of the cam followers. The other end is machined so the pushrod can be inserted. Several designs are used. One of the most common is shown in *Figure 10-28*. In this design, the roller rides directly on the camshaft. The pushrod sits in the cam follower. Engines that use cam followers of this design also have an adjustable valve train clearance.

Hydraulic Lifters

Hydraulic lifters are used to reduce noise and control the valve clearance. Hydraulic lifters use oil pressure to keep the valve clearance at zero. The main parts of the hydraulic lifter are shown in *Figure 10-29*. The body houses the internal parts of the hydraulic lifter. The plunger moves up and

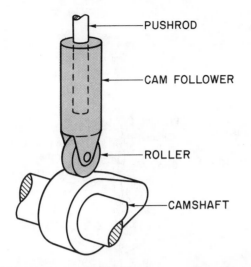

FIGURE 10-28. Cam followers are used in place of solid lifters. They have a roller that rides on the camshaft.

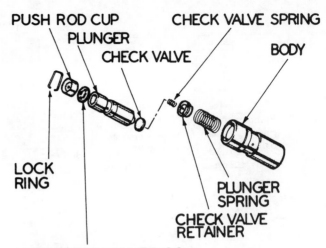

FIGURE 10-29. The parts of the hydraulic lifter are shown. The body of the lifter holds the other parts inside. *(Courtesy of Motor Publications, Auto Repair Manual, 1981-1987)*

down inside the body. *Figure 10–30* shows the hydraulic lifter operation for an overhead cam engine. The rocker arms push directly on the lifter.

The hydraulic lifter operates as shown in *Figure 10–31*. Referring to Figure 10–31 (A), before the cam lobe starts to lift, the plunger is pushed upward by the internal return spring. Oil is fed into the recessed area called the *oil relief*. The oil pressure comes from the oil system in the engine. During this time, the high pressure chamber is filled

with oil. Any clearance within the valve mechanism is now taken up by the oil pressure. Note that the oil passes by the check ball. The check ball allows oil to pass in one direction only. If oil tries to flow from the high pressure chamber into the plunger, it will be stopped. Some lifters use a check valve rather than a check ball, but the principle is still the same.

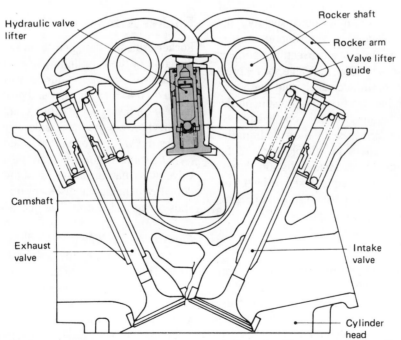

FIGURE 10-30. On an overhead cam engine, the lifters are set directly above the camshaft. No pushrod is needed. *(Courtesy of Nissan Motor Corporation in USA)*

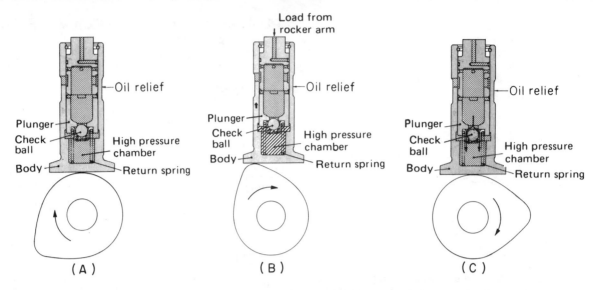

FIGURE 10–31. The operation of a hydraulic lifter is shown. (A) shows the operation before the cam lobe lifts. (B) shows the operation during cam lobe lifting, and (C) shows the operation after the cam lobe has been lifted. *(Courtesy of Nissan Motor Corporation in USA)*

When the cam lobe starts to raise the lifter, Figure 10–31 (B), the valve spring pressure, which is felt through the rocker arm, keeps the plunger from moving upward. The body of the lifter, however, is raised by the cam lobe. This causes the pressure in the high pressure chamber inside the lifter to be increased. As the cam lobe continues to lift the body of the lifter, the high pressure locks the system, and the valve is forced to open. A small amount of oil then leaks from the clearance between the plunger and the body, causing the plunger to move down slightly. This is called "leak-down." Leak-down is very important. It is controlled precisely by the clearance between the plunger and the body. Its purpose is to allow the valve spring pressure to push the plunger down adequately into the body.

As the cam lobe continues to turn, the pressure on the body is reduced. This is because the cam lobe has passed the lifter. As the pressure decreases, the pressure from the valve spring forces the valve to close. The plunger and the body of the lifter now return to the original position. Again, any clearance in the valve mechanism is taken up by the oil pressure. When the lifter is at the position shown in Figure 10–31 (C), it is ready to repeat the cycle.

Pushrods

Pushrods are designed to be the connecting link between the rocker arm and the valve lifter. They are made as light as possible. Some are designed to have small convex balls on the end. These small balls ride inside the lifter and the rocker arm. Pushrods are used only on engines that have the camshaft placed within the block. Overhead camshafts do not need pushrods.

Pushrods are either solid or hollow. On certain engines, the pushrods have a hole in the center to allow oil to pass from the hydraulic lifter to the upper portion of the cylinder head. Rather than having a convex ball, the end of the pushrod on solid lifters is concave. This end then fits into a ball on the rocker arm. *Figure 10–32* shows examples of different types of ends on pushrods.

Rocker Arms

Rocker arms are designed to do two things: 1) change the direction of the cam lifting force, and 2) provide a certain *mechanical advantage* during valve lifting. Referring to *Figure 10–33*, as the lifter and pushrod move upward, the rocker arm pivots at the center point. This causes a change in direction on the valve side. This change in direction causes the valve to open downward.

On some engines, it may be important to open the valve more than the actual lift of the cam lobe. This can be done by changing the distances from the *pivot point* to the ends of the rocker arm. The distance from the ends of the rocker arm to the pivot point is not the same. Refer to *Figure 10–34*. Note that the distance from point A to the pivot point is less than the distance from point B to the pivot point. The ratio between these two measurements is called "rocker arm ratio." In this example, the ratio is 1.5 to 1. This means that the valve will open 1.5 times more than the actual lift on the cam lobe.

Rocker arms are designed and mounted in several ways. Some are designed to fit on a rocker arm shaft. Springs, washers, individual rocker arms, and bolts are used in this

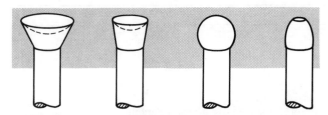

FIGURE 10–32. Several types of ends are used on pushrods.

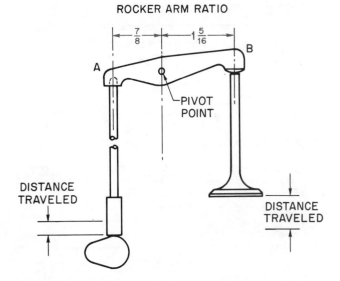

ROCKER ARM RATIO

FIGURE 10–34. Rocker arms have different lengths from the center to the ends. This produces a "rocker arm ratio," which is used to open the valve more than the cam lift.

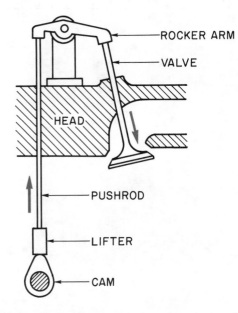

FIGURE 10–33. Rocker arms are used to change the direction of motion on the valve operating mechanism.

type of assembly. Other rocker arms are placed on studs that are mounted directly in the cylinder head.

Some overhead camshaft engines use rocker arms in such a way that the camshaft rides directly on top of the rocker arm. See *Figure 10–35*. Other overhead camshaft engines do not use rocker arms. In this type of engine, the camshaft rides directly on top of the valves.

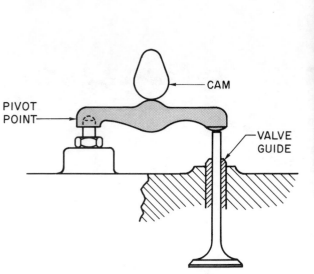

FIGURE 10–35. Certain engines have rocker arms that ride directly under the camshaft. These are used on overhead camshaft engines.

DRAGSTERS

This vehicle is called a top fuel class dragster. It uses a special frame, engine, and wheels to obtain top speeds of 265–278 mph. (Courtesy of Jim Rennich, Bloomington, MN)

DIAGNOSIS AND SERVICE

1. Always look between the combustion chamber and the valve guides for cracks in the cylinder head.

2. Check the *cylinder head* for *warpage* by using a straightedge and a feeler gauge.
   a. Using a steel straightedge, lay the straightedge across the length of the head.
   b. Try to slip a feeler gauge between the head and the straightedge. These spots show that the head is warped.
   c. Start with a 0.003 inch feeler gauge, then go to a larger feeler gauge, if possible.

   d. Check several spots on the length as well as the width of the head.
   e. A typical specification is that maximum warp is 0.003 inch for any 6-inch length or 0.007 inch overall.

3. When *cleaning the valve*, clean off the carbon buildup.
   a. Chip off the large particles of carbon with a steel punch. Be careful not to chip the valve surface, especially at the fillet.
   b. Using a wire wheel on a power grinder, clean off the remaining valve deposits.
   c. Always remember to wear appropriate safety glasses and face mask.
   d. Do not clean the stem of the valve with the wire wheel.
   e. Soak the valve in cleaning solvent to soften the remaining carbon and varnish. Use a polishing grade of emery cloth (not coarser than 300 grit) to clean the remaining varnish.

4. *Check the valve guides* for wear according to the manufacturer's specifications. These specifications are listed in many of the auto maintenance manuals.
   a. Check the valve guide by determining the maximum dimension at the port end or top end of the guide.
   b. Place a telescoping or small hole gauge inside the valve guide.
   c. After removing the gauge, measure it with a micrometer.
   d. Measure the smallest valve stem diameter on the valve. This is usually read where the valve stem touches the top of the valve guide.
   e. Subtract the two readings. This value is called the valve clearance.
   f. Average specifications are: intake, 0.001–0.003″; exhaust, 0.0015–0.0035″.

5. After the valve has been ground, check the *valve margin* and compare it to the manufacturer's recommended specifications.
   a. Use a vernier caliper to check the margin.
   b. The minimum width of valve margins is 1/32". If the width is less than 1/32", the valve should be discarded and replaced.

6. Check the valve face for indications of burned valves.

7. Always replace the valve seals when rebuilding the cylinder head.

8. Check the tip of the valve stem for wear from the rocker arms.
   a. If the stem is not square or has nicks and is rounded off, reconditioning is needed.
   b. Use the valve grinding machine by first dressing the stone correctly.
   c. Place the valve in the holding clamp and tighten the clamp.
   d. Turn on the machine and move the valve closer to the wheel. Make sure the coolant is on the valve grinder.
   e. Advance and grind the stem slowly, making sure the valve is not turning a blue color. This means that the valve is too hot and too much metal is being ground off at one time.
   f. Grind the stem until all pits and signs of wear have been removed.
   g. If the chamfer has been removed on the valve stem, it may have to be reground. Make sure the chamfer is no wider than 1/32".

9. Completely remove all carbon deposits from the cylinder head and combustion chamber.
   a. Use the valve removal tool to remove the valves.
   b. Using a small punch, chip away the larger particles of carbon.
   c. Using an electric drill with a coarse wire wheel, clean the carbon out of the combustion chamber. Be sure to wear a face mask and safety glasses.
   d. Be careful not to damage the seat area of the valves.

10. Check the valve springs for *alignment* and squareness.
    a. Stand the valve spring on a flat surface.
    b. Hold the bottom of the spring against a square.
    c. Using a feeler gauge or vernier caliper, measure the distance away from the square, the gap, from the spring top to the square.
    d. Compare the distance measured to the manufacturer's specifications. The maximum

distance allowed is usually 1/16" or 5/64". Discard the spring if the measurement is greater than the manufacturer's specifications.

11. Check all valve springs for breakage.

12. When rebuilding the cylinder head, use the correct tester to check the *valve spring tension*.
    a. Use a standard spring tester to measure the spring tension.
    b. After placing the spring in the tester, compress each spring the same amount. There is usually a stop on the handle of the spring tension tester.
    c. Read the tension on the dial and compare the reading to the manufacturer's specifications. The tension should usually be within 10% of the specification.

13. Check each pushrod for straightness. This can be done by rolling the pushrod along a flat surface.

14. Check the rocker arms for wear on the valve and the pushrod end. If signs of wear are evident, the rocker arm can be ground to eliminate the wear. Maximum metal removal is approximately 0.010".

15. When the camshaft is removed, check each cam lobe for excessive wear by *checking the lift* with a micrometer.
    a. Remove the camshaft from the engine.
    b. Place the camshaft in a set of V blocks so it can be rotated.
    c. Using a dial indicator, rotate the cam to determine the height of the cam lobe.
    d. If a dial indicator is not available, a micrometer can be used. Check the smallest diameter of each cam and subtract this dimension from the largest diameter at the cam height. Compare each cam lobe and the lift. Compare the wear to the manufacturer's specifications. The maximum lobe wear allowed is usually about 0.005".

16. When installing the camshaft, make sure the timing marks are aligned according to the service manual specifications. Automotive maintenance manuals have drawings showing how the timing marks should look. Although each manufacturer's procedure may be different, the camshaft must be timed correctly to the crankshaft.

17. When the camshaft is removed, check for excessive wear or scraping on the thrust plate.

18. Check each lifter for wear at the bottom. Lifter bottoms should show no signs of wearing, scuffing, scoring, or pitting. Lifters that are concave or flat are worn. Lifters that are slightly convex are not worn. Replace lifters if wear is evident.

19. The outer surface of the lifter should have no signs of wearing. If there is wear on the outside of the lifter, also check the lifter bores for wear. Replace the lifter if wear is observable.

20. If in doubt about lifter quality, *leak-down tests* can be performed.

   a. Two systems can be used to complete a leak-down test. One system uses a special testing fixture, the other can be done on a workbench. If a fixture is available, follow the recommended procedure in the operator's manual.

   b. If a testing fixture is not available, disassemble the lifter, removing the plunger spring and oil.

   c. Reassemble the lifter without the plunger spring and snap ring.

   d. Depress the plunger until a light resistance is felt. Note the position of the plunger. Now rapidly depress and release the plunger.

   e. Observe the position the plunger returns to. If it stays down or does not return to the previously noted position (in step d above), the leak-down rate is excessive. The lifter should be replaced.

21. Always use the correct tools to remove and replace valves. Valve springs are under a high-compression pressure and may spring loose, causing damage. *Figure 10–36* shows several valve removal tools. Note that various end adaptors are used to fit the tools to different valve spring sizes.

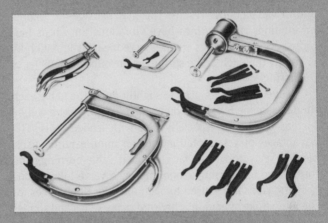

FIGURE 10–36. These valve removing tools should be used to aid in valve removal and installation. *(Courtesy of Snap-On Tools Company)*

SUMMARY

The purpose of this chapter was to investigate the design and operation of the cylinder heads and valve assemblies. These parts include the cylinder head, the valve assembly, the camshaft assembly, and the valve operating mechanism. Several service tips and diagnosis checks were also studied.

The cylinder head acts as a cap or seal for the top of the engine. The cylinder head can be made of cast iron or aluminum. Once cast, various surfaces must be machined so that other parts can fit on the cylinder head.

The cylinder head has both intake and exhaust ports to allow fuel and air to enter and leave the engine. In addition, the cylinder head has cooling passages to allow the coolant to flow through the head.

One of the more important parts of the cylinder head is the combustion chamber. Two types of combustion chamber are commonly used. They are the wedge-shaped combustion chamber and the hemispherical combustion chamber. Both are designed to improve the combustion process under certain engine conditions. Some other types of combustion chambers are used for diesel engines only. They include the open, precombustion, and turbulence combustion chambers. These chambers are needed because diesel fuel burns differently than gasoline.

The valve assembly consists of the valves, valve seats, seals, guides, springs, and retainers. There are many parts to the valve. They include the valve face, stem, fillet, margin, and valve head.

Valves can be made of several materials. The selection of material is usually based on the idea that valves must conduct heat rapidly. Nickel, chromium, and manganese are used. Some valves use a liquid sodium inside the valve to help conduct the heat.

Valve guides are used to hold the valve within the cylinder head. They also act as bushings for the valves to slide open and closed. There are two types of valve guides: integral and insert guides. Both have advantages and disadvantages for correct operation.

Valve seals are used to stop oil from seeping past the valve guides. Several types of valve seals are used. These include oil deflectors, positive guide seals, and passive guide seals. Without these seals, oil would flow past the guides into the intake and exhaust stream of gases. This might cause excessive oil consumption.

The valve seat is machined into the cylinder head to allow the valve face to seal against it. The valve seats can be either integral or insert type. These seats should always be checked for cracks and excessive wear. Seats can be ground or resurfaced. An interference angle is machined on the seat when grinding. The interference angle means that the valve is ground at 45 degrees, while the seat is ground at 46 degrees. This improves the initial seating of the valve.

Valve springs are used to keep the valves closed. They are held in place by retainers and keepers. When the springs are broken or do not have the right tension, valve float and/or valve bounce results. This may damage the engine. In addition, valve springs have vibrations known as harmonics. Special springs are used to reduce this vibration by changing the pitch of the springs or the outside diameter of the spring.

Valve rotators are used to slightly rotate the valve each time it is opened. If this is done, the life of the valve will be increased. This is especially true with leaded fuels. The advantages of valve rotation include minimized deposits, cleaner seats and valve face, reduced valve burning, uniform valve temperatures, and improved valve lubrication.

The camshaft is used to open and close the valves for the four-stroke operation. There are many styles and shapes of cam lobes. The shape determines how quickly the valves open and close. The camshaft is driven by the crankshaft. This is done with timing gears, belts, or chains. Always make sure that the camshaft and crankshaft are timed according to the manufacturer's specifications. The camshaft is held in place with the thrust plate bolted to the cylinder block.

The valve operating mechanism includes lifters, pushrods, and rocker arms. This mechanism is used to change the rotary motion of the camshaft to reciprocating motion to open and close the valves. Because of the heat of expansion, there must be a valve train clearance. This clearance is reduced as the parts heat up.

Several types of lifters are used on the valve mechanism. Solid lifters transfer motion through a solid piece of metal. All solid lifters have some means of adjusting the valve train clearance. This is usually done by moving a small adjusting screw on the rocker arm.

Cam followers are also considered solid lifters. Cam followers use a roller that rides directly on the camshaft. They must also have some means of adjusting the valve train clearance. Cam followers wear less, but they are used in applications requiring stronger valves.

Hydraulic lifters are used on many engines. Hydraulic lifters absorb the valve train clearance by using oil pressure inside the lifter body. As the camshaft lobe starts to raise the lifter, oil inside the lifter locks up and lifts the pushrod, rocker arms, and valves.

Pushrods are defined as the connecting link between the rocker arm and the lifters. Pushrods usually have small convex balls inserted on each of their ends. These convex balls fit into the lifter and rocker arm to allow proper movement. Overhead camshaft engines do not need pushrods.

Rocker arms are designed to change the direction of the cam lifting force and to provide a mechanical advantage for lifting the valve. The purpose of the mechanical advantage is to open the valve more than the actual lift of the camshaft lobe. This is done by making the distance from the center of the rocker arm to each end different lengths.

Service is very important when dealing with the cylinder head and valves. The cylinder head can be checked for flatness, cracks, and warpage. The valve seats can be ground during an overhaul. The valves can also be ground to either a 30 or 45 degree angle, depending upon the manufacturer's specifications. Other important service checks include valve spring tension and alignment, valve guide wear, broken springs, camshaft lobe wear, and lifter wear.

TERMS TO KNOW

Can you explain each of the following terms? Review the chapter until you can use each term correctly.

Siamese ports
Wedge-shaped combustion chamber
Hemispherical combustion chamber
Turbulence
Quenching
Shrouding
Precombustion chamber
Poppet-type valves
Valve face
Valve guide
Seating
Integral guide
Insert guide
Passive seal
Stellite

Interference angle
Line contact
Keeper
Valve float
Valve bounce
Harmonics
Pitch
Inclined surface
Lobe
Thrust plate
Keyway
Valve train clearance
Tappet
Oil relief
Mechanical advantage
Pivot point

REVIEW QUESTIONS

Multiple Choice

1. Which of the following is used to seal the top of the cylinder block and hold the valves?
   a. Tappets
   b. Cylinder head
   c. Lifters
   d. Camshaft
   e. Crankshaft

2. Two cylinders drawing air and fuel from the same port are called:
   a. Double porting
   b. Multiple porting
   c. Siamese ports
   d. Single ports
   e. All of the above

3. Which type of combustion chamber is designed as a half circle?
   a. Hemispherical combustion chamber
   b. Wedge-shaped combustion chamber
   c. Open combustion chamber
   d. Precombustion chamber
   e. Closed combustion chamber

4. When gases are cooled by pressing them in the cylinder area, _____ has occurred.
   a. Turbulence
   b. Quenching
   c. Combustion
   d. Cool down
   e. Convection

5. Which type of combustion chamber is called a turbulence-type combustion chamber?
   a. Hemispherical
   b. Wedge
   c. Precombustion
   d. Open
   e. Closed

6. The _____ combustion chamber is a second small chamber used to ignite a rich mixture of fuel.
   a. Hemispherical
   b. Wedge
   c. Precombustion
   d. Closed
   e. Open

7. The valve assembly consists of:
   a. The valve guides
   b. The valve seats and valves
   c. The springs, retainers, and seals
   d. All of the above
   e. None of the above

8. The _____ on the valve makes direct contact with the valve seat.
   a. Valve stem
   b. Valve face
   c. Valve margin
   d. Valve bottom
   e. Valve side

9. The distance between the valve face and the valve head is called the:
   a. Stem
   b. Fillet
   c. Margin
   d. Seat
   e. Guide

10. Which of the following valve parts are worn when oil leaks into the exhaust stream?
    a. Valve guides
    b. Valve margin
    c. Valve fillet
    d. Valve seat
    e. Valve head

11. Which type of valve seat can be removed and replaced if it is defective?
    a. Insert type
    b. Integral type
    c. Ground seats
    d. Hemi seat
    e. Wedge seat

12. When the valve is ground at a different angle than the seat by one degree, a/an _____ is obtained.
    a. Wrong angle
    b. Interference angle
    c. Poor contact
    d. Side angle
    e. Offset angle

13. Which of the following are used to reduce harmonics in valve springs?
    a. Variable spring pitch
    b. More than one spring
    c. Small spring vibration damper
    d. All of the above
    e. None of the above

14. Camshaft thrust is absorbed by the use of a:
    a. Thrust plate
    b. Thrust gear
    c. Center cam thrust bearing
    d. Special washer
    e. Thrust bearing

15. Camshafts are driven from the crankshaft by:
    a. Timing belts
    b. Timing chains
    c. Timing gears
    d. All of the above
    e. None of the above

16. Which of the following lifters has "leak-down"?
    a. Hydraulic lifters
    b. Cam followers
    c. Solid lifters
    d. Rollers
    e. Solid rollers

17. Which of the following lifters will produce the least wear on the camshaft?
    a. Hydraulic lifters
    b. Cam followers
    c. Solid lifters
    d. Hollow lifters
    e. Slide lifters

18. Rocker arms are designed in an I-head engine to do which of the following?
    a. Change the rotary motion of the camshaft to reciprocating motion for the valves
    b. Reduce wear on the valve face
    c. Hold the valve in place in the cylinder head
    d. Control the speed of the camshaft
    e. All of the above

19. Rocker arm ratio is used on valve mechanisms to:
    a. Open the valve a certain distance
    b. Reduce wear on the valve stem
    c. Increase oil sealing on the valve stem
    d. Increase the valve size
    e. Decrease the valve size

20. When installing the timing gear and camshaft:
    a. Always time the gears according to the manufacturer's specification
    b. Never worry about correct timing of the two gears
    c. Always force the gears together with a rubber hammer
    d. Put the gears 90 degrees apart from each other
    e. Bolt the gears together

21. Valve mechanisms can wear:
    a. On the valve stem
    b. On the valve face and seat
    c. On the valve guide
    d. All of the above
    e. None of the above

The following questions are similar in format to ASE (Automotive Service Excellence) test questions.

22. Technician A says a leak-down test can be performed on pushrods. Technician B says a leak-down test can be performed on lifters. Who is right?
    a. A only
    b. B only
    c. Both A and B
    d. Neither A nor B

23. After checking the lift on the lobes of a camshaft, one lobe is found to be 0.002 inch less than the other lobes. Technician A says the camshaft need not be replaced. Technician B says the camshaft should be replaced. Who is right?
    a. A only
    b. B only
    c. Both A and B
    d. Neither A nor B

24. Technician A says valve springs should be checked for weight. Technician B says valve springs should be checked for color. Who is right?
    a. A only
    b. B only
    c. Both A and B
    d. Neither A nor B

25. Technician A says the camshaft cannot wear and thus need not be checked. Technician B says the camshaft lobes will wear and thus should be checked for lift. Who is right?
    a. A only
    b. B only
    c. Both A and B
    d. Neither A nor B

Essay

26. What is quenching in reference to the combustion chamber?

27. What is the purpose of a precombustion chamber?

28. What should be done to the valve if the valve face is damaged?

29. What will be the result if the valve stem-to-guide clearance is too large?

30. What is the purpose of the interference angle on valves?

31. What is the purpose of valve rotators?

32. What is the definition of rocker arm ratio?

CHAPTER 11

Lubricating Systems

INTRODUCTION

The automobile engine has many moving parts. These parts must be lubricated adequately in order for the engine to operate correctly. During operation many contaminants enter and are produced within the engine. The lubrication system must also clean out these contaminants. To keep the engine well lubricated and able to operate for long periods of time, high-quality lubrication oil is used. The goal of this chapter is to define the purpose, operation, and characteristics of the lubrication system.

OBJECTIVES

After reading this chapter, you will be able to:
■ Define the purposes of the lubrication system.
■ Identify the contaminants within the engine that must be removed by the lubrication system.
■ Analyze the characteristics of lubricating oil.
■ Compare the different ways oil can be classified.
■ Compare the advantages and disadvantages of using synthetic oils.
■ Follow the flow of oil through an engine.
■ Examine the parts of the lubrication system in both gasoline and diesel engines.
■ Identify service procedures and suggestions for the lubrication system.

CHAPTER HIGHLIGHTS

11.1 PURPOSES OF LUBRICATION
    A. Oil Cools
    B. Oil Seals
    C. Oil Cleans
    D. Oil Lubricates
    E. Fluid Friction

11.2 CONTAMINANTS IN THE ENGINE
    A. Road Dust and Dirt
    B. Carbon and Fuel Soot
    C. Water Contamination
    D. Fuel Contamination
    E. Oil Oxidation
    F. Acids in the Engine

11.3 OIL CHARACTERISTICS
    A. Viscosity
    B. Viscosity Index
    C. Pour Point
    D. Oxidation Inhibitors
    E. Detergent-dispersants
    F. Other Additives

11.4 OIL CLASSIFICATIONS
    A. SAE Ratings
    B. API Ratings
    C. Military Ratings

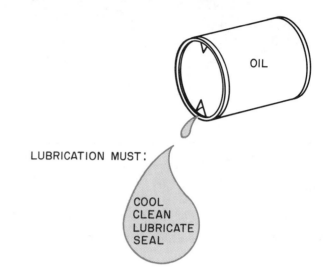

LUBRICATION MUST:

COOL
CLEAN
LUBRICATE
SEAL

FIGURE 11–1. The purpose of any lubrication system is to cool, clean, seal, and lubricate within the engine.

11.1 PURPOSES OF LUBRICATION

Lubricating oil is used in gasoline and diesel engines for many purposes. Lubricating oil is designed to do four major things within the engine. These are to cool, clean, lubricate, and seal various parts within the engine. See *Figure 11–1*.

Oil Cools

Lubricating oil comes directly in contact with vital internal moving engine parts. Lubricating oil is designed to effectively carry away the excess heat being produced. As the lubricating oil passes over a hot engine part, it removes a certain amount of heat. The heat is transferred to the crankcase oil reservoir located in the oil pan. Here the heat is dissipated into the oil. If engine lubricating oil is lacking within the engine, the engine may overheat. This is because there is not enough oil to remove the excess heat.

Oil Seals

Lubricating oil inside the engine produces a film on the moving parts. This film acts as a protective sealing agent in the vital ring zone area. Oil located on the cylinder walls and rings helps to produce a seal between the two. The oil actually takes up space between the two moving parts. Without this oil, compression gases would escape into the

crankcase area. As was discussed in an earlier chapter, cylinders and rings are deglazed or honed. Honing helps keep the oil on the walls so that sealing can be produced. It is necessary to keep all foreign materials out of this area so that proper sealing can be obtained.

Oil Cleans

It is also very important to keep all engine parts clean and free of carbon and dirt. Clean engine parts will assure proper circulation and cooling of the engine. As particles of dirt get into the engine, the lubricating oil helps to remove these parts and send them to an oil filter. Most of the physical dirt that gets into the engine comes from blow-by. Other sources of dust and dirt come in through the air filter. This dirt must be removed from the engine and sent to the filter.

Oil Lubricates

The most important function of the oil is to lubricate the parts that are moving close together. As engine parts move close together, they produce friction. *Friction* is defined as resistance to motion between two bodies in contact with each other. In an earlier chapter, frictional horsepower (the horsepower used to overcome friction) was defined. Lubricating oil helps to reduce frictional horsepower. If used correctly, lubricating oil can substantially reduce engine wear between two parts.

Fluid Friction

Oil, like other matter, is composed of molecules. These oil molecules exhibit two important characteristics for engine lubrication.

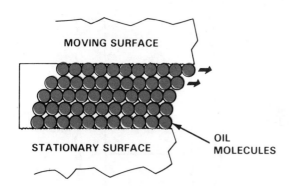

FIGURE 11–2. Oil acts as many small ball bearings. When movement occurs, the internal layers of oil will slide against each other, reducing friction. (Courtesy of Federal-Mogul Corporation)

1. Oil molecules will stick to metal surfaces more readily than to other oil molecules.
2. Oil molecules will slide against each other freely.

Referring to **Figure 11–2**, if a film of oil is flowing between two metal surfaces, the top layer of oil molecules will stick to the top piece of metal. The bottom layer of oil molecules will stick to the bottom piece. As the top piece of metal is moved, the internal layers of oil molecules will slide against each other. This action will reduce friction as compared to metal pieces that slide against each other without the oil film.

CAR CLINIC ▬▬▬

PROBLEM: CONSTANT OIL LEAK

An older foreign car is being restored. The engine has been rebuilt and is running well. An oil leak around the oil pan persists, however, even though all of the gaskets have been replaced. What other checks could be made?

SOLUTION:

Many older engines, especially foreign engines, leak oil. The most common reason is that the oil pan bolts (or other bolts) index directly with the crankcase area. Oil vapors carrying oil seep around the threads of the bolts (because of crankcase pressure), and leak out of the engine. To cure this problem, all of the oil pan bolts will have to be sealed with an epoxy or other sealant to stop the leaking. Some engines use copper washers to help seal bolts indexing with an oil area. The rule to remember is that any bolt that indexes with an oil area should be sealed to prevent leaking.

11.2 CONTAMINANTS IN THE ENGINE

Many *contaminants* get into the engine oil. These contaminants must be controlled. This means that either the oil must be designed to reduce the amount of contaminants in the engine or the oil must be filtered to reduce the amount of contaminants. To understand the lubricating system better, several contaminants should be defined. These include dust and dirt particles, carbon, water, fuel, oil, oxidation, and acid buildup.

Road Dust and Dirt

Road dust and dirt can get into gasoline and diesel engines from the outside. Road dust and dirt enter the engine through the air cleaner and travel into the combustion chamber. They then pass by the rings during blow-by and into the crankcase area. After the dirt gets into the crankcase, it settles in the oil. The typical gasoline engine takes in an excess of 10,000 gallons of dust-carrying air for each gallon of gasoline it consumes. This dust and dirt can be very harmful to the engine. It is the job of the oil system to remove these contaminants.

Carbon and Fuel Soot

Many particles of carbon and soot are not burned completely in the combustion process. Pressures produced by combustion forces some of this unburned fuel and carbon past the rings and into the crankcase area. This process is known as blow-by. Again, if carbon and soot particles were to remain in the oil, the life of the engine could be reduced drastically. The oil must be filtered to remove these particles.

Water Contamination

Water can also contaminate the engine lubrication system. Water can contaminate the lubricating oil because of leaks in the cooling system or from water vapor. Water vapor is part of the combustion by-products. The blow-by process described earlier forces the water vapor into the crankcase area. As the water vapor enters the crankcase, it cools and condenses into a liquid. Water vapor can also enter the crankcase through several pollution control devices. Air usually has moisture in it. As outside air flows through the crankcase, the moisture from the air is condensed and forms a liquid. Water within the crankcase can produce problems with *sludges* and acids.

Fuel Contamination

The performance of motor oil is seriously affected when unburned or partially burned fuel enters the crankcase. Faulty operation of the choke (rich mixture), the fuel pump, poor combustion, bad timing, or worn pistons and rings can all cause fuel to enter the crankcase. Faulty injection and

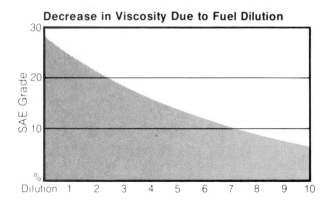

FIGURE 11-3. As fuel contaminates the oil, the oil becomes thinner. This reduces the viscosity and makes the oil less desirable. *(Courtesy of Dana Corporation)*

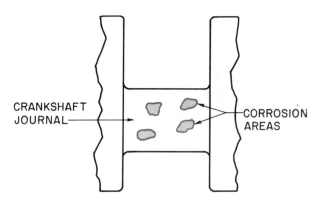

FIGURE 11-4. As acids get into the engine oil, they tend to cause corrosion and pitting. Note the pitting caused by the acidic oil.

worn engine parts produce the same condition in diesel engines.

Fuel *dilution* of crankcase oil may occur in any engine. The amount may vary from a mere trace to as much as 50%. The presence of fuel in excess of 5% gasoline or 7% diesel fuel may lead to rapid engine wear and deterioration of oil. Fuel contamination of 5% changes the qualities of motor oil. It makes it thinner and less able to stick to the moving parts. This reduces the ability of the oil to lubricate. See *Figure 11-3*.

Oil Oxidation

Oil *oxidation* occurs within the engine during normal operation. Oil oxidation is defined as the combining of hydrocarbons and other combustion products with oxygen. This is a normal process or by-product of combustion. The result is the actual production of organic acids within the crankcase. This produces oil with high *acidity*. Oil oxidation is enhanced or increased in the presence of certain types of metals used in the engine. In fact, certain metals act as a *catalyst* to promote further oxidation and add more corrosive acids to the oil. Increases in oil temperature and in water in the crankcase also enhance oil oxidation. Once the process of oil oxidation starts, the situation becomes progressively worse because it is self-accelerating.

Acids in the Engine

Certain *organic* acids are highly corrosive. Other organic acids tend to form gums and lacquers within the engine crankcase. If allowed to become concentrated from not changing oil regularly, the organic acids will attack certain bearing metals. This causes pitting and failure. See *Figure 11-4*.

The acids also react with the remainder of the oil to form soft masses. These are referred to as "sludges." Sludges cause engine trouble by settling in the oil passageways, in the oil pan, in the filters, and in the oil coolers. The heavier oxidation products form hard *varnish* deposits on pistons, valve stems, and other engine parts. The sludges, acids, and varnish obviously reduce engine life. Therefore, it is very important that oxidation be reduced as much as possible.

11.3 OIL CHARACTERISTICS

As has been discussed, oil has several major jobs to do within the engine. Oil must reduce contaminants within the engine, and cool, clean, lubricate, and seal under various conditions.

Viscosity

The *viscosity* of lubrication oil is defined as the oil's *fluidity* or thickness at a specific temperature. Viscosity is also defined as "resistance to flow." Viscosity is measured by a device known as the *Saybolt Universal Viscosimeter*. See *Figure 11-5*. Viscosity is determined at a specific oil temperature. The most common temperatures are 0, 150, and 210 degrees F. A sample of oil is drained from the viscosimeter into a receiving flask. This is done at a specific temperature. The time required for the sample to completely drain is recorded in seconds. The viscosity is given in Saybolt Universal Seconds (SUS) at a particular temperature.

Viscosity Index

As an oil becomes cooler, it also becomes less fluid. As an oil becomes hotter, it becomes thinner. The *viscosity index* (VI) is used to control oil thickness at different temperatures. The viscosity index is a measure of how much the viscosity of an oil changes with a given temperature. Chemicals are

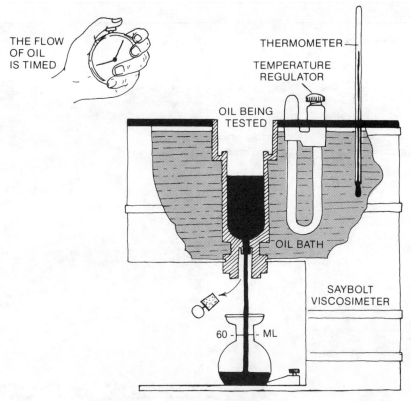

FIGURE 11–5. A Saybolt Universal Viscosimeter measures the thickness of oil. *(Courtesy of Davis Publications)*

added to oil so that as the temperature of the oil changes, the viscosity will not change as much. Usually the higher the viscosity index, the smaller the relative change in viscosity with temperature.

Pour Point

Oil must flow as a liquid in order to be used. Extreme cold conditions may increase oil viscosity until it cannot flow. The temperature at which the oil ceases to flow is defined as the *pour point*. As the temperature of the oil approaches its pour point, it becomes thicker and more difficult to pump. *Additives* are put into the oil to control the pour point.

Oxidation Inhibitors

Oxidation *inhibitors* reduce oxygen, which reduces the formation of sludges and varnish produced within the engine. As indicated earlier, any oxygen in the crankcase area may increase oil oxidation in the engine and thus produce acids. Chemicals that help eliminate or absorb oxygen are added to oil during the refining process. If the oxygen is absorbed, less is available to produce oil oxidation. In turn, fewer acids are produced. Note that if oil is not changed regularly, oxidation may increase.

Detergent-dispersants

Detergents are added to engine oil to help clean particles, dust, dirt, and other foreign materials from the engine. Detergents in the oil loosen and detach deposits of carbon, gum, and dirt. The oil suspends these particles. This characteristic is called *suspendability*. The oil then carries the loosened materials to the filter. Particles that are smaller than the filter size stay in the oil until the oil is changed. Note, however, that detergents also cause a certain amount of increased wear on the engine.

Dispersants are also added to oil. A chemical is added to the oil to keep the small dirt particles dispersed or separated from each other. Without dispersants, particles tend to collect and form larger particles. These particles, which are heavier and harder to suspend in oil, can block oil passageways and filters. Dispersants greatly increase the amount of contaminants the oil can carry while functioning correctly.

Petroleum manufacturers are now placing more emphasis on dispersants than on detergents. It is felt that if the contaminants can be kept suspended in the oil, they will not deposit on engine parts. This means there is less need for detergent chemicals in the oil.

ROAD TESTS

Car manufacturers test their vehicles under many adverse conditions. These range from super highways to dirt roads, from banked, high-speed test tracks to punishing bumps, from powdery sand to soupy mud, from hairpin curves to seemingly impossible grades. Manufacturers have developed their own test tracts, as shown here at the Milford, Michigan Proving Grounds. (Courtesy of General Motors Milford Proving Grounds)

Other Additives

Antifoaming additives are added to the lubricating oil to reduce foaming of the oil. During normal operation, the oil is pressurized and pumped through the engine. This constant pumping may cause the oil to foam. Any foam produced in the engine reduces lubricating, sealing, cooling, and cleaning. Hydraulic lifters also operate poorly with oil that has foam mixed in it. Chemicals are added to the oil to reduce foaming.

Corrosion and *rust inhibitors* are added to oil to reduce or *inhibit* corrosion of engine parts. Chemicals are added to help neutralize any acids produced. Rust inhibitors are also used to help remove water vapors from engine parts. Oil can then coat the engine parts correctly.

Antiscuff additives are put in the oil to help polish moving parts. This is especially important during new engine break-in periods.

Extreme-pressure resistance additives are added to the lubricating oil to keep oil molecules from splitting apart under heavy pressures. During operation, oil is constantly being pressurized, and squeezed, and so on. This may cause the oil molecules to separate, and reduce the oil's lubricating properties. The added chemicals react with metal surfaces to form very strong, slippery films, which may be only a molecule or so thick. With these additives, protection is increased during moments of extreme pressure.

CAR CLINIC

PROBLEM: OIL LEAK WHEN CAR SITS OVERNIGHT

Oil leaks from the vehicle when it sits overnight in a driveway. The valve cover gaskets have been replaced several times, but the leak always comes back.

SOLUTION:

A chronic oil leak like this is usually caused by too much crankcase pressure. Check the crankcase pressure and compare it to the manufacturer's specifications. If it is above specifications, perform a cylinder leak-down test. This will check for leaky piston rings. If the piston rings are in good condition, check the condition of the PCV system for correct venting. Fixing one of these two problems should solve leaking valve cover gaskets.

11.4 OIL CLASSIFICATIONS

Lubricating oils are rated by various agencies. These ratings are used so that the correct oil can be selected for the correct application. Three organizations rate oil: Society of Automotive Engineers (SAE ratings), the American Petroleum Institute (API ratings), and the federal government (Military ratings).

SAE Ratings

SAE ratings are given in terms of the viscosity or thickness of the oil. The SAE has applied the Saybolt Universal Seconds test to their ratings. SAE ratings are determined for different temperatures. Oil is tested for viscosity at 0 degrees, 150 degrees, and 210 degrees F. The most common are 0 and 210 degrees. A rating that has a letter *W* after it means that the viscosity is tested at 0 degrees. If there is no *W* after the rating, the oil has been tested at 210 degrees. For example, referring to **Figure 11–6**, a 10W oil has a rating of

	SECONDS AT 0°F		SECONDS AT 210°F	
SAE NUMBER	MIN.	MAX.	MIN.	MAX.
5W		4,000	39	
10W	6,000	12,000	39	
20W	12,000	48,000	39	
20			45	58
30			58	70
40			70	85
50			85	110

VISCOSITY RANGE, SAYBOLT UNIVERSAL SECONDS

FIGURE 11–6. Viscosity is measured in seconds. Oil must have a certain flow rate in seconds for each SAE rating.

Viscosity: (stable ambient temperatures)

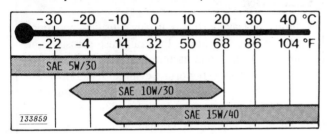

FIGURE 11–7. Car manufacturers recommend certain viscosity ratings for different temperatures. *(Courtesy of Volvo of North America)*

6,000 to 12,000 seconds at 0 degrees. When the oil temperature is increased to 210 degrees, the oil has a rating of 39 seconds. Multigrade oils that have ratings such as 10W30 meet the viscosity requirements for both the 0 degree and 210 degree ranges. This means that the oil is less affected by temperature changes. Viscosity Index improvers are added to make this possible.

It is sometimes difficult to determine exactly which type of oil viscosity to use. Manufacturers include various charts to help determine the best viscosity for the best service. *Figure 11–7* shows such a chart. For example, SAE 10W30 oils should be used in temperatures from about −10 degrees to about 68 degrees F. Other oils are shown for different temperature conditions. The manufacturer's recommendations should always be followed to determine which oil is correct for their engine. Failure to do so will void the new car warranty.

API Ratings

The American Petroleum Institute (*API*) rates oil based on the type of engine service the oil should be used in. API ratings for gasoline engines include SA, SB, SC, SD, SE, and SF. API diesel ratings are CA, CB, CC, and CD. SA and SB ratings are for engine operations that are light duty. There is very little protection against corrosion, scuffing, and deposits. Oils rated SC meet the requirements for vehicles manufactured from 1964 to 1967. SD oils meet the requirements for vehicles manufactured from 1968 to 1979. The SE rating means that the oil can be used in vehicles built in 1972 and in certain 1971 vehicles. Oils rated SF are used in gasoline engines built in 1981 and later cars. Only SF oils should be used in vehicles built in recent years.

As the automotive manufacturers design engines with higher temperatures and closer clearances, the type of oil used will have to be changed. Other ratings will eventually be used for vehicles that require even more protection against sludge, varnish, wear, and engine deposits.

CA oils are designed for mild- to moderate-duty operation with high-quality fuel used with diesel engines. CB oils are for mild- to moderate-duty operation of diesel engines with lower-quality fuels. More protection is needed against wear deposits and bearing corrosion. CC oils are intended for moderate- to severe-duty supercharged or turbocharged diesel engines. There is more protection against rust, corrosion, and low-temperature deposits. CD oils are designed for protection with supercharged or turbocharged diesel engines in severe-duty service. Protection is given against high-temperature deposits, bearing corrosion, and use of various fuels. Automotive manufacturers usually require the use of CD oils in automotive diesel applications.

Military Ratings

The third rating is associated with military testing and performance. The federal government contracts with an engine manufacturer to build engines for a particular application. When this is done, a military specification is issued for the type of oil recommended in that specific application. Some of the tested ratings are:

1. MIL-L2104 B, MIL-L46152, or MIL-L45199B
2. Supplement 1
3. Series 3

FIGURE 11–8. Each oil can has different ratings. SAE, API, and MIL ratings are printed on the can. *(Courtesy of Davis Publications)*

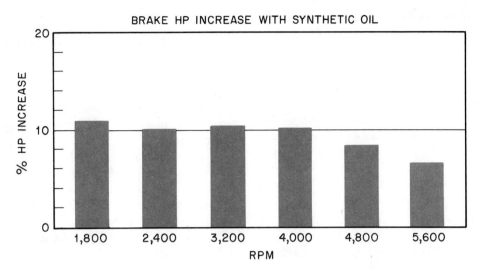

FIGURE 11-9. Use of synthetic oils changes horsepower.

From these tests, an engine manufacturer then recommends a specific type of oil for its engines in the military application. SAE and API ratings are usually included with this recommendation. If there is a military rating stamped on the can, it can be assumed that this oil has been tested for a specific application with military equipment.

These three ratings are normally stated on the oil can. *Figure 11-8* shows a typical can with the ratings stamped in the cover. If the oil can does not have an API rating, the oil probably won't meet the API specifications.

11.5 SYNTHETIC LUBRICANTS

In recent years the use of *synthetic* lubricants has increased. *Synthesize* means to combine parts into a whole or to make complex compounds from a series of individual molecules. Synthetic lubricants are made chemically by mixing alcohol, various acids, other chemicals, and hydrocarbons together. The hydrocarbons can be taken from coal, oil, natural gas, wood, or any agricultural resource. The result is a synthesized product that is capable of meeting and exceeding the lubrication needs of various engines. Synthetic oils are manufactured by combining specific molecules into an end product tailored to do a specific job. They are designed to meet or exceed the SAE and API recommendations.

Advantages of Synthetic Oils

Synthetic oils were initially designed for jet engine use because of the higher temperatures and pressures. They are now used in automobile applications because of their ability to withstand high temperatures with little change in the viscosity of the oil. There are several advantages and disad-

vantages of using synthetic oil in automotive engines. Some advantages include:

1. *Increased thermal and oxidation stability.* For example, synthetic oils can operate effectively from −60 degrees F to +400 degrees F. Oxidation occurs at about 50 degrees higher as compared to standard oils.

2. *Less evaporation.* Only about 1% of the oil is evaporated over a standard period of time as compared to about 25% for standard lubrication oils.

3. *Less viscosity change with temperature.* This improves cold starting ability and fuel mileage during cold weather.

4. *Improved fuel mileage.* This is because synthetic oils have increased lubricating properties. The result is less frictional resistance and more horsepower. *Figure 11-9* shows horsepower increases at different rpm.

5. *Reduced oil consumption.* Synthetic oils increase sealing characteristics.

6. *Cleaner engine parts.* Less maintenance is required when engine parts remain cleaner.

7. *Not affected by fuel contamination.*

Disadvantages of Synthetic Oils

Although synthetic oils seem to be better than conventional oils, there are several disadvantages that should be mentioned. These include:

1. *Poor break-in characteristics.* Because of the increased lubrication property, break-in will not

occur correctly. Synthetic oils should not be used until the vehicle has reached about 6,000 miles.

2. *Ineffectiveness in older engines.* Synthetic oils should not be used in engines with bad rings, valve seals, or seals. If used on such engines, the oil may be consumed more rapidly.

3. *Higher cost.* The cost per quart is still higher for synthetic oil than for conventional motor oil. Over the life of the vehicle, however, the total cost for lubrication will be about the same. The savings come from cleaner engines and better fuel mileage. Oil change periods range in the area of 25,000 miles.

4. *Inability to be mixed.* It is not advisable to mix different brands of synthetic oils. This may cause some inconvenience when trying to buy a particular brand.

A great deal of research is being done on synthetic oils. As the temperatures of engines increase in the future, synthetic lubricants will have to be reevaluated to see if they can be used more effectively than conventional oils. It is important to note that each person will have to decide about the use of synthetic lubricants.

11.6 GREASES

The term *grease* has long been employed to describe lubricants that are in a semi-fluid state and are used to reduce friction. Grease is made by adding various thickening agents such as soap to a liquid lubricant such as oil. Greases are commonly used on automotive chassis parts such as the universal joints, wheel bearings, and front end steering parts.

Although greases are suitable for many lubrication jobs, several modifying agents are often added to greases to improve certain properties. Examples of properties that may be improved include: tackiness, oxidation stability, consistency stability, load-carrying capacity, and rust prevention. Always follow the manufacturer's recommendation when selecting greases.

11.7 LUBRICATING SYSTEM OPERATION

The lubricating system is composed of several parts. These include the oil pan, oil pump, main oil galleries, oil filters, oil pressure regulators, oil coolers, and oil sensors. These parts are needed to make the lubricating system operate correctly.

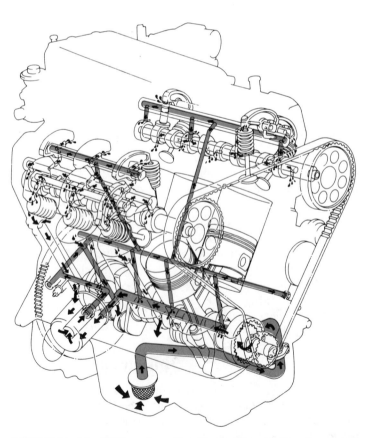

FIGURE 11–10. Oil flows from the oil pan, through the pump, and into the main oil gallery. From there the oil is sent to all of the parts of the engine. *(Courtesy of Nissan Motor Corporation in USA)*

Oil Pan

The purpose of the oil pan is to hold the excess oil during operation and nonrunning conditions. This reservoir for the oil is located on the bottom of the engine. After the parts of the engine are lubricated, gravity causes the oil to flow back to the oil pan. A plug in the bottom of the oil pan is used to drain the oil. The cooling of the oil takes place within the oil pan.

Oil Flow

Each engine manufactured has a certain flow pattern to get the oil from the oil pan or *sump* to the various parts that need lubrication. Flow diagrams are different for each type of engine and for each year manufactured. A standard flow diagram is shown in *Figure 11–10*. The flow of oil starts at the oil pan where the oil is stored and cooled. From there the oil is drawn into the oil pump through the oil screen. In this case, the oil pump is located in the front of the engine on the crankshaft. Other oil pumps are driven from the camshaft or through the distributor shaft. After the oil is pressurized, it is sent to the oil filter to be cleaned. From the oil filter, the oil is sent into passages in the block of the engine. These are referred to as the main oil gallery. From the main oil gallery, oil is sent to all other parts of the engine. There is an oil pressure sensor on the main oil gallery. The oil pressure sensor is used to sense the oil pressure. It is attached to the oil light or gauge on the dashboard.

Oil is sent through drilled passages that index with the main oil gallery to the crankshaft. Here the oil is used to lubricate the main and connecting rod bearings. *Figure 11–11* shows oil holes drilled in a crankshaft. Oil from the main bearings feeds to the connecting rod bearings and then drops back to the oil pan and sump.

From the main oil gallery, the oil is also sent through drilled passages to the head. On some engines, the oil flows from the main oil gallery, through the lifters, through the pushrods, and up to the rocker arms. Here the oil is used to lubricate the camshafts, rocker arms, and valves. The oil then drains back to the oil pan and sump through holes in the block and cylinder heads. An oil flow diagram for a diesel engine is shown in *Figure 11–12*. Some differences include oil sent to the turbocharger and to the oil cooler.

Oil Pumps

To distribute the oil throughout the engine, oil must be pressurized. This is done with an oil pump. The oil pump is located in the crankcase area so that oil can be drawn from the oil pan and sent into the engine. Oil pumps are considered positive displacement pumps. This means that for every revolution on the pump, a certain volume of oil is pumped. Therefore, as rpm increases, oil pressure also increases.

There are several types of pumps. *Figure 11–13* shows how a standard gear-type pump operates. In this case, the oil is drawn in on one side as if into a vacuum. It is then pressurized on the other side. The *eccentric* gear-type pump operates on the same principle. (Eccentric means not having the same center.) *Figure 11–14* shows two gears. The centers of the gears are not at the same position. As the inside gear is turned, a suction is produced on one side and a pressure is produced on the other side.

Figure 11–15 shows a typical oil pump driven by the crankshaft. The pump is an eccentric type. As the inside gear is turned, the vacuum produced draws oil from the oil pan, through the oil screen, through the tube, and into the oil pump. The oil is then pressurized and sent into the engine.

Oil Pressure Regulating Valve

Because all oil pumps are positive displacement types, a pressure regulator valve must be used. A pressure regulator valve is used to keep the pressure within the oil system at a constant maximum value. As the rpm of the engine

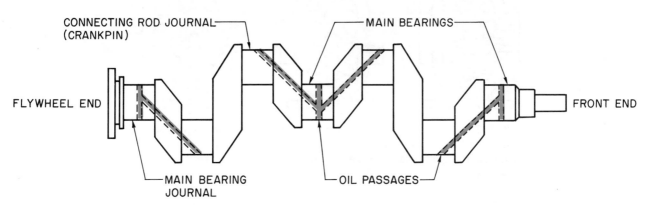

FIGURE 11–11. Oil from the main oil gallery is sent through passages into the crankshaft. The oil at the main bearings is then sent to the connecting rod bearings.

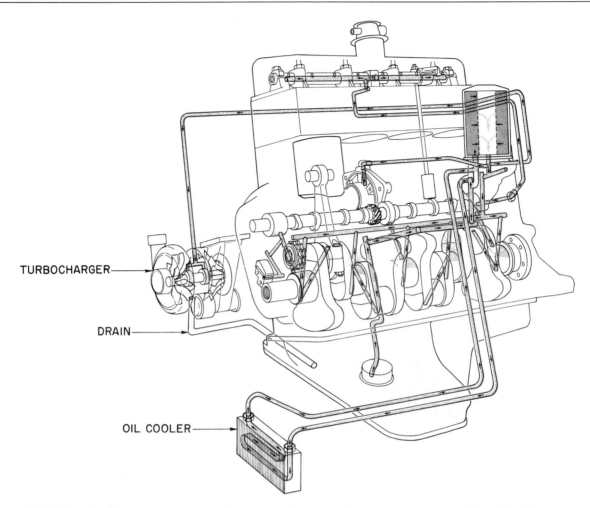

TURBOCHARGER

DRAIN

OIL COOLER

FIGURE 11-12. On certain engines, oil must also flow to the turbocharger and the oil cooler. (Courtesy of Peugeot Motors of America, Inc.)

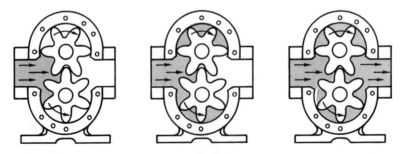

FIGURE 11-13. As the gears turn on the oil pump, a suction is produced on the left side, drawing in oil. The oil is carried around to the other side. As the gears mesh with each other, the oil is squeezed out and pressurized.

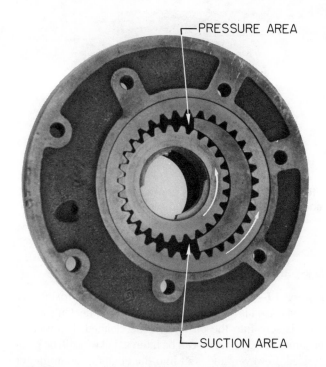

FIGURE 11-14. An eccentric gear pump has two gears. Each has a different center point (eccentric). As they are turned, a suction and pressure are produced.

changes, the amount of pressure produced by the oil pump also changes. In addition, as the oil gets thicker because of cold weather, oil pressure may increase. The pressure regulator valve maintains a constant maximum pressure. Whenever the pressure exceeds this maximum, the regulator valve opens to reduce the pressure. Normal oil pressures vary according to the engine manufacturer. Check the engine manufacturer's specifications to determine the oil pressure for a particular engine.

Figure 11-16 shows how an oil regulator valve works. As the oil pressure increases in the pump, the pressure pushes against a ball or valve held in place by a spring. When the oil pressure is greater than the spring tension, the ball lifts off its seat. At this point, some of the oil is returned back to the suction side of the oil system or to the oil pan. This reduces the pressure, which seats the ball again. The spring tension is designed to set the oil pressure at the manufacturer's specifications. If a stronger spring is used, the oil pressure will increase. If the spring pressure is less, or the spring is broken, the oil pressure will be less. Pressures are normally about 40-60 psi for passenger cars and light trucks. Larger engines may have pressures in the range of 40-70 psi.

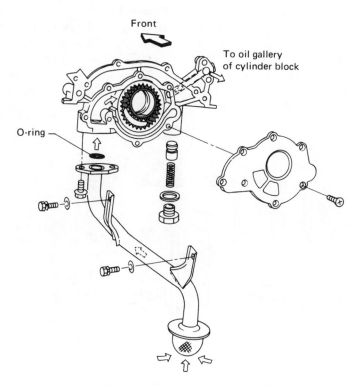

FIGURE 11-15. This eccentric pump is driven directly by the crankshaft. *(Courtesy of Nissan Motor Corporation in USA)*

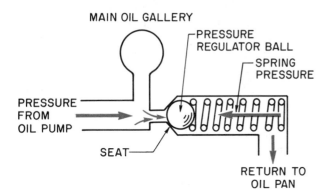

FIGURE 11-16. During normal operation, the oil is sent into the main oil gallery. As oil pressure increases, the ball is lifted off its seat (against spring pressure). Some of the oil is then returned to the oil pan, reducing the oil pressure.

FIGURE 11-17. Oil filters are used to remove dust, dirt, and sludge particles from the oil.

Oil Filters

Oil must be filtered and cleaned constantly within the engine. As was mentioned earlier in this chapter, there are several contaminants that get into the oil. Oil filters are used to clean the dirt particles out of the oil. *Figure 11-17* shows a typical oil filter on an engine.

There are two types of oil filtering systems. One is called the "full-flow" system and one is called the "by-pass" system. Years ago only the by-pass system was used. However, because of a greater need to filter the oil, full-flow systems are now being used. Full-flow systems filter all of the oil before it enters the engine. By-pass systems filter only a part of the oil during operation.

By-pass Filter

Figure 11-18 shows the by-pass system. Approximately 90% of the oil is pumped directly to the engine. Only about 10% is sent into the oil filter to be cleaned. If this filter becomes plugged, no oil can be filtered. Oil will still be pressurized, however, and sent into the engine. Certain diesel applications use this system along with a full-flow filter.

Full-flow Filter

Figure 11-19 shows the full-flow system. In this system, all of the oil must pass through the oil filter before entering the engine. If dirt plugs the oil filter, the oil pressure will increase before the filter, causing all of the oil to be returned to the crankcase through the regulator valve. To prevent this problem, a relief valve is used. The relief valve is designed

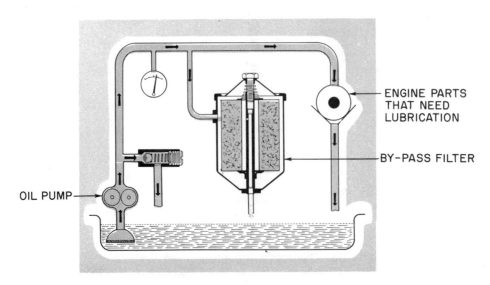

FIGURE 11-18. A by-pass oil filter is used on some engines. In this system, only a small amount of oil is filtered. The filter is a by-pass line around the oil system. *(Courtesy of Dana Corporation)*

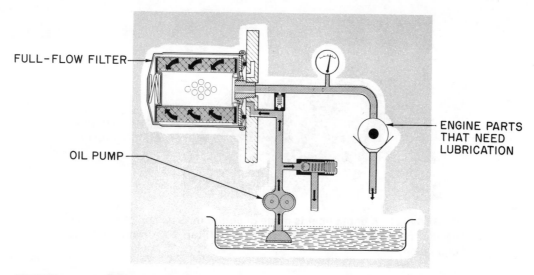

FIGURE 11–19. When a full-flow oil filter is used, all of the oil must go through the filter before entering the engine passages. *(Courtesy of Dana Corporation)*

to open at about 5–40 psi difference in pressure across the filter. This is called *differential pressure*. Differential pressure is the difference in pressure between the inlet and outlet of the filter. The inlet pressure will always be controlled by the regulator valve. As the oil filter plugs up, the pressure on the other side of the filter will drop. When the difference is equal to the relief valve setting, a certain amount of oil will pass through the filter and go into the engine. This means that even if the filter gets plugged, the engine will still receive oil. If the filter is totally plugged, the oil pressure will be slightly less than normal pressure.

Filter Design

Oil filters come in a variety of shapes and sizes. Two common types are the surface and depth filters. The surface filter is shown in *Figure 11–20*. The oil flows over the surface of the paper material. The contaminants are trapped on the surface of the paper, but the oil flows through microscopic pores in the paper.

The depth filter material is a blend of cotton thread and various fibers. *Figure 11–21* shows a depth filter. As the oil and dirt flow through the filter material, contaminants are trapped inside the filter material. Depth filters are used less today because they produce more restriction to the oil flow. Today's engines need more oil flow to aid in cooling the engine.

Oil Coolers

Oil coolers are used on certain heavy-duty gasoline engines and many diesel engines. An oil cooler is a device that helps

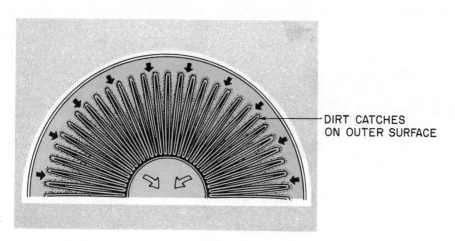

FIGURE 11–20. A surface filter catches the dirt particles on the surface of the filter material. *(Courtesy of Dana Corporation)*

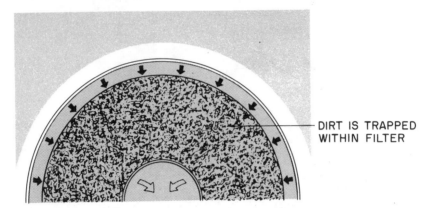

FIGURE 11-21. A depth filter is made of cotton threads and other fibers. The dirt particles are trapped within the filter. (Courtesy of Dana Corporation)

keep the oil cool. Oil temperature should be in the range of 180–250 degrees F. Under normal conditions, oil is cooled by having the right amount of oil in the oil pan. When excess temperatures occur, however, this cooling may not be enough. Oil coolers are used then.

Common oil coolers are designed with many copper tubes which are sealed together. The assembly is then sealed in a shell-type housing. This type of device is called a liquid-to-liquid heat exchanger. This means that the hotter liquid transfers some of its heat to a cooler liquid. Coolant from the engine cooling system passes from one end to the other through the copper tubes. The higher temperature lubricating oil passes through the shell. Although the two fluids do not mix, the heat in the lubricating oil is transferred to the coolant. The coolant then expels the excess heat through the radiator. *Figure 11–22* shows an example of an oil cooler used on a diesel engine.

Oil Sensors and Gauges

Oil pressure sensors are used to indicate if the oil system has the right amount of pressure. Oil pressure is usually sensed or measured directly from the main oil gallery. This information is sent to the dashboard on the vehicle so the operator can read the pressure. Two types of systems are commonly used. These are the pressure gauge and the oil indicator light. The pressure gauge reads the pressure of the oil within the system. The oil indicator light system has a light that goes on when oil pressure is low, usually at 7 psi or below.

Electric Oil Gauge

Figure 11–23 shows how an electric oil pressure gauge works. As oil pressure is sensed from the main oil gallery, the pressure moves an arm in such a way that it changes the

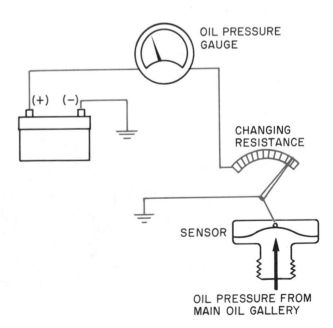

FIGURE 11-23. An electric oil pressure gauge senses oil pressure. As pressure changes, resistance within the circuit also changes, causing the needle to read differently.

FIGURE 11-22. Oil coolers are used to keep the oil cool during operation. Oil coolers use a series of metal tubes connected together. Oil flows on one side, while engine coolant flows on the other.

resistance of the circuit. This change in resistance causes the oil pressure gauge to read differently for different pressures. The oil sensor is connected to the main oil gallery. Wires are then connected from the sensor to the electric oil gauge on the dashboard.

Bourdon Gauge

The Bourdon pressure gauge is also used to measure and read oil pressure. *Figure 11–24* shows a typical Bourdon gauge. The tube is made of thin brass. The free end is connected to an indicating needle on the gauge dial. As pressure in the oil system increases, the gauge tends to become straighter. This causes the needle to read differently

on the dial. This type of gauge reading requires the oil to be sent directly to the gauge. A small copper or plastic tube is usually connected from the engine main oil gallery to the gauge on the dashboard.

Indicator Light

The oil indicator light simply goes on when the oil pressure is low. The light is connected to the ignition switch. There is also a pressure switch on the main oil gallery. When the ignition switch is turned on, the pressure switch is still closed because there is no oil pressure. The light is on then. When the engine starts, oil pressure builds up and opens the switch on the main oil gallery. Opening this switch turns off the oil light. *Figure 11–25* shows two types of oil sensors.

FIGURE 11–24. In a Bourdon pressure gauge, as pressure increases, the tube becomes straighter, giving a different reading.

FIGURE 11–25. Oil sensors are used to sense the pressure in the main oil gallery. As pressure increases, a switch in the sensor opens. When pressure is low, the switch closes and turns on the oil indicator light.

SAFETY TIPS
1. *The vehicle must be lifted up whenever the oil is changed. When jacking up a vehicle with a hydraulic lift, always use jack stands to support the vehicle if the hydraulic lift should fail.*
2. *When changing oil or oil filters, oil often spills on the floor. If oil spills, clean up the spill immediately so that you won't slip and fall.*
3. *Always use proper clothing (lab coats, etc.) when working on the lubrication system. Oil spilled on clothing, especially diesel oil, may damage the clothing.*
4. *When running the engine to check oil pressure, and so on, make sure the exhaust fumes are adequately removed by the exhaust system in the building.*

1. *Change oil* at the manufacturer's recommended time periods, using the correct quantity of oil. When the oil filter is changed, an additional quart of oil typically must be added.
   a. Jack up the vehicle and place the stands securely under the car.
   b. Remove the oil plug.
   c. Let the oil drain into an appropriate oil pan.
   d. Make sure all of the oil has been drained before tightening the oil plug.
   e. Be careful not to strip the threads on the oil plug.
   f. Remove the vehicle from the jacks.
   g. Add the correct quantity of oil to the engine.

2. Oil that sits in an unused engine may have a buildup of acids. Always change oil in an engine that has been stored (three months or longer) before running.

3. Make sure the oil viscosity is correct for the application and weather conditions. Again, always refer to the manufacturer's specifications, as use of incorrect oil will void the manufacturer's warranty.

4. Make sure the API ratings on oil are correct for the application.

5. Fuel can be detected in oil by a gaseous smell or by reduced viscosity. Fuel can get into the oil in several ways.
   a. If the fuel pump is damaged, oil may seep through the pump and into the crankcase.
   b. If the engine is flooded continuously when starting, or if the engine carburetion is set rich, fuel may get into the oil.
   c. If fuel is detected in the oil, change the oil immediately.

6. Water can be detected in oil by observing a light brown or white color in the oil. Water can get into the oil from the cooling system. This usually comes from a cracked block.

7. Never use a synthetic oil in an older engine that has an excess amount of wear on the rings.

8. Never mix synthetic oils.

9. Never use synthetic oil for break-in periods.

10. A broken spring on the regulator valve could be a cause of lower oil pressure.

11. On diesel engines that have both the full-flow and by-pass filters, make sure the filters are the correct type. Never replace a by-pass filter with a full-flow filter.

12. If the oil pressure drops suddenly, check the oil level first, then check the oil pressure sensor and the gauge.

13. Oil leaks occur througout the engine. When working on the lubrication system, *check for leaks* at these possible trouble spots:
    a. the oil plug, which might be loose.
    b. the gaskets around valve covers.
    c. all of the bolts that index with the crankcase. Oil can leak around the threads of the bolts. These bolts should have copper washers to stop leaks. These are common on older engines.
    d. the front and/or rear main seals on the crankshaft, which might leak.
    e. the filter, which might be loose or not sealed correctly.

14. *Oil consumption* in the engine can be caused by several conditions.
    a. Oil can be consumed because the valve guides are worn. Oil then seeps down into the exhaust stream and is burned. A bluish-colored exhaust indicates that oil is being burned.
    b. Oil can be consumed because the oil control rings are worn. When rings are worn, the oil is not scraped completely from the cylinder walls. This oil is then burned in the combustion chamber and exhausted. Again, a bluish exhaust is evidence of oil burning in the combustion.

15. Always check the oil pressure in the engine and compare it to the manufacturer's specifications. *Low oil pressure* is normally caused by:
    a. worn main bearings.
    b. worn connecting rod bearings.

16. *Figure 11–26* shows a complete listing of various lubrication problems. Identify the condition on the left side of the chart. Then move to the right to see possible causes and the corrections.

17. Change oil filters at the manufacturer's recommended time periods. Oil filters can be removed easily, using the oil filter wrenches shown in *Figure 11–27*.

Condition	Possible Cause	Correction
EXTERNAL OIL LEAKS	(1) Fuel pump gasket broken or improperly seated.	(1) Replace gasket.
	(2) Cylinder head cover RTV sealant broken or improperly seated.	(2) Replace sealant; inspect cylinder head cover sealant flange and cylinder head sealant surface for distortion and cracks.
	(3) Oil filler cap leaking or missing.	(3) Replace cap.
	(4) Oil filter gasket broken or improperly seated.	(4) Replace oil filter.
	(5) Oil pan side gasket broken, improperly seated or opening in RTV sealant.	(5) Replace gasket or repair opening in sealant; inspect oil pan gasket flange for distortion.
	(6) Oil pan front oil seal broken or improperly seated.	(6) Replace seal; inspect timing case cover and oil pan seal flange for distortion.
	(7) Oil pan rear oil seal broken or improperly seated.	(7) Replace seal; inspect oil pan rear oil seal flange; inspect rear main bearing cap for cracks, plugged oil return channels, or distortion in seal groove.
	(8) Timing case cover oil seal broken or improperly seated.	(8) Replace seal.
	(9) Excess oil pressure because of restricted PCV valve.	(9) Replace PCV valve.
	(10) Oil pan drain plug loose or has stripped threads.	(10) Repair as necessary and tighten.
	(11) Rear oil gallery plug loose.	(11) Use appropriate sealant on gallery plug and tighten.
	(12) Rear camshaft plug loose or improperly seated.	(12) Seat camshaft plug or replace and seal, as necessary.
	(13) Distributor Base Gasket damaged.	(13) Replace Gasket.

FIGURE 11–26. The troubleshooting chart shows many of the common problems, possible causes, and corrections for trouble in the lubrication system. *(Courtesy of Motor Publications, Auto Repair Manual, 1981–1987)*

Condition	Possible Cause	Correction
EXCESSIVE OIL CONSUMPTION	(1) Oil level too high.	(1) Drain oil to specified level.
	(2) Oil with wrong viscosity being used.	(2) Replace with specified oil.
	(3) PCV valve stuck closed.	(3) Replace PCV valve.
	(4) Valve stem oil deflectors (or seals) are damaged, missing, or incorrect type.	(4) Replace valve stem oil deflectors.
	(5) Valve stems or valve guides worn.	(5) Measure stem-to-guide clearance and repair as necessary.
	(6) Poorly fitted or missing valve cover baffles.	(6) Replace valve cover.
	(7) Piston rings broken or missing.	(7) Replace broken or missing rings.
	(8) Scuffed piston.	(8) Replace piston.
	(9) Incorrect piston ring gap.	(9) Measure ring gap, repair as necessary.
	(10) Piston rings sticking or excessively loose in grooves.	(10) Measure ring side clearance, repair as necessary.
	(11) Compression rings installed upside down.	(11) Repair as necessary.
	(12) Cylinder walls worn, scored, or glazed.	(12) Repair as necessary.
	(13) Piston ring gaps not properly staggered.	(13) Repair as necessary.
	(14) Excessive main or connecting rod bearing clearance.	(14) Measure bearing clearance, repair as necessary.

FIGURE 11–26. Continued

NO OIL PRESSURE	(1) Low oil level.	(1) Add oil to correct level.
	(2) Oil pressure gauge, warning lamp or sending unit inaccurate.	(2) Inspect and replace as necessary.
	(3) Oil pump malfunction.	(3) Repair or replace oil pump.
	(4) Oil pressure relief valve sticking.	(4) Remove and inspect oil pressure relief valve assembly.
	(5) Oil passages on pressure side of pump obstructed.	(5) Inspect oil passages for obstructions.
	(6) Oil pickup screen or tube obstructed.	(6) Inspect oil pickup for obstructions.
	(7) Loose oil inlet tube.	(7) Tighten or seal inlet tube.
LOW OIL PRESSURE	(1) Low oil level.	(1) Add oil to correct level.
	(2) Inaccurate gauge, warning lamp or sending unit.	(2) Inspect and replace as necessary.
	(3) Oil excessively thin because of dilution, poor quality, or improper grade.	(3) Drain and refill crankcase with recommended oil.
	(4) Excessive oil temperature.	(4) Correct cause of overheating engine.
	(5) Oil pressure relief spring weak or sticking.	(5) Remove and inspect oil pressure relief valve assembly.
	(6) Oil inlet tube and screen assembly has restriction or air leak.	(6) Remove and inspect oil inlet tube and screen assembly. (Fill inlet tube with lacquer thinner to locate leaks.)
	(7) Excessive oil pump clearance.	(7) Inspect and replace as necessary.
	(8) Excessive main, rod, or camshaft bearing clearance.	(8) Measure bearing clearances, repair as necessary.
HIGH OIL PRESSURE	(1) Improper oil viscosity.	(1) Drain and refill crankcase with correct viscosity oil.
	(2) Oil pressure gauge or sending unit inaccurate.	(2) Inspect and replace as necessary.
	(3) Oil pressure relief valve sticking closed.	(3) Remove and inspect oil pressure relief valve assembly.

FIGURE 11-26. Continued

FIGURE 11–27. An oil filter wrench such as this is used to remove oil filters from the engine. *(Courtesy of Owatonna Tool Company)*

SUMMARY

The lubrication system on both gas and diesel engines is designed to cool, clean, lubricate, and seal the internal parts of the engine. Oil is cooled in the oil pan. As the oil passes through the filter, it is cleaned. Oil must lubricate the many parts of the engine. Oil must also act as a seal around the compression rings.

Gasoline and diesel engines have many contaminants that must be removed. These include road dust and dirt, carbon and fuel soot from combustion, water from water vapor and the cooling system, fuel, and oil oxidation. All contaminants reduce the oil's ability to cool, clean, seal, and lubricate. The oil oxidation process also produces acids in the oil. These acids produce corrosion and damage the engine.

Because of these contaminants, oil must be designed with various characteristics. Viscosity is defined as the thickness of the oil. As temperatures change, the oil viscosity also changes. The Viscosity Index is a measure of how much the oil viscosity changes with temperature. Pour point is a measure of the temperature at which the oil is too thick to pour. Detergents and dispersants are added to oil to help with the cleaning process. Other chemicals such as antifoaming additives, corrosion and rust inhibitors, antiscuff additives, and extreme-pressure resistance additives are included to improve the quality of the oil during operation.

The SAE (Society of Automotive Engineers) classifies oil according to its thickness or viscosity. API (American Petroleum Institute) rates oil according to the type of service it is used in. The military also rates certain oils for use in specific applications.

The use of synthetic lubricants is increasing. Various chemicals are combined to make synthetic lubricants, which are very stable. Synthetic lubricants have many advantages and disadvantages. These must be studied before deciding to use a synthetic lubricant.

The lubricating system has many parts. Oil is held in an oil pan. This is called the sump. After being pressurized by a gear-type oil pump, the oil passes under pressure to the oil filter. From the oil filter, it is sent into the main oil gallery, which feeds the remaining parts of the engine.

The oil pump is a positive displacement pump that increases pressure with rpm. Several types of pumps are used, including the eccentric gear-type pump. As the pump turns, a suction is created on one side, while a pressure is produced on the other side. The pressure from the oil pump is controlled by the oil pressure regulator. When the oil pressure is greater than the spring pressure of the regulator, oil is returned to the suction side of the pump.

Two types of filters are used on lubricating systems: the full-flow and by-pass filter systems. Depending upon the engine, either one or both may be used. There are also two types of filters. The surface filter catches the dirt particles as the oil flows through the paper. The depth filter uses cotton and other fibers to catch the dirt particles within the filter.

Oil coolers are used on some engines for cooling the oil. During operation, the oil passes on one side of a series of baffles. Engine coolant passes on the other side. The heat from the lubricating oil is transferred to the cooling system and removed through the radiator.

Oil sensors and gauges are used to monitor the oil pressure being produced within the lubricating system. Electric sensors and gauges sense the pressure in the main oil gallery. Based on this pressure, either a gauge or an oil light is used. Other mechanical gauges are used as well. A Bourdon gauge straightens as the pressure increases. This motion is then read on a scale, indicating the correct oil pressure within the engine.

TERMS TO KNOW

Can you explain each of the following terms? Review the chapter until you can use each term correctly.

Contaminants   Varnish

Dilution   Viscosity

Oxidation

Acidity

Catalyst

Organic

Additive

Corrosion

Detergent

Dispersant

Suspendability

Inhibitor

Inhibit

Sludge

Society of Automotive
    Engineers (SAE)

American Petroleum
    Institute (API)

Viscosity Index

Saybolt Universal
    Viscosimeter

Fluidity

Synthetic

Eccentric

Differential pressure

Sump

REVIEW QUESTIONS

Multiple Choice

1. An engine lubricating system is designed to:
   a. Cool the engine
   b. Clean and seal the engine
   c. Lubricate the engine
   d. All of the above
   e. None of the above

2. Which of the following is not an oil contaminant in an engine?
   a. Road dust
   b. Carbon from combustion
   c. Wheel bearing grease
   d. Oil oxidation
   e. Road dirt

3. The viscosity of oil will _____ as fuel enters the oil.
   a. Increase
   b. Decrease
   c. Remain the same
   d. Heat up
   e. Thicken

4. Acids _____ in an engine's lubricating system.
   a. Cannot be produced
   b. Can be produced
   c. Form from the bearing noise
   d. Form from vibration
   e. None of the above

5. Viscosity is defined as the _____ of oil.
   a. Acidity
   b. Pour point
   c. Oxidation
   d. Thickness
   e. Temperature

6. At what temperatures are oils normally tested for viscosity?
   a. 0 and 100 degrees F
   b. 0, 150, and 210 degrees F
   c. 210 degrees F only
   d. 80 and 100 degrees F
   e. 1,000 to 1,005 degrees F

7. Chemicals added to oil to keep small particles of dirt separated from each other are called:
   a. Detergents
   b. Dispersants
   c. Viscosity index improvers
   d. Filters
   e. Acids

8. Which organization classifies oil by its viscosity?
   a. American Petroleum Institute
   b. Military
   c. Society of Automotive Engineers
   d. All of the above
   e. None of the above

9. An oil with a 10W30 label means the oil has a/an _____ at 0 degrees.
   a. API 30 rating
   b. SAE rating of 10
   c. SAE rating of 30
   d. API CD rating
   e. API SD rating

10. When an oil is manufactured by combining specific molecules into an end product, the oil is said to be:
    a. Synthetic
    b. Oxidized
    c. High in viscosity
    d. Low in viscosity
    e. High in acidity

11. Which type of oil pump has two gears that have different center points?
    a. All oil pumps
    b. Eccentric oil pumps
    c. Differential pressure pumps
    d. Centrifugal
    e. None of the above

12. Which type of filter forces only a small portion of the oil through the filter?
    a. Full-flow filter
    b. By-pass filter
    c. Differential pressure filter
    d. Easy-flow filter
    e. None of the above

13. Which device controls the pressure of the oil directly from the oil pump?
    a. Oil relief valve
    b. Oil regulator valve
    c. Oil screen
    d. Oil pump
    e. Oil pan

14. The difference between the input pressure of the oil filter and the output pressure of the filter is called:
    a. Pressure equalness
    b. Filter difference
    c. Pressure differential
    d. Input vs. output pressure
    e. None of the above

15. Which type of filter catches the dirt and dust particles on the outside of the filter surface?
    a. Depth filter
    b. Round filter
    c. By-pass filter
    d. Surface filter
    e. None of the above

16. The oil cooler takes heat from the oil and transfers it to the _____.
    a. Cooling system
    b. Engine block
    c. Oil sump
    d. All of the above
    e. None of the above

17. Which type of oil gauge straightens a curved tube as the oil pressure increases?
    a. Electric oil gauge
    b. Bourdon gauge
    c. By-pass filter gauge
    d. Curved-tube gauge
    e. Low-pressure differential gauge

18. If the oil pressure drops off suddenly, the problem may be the:
    a. Oil crankshaft
    b. Oil regulator valve
    c. Oil pan
    d. Oil passages
    e. Oil seals

19. There is no problem when using a full-flow filter for a by-pass filter, or a by-pass filter for a full-flow filter.
    a. False, the two should never be mixed
    b. True, it is OK to interchange the filters
    c. True, but always use a clean filter if they are changed
    d. False, the two can only be mixed after 5,000 miles
    e. False, the two can only be mixed after 10,000 miles

The following questions are similar in format to ASE (Automotive Service Excellence) test questions.

20. A car must be put away for a four-month period for storage. Technician A says the oil should be changed before storage because of acid buildup in the oil. Technician B says keep the old oil in the engine and then change it after the car is taken out of storage. Who is right?
    a. A only
    b. B only
    c. Both A and B
    d. Neither A nor B

21. Fuel has been found in the oil. Technician A says it could be because of too rich a carburetor mixture. Technician B says it could be caused by a bad diaphragm in the mechanical fuel pump. Who is right?
    a. A only
    b. B only
    c. Both A and B
    d. Neither A nor B

22. Technician A says excessive oil consumption can be caused by bad rings. Technician B says excessive oil consumption is caused by bad valve guides. Who is right?
    a. A only
    b. B only
    c. Both A and B
    d. Neither A nor B

23. Oil pressure in a car engine has dropped. Technician A says the first thing to check is the oil sensor. Technician B says the oil pump is bad and should be replaced immediately. Who is right?
    a. A only
    b. B only
    c. Both A and B
    d. Neither A nor B

Essay

24. What are the four things that an oil lubricant must do in an engine?

25. List at least three contaminants that can make the oil dirty.

26. How can fuel contaminate oil?

27. What is oil oxidation and what are the results of oil oxidation?

28. Define two methods used to classify oil.

29. What is an oil dispersant?

30. List three advantages and disadvantages of using a synthetic oil.

31. What is the difference between a by-pass filter and a full-flow filter?

CHAPTER 12

Cooling Systems

INTRODUCTION

The cooling system is one of the more important systems on the automotive engine. If the cooling system is not operating correctly, the engine may be severely damaged. To understand the cooling system, it is important to study coolant characteristics. The various parts of the cooling system, including the water pump, thermostats, radiators, hoses, pressure caps, fans, and temperature indicators must also be studied.

OBJECTIVES

After reading this chapter, you will be able to:
- Identify the purposes of the cooling system.
- Compare the ways in which heat can be transferred.
- Compare the different types of cooling systems.
- Define the characteristics of coolant and antifreeze.
- Describe the operation of water pumps.
- State the purpose and operation of thermostats and pressure caps.
- Describe the operation of expansion tanks.
- State the purpose and operation of radiators.
- Compare the operation and design of fans, shrouds, and belts.
- Describe the operation of temperature indicators.
- Identify various diagnosis and service procedures on the cooling system.

CHAPTER HIGHLIGHTS

12.1 PRINCIPLES OF THE COOLING SYSTEM
    A. Purpose of the Cooling System
    B. Heat Removal
    C. Heat Transfer

12.2 TYPES OF COOLING SYSTEMS
    A. Air-cooled Engines
    B. Liquid-cooled Engines
    C. Liquid Coolant Flow

12.3 COOLANT CHARACTERISTICS
    A. Antifreeze
    B. Freezing Points
    C. Boiling Points
    D. Corrosion
    E. Cooling System Leaks

12.4 COOLING SYSTEM PARTS AND OPERATION
    A. Water Pump

CHAPTER HIGHLIGHTS (CONTINUED)

DIAGNOSIS AND SERVICE

SUMMARY

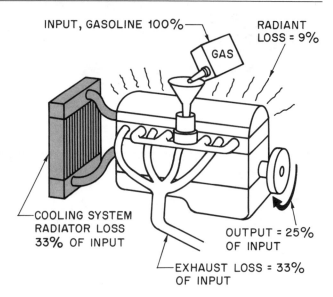

FIGURE 12–1. Approximately 33% of the heat generated by the engine must be removed by the cooling system. *(Courtesy of DCA Educational Products)*

12.1 PRINCIPLES OF THE COOLING SYSTEM

Purpose of the Cooling System

The purpose of the cooling system is to do three things. The first is to maintain the highest and most efficient operating temperature within the engine. The second is to remove excess heat from the engine. The third is to bring the engine up to operating temperature as quickly as possible. If the engine is not at the highest operating temperature, it will not run efficiently. Fuel mileage will decrease and wear on the engine components will increase.

In heavy-duty driving, an engine could theoretically produce enough heat to melt an average 200 pound engine block in 20 minutes. Even in normal driving conditions, combustion gas temperatures may be as high as 4,500 degrees F. Lubricated parts such as pistons may even run 200 degrees F or more above the boiling point of water (212 degrees F).

Heat Removal

Within the gasoline or diesel engine, energy from the fuel is converted to power for moving the vehicle. Not all of the energy, however, is converted to power. Referring to *Figure*

12–1, of the energy going into the engine, about 25% is used to push the vehicle. About 9% of the heat generated by the fuel is lost through radiation and 33% is sent out through the exhaust system. The remaining 33% must be removed by the cooling system. If this is done correctly, the temperature of the engine will be at its highest efficiency.

If the engine temperature is too high, various problems will occur. These include:

1. *Overheating of lubricating oil*. This will result in the lubricating oil breaking down.
2. *Overheating of the parts*. This may cause loss of strength of the metal.
3. *Excessive stresses between engine parts*. This may cause increases in friction, which may cause excessive wear.

If the engine temperature is too low, various problems will occur. These include:

1. *Poorer fuel mileage*. The combustion process will be less efficient.
2. *Increases in carbon buildup*. As the fuel enters the engine, it will condense and cause excessive buildup on the intake valves.
3. *Increases in varnish and sludges within the lubrication system*. Cooler engines enhance the buildup of sludges and varnishes.
4. *Loss of power*. If the combustion process is less efficient, the power output will be reduced.

WAKE IMAGE TEST

It is known that the smoother a vehicle cuts through the air, the more efficient and stable it is. The wake image test uses computers to help establish aerodynamic disturbances as the vehicle cuts through the air. Air currents are sent across the vehicle in the upper photo. The probes on the right of the vehicle sense the air flow at different positions. The data are then sent to the computer for a graphic display. The photos below show various graphic displays of aerodynamic disturbances created by the vehicle shape. (Courtesy of Ford Motor Company)

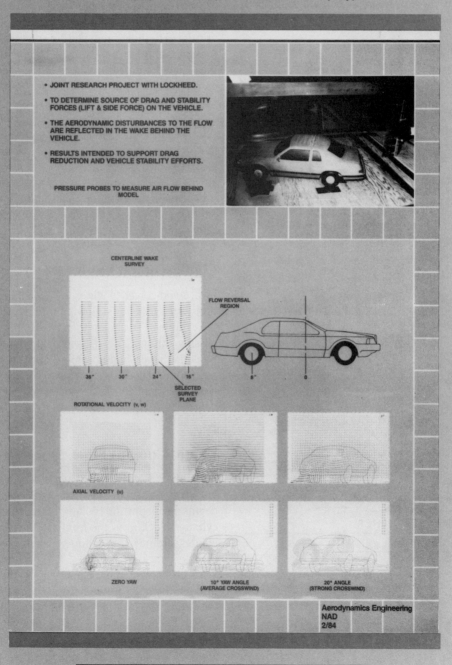

5. *Fuel not being burned completely.* This will cause fuel to dilute the oil and cause excessive engine wear.

Heat Transfer

The cooling system works on the principles of *heat transfer.* Heat will always travel from a hotter to a cooler object. Heat transfers in three ways: conduction, convection, and radiation. *Conduction* is defined as transfer of heat between two solid objects. For example, referring to *Figure 12–2*, heat must be transferred from the valve stem to the valve guide. Since both objects are solid, heat is transferred from the hotter valve stem to the cooler valve guide by conduction. Heat is also transferred from the valve guide to the cylinder head by conduction.

Heat can be transferred by *convection.* Convection is defined as the transfer of heat by the circulation of heated parts of a liquid or gas. When the hot cylinder block transfers heat to the coolant, it is done by convection. Convection also occurs when the hot radiator parts transfer heat to the cooler air surrounding the radiator.

Radiation is another way that heat is transferred. Radiation is defined as the transfer of heat by converting heat energy to radiant energy. Any hot object will give off radiation. The hotter the object, the greater the amount of radiant energy. When the engine is hot, some of the heat is converted to radiation (about 9% in Figure 12–1). The cooling system relies on these principles to remove the excess heat within an engine.

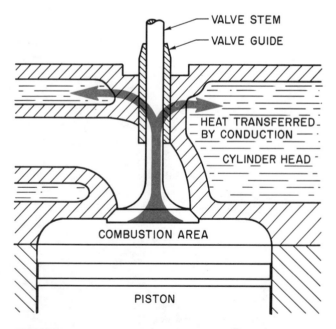

FIGURE 12–2. Heat is transferred by conduction from the valve stem to the valve guide. Both objects are solid.

12.2 TYPES OF COOLING SYSTEMS

Engine manufacturers today commonly use two types of cooling systems. These are the air-cooled and liquid-cooled systems.

Air-cooled Engines

Several manufacturers have designed engines that are *air cooled.* Certain foreign manufacturers still use air-cooled engines. Air-cooled engines have fins or ribs on the outer surfaces of the cylinders and cylinder heads. These fins are cast directly into the cylinders and heads. The fins increase the surface area of the object which, in turn, increases the amount of convection and radiation available for heat transfer. The heat produced by combustion transfers from the internal parts of the engine by conduction to the outer fins. Here the heat is dissipated to the passing air. In some cases, individual cylinders are used to increase air circulation around the cylinders.

Air-cooled engines require air circulation around the cylinder block and heads. Some sort of fan is usually used to move the air across the engine. A shroud is also used in some cases to direct or control the flow of air across the engine. Air-cooled engines usually do not have exact control over engine temperature; however, they do not use a radiator and water pump. This may reduce maintenance on the engine over a long period of time.

Liquid-cooled Engines

In a *liquid-cooled* engine, the heat from the cylinders is transferred to a liquid flowing through jackets surrounding the cylinders. The liquid then passes through a radiator. Air passing through the radiator removes the heat from the liquid to the air. Liquid-cooling systems usually have better temperature control than air-cooled engines. They are designed to maintain a coolant temperature of 180–205 degrees F.

Liquid Coolant Flow

Referring to *Figure 12–3*, a liquid-cooling system has several parts. When the vehicle is started, the coolant pump begins circulating the coolant. The coolant goes through the cylinder block from the front to the rear. The coolant circulates around the cylinders as it passes through the cylinder block.

The coolant then passes up into the cylinder head through the holes in the head gasket. From there, it moves forward to the front of the cylinder head through internal passages. These passages permit cooling of high-heat areas like the spark plug and exhaust valve areas.

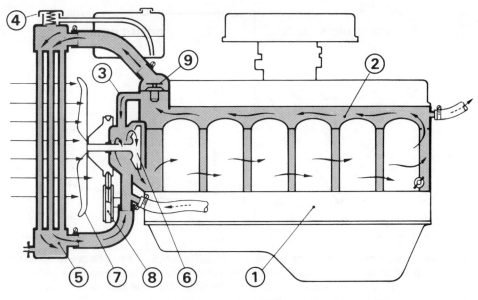

1. CYLINDER BLOCK
2. CYLINDER HEAD
3. BYPASS
4. RADIATOR PRESSURE CAP
5. RADIATOR
6. COOLANT PUMP
7. FAN
8. FAN BELT
9. THERMOSTAT

FIGURE 12–3. Coolant is pumped from the water pump, through the cylinder block and heads, through the thermostat into the radiator, and back to the water pump. *(Courtesy of Union Carbide Corporation)*

As the coolant leaves the cylinder head, it passes through a thermostat on the way to the radiator. As long as the coolant temperature remains low, the thermostat stays closed. Under these conditions the coolant flows through the *by-pass tube* and returns to the pump for recirculation through the engine. As the coolant heats up, the thermostat gradually opens to allow enough hot coolant to pass through the radiator. This will maintain the engine's highest operating temperature.

From the thermostat, the coolant flows to the internal passages in the radiator. These are tubes in the core with small fins on them. The coolant is now being cooled by the air passing through the radiator. From there it returns to the outlet of the radiator and back to the pump. It then continues its circulation through the engine.

CAR CLINIC

PROBLEM: HOT ENGINE

A car with 45,000 miles on it seems to be overheating. There is a "hot" light on the dashboard. Driving down the highway at 55 mph, the light goes on and then off. Does this mean the temperature of the engine is very hot? The engine doesn't seem to lose power, and coolant never has to be added.

SOLUTION:

The first and easiest component to replace is the coolant temperature sensor. At times, these coolant sensors will operate incorrectly. Although the sensor is reading "too hot," the engine may be at the correct temperature. Try replacing the coolant sensor to eliminate the problem.

12.3 COOLANT CHARACTERISTICS

Antifreeze

Water has been the most commonly used engine coolant. This is because it has good ability to transfer heat and can be readily obtained. Water alone, however, is not suitable for today's engines for a number of reasons. Water has a freezing point of 32 degrees F. Engines must operate in colder climates. Also, water has a boiling point of 212 degrees F. Engine coolant temperature often exceeds this point. In addition, water can be very *corrosive* and produce rust within a coolant system.

To overcome these problems, *antifreeze* is added to the coolant. An ethylene-glycol-type antifreeze coolant is the most common type used. When purchased on the market,

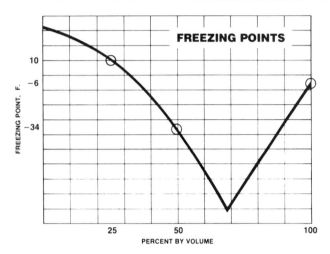

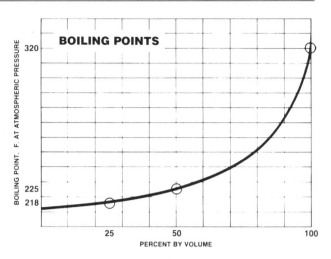

FIGURE 12–4. As the percentage of antifreeze increases, the freezing point of the solution is lowered. This is true up to the point where there is 68% antifreeze and 32% water. Beyond this point, the freezing temperature starts to increase. *(Courtesy of Chevrolet Division, General Motors Corporation)*

FIGURE 12–5. As antifreeze is added to the coolant, its boiling point also increases. *(Courtesy of Chevrolet Division, General Motors Corporation)*

this antifreeze includes suitable corrosion inhibitors. The best percentage of antifreeze to water to use is about 50% antifreeze mixed with 50% water.

Freezing Points

Figure 12–4 shows what happens to the freezing point of a coolant when different percentages of antifreeze are used. For example, when 100% water is used, the freezing point is 32 degrees F. When 25% antifreeze and 75% water is used, the freezing point of the coolant is about 10 degrees F. At 68% antifreeze, the freezing point of the coolant is about −92 degrees F. As the amount of antifreeze percentage increases from this point, the freezing point goes back toward 0 degrees F.

Boiling Points

The addition of antifreeze in the cooling system increases the boiling point. The *boiling point* of a fluid is the temperature at which a liquid becomes a vapor. Any coolant that becomes a vapor has very poor conduction and convection properties; therefore, it is necessary to protect it from boiling. This protection provides a greater margin of safety against engine cooling system overheating failure.

Figure 12–5 shows how boiling points increase with increases in percentages of antifreeze. For example, when there is no antifreeze, the boiling point is at 212 degrees F. When there is 50% antifreeze in the coolant, the boiling point increases to 226 degrees F. If there is 70% antifreeze, the boiling point increases to about 238 degrees F. It can be

seen that antifreeze protects the coolant during summer operation and winter operation.

Corrosion

Corrosion in the cooling system can be very damaging to the engine. Corrosion can be produced in several ways. *Direct attack* means the water in the coolant is mixed with oxygen from the air. This process can produce rust particles, which can damage water pump seals and cause increased leakage. *Electromechanical attack* is a result of using different metals in an engine. In the presence of the coolant, different metals may set up an electrical current in the coolant. If this occurs, one metal may deteriorate and deposit itself on the other metal. For example, a core plug may deteriorate to a point of causing leakage. *Cavitation* is defined as high shock pressure developed by collapsing vapor bubbles in the coolant. These bubbles are produced by the rapid spinning of the water pump impeller. The shock waves erode nearby metal surfaces such as the pump impeller.

Mineral deposits such as *calcium* and *silicate deposits* are produced when a hard water is used in the cooling system. Both deposits restrict the conduction of heat out of the cooling system. See *Figure 12–6*. The deposits cover the internal passages of the cooling system, causing uneven heat transfer out of the engine. See *Figure 12–7*. When an ethylene-glycol antifreeze solution is added to the coolant, many of these corrosion problems are eliminated. Chemicals are added to the antifreeze to reduce corrosion. It is usually not necessary to add any corrosion inhibitor to the cooling system. In some cases, mixing different corrosion inhibitors produces unwanted sludges within the cooling system.

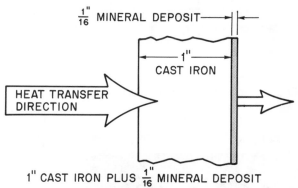

FIGURE 12–6. Mineral deposits inside the cooling system restrict the conduction of heat. Here, a 1/16-inch deposit makes the cast iron equivalent to 4 1/4 inches for heat transfer.

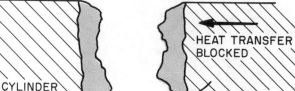

FIGURE 12–7. When minerals like calcium and silicate deposits build up inside the cooling system, certain areas may not transfer heat, while other areas may transfer heat more rapidly. This causes hot spots inside the engine.

Cooling System Leaks

While a full cooling system is essential, leakage ranks very high on the list of cooling system problems. There are numerous sources of leaks. Some leaks can be corrected by tightening hose clamps, but a large percentage of leaks are from small pinholes. These are commonly found in the radiator, freeze plugs, heater core, or around gaskets. To omit leaks, certain antifreezes have stop leak protection. These products are designed to seal the common pinhole leaks. This will prevent inconvenient breakdown and costly repair bills.

CAR CLINIC

PROBLEM: RUST IN THE COOLING SYSTEM

An engine has rust in the antifreeze. The antifreeze looks very brown and water pump bearings keep going out. There is not an overflow tank on this vehicle. What could be the problem?

SOLUTION:

Rust in the cooling system and antifreeze is normally caused by air getting into the radiator when the engine cools. As the coolant cools, it shrinks in size. An overflow tank normally has sufficient coolant in it to keep the radiator full even at cold temperatures. Without an overflow tank, air will enter the cooling system through the radiator cap. The rust in the coolant also causes wear on the water pump seal and the bearings. The best solution is to purchase an overflow tank and install it on the cooling system. Keep the overflow tank filled to the mark shown on the tank.

12.4 COOLING SYSTEM PARTS AND OPERATION

Various parts are used to operate the cooling system correctly. These include the water pump, water jackets, thermostats, radiators, transmission coolers, pressure caps, expansion tanks, fans, shrouds, belts, and temperature indicators.

Water Pump

The purpose of the water pump is to circulate the water through the cooling system. The pump is located on the front of the engine. It is driven by a belt that is attached to the crankshaft. As the crankshaft turns, the fan belt turns the pump, causing coolant to be circulated. See *Figure 12–8*.

The coolant pump is called a *centrifugal pump*. This means that as coolant is drawn into the center of the pump, centrifugal forces pressurize water and send it into the cooling system. See *Figure 12–9*. Centrifugal forces throw the coolant outward from the impeller tips. This type of pump will only pump the coolant that is required by the system. Therefore, it does not require a regulator or relief valve.

The pump consists of a housing, a bearing on a shaft, the impeller, and seal. The housing has a coolant inlet and outlet. The seal is used to keep the water inside the pump. The bearing is used to support the shaft on which the impeller rides. See *Figure 12–10*. This assembly is bolted to the cylinder block. A gasket is used to keep the coolant from leaking. On some engines, there is also a hub where a fan and pulley can be connected to the shaft. Other engines have the fan mounted directly on the radiator. In this case, the fan is operated electrically. Only the drive belt pulley is connected to the water pump shaft.

There is a small hole on the bottom of the water pump housing. This hole allows coolant to escape if the seal leaks. Today it is easier to simply replace the damaged water pump with a rebuilt or new pump. The old pump can then be turn-ed in to get a "core charge" and possibly be rebuilt by a manufacturer.

Water Jackets

Water jackets are defined as the open spaces within the cylinder block and cylinder head where coolant flows.

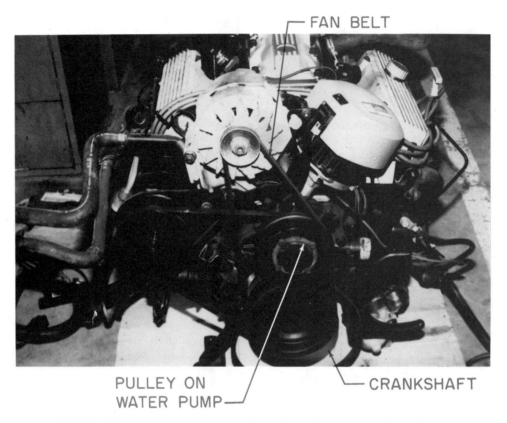

FAN BELT

PULLEY ON WATER PUMP

CRANKSHAFT

FIGURE 12–8. The water pump is driven from the crankshaft by V belts.

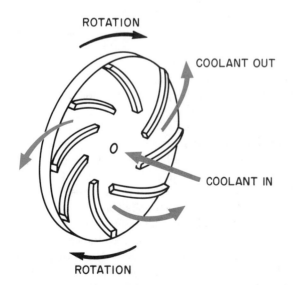

ROTATION

COOLANT OUT

COOLANT IN

ROTATION

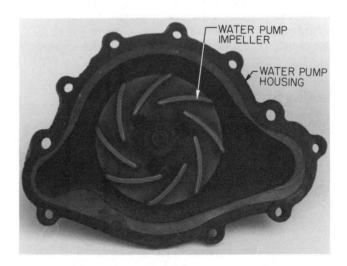

WATER PUMP IMPELLER

WATER PUMP HOUSING

FIGURE 12–9. A centrifugal pump takes the fluid into the center of the vanes, and centrifugal forces throw the fluid outward.

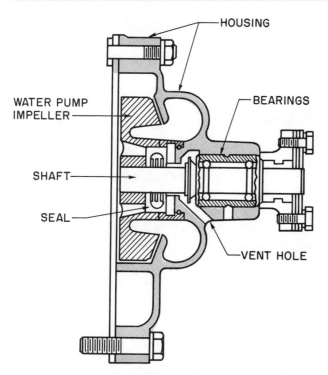

FIGURE 12–10. The water pump consists of a housing, bearings, shaft, and seals.

These water jackets are designed to allow coolant to flow to the right spots so that maximum cooling can be obtained. *Figure 12–11* shows internal water jackets used on an engine.

Thermostats

The thermostat is one of the most important parts of the cooling system. It is designed to sense the temperature of the coolant. If the temperature of the coolant remains cold, the thermostat will be closed. See *Figure 12–12*. The coolant then goes to the by-pass tube. This allows a small

amount of coolant to pass into the radiator to be cooled. The remaining coolant flows through the by-pass tube. This coolant is recirculated without being cooled. If the engine is under a heavier load, more cooling will be necessary. If the temperature of the coolant increases to the opening temperature, the thermostat will open slightly. As the temperature of the coolant increases further, the thermostat opens more. This allows more coolant to reduce its temperature through the radiator. When the engine is under full load, the thermostat will be fully open. The maximum amount of coolant will be sent to the radiator for cooling, and a small amount of coolant will continue to flow through the by-pass tube.

Thermostats operate on a very simple principle. A wax pellet material within the thermostat expands and causes the mechanical motion that opens the thermostat. See *Figure 12–13*. This allows coolant to pass through to the radiator. It should be noted that the thermostat is opened only partially when the temperature reaches its opening point. As the coolant temperature increases, the thermostat opens further. Eventually, the coolant is hot enough to cause the thermostat to open fully to get maximum cooling.

Thermostats are designed to open at different temperatures. Common thermostat temperatures are 180 and 195 degrees. Always follow the manufacturer's recommendations for determining the correct thermostat temperature.

Radiators

The purpose of the radiator is to allow fresh air to reduce the temperature of the coolant. This is done by flowing the coolant through tubes. As the coolant passes through the tubes, air is forced around the tubes. This causes a transfer of heat from the hot coolant to the cooler air. This process is called heat exchange. In this case, heat is exchanged from a liquid, the coolant, to air. This is called a liquid-to-air *heat*

FIGURE 12–11. Water jackets are used to allow the coolant to flow to the right spots within the engine.

exchanger. ***Figure 12–14*** shows an internal diagram of a typical radiator. Note that the coolant flows through the tubes and air flows through the air fins.

Down-flow and Cross-flow Radiators

Two types of radiators are commonly used in the automobile: the down-flow radiator and the cross-flow radiator. In the down-flow radiator, coolant flows from the top of the radiator to the bottom. In the cross-flow radiator, the coolant flows from one side of the radiator to the other side. ***Figure 12–15*** shows a diagram of both the down-flow radiator and

the cross-flow radiator. The down-flow radiator is found on most older vehicles. Because newer vehicles are lower in front, the cross-flow radiator has been used on most vehicles manufactured after 1970. Some American and foreign manufacturers, however, still use the down-flow radiator.

Radiator Parts

Radiators are made of several parts. The fins and tubes mentioned earlier are called the *core.* Depending upon the size of the engine and the cooling requirements, one, two, three, or four cores can be used. There is also an inlet and outlet

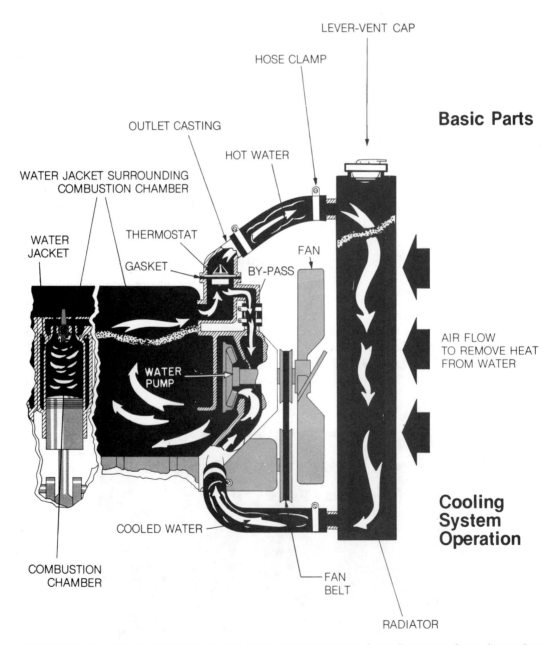

FIGURE 12–12.   The thermostat is used to open a passageway to the radiator or to force the coolant through a by-pass tube. (Courtesy of Robertshaw Controls Company)

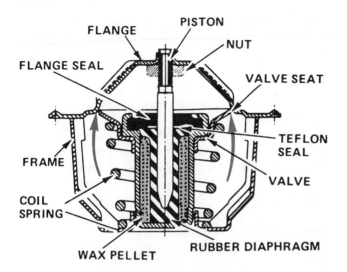

FLANGE PISTON NUT FLANGE SEAL VALVE SEAT FRAME TEFLON SEAL VALVE COIL SPRING WAX PELLET RUBBER DIAPHRAGM

FIGURE 12-13. As the wax pellet increases in temperature, the coil spring and valve lift. This causes the valve to lift off its seat, allowing coolant to pass through the thermostat. *(Courtesy of Chevrolet Division, General Motors Corporation)*

tank. These tanks hold the coolant before it goes into the radiator or into the block. The inlet tank also has a hose connection to allow coolant to flow from the engine into the radiator. The outlet tank has a hose connection to allow coolant to pass back to the engine. In addition, there is a filler neck attached to one of the tanks. The radiator pressure cap is placed here. *Figure 12–16* shows a breakdown of the common parts of the radiator.

Transmission Coolers

Cars that have automatic transmissions must have some means of cooling the transmission fluid. If the transmission fluid gets too hot, the transmission may be severely damaged. Transmission fluid is cooled by passing the fluid out of the transmission, into a tube in the radiator outlet tank, then back to the transmission. The liquid in the radiator is cool enough to lower the temperature of the transmission fluid. This heat transfer is done by using a liquid (transmission fluid) to liquid (engine coolant) heat exchanger. See *Figure 12–17*.

Pressure Caps

Pressure caps are placed on the radiator to do several things. They are designed to:

1. Increase the pressure on the cooling system
2. Reduce cavitation
3. Protect the radiator hoses
4. Prevent or reduce surging

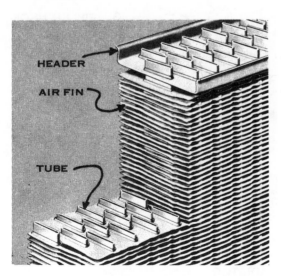

HEADER AIR FIN TUBE

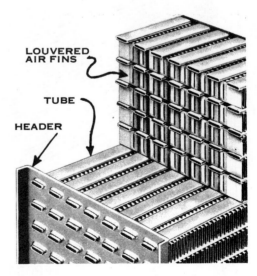

LOUVERED AIR FINS TUBE HEADER

FIGURE 12-14. Inside the radiator there are small tubes through which the coolant flows. The air fins help to remove the heat into the air. *(Courtesy of Harrison Radiator Division of General Motors Corporations)*

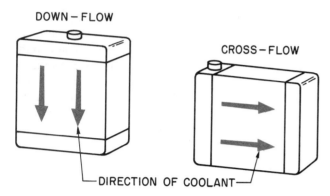

FIGURE 12-15. The down-flow radiator moves the coolant from the top to the bottom. The cross-flow radiator moves coolant from one side to the other.

It is very important to maintain a constant *pressure* on the cooling system. The pressure should be near 15 pounds per square inch. Pressure caps are placed on the radiator to maintain the correct pressure on the cooling system.

Pressure on the cooling system changes the boiling point. As pressure is increased, the boiling point of the coolant also increases. This is shown in **Figure 12-18**. The bottom axis shows pressure. The left vertical axis shows the boiling point. Different solutions of antifreeze are also

shown. For example, using water, the boiling point at 0 *psig* (pressure per square inch, on a gauge) is 212 degrees F. If the pressure is increased to 15 psig, the boiling point increases to about 250 degrees F.

Figure 12-19 shows how a pressure cap maintains the constant pressure. As the coolant increases in temperature, it begins to expand. As it expands, the coolant cannot escape. The spring holds a rubber washer against the filler neck. This keeps the fluid in the cooling system and increases the pressure. When the pressure reaches 15 psig, the rubber seal is lifted off the filler neck against spring pressure. The coolant then passes through the pressure cap to a tube that is connected to a recovery bottle. This type of system is called a closed system. An open system allows the coolant to pass through the pressure cap directly to the road surface.

The pressure cap also protects the hoses from expanding and collapsing. When the engine is shut down, the coolant starts to cool. As it cools, the coolant shrinks. Eventually, a *vacuum* is created in the cooling system. This means that the pressure outside the radiator is greater than the pressure inside the radiator. This causes the hoses to collapse. Continued expanding and collapsing of the hoses causes them to crack and eventually leak. The pressure cap has a vacuum valve which allows atmospheric pressure to seep into the cooling system when there is slight vacuum.

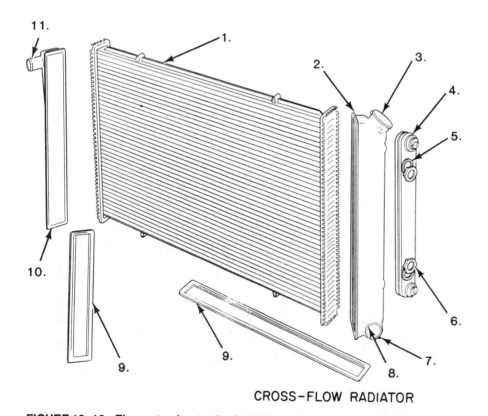

1. RADIATOR CORE ASSEMBLY
2. OUTLET TANK
3. FILLER NECK
4. TRANSMISSION OIL COOLER
5. OIL COOLER GASKETS
6. OIL COOLER ATTACHING NUTS
7. DRAIN COCK
8. OUTLET PIPE
9. GASKETS
10. INLET TANK
11. INLET PIPE

CROSS-FLOW RADIATOR

FIGURE 12-16. The parts of a standard radiator are shown. *(Courtesy of Chevrolet Motor Division, General Motors Corporation)*

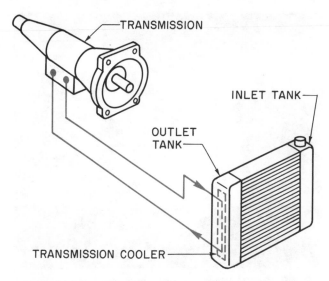

FIGURE 12-17. Fluid from the automatic transmission is sent through metal tubes to the radiator to be cooled.

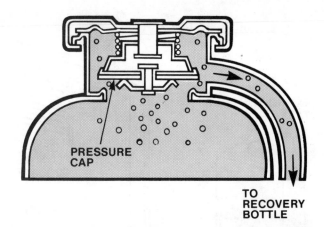

FIGURE 12-19. As the pressure increases on the cooling system, the large spring will eventually be lifted off its seat. This action releases any pressure over 15 pounds per square inch. *(Courtesy of Chevrolet Division, General Motors Corporation)*

During operation, a small spring holds the vacuum valve closed. When there is a vacuum inside the cooling system, the vacuum valve is pulled down and opened. The vacuum is then reduced within the cooling system. See *Figure 12-20*.

Increasing the pressure also reduces cavitation. Cavitation was defined earlier as small vacuum bubbles produced by the water pump action. Increased pressure reduces this action.

Pressure on the cooling system also reduces *surging*. Surging is defined as a sudden rush of water from the water pump. This could be caused by rapidly increasing the rpm of the engine. Surging can produce air bubbles and agitation of the coolant. Pressure on the cooling system tends to reduce this action.

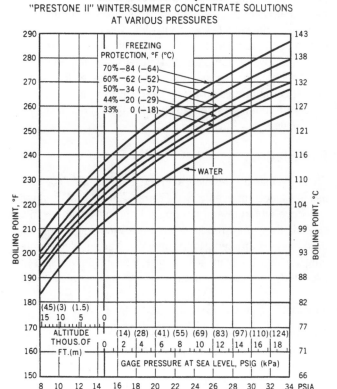

FIGURE 12-18. As the pressure is increased on the cooling system, the boiling point also increases. *(Courtesy of Union Carbide Corporation)*

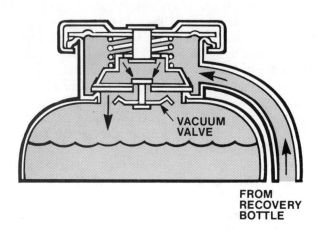

FIGURE 12-20. When the cooling system cools down, a vacuum is produced in the system. The vacuum spring is opened and the system equalizes the pressure. *(Courtesy of Chevrolet Division, General Motors Corporation)*

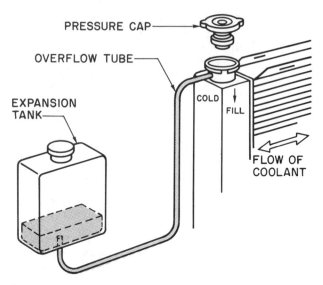

FIGURE 12–21. When the pressure cap releases coolant, it is sent to the expansion tank. When the engine cools, the fluid in the expansion tank is drawn back into the radiator.

FIGURE 12–22. When a closed system is used, coolant can be added directly to the expansion tank.

Expansion Tank (Closed System)

Many cooling systems use an expansion or recovery tank. Cooling systems with expansion tanks are also called "closed-cooling systems." They are designed to hold any coolant that passes through the pressure cap when the engine is hot. *Figure 12–21* shows a typical expansion tank.

As the engine warms up, the coolant expands. This eventually causes the pressure cap to release. The coolant that normally passes to the atmosphere is now sent to an expansion tank. When the engine is shut down, the coolant begins to shrink. Eventually, the vacuum spring inside the pressure cap opens. When this happens, the coolant from the expansion tank is drawn back into the cooling system. (On open systems without an expansion tank, air is drawn into the cooling system.) The major advantage of using this system is that air never gets into the cooling system. The presence of air in the cooling system can cause rust and corrosion to be produced, which can damage the cooling system components.

Coolant is added to the expansion tank. There is a small plastic cap on the expansion tank that can be removed. See *Figure 12–22*. Coolant can be added when the engine is hot because the pressure cap does not have to be removed to add the coolant.

Fan Designs

The purpose of the fan is to draw air through the radiator for cooling during low-speed and idle operation. A fan is not needed at faster speeds because air is pushed through the radiator by the vehicle speed. On older vehicles, the fan was driven on the same shaft as the water pump. A belt was used to turn the water pump and fan. Because the fan produces frictional horsepower losses, it has been designed to operate only at certain times on newer cars. It is estimated that the fan on the cooling system can absorb up to 6% of the engine horsepower.

Fan Blades

There are usually 4–6 blades on the fan. The fan blades are usually spaced unevenly. See *Figure 12–23*. This is done to reduce vibration and fan noise. If the blades are evenly spaced, they may produce fan noise and vibration that becomes very annoying to the operator. The fan blades slapping the air actually make the noise. If the blades are spaced irregularly, the noise is broken up and reduced.

Variable Speed Fan

The fan does not need to be turned when the engine is cold or at high speeds. Variable speed fans are used to control the time when the fan should turn. *Figure 12–24* shows a typical variable speed fan. As air passes through the radiator, its temperature is sensed on the front of the fan clutch. During cold engine operation, the fan is allowed to

FIGURE 12–23. Fan blades are unevenly spaced to reduce vibration and fan noise.

TEMPERATURE SENSOR

TEMPERATURE SENSOR (BIMETAL STRIP)   COOLING FINS

FIGURE 12–24. The variable speed fan senses the temperature of the air coming through the radiator. When the air is hot enough, the inside clutch engages and causes the fan to turn.

slip. When the temperature increases to the correct point, a *bimetal strip* on the front of the clutch expands. This causes a shaft inside the clutch to turn slightly. As the shaft turns, it opens a hole to allow silicon oil to enter a fluid coupling. This oil causes the fan to lock up and start turning. Note that there are several variations of this design; however, the result is the same.

Electric Fans

Another method used to turn the fan involves using a small motor. See *Figure 12–25*. In this system, the fan is turned on and off according to the engine coolant temperature. When the coolant reaches a certain temperature, usually around 200 degrees F, a sensor tells the electric fan to turn on. If the temperature is less than 200 degrees F, the fan remains off. The advantage of this system is that there are no fan belts. There is no frictional horsepower loss, as with belt driven fans. There is, however, a small power loss because the alternator must charge the battery more often. This is because the power used to turn the electric motor comes from the electrical system of the vehicle.

Flexible-blade Fans

Flexible blades are also used to reduce frictional horsepower loss from the fan. See *Figure 12–26*. These flexible blades are made of fiberglass or metal. As the speed of the engine and vehicle increases, the blades flatten. The blades move less air, causing less frictional horsepower to be lost.

Fan Shrouds

Certain vehicles use a fan *shroud*. A fan shroud is used to make sure the fan pulls air through the entire radiator. If a fan shroud is not used, there may be hot spots in the radiator. See *Figure 12–27*. For example, the fan normally

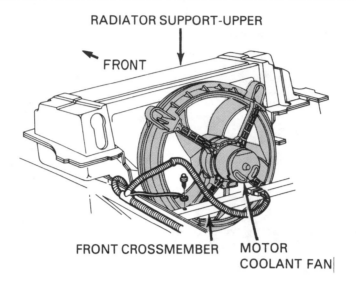

RADIATOR SUPPORT-UPPER

FRONT

FRONT CROSSMEMBER   MOTOR
COOLANT FAN

FIGURE 12–25. On certain vehicles, the fan is driven by an electric motor. The motor turns the fan off and on whenever it is needed. This system reduces frictional horsepower losses. *(Courtesy of Chevrolet Division General Motors Corporation)*

FLEXIBLE
FAN BLADE

FIGURE 12–26. Flexible-blade fans are used to reduce frictional horsepower losses. As the blade speed increases, the blades bend and cut through less air.

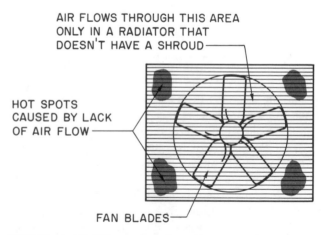

AIR FLOWS THROUGH THIS AREA
ONLY IN A RADIATOR THAT
DOESN'T HAVE A SHROUD

HOT SPOTS
CAUSED BY LACK
OF AIR FLOW

FAN BLADES

FIGURE 12–27. When a fan shroud is not used, hot spots can be created in the corners of the radiator. Using a fan shroud reduces this possibility.

pulls air through the radiator directly in front of the blades. There is very little air moving through the corners of the radiator. Using a fan shroud causes air to be pulled evenly through the entire surface area of the radiator.

Fan Belts

The fan belt is used to turn the water pump and the fan. Most fan belts are V-type belts. Friction is produced between the sides of the belt and the pulley. Using the V-type belt produces a larger area of contact. This means that a more positive connection can be made between the belt and the pulley.

Other types of belts are also being used today. One popular type is called the flat serpentine belt. It is about 1 1/2 inches wide and has several grooves on one side. It is flat on the other side. On some engines, this belt is used to reverse the direction of a water pump. This is done because objects can be driven from either side of the belt.

Temperature Indicators

It is important for the vehicle driver to know the engine temperature. Several systems, including a gauge showing the temperature or a warning light, can be used.

The warning light is the simplest system. An indicator is placed in the coolant. As the temperature of the coolant increases to a certain level, the heat from the engine causes an electrical circuit to close inside the sensor. This sensor connects the battery to the warning light on the dashboard. See *Figure 12–28*.

A second type of temperature indicator reads the actual coolant temperature and displays this on the dashboard of the vehicle. See *Figure 12–29*. The electrical indicator and the *Bourdon tube* gauge operate in this manner.

The electrical unit has a sensor placed in the coolant. As the temperature of the engine coolant increases, the electrical resistance of the sensor decreases. As the temperature of the coolant decreases, the electrical resistance of the sensor increases. This increase or decrease in resistance causes the needle in the temperature gauge to read differently with different coolant temperatures.

As pressure changes within the Bourdon tube, the needle reads differently on the dial. Pressure is produced in the tube inside the sensor, which is immersed in the coolant. As the temperature of the coolant increases, a liquid inside the sensor vaporizes. The vapor produces a pressure that is transmitted to the gauge through a tube so a dial can be moved.

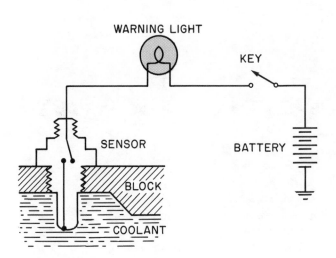

FIGURE 12–28. A warning light operates when the temperature of the coolant reaches a certain temperature.

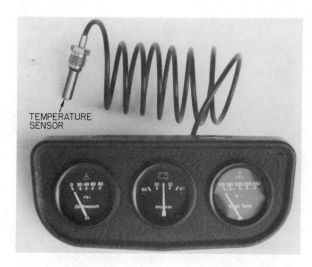

FIGURE 12–29. Electrical temperature gauges use a sensor and a gauge to read the exact temperature inside the engine.

WIND TUNNEL

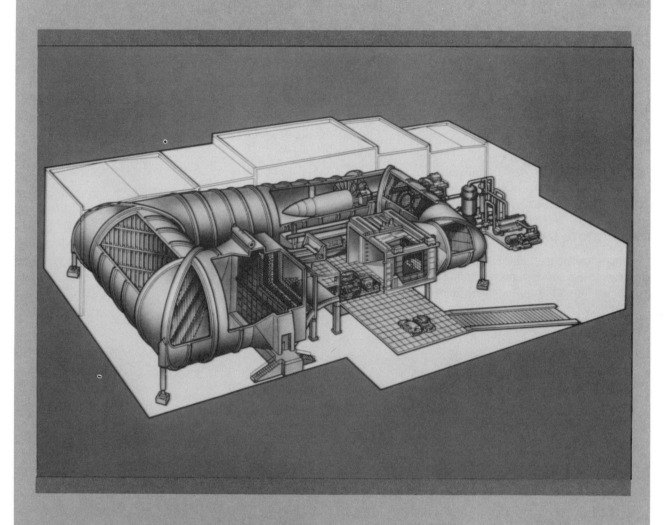

Wind tunnels are used to test the ease at which a car moves through the air at high speed. The wind tunnel produces a high velocity wind. Smoke can be introduced into the air stream to help observe the aerodynamics of the vehicle. Based upon information derived from the wind tunnel test, a coefficient of drag can be determined. The coefficient of drag is a number used to indicate how easy a car (and its shape) slices through the air as it moves down the road. The lower the coefficient of drag, the less wind resistance to the vehicle. Fuel mileage will increase with a lower coefficient of drag. (Courtesy of Volkswagen United States, Inc.)

DIAGNOSIS AND SERVICE

SAFETY TIPS

1. *When working on the cooling system, remember that at operating temperature the coolant is extremely hot. Touching the coolant or spilling the coolant on the body may cause serious injury.*

2. *When working on the coolant system e.g., replacing the water pump or thermostat, a certain amount of coolant will spill on the floor. The antifreeze in the coolant causes it to be very slippery. Always immediately wipe up any coolant that spills to reduce or eliminate the chance of injury.*

3. *Always wear proper clothing and eye protection when using coolant additives to remove silicate and calcium deposits, as they may be very corrosive.*

4. *When working on the fan, water pump, or belts, make sure the engine ignition system is off and/or the battery is disconnected.*

5. *Whenever the vehicle is running, always keep all fingers, tools, and clothing away from the moving fan.*

6. *When working on the water pump, fan, or belts, be careful not to scrape your knuckles against the radiator, as injury may result.*

CAUTION: Never remove the pressure cap when the engine is hot. Removing the pressure cap removes the pressure on the cooling system. This reduces the boiling point of the coolant. The coolant could then boil violently and burn or injure a person's hands or face.

1. Always use antifreeze in the cooling system for summer and winter protection. The *correct amount of antifreeze* can be checked by using a cooling system hydrometer. See *Figure 12–30*.

 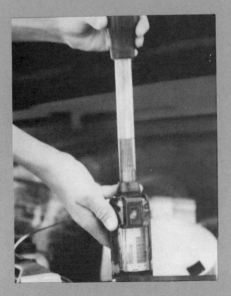

FIGURE 12–30. A hydrometer is used to check the coolant for protection against freezing.

a. Never remove the radiator cap when the engine is at operating temperature.

b. Remove the cap and, using the hydrometer, draw coolant up into the hydrometer.

c. Because there are several types of hydrometers, read the directions as to what the reading means. Usually, the hydrometer indicates the lowest temperature to which the coolant is protected.

2. If the coolant has a rusty color to it, corrosion may be building up in the cooling system. This is usually caused by the presence of oxygen in the coolant. Change the coolant completely, flush the system, and add the correct ratio of antifreeze. The coolant should be replaced at specific intervals. Check the manufacturer's recommendation for the correct time or mileage interval.

3. Check for silicate and calcium buildup by looking down into the radiator and observing the tubes. See *Figure 12–31*. Deposits can be seen building up around each tube. If there appears to be silicate and calcium buildup, the radiator may have to be cleaned. This is done using special equipment and tools. Radiator shops are often fully equipped to clean the internal passageways of the radiator. If there is equipment in the shop, follow the manufacturer's suggested procedure for cleaning radiators.

4. Always check fan belts for correct tension. See *Figure 12–32*. Tension can be checked by using this gauge. Position the gauge between the two pulleys as shown. The tigher the belt, the less it will bend. Read the gauge to determine the tightness of the belt.

5. Cooling systems can be checked for leaks by using a *pressure tester*. See *Figure 12–33*.

a. Remove the radiator cap. *CAUTION: Be sure the engine is not hot or at operating temperature.*

FIGURE 12–32. Fan belts are checked for tension by using this gauge. If the belts are too tight, the bearings may be damaged.

b. Place the pressure tester on the radiator and seal it like the radiator cap.

c. Pump up the pressure tester to the pressure reading on the cap.

d. With pressure on the system, observe the hoses, water pump, radiator, and so on for small leaks.

e. If no external leaks are evident but the gauge refuses to hold pressure for at least two minutes, there may be an internal coolant leak. If this is the case, the engine must be disassembled and the block and heads checked for cracks.

f. When finished, remove the pressure tester and replace the pressure cap.

6. *Check the pressure cap* by using the pressure tester shown in *Figure 12–34*.

a. Remove the radiator cap. *CAUTION: Be sure the engine is not hot or at operating temperature.*

b. Place the pressure cap on the pressure tester and seal by turning, as if on a radiator.

c. Increase the pressure on the cap, using the pump on the pressure tester.

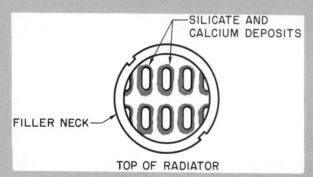

FIGURE 12–31. Deposits of calcium and silicate can be seen on down-flow radiators by looking directly into the filler neck. There will be a buildup of deposits around each tube.

FIGURE 12-33. The cooling system can be checked for small leaks by putting the system under pressure.

FIGURE 12-34. The pressure cap tester is used to determine exactly what pressure will open the pressure cap.

d. Note the pressure at which the pressure cap releases. (As the pressure is increased by pumping, the pressure will not increase any higher.)
e. Make sure the cap releases pressure at the psi setting shown on the cap.
f. If the pressure is released at a lower pressure, the cap is defective. Replace the cap if this occurs.

7. If coolant is leaking from the water pump, the water pump seal is damaged. If leakage occurs, replace the water pump. To *replace the water pump*:
a. Make sure the engine is cool.
b. Remove the radiator cap.

c. Drain the coolant out of the radiator and block.
d. Remove the belts by loosening the appropriate component on the belts. This may include the alternator, air conditioning compressor, and so on.
e. On some engines, the radiator may have to be removed to get at the water pump.
f. Remove the fan if it is driven by the water pump.
g. Remove the bolts holding the water pump to the engine block.
h. Scrape the old gasket completely off of the block.
i. Replace the old gasket with a new gasket.
j. Replace the water pump with a new or rebuilt pump.
k. Reverse the previous procedure.
l. After all of the parts have been replaced, add the coolant back into the cooling system. Extra antifreeze may have to be added.
m. Run the engine and pressure test for leaks.

8. *Check the bearings* on the water pump by moving the fan back and forth to observe movement. If any movement is noticed, the bearings are wearing. The water pump must then be replaced.

9. If the engine is overheating, *check the thermostat*, using the following procedure:
a. Suspend the thermostat in a pan of hot water with a thermometer in the water.
b. Increase the temperature of the water in the pan until the thermostat starts to open. A thermostat should still be closed about 10 degrees F below the number stamped on the thermostat.

c. The thermostat should be fully open about 25 degrees F above the number stamped on it.

10. Always check the radiator hoses for leaks, cracks in the rubber, and damaged clamps.

11. If there is evidence of corrosion buildup, the temperature sensors may read the wrong temperature. Cooling systems can be flushed to clean out these silicate and calcium deposits.

12. Always check fan belts for cracks and wear.

13. Keep the front of the radiator clean and free of dirt, bugs, and other debris.

14. Exhaust can leak into the cooling system if the cylinder head is cracked or if the head gasket is damaged. This can be checked by observing small air bubbles in the coolant.

15. Many vehicles have reinforcement springs inside the lower radiator hose. The spring is there to prevent the rubber hose from collapsing due to water pump suction. If the lower hose is replaced, make sure the spring is not deformed, missing, or out of position. Without the spring, the vehicle is likely to overheat at cruising speed. See **Figure 12–35**.

16. If the upper radiator hose collapses after the engine has cooled down, check the radiator cap's vent valve. It may be clogged. When clogged or damaged, the valve will not allow pressure in the cooling system to equalize when being cooled down. Replace or clean the cap vent to remedy the problem. **Figure 12–36**.

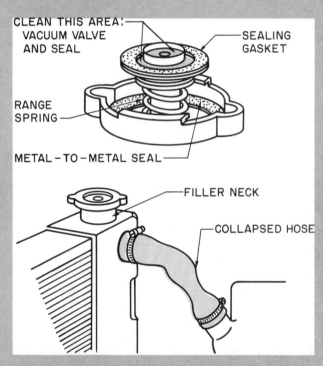

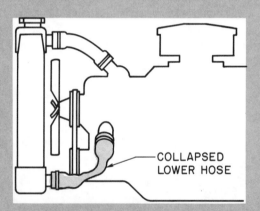

FIGURE 12–35. The lower radiator hose may collapse at cruising speed due to the suction of the water pump. A spring is placed inside the hose to prevent the hose from collapsing.

FIGURE 12–36. If the pressure cap vacuum valve is defective or clogged, the upper radiator hose will collapse when the engine cools down. Clean the valve or replace the pressure cap.

SUMMARY

The cooling system is designed to maintain the highest operating temperature within the engine. To do this, the cooling system must remove large quantities of heat from the engine. The cooling system uses three means of heat transfer to accomplish this: conduction, convection, and radiation. It is designed to maximize the principles of transferring heat by these three means.

Several types of cooling systems are now being used on the automobile. Air-cooled engines use air passing over fins to cool the engine block and cylinder heads. Heat is dissipated to the air passing around the engine. Liquid-cooled engines circulate a mixture of antifreeze and water through the engine to remove heat. The coolant is pumped with a water pump into the engine block, through internal passages, into the cylinder head, and into a radiator. Air flows through the radiator to remove the excess heat captured by the coolant. A thermostat is used to direct the coolant, which is either recirculated without being cooled or sent to the radiator to be cooled.

To understand the design and operation of the cooling system, it is important to identify coolant characteristics. Antifreeze is added to the coolant to protect it against freezing. The addition of antifreeze also increases the coolant's boiling point. Antifreeze also reduces corrosion within the cooling system. Corrosion can be caused by oxygen in the system, electromechanical attacks, and the buildup of calcium and silicate deposits. Antifreeze can also reduce leaks. Chemicals are added to the antifreeze to stop pinhole leaks in the cooling system.

The cooling system has many parts that enable it to operate effectively. These include the water pump, water jackets, thermostat, radiator, pressure cap, expansion tanks, fan, shrouds, belts, and temperature indicators.

The water pump is used to circulate the coolant throughout the cooling system. It is called a centrifugal pump. It is usually driven by a belt from the crankshaft. The water jackets are open spaces within the cylinder block. The coolant passes through the water jackets, picking up heat to be removed from the engine. The thermostat is used to control the temperature of the engine. If the coolant is cold, the thermostat remains closed. This directs the coolant back into the water pump to be recirculated. If the coolant is hot, the thermostat opens and sends the coolant to the radiator to be cooled.

The radiator is used to remove the heat from the coolant. The coolant passes through small tubes inside the radiator. Air flows across the tubes and removes the heat. Both down-flow and cross-flow radiators are used in vehicles today. Automatic transmission fluid is also passed through a tube in the bottom of the radiator to be cooled. A pressure cap is placed on the radiator to maintain approximately 15 pounds of pressure on the cooling system. The pressure increases the boiling point of the coolant, reduces cavitation, protects radiator hoses, and reduces surging.

An expansion tank is used on some vehicles. When the engine heats up and the coolant increases in temperature, the coolant expands. At times the pressure cap will release excess fluid into the air. An expansion tank collects the excess coolant. Using an expansion tank reduces the amount of oxygen allowed into the system. Corrosion is then greatly reduced.

Several types of fans and fan blades are used on the cooling system. The fan is normally a source of frictional horsepower loss; therefore, it should only be operated when necessary. The variable speed fan uses silicon oil inside a fan clutch to engage and disengage the clutch. It does this based on coolant temperature. Fans are driven by an electric motor or by using V belts. On electric fans, a sensor tells the fan when to operate. Some fan blades are made of fiberglass. As the speed of the fan increases, the fan blades flatten, reducing frictional horsepower losses. Fan shrouds are used to make sure air flows evenly through the radiator.

Temperature indicators are used to tell the operator the condition of the cooling system. Several types are used. The simplest type turns on a light when the temperature gets too high. A second style reads the engine coolant temperature

on a gauge. Two gauges are available: the Bourdon tube gauge and the electrical gauge.

Several important diagnosis and service tips about the cooling system should be known. Some of the more important ones include always checking fan belt tension, checking the pressure cap, making sure the thermostat opens at the correct temperature, checking hoses that may be damaged, checking the water pump for leakage and damaged bearings, and checking for calcium and silicate buildup within the radiator.

TERMS TO KNOW

Can you explain each of the following terms? Review the chapter until you can use each term correctly.

Heat transfer	Centrifugal pump
Conduction	Heat exchanger
Convection	Core
Radiation	Pressure
Air-cooled	PSIG
Liquid-cooled	Vacuum
By-pass tube	Surging
Corrosive	Bimetallic strip
Antifreeze	Shroud
Boiling point	Bourdon tube

REVIEW QUESTIONS

Multiple Choice

1. Which of the following is *not* a purpose of the cooling system?
   a. Keeping the engine temperature as low as possible
   b. Removing excess heat from the engine
   c. Bringing the temperature to operating range as quickly as possible
   d. Operating the engine at the best operating temperature for highest efficiency
   e. Protecting the internal parts from overheating

2. Which of the following is transfer of heat when both objects are solid?
   a. Conduction
   b. Convection
   c. Radiation
   d. Surging
   e. Vacuum

3. Heat transfer by _____ means to move heat by circulation of air or liquid.
   a. Conduction
   b. Convection
   c. Radiation
   d. Surging
   e. Vacuum

4. The _____ engine uses air passing over fins to cool the engine.
   a. Liquid-cooled
   b. Air-cooled
   c. Conduction
   d. Radiant
   e. Closed-loop

5. If the coolant is not hot enough to be cooled, it is sent to the _____.
   a. Radiator
   b. Pressure cap
   c. By-pass tube
   d. Differential regulator
   e. Coolant sensor

6. Antifreeze is used to protect the coolant against:
   a. Freezing
   b. Boiling
   c. Corrosion
   d. All of the above
   e. None of the above

7. Which two deposits build up inside the cooling system?
   a. Calcium and water
   b. Silicate and rust
   c. Calcium and silicate
   d. Water and rust
   e. Nitrogen and water

8. What is the most common mixture for antifreeze and water?
   a. 20% water, 80% antifreeze
   b. 50% water, 50% antifreeze
   c. 90% water, 10% antifreeze
   d. 30% water, 70% antifreeze
   e. None of the above

9. The water pump is considered a/an _____ type pump.
   a. Positive displacement
   b. Centrifugal
   c. Eccentric
   d. All of the above
   e. None of the above

10. The water pump is driven by:
    a. Fan belts from the starter
    b. Fan belts from the oil pump
    c. Fan belts from the crankshaft
    d. Fan bents from the transmission
    e. None of the above

11. When the temperature of the engine gets high enough, the automotive thermostat:
    a. Closes
    b. Opens
    c. Blocks off the radiator
    d. Tells the radiator to shut down
    e. None of the above

12. The thermostat is _____.
    a. Open all of the time
    b. Closed all of the time
    c. Opened more and more as the temperature increases
    d. Opened to allow coolant to flow to the transmission
    e. None of the above

13. The radiator is considered a _____ type of heat exchanger.
    a. Air-to-liquid
    b. Liquid-to-air
    c. Air-to-air
    d. All of the above
    e. None of the above

14. Which type of radiator is used to make the vehicle lower in front?
    a. Down-flow
    b. Cross-flow
    c. Centrifugal-flow
    d. Up-flow
    e. Side-flow

15. Which of the following is *not* a part of the radiator?
    a. Inlet tank
    b. Outlet tank
    c. Core
    d. Water pump
    e. Cooling baffles, fins, and tubes

16. Pressure on the cooling system will protect the cooling system from:
    a. Cavitation
    b. Increased operating temperatures
    c. Cold weather operation
    d. All of the above
    e. A and B

17. What is the standard pressure on a cooling system?
    a. 60–70 pounds
    b. 10 pounds
    c. 15 pounds
    d. 2–4 pounds
    e. None of the above

18. When an expansion tank is used on the cooling system:
    a. Overflow coolant from the pressure cap enters the expansion tank
    b. Coolant is drawn back into the radiator when the engine cools down
    c. Coolant can be added to the expansion tank
    d. All of the above
    e. None of the above

19. Which system will produce more rust and corrosion in a cooling system?
    a. A cooling system without an expansion tank
    b. A cooling system with an expansion tank
    c. A cooling system without a cross-flow radiator
    d. A cooling system with a radiator cap
    e. A cooling system with a thermostat

20. What is a big disadvantage of using a fan to draw air through the radiator?
    a. Fans make too much noise
    b. Fans vibrate too much
    c. Fans consume frictional horsepower
    d. All of the above
    e. None of the above

21. The electric-driven fan is turned off and on based on the:
    a. Temperature of the engine
    b. Pressure of the coolant
    c. Speed of the engine
    d. Pressure of the lubrication
    e. Amount of coolant flow

22. Which method of temperature indicators uses a curved tube that straightens when pressure increases?
    a. The electrical sensor
    b. The Bourdon tube gauge
    c. The silicon oil fan clutch
    d. All of the above
    e. None of the above

23. A fan shroud is used:
    a. To reduce any hot spots in the corners of the radiator
    b. To protect the fan from damage
    c. To protect the radiator from damage
    d. To contain leaks in the cooling system
    e. To protect the hoses

The following questions are similar in format to ASE (Automotive Service Excellence) test questions.

24. Technician A says that brownish colored coolant is a sign of rust in the cooling system. Technician B says that brownish colored coolant is caused by the addition of antifreeze in the coolant. Who is right?
    a. A only
    b. B only
    c. Both A and B
    d. Neither A nor B

25. Technician A says the thermostat can only be checked by keeping it in antifreeze. Technician B says to check the thermostat, remove it, and suspend it in water with a thermometer while heating up the water. Who is right?
    a. A only
    b. B only
    c. Both A and B
    d. Neither A nor B

26. After checking the engine, it was found that the coolant was boiling at 212 degrees F. Technician A says this is normal. Technician B says the thermostat should be checked with a pressure tester. Who is right?
    a. A only
    b. B only
    c. Both A and B
    d. Neither A nor B

27. An engine has a large amount of silicon and calcium in the cooling system. Technician A says the silicon is there because of the high quality of the antifreeze. Technician B says the silicon is there because of the aluminum cylinder block. Who is right?
    a. A only
    b. B only
    c. Both A and B
    d. Neither A nor B

Essay

28. What problems can occur when the engine temperature is too high?

29. What problems can occur when the engine temperature is too low?

30. What is the difference between convection, conduction, and radiation forms of heat transfer?

31. What is the purpose of the by-pass tube on the cooling system?

32. What happens to the freezing point of coolant if the percentage of antifreeze is too high?

33. Describe the purpose and operation of a thermostat.

34. Describe the purpose and operation of a pressure cap.

CHAPTER 13

Fuel Characteristics

INTRODUCTION

The correct operation and driving characteristics of automotive engines are directly related to the type of fuel that is used. Today's automotive engines use gasoline as the primary fuel. However, diesel fuel is available for diesel engines. In addition, gasohol and other fuels are constantly being tried to see if they are efficient and clean burning. This chapter is related to the study of these fuels.

OBJECTIVES

After reading this chapter, you will be able to:
■ Define the refining process and determine the heating values of different fuels.
■ State the characteristics and properties of gasoline.
■ Identify the characteristics and properties of diesel fuels.
■ Recognize the characteristics of gasohol as a fuel.
■ Compare liquid petroleum gas as a fuel to diesel fuel and gasohol fuel.

CHAPTER HIGHLIGHTS

13.1 REFINING PROCESSES
    A. Hydrocarbons and Crude Oil
    B. Boiling Points
    C. Distillation
    D. Heating Values

13.2 GASOLINE PROPERTIES
    A. Knocking
    B. Octane Number
    C. Types of Octane Ratings
    D. Octane Requirements
    E. Adding Lead to Gasoline
    F. Removing Lead from Gasoline
    G. Volatility
    H. Gasoline Additives
    I. Engine Run-on

13.3 DIESEL FUEL PROPERTIES
    A. Cetane Number

    B. Distillation Endpoint
    C. Sulfur Content
    D. Pour Point
    E. Cloud Point
    F. Viscosity
    G. Flash Point
    H. Ash Content
    I. ASTM Diesel Fuel Classification

13.4 GASOHOL AS A FUEL
    A. Gasohol Characteristics
    B. Questions About Gasohol

13.5 LIQUID PETROLEUM GAS
    A. LPG Characteristics
    B. Advantages and Disadvantages of LPG

SUMMARY

13.1 REFINING PROCESSES

Hydrocarbons and Crude Oil

Fuels used in vehicles today are called *hydrocarbons*. In simple terms, hydrogen and carbon molecules are combined chemically to make different fuels. These combinations of molecules are called hydrocarbons. The oil industry begins the manufacture of automotive fuels by first exploring for crude oil. Crude oil is the thick brown and black, slippery liquid that all fuels are made from. It contains thousands of different chemical combinations of hydrogen and carbon. See *Figure 13–1*.

Crude oil is usually located between 5,000 and 20,000 feet under the surface of the earth. It is the oil company's job to find the oil, then get it out of the ground. Large offshore and land oil rigs are used to pump the crude oil to the surface. See *Figure 13–2*.

After the crude oil has been pumped from the ground, it is transported to the refinery to be processed into different fuels. After the fuels have been refined, they are shipped to the proper distributor for sale to the public.

Boiling Points

The *boiling point* of a hydrocarbon is defined as the temperature at which the hydrocarbon will start to boil. It is at this temperature that the hydrocarbon becomes a vapor. If the vapor is cooled below its boiling point, the vapor will *condense* or become a liquid again. Hydrocarbon boiling points range from −250 degrees F to more than 1,300 degrees F. Most hydrocarbons have a range of boiling points. For example, gasoline molecules start to boil from about 195 degrees F to 390 degrees F. Other fuels have different boiling points.

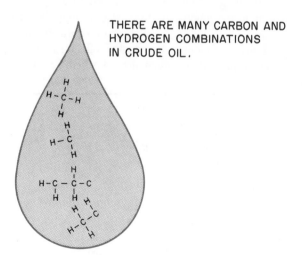

THERE ARE MANY CARBON AND HYDROGEN COMBINATIONS IN CRUDE OIL.

FIGURE 13–1. Crude oil contains many combinations of hydrogen and carbon. Different combinations make different fuels.

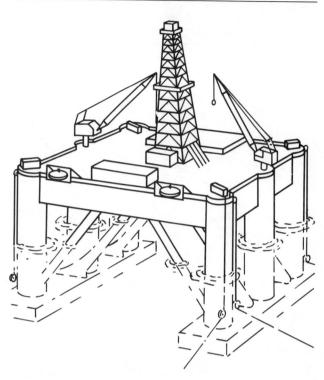

FIGURE 13–2. Large oil rigs such as this example are used to remove the crude oil from the ground.

Distillation

Distillation is the process in a refinery that separates the different hydrocarbons in the crude oil. Certain hydrocarbons from the crude oil will be used to make gasoline. Other hydrocarbons will be used to make diesel fuels. Others will be used to make many other products. *Figure 13–3* shows some of the products that can be made from crude oil.

Distillation is done by heating all of the crude oil and the hydrocarbons to the highest boiling point. Referring to *Figure 13–4*, at the bottom of the column all hydrocarbons are vaporized. As the vapors rise to the top of the column, they start to cool. The hot vapors are cooled near their boiling points. At this point, they are condensed back to a liquid. For example, on the right side of the column, vapors cool to form gasolines, diesel fuels, heavier oils, and different residues. This column is also called a "fractionating column."

From this point, the fuels are cleaned and chemicals are added to improve the fuels. They are now ready to be transported to the distributors and service stations.

Heating Values

Each of the fuels that are refined has a different heating value. Heating values are measured in Btus. The term *Btu* means "British Thermal Unit." It is defined as the amount of heat needed to raise one pound of water one degree F. For example, one Btu is about equal to the heat given off when one wooden match burns completely. A home on a cold winter day may require 30,000 Btus per hour to be heated.

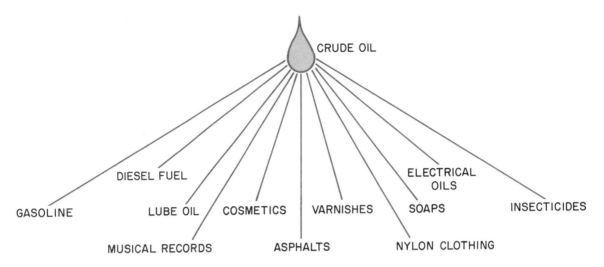

FIGURE 13-3. Many products, including fuels, can be extracted from crude oil. These are just a few of the many products.

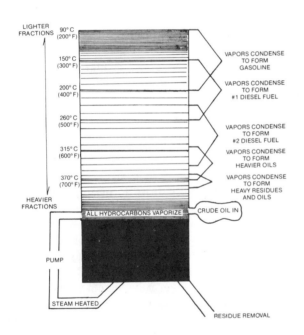

FIGURE 13-4. The process of distillation vaporizes all hydrocarbons. Then as they condense at different temperatures, they are separated and removed as specific products. (Courtesy of Davis Publications, Inc.)

a hydrocarbon, the higher the heating value of the fuel. Diesel fuels have a higher boiling point than gasoline. This means that diesel fuels have more Btus per gallon than gasoline.

CAR CLINIC

PROBLEM: ALCOHOL IN THE FUEL

Many car owners are concerned about the use of alcohol in fuel such as gasohol. Is there any problem with running high levels of alcohol in the fuel?

SOLUTION:

Alcohol will generally cause engine damage. The amount of damage depends upon the percentage of alcohol in the gasoline. Anything above 10% could be damaging to the engine. At 15%, alcohol may damage aluminum parts, and cause steel to begin rusting. Alcohol may also dilute the engine oil considerably. Always be aware of how much alcohol is being added to the fuel.

Gasoline and diesel fuels are compared by the amount of Btus they contain if converted to heat. The higher the heating value of a fuel, the better the combustion and thus, the better the fuel mileage. For example, one gallon of gasoline contains about 110,000–120,000 Btus. Diesel fuels typically contain about 130,000 to 140,000 Btus per gallon.

The heating value of a fuel is related to its boiling point during distillation. Normally, the higher the boiling point of

13.2 GASOLINE PROPERTIES

Knocking

The term *knocking* is used when studying fuel and different gasoline qualities. Knocking is a process that happens within the combustion chamber. It sounds like a small ticking or rattling noise within the engine. It can be very damaging to the pistons and rings, as well as to the spark plug and valves. Another name for knocking is detonation.

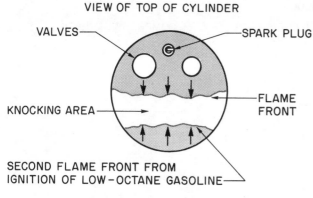

VIEW OF TOP OF CYLINDER

VALVES

SPARK PLUG

KNOCKING AREA

FLAME
FRONT

SECOND FLAME FRONT FROM
IGNITION OF LOW-OCTANE GASOLINE

TWO FLAME FRONTS COLLIDE

FIGURE 13-5. Detonation is a result of having two flame fronts hit each other. One is from the spark plug combustion and one is from the fuel ignition because of poor quality.

Figure 13-5 shows what happens inside a combustion chamber when detonation occurs. When the spark plug fires to ignite the air-fuel mixture, it produces a flame front. Shortly after ignition, a second explosion (flame front) and ignition of fuel take place on the other side of the combustion chamber. When these two energy fronts hit each other, they cause a *pinging* or knocking within the engine. Knocking is usually caused by poor quality fuel. The temperatures within the combustion chamber are high enough to cause the air and fuel mixture to ignite without the spark plug. Fuels are made to have antiknocking characteristics. Actually, antiknocking fuels need higher temperatures to start burning.

Octane Number

Different types of gasoline are identified by the *octane number*. The octane number is defined as the resistance to burning. For example, the higher the octane number of a gasoline, the higher the temperature needed to ignite the fuel. The lower the octane number, the easier the fuel can start burning. Standard octane ratings range from 85–90 for regular gasoline. Premium gasoline or hi-test gasoline has an octane range from 90–95. Gasolines that have higher octane ratings tend to reduce knocking and detonation.

Types of Octane Ratings

Three types of octane ratings are used in the petroleum industry. They include the research octane, motor octane, and road octane. Research octane is a laboratory measure of gasoline and its antiknock characteristics. The octane is tested under mild engine operating conditions, low speed, and low temperatures. Motor octane is a laboratory measure

of the antiknock characteristics of the fuel under severe engine operation. This includes high speed and high temperatures within the engine. Road octane represents actual road driving conditions. Road octane is the one posted on gasoline pumps. It is calculated by taking the average of the research and motor octanes. On gasoline pumps, road octane is stated as research octane plus motor octane divided by two. This is illustrated on the pump as:

$$\text{Octane} = \frac{R + M}{2}$$

where:

$$R = \text{Research octane}$$
$$M = \text{Motor octane}$$

Until about the 1970s, the octane numbers of gasoline were increasing. This was because the automotive manufacturers were constantly increasing compression ratios. Higher compression ratios usually meant more efficient combustion. As the compression ratios increased, so did the compression temperatures. If a lower octane fuel was used in these higher compression ratio engines, knocking always occurred. *Regular gasoline* was used for low-compression engines. *Hi-test* was used for higher compression engines.

Octane Requirements

The octane requirements of automobiles vary a great deal. In fact, as an engine gets older, the octane requirement may increase somewhat. This is because when the engine is new, there is a set compression ratio. As the engine gets older, however, carbon buildup may increase the compression ratio. This means that the engine will need a higher octane gasoline. See *Figure 13-6*.

Many other factors also influence the octane needed in an engine. These include:

1. *Air temperature* — The higher the air or engine temperature, the greater the octane requirements.
2. *Altitude* — The higher the altitude, the lower the octane.
3. *Humidity* — The lower the humidity, the greater the octane needed in the engine.
4. *Spark timing* — The more advanced the spark timing, the greater the octane needed.
5. *Carburetor settings* — The leaner the carburetor, the higher the octane needed.
6. *Method of driving* — If the vehicle is accelerated rapidly, higher octane is needed. Start and stop driving also increases combustion deposits and buildup. Higher octane will then be needed.

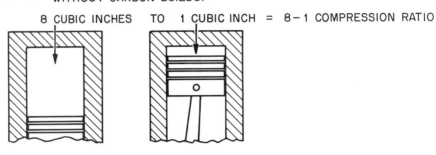

WITHOUT CARBON BUILDUP

8 CUBIC INCHES   TO   1 CUBIC INCH = 8 – 1 COMPRESSION RATIO

WITH CARBON BUILDUP

7.7 CUBIC INCHES   TO   0.7 CUBIC INCH = 11 – 1 COMPRESSION RATIO

0.3 CUBIC INCH BUILDUP

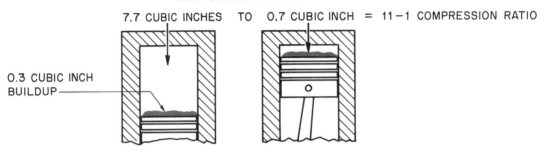

FIGURE 13–6. As an engine gets older, the compression ratio may increase. As carbon builds up on top of the piston, the compression ratio may change.

CARBON BUILD UP

Adding Lead to Gasoline

One method used to increase the octane of gasoline is by adding a chemical called TEL. This stands for *tetraethyl lead*. Tetramethyl lead, called TML, is also used to increase the octane rating. These chemicals were added during the refining process. They also act as a lubricant on the valves and valve guides. It was found, however, that TEL and TML are dangerous pollutants. The exhaust gases from cars were carrying lead pollution. Today many pollution control devices, particularly catalytic converters, should not be used with leaded fuels.

Removing Lead from Gasoline

In the early 1970s, TEL and TML were removed from gasoline because of pollution. The result was called low-lead or nonleaded gasoline. Other chemicals were added to the gasoline to aid in lubricating the valves.

When the lead was removed from gasoline, the octane ratings also fell. Therefore, the automobile manufacturing companies had to reduce their compression ratios. Compression ratios dropped from an average of 12 to 1, down to 8.5 to 1. This made the engine less efficient. Recent engines still have compression ratios from 8 to 1 up to about 9 to 1.

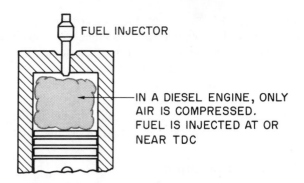

FIGURE 13-7. In a diesel engine, only air is compressed. This means there is no chance of having the fuel ignite too soon.

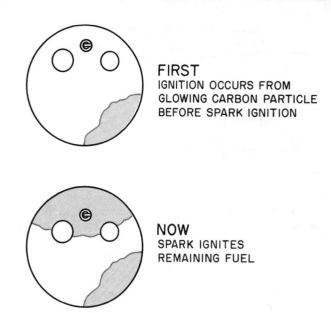

FIGURE 13-8. Preignition is caused by having another ignition source ignite the fuel before the spark plug does. This also causes pinging.

One advantage to lowering compression ratios is that the internal combustion temperatures are lower. This means fewer *nitrogen oxides* are produced within the exhaust. Nitrogen oxides are also a source of pollution. Nitrogen oxides are referenced as NO_x.

The problem with compression ratios and octane ratings exists only in engines that compress an air and fuel mixture. Diesel engines do not have this problem. Diesel engines have increased their compression ratios to as high as 22.5 to 1. This can be done because the compression stroke on a diesel engine compresses only air. Fuel is then injected near top dead center. It is nearly impossible to have a diesel engine detonate because fuel does not enter the cylinder until the piston is near top dead center. See *Figure 13-7*.

Volatility

Gasoline must vaporize in order to burn effectively. If a fuel does not vaporize completely before going into the combustion chamber, combustion efficiency and fuel efficiency, in turn, will decrease. *Volatility* is defined as the ease with which a gasoline vaporizes. Highly volatile fuels *vaporize* very easily. Fuels that are used in cold weather must be highly volatile. In warm weather, however, gasoline should be less volatile to prevent formation of excess vapor. Excess vapor causes a loss of power or stalling due to *vapor lock*. Gasolines vary in volatility seasonally.

Gasoline Additives

Additives other than the chemicals that change octane are found in gasoline. Fuel detergents are added to keep the carburetor or fuel injector clean. If the carburetor is dirty, the air-fuel mixture may not be accurate. Other additives help to keep intake valves and other internal engine parts clean. Winter additives, including *anti-icers*, stop fuel lines and

carburetors from icing up in cold weather. For example, the carburetor anti-icers help keep ice from building up on the throttle plates. If ice does build up, missing and stalling will result. The ice is formed when moist air hits a throttle plate that has been cooled by the vaporization of gasoline. The ice then restricts the flow of air into the carburetor.

Ignition control and combustion modifiers are also added to gasoline. These additives help prevent spark plug fouling and *preignition*. Preignition is caused by glowing deposits in the combustion chamber. These deposits ignite the air-fuel mixture before the spark plug fires. See *Figure 13-8*. This causes a pinging or knocking sound as with detonation. Other additives include antirust, antigum, and antiwear chemicals.

Engine Run-on

Another characteristic of combustion caused by poor-quality gasoline is called *postignition*. Postignition is defined as having the engine run after the ignition has been shut off. See *Figure 13-9*. For example, a poor-quality gasoline may have many glowing deposits inside the combustion chamber during operation. When the ignition is shut off, these glowing deposits act as the ignition source for the air and fuel still in the combustion chamber. This is also called "dieseling" or "run-on." The only method of stopping this is to remove the fuel completely, or to increase the quality of gasoline being used. Both methods are used today.

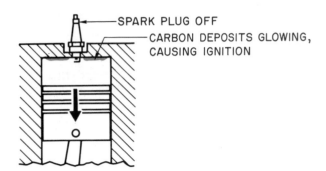

FIGURE 13–9. Postignition occurs after the ignition has been shut off. A glowing particle of carbon ignites the air-fuel mixture.

CAR CLINIC

PROBLEM: CAR SMOKES WHEN STARTED

A vehicle produces blue smoke when the engine is first started. Once the engine is running and at operating temperature, the smoking stops. What could be the problem?

SOLUTION:

The most common cause of this problem is oil seeping past the exhaust valves when the engine is shut down. The oil then sits in the exhaust manifold until the engine is started. When the engine is started, the oil is burned, producing the blue smoke. Replacing the valve stem seals should solve the problem.

13.3 DIESEL FUEL PROPERTIES

Diesel fuels are also a mixture of hydrogen and carbon molecules. In comparison to gasoline, however, diesel fuels have more energy per gallon than gasoline. In addition, diesel fuels burn with higher efficiency during combustion.

Diesel fuel is part of a group of fuel oils called distillates. This group of fuel oils make up such products as jet fuels, kerosene, home fuel oil, and diesel fuels.

Three properties are important in the selection of diesel fuel. These include cetane number, distillation endpoint, and sulfur content. Other characteristics, including pour point, flash point, viscosity, and ash content, are also considered.

Cetane Number

There is a delay between the time that fuel is injected into the cylinder and the time that ignition by hot gases takes place. This time period or delay is expressed as a *cetane number*. Cetane numbers range from 30 to 60 on diesel fuels. The cetane number is an indication of the ignition quality of diesel fuel. The higher the cetane number, the better the ignition quality of the fuel. High cetane numbers should be used for cold weather starting. *Ether*, with a cetane number of 85–96, is often used for starting diesel engines in cold weather. If a low cetane number is used in a diesel engine, some of the fuel might not ignite. The fuel will then accumulate within the cylinder. When combustion finally does occur, this excess fuel will explode suddenly. This may result in a knocking sound as in a gasoline engine.

Distillation Endpoint

Fuels can be burned in an engine only when they are in vaporized form. The temperature at which a fuel is completely vaporized is called the *distillation endpoint* temperature. The distillation endpoint should be low enough to permit complete vaporization at the temperatures encountered in the engine. This means for engines operating at reduced speeds and loads or in cold weather, lower distillation endpoints will give better performance.

Sulfur Content

The *sulfur* content in a diesel fuel should be as low as possible. Sulfur is part of the crude oil when it is taken out of the ground. Refining is designed to remove as much sulfur as is practical. There is a direct link between the amount of sulfur present in the fuel and the amount of corrosion and deposit formation within the engine. Sulfur is also contained in the exhaust gases. These gases can pollute and cause a significant amount of *acid rain*. Certain tests have shown that increasing sulfur content from 0.25% to 1.25% increases engine wear by 135%. Engine wear is most noticeable on the cylinder walls and piston rings.

Pour Point

Diesel fuels are also rated as to their pour point. Pour point is defined as the temperature at which fuel stops flowing. For cold weather operation, the pour point should be about 10 degrees F below the *ambient* (surrounding air) temperature at which the engine is run.

Cloud Point

The *cloud point* of a diesel fuel is defined as the temperature at which wax crystals start to form in the fuel. Diesel fuels tend to produce wax crystals in very cold weather. The wax crystals plug up fuel filters and injectors.

Viscosity

As we saw in the lubrication system, viscosity is a measure of the resistance to flow. The fuel should have a viscosity that allows it to flow freely in the coldest operation. The viscosity of fuel oil also affects the size of the fuel spray droplets from the fuel injection nozzles. The higher the viscosity, the larger the droplets. This, in turn, affects the *atomization* qualities of the fuel spray.

Fuel oil viscosity is normally checked at 100 degrees F. It is measured in centistokes. This is another unit used to measure viscosity. It is used instead of the Saybolt Universal Second. Additives are incorporated into the fuel during refining to keep the viscosity at the correct level.

Flash Point

The *flash point* of a diesel fuel is defined as the fuel's ignition point when exposed to an open flame. It is determined by heating the fuel in a small enclosed chamber. The temperature at which vapors ignite from a small flame passed over the surface of the liquid is the flash point temperature. The flash point should be high enough so that the fuel can be handled safely and stored without danger of explosion.

Ash Content

The ash content of diesel fuels is a measure of the impurities, which include metallic oxides and sand. These impurities cause an *abrasive* action on the moving parts of the engine. The amount of ash content should be kept to a minimum.

ASTM Diesel Fuel Classification

Diesel fuels are classified as either No. 1-D or No. 2-D. The characteristics of each are shown in *Figure 13–10*. This is the American Society for Testing of Materials (ASTM) classifications.

Grade No. 1-D fuels have the lowest boiling ranges (most volatile). These fuels also have the lowest cloud and pour points. Fuels within this classification are suitable for use in high-speed diesel engines in services involving frequent and relatively wide variations in loads and speeds. This may include stop-and-go bus and door-to-door operations. They are also used in cases where abnormally low fuel temperatures are encountered.

Grade No. 2-D fuels have higher and wider boiling ranges than 1-D fuels. These fuels normally have higher cloud and pour points than 1-D fuels. They are used in diesel engines in services with high loads and uniform speeds. They are also used in climates where cold starting and cold fuel handling are not severe problems. These fuels satisfy the majority of automotive diesel applications.

13.4 GASOHOL AS A FUEL

Gasohol is a term used to describe a motor fuel that blends 90% gasoline and 10% alcohol. See *Figure 13–11*. The alcohol in gasohol is called *ethanol*. Ethanol is produced by distilling agricultural crops such as corn, wheat, timber, and sugar cane. The purpose of using gasohol is that for every

gallon of gasohol sold, 10% of the fuel is renewable, whereas gasoline is considered a nonrenewable source of fuel. The study and use of gasohol grew out of the energy shortage of 1973. There is still a rather large debate going on as to its effectiveness and use in the automobile.

Gasohol Characteristics

Overall, the Btus in a gallon of gasohol are slightly less than in gasoline. Gasoline has about 115,000 Btus per gallon.

ASTM CLASSIFICATION OF DIESEL FUEL OILS		
	NO. 1-D	NO. 2-D
Flash Pt.; °F Min.	100	125
Carbon Residue; %	0.15	0.35
Water and Sediment; (% by volume) Max.	Trace	0.10
Ash; % by Wt.; Max.	0.01	0.02
Distillation, °F 90% Pt.; Max. Min.	550 —	640 540
Viscosity at 100°F; centistokes Min. Max.	1.4 2.5	2.0 4.3
Sulfur; % Max.	0.5	0.7
Cetane No; Min.	40	40

FIGURE 13–10. This chart shows the ASTM classifications for different types of diesel fuels. *(Courtesy of American Society for Testing Materials)*

FIGURE 13–11. Gasohol is made from 90% gasoline and 10% alcohol.

LABORATORY ANALYSIS

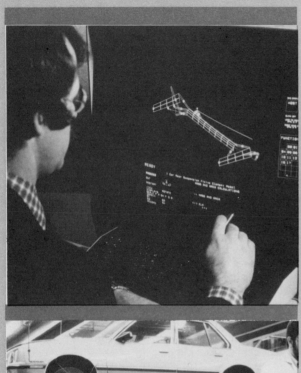

Laboratory tests and road tests are important in evaluating component design and material. Various parts that are under development are sent to the labs for analysis. Here they are studied under precise scientific conditions. Material composition, strength, design, chemical properties, reaction to temperature changes, and other factors are evaluated. (Courtesy of General Motors Proving Grounds)

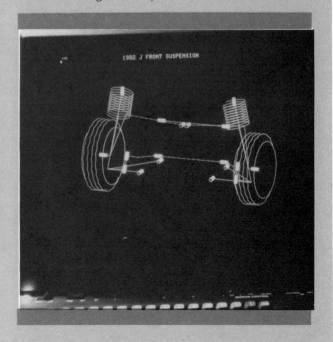

Ethanol contains only 75,000 Btus per gallon. By adding ethanol to gasoline, the total Btu content is reduced slightly as compared to gasoline. Alcohol, however, has a slightly higher octane than gasoline. The octane of gasohol is normally about 3–4 points higher than that of gasoline. The increase in octane reduces or eliminates engine knock and ping. Gasohol's higher octane also reduces engine postignition and run-on. Testing has shown that when gasohol is used, cleaner burning occurs in the combustion chamber. This tends to reduce harmful carbon deposits and soot buildup, which, in turn, promotes longer engine life. The alcohol in gasohol is also a de-icer (it prevents freezing in cold weather). Gasohol has been used in many parts of the country. Car manufacturers vary in their recommendations concerning the use of gasohol.

Questions About Gasohol

Questions still need to be answered about the use of gasohol. Is the reduced gasoline mileage when using gasohol sufficient to warrant its continued use? Should the U. S. grow grain to run our automobiles when millions of people are starving in other countries? What effect would gasohol production have on farming and other farm product prices? Can the price of gasohol be competitive with the price of gasoline? Before gasohol becomes widely accepted, these and other questions will have to be addressed in detail.

13.5 LIQUID PETROLEUM GAS

LPG Characteristics

Another type of fuel that can be used in the automobile is called liquid petroleum gas or LPG. LPG is also a hydrocarbon fuel and a by-product of the distillation process discussed earlier. It is made primarily of *propane* and butane. One major difference of LPG is its boiling point. See *Figure 13–12*. LPG has a very low boiling point. In normal ambient pressures, LPG has boiled and is in the vapor form. If the fuel is put under a pressure, the boiling point can be increased. The fuel can then be stored in a pressurized container as a liquid. See *Figure 13–13*. This makes storage much easier but possibly more dangerous in accidents. When the fuel is allowed to come out of the container, it turns back to a vapor.

LPG also has a higher octane rating (about 100 octane) than gasoline. This means that compression ratios could be increased slightly to improve efficiency. However, because the Btu content of LPG per gallon is less, there is a slight reduction in power as compared to gasoline.

Advantages and Disadvantages of LPG

One of the big advantages of using LPG is that it is extremely clean and pollution free. This is the reason many lift trucks use LPG fuels inside buildings. The exhaust is not as dangerous as the carbon monoxide that is produced by

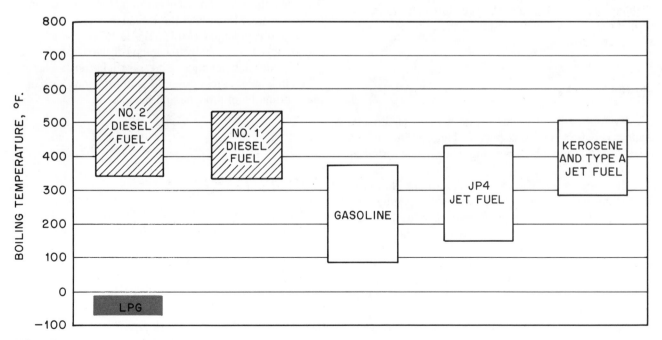

FIGURE 13–12. The boiling point of LPG is considerably lower than that of other fuels. This means that in ambient pressure and temperatures, LPG fuel has boiled and is in a vaporized form.

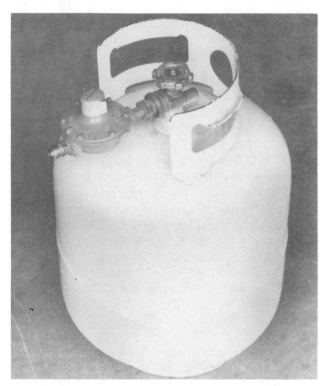

FIGURE 13–13. LPG is stored in a pressurized container to keep it in liquid form.

gasoline. Large fleet operations are probably the biggest users of LPG fuels in the automotive industry. In these operations, the fuel can be containerized and sold in larger quantities. However, the fuel systems must be altered slightly to allow LPG to be used. Regulators, converters, and several other components must be installed to operate the system correctly.

SUMMARY

Several types of fuels are used today in the automotive engine. The most popular is gasoline, but diesel fuel is also used. Gasohol and LPG are being studied and tested as possible fuels.

These fuels are all considered hydrocarbons. Normally, oil companies explore and locate crude oil under the ground. When it is found, the refinery separates the crude oil into many products. Gasoline, diesel fuel, and LPG all come from the crude oil. Distillation is the process that separates the crude oil into different fuels. The biggest difference between the fuels is that each has a different boiling point. The heating value of each fuel is also different. Heating values are measured in Btus.

Gasoline has many properties. Knocking occurs when a poor-quality fuel ignites on one side of the combustion chamber, and the spark plug ignites the other side. The two flame fronts hitting each other causes knocking. Octane is added to the fuel to reduce knocking. It is important to use the right octane. Temperatures, altitude, humidity, spark timing, carburetor setting, and driving methods all influence the octane rating needed.

Tetraethyl lead was used to increase the octane, but it also produced pollution. In addition, many new emission control devices would not work with lead in the fuel. The lead was removed and no-lead fuels were put in cars. This meant the compression ratios had to be reduced. Reducing compression ratios has had a significant effect on the efficiency of combustion in engines.

Gasoline has many additives. These include anti-icers, carburetor detergents, lubricants for valves, and combustion modifiers.

Poor-quality gasoline can cause postignition. When the ignition is shut off, the vehicle's engine continues to run. Postignition is produced by glowing carbon deposits inside the combustion chamber.

Diesel fuels are also used in the automotive market. The primary rating of diesel fuel is the cetane number. Cetane is a measure of the delay time between injection and combustion of the fuel. Other characteristics of diesel fuels include distillation endpoint, sulfur content, pour point, cloud point, flash point, and ash content.

Diesel fuels come in two classifications: No. 1-D and No. 2-D. Each has been designed to have the right cetane number, distillation endpoint, cloud, pour and flash points, and ash content.

Gasohol is being used in some vehicles. It is made from a mixture of 90% gasoline and 10% alcohol. The alcohol comes from distilling corn, wheat, and other farm products. Although many questions still surround the use of gasohol, it is being used in parts of the United States.

Liquid Petroleum Gas (LPG) is also considered a hydrocarbon. It is made from hydrocarbons distilled from natural gas or crude oil. If put under a pressure, it becomes a liquid. When in ambient pressures, LPG is in a gaseous form. It is made primarily of propane and butane.

TERMS TO KNOW

Can you explain each of the following terms? Review the chapter until you can use each term correctly.

Hydrocarbon	Vapor lock
Molecule	Anti-icer
Boiling point	Preignition
Condense	Cetane number
Distillation	Ether
Btu	Sulfur

Pinging

Octane number

Regular gasoline

Hi-test

Tetraethyl lead

Nitrogen oxide

Vaporize

Acid rain

Ambient

Atomization

Abrasive

Ethanol

Propane

REVIEW QUESTIONS

Multiple Choice

1. Crude oil is primarily made of what molecules?
   a. Nitrogen and oxygen
   b. Carbon and hydrogen
   c. Sulfur and nitrogen
   d. Nitrogen and sulfur
   e. Sulfur and carbon

2. What is the major difference between different fuels?
   a. Boiling points
   b. Acid levels
   c. Degree of lubrication
   d. Density
   e. Thickness

3. The process of separating crude oil into different fuels is called _____.
   a. Combustion
   b. Condensing
   c. Distillation
   d. Departing
   e. None of the above

4. When two flame fronts hit each other, a _____ sound is heard.
   a. Knocking
   b. Soft sound
   c. Cracking
   d. Hissing
   e. None of the above

5. The resistance to burning of gasoline is defined as:
   a. Cetane
   b. Octane
   c. Butane
   d. Methane
   e. Ethane

6. Which octane is a result of calculating (R + M)/2?
   a. Road octane
   b. Motor octane
   c. Research octane
   d. All of the above
   e. None of the above

7. Which of the following will influence the octane number needed in an engine?
   a. Carburetor settings
   b. Air temperature
   c. Altitude
   d. All of the above
   e. None of the above

8. Which chemical is added to gasoline to increase the octane number?
   a. Sulfur lead
   b. Distillation chemicals
   c. Tetraethyl lead
   d. Nitrogen
   e. Oxygen

9. When lead was removed from gasoline, what had to be done with the compression ratios?
   a. They were increased
   b. They were kept the same
   c. They were reduced or lowered
   d. They were doubled
   e. None of the above

10. The condition that occurs when the engine ignition is shut off and the engine continues to run is called _____.
   a. Pinging
   b. Knocking
   c. Postfiring
   d. Preignition
   e. Run-on

11. The measure of a diesel fuel's quality is called:
   a. Ash point
   b. Cetane number
   c. Sulfur content
   d. Methane number
   e. Octane number

12. The temperature at which a diesel fuel is completely vaporized is called the:
   a. Distillation endpoint
   b. Flash point
   c. Pour point
   d. Fire point
   e. Boiling point

13. The term *ambient* is defined as:
   a. Surrounding
   b. Closed in
   c. Open
   d. Endpoint
   e. Cetane point

14. A diesel fuel's ignition point when exposed to open flame is defined as:
    a. Pour point
    b. Flash point
    c. Distillation point
    d. Burning point
    e. Firing point

15. What are the two types of diesel fuel classifications?
    a. 1 D and 2 D
    b. 2 D and 4 D
    c. 3 D and 4 D
    d. 4 D and 5 D
    e. 4 D and 1 D

16. Gasohol is a mixture of what percentages of gasoline and alcohol?
    a. 10% gasoline and 90% alcohol
    b. 20% gasoline and 80% alcohol
    c. 30% gasoline and 70% alcohol
    d. 40% gasoline and 60% alcohol
    e. None of the above

17. Gasohol has _____ ratings than gasoline.
    a. Lower octane
    b. Higher octane
    c. 17 higher octane
    d. Higher cetane
    e. Lower cetane

18. If put under a pressure, LPG will change to a _____.
    a. Liquid
    b. Gas
    c. Solid
    d. Semi-solid
    e. Semi-liquid

The following questions are similar in format to ASE (Automotive Service Excellence) test questions.

19. Technician A says gasoline has an octane rating. Technician B says gasoline has a cetane rating. Who is right?
    a. A only
    b. B only
    c. Both A and B
    d. Neither A nor B

20. Technician A says that resistance to burning is measured with a cetane rating on gasoline. Technician B says that resistance to burning has no type of measurement for gasoline. Who is right?
    a. A only
    b. B only
    c. Both A and B
    d. Neither A nor B

21. Technician A says that preignition is caused by glowing deposits inside the combustion chamber. Technician B says that preignition is caused by bad lubrication. Who is right?
    a. A only
    b. B only
    c. Both A and B
    d. Neither A nor B

22. Technician A says that lead has recently been added to fuel to improve nitrogen oxides. Technician B says that lead has recently been added to fuel to lower the compression ratio. Who is right?
    a. A only
    b. B only
    c. Both A and B
    d. Neither A nor B

Essay

23. What does the term *distillation* mean?

24. What is knocking?

25. How does octane of a fuel relate to the burning of fuel?

26. Why are anti-icers added to fuel?

27. State several advantages and disadvantages of using gasohol.

CHAPTER 14

Basic Fuel Systems

INTRODUCTION

The fuel system is very critical to engine operation. This chapter is about the fuel system components and their operation. The fuel system consists of the fuel tank, lines, pumps, filters, carburetors, and injectors. However, because carburetors and injectors have become rather complex, a separate chapter is devoted to each. This chapter deals only with the basic components of the fuel system needed to get the fuel to the carburetor or injectors.

OBJECTIVES

After reading this chapter, you will be able to:
- Identify the total fuel flow on an automobile engine.
- Analyze the parts and operation of the fuel tank.
- Recognize the parts and operation of fuel pumps.
- State the purpose and operation of fuel filters.
- Identify common diagnosis and service procedures in the basic fuel system components.

CHAPTER HIGHLIGHTS

14.1 TOTAL FUEL FLOW

14.2 FUEL TANK
    A. Fuel Cap and Filler Neck
    B. Fuel Metering Unit
    C. Diesel Fuel Pickup and Sending Unit

14.3 FUEL PUMPS
    A. Mechanical Fuel Pumps
    B. Vapor Return Line
    C. Electric Fuel Pumps
    D. Bellows-type Electric Fuel Pump
    E. Roller Vane Fuel Pump
    F. Fuel Pressure Regulators
    G. Computers and Fuel Control

14.4 FUEL FILTERS
    A. Purpose of Fuel Filters
    B. Sources of Gasoline Contaminants
    C. Gasoline Fuel Filters
    D. Diesel Fuel Contaminants
    E. Diesel Fuel Filters

DIAGNOSIS AND SERVICE

SUMMARY

14.1 TOTAL FUEL FLOW

The fuel system is made so that fuel can be stored in a fuel tank and be ready for delivery to the engine. Referring to **Figure 14–1**, the fuel in the fuel tank is ready to be used when the engine needs it. The fuel pump, which is electrically or mechanically driven, draws fuel from the tank and sends it to the carburetor. The purpose of the carburetor is to keep the fuel at the right air-fuel ratio. It is very important to maintain a ratio of 15 parts air to 1 part fuel. The air-fuel ratio must be accurate throughout engine operation. *Fuel injectors* can also be used.

Figure 14–2 shows a diesel fuel system. Here the fuel goes to the fuel injector pump, rather than to the carburetor. On all fuel injector systems on diesel engines, there must be a fuel return line to the fuel tank. This is also shown. Any fuel not used by the engine is returned to the fuel tank or to the suction side of the fuel pump to be used again.

The parts of the fuel sytem include the vapor separator, the carbon canister, and the fuel filter. The vapor separator is used to separate any *fuel vapors* and send them back to the fuel tank. The carbon canister is a pollution control device that holds excess vapors from the tank and carburetor. The carbon canister is studied in detail in the chapter on pollution. The fuel filter, which is located directly before the carburetor or fuel injectors, is used to keep the fuel clean and free of contaminants.

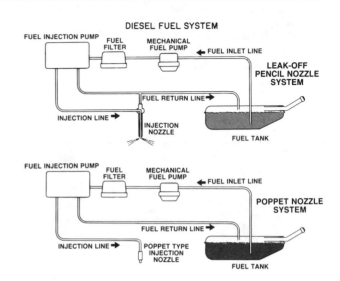

FIGURE 14–2. A diesel fuel system has a tank, mechanical pump, fuel filter, and fuel injection pump with injectors. *(Courtesy of General Motors Product Service Training)*

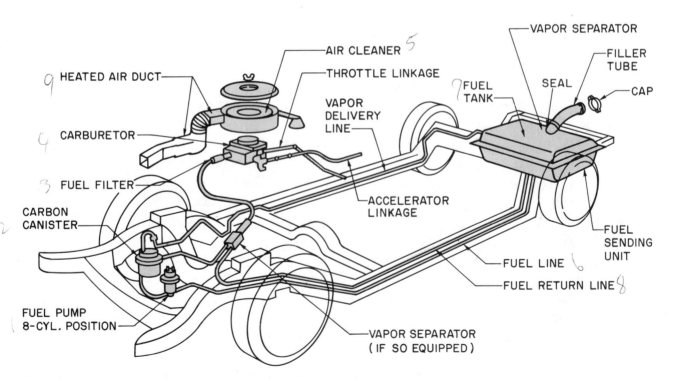

FIGURE 14–1. The fuel system consists of the fuel tank, fuel lines, fuel pump, carburetor or fuel injectors, fuel filter, and vapor separator. The carbon canister and heated air duct also play a part in the fuel system. *(Courtesy of Ford Motor Company)*

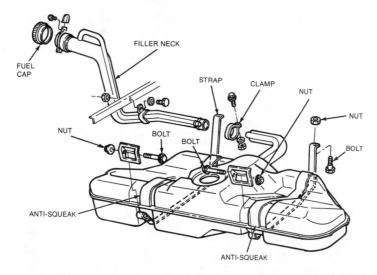

FIGURE 14–3. The fuel tank consists of many parts. They are designed to fit around the frame of the vehicle. (Courtesy of Chevrolet Division, General Motors Corporation)

CAR CLINIC

PROBLEM: BLACK SMOKE DURING STARTING

An engine has become hard to start and produces large amounts of black smoke when it is cold. Also, the fuel mileage has been decreasing each month. What could be the problem?

SOLUTION:

Black smoke is always a sign of an extremely rich mixture in the carburetor. Make sure the choke is adjusted properly. The choke should be set so that the choke plate is just closed. Make sure the bimetal strip is not holding the choke closed under high spring pressure. Also, the choke should be fully open when the engine is at operating temperature. Often, the choke must be readjusted by bending the choke rod. In some cases, if this doesn't solve the problem, the bimetal spring controlling the choke valve must be replaced.

14.2 FUEL TANK

The fuel tank is made to hold or store the excess fuel needed by the automobile. It is usually located in the rear of the vehicle, but in cars that have rear-mounted engines, it is located in the front part of the vehicle. *Figure 14–3* shows a fuel tank. The shape of the fuel tank depends upon the physical design of the vehicle. The fuel tank must fit around the frame and still be protected from impacts.

The filler neck is used to fill the tank. The filler neck must be attached to the outside of the vehicle so the owner can fill the tank. Straps are used to hold the tank in place within the frame.

Fuel Cap and Filler Neck

The fuel cap on the fuel tank is used for several reasons. First, the fuel cap keeps the fuel from splashing out of the tank. Second, the cap releases the vacuum created when the fuel is removed by the engine. Third, the cap releases pressure when the fuel tank heats up and expands in hot weather. Fourth, the cap passes vapors to the carbon canister when needed. The filler neck on newer vehicles has a restrictor door so that only unleaded fuel nozzles can be used.

Many designs are used for fuel caps. One common type seals the tank with a cap that has threads, *Figure 14–4*. A ratchet tightening device on the threaded filler cap reduces the chances of incorrect installation. If the wrong cap is used, vapor control will be lost.

The filler neck has internal threads. The cap contains a plastic center extension. This acts as a guide when inserted into the filler neck. As the cap is turned, a large O ring is seated upon the filler neck flange. After the seal and flange make contact, the ratchet produces a clicking noise. This indicates that the seal has been set. Now the vacuum and pressure lines will operate correctly.

FIGURE 14–4. The filler cap is used to seal the fuel tank so pressure and vapors can escape.

Fuel Metering Unit

The operator must know how much fuel is in the tank. This is done with a fuel metering unit. Many styles are used. One type has a hinged float inside the tank. As the float changes position with different levels of fuel, the sensing unit changes its resistance. The resistance then changes the current in the circuit and the position of the needle on the dashboard gauge. *Figure 14–5* shows a fuel metering unit.

Diesel Fuel Pickup and Sending Unit

Diesel engines are very sensitive to water contamination. Water can get into the fuel by *condensation*. When a fuel tank is only partially full, water tends to form within the tank. The moisture in the air within the tank condenses on the inside tank wall. This occurs in the fuel tank or in any storage tank between the refinery and the vehicle.

By law in many states, water in diesel fuel should be no more than 1/2 of 1%. That quantity of water will be absorbed by the fuel. Higher levels may have a damaging effect on the injectors.

Hydrocarbon fuels are lighter than water. This means that water will be on the bottom of the tank and not on the top. Pickup systems are made to keep water out of the fuel. Some use a "sock" to absorb the water. Others use a detector to determine if water should be drained from the system. *Figure 14–6* shows a common type of fuel pickup with

a water warning system. This unit uses a capacitive probe to detect the presence of water when it reaches the 1–2 gallon level. An electronic module provides a ground through a wire to a light in the instrument panel. The light reads "water in fuel."

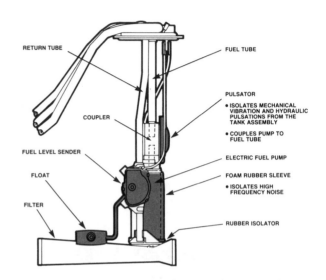

FIGURE 14–5. The fuel metering unit has a float that changes resistance as the fuel level changes. *(Courtesy of General Motors Product Service Training)*

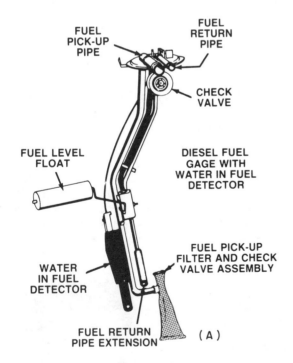

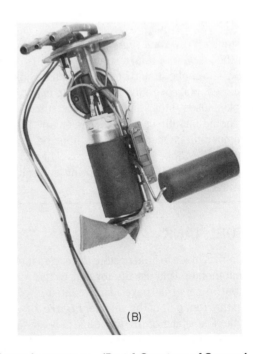

FIGURE 14–6. This fuel pickup/sending unit has a water-in-fuel warning system. *(Part A Courtesy of General Motors Product Service Training)*

BASIC FUEL SYSTEMS   253

PROBLEM: USING UNLEADED FUEL IN OLDER CARS

An older vehicle has always used a leaded fuel. Is it OK to run unleaded fuel in such an engine? Could there be any damage to the engine from unleaded fuels?

SOLUTION:

The biggest problem in using an unleaded fuel in older engines is that it shortens the life of the exhaust valves. The lead in leaded fuels lubricates the valves. In engines designed to use unleaded fuel, the valves are harder, so they don't need as much lubrication. If the engine is going to be used for a long time, try changing the valves to a harder type if they are available. Also, an oiler kit that oils the valves from the top could improve the situation. It is important to get used to the idea that there will eventually be no leaded fuels on the market.

FIGURE 14–7. The mechanical fuel pump is used on many vehicles that use carburetors. It is driven by the camshaft.

14.3 FUEL PUMPS

The purpose of the fuel pump is to transfer the fuel from the tank to the carburetor or fuel injectors. Gasoline engine fuel pumps are either mechanical or electrical. Both will be studied. Diesel engines have a complex fuel pump. This unit will be studied further in the chapter on fuel injection.

Mechanical Fuel Pumps

A typical mechanical fuel pump is shown in *Figure 14–7*. This unit is driven by the camshaft. There is a cam or an eccentric lobe on the camshaft. As the camshaft turns, the lobe lifts a lever up and down, causing a pumping action. Fuel is drawn from the tank by a vacuum and is sent to the carburetor.

This pump is called a *diaphragm*-type pump. A diaphragm is used internally to cause the suction and pressure needed to move the fuel. Check valves are used to keep the fuel moving in the right direction.

Check valve operation is shown in *Figure 14–8*. The check valve is made of a small disc held on a seat by a spring. If a suction is produced on the right side of the valve, the disc will lift off the seat and draw fuel into the pump, Figure 14–8 (A). When a pressure is produced on the back side of the disc, it is forced against the seat. Also, when a pressure is produced on the left side of the valve, the disc will open to allow flow, Figure 14–8 (B).

The same action occurs inside the fuel pump. Referring to *Figure 14–9*, when the eccentric cam lifts the diaphragm, a suction is produced. This suction causes the left check valve to open and draw in fuel. When the diaphragm moves in the opposite direction, a pressure is produced

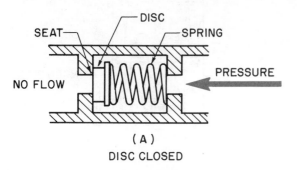
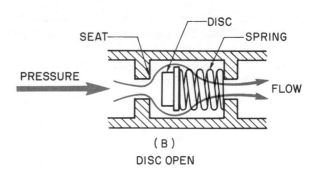

FIGURE 14–8. Check valves are used inside the mechanical fuel pump to control the suction and pressure of the fuel. When a suction is on the left side, the disc lifts off the seat and brings in fuel. Pressure on the right side causes the disc to seat again.

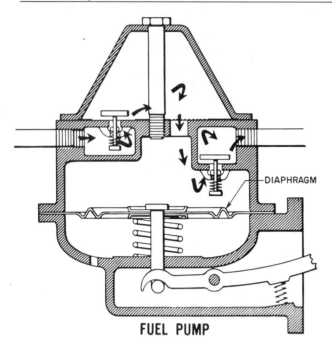

FIGURE 14–9. A diaphragm is used to cause a vacuum. The fuel is drawn in and pressurized. It is then sent through the check valve to the carburetor.

advantage is that electric fuel pumps can be located farther away from the engine. This means that heat from the engine will not produce vapors in the fuel pump. Therefore, there is less risk of vapor lock. Electric fuel pumps also work well in conjunction with computer-controlled vehicles. In addition, electric fuel pumps consume less frictional horsepower.

Several styles of the electric fuel pump are used, including the bellows type and the impeller or roller vane type. The year of manufacture and the type of vehicle will determine which type is used.

Bellows-type Electric Fuel Pump

In the bellows-type electric fuel pump, a metal *bellows* is used instead of a diaphragm. A bellows can cause a vacuum or pressure when stretched or compressed. The bellows is moved back and forth by the action of a solenoid. See *Figure 14–10*. When the electric current is sent to the magnetic coil, the *armature* is drawn downward. This action causes the metal bellows to expand and stretch. Fuel is drawn in by a vacuum. When the armature reaches its lowest point, the electric current on the magnetic coil is removed. The return spring then pushes the bellows upward. This causes a pressure that forces the fuel out of the fuel pump and into the carburetor.

inside the fuel pump. This pressure closes the left check valve and forces the fuel through the right check valve. This action causes a suction, which draws fuel from the tank. This creates a pressure, which forces the fuel to the carburetor.

Vapor Return Line

Some mechanical fuel pumps have a small connection on the fuel pump called a vapor return. The purpose of this line is to return fuel vapor that has built up in the fuel pump to the fuel tank. As engine and underhood temperatures increase, vapors may develop in the fuel lines. Vapors are removed at this point.

Electric Fuel Pumps

Electric fuel pumps can be used instead of mechanical fuel pumps. Electric fuel pumps have certain advantages over the mechanical type. Electric fuel pumps work independently on electric current. They provide fuel when the switch is turned on. Mechanical fuel pumps must have the engine cranking or turning before they deliver fuel. Another

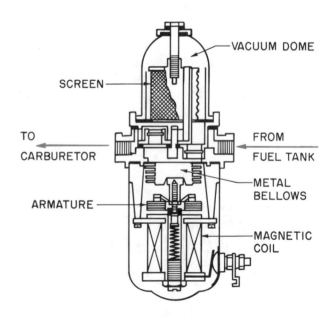

FIGURE 14–10. A bellows-type fuel pump operates from the movement of the armature in the magnetic coil. The bellows is stretched and squeezed to produce the suction and pressure needed to move the fuel.

PRO RALLYING

PRO Rally, a registered service mark of the Sports Car Club of America, is another term for performance rallying. It has its roots in such historic events as the Monte Carlo Rally, the London to Sidney Marathon, and the renowned East African Safari. A PRO Rally is a race over closed sections of public roads by modified production automobiles. A typical PRO Rally lasts from 12 to 16 hours and covers a total distance of 300 to 600 miles. Other characteristics of PRO Rallys include:

1. The cars race one at a time against the clock.
2. The racing is done on roads rather than on prepared tracks.
3. PRO Rallies may last several days.
4. Each car has two people: a driver and a codriver or navigator.
5. Between the racing sections (called stages), the cars travel over public roads and must obey all traffic laws. (Courtesy of Robert B. Nielsen, St. Paul, Minnesota)

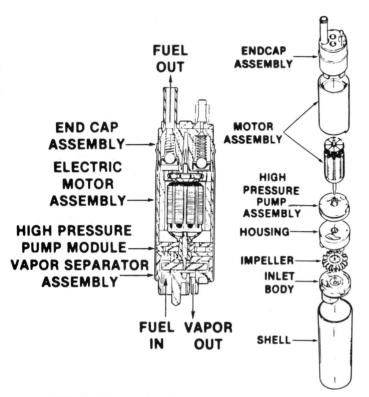

FIGURE 14-11. The roller vane fuel pump has a motor that rotates an impeller to draw fuel in and pressurize it. *(Courtesy of General Motors Corporation)*

FIGURE 14-12. A fuel pressure regulator keeps fuel at the right pressure for the system.

Roller Vane Fuel Pump

A second type of fuel pump used on newer vehicles is called the roller vane electric fuel pump. There are several variations of this type of pump. *Figure 14-11* shows an example of such a pump. The electric motor assembly is operated by electric current. As the motor turns an impeller, fuel is drawn in by the lower section of the pump. It is pressurized and sent out the top of the pump for delivery to the engine. The impeller at the inlet end serves as a vapor separator. The unit operates at approximately 3,500 rpm. A pressure relief valve keeps fuel pump pressure at a constant pressure such as 60–90 psi. The fuel pump delivers more fuel than the engine can consume even under the most extreme conditions.

Fuel Pressure Regulators

Fuel systems that have electric fuel pumps and fuel injectors may use a fuel pressure regulator. *Figure 14-12* shows such a regulator. It is located as shown in *Figure 14-13*. The fuel pressure regulator regulates the fuel that comes out of the fuel pump, through the fuel filter, and into the injector. It controls the fuel pressure at the fuel injectors. The purpose of the regulator is to keep the pressure of the fuel at the injectors between 60–90 psi. Some manufacturers have lower fuel injector pressure.

The fuel pressure regulator operates internally as shown in *Figure 14-14*. The fuel pressure regulator contains a pressure chamber that is separated by a diaphragm. A *calibrated* spring is placed in the vacuum chamber side. Fuel pressure is regulated when the fuel pump pressure overcomes the spring pressure. At this point, a small relief valve in the center opens. This action passes excess fuel back to the fuel tank. Vacuum action on the top side of the diaphragm, along with spring pressure, controls fuel pressure. A decrease in vacuum creates an increase in fuel pressure. A small increase in the intake manifold vacuum creates a decrease in fuel pressure. As an example, when the engine is under heavy load, vacuum is decreased. This condition requires more fuel. A decrease in vacuum allows more pressure to pass the top side of the pressure relief valve. This increases fuel pressure, and thus more fuel is pumped.

Computers and Fuel Control

Many manufacturers are using computers to control various functions of the engine. Fuel systems are also being controlled by computers. One such system is diagrammed in *Figure 14-15*. This figure shows a diagram of a computer used to control several systems. The left side of the diagram shows the signals being fed into the Control Unit. The right side of the diagram shows what systems are being controlled by the control unit. In this case, the control unit is operating the fuel pump, among other systems.

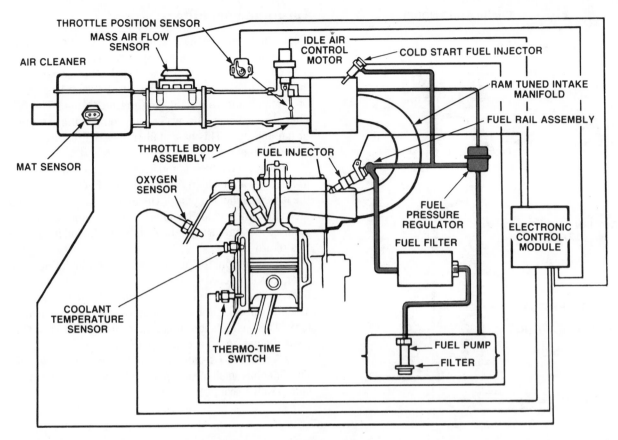

FIGURE 14–13. **The fuel pressure regulator is located in a position where it can regulate the fuel going to the injector.** *(Courtesy of General Motors Product Service Training)*

Operation of the fuel pump is controlled by the computer control unit. This particular manufacturer calls the computer control unit an E.C.C.S. (Electronic Concentrated Engine Control System). Several signals are fed into the computer to control the fuel pump. These include the engine speed, temperature, a start signal, the throttle valve idle position, and the battery voltage. The E.C.C.S. unit then takes each of these signals and operates the fuel pump accordingly. More information will be given about computer-controlled systems in later chapters.

14.4 FUEL FILTERS

Purpose of Fuel Filters

Most internal combustion engines consume a mixture of fuel and air to produce power. The key word in the operation of any such engine is cleanliness. Any fuel system using a carburetor or fuel injectors has many small passages and delicate parts which can be damaged by dirt particles. A dirty carburetor or fuel injector can cause erratic performance or complete engine shutdown.

Diesel engines use a fuel injector under very high pressure. Many small openings in the tip of the injectors will be damaged if dirt particles get into the system.

Sources of Gasoline Contaminants

Contaminants may enter the fuel system from various sources. These include:

1. unfiltered fuel that is pumped into the vehicle tank.
2. loose tank caps or faulty sealing gaskets.
3. rust, a powerful abrasive, that flakes off from the fuel tank and lines.
4. contaminants or dirt particles left in the tanks or lines during manufacturing and assembly.

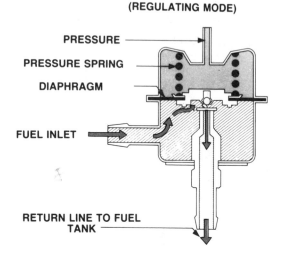

(REGULATING MODE)

PRESSURE

PRESSURE SPRING

DIAPHRAGM

FUEL INLET

RETURN LINE TO FUEL TANK

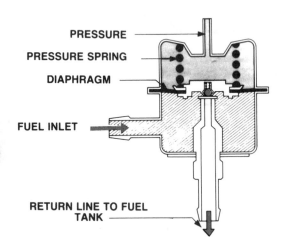

PRESSURE

PRESSURE SPRING

DIAPHRAGM

FUEL INLET

RETURN LINE TO FUEL TANK

FIGURE 14–14. Fuel pressure is regulated by using a diaphragm and spring. As fuel pressure increases, the spring lifts and allows fuel to pass through the check valve, back to the tank. *(Courtesy of General Motors Product Service Training)*

Gasoline Fuel Filters

Depending upon the manufacturer, some engines may have one or two filters in the fuel system. Referring to *Figure 14–16*, the first filter, which is located in the gasoline tank, is made of fine woven fabric. This filter prevents large pieces of contaminant from damaging the fuel pump. The tank filter also prevents most water from going to the carburetor or fuel injectors. This type of filter does not require service or replacement.

The second filter is found in one of several locations. An in-line filter is shown in *Figure 14–17*. Fuel is drawn from the tank into the fuel pump. The fuel pump pressurizes the fuel and sends it to the filter. A return line is used on vehicles that have a vapor lock problem. (Not all filters have this return line.)

Gasoline from the pump flows to the inlet fitting of the filter. Gasoline used by the carburetor passes through the filter *media* and out the center fitting. A small amount of gasoline will exit through the second outlet and return to the tank. The recirculation of gasoline through the vapor line cools the gas and prevents vapor lock.

Figure 14–18 shows the in-carburetor type fuel filter. This filter has small pleated paper filters with a gasket to provide positive sealing.

Diesel Fuel Contaminants

When diesel fuel is being shipped from the refinery, it is possible to pick up certain fuel contaminants. These include rust, dirt, and water.

Rust usually comes from large storage tanks or vehicle tanks. Rust occurs when there are low fuel levels in the tank

FUEL PUMP CONTROL

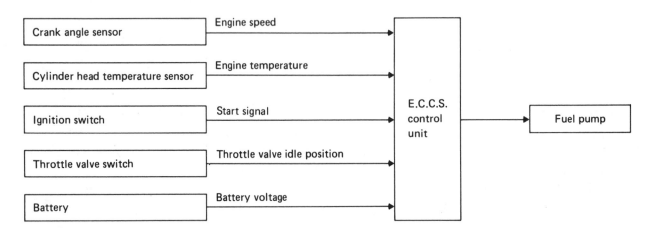

FIGURE 14–15. On computer-controlled engines, the fuel pump is controlled by the computer. Various inputs help determine the operation of the fuel pump. *(Courtesy of Nissan Motor Corporation in USA)*

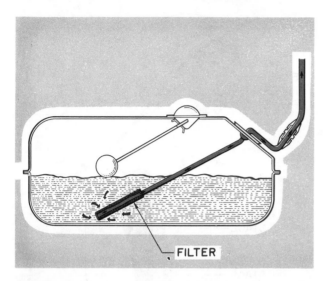

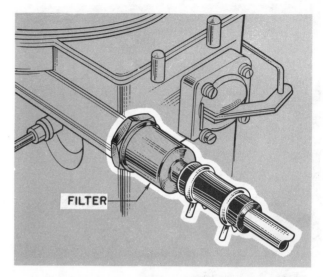

FIGURE 14–16. The first filter in a gasoline engine is located directly in the tank. *(Courtesy of Dana Corporation)*

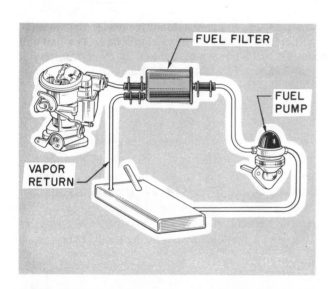

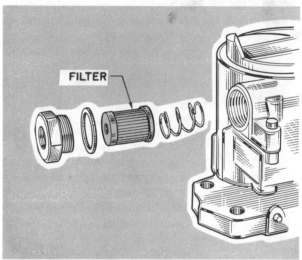

FIGURE 14–18. The filter is attached directly to the carburetor on some engines. *(Courtesy of Dana Corporation)*

FIGURE 14–17. An in-line filter is used to filter out small particles of dirt and rust. A vapor return line helps to reduce vapor buildup in the system. *(Courtesy of Dana Corporation)*

over a long period of time. Rust is an abrasive and can cause damage to the injection system.

Water can enter the fuel when it is held in underground storage tanks, and when a vehicle tank is being filled on a wet, rainy day. The most common source of water is condensation in the fuel tank. If the fuel tank is not kept filled, warm, moisture-filled air condenses on the cooler inside metal wall of the fuel tank.

Dirt can find its way into the fuel several ways. Dirt can be found on the tank spouts and dispensing nozzles. The vent system must be in good condition and checked as a source of dirt.

Two other contaminants are sometimes found in fuel. These are bacteria and wax crystals. Bacteria growth takes place in diesel fuel when certain microorganisms begin to grow. They grow at the point where water and diesel fuel meet in the tank. Since the tank is very dark inside and if the fuel moves very little, conditions are ideal for bacteria growth. As the bacteria grows, it forms a "slime" which will eventually be carried to the filter. This causes it to plug prematurely.

Wax crystals form when the temperature of the diesel fuel reaches its cloud point. As the wax crystals begin to form, they plug the filters. This means there will be less fuel sent to the engine, which could cause stalling or complete shut down.

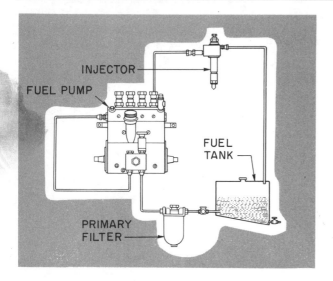

FIGURE 14–19. It is desirable to use two filters in the diesel fuel system. One is a primary filter and one is a secondary filter. *(Courtesy of Dana Corporation)*

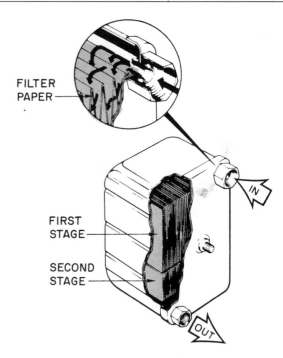

FIGURE 14–20. In some automotive diesel fuel systems, the filters are combined into a single filter. The first and second stage filters are built into this filter. *(Courtesy of General Motors Product Service Training)*

Diesel Fuel Filters

The most desirable arrangement for diesel engines is the two-filter system. However, many diesel engines used in automotive applications use only one filter, which is called the primary filter shown in *Figure 14–19*. The primary filter will catch most of the solid contaminants and remove small amounts of water. The secondary filter is much more efficient. It removes all remaining solid particles from the fuel.

On certain engines, the primary and secondary filters are combined, *Figure 14–20*. This is a surface-type filter with pleated paper. The first stage consists of about 400

square inches of filtering area. It removes 94% of the particles 10 *microns* and larger. One micron is equal to 0.000039 inch. See *Figure 14–21*. The second stage filter is made of the same paper material. It consists of about 200 square inches of filtering surface. This stage is 98% effective in filtering the fuel already filtered by the first stage.

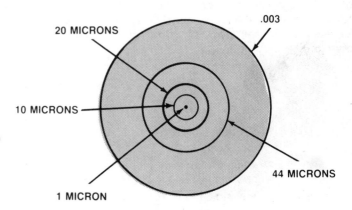

RELATIVE SIZE OF MICRON PARTICLES MAGNIFICATION 1000 TIMES

1 MICRON = .000039
LOWEST VISIBILITY RANGE = 44 MICRONS (.0017)
HUMAN HAIR = .003

FIGURE 14–21. One micron is equal to 0.000039 inch. It is a term used to describe filter sizes. *(Courtesy of General Motors Product Service Training)*

DIAGNOSIS and SERVICE

SAFETY TIPS
1. *Gasoline is very toxic and dangerous to the skin and eyes. Always wear safety glasses and correct clothing to protect your skin from gasoline. If gasoline does get into the eyes, wash with warm water immediately.*
2. *Before working with the fuel system, disconnect the battery so there is no possibility of producing a spark that would ignite gasoline.*
3. *Never remove a fuel filter when the engine is at operating temperature. When removing the filter, gasoline will spill onto the hot manifold, possibly causing a fire.*
4. *When checking the fuel pump pressure as indicated in the diagnosis and service section, make sure there is no electrical spark that could ignite the fumes of the open container of gasoline.*

1. When working with gasoline, always use approved containers and maintain all safety rules. Because it is possible that fuel will be spilled, disconnect the negative cable on the battery. This will eliminate the possibility of causing a spark.
2. If water is suspected to be in diesel fuel, drain the water out before operation. Fuel is lighter than water so the water will settle to the bottom. A bottom drain can be used to drain the water out of the system. On some diesel fuel systems, there are valves called "pet cocks" at the bottom of the fuel tank and at the bottom of the primary and secondary filter canisters. These are used to remove the water from the fuel system. They should be drained regularly.
3. When removing the mechanical fuel pump, avoid spilling the fuel. Any fuel spilled could be a fire hazard. Remove the *mechanical fuel pump* as follows:
   a. Disconnect the input and output fuel lines from the fuel pump. Use an open end (flare) wrench to loosen the fuel line connections. On some flexible lines, the clamp must be removed before the line can be removed.
   b. Remove any emission control lines that are attached.
   c. Using the correct size socket and ratchet wrench, remove the two bolts holding the fuel pump to the block.
   d. After the fuel pump has been removed, make sure the gasket is also removed. In some cases, the gasket must be scraped from the block.
4. Avoid welding on or near the fuel tank. Send the fuel tank to a repair shop equipped to handle this service.
5. When removing the in-line fuel filter on a gasoline engine, use two open ended or flare nut wrenches to disconnect the filter. See *Figure 14-22*.

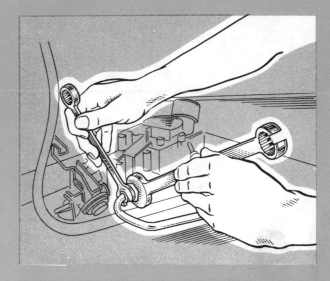

FIGURE 14-22. **When removing an in-line filter, always use two wrenches. (Courtesy of Dana Corporation)**

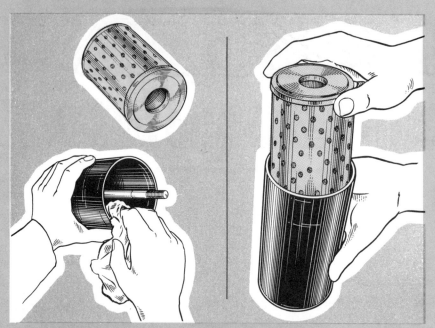

FIGURE 14–23. Always clean the inside of the container before replacing diesel fuel filters. *(Courtesy of Dana Corporation)*

6. When replacing diesel fuel filters, always clean the filter container before reinstalling. See **Figure 14–23**. It is best to fill the fuel filter canisters with fuel before reinstalling them on the engine. It could save having to prime the engine later.

7. An increased level of oil in the lubrication system may indicate a bad mechanical fuel pump. If the diaphragm is cracked, fuel will leak through the diaphragm and into the crankcase. This can be detected by smelling the oil. If it has a gasoline smell, the fuel pump may be damaged and should be replaced.

8. *Fuel pump pressure* and capacity can be checked on the car with a pressure gauge, a hose, and a quart measuring container. This is shown in **Figure 14–24**. Use the following procedure to check fuel pump pressure.
    a. Disconnect the fuel pipe at the carburetor inlet.
    b. Attach the correct pressure gauge and hose between the carburetor inlet and the disconnected fuel pipe.
    c. Start and run the engine.
    d. Check the automotive repair manual to see if the pressure is within the limits in the manufacturer's specifications for that particular vehicle.
    e. The pressure should remain constant or return very slowly to zero when the engine is shut off.
    f. Make sure the hose is placed inside the pint container to check the capacity.

g. Run the engine at idle speed and note the time it takes to fill the container.
h. Depending upon the pump being tested, it should take about 20–30 seconds to fill the container.

9. Never operate an electric fuel pump out of the gasoline container for more than 30 seconds.

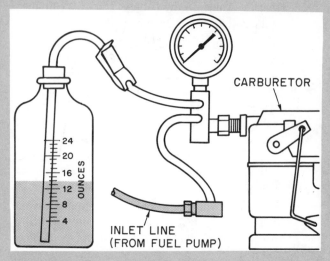

FIGURE 14–24. Fuel pump pressure can be tested by using a gauge and a pint size container. The engine is run and the amount of fuel measured in the container is compared to the engine specifications.

10. When an electric fuel pump that is an integral part of the tank is faulty, replace the entire unit. No service can be performed on this type of unit. Use the following general procedure to remove and install the *electric fuel pump.*

   a. Remove the negative battery cable.
   b. Relieve any fuel pressure in the system.
   c. Lift the vehicle up on an appropriate hoist.
   d. Drain and remove the fuel tank. This procedure may be different and will vary with different vehicles.
   e. Loosen and remove the assembly that holds the fuel pump in the fuel tank.
   f. Remove the fuel tank sending unit from the tank, being careful not to produce any electrical sparks.
   g. Inspect hoses, sender, and pump assembly for signs of deterioration.
   h. Reverse the procedure to install the assembly, making sure the O rings and locking assembly are correctly in place.

11. If the fuel filter element becomes plugged, the engine will stop. An engine stoppage caused by a plugged filter element will be preceded by a hesitation or sluggish operation.

12. The engine *fuel filter* can be replaced using the following general procedure.

   a. Disconnect the fuel lines at the fuel inlet nut. This is shown in *Figure 14–25*.
   b. Remove fuel inlet nut from carburetor.
   c. Remove the filter and spring.
   d. Some filters have a filter check valve. Make sure the check valve is replaced correctly inside the filter.
   e. Install the spring, filter, gasket, and fuel inlet nut.
   f. Be careful not to strip the threads when inserting the fuel inlet nut. It is very easy to misalign the threads, causing them to be damaged. Always start the fuel inlet nut with your fingers. Use a wrench to tighten the nut to the correct torque specifications.

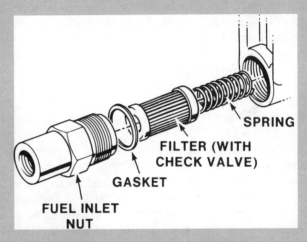

FIGURE 14–25. The fuel filter can be replaced by removing the assembly from the side of the carburetor. *(Courtesy of Oldsmobile Division, General Motors Corporation)*

SUMMARY

The fuel system is critical to the operation of the automobile. The main purpose of the fuel system is to supply fuel to the carburetor or fuel injectors so that a constant 15 parts of air can mix with 1 part of fuel. Several components are used to accomplish this task. These include the fuel tank, fuel pump, and fuel filters. Carburetors and fuel injectors are more complex and a separate chapter is devoted to each major area.

The fuel tank is made to hold excess fuel where it is ready to be used by the carburetor. The fuel tank cap is made to seal the fuel tank. It must operate so that both a vacuum and a pressure are released inside the tank.

A fuel metering unit is located inside the fuel tank. This unit is made to measure the amount of fuel inside the tank. It uses a float device that changes the resistance of an electrical circuit as it moves. Diesel fuel tanks also have a water sensor to determine if there is too much water in the tank. On newer vehicles, the electric fuel pump is also part of the total unit inside the fuel tank.

Fuel pumps are used to transfer the gasoline or diesel fuel from the tank to the carburetor or fuel injectors. Several types are used, including the mechanical and the electrical fuel pump. The mechanical fuel pump uses a diaphragm that is moved up and down by a lever connected to a cam device on the camshaft. The up and down movement causes a suction and pressure to be created inside the fuel pump. Check valves are used to control the fuel within the fuel pump.

Several types of fuel systems use a vapor return line. As the fuel is heated near the engine, vapor forms. This can cause vapor lock in the fuel system. Vapor return lines eliminate vapor lock.

Electric fuel pumps are used on many vehicles today. They are located in the rear of the vehicle or inside the fuel tank, so there is less risk of vapor lock. Two types of electric fuel pumps are used. The bellows type uses a bellows that is stretched and squeezed. This action creates a suction and pressure. The second type is called a roller vane fuel pump.

An electric motor rotates an impeller to pump the fuel from the tank to the engine.

Fuel systems that have electric fuel pumps with fuel injectors use a pressure regulator. The pressure regulator keeps the fuel pressure at a constant pressure. It is controlled by the intake manifold vacuum.

Computers are also being used to control fuel pumps. Several signals are fed into the computer. The computer then controls the fuel pump accordingly.

Fuel filters are used to stop any contamination from getting into the fuel system. Both gasoline and diesel fuel systems use filters. There are several types used on the gasoline engine. One type of filter is usually placed inside the fuel tank. This filter is made of a fine woven fabric.

The second filter is an in-line filter that can be found in several locations. The filter can be placed directly in the pressure fuel line or on the input to the carburetor. This filter is considered to be a surface filter.

Diesel fuel systems use either one or two filters. One is called the primary, and one is called the secondary. Each filter has a certain micron size to allow a certain size particle through the filter.

Several service tips are important when working on the fuel system. Always refer to the maintenance manual before working on the fuel system. Several important service suggestions include: 1) Be careful when handling gasoline, as it is very explosive. 2) Water can be very damaging to a diesel fuel system. 3) Always use two wrenches when removing filters connected to the input of the carburetor. 4) Fuel pressure can be checked with a pressure gauge and a pint size container. 5) Oil that has been mixed with fuel (gasoline) may indicate a bad diaphragm on the fuel pump.

TERMS TO KNOW

Can you explain each of the following terms? Review the chapter until you can use each term correctly.

Fuel injector	Armature
Fuel vapor	Calibrated
Contaminant	Media
Condensation	Microorganism
Diaphragm	Prematurely
Eccentric	Micron
Bellows	

REVIEW QUESTIONS

Multiple Choice

1. Which of the following is *not* part of the basic fuel system flow?
   a. Fuel tank
   b. Catalytic converter
   c. Fuel pump
   d. Fuel filter
   e. Vapor separator

2. Which system always has a return line back to the fuel pump or tank?
   a. Diesel fuel system
   b. Gasoline fuel system
   c. Contaminated fuel system
   d. All of the above
   e. None of the above

3. The fuel cap is used to:
   a. Help produce a vacuum on the fuel tank
   b. Help produce a pressure on the fuel tank
   c. Relieve the pressure and vacuum on the fuel tank
   d. Clean the fuel
   e. Filter the fuel

4. The fuel metering unit changes the _____ in an electrical circuit when the float changes position.
   a. Resistance
   b. Voltage
   c. Calibration
   d. Filters
   e. Wattage

5. Water in diesel fuel will _____.
   a. Increase the horsepower
   b. Damage the fuel system
   c. Cause vaporization
   d. All of the above
   e. None of the above

6. The mechanical fuel pump is driven from the:
   a. Crankshaft
   b. Distributor shaft
   c. Camshaft
   d. Alternator
   e. Valve system

7. Vapor in the gasoline fuel system is caused by:
   a. Rapidly cooling down the temperature of gasoline
   b. Heating gasoline to its boiling temperature
   c. Pumping gasoline through small holes
   d. Mixing it with nitrogen
   e. None of the above

8. Which of the following is a type of electric fuel pump?
   a. Bellows type
   b. Roller vane type
   c. Mechanical diaphragm
   d. All of the above
   e. A and B

9. What is the purpose of using a fuel pressure regulator?
   a. It regulates the oil pressure
   b. It regulates the fuel pressure at the injectors
   c. It controls the thickness of the fuel in cold weather
   d. It controls the temperature of the fuel
   e. None of the above

10. Which of the following is a contaminant that is of concern in diesel fuel?
    a. Wax crystals
    b. Microorganisms
    c. Rust
    d. All of the above
    e. None of the above

11. Fuel filters are _____ inside of the fuel tank.
    a. Never placed
    b. Always placed
    c. Sometimes placed
    d. Dissolved
    e. Expanded

12. What is the most desirable arrangement of fuel filters on a diesel engine?
    a. Only one filter should be used
    b. Two filters should be used
    c. Three filters should be used
    d. No filters should be used
    e. Five or more filters should be used

13. One micron is equal to _____ inch(es).
    a. 0.0045
    b. 0.000039
    c. 1.3
    d. 1.35
    e. 0.0039

14. Which filter has the smallest micron size holes?
    a. The primary filter
    b. The secondary filter
    c. Both filters have the same micron size
    d. The combining filter
    e. Filters are not rated by micron size

The following questions are similar in format to ASE (Automotive Service Excellence) test questions.

15. Technician A says to check fuel pump pressure, use a compression gauge. Technician B says to check fuel pump pressure, disconnect the fuel line to the carburetor, and use a pint jar and a pressure gauge. Who is right?
    a. A only
    b. B only
    c. Both A and B
    d. Neither A nor B

16. An engine seems to be running irregularly and stalls often. Technician A says the fuel filter needs to be replaced. Technician B says the fuel lines are broken. Who is right?
    a. A only
    b. B only
    c. Both A and B
    d. Neither A nor B

17. A gasoline engine hesitates and is very sluggish. Technician A says the problem could be the use of high octane fuel. Technician B says the problem could be a dirty fuel filter. Who is right?
    a. A only
    b. B only
    c. Both A and B
    d. Neither A nor B

18. There is an increase in the oil level in a gasoline engine. Technician A says fuel could be leaking into the oil through the fuel pump. Technician B says fuel cannot get into the oil, and probably too much oil was put into the crankcase. Who is right?
    a. A only
    b. B only
    c. Both A and B
    d. Neither A nor B

Essay

19. Why is there a vacuum release on the fuel cap?

20. Describe the operation of the diaphragm-type fuel pump.

21. What are some advantages of using an electric fuel pump?

22. Why do some cars have a fuel pressure regulator?

23. List the type of contaminants that can enter the fuel system.

24. What is a micron?

25. Describe the procedure used to check fuel pump pressure.

CHAPTER 15

Carburetor Systems

INTRODUCTION

Carburetors used on gasoline engines are designed to mix the air and fuel at the correct ratio. The most correct air-fuel ratio is 14.7 parts of air to 1 part of fuel. A carburetor's job is to maintain this ratio during all engine operations. There are many types of carburetors and accessories. Over the years many changes have been incorporated into the carburetor. In fact, carburetors are now being controlled by computers. Also, there are many variations to the basic carburetor. This chapter is designed to study carburetor principles, types of carburetors, circuits, and electronic controls.

OBJECTIVES

After reading this chapter, you will be able to:
- Define the basic principles of carburetion.
- Compare the different types of carburetors used on automotive engines.
- Analyze carburetor circuits, including the float, idle, low-speed, main metering, power, acceleration, and choke circuits.
- State the design and purpose of common carburetor accessories.
- Identify how carburetors are controlled by electronic computers.
- State various diagnosis and service procedures regarding carburetors.

CHAPTER HIGHLIGHTS

15.1 CARBURETOR PRINCIPLES
    A. Air-fuel Ratio
    B. Atomization
    C. Venturi
    D. Throttle Plate

15.2 TYPES OF CARBURETORS
    A. Carburetor Draft
    B. Carburetor Barrels
    C. Venturi Types

15.3 CARBURETOR CIRCUITS
    A. Float Circuit
    B. Float Bowl Venting

    C. Idle Circuit
    D. Low-speed Circuit
    E. Main Metering Circuit
    F. Power Circuit
    G. Mechanically Controlled Metering Rod
    H. Vacuum-controlled Metering Rod
    I. Acceleration Circuit
    J. Duration Spring
    K. Choke Circuit
    L. Manual Choke Control
    M. Automatic Choke Control

CHAPTER HIGHLIGHTS (CONTINUED)

15.1 CARBURETOR PRINCIPLES

Air-fuel Ratio

Research has shown that if the gasoline engine operates on an accurate and precise air-fuel ratio, engine efficiency will be improved. However, many things cause the air-fuel ratio to be upset or changed from the optimum 14.7–1. The following factors may change the air-fuel ratio:

1. air density and altitude
2. acceleration of the vehicle
3. deceleration of the vehicle
4. temperature of the air
5. moisture content of the air
6. speed of the engine or vehicle
7. load on the engine
8. overall condition and efficiency of the engine

The carburetor must be designed to operate under different conditions. *Figure 15–1* shows how the air-fuel ratio changes with different speeds. Air-fuel ratios range from 9 to 1 during idle to 18 or 19 to 1 during deceleration. When the vehicle is started and at idle, air-fuel ratio is about 12 to 1. This is a very rich mixture and produces poor fuel economy. As the engine speed is increased to move the vehicle to 40 mph, the air-fuel ratio settles at about 15 to 1. As the speed of the vehicle increases to 60 mph or above, the air-fuel ratio drops off again to about 12 to 1. Also, when the vehicle is accelerated or decelerated, the air-fuel ratio changes accordingly. Deceleration produces a leaner mixture, while acceleration produces a richer mixture.

Atomization

Carburetors operate on the principle that as more air flows through the carburetor, more fuel is added. However, the

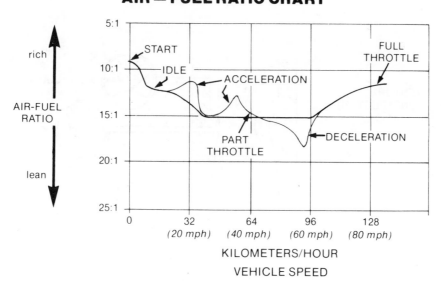

FIGURE 15–1. Air-fuel ratios will change with different driving speeds. Acceleration and deceleration points are also shown. (Courtesy of Davis Publications, Inc.)

A = AIR
F = ATOMIZED FUEL

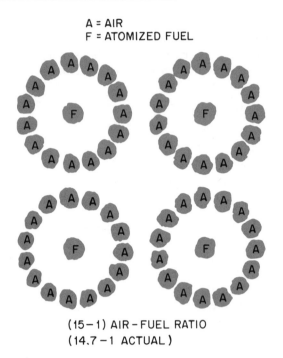

(15—1) AIR—FUEL RATIO
(14.7—1 ACTUAL)

FIGURE 15–2. When air and fuel are mixed correctly, 15 particles of air will surround 1 part of fuel. The fuel mixes best in the atomized state.

fuel must be in an *atomized* state. Fuel that is atomized is in very small mist-like droplets. The fuel is then *vaporized*. When a liquid is changed to a vapor, it is vaporized. Vaporization occurs after the fuel is atomized. This can be done by increasing the temperature of the fuel. Fuel in a vaporized state mixes very well with the air that is passing through the carburetor. *Figure 15–2* shows how atomized particles of fuel are surrounded by 15 parts of air for most

efficient combustion. Fuel that is in a liquid state, not vaporized or atomized, mixes poorly with the air passing through the carburetor.

Venturi

Air is drawn into the engine by the action of the piston moving downward on the intake stroke. When the piston moves down and the intake valve is opened, a *vacuum* is produced. This vacuum causes air to be drawn or pushed into the engine. The air, however, passes through the carburetor and *venturi* as it goes into the engine.

A venturi is a streamlined restriction that partly closes the carburetor bore. See *Figure 15–3*. Air is drawn into the engine by the intake manifold vacuum. As the air enters the venturi, it is forced to speed up or increase in velocity in order to pass through the restriction. This restriction causes an increase in vacuum by the venturi. The vacuum is also felt slightly below the major restricted area, and it continues to be reduced farther down the bore.

As the engine speed increases during acceleration, more air goes into the carburetor. This causes the venturi vacuum to increase because the greater the velocity of air passed through the venturi, the greater the vacuum. The vacuum produced at the venturi is used to draw in the correct amount of fuel. As the vacuum increases, more fuel is drawn in. As it decreases, less fuel is drawn in.

The venturi also aids fuel atomization and vaporization by exposing the fuel to air. The fuel is added in the center of the strongest vacuum point of the venturi. *Figure 15–4* shows fuel being added to the air flow through the carburetor. A discharge tube is located near the venturi. As the air flows through the venturi, the vacuum draws the fuel from a bowl into the stream of air going into the engine.

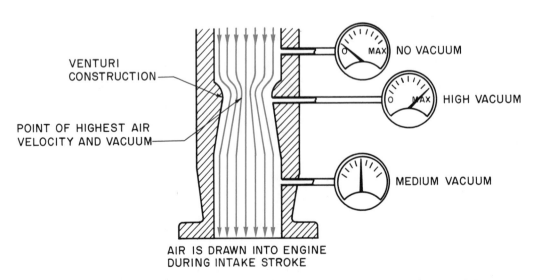

FIGURE 15–3. A venturi is a restriction in the path of air flow. A vacuum is produced at the point of greatest restriction.

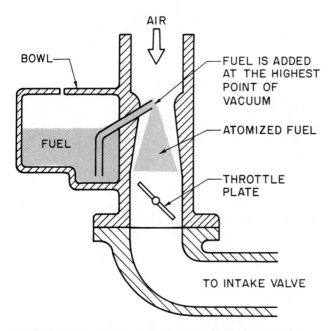

FIGURE 15-4. The vacuum that is produced at the venturi is used to draw in the fuel from the carburetor.

FIGURE 15-6. The throttle plate is a circular disk attached to a center rod. As the rod is turned, the throttle plate opens and closes.

Throttle Plate

The flow of air and fuel through the carburetor is controlled by the throttle plate, *Figure 15–5*. The throttle plate, which is made of a circular disk, is placed directly in the flow of air and fuel, below the venturi. Its purpose is to control the amount of air and fuel that enters the engine. The throttle plate is connected to the driver's throttle. As the driver's foot is depressed, the throttle plate opens to a vertical position. During this condition, there is very little restriction of air and fuel. This is a maximum load and speed condition. As the driver's foot is removed, a spring closes the throttle plate and restricts the amount of air and fuel going into the engine. This is a low speed and load condition. *Figure 15–6* shows a throttle plate from a carburetor.

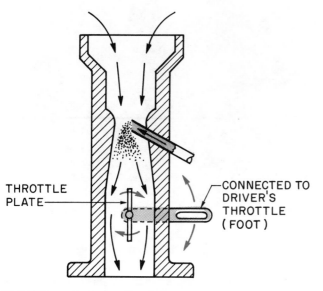

FIGURE 15-5. A throttle plate is put in the base of the carburetor to control the amount of air and fuel flowing into the engine.

CAR CLINIC

PROBLEM: CAR STALLS ON TURNS

After about 15,000 miles, the customer notices that the car seems to stall on right turns. The carburetor has recently been rebuilt.

SOLUTION:

A vehicle that stalls on turns usually has fuel-level problems in the carburetor. Start by checking the float levels. When the float level is too low, the fuel may slosh to one side when the car turns. Too high a float level may interfere with the carburetor venting system. Also, on certain carburetors the lower section of the carburetor may become loose. If this is the case, centrifugal forces during turning will cause the carburetor to tilt to one side. This would create a severe vacuum leak which could cause the engine to stall. To check, simply move the top part of the carburetor back and forth to observe if it is loose.

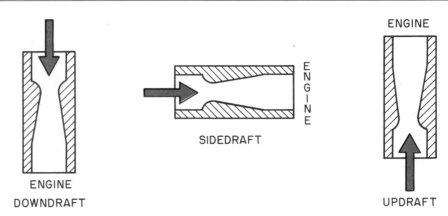

FIGURE 15–7. Carburetors are classified by the type of draft used. Downdraft, sidedraft, and updraft carburetors are used.

15.2 TYPES OF CARBURETORS

Many types of carburetors have been built in the past. Different load conditions, engine shapes, and air-fuel requirements mean that different types of carburetors must be built. Carburetors are designed with different drafts, different numbers of barrels, different types of venturi, and different flow rates.

Carburetor Draft

Draft is defined as the act of drawing or pulling air. Carburetors are classified as having different directions of draft. For example, a downdraft carburetor has air flowing down vertically into the engine. Most engines today have a downdraft carburetor. *Figure 15–7* shows a schematic of the different drafts used on carburetors.

The sidedraft carburetor is designed so that air flows through the carburetor in a horizontal direction. Many foreign sports cars use a sidedraft carburetor because it is placed on the side of the engine rather than on the top. The engine compartment can then be more streamlined.

An updraft carburetor brings the air and fuel into the engine in an upward direction. Not many automobiles use this type. Updraft carburetors are used in forklifts and other industrial engine applications.

Carburetor Barrels

A *carburetor barrel* is defined as the passageway or bore used to mix the air and fuel. One barrel consists of the throttle plate, venturi, and air horn. This design is used on small engines that do not require large quantities of fuel.

A two-barrel carburetor has two throttle plates and two venturis. The area where the air comes into the carburetor is common on both barrels. *Figure 15–8* shows the throttle plates from a two-barrel carburetor. A two-barrel carburetor is used on many intermediate load applications. Many newer small vehicles also use two-barrel carburetors. These carburetors, however, have one barrel that is very small in diameter and one that is larger.

A four-barrel carburetor has four barrels to mix the air and fuel. In most driving conditions, the engine operates on two of the barrels. When additional power is needed, the other two barrels add fuel to increase the amount of horsepower and torque produced by the engine. *Figure 15–9* shows a four-barrel carburetor.

FIGURE 15–8. Carburetors are also classified by stating the number of barrels. This is a two-barrel carburetor. There is a throttle plate for each barrel.

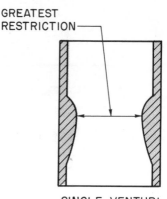

FIGURE 15–9. A four-barrel carburetor has four barrels and four throttle plates.

Modern engines are designed to be fuel efficient. In the past, when fuel efficiency was not a concern, many four-barrel carburetors were used. As the need to increase fuel mileage increased, however, the use of four-barrel carburetors decreased. Today, four-barrel carburetors are used on engines that require high horsepower and torque loads.

Venturi Types

Carburetors are also classified by the type of venturi used. *Figure 15–10* shows the different types used. The single venturi is used on many small single-cylinder engines such as those used in gardening equipment.

The double (dual) venturi is called the secondary or boost venturi. This design increases the venturi effect which, in turn, increases the efficiency of the carburetor. The bottom of the center venturi is located at the greatest restriction area of the next larger venturi. In this arrangement, the vacuum developed in the venturi is multiplied. This causes better vaporization and atomization and more control of fuel entering the carburetor.

The third type of venturi is called the triple venturi. Even more control and atomization occurs with this design. The fuel discharge tube is located inside the smallest venturi for maximum control and atomization.

Some carburetors are built with a variable or changing venturi. As the throttle is depressed, the venturi increases in size. As the throttle is released, the venturi decreases in size. Some American cars and several foreign cars use variable venturi carburetors.

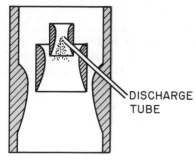

FIGURE 15–10. Carburetors have several types of venturis. One, two, and three venturis can be used. These can also be called "boost venturis."

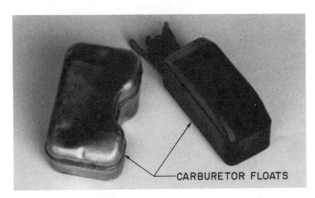

FIGURE 15–11. Brass or plastic floats are used to keep the right amount of fuel inside the bowl area.

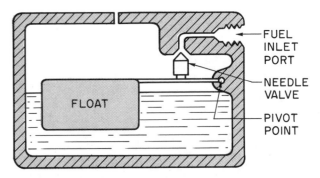

FIGURE 15–12. The float system keeps the correct amount of fuel in the float bowl. When fuel is used, the float drops slightly. This opens the needle valve and more fuel enters. As the fuel enters, the float rises and shuts off the fuel inlet.

15.3 CARBURETOR CIRCUITS

Carburetors are studied by analyzing different circuits of operation. There are several carburetor circuits. These include the float, idle, low-speed, main metering, power, acceleration, and choke circuits. Most carburetors use these circuits for operation.

There has been an increase in the number of changes made on the carburetor. Emission controls and standards have had an impact on carburetor designs. Carburetors have been changed and altered drastically to meet these standards. This section looks at the basics of each of these circuits.

Float Circuit

The float circuit is made to provide a source of fuel in a bowl for the carburetor. The carburetor float circuit maintains the correct fuel level for all conditions of operation. The circuit includes a fuel bowl that contains a supply of fuel, a float, and a float-operated valve to control the level of fuel in the bowl. There are also bowl vents to maintain correct fuel bowl pressure.

The bowl is a cast part of the carburetor. It holds the fuel to be used in the carburetor. It is like a reservoir of fuel. There is a float inside the float bowl, *Figure 15–11*. The float, which is made of a brass or plastic bulb, floats on top of the fuel. The float is connected to a needle valve. The needle valve fits into the fuel inlet.

Referring to *Figure 15–12*, as the fuel in the bowl is used, the float drops slightly. As the float drops, it opens the needle valve to allow more fuel to enter the float bowl chamber. As the fuel enters, the float lifts and causes the needle valve to close off the fuel inlet port. During operation, the fuel is kept at almost a constant level. The float tries to hold the needle valve partly closed. The incoming fuel makes up for the fuel being used at the venturi.

It is very important to maintain the exact amount of fuel inside the bowl. The level of fuel in the bowl controls the amount of fuel entering the venturi area. If the fuel level is too high, the extra fuel adds pressure. This pressure increases the amount of fuel entering the venturi, causing a rich condition. See *Figure 15–13*. If the fuel level is too low, less fuel will enter the venturi. This will cause a lean condition. Both conditions are damaging to the engine.

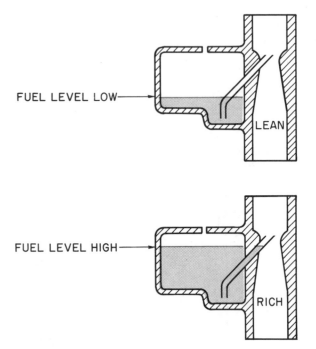

FIGURE 15–13. If the float level is too low, the mixture will be too lean. If the float level is too high, the mixture will be too rich.

Float Bowl Venting

As fuel is removed from the float chamber or bowl, air must replace the fuel. If there is no air for replacement, a vacuum will develop inside the float chamber. This will upset the air-fuel ratio. Two vents are used. One vent is connected to the carbon canister, which is a pollution control device. Fumes from the carburetor float bowl area are sent to the carbon canister. The carbon canister is explained in detail in the chapter entitled "Pollution Control."

A second tube is connected from the float bowl to the top of the air horn, *Figure 15–14*. This is called the internal vent. Almost all carburetors have this vent. Its purpose is to keep the pressure on top of the fuel the same as that going into the carburetor air horn. For example, if the air cleaner is dirty and produces a drop in pressure (an increase in vacuum), less air will be going into the air horn and through the venturi. This means that less fuel should be going into the carburetor for normal operation. This lower pressure is also felt inside the bowl area through the internal vent. However, if atmospheric pressure were felt inside the bowl, it would be higher than that before the venturi. This would cause a higher pressure (as compared to the top of the air horn) on the float area. The air-fuel mixture would now be richer. The internal vent equalizes these pressures.

Idle Circuit

During idle conditions, there is not enough air entering the venturi to cause a vacuum to move the fuel. The throttle

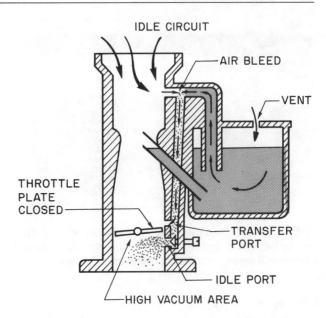

FIGURE 15–15. During idle conditions, the throttle plate is closed. Fuel is drawn in from the intake manifold vacuum through the idle port. *(Courtesy of Ignition Manufacturers Institute)*

plate is almost all the way closed as shown in *Figure 15–15*. During this condition, there is a large vacuum below the throttle valve. This vacuum causes fuel to be drawn from the carburetor float bowl through internal passages to the idle port. The idle port is below the throttle plate. As fuel is drawn from the float bowl to the idle port, air is drawn in through an air-bleed passageway near the top of the carburetor. Only a small amount of air passes by the throttle plate. During this condition, the air-fuel mixture needs to be rich to keep the engine idling.

On older engines there is an idle mixture needle valve. This valve is used to control or adjust the amount of air and fuel at idle. See *Figure 15–16*. There is a certain amount of adjustment available on these carburetors. More current carburetors, however, have limiting caps on the idle screws. This limits the amount of adjustment available on the idle mixture screws. On newer carburetors, the idle mixture screws are sealed with steel plugs to eliminate all adjustment.

Single-barrel carburetors have only one idle circuit. Most two-barrel carburetors have two idle circuits. On four-barrel engines, only the two primary barrels use an idle system. The secondary barrels do not have an idle circuit.

Some carburetors have a transfer slot between the transfer port and idle port. This slot helps to maintain the correct air-fuel ratio during low-speed conditions. There are times during idle when more air is needed. Additional air is transferred from above the throttle plate to below it through the transfer slot.

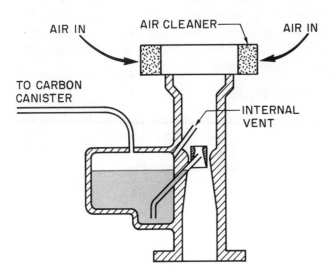

FIGURE 15–14. An internal vent is used to keep the pressure in the float bowl the same as that going into the engine.

FIGURE 15–16. On older carburetors the air-fuel mixture can be adjusted during idle by using the idle mixture adjustment screws located at the base of the carburetor.

Low-speed Circuit

The low-speed circuit is used to provide the correct air-fuel mixture during very low speeds from idle to 15–25 mph. (Speeds will vary with each carburetor.) During this condition, the throttle port is open slightly. *Figure 15–17* shows the position of the throttle plate during low-speed operation. Now a small amount of air flows past the throttle plate. This is still not enough air, however, to cause a vacuum at the venturi. Therefore, fuel must enter through the idle port and the transfer port. When the throttle plate is opened, it gradually exposes the transfer port to the intake manifold vacuum. This causes the fuel to be discharged through the transfer and idle ports. Some manufacturers call the transfer port operation the "off idle operation." Carburetors have 1, 2, or 3 off idle ports, depending upon the manufacturer.

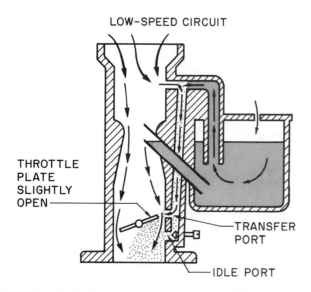

FIGURE 15–17. During low speeds, the throttle plate is slightly open. Fuel enters the carburetor through the idle and transfer ports.

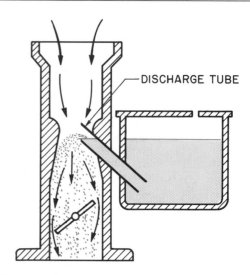

FIGURE 15–18. During normal medium-speed driving, the throttle plate is open enough for the venturi to create a vacuum. Fuel is drawn in by this vacuum through the discharge tube.

Main Metering Circuit

As the operator demands more speed, the carburetor transfers from the idle or low-speed circuit to the main metering circuit. This circuit operates the vehicle from about 25–60 mph. This varies, depending upon the type of vehicle and manufacturer.

Referring to *Figure 15–18*, during main metering, the throttle plate is opened enough so that air moves rapidly past the venturi. As the air passes, a vacuum is created. The vacuum draws fuel from the float bowl area, through a tube, and into the flow of air. The tube is called the main nozzle, high-speed nozzle, or discharge tube. Very little vacuum is felt inside the idle and transfer ports. During this time, the operator opens the throttle plate to increase speed, and more air enters the venturi. More air means more vacuum, and thus, more fuel.

Power Circuit

The size of the discharge tube controls the maximum amount of fuel allowed into the engine during main metering operation. When more fuel is required, for example, when pulling heavy loads, the power circuit supplies it. *Figure 15–19* shows the addition of a metering rod in the circuit. The metering rod is located in the hole that leads to the discharge tube. This hole is called a main jet.

The metering rod has different diameters on the end. When the thickest diameter of the metering rod is inserted in the jet, only a certain amount of fuel can pass through the jet. When the metering rod is lifted, the thinner diameter is in the main jet. Now more fuel can pass through the jet and into the engine. The carburetor is operating on the main

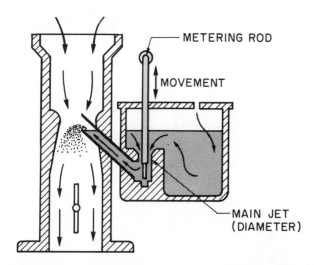

FIGURE 15-19. The metering rod is used to increase the size of the main jet during high-power conditions. The metering rod is lifted so that the smaller diameter end fills the main jet. Now more fuel can enter into the carburetor for increased loads. *(Courtesy of Ignition Manufacturers Institute)*

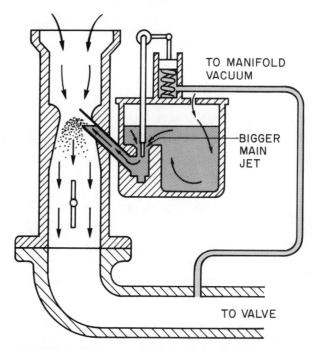

FIGURE 15-20. The metering rod can be controlled by the engine vacuum. As the vacuum decreases under heavy load, the vacuum is reduced and the metering rod is lifted.

metering circuit when the thickest diameter is inserted. When more fuel is needed under heavier loads, the metering rod is lifted, allowing more fuel to enter the venturi.

Mechanically Controlled Metering Rod

The metering rod can be controlled in several ways. One method is to control the metering rod mechanically. The metering rod movement is connected to the movement of the throttle. As the throttle is depressed during medium loads, the metering rod is not moved. Then as the operator pushes the throttle plate fully open, the metering rod is lifted. As it lifts upward, the smaller diameter end of the metering rod is moved into the main jet. This causes more fuel to pass through the main jet and into the discharge tube. When the operator decelerates, the metering rod returns to its original position.

Vacuum-controlled Metering Rod

A second method used to control the movement of the metering rod is to use vacuum. Compared to the mechanical metering rod, the vacuum-controlled metering rod provides more metering rod movement with less throttle action. When the throttle is nearly closed, engine manifold vacuum is high. This vacuum pulls a spring-loaded piston downward. The metering rod now provides maximum restriction in the main jet.

When the throttle is opened, there is a corresponding decrease in manifold vacuum. The spring pressure then raises the piston and metering rod. See *Figure 15-20*. This

permits an increased flow of fuel through the main jet. As the throttle opens further, the spring pressure pushes the piston and metering rod up even further. This produces less restriction in the main jet. The maximum amount of fuel will now flow through the carburetor. The vacuum-operated metering rod also operates when the throttle is opened quickly during acceleration.

Certain carburetors use both the mechanically and vacuum-operated systems. New carburetion systems use computers to control the movement of the metering rod. Computer-controlled carburetors will be discussed later in this chapter.

Acceleration Circuit

When the throttle is opened quickly during acceleration, a large volume of air enters the engine. This causes the air-fuel ratio to be very lean, and the engine may stall. An acceleration pump is used to supply extra fuel during acceleration. The acceleration pump is connected to the linkage of the throttle. As the throttle is depressed, the acceleration pump forces fuel under pressure into the stream of air entering the carburetor.

There are many pump designs. One system is shown in *Figure 15-21*. This system has the pump lever connected to the throttle movement. When the throttle is depressed, the pump plunger is forced down. The cup seal forces fuel through a discharge passage, through a pump discharge

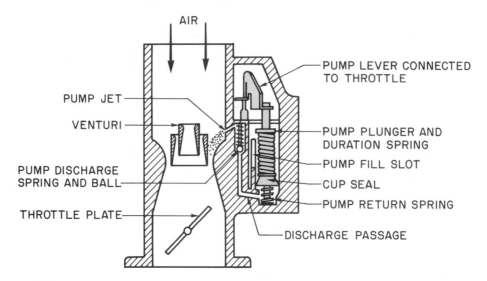

AIR

PUMP JET

VENTURI

PUMP DISCHARGE
SPRING AND BALL

THROTTLE PLATE

PUMP LEVER CONNECTED
TO THROTTLE

PUMP PLUNGER AND
DURATION SPRING

PUMP FILL SLOT

CUP SEAL

PUMP RETURN SPRING

DISCHARGE PASSAGE

FIGURE 15-21. An acceleration pump is used to force the fuel into the air stream during acceleration. The pump is connected to the mechanical motion of the throttle.

spring and ball, and into the flow of air just before the venturi. When the operator decelerates, the pump return spring forces the pump plunger upward. During this upward motion, more fuel is brought into the area below the cup seal.

The discharge spring and ball allow the fuel to flow in only one direction. If the engine backfired, pressure would build up inside the venturi. The spring and ball would now seat. This action would stop any backfire pollutants from getting into the acceleration system.

This particular system also has a pump fill slot. The plunger is located within the fuel bowl. A slot is cut into the plunger well to allow fuel to enter the area below the plunger. Certain systems use an intake *check ball* (valve) to allow fuel to flow into the plunger well. When the plunger moves down during acceleration, the ball is seated and fuel cannot pass back into the bowl area.

Duration Spring

On many acceleration pumps, there is a duration spring. This spring is used to lengthen the time of the pump downstroke. It *meters* the fuel with the opening time of the throttle valve. The duration spring also prevents damage to the diaphragms on the accelerator pump caused by trying to compress liquid fuel. During operation, the duration spring delays the plunger motion slightly. The movement of the plunger by the throttle motion pushes down on the duration spring. The duration spring then pushes down on the plunger to force fuel into the carburetor. *Figure 15–22* shows the plunger and duration spring.

Choke Circuit

When the engine is cold, a richer air-fuel mixture is needed to keep the engine running. A rich mixture is required for starting because fuel atomization is very poor in a cold engine. Much of the gasoline condenses on the cold engine parts before entering the cylinder area. A choke valve is used to provide the richer mixture. The choke valve is located directly on top of the air horn. See *Figure 15–23*. It is used to block the air going into the carburetor. As the valve closes, intake manifold vacuum is felt up through the venturi area. This higher vacuum draws increased amounts of fuel through the discharge tube and into the engine.

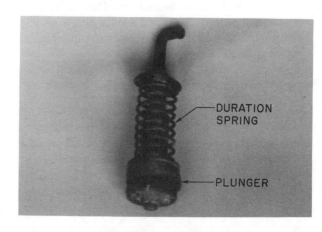

DURATION
SPRING

PLUNGER

FIGURE 15-22. The duration spring is used to lengthen the time of the acceleration pump downstroke.

FIGURE 15–23. The choke valve is used to restrict the air flow into the carburetor during cold starting conditions.

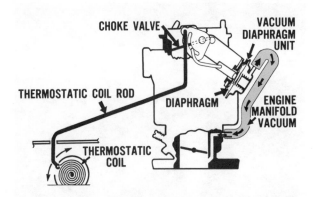

FIGURE 15–24. The choke plate is controlled by a vacuum diaphragm unit. *(Courtesy of Ignition Manufacturers Institute)*

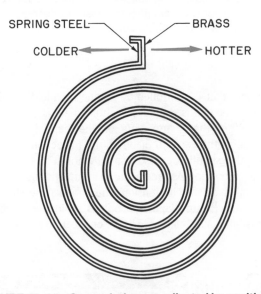

FIGURE 15–25. Some chokes are adjusted by positioning a bimetallic strip with a slight tension. As the spring heats up, it decreases its tension and opens the choke plate.

Manual Choke Control

Several methods are used to control the opening and closing of the choke plate. The simplest method is to manually control the choke. When the engine is cold, the operator pulls a choke control knob on the dashboard. A wire is connected from this knob to the choke plate. This action closes the choke plate. As the engine warms up, the operator must slowly push the knob back in or the air-fuel ratio will be too rich. This system was used on many older vehicles and many foreign sports cars.

Automatic Choke Control

The choke can also be controlled automatically. As the temperature of the engine increases after starting, the choke should be opened accordingly. When the engine is at operating temperature, the choke should be fully open. Automatic chokes operate on vacuum diaphragms, small pistons, and electric controls.

A vacuum diaphragm system is shown in *Figure 15–24*. The choke valve is connected to a thermostatic coil rod. This rod is connected to a thermostatic coil that is made of a *bimetallic strip*. The coil is located close to the intake manifold. As the coil cools, it increases in tension. When the coil warms up, the tension is reduced. When the engine is cold, the thermostatic coil has greater tension. The increased tension closes the choke plate, putting the choke circuit into operation. As the engine warms up, the thermostatic coil decreases its tension and the choke valve slowly opens. See *Figure 15–25*.

Thermostatic coils are adjusted in many ways. On some systems, the coil can be tightened or loosened for adjustment. A plastic plate on the outside of the thermostatic coil shows the direction for a leaner or richer mixture. See *Figure 15–26*. On other carburetors, the thermostatic coil rod may have to be bent to get the correct adjustment.

The system shown in Figure 15–24 also has a *vacuum diaphragm* unit. Vacuum is used to open the choke slightly when the engine is cold. A vacuum-operated piston is used on some carburetors. If the choke plate is closed completely, the engine will not get any air. This may cause an over-choked or rich condition. The choke plate needs to be slightly open when at full choke. This is done by using a

PLASTIC PLATE FOR CHOKE ADJUSTMENT

FIGURE 15–26. Some carburetors have a plastic adjustment cover on the outside housing of the thermostatic coil. Turning the plastic cover either increases or decreases the tension of the thermostatic coil.

diaphragm that is controlled by engine manifold vacuum. Before the engine is started, the choke plate is completely closed. When the engine starts, engine manifold vacuum pulls against the vacuum diaphragm. This action pulls the choke valve slightly open so a small amount of air can get into the carburetor. As the engine heats up and the thermostatic coil decreases its tension, the choke plate starts to open. The vacuum diaphragm unit is no longer in effect. This system is also referred to as the choke unloader system.

CAR CLINIC

PROBLEM: OIL LEVEL INCREASES

A customer has noticed when checking the oil that the level has increased. How can the level of oil be increased without adding oil to the crankcase?

SOLUTION:

The level cannot increase without adding some type of fluid. The most likely problem is that fuel (gasoline) is being added to the oil. A quick check is to smell the oil. It should have a gasoline odor. As fuel is added, the oil level appears to increase. Fuel can be added to the oil in two major ways. The fuel pump (if mechanical and located on the side of the engine) can have a leak in the diaphragm. Fuel then passes through the diaphragm into the crankcase area. A second way in which fuel can be added to oil is by having the carburetor or fuel injectors extremely rich during starting. A heavily choked engine will cause raw fuel to slip past the rings and into the

crankcase area. To eliminate the problem, either replace the fuel pump or readjust the choke according to the manufacturer's specifications. If the vehicle uses injectors, the computer controls will have to be checked for a rich mixture.

15.4 CARBURETOR ACCESSORIES

There have been many changes in carburetor designs. Many accessories have been used during different years on carburetors. Certain accessories are used on cars with air conditioning and with manual and/or automatic transmissions. The increased emphasis on emission and pollution control has also produced more accessories. This section introduces some of the more common accessories used on carburetors.

Hot Idle Compensator

During times when the engine is very hot and under-hood temperatures are very high, fuel vapors have a tendency to collect in the intake manifold. The addition of these fuel vapors can cause the air-fuel mixture to be too rich. This may cause the engine to have a rough idle and a tendency to stall at low or idle speeds. The hot idle compensator is used to open a bleed circuit. This permits more air to enter the manifold below the throttle plate. The hot idle compensator helps to dilute the rich air-fuel mixture.

Figure 15–27 shows the hot idle compensator valve. It is made from a bimetallic strip attached to the valve on top of the bleed circuit. It is usually located directly on top of the carburetor. As the temperature of the engine increases to about 120 degrees F, the bimetallic strip starts to bend and open the valve. This permits additional air to enter the intake manifold. When operating temperatures go below the 120 degrees F, the valve is closed and the extra air flow is stopped below the throttle plate.

Idle Speed Adjusting Screw

On most carburetors there is an idle speed adjusting screw. Its purpose is to adjust the speed of the idle when the engine is warm. A metal bracket is attached to the throttle plate. The operator's foot pedal is attached to this plate. There is also a small adjusting screw located either on the frame of the carburetor or the bracket. As the screw is adjusted, the position of the throttle plate is opened or closed. This action will either increase or decrease the idle speed. Once the operator accelerates, the idle speed adjusting screw does not affect the operation of the throttle plate.

Fast Idle Cam

When the engine is operating on the choke circuit, the engine idle must be increased or the engine will stall. This

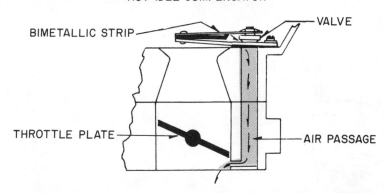

FIGURE 15-27. The hot idle compensator valve uses a bimetallic strip to open an air bleed circuit. The extra air dilutes the rich air-fuel mixture. *(Courtesy of Ignition Manufacturers Institute)*

is done mechanically with a fast idle cam. *Figure 15–28* shows a fast idle cam. There is a connecting rod linkage attached from the choke plate rod to the fast idle cam. The fast idle cam is a small, stepped cam. When the choke plate closes, the fast idle cam is lifted. The idle speed adjusting screw now rides on a higher spot on the cam. This action causes the idle speed to be increased during choked conditions. When the engine heats up and the choke is opened, the fast idle cam drops down from gravity. Now the idle speed is reduced to normal conditions.

When an engine is started during cold conditions, the throttle must be depressed in order for the choke to close. If it is not depressed, the choke will not engage and the high-speed cam will not increase the idle. Also, once the engine

is running at fast idle (on the cam), the throttle must be depressed or opened in order for the fast idle cam and the choke plate to be released.

Idle Stop Solenoid

Most current engines with emission controls idle at higher speeds, use leaner air-fuel mixtures, have slightly retarded timing, and have higher cooling temperatures. These conditions increase the tendency of the engine to "run on" or "diesel" after the ignition switch has been turned off. One way to stop this condition is to remove the air and fuel completely when the engine is shut down.

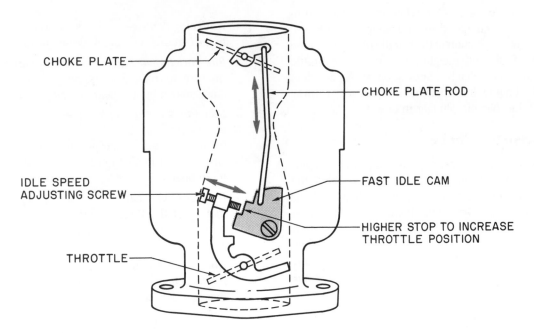

FIGURE 15-28. The fast idle cam is used to increase the idle speed during choked conditions.

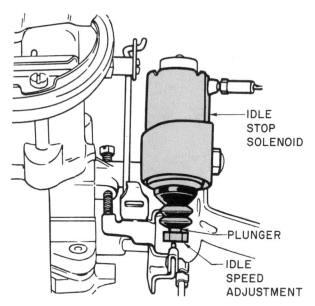

FIGURE 15-29. An idle stop solenoid is used to control the position of the idle setting when the ignition is turned off and on. *(Courtesy of Echlin Manufacturing Company)*

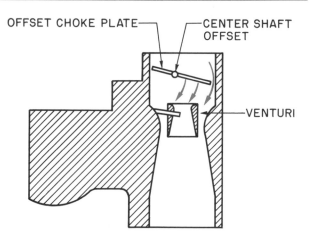

FIGURE 15-30. The choke plate is offset on certain carburetors. Air trying to enter the engine can easily push the long side down, allowing air to flow into the carburetor.

In the past, the position of the idle screw always kept the throttle plate slightly open. Air and fuel could enter the engine as it slowed to a stop. With this system, the throttle plate is completely closed when the engine is shut down. Now there is less chance for run-on. This is done by using an idle stop *solenoid*, also called a TSCV (Throttle Solenoid Control Valve).

Referring to ***Figure 15-29***, when the engine is running, the solenoid is energized and the plunger is extended. The plunger sets the position of the throttle plate. When the ignition switch is shut off, the plunger retracts. The throttle plate now closes completely, shutting off all air and fuel going into the engine. Idle adjustment is made by turning the center of the solenoid plunger in or out.

Electric-assisted Choke

Certain carburetors have an electric-assisted choke. Its purpose is to decrease the time the choke is closed. While the choke is closed, the engine's production of hydrocarbons and carbon monoxide emissions increases. This can be reduced by decreasing the amount of time the choke is closed. A heating element is placed in the thermostatic coil assembly. The addition of the heat from the electric heating element plus the heat from the intake manifold causes the choke to open earlier.

Offset Choke Plate

An offset choke plate is used when the choke is opened and closed by a thermostatic spring. In ***Figure 15-30***, the plate

is offset on the center shaft of the choke. Air trying to enter the engine can easily push the long side down, allowing air to flow into the carburetor. During starting, the offset choke plate allows a small amount of air to enter the carburetor during each vacuum pulse. As the engine warms up, changing speeds cause the opening of the choke plate to vary. This action helps to regulate the position of the choke.

Other Accessories

A variety of small carburetor accessories is used to improve the operation of the engine. Each has a special purpose and operation. They include:

1. *Anti-icing vents* — used to send warm air to a spot in the carburetor that may ice up due to evaporation.

2. *Idle enrichment valves* — used to enrich the idle mixture to reduce emissions. A small vacuum diaphragm mounted near the carburetor top is used.

3. *Secondary throttle linkage* — used to engage the secondary barrels of a carburetor. Mechanical linkage is used to engage the secondary throttles when the primary throttles are about one half or more open.

4. *Deceleration dashpots* — used to slow down the motion of the throttle during deceleration. If the throttle closes too fast, a very rich condition could cause the engine to stall.

5. *Hot water chokes* — used to apply heat to the thermostatic housing. The choke is released earlier so that less pollution is produced.

6. *Altitude compensators* — used to keep the air-fuel mixture correct at different altitudes. A bellows is usually used to sense altitude differences, causing the air-fuel ratio to be adjusted.

7. *Poppet valve choke plates* — used on some manual chokes. A small poppet valve and spring are placed on the choke plate. When the choke is fully closed, a small amount of air can still get through the poppet valve.

Other connections are also made to the carburetor. These connections are used to sense or control other systems. These systems will be discussed in many of the following chapters.

15.5 ELECTRONIC CARBURETOR CONTROLS

In recent years, computers have been used to control various components on the engine. This is also true in the fuel system. Control of fuel by computers is referred to as "fuel management." A small computer is placed on board the vehicle. During normal operation, many sensors electronically feed signals into the computer, which acts as a brain. Based on these input signals, the computer sends out a signal to operate a certain component.

One such system is called the ECM (Electronic Control Module). See *Figure 15–31*. This computer is part of a total system called CCC (Computer Command Control), which is manufactured by General Motors. Other manufacturers use similar systems. Computers are designed to manage fuel so that higher gasoline mileage can be obtained while produc-

FIGURE 15–31. A computer is used to control or manage fuel. The computer is part of a system called "Computer Command Control," manufactured by General Motors.

ing less emissions. *Figure 15–32* shows a schematic of a fuel management system using the ECM.

Closed and Open Loop

Within the CCC system, there are two modes of operation. When the temperature of the engine is below 150 degrees F, for example, during a cold engine start, the computer operates in an "open loop." This means that the computer

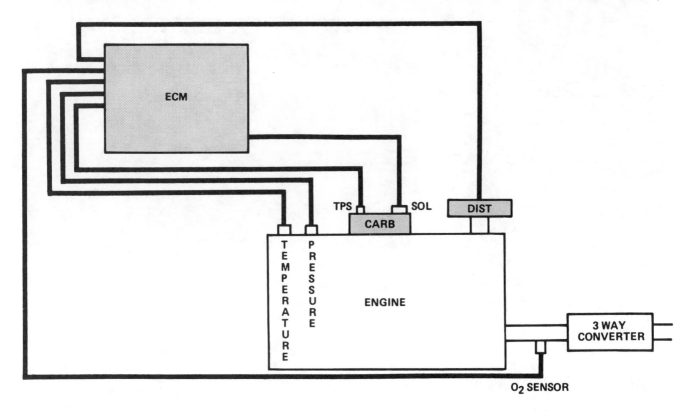

FIGURE 15–32. This schematic shows the inputs and output on a fuel management system. *(Courtesy of General Motors Product Service Training)*

PERFORMANCE RACING

Drag racing is a very popular automotive sport within the United States. Many types of engines are used to produce the required power. Normally, a drag race consists of two vehicles that start at the same point and race to the end of a track of a predetermined length. Both time and speed are monitored, but the first vehicle to cover the predetermined length wins.

Various components are designed to give maximum torque and speed to the vehicle. All of the vehicles use some form of supercharging, normally a blower. Other components that are changed include the gear ratios, horsepower and torque of the engine, size of the tires, weight of the vehicle, and aerodynamic design. These factors are designed to produce the quickest acceleration along with the maximum top speed. (Courtesy of Anthony Gilberti)

will not adjust the fuel measurement; it will allow engine operation to be governed by a preprogrammed set of instructions.

When the temperature of the engine is about 150 degrees F, the computer operates in a "closed loop." This means that the computer is constantly monitoring the information it receives from the various sensors and adjusting the fuel accordingly.

Carburetor Solenoid

An electric solenoid in the carburetor controls the air-fuel ratio. The solenoid is connected to the ECM. The ECM sends a signal to the solenoid. The solenoid, in turn, controls a metering rod and an idle air bleed valve. With the use of the computer, the air-fuel ratio can be accurately controlled throughout the entire operating range of the engine.

Figure 15–33 shows a mixture control (M/C) solenoid. The mixture control solenoid is an electrical unit that controls fuel flow from the bowl to the main discharge tube or nozzle. The idle circuit air bleed is also controlled by the solenoid. The solenoid coil and plunger are mounted vertically in the carburetor. The stem and valve end reach to the bowl floor where the solenoid controls a passage between the bowl and the discharge tube. It acts as a metering valve that opens and closes. This happens at a rapid rate of ten times per second.

The upper end of the solenoid rod plunger opens and closes the idle air bleed. When the solenoid is energized, the plunger moves down. This opens the idle air bleed and closes the discharge tube passage. Both the idle and main metering systems become lean or rich together.

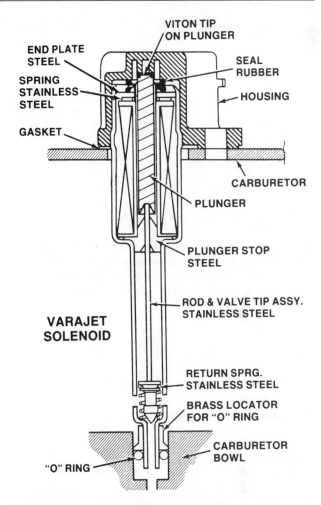

FIGURE 15–33. The M/C (mixture control) solenoid is controlled by the computer. It keeps adjusting the size of the main jet for different air-fuel conditions. *(Courtesy of General Motors Product Service Training)*

O₂ Sensor (Oxygen Sensor)

A sensor is located in the exhaust stream close to the engine. It is known as an oxygen sensor. It measures the amount of oxygen in the exhaust gas. *Figure 15–34* shows an oxygen sensor. There is a direct relationship between the air-fuel mixture and the amount of oxygen in the exhaust gas. The oxygen sensor determines whether the exhaust is too rich or too lean. It sends a low-voltage (below 450 millivolts) signal to the ECM when the mixture is lean. A high-voltage (above 450 millivolts) signal is sent to the ECM when the mixture is rich.

The ECM then signals the mixture control (M/C) solenoid to deliver a richer or leaner mixture. As the carburetor makes an air-fuel change, the oxygen sensor immediately senses that change and again signals the ECM. This goes on continually during the engine operation.

This process is a closed loop operation. Closed loop operations deliver an accurate 14.7 to 1 air-fuel ratio to the engine.

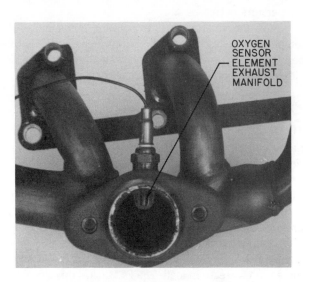

FIGURE 15–34. An oxygen sensor is used to determine the amount of oxygen in the exhaust. This data is sent to the computer to help control the air-fuel ratio.

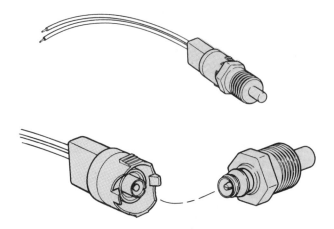

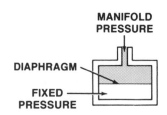

FIGURE 15–36. A pressure sensor located in the intake manifold detects changes in manifold pressure or vacuum. *(Courtesy of General Motors Product Service Training)*

FIGURE 15–35. A temperature sensor is also used to sense the temperature of the engine coolant. This data is sent to the computer to help control the air-fuel ratio. *(Courtesy of General Motors Product Service Training)*

Temperature Sensor

A temperature sensor is also used with the ECM, *Figure 15–35*. The sensor is located in the cooling system and is connected to the ECM. Whenever the engine temperature is cold, there is no need for the oxygen sensor to control the air-fuel ratio. Under these conditions, the ECM tells the carburetor to deliver a richer mixture. The mixture is based on what has been programmed into the ECM and what other sensors are telling the computer. This is an open loop operation. The sensor's resistance is lowered as coolant temperature increases. The resistance is raised as the temperature decreases.

After the engine reaches operating temperature, the temperature sensor signals the ECM to read what the oxygen sensor is providing. If other requirements are met, closed loop operation begins.

Pressure Sensor

The load on the engine also affects the air-fuel mixture needed in the engine. As a load is placed on the engine, a richer air-fuel mixture is needed. This can be measured by sensing the intake manifold vacuum. This is shown in *Figure 15–36*. A pressure sensor located in the intake manifold detects changes in the manifold pressure. As the pressure changes, a flexible resistor attached to a diaphragm also changes its resistance. This change causes a voltage change that the ECM can read. This signals the ECM that there is an increase in load. The ECM takes this signal into account to determine the exact air-fuel ratio required.

Throttle Position Sensor

The position of the throttle opening is another factor in determining what air-fuel ratio is needed. The more the throttle is open, the richer the mixture required by the engine. *Figure 15–37* shows the TPS (Throttle Position Sensor). It is a variable resistor that sends a signal to the ECM. Depending upon the position of the throttle, the ECM will signal the carburetor solenoid to increase or decrease the air-fuel mixture.

The throttle position sensor is a variable resistor that is mounted in the float bowl. As the position of the throttle changes, the voltage also changes. At closed throttle, the voltage is about one volt or less. As the throttle opening increases, the voltage increases to about five volts at wide open throttle.

Engine Speed Sensor

Engine speed also has a direct bearing on the air-fuel mixture. When the engine is operating at a low speed, less fuel is needed. When the engine operates at a higher rpm, more fuel is needed. This is done by using an engine speed sensor. A tachometer signal from the distributor is sent to the ECM. Refer back to the schematic in Figure 15–32. This tells the ECM the rpm of the engine. The computer considers this signal when setting the exact air-fuel ratio needed.

Idle Speed Control

Since the increase in emission standards, idle speed must be controlled more precisely. The CCC system controls the idle speed for drivability as well as for fuel economy and emission control. This is done by using the ECM and a reversible electric motor. The ECM maintains a selected idle speed regardless of the load imposed on the engine. A plunger that acts as a movable idle stop changes the idle speed. The plunger is positioned by a small electric motor. A throttle contact switch tells the ECM to operate only when the throttle lever is closed. When the throttle lever moves away from the throttle contact switch, the ECM is instructed not to

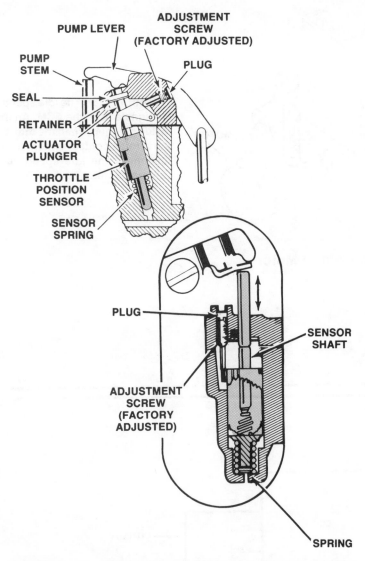

FIGURE 15-37. A throttle position sensor is used to determine the exact position of the throttle. This data then helps the computer determine the exact air-fuel ratio needed. *(Courtesy of General Motors Product Service Training)*

operate. The driver now has control of the engine speed. See *Figure 15-38*.

Figure 15-39 shows a schematic of the ECM and identifies the inputs and the outputs to the electric motor. Note that the throttle switch is considered an input to the ECM. Also, many of the inputs are the same as those used with the carburetor solenoid. The following controls are considered input signals to the ECM:

1. *Distributor* — sensing rpm.

2. *Oxygen sensor* — sensing oxygen in exhaust.

3. *Temperature sensor* — sensing coolant.

4. *Pressure sensor* — sensing manifold pressure and load.

5. *Throttle position sensor* — sensing throttle position.

The *battery signal* is used to sense the system's operating voltage. If the voltage signal falls below a predetermined level, the ECM will instruct the idle speed control plunger to extend. This will increase the engine speed, which will, in turn, increase the generator speed so that generator output will increase.

The *park-neutral switch* tells the ECM when the transmission has shifted. When the transmission shifts, the load on the engine changes. The idle speed must then be changed. This prevents different idle speeds at neutral, drive, and reverse conditions.

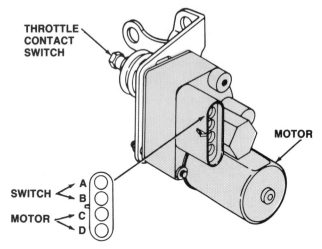

THROTTLE
CONTACT
SWITCH

MOTOR

SWITCH
A
B
MOTOR
C
D

Idle Speed Control Motor

FIGURE 15–38. The idle speed control motor is operated by signals from the ECM. An internal motor positions an inside plunger. When the throttle lever is moved away from the throttle contact switch, the computer is instructed not to control the idle speed control motor. *(Courtesy of General Motors Product Service Training)*

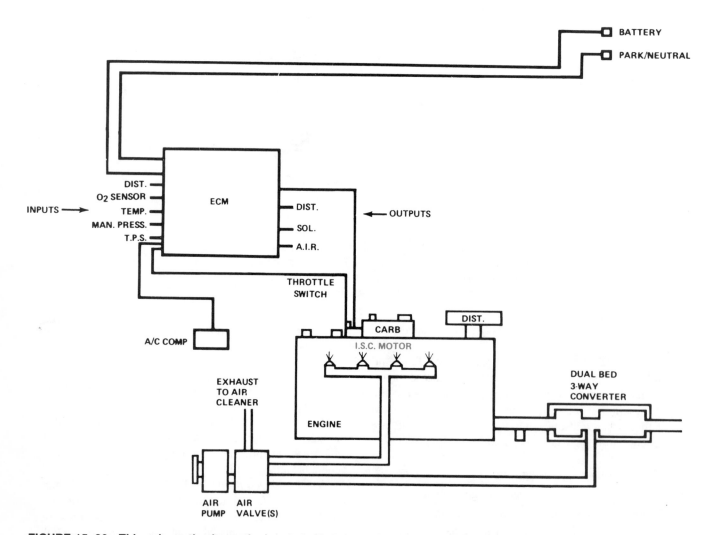

FIGURE 15–39. This schematic shows the inputs and outputs used to control the idle speed control motor. *(Courtesy of General Motors Product Service Training)*

DIAGNOSIS and SERVICE

SAFETY TIPS
1. *When checking the color of exhaust when the engine is running, make sure there is adequate ventilation for the exhaust fumes.*
2. *The cleaner used to clean carburetors is extremely toxic. When cleaning carburetors, always wear protective gloves, safety glasses, and protective gowns to eliminate any possibility of injury.*
3. *When adjusting the carburetor with a screwdriver or other tool, always approach the carburetor from the side. Approaching from the front may cause injury if the tool is accidentally dropped into the spinning fan.*
4. *When setting any adjustment on the carburetor, always put the car in park, with the emergency brake on, and the wheels blocked.*

1. A rich air-fuel mixture can be detected by observing the color of the exhaust. The exhaust can be observed when the engine is accelerated. A rich mixture is usually identified by observing black smoke. In addition, black carbon will build up inside the exhaust pipe.

2. *Floats* must be *adjusted* according to the manufacturer's specifications. Many automotive repair manuals and tune-up manuals have such carburetor specifications. Too high a float level will cause a rich mixture and poor gasoline mileage. Too low a float level will cause a lean mixture and possibly damage the internal parts of the engine. Floats are commonly checked by removing the top of the carburetor and holding the floats and air horn in a set position. A measurement is then taken to determine if the floats are too high or too low. The measurement is usually taken from the top of the floats to the gasket on the air horn. Floats can be adjusted by bending small float tabs to reposition them at the right level.

3. Check the float for small pinhole leaks by shaking the float to see if there is fuel inside it. Some floats can be weighed with a scale to check for fuel absorption.

4. If the *float needle valve* is stuck open with a piece of dirt, fuel will spill over through the bowl vent. To check the valve:
   a. Remove the main fuel line to the carburetor.
   b. Remove the accelerator linkage to the carburetor.
   c. Remove any choke or other linkage that will restrict the removal of the air horn from the carburetor.
   d. Remove the top of the carburetor by loosening and removing all screws.
   e. Remove the float and float pins.
   f. At this point, the needle valve will be observable.
   g. Use a small needle pliers to remove the valve.
   h. Check for dirt particles and clean the needle valve and passageways.
   i. Make sure the needle valve moves in the hole without sticking.
   j. After cleaning, assemble the carburetor in the reverse order.

5. Adjust the *idle mixture adjustment screws* as far as possible with the highest idle. Usually 1 to 1 1/2 turns out is the standard, but always check the manufacturer's specifications. When starting, make sure the needle is turned in (seated) lightly so as not to damage the needle and seat.
   a. Begin by turning the idle adjustment screws 1 and 1/2 turns out. The engine should be off during this adjustment. ***CAUTION:*** *Set the parking brake for safety*.
   b. Start the engine and run it to operating temperature, making sure the choke is open.
   c. Turn the idle screws in until a slight drop in engine rpm is observed. This can be done by using a tachometer attached to the engine's electrical system.

d. Now turn the screws out until a slight drop is observed in rpm.

e. Adjust the idle mixture screws between these two points to get the smoothest idle.

f. Note that this procedure may vary with each vehicle and manufacturer. More accurate adjustments can be made using an exhaust gas analyzer to measure emissions.

6. Irregular operation may be caused by the idle transfer port being plugged with dirt or varnish deposit from the fuel. This is sometimes called "off idle stumble." This can be fixed by cleaning the carburetor completely.

7. A carburetor rebuild kit that replaces the seals, needle valves, gaskets, and so on can be purchased. Always clean the internal parts of the carburetor with carburetor cleaner. Carburetor cleaner will also clean off varnish on the internal parts of the carburetor circuits. *CAUTION: Carburetor cleaner is very toxic and harmful to human skin. Always wear safety glasses, gloves, and use an approved container to hold the carburetor cleaner solution.*

8. When *adjusting the choke*, check that it is fully open when the engine is at operating temperature and fully closed when the engine is cold, before starting. Because there are so many variations and accessories on carburetors, always refer to the manufacturer's specifications and procedures. These procedures can be found in automotive repair manuals.

9. A vacuum/pressure pump can be used to check for leaks and correct operation on all diaphragms.

10. Main jets on carburetors can be purchased and changed for high-altitude operation.

11. Flat spots during acceleration can be caused by a faulty acceleration pump. When the carburetor is rebuilt, the acceleration pump should be replaced.

12. Over a period of time, the bimetallic strip on the choke may change its characteristics. This may cause the choke to open and close at different temperatures. If this is the case and the choke cannot be adjusted to compensate for the change, the bimetallic strip may have to be replaced.

13. The *idle speed* on the idle stop solenoid is adjusted by turning the inside hex nut on the plunger of the solenoid or by repositioning the solenoid. Use the following general procedure to check the solenoid:

a. Check the solenoid by holding the throttle about one-quarter open with the engine off.

b. Apply battery voltage to the solenoid with a jumper wire. The plunger should extend.

c. Remove the jumper wire and the solenoid should retract.

To check the idle and shut-down adjustments:

a. Connect a tachometer to the engine. Set the brake and block the drive wheels.

b. Start the engine and warm it up to normal operating temperature.

c. Check the manufacturer's specifications for transmission position (drive or park) and whether the air conditioning should be on or off.

d. Make sure the plunger on the solenoid is fully extended.

e. Adjust the solenoid body or plunger screw to obtain the specified slow idle speed.

f. Disconnect the solenoid lead wire to deenergize the solenoid and retract the plunger.

g. Adjust the carburetor idle speed screw of the solenoid to obtain the specific shutdown idle. Reconnect the solenoid and return to the slow idle position.

14. Always hold the hot idle compensator valve closed when adjusting the idle mixture screws.

15. CCC systems have a built-in self-diagnostic system that catches the problems most likely to occur. The diagnostic system lights a "CHECK ENGINE" light in the instrument panel when a fault is detected. By grounding a trouble code "TEST" lead terminal under the dash, the "CHECK ENGINE" light will flash a trouble code indicating the problem area.

16. A considerable amount of information is available to help the service technician diagnose problems in carburetors and computer systems. Automotive repair manuals and tune-up manuals have complete listings of the procedures used to troubleshoot carburetor systems. These include trouble codes, pinpoint test, and diagnostic codes. Because of the complexity of engines and systems today, it is important to be able to locate this information.

SUMMARY

The purpose of this chapter was to introduce basic carburetor principles, study carburetor circuits, analyze carburetor accessories, and study computer-controlled carburetors.

The carburetor works on several main principles. It is designed to keep the air-fuel ratio close to 14.7 parts of air to 1 part of fuel. The carburetor must also change the fuel into tiny droplets or spray for best atomization. The venturi is the main component that is used to draw fuel into the air flow. The throttle plate is used to control the amount of air and fuel flowing into the engine.

Three methods are used to identify types of carburetors. The carburetor draft refers to the direction in which air flows through the carburetor. The number of barrels refers to the number of throttle plates, idle circuits, and choke plates used. The number of venturis also helps determine the type of carburetor used. One, two, and three venturis may be used in any one carburetor.

Several circuits are built into the carburetor to obtain the best performance. The float circuit keeps the level of fuel in the float bowl at the correct level. It uses brass or plastic floats and a needle valve for operation. The idle circuit controls the fuel during idle and very low speed operation. When the throttle plate is closed, fuel is brought in through an idle port below the throttle by the intake manifold vacuum. The low-speed circuit uses an additional transfer port to increase fuel flow during low speeds. The main metering circuit uses a stepped metering rod to control the amount of fuel during intermediate driving conditions. The power circuit is used for increased power. The metering rod is lifted so that more fuel can enter the main jet. The acceleration circuit uses a mechanical pump to squirt extra fuel for acceleration. The choke circuit increases the amount of fuel during cold starting conditions.

Various carburetor accessories are used to reduce emissions and to make the carburetor more efficient. The hot idle compensator bleeds air into the intake manifold to dilute the air-fuel ratio during hot idle conditions. The idle speed adjusting screw adjusts the speed of the idle by mechanically adjusting the position of the throttle plate. The fast idle cam increases the idle when the choke is engaged. The idle stop solenoid completely shuts off the fuel and air when the engine ignition is shut off. The electric-assisted choke uses an electric heater in the choke to heat up the thermostatic spring quickly. The offset choke plate allows a small amount of air to pulse into the carburetor during fully choked conditions. Other accessories used on carburetors include anti-icing vents, idle enrichment valves, secondary throttle linkage, dashpots, hot water chokes, altitude compensators, and poppet valve choke plates. Each accessory has a special purpose and improves carburetor performance.

Carburetors are now being controlled by computers. Fuel management is needed as a result of stricter emission standards. A common type of fuel management system is called the CCC (Computer Command Control). This system uses an ECM (Electronic Control Module). Various data are sent into the computer. These include oxygen in the exhaust, coolant temperature, intake manifold pressure/vacuum, engine speed, battery voltage, and park-neutral conditions. Based on these inputs, the air-fuel ratio is controlled at idle and during normal driving. This is done by operating a metering control solenoid and an idle speed control motor.

Many service checks can be performed on the carburetor. A carburetor that is operating correctly will deliver a 14.7 to 1 air-fuel ratio throughout its operation. If this changes drastically, poor fuel mileage, increased emissions, and irregular operation will be the result.

TERMS TO KNOW

Can you explain each of the following terms? Review the chapter until you can use each term correctly.

Atomized	Meter
Vaporized	Bimetallic strip
Vacuum	Vacuum diaphragm
Venturi	Solenoid
Draft	Dashpot
Check ball	

REVIEW QUESTIONS

Multiple Choice

1. The main purpose of the carburetor is to mix the air-fuel ratio to:
   a. 12.3 to 1
   b. 14.7 to 1
   c. 20 to 1
   d. 25.3 to 1
   e. 30.2 to 1

2. Atomization is used to help _____ the fuel.
   a. Cool
   b. Vaporize
   c. Weigh
   d. Pressurize
   e. Reduce the weight of

3. A restriction in the flow of air in a carburetor is called a _____.
   a. Barrel
   b. Pressure
   c. Venturi
   d. Dashpot
   e. None of the above

4. Fuel is drawn into the carburetor by a _____ produced by a venturi.
   a. Pressure
   b. Vacuum
   c. Temperature difference
   d. Vacuum pump
   e. Hydraulic pressure

5. What controls the amount of air and fuel going into the carburetor during normal operation when the engine is warm?
   a. Throttle plate
   b. Air horn
   c. Bowl
   d. Position of the choke
   e. Position of the dashpot

6. Which of the following is a method used to identify a type of carburetor?
   a. Number of barrels
   b. Number of venturis
   c. Type of draft
   d. All of the above
   e. None of the above

7. A carburetor that has two throttle plates is called a:
   a. Two-barrel carburetor
   b. Two-venturi carburetor
   c. Two-draft carburetor
   d. All of the above
   e. None of the above

8. During idle, what causes the fuel to enter into the air flow?
   a. Fuel pressure
   b. Intake manifold vacuum
   c. Air horn pressure
   d. Injector valve
   e. Accelerator pump

9. The idle port is positioned _____ the throttle plate.
   a. Below
   b. Above
   c. Even with
   d. To the right of
   e. To the left of

10. Too high a float level will produce a _____ mixture.
    a. Lean
    b. 14.7 to 1
    c. Rich
    d. High pressure in the venturi
    e. Low pressure in the venturi

11. Which vent on the float bowl is used to equalize pressure between the bowl and the air horn?
    a. External vent
    b. Internal vent
    c. Bypass vent
    d. All of the above
    e. None of the above

12. Idle mixture adjustment screws are used on _____.
    a. Older carburetors
    b. Newer carburetors
    c. All carburetors
    d. Electronic carburetors
    e. Fuel injectors

13. When the carburetor is operating on the low-speed circuit, fuel enters the air flow:
    a. Two inches above the throttle
    b. Below the throttle
    c. By accelerator pump pressure
    d. All of the above
    e. None of the above

14. When the carburetor is operating on the low-speed circuit, fuel is drawn into the air stream by the:
    a. Venturi
    b. Intake manifold vacuum
    c. Fuel pump pressure
    d. Air cleaner
    e. Fuel pump

15. Which of the following carburetor circuits uses a metering rod?
    a. Idle
    b. Low-speed
    c. Main metering
    d. Choke
    e. Acceleration

16. Which of the following carburetor circuits uses the small diameter end of a metering rod?
    a. Main metering
    b. Low-speed
    c. Idle
    d. Power
    e. Choke

17. The metering rod can be controlled by:
    a. Vacuum
    b. Mechanical linkage
    c. The computer
    d. All of the above
    e. None of the above

18. Which circuit is used for high speed and heavy loads?
    a. Power circuit
    b. Acceleration circuit
    c. Main metering circuit
    d. Idle circuit
    e. Choke

19. The acceleration circuit operates:
    a. Only when the choke is on
    b. Only during idle
    c. Only when the throttle linkage moves to increase fuel
    d. When the throttle closes
    e. Only when the engine is cold

20. What is used to lengthen the time of the downstroke on the acceleration pump?
    a. Vacuum diaphragm
    b. Duration spring
    c. Bimetallic strip
    d. Fuel pump
    e. Choke plate

21. Chokes can be controlled by:
    a. Manual means
    b. Thermostatic coils
    c. Vacuum diaphragms
    d. All of the above
    e. None of the above

22. When brass and spring metal are made into a strip, it is called a _____.
    a. Two-metal strip
    b. Diaphragm strip
    c. Bimetallic strip
    d. Cold spring
    e. Bent spring

23. What circuit is used to increase the air during idle and when the engine is hot?
    a. Idle speed adjustment
    b. Fast idle cam
    c. Hot idle compensator
    d. Choke
    e. Main

24. The fast idle cam is used only during:
    a. Choked conditions
    b. Main metering conditions
    c. Power conditions
    d. High-speed conditions
    e. Low-speed conditions

25. Which accessory helps to eliminate run-on and dieseling?
    a. Fast idle cam
    b. Idle stop solenoid
    c. Hot idle compensator
    d. Anti-choke valve
    e. None of the above

26. The electric-assisted choke is used to:
    a. Heat up the thermostatic coil more quickly
    b. Shut off the choke immediately with a solenoid
    c. Sense the temperature of the choke for the computer
    d. Increase the engine speed
    e. Decrease the engine speed

27. When the computer sends a set of programmed instructions to the carburetor to tell it what to do, it is said to be in:
    a. Open loop operation
    b. Closed loop operation
    c. Intermediate loop operation
    d. High-speed loop
    e. Low-speed loop

28. The computer on the CCC system controls:
    a. The metering control solenoid
    b. The idle speed control motor
    c. The choke plate
    d. All of the above
    e. A and B

29. Which of the following is *not* used as an input on the ECM when it is used for fuel management?
    a. Coolant temperature
    b. Rpm
    c. Throttle position
    d. Coolant fan speed
    e. Exhaust oxygen

30. Idle mixture adjustment screws should be turned:
    a. Out about 1½ turns
    b. All the way in
    c. Out at least 4½ turns
    d. Out 3½ turns
    e. Out 5 turns

31. The choke should be adjusted so it is _____ when the engine is at operating temperature.
    a. Fully open
    b. Fully closed
    c. Partly open
    d. Two-thirds closed
    e. None of the above

The following questions are similar in format to ASE (Automotive Service Excellence) test questions.

32. Technician A says that black smoke coming from the exhaust is an indication of a lean mixture. Technician B says that black smoke coming from the exhaust is an indication of a bad fuel pump. Who is right?
    a. A only
    b. B only
    c. Both A and B
    d. Neither A nor B

33. Technician A says that when the floats are set too high the engine will run rich. Technician B says that when the floats are set too high the engine will run lean. Who is right?
    a. A only
    b. B only
    c. Both A and B
    d. Neither A nor B

34. Technician A says the choke should be checked only when the engine is cold. Technician B says the choke should be checked only when the engine is warm. Who is right?
    a. A only
    b. B only
    c. Both A and B
    d. Neither A nor B

35. Technician A says the choke should be fully open when the engine is cold. Technician B says the choke should be fully closed when the engine is at operating temperature. Who is right?
    a. A only
    b. B only
    c. Both A and B
    d. Neither A nor B

Essay

36. Describe at least three things that may change the air-fuel ratio.

37. What is the difference between atomizing and vaporizing?

38. What is the purpose of a venturi?

39. What is the difference between the main metering circuit and the power circuit on a carburetor?

40. What is the purpose of the fast idle cam?

41. What is the purpose of the oxygen sensor?

42. Describe the purpose and operation of the throttle position sensor.

43. What are some of the input signals to the ECM on computer-controlled carburetors?

CHAPTER 16

Injection Systems

INTRODUCTION

In the past few years, many changes have occurred in the design of fuel systems. One change is to use fuel injection rather than carburetion to mix the fuel. Fuel injection has always been used in diesel engines. It has only recently been used on gasoline engines. The reason for using fuel injection is to control the air-fuel ratio of the engine more precisely. The purpose of this chapter is to study different fuel injection systems used on automobile engines.

OBJECTIVES

After reading this chapter, you will be able to:
- Define the purposes of using fuel injection systems on engines.
- State the different types of fuel injection systems used.
- Analyze the throttle body fuel injection system.
- Analyze the port injection system.
- Analyze high-pressure fuel injection systems used in diesel engines.
- Describe the operation of the injector nozzles used on high-pressure injection systems.
- Identify the operation of governor systems used on high-pressure diesel injection systems.
- Explain how electronic controls are used on diesel fuel injection systems.
- State various diagnosis and service procedures used on injection systems.

CHAPTER HIGHLIGHTS

16.1 CLASSIFICATIONS OF
FUEL INJECTION SYSTEMS
   A. Direct and Indirect Fuel Injection
   B. Port and Throttle Body Fuel Injection
   C. Timed and Continuous Fuel Injection

16.2 GASOLINE ELECTRONIC
FUEL INJECTION SYSTEMS
   A. Fuel Injection Defined
   B. Fuel Injection and Air-fuel Ratio
   C. Nonstoichiometric Air-fuel Ratio Conditions

D. Throttle Body Injection
E. Fuel Injector
F. Injector Pulse Width
G. Fuel Pressure Regulator
H. Vacuum Assist
I. Throttle Position Sensor (TPS)
J. Idle Air Control (IAC)
K. Throttle Body Injector (TBI) Controls
L. Manifold Absolute Pressure (MAP)
M. Crossfire Injection

16.1 CLASSIFICATIONS OF FUEL INJECTION SYSTEMS

Direct and Indirect Fuel Injection

Fuel injection systems can be divided into two major types: the high-pressure (direct) injection system used on diesel engines and the low-pressure (indirect) systems used on gasoline engines. See *Figure 16–1*. Direct fuel injection means the fuel is injected directly into the combustion chamber. Indirect fuel injection means the fuel is injected either into the port before the intake valve or into the intake manifold by a *throttle body* (throttle body injection — TBI) injector. When fuel is injected into the combustion chamber, the pressure of the injection must be increased to high values. High-pressure injection must be used because the injection occurs during the compression stroke. The injection pressure must be much higher than the compression

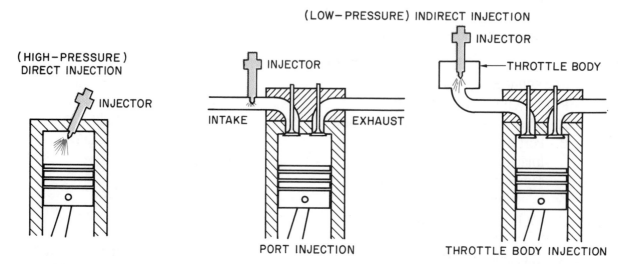

FIGURE 16–1. There are several types of fuel injection. Fuel can be injected either into the combustion chamber, the intake port, or into a throttle body.

pressure for correct atomization. Indirect pressure injection systems inject fuel either into the port before the intake valve or into the throttle body. The area of injection is low in pressure and thus low-pressure injection can be used.

Port and Throttle Body Fuel Injection

Two types of indirect fuel injection are used. When the fuel is injected into the port, it is called port fuel injection (PFI). In this case, there is one fuel injector for each cylinder and set of valves. When the fuel is injected into the center of the throttle body where the carburetor used to be, it is called throttle body injection. In this case, only one fuel injector is used on the system. The fuel injector feeds all of the cylinders of the engine. These two systems are also called multiple-point (port) and single-point (throttle body) fuel injection.

Timed and Continuous Fuel Injection

Fuel injection systems are also classified by the type of injection action. Some injection systems are defined as timed fuel injection. This means that fuel injection occurs at a precise time. Both gasoline and diesel engines use the timed injection system. Diesel engines have high-pressure timed injection systems. Gasoline engines have low-pressure timed injection systems.

Continuous injection is another method of defining the type of injection. Certain gasoline engines have used continuous fuel injection in the past. They are designed to spray a continuous flow of fuel into the engine. Diesel engines do not have a continuous injection system.

In addition to these classifications, fuel injection can be either mechanical or electronic. Diesel engines utilize mechanical fuel injection. Initially, gasoline engines used mechanical fuel injection. Due to stricter *emission standards*, and the increased use of computers on gasoline engines, however, today's fuel injection systems are almost totally electronic.

16.2 GASOLINE ELECTRONIC FUEL INJECTION SYSTEMS

Fuel Injection Defined

Fuel injection is defined as the process of injecting fuel before the valves so that the 14.7 to 1 air-fuel ratio can be maintained in the combustion chamber. Carburetors have been used in the past to mix the correct amount of fuel with the air. With the increased emphasis on pollution and emissions, however, more precise methods are needed. Fuel injection can precisely measure the amount of fuel to maintain the exact and most perfect air-fuel ratio.

Fuel injection systems operate in conjunction with electronic computers. As discussed in an earlier chapter, the electronic control module (ECM) takes in data from many sensors. Based on this information the ECM tells the fuel injector to inject the exact amount of fuel at the correct time.

Fuel Injection and Air-fuel Ratio

When the air-fuel ratio is at 14.7 to 1, conditions are ideal for complete combustion. Complete combustion helps to ignite the mixture, assuring release of all the heat energy in the fuel. If this is done, very little unburned fuel is left. The 14.7 to 1 air-fuel ratio is known as the *stoichiometric ratio*. This is called the best ratio for achieving both optimum fuel efficiency and optimum emission control under ideal conditions.

There are three primary pollutants caused by poor combustion. These include carbon monoxide (CO), hydrocarbons (HC), and nitrogen oxides (NO_x). These pollutants will be discussed in detail later in this textbook. However, as air-fuel ratios are changed, these pollutants increase or decrease. *Figure 16–2* shows the relationship between air-fuel ratios and pollution characteristics. Referring to this chart, it is easy to understand why it is very important to maintain a precise 14.7 to 1 air-fuel ratio. The stoichiometric ratio is the optimum ratio to minimize undesirable emissions.

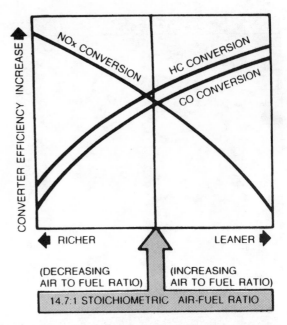

FIGURE 16–2. When the air-fuel mixture is lean, there is an increase in NO_x. When it is rich, HC and CO increase. Fuel injection can be designed to increase control over these pollutants. *(Courtesy of Chrysler Corporation)*

SAFETY TESTING

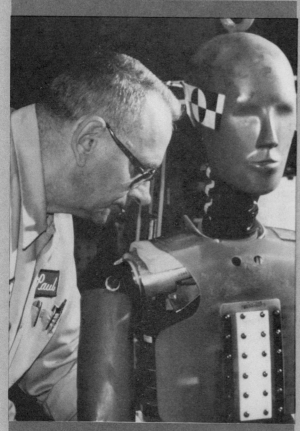

Car manufacturers are very concerned with vehicle crashworthiness and occupant safety. These photos show safety devices being evaluated. Accident data from various sources throughout the U. S. are studied. In addition, collisions are simulated and hundreds of cars are demolished each year in carefully planned "accidents." In crash tests, dummy occupants with sensors at many critical points ride remote-controlled vehicles right into the impact. Crashes are also recorded in slow motion so that the actions of the dummy occupants can be studied.

Many improvements have been developed from this type of testing. These include energy-absorbing steering columns, sideguard door beams, less-aggressive instrument panel controls, and injury-mitigating vehicle structures. (Courtesy of General Motors Proving Grounds)

Nonstoichiometric Air-fuel Ratio Conditions

There are certain operating conditions when special air-fuel ratio requirements have priority over those of emission control. These conditions include times when the engine is cold and times when there is low manifold vacuum. This is a result of an increased load put on the engine.

Cold engine operation occurs whenever the engine is below normal operating temperature and is started or operated. During this condition, the cold inner surfaces of the intake manifold cause some of the fuel in the air-fuel mixture to condense. If the intake air is also below the desirable temperature of 70 degrees F, its ability to vaporize the fuel and mix with it is reduced. During cold engine operation, a richer mixture is needed. The rich mixture replaces the fuel that is lost through condensation or poor vaporization. The rich mixture also aids prompt starting and smooth responsive performance during warm-up periods.

Low manifold vacuum is produced when the load on the engine is increased. When the operator needs more power from the engine, the throttle is pushed down. This opens the throttle plate. When the throttle plate is open, the *intake manifold vacuum* is reduced. Any reduction in manifold vacuum is a positive indication that the engine is being asked to take on an added load. During this time, a richer mixture is needed. A richer-than-stoichiometric ratio helps provide excess fuel for the increased load. This condition is similar to acceleration. Once the engine and load stabilize, the air-fuel mixture is returned to the optimum ratio. Although carburetion systems provide for this increased air-fuel ratio, electronic fuel injection can more precisely limit the enrichment to the degree needed and for the exact duration required.

Throttle Body Injection

The throttle body injection (TBI) system uses a computer to control the amount of fuel injected into the manifold. It is considered an indirect type of injection. Air is drawn into the engine and passes by the injector nozzle. The exact amount of fuel is added for the conditions of operation, *Figure 16–3*.

Fuel Injector

A typical TBI system uses a solenoid-operated fuel injector controlled by the computer. The throttle body injector is centrally located on the intake manifold, usually near the top center of the fuel charging assembly. Here air and fuel are mixed correctly. Fuel is supplied to the injector from the electric fuel pump located in the fuel tank. The incoming fuel is directed to the low end of the injector assembly. An electrically operated solenoid valve in the injector is used to control fuel delivery from the injector nozzle. Referring to

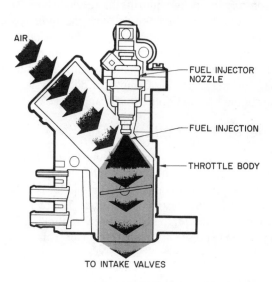

FIGURE 16–3. The throttle body injector injects fuel into an air stream going into the engine. *(Courtesy of Chrysler Corporation)*

Figure 16–4, when there is no electric current from the electronic control module (ECM) to the solenoid, a spring closes a ball-type metering valve inside the injector nozzle. This prevents fuel from flowing through the nozzle.

When the solenoid valve is energized by the computer, the spring-loaded metering valve moves to its full open position. Fuel under pressure from the fuel pump is injected in

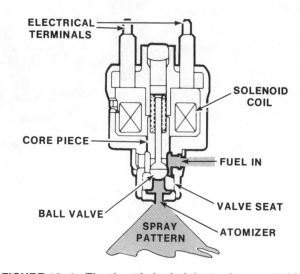

FIGURE 16–4. The throttle body injector is operated by a solenoid. When energized, the ball valve is lifted off its seat and fuel flows into the engine. A spring keeps the ball on the seat when the injector is not energized. *(Courtesy of Rochester Products, Division of General Motors Corporation)*

a *conical* spray pattern into the throttle body bore. The throttle body bore is located directly above the *throttle plate*. The volume of fuel flow is changed by varying the length of time the injector is held open by the ECM. Note that manufacturers may refer to the computer systems differently. Examples include Electronic Computer Control System (ECCS), Electronic Engine Control (EEC), and Logic Module.

Injector Pulse Width

The length of time the injector is open (turned on) and emitting fuel is called the *pulse width*. The pulse width is measured in milliseconds (ms). The injector is pulsed electronically for each piston intake stroke. The correct amount of fuel is *metered* into the engine by controlling how long and when to pulse or turn on the injector. This is controlled by the computer.

Fuel Pressure Regulator

The fuel pressure must be regulated in order for the TBI to work correctly. This is done by a fuel pressure regulator. *Figure 16–5* shows how the fuel pressure regulator is built into the fuel circuit. The throttle body injector is located directly above the throttle body bore. Fuel from the fuel tank is pressurized and sent into the fuel inlet. From the fuel inlet the fuel is sent to the TBI. There is usually an excessive amount of fuel available at the injector. Excess fuel passes through the injector and into the fuel pressure regulator assembly. The fuel pressure regulator is an integral part of the throttle body injection unit. *Figure 16–6* shows such a unit.

The fuel pressure regulator is a mechanical device that maintains approximately 11 psi across the tip of the injector. The exact pressure depends upon the manufacturer. Some are as high as 36 psi. When the fuel pressure exceeds the regulator setting, it pushes the spring-loaded diaphragm down. This uncovers the fuel return port. As the fuel pressure drops below the regulator setting, the spring tension pushes the diaphragm up. This closes off the fuel return port to maintain the pressure of the fuel.

Vacuum Assist

A vacuum assist is used on some, but not all, fuel pressure regulators. The vacuum assist is used to change the fuel pressure during different operating conditions. For example,

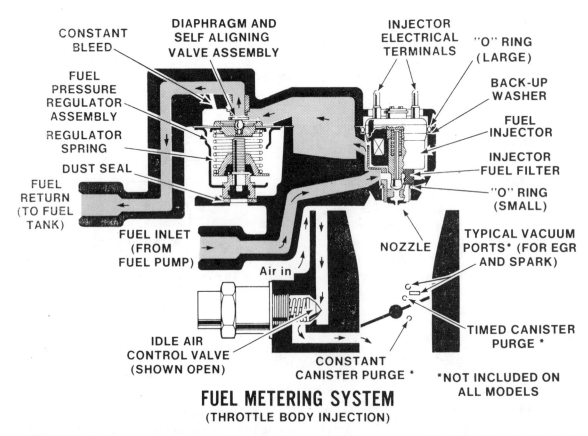

FUEL METERING SYSTEM
(THROTTLE BODY INJECTION)

FIGURE 16–5. The fuel pressure regulator is used to control the pressure of the fuel at the throttle body injector. If fuel pressure increases too much, the fuel will pass by the regulator and back to the fuel tank. *(Courtesy of Chevrolet Division, General Motors Corporation)*

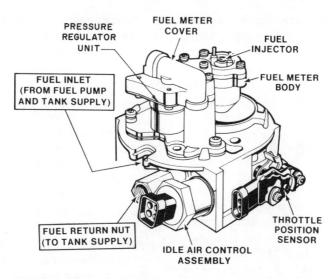

FIGURE 16-6. The fuel pressure regulator can be designed as an integral part of the throttle body injection unit. *(Courtesy of Chevrolet Division, General Motors Corporation)*

when the throttle plate is open, a greater vacuum is felt on the injector tip. Therefore, less fuel pressure is needed to make the fuel flow at the same rate. A vacuum line is connected from above the throttle plate to the bottom of the fuel pressure regulator. This vacuum line reduces the spring pressure, which causes the fuel pressure regulator to open sooner, or at a lower pressure.

Throttle Position Sensor (TPS)

Throttle position sensors were introduced in the carburetion section. The TPS contains a variable resistor that is used to regulate an input voltage, based on the angle of the throttle valve. The ECM uses this signal as a reference to determine idle speeds and air-fuel ratios.

Idle Air Control (IAC)

On certain throttle body injection systems, an idle air control (IAC) is used. Some manufacturers call this system the automatic idle speed motor (AIS). The purpose of the idle air control system is to control engine rpm at idle, while preventing stalls due to changes in engine load. Changes in engine load may be caused from accessory loads, such as air conditioning, during idle.

An IAC assembly motor is mounted on the throttle body unit. It provides control of bypass air around the throttle valve. By extending or retracting a *pintle*, a controlled amount of air is routed around the throttle valve. See *Figure 16-7*.

If the rpm of the engine is lower than desired during operation, more air is diverted around the throttle valve. This increases the rpm. If the rpm is higher than desired, less air is diverted around the throttle valve. This decreases rpm. The ECM monitors the manifold vacuum and adjusts the fuel delivery as idle requirements change.

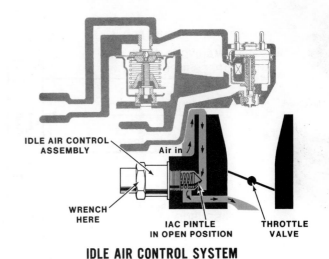

IDLE AIR CONTROL SYSTEM

FIGURE 16-7. Idle air control (IAC) is achieved by controlling the air passing around the throttle valve. The IAC is attached directly to the throttle body. *(Courtesy of Rochester Products, Division of General Motors Corporation)*

During idle, the ECM uses the information from several input signals to calculate the desired pintle position. If the rpm drops below a value stored in the ECM's memory, and the throttle position sensor indicates that the throttle is closed, the ECM calculates the desired pintle position. The ECM will increase or decrease rpm to prevent stalling.

Throttle Body Injector (TBI) Controls

Throttle body injection systems utilize several sensors that send information to the ECM. These sensors include manifold absolute pressure (MAP), oxygen in the exhaust (oxygen sensor), coolant temperature (coolant sensor), engine speed (rpm), and throttle position sensor (TPS). See *Figure 16–8*. Information sent to the ECM from these sensors tells the TBI unit exactly how much fuel should be metered at a specific time. Other inputs are also used to provide information to the ECM. These include a park/neutral safety switch, brake switch, and air conditioning switch.

Manifold Absolute Pressure (MAP)

Manifold absolute pressure (MAP) is a sensor used to measure the absolute pressure (vacuum) inside the intake manifold. In order to understand MAP, absolute pressure and gauge pressure must be studied.

Gauge pressure is defined as pressure on a scale (psig) starting with zero at atmospheric pressure or 14.7 atmospheres. *Absolute pressure* is defined as pressure on a scale (psia) starting with zero at zero atmospheric pressure. *Figure 16–9* shows the differences between absolute pressure and gauge pressure. Zero psig is the same as 14.7 psia. One advantage of using the absolute scale is that there are no vacuum readings. A vacuum on the psig scale is a pressure on the psia scale.

Intake manifold vacuum can now be stated as a pressure. Manifold absolute pressure (MAP) is a pressure reading on the psia scale. It would be considered a vacuum on the psig scale. Note that as manifold pressure increases, vacuum decreases. Manifold absolute pressures are used in the study of computer-controlled combustion (CCC) systems.

The MAP sensor uses a flexible-type resistor. When the resistor is flexed because of an increase in manifold pressure, its resistance changes. This change causes a voltage change that the ECM uses to control the TBI.

Crossfire Injection

Certain engines utilize a pair of throttle body injection units. These are mounted on the front and rear of a single manifold cover. This arrangement allows each TBI unit to supply the correct air-fuel mixture through a crossover port. The port is located inside the intake manifold and feeds the cylinders on the opposite side of the engine, thus the term *crossfire injection*.

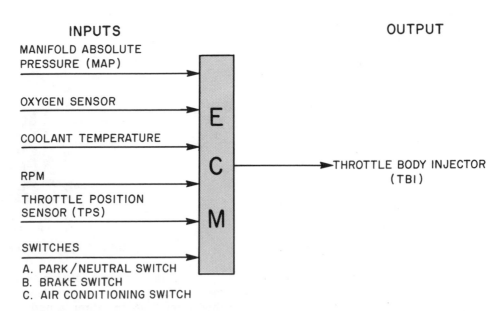

FIGURE 16–8. The throttle body injector is controlled by the ECM. Several inputs are needed to obtain the correct control on the TBI.

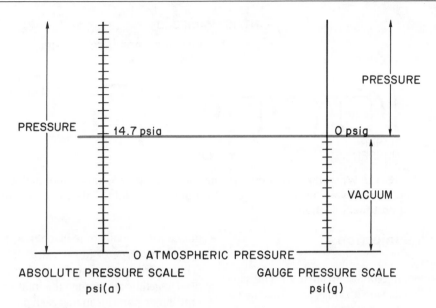

FIGURE 16-9. Manifold absolute pressure can be understood by comparing different types of pressure scales. Psig and psia scales are used. A vacuum reading on the psig scale is a pressure on the psia scale.

CAR CLINIC ▰▰▰

PROBLEM: POOR MILEAGE

A customer complains that the fuel mileage has been decreasing steadily. What could be causing such a condition?

SOLUTION:

The most common causes of increased fuel consumption are the spark or fuel settings. First make sure the timing and the advance are set correctly. Then use a carbon monoxide (CO) meter to check the air-fuel mixture. If these two checks don't eliminate the problem, determine if the front or rear brakes are dragging. This can be done by placing the car on a jack and turning the tires by hand to check for dragging.

16.3 PORT FUEL INJECTION SYSTEMS

Port fuel injection is another way of electronically injecting fuel into a gasoline engine. Port fuel injection is designed to have a small fuel injector placed near the intake port of each cylinder. See *Figure 16-10*. Each injector is controlled by the ECM, and the metering is based on several inputs, as with throttle body injection.

One of the biggest advantages of using port injection over TBI is the ability to get the same amount of fuel to each cylinder. When a throttle body injector or carburetor is

used, the intake manifold acts as a sorting device to send fuel to each cylinder. As air flows through the curved ports in the intake manifold, air, being lighter than fuel, is sorted evenly. However, fuel, which is heavier than air, tends to collect at certain spots inside the intake manifold. This causes some cylinders to be richer than others during operation, *Figure 16-11*.

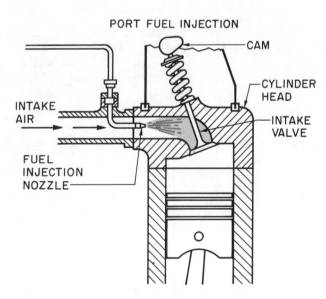

FIGURE 16-10. Port fuel injection has an injector located near the intake port. The fuel is injected directly into the port before the intake valve.

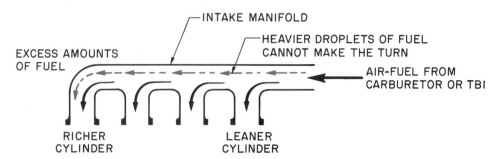

FIGURE 16–11. When throttle body injection or carburetion is used, fuel delivered to the cylinders may vary because of the design of the manifold. This can be overcome by using port fuel injection.

Types of Port Fuel Injection

Several types of port fuel injection (PFI) are used. The standard type of PFI has a double-fire fuel injection. This means that all injectors are pulsed one time each engine revolution. Two injections of fuel are mixed with incoming air to produce the charge for each combustion cycle. On some engines, there are groups of injectors that are fired at the same time.

A second type of PFI is referred to as sequential fuel injection (SFI). On certain applications, the injectors are pulsed sequentially (one-by-one) in spark plug firing order.

A third type of PFI is called the tuned port injection (TPI). Tuned ports are a way of identifying the type of ports used to get the air through the intake manifold to the intake port. Tuned ports mean that each port is designed with equal and minimum restriction. This assures that the same amount of air will be delivered to each cylinder.

Port Fuel Injection Flow Diagram

Figure 16–12 shows a complete schematic of a port fuel injection system. Various components of this system that dif-

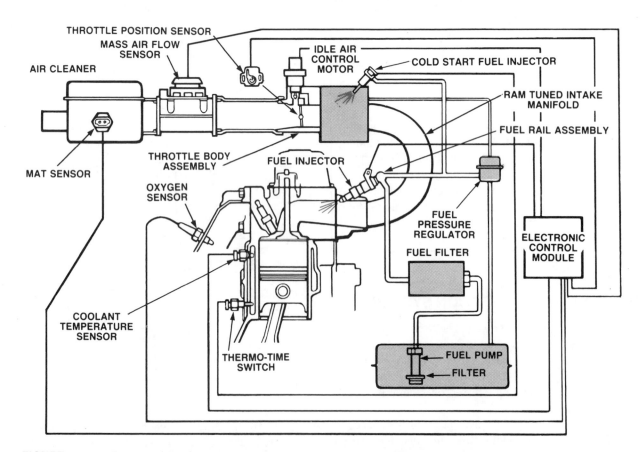

FIGURE 16–12. This schematic shows a port fuel injection system. Many of the sensors were also used in the TBI and carburetion systems. *(Courtesy of General Motors Product Service Training)*

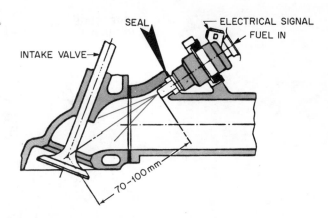

FIGURE 16–13. Fuel is injected at the correct angle in the intake port. *(Courtesy of General Motors Product Service Training)*

fer from carburetion and the TBI system will be discussed. Although port fuel injection systems may be designed differently by other manufacturers, the basic principles remain the same.

The port fuel injector is located just in front of the intake valve. It is controlled by the ECM. Fuel from the fuel pump and filter in the tank is sent through a fuel filter into the injector. As with the TBI system, a fuel pressure regulator controls the pressure of the fuel being sent to the injector. Other components that have already been studied as part of the carburetion or TBI section include the oxygen sensor, coolant temperature sensor, throttle position sensor, and the idle air control motor.

Manifold Air Temperature (MAT)

In order for the ECM to meter the correct amount of fuel during all driving conditions, the temperature of the air coming into the intake manifold is sensed. As the temperature of the air changes, the amount of oxygen per cubic foot also changes. Referring to Figure 16–12, the MAT sensor is located on the air cleaner. The temperature of the incoming air is used as a signal and sent to the ECM.

Port Fuel Injectors

The fuel injectors used with PFI inject fuel directly before the intake valve, *Figure 16–13*. Nozzle spray angle is 25 degrees F. Two O rings are used for installation. One O ring is used to seal the injector nozzle to the intake manifold. The second O ring is used to seal the injector with the fuel inlet connection. Both O rings also prevent excessive injector vibration.

The injector is designed as shown in *Figure 16–14*. It is a solenoid-operated injector, consisting of a valve body and a *needle valve* that has a specially ground pintle There is a solenoid winding that is operated electrically by the

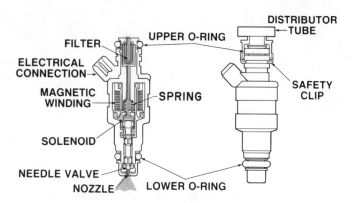

FIGURE 16-14. The solenoid for port fuel injection is a solenoid-operated needle valve. When the solenoid is energized, the needle valve lifts to inject fuel. *(Courtesy of General Motors Product Service Training)*

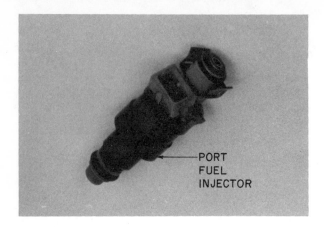

FIGURE 16–15. A port fuel injector.

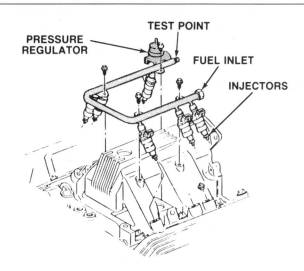

FIGURE 16–17. The fuel pressure regulator is located directly on the fuel rail. *(Courtesy of General Motors Product Service Training)*

ECM. When an electric pulse is sent to the injector, the magnetic field inside the solenoid lifts the needle valve from its seat. The ECM controls the length of the pulse, which establishes the pulse width or the amount of injection. A small helical spring closes the needle valve when the electrical signal is removed. *Figure 16–15* shows a port fuel injector.

Fuel Rails

Fuel must be sent to each injector from the fuel pump. The fuel is sent into a fuel rail, *Figure 16–16*. Here the fuel

is distributed to each injector. The pressure regulator shown in *Figure 16–17* keeps the pressure inside the fuel rail at approximately 350 psi on this particular system. The injectors are locked in place using retainer clips.

Throttle Blade

The throttle blade controls the volume of air that enters the engine, *Figure 16–18*. The throttle blade is controlled by the position of the operator's foot. As the foot is depressed,

FIGURE 16–16. The fuel rail is used to send the fuel to each port fuel injector.

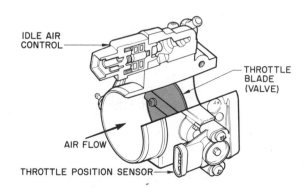

FIGURE 16–18. The throttle body consists of the throttle valve, which controls the air flow, and the throttle position sensor. *(Courtesy of General Motors Product Service Training)*

1984 Victory Circle – Rick Mears Winner. (*Indianapolis Motor Speedway, Photo by Ron McQueeney*)

1986 Winner Bobby Rahal – Pit Stop. (*Indianapolis Motor Speedway, Photo by Ron McQueeney*)

1985 Danny Sullivan – Spin. (*Indianapolis Motor Speedway, Photo by Ron McQueeney*)

1986 Remote Start. (*Indianapolis Motor Speedway, Photo by Ron McQueeney and Ed Kackley*)

1986 Pace Lap – Three Corvettes. (*Indianapolis Motor Speedway, Photo by Ron McQueeney*)

An aerial view of General Motors Proving Grounds in Milford, Michigan. (*Courtesy of General Motors Proving Grounds*)

Testing of vehicle design has taken on a very important part of automotive design. (*Courtesy of General Motors Proving Grounds*)

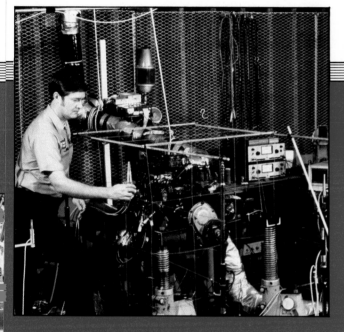

Testing an engine for sound and vibration. (*General Motors Proving Grounds*)

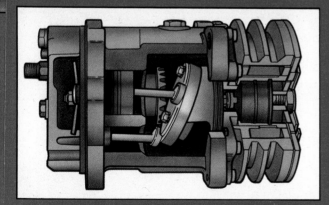

An air compressor pump uses a rotating "swash plate" to cause pistons to move up and down. (*Courtesy of Volkswagen United States, Inc.*)

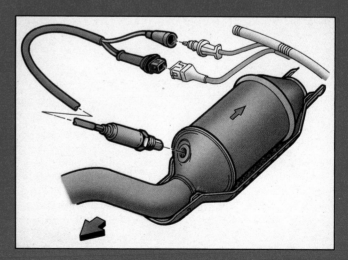

This oxygen sensor is used to measure the oxygen in the exhaust flow and feed this data back to an on-board computer for engine control. (*Courtesy of Volkswagen United States, Inc.*)

Bob Tullius GTP (Grand Tourney Protype) Jaguar at a 1984 Trans Am Race at Brainard International Raceway. (*Courtesy of Jim Rennich*)

One of the top Eliminator Street Rods at the "1986 World of Wheels." (*Courtesy of Jim Rennich*)

This vehicle is being tested on a chassis dynamometer that is placed under the rear wheels of the vehicle. (*Courtesy of Clayton Dynamometers*)

This computerized engine analyzer not only can check computer and electrical systems on the automobile but also gives one printout for the customer and a different printout for the technician. (*Courtesy of Sun Electric Corporation*)

Sophisticated engine analyzers are used to check emissions, electrical circuits, computer systems, timing, and ignition systems. The analyzer is able to interface directly with computer controlled engine codes used today for trouble-shooting. (*Courtesy of Sun Electric Corporation*)

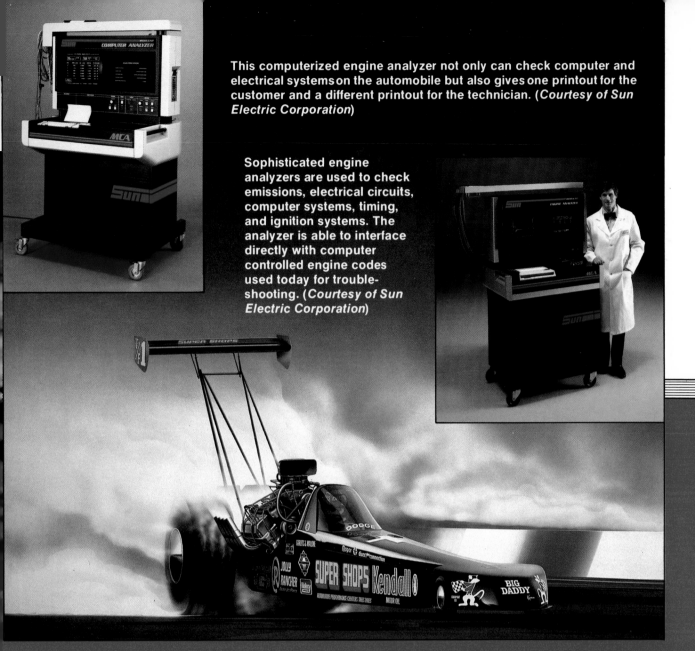

This dragster uses a blower to force excessive amounts of air into the engine. The airfoil on the back is to stabilize the vehicle at high speeds. Some dragsters are capable of horsepower ranges between 900 – 1100 hp or more. (*Courtesy of Mopar Division, Chrysler Corporation*)

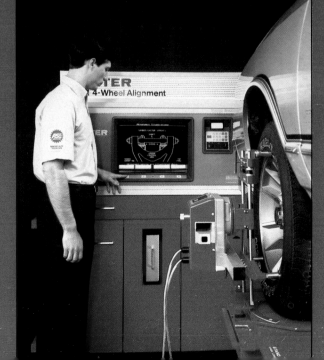

Sophisticated equipment is used to balance tires on automobiles. (*Courtesy of Hunter Engineering Company*)

Computers are used on these machines to help the service technician accurately adjust the vehicle's alignment. (*Courtesy of Hunter Engineering Company*)

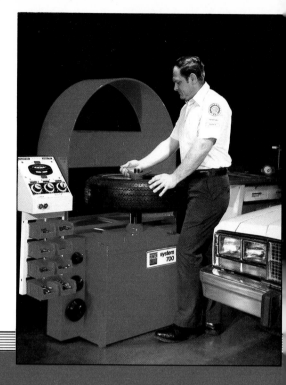

Al Unser Jr.'s car takes on a new set of tires at a pit stop during a Long Beach, California race. (*Courtesy of Goodyear Tire & Rubber Co.*)

Designers and engineers use the CAD (computer-aided-design) system to create bodies of vehicles, including the underbody and panels, steering geometry, suspension and other systems. Shown here is a Chrysler LeBaron GTS/Dodge Lancer that can be fitted with components, passengers, and cargo, all done by the computer program. (*Courtesy of Chrysler Motors*)

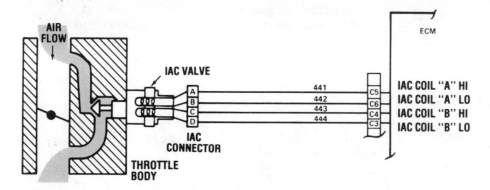

FIGURE 16-19. The idle air control is attached to the throttle body. The ECM controls the motor windings to open or close the IAC valve. *(Courtesy of General Motors Product Service Training)*

the blade opens and allows more air into the engine. As the foot is released, the blade closes and reduces the amount of air allowed into the engine. On some engines, there is also a coolant passage that allows engine coolant to flow through the throttle body unit. The purpose of this coolant flow is to increase the temperature of the incoming air to assist in preventing throttle blade icing in cold weather operation.

The throttle body also supports and controls the movement of the throttle position sensor (TPS). As was stated earlier, the TPS is used to send a signal to the ECM. This signal gives the position of the throttle under all operating conditions.

Idle Air Control (IAC)

Attached to the throttle body is an idle air control (IAC). The IAC is used to control the amount of air during idle conditions. The ECM controls idle speed by moving the IAC valve in and out as shown in *Figure 16-19*. The ECM does this by sending the IAC voltage pulses called "counts" to the proper motor winding. The motor shaft and valve move a given distance for each count received.

For example, to increase the idle speed the ECM sends enough counts to retract the IAC valve and allow more air to flow around the throttle blade. The increase in air flow into the engine causes the idle to increase. (A corresponding increase in fuel metered by the ECM must also occur.) To decrease the idle speed, the ECM sends the correct number

of counts to the IAC to extend the valve and reduce air flow. *Figure 16-20* shows the throttle body unit with the idle air control valve and the throttle position sensor.

Cold Start Injector

When starting a cold engine, additional fuel is required to operate the engine. If there is sufficient fuel, correct fuel vaporization and atomization for combustion will occur. The cold start injector is used to provide additional fuel during cranking. This circuit is important when engine coolant temperatures are low. During this time, the main injectors are not pulsed "ON" long enough to provide the amount of fuel needed to start the engine.

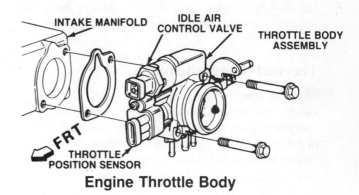

Engine Throttle Body

FIGURE 16-20. The throttle body also has the throttle position sensor and idle air control valve attached. *(Courtesy of General Motors Product Service Training)*

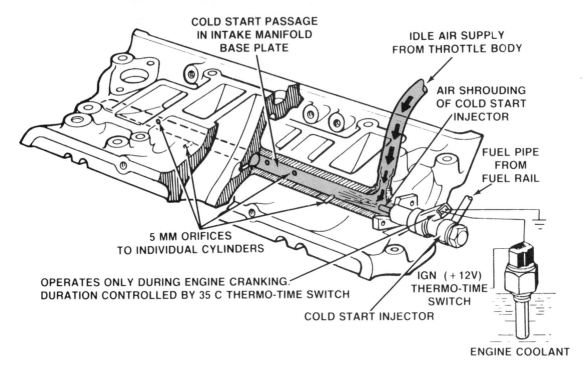

COLD START PASSAGE
IN INTAKE MANIFOLD
BASE PLATE

IDLE AIR SUPPLY
FROM THROTTLE BODY

AIR SHROUDING
OF COLD START
INJECTOR

FUEL PIPE
FROM
FUEL RAIL

5 MM ORIFICES
TO INDIVIDUAL CYLINDERS

OPERATES ONLY DURING ENGINE CRANKING.
DURATION CONTROLLED BY 35 C THERMO-TIME SWITCH

IGN (+12V)
THERMO-TIME
SWITCH

COLD START INJECTOR

ENGINE COOLANT

FIGURE 16–21. The cold start injector is placed on the intake manifold to help in cold starting conditions. *(Courtesy of General Motors Product Service Training)*

Figure 16–21 shows how the cold start system operates. During cranking, fuel is injected into the cylinder ports through the individual fuel injectors. Additional fuel is injected into a separate passage within the inlet manifold by the cold start injector. This passage has small individual orifices by each cylinder. These small passages send extra fuel to each cylinder. Air is also sent through the passage from the throttle body.

Thermal-time Switch

The cold start injector is controlled by the thermal-time switch. This switch is connected to the starter solenoid. The cold start injector operates only during cranking or starting of the engine. The cold start injector also operates only at coolant temperatures below 95 degrees F (35 degrees C) and can operate only a certain length of time.

The switch is made of a bimetal material which opens at a specified coolant temperature. The bimetal material is heated by a winding inside the thermal switch. This allows the cold start injector to stay on for a specific amount of time (8 seconds). The time the thermal switch stays closed varies inversely with the coolant temperature. As the coolant temperature goes up, the cold start valve on-time goes down. *Figure 16–22* shows an example of the cold start injector circuit and the thermal-time switch.

Mass Air Flow (MAF)

Mass air flow is defined as the total amount (mass) of air going into the engine. This depends upon the altitude of the vehicle, temperature of the air, density of the air, and moisture of the air. All of these variables cause the air-fuel ratio to change. In order to maintain a stoichiometric air-fuel ratio of 14.7 to 1, these variables must be known. This is done by using a mass air flow sensor.

Figure 16–23 shows a MAF sensor. It consists of the following:

a. a flow tube that houses the parts.
b. a sample tube that directs air to a sensor.
c. a screen that breaks up the air flow.
d. a ceramic resistor that measures the temperature of the incoming air (Air Temperature Sending Resistor).

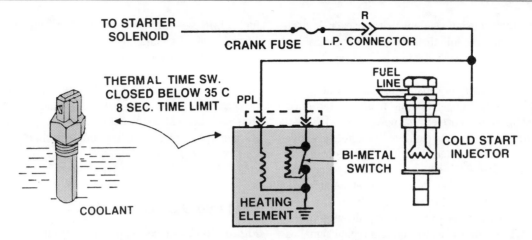

FIGURE 16–22. The thermal-time switch controls the time the cold start injector is on. The switch operates only during cranking or starting the engine. A small heating element also heats the bimetal switch to limit the "ON" to 8 seconds. *(Courtesy of General Motors Product Service Training)*

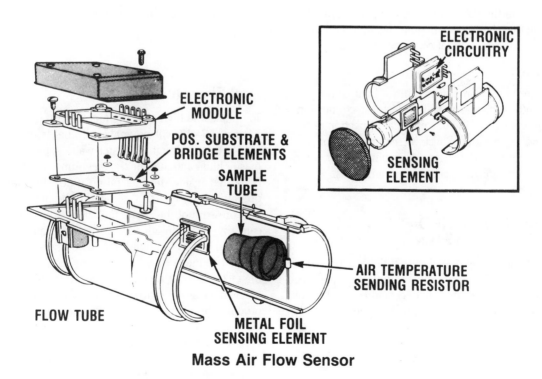

Mass Air Flow Sensor

FIGURE 16–23. The mass air flow (MAF) determines the exact amount (mass) of air flowing into the engine. *(Courtesy of General Motors Product Service Training)*

e. a metal foil sensing element that senses air mass.

f. an electronic module that determines the mass air flow from the sensing element.

This type of sensor is capable of compensating for altitude and humidity. In operation, air mass is determined by measuring the amount of electrical power needed to keep the temperature of the sensing element 75 degrees C above the incoming air. As air enters the unit, it passes over and cools the sensing element. When the element is cooler, it requires more electrical power to keep the element at 75 degrees C above the incoming air temperature. The electrical power requirement is a measure or indication of the mass air flow. The power is converted to a digital signal as a frequency. This signal is sent to the ECM and used to calculate engine load. Using mass air flow, engine temperature, and rpm, the ECM can calculate the exact amount of fuel to be metered to provide a stoichiometric ratio of 14.7 to 1.

CAR CLINIC

PROBLEM: "CHECK ENGINE" LIGHT IS ERRATIC

It is noticed on the dashboard of a computer-controlled GM engine that the "Check Engine" light goes on intermittently or is erratic. What could be the problem?

SOLUTION:

Most of the time a component fault or light coming on intermittently indicates a faulty electrical connection. The diagnosis should include physical inspection of the wiring and connectors. Check all connections on the electrical circuits on the engine. Also, physically observe the connections for damaged or bent terminals.

16.4 HIGH-PRESSURE DIESEL FUEL INJECTION SYSTEMS

All diesel engines being manufactured today use injection systems. The injection of diesel fuel is considered mechanical, direct (into the cylinder), and timed. In other words, diesel fuel is injected at a precise time into the top of the cylinder, during the compression stroke of the diesel engine. In order to do this, the pressure to produce atomized fuel must be very high. Injection systems inject fuel in the range of 4,000–10,000 psi.

This type of injection system requires study in several major areas. These include: 1) pressurizing the fuel, 2) timing the fuel, 3) atomizing the fuel, and 4) metering the fuel. *TAMP* is a term used to remember *T*iming, *A*tomizing, *M*etering, and *P*ressurizing. In addition, diesel fuel injection systems also use governors for speed control.

There have been many changes through the years to diesel fuel injection design. Each of the TAMP principles can be accomplished in many ways. In addition, electronic controls have been added to get precise fuel control. However, the basic principles of fuel injection remain much the same. One of the more common types of fuel injection pump is called the rotor-distributor-type injection pump.

Fuel Pump Components

Figure 16–24 shows a cutaway of a standard high-pressure fuel injection pump. The pump has several components including the:

1. drive shaft
2. housing
3. metering valve
4. hydraulic head assembly
5. transfer pump
6. pressure regulator assembly
7. distributor rotor
8. internal cam ring
9. automatic advance (optional)
10. pumping plungers
11. governor

Fuel Flow Diagram

To understand high-pressure fuel systems, the schematic in *Figure 16–25* will be analyzed. Fuel is drawn from the supply tank, through a master water separator, through a master filter, and into the transfer pump. The transfer pump is located inside the fuel pump unit. The transfer pump pressurizes the fuel and sends it to the head passage area. The pressure of the diesel fuel at this point acts upon the cam ring to aid in advancing the fuel timing.

Fuel is sent through the head passage to the metering valve. The metering valve is positioned by the governor to meter the correct amount of fuel to the distributor rotor. Inside the distributor rotor, there are pumping plungers that produce the high pressure needed for injection. When the high pressure is produced, fuel is sent through a delivery valve to each injector nozzle. There is an injector nozzle for each cylinder. Any excess fuel from the nozzle or governor area is returned to the fuel tank. Timing occurs in the distributor rotor, atomizing occurs at the nozzle, metering occurs at the metering valve, and pressurizing occurs at the distributor rotor.

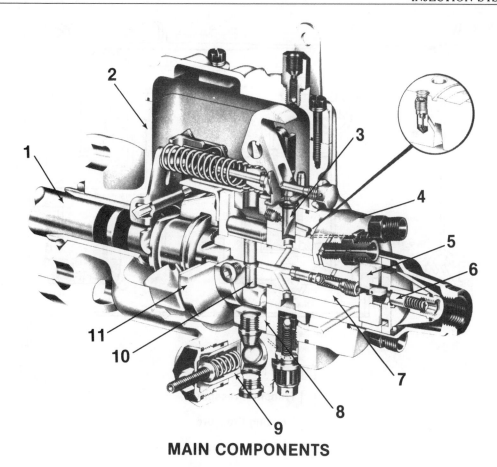

MAIN COMPONENTS

1. DRIVE SHAFT
2. HOUSING
3. METERING VALVE
4. HYDRAULIC HEAD ASSEMBLY
5. TRANSFER PUMP BLADES
6. PRESSURE REGULATOR ASSEMBLY
7. DISTRIBUTOR ROTOR
8. INTERNAL CAM RING
9. AUTOMATIC ADVANCE (OPTIONAL)
10. PUMPING PLUNGERS
11. GOVERNOR

FIGURE 16–24. This high-pressure fuel injection pump is used on automotive diesel engines. It is designed to inject high-pressure fuel into the combustion chamber during the compression stroke. (Courtesy of Society of Automotive Engineers)

Transfer Pump

The purpose of the transfer pump is to transfer the diesel fuel from the fuel tank to the fuel pump. The transfer pump is located on the main shaft of the fuel injection pump. As the fuel injection pump is rotated by the engine, the transfer pump also rotates.

The transfer pump is a positive displacement vane-type pump. *Figure 16–26* shows a transfer pump. The inside diameter of the liner is eccentric to the rotor axis. This causes the blades to move in the rotor slots. This blade movement changes the volume between the blade segments. As the blade is rotated, a suction and pressure are created in the inlet slot and outlet groove.

Pressure Regulator Assembly

The pressure of the transfer pump increases with the rpm of the main shaft. When too much pressure is created, some of the fuel is bypassed back to the inlet side of the transfer

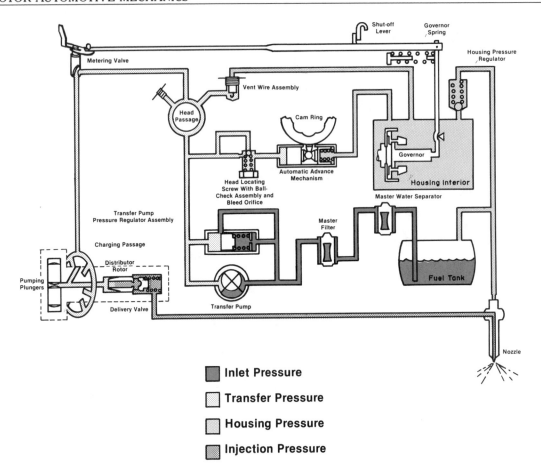

Inlet Pressure

Transfer Pressure

Housing Pressure

Injection Pressure

FIGURE 16–25. This flow diagram shows the flow of fuel from the fuel tank to the injector. (Courtesy of Stanadyne, Inc.)

pump. This is done by the regulator assembly. As flow increases, the regulating spring is compressed. Eventually the regulating spring is compressed until the edge of the regulating piston uncovers the pressure regulating slot. Diesel fuel is now transferred back to the inlet side of the pump. See *Figure 16–27*.

Charging Cycle

After the fuel is pressurized by the transfer pump, it must be charged. The charging cycle is defined as that part that brings in the correct amount of fuel to be pressurized. Referring to *Figure 16–28*, pressurized fuel from the transfer pump is sent through internal passages into a circular fuel passage. From the circular fuel passage the fuel is sent past the metering valve. The position of the metering valve determines exactly how much fuel will be sent to the injectors. From the metering valve the fuel is sent into a charging passage. This is a circular passage with holes that line up with holes in the rotor. When the holes line up during rotation, the metered fuel is sent to the center of the rotor.

Figure 16–29 shows a breakdown of the rotor. Remember that the rotor is being turned by the engine.

When the fuel is sent to the center of the rotor, it forces the small plungers outward. The metered fuel is now in a pumping chamber. If only a small amount of fuel is admitted to the pumping chamber, as in idle, the plungers move out only a short distance. If more fuel is admitted to the pumping chamber, the plungers move out further. The fuel injection pump is now ready for the discharge cycle.

Discharging Cycle

Figure 16–30 shows a circular cam with the plungers and rollers placed in the center. As the rotor turns, the ports from the circular passages are closed off. The fuel inside the rotor is captured. As the center rotor continues to turn, the rollers are forced inward by the shape of the cam. This produces a high pressure (injection pressure) because the fuel is being squeezed inside the rotor.

Figure 16–31 shows the flow of fuel during the discharge cycle. When the fuel is fully pressurized, it flows

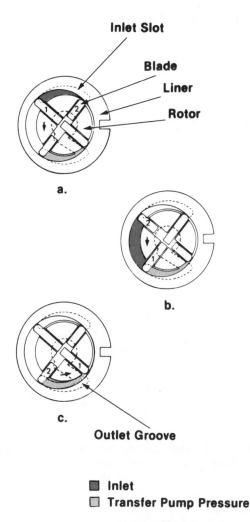

FIGURE 16–26. This vane-type pump is used to transfer fuel from the fuel tank to the injection pump. As the vanes rotate in an eccentric housing, a suction and pressure are created. *(Courtesy of Stanadyne, Inc.)*

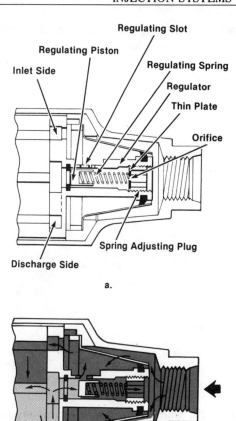

FIGURE 16–27. The pressure regulator is used to maintain the correct transfer pump pressure. As the pressure increases with rpm, the regulating spring will compress. Eventually the regulating piston moves to open a regulating slot. Fuel is bypassed back to the suction side of the pump. *(Courtesy of Stanadyne, Inc.)*

past a delivery valve to a discharge port. As the rotor continues to turn, the discharge port lines up with a port that goes to each discharge fitting and finally to the injector. The high-pressure fuel is now being distributed to each cylinder. Delivery of the fuel continues until the cam rollers pass the innermost point on the cam lobe and begin to move outward.

Delivery Valve

The purpose of the delivery valve is to shut off the flow of fuel during injection. It is important to end injection rapidly. If the injection does not end rapidly, the injector will dribble fuel into the combustion chamber. The result will be poorly timed and atomized fuel.

The delivery valve operates in a bore in the center of the distributor rotor. When injection starts, fuel pressure created by the rollers and cam moves the delivery valve slightly out of its bore. *Figure 16–32* shows this action. Pressure from the cam and rollers pushes the delivery valve to the right. Fuel now passes to the discharge port and to the injectors. Delivery ends when the pressure on the plunger side is reduced quickly. This is due to the cam rollers passing the highest point on the cam lobe.

At this point, there is a drop in pressure on the left side of the delivery valve, while a high pressure still exists on the right side of the delivery valve. This difference in pressure causes the delivery valve to close rapidly. This action reduces the high pressure to the injector very rapidly. Following this, the discharge port closes completely, and the fuel that remains is maintained until the next injection.

Return Fuel Oil Circuit

Excess fuel from the transfer pump is discharged into a vent passage. It is returned to the interior housing of the governor. See **Figure 16–33**. The amount of fuel sent back to the housing depends upon the specifications. The amount of return fuel oil is controlled by the size of a wire used in the vent assembly. The smaller the wire, the more return flow, and vice versa. The vent wire is available in several sizes.

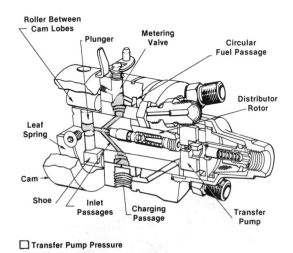

☐ Transfer Pump Pressure

FIGURE 16–28. During the charging cycle, fuel passes the metering valve and enters the area below the plunger. Fuel is now ready to be pressurized in the discharge cycle. *(Courtesy of Stanadyne, Inc.)*

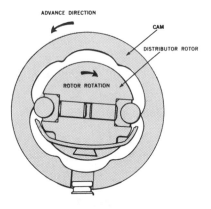

FIGURE 16–30. This circular cam and rotor is used to produce the high-pressure fuel injection. As the rotor and rollers turn inside the cam, the shape of the cam pushes the rollers in to produce the high pressure. *(Courtesy of Stanadyne, Inc.)*

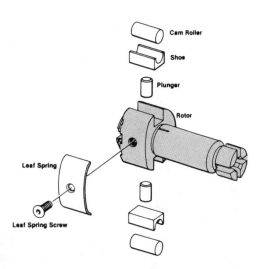

FIGURE 16–29. A breakdown of the distributor rotor is shown. The right side is the transfer pump and the left side is the plunger and cam rollers are used to produce the high pressure. *(Courtesy of Stanadyne, Inc.)*

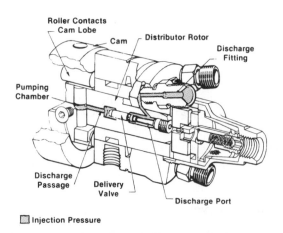

▨ Injection Pressure

FIGURE 16–31. Fuel flow during the discharge cycle is shown. Fuel pressure created from the cam and rollers is sent through a delivery valve to each discharge fitting. From there the fuel is sent to each injector. *(Courtesy of Stanadyne, Inc.)*

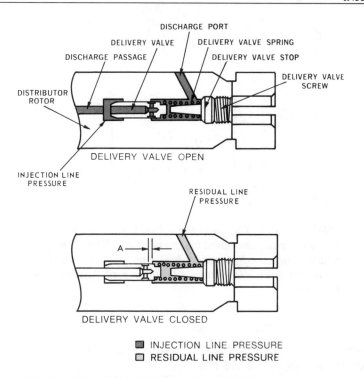

FIGURE 16–32. The delivery valve is used to rapidly shut off the fuel to the injectors. *(Courtesy of Stanadyne, Inc.)*

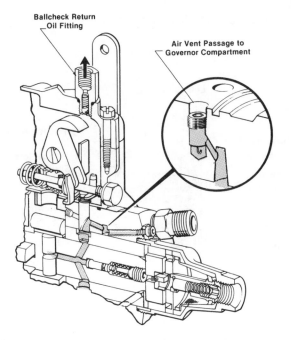

FIGURE 16–33. Fuel flow back to the housing is controlled by the size of the vent wire in the passage. *(Courtesy of Stanadyne, Inc.)*

Metering Valve

The metering valve is used to control the exact amount of fuel being sent into the circular charging passage. *Figure 16–34* shows examples of two metering valves. As the valves are turned, the position of the metering edge changes with respect to the holes in the housing. This means either a larger quantity or lesser quantity of fuel will be delivered to the circular charging head.

Governor Principles

A governor is a speed-sensing device that tells the metering valve how much fuel to put into the engine. *Figure 16–35* shows the principles of any governor operation. Two forces are working against each other. A spring force pushes in one direction. A centrifugal force from several weights produces the second force. The weights are turned by the engine. The center point is connected to the engine throttle or metering valve.

The governor tries to keep the rpm of the engine the same. The speed of the engine is controlled by the tension adjusted on the spring. If the engine is running at 3,000 rpm under no load, a balance has been established between the

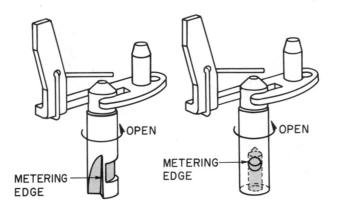

FIGURE 16-34. The metering valve is used to control or meter the exact amount of fuel for each injection. One uses a helix (angled cut) to meter fuel. The other uses a hole to meter fuel. *(Courtesy of Society of Automotive Engineers)*

centrifugal force and the weight force. When the engine encounters a load, the rpm drops. This causes the centrifugal forces to be reduced. There is now an unbalanced condition between the two forces. The spring will now push the metering rod to add more fuel. As more fuel is added to the engine, the rpm will increase back to 3,000 to handle or carry the load.

If the load is removed, the reverse happens. As the load is removed, the engine rpm increases. This causes the centrifugal forces to be greater than the spring forces. This action moves the metering valve to reduce the fuel setting. The rpm is subsequently reduced to the 3,000 originally set.

Governor Operation

Figure 16-36 shows the mechanical system and how it operates. The governor weights are located inside the weight retainer. The governor spring is connected to the throttle shaft. The throttle shaft controls the position of the metering valve. As the weights move outward from a decrease in load, they move the thrust sleeve and washer to the left. This action pushes the lower section of the governor arm to the left. The governor arm is supported on a pivot shaft. Movement to the left on the bottom of the governor arm causes movement to the right on the top of the governor

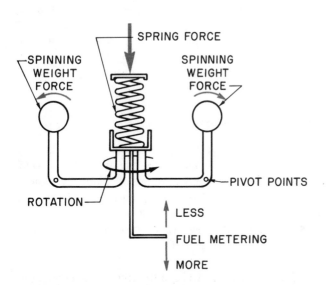

FIGURE 16-35. Any governor is a speed-sensing device. Centrifugal forces from spinning weights oppose spring forces. The resultant force controls the metering of the fuel.

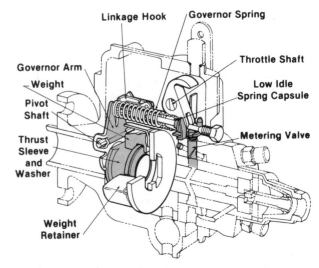

FIGURE 16-36. On a fuel injection pump, weights move a thrust sleeve. The sleeve is connected through a linkage to the governor spring. This sets the metering rod in the correct position. *(Courtesy of Stanadyne, Inc.)*

arm. This movement pushes against the governor spring. The throttle shaft also moves to the right. This action reduces the amount of fuel sent through the metering valve.

If the load is increased, the reverse occurs: 1) the governor weights drop in, 2) the governor spring pushes the governor arm to the left, 3) the throttle shaft moves to the left, 4) the metering valve increases its flow, 5) rpm increases to handle the extra load.

Automatic Fuel Injection Advance

The advance mechanism will advance or retard the delivery of fuel to each injector. As the rpm of the engine increases, the fuel should be advanced. Advancing the fuel assures that combustion will take place when the piston is at the most effective position during all rpm and loads.

Figure 16–37 shows the operation of an advance mechanism. Advance is accomplished by moving the cam ring clockwise. When the cam is moved counterclockwise, the injection is retarded. The cam ring is controlled by the position of two pistons. One is called the spring piston and the other is called the power piston. The pressure from these two pistons positions the advance cam screw to either advance or retard the injection.

Governor housing pressure is applied to the spring piston. Governor housing pressure is controlled by the pressure regulator on the housing. This uses a ball check valve to control the pressure inside the housing.

Transfer pump pressure is applied to the power piston. As the engine speed increases, the transfer pump pressure increases. This causes an unbalanced pressure, which

moves the cam ring clockwise. The fuel injection is now being advanced. When the engine speed decreases and transfer pump pressure is reduced, the advance spring pushes the advance cam screw to the right. The fuel injection is now being retarded. A trimmer screw is provided to adjust the advance spring. This controls the start of cam movement.

Fuel Injector Nozzles

The fuel injector nozzle is used to inject the fuel directly into the combustion chamber. See *Figure 16–38*. During the injection process, the fuel must be finely atomized. The injector nozzle is designed to atomize the fuel correctly.

There are several types of injector nozzles. The most common are the pintle nozzle and the hole nozzle. *Figure 16–39* shows both types of nozzles. The pintle nozzle has a small needle valve located in the center of the tip protruding into the combustion chamber. The needle valve or stem is used to open and close a small hole for injecting fuel. The hole nozzle has a valve stem that opens and closes several holes leading to the combustion chamber. The most common type of nozzle in the automotive market is the hole nozzle. Although there are many designs used to accomplish atomization, the principles are much the same.

Nozzle Operation

In operation, high-pressure fuel delivered from the fuel injection pump is fed through steel lines to the injector. Once at the injector, the fuel is fed through internal passages

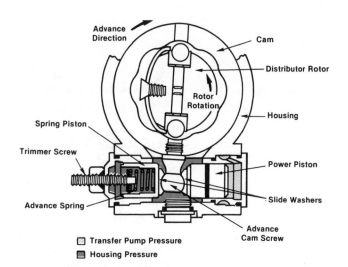

FIGURE 16–37. Advance on a fuel injection system is accomplished by moving the cam ring. This is done by using two pressures working against each other. *(Courtesy of Stanadyne, Inc.)*

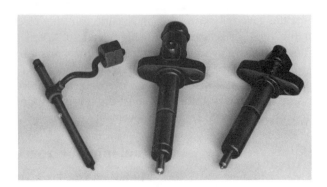

FIGURE 16–38. There are many types of fuel injector nozzles. The nozzles are used to inject and atomize the fuel inside the combustion chamber.

to the nozzle assembly. See **Figure 16–40**. The nozzle assembly has a stem that keeps the holes closed off when there is no pressure from the injection pump. A spindle assembly and large spring are used to hold the nozzle closed.

Referring to **Figure 16–41**, as fuel pressure is increased by the injection pump, this pressure is felt on the stem or needle valve of the nozzle. As the pressure continues to increase, the stem or needle valve will lift against spring tension. When this happens, fuel is injected rapidly into the cylinder. As the fuel is injected, the pressure drops. When the pressure drops below the spring pressure, the stem or needle valve is rapidly forced closed by the injector spring. This action "pops" the injector. This means that the injection stroke is started and stopped very rapidly. This improves the atomization and timing of fuel into the combustion chamber.

16.5 DIESEL ELECTRONIC CONTROL SYSTEMS (DECS)

In order to meet emission standards, diesel engines also use computers to control the fuel and emission systems. One type of computer system is called the diesel electronic control system (DECS). This system uses an electronic control module to control various inputs and outputs. **Figure 16–42** shows a schematic of the inputs and outputs on the fuel system.

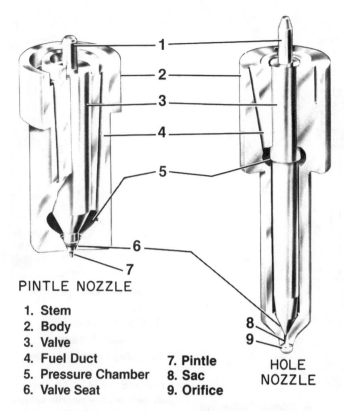

PINTLE NOZZLE

1. Stem
2. Body
3. Valve
4. Fuel Duct   7. Pintle
5. Pressure Chamber   8. Sac
6. Valve Seat   9. Orifice

HOLE NOZZLE

FIGURE 16–39. Several nozzle tips are used to inject the fuel into the cylinder. Both the hole and pintle nozzles are used. *(Courtesy of United Technologies, formerly American Bosch)*

Fast Idle Solenoid

The fast idle solenoid is a plunger-type solenoid. When extended, it pushes on the throttle linkage of the fuel injection pump. This action increases the spring pressure on the governor and slightly increases the engine speed. It is energized with temperatures below 100 degrees F, and above 248 degrees F for overheating. The fast idle solenoid is operated whenever the air conditioning is on.

Housing Pressure Cold and Altitude Advance (HPCA and HPAA)

It is desirable to advance the fuel injection pump on a cold engine and at higher altitudes. This helps to reduce emissions, white smoke, and noise. Cold starting is also improved. The HPCA and HPAA solenoid operates with the fuel injection pump return line pressure regulator. The solenoid operates at cooling temperatures below 98.6 degrees F and above 4,000 feet above sea level.

When energized, the solenoid pushes the injector housing check ball off its seat. This causes the housing pressure to drop, which results in advanced fuel injection pump timing.

Altitude Fuel Limiter (AFL)

At altitudes above 4,000 feet, the fuel must be limited. The AFL is a solenoid that, when activated, limits the travel of the metering valve at wide open throttle.

Metering Valve Sensor (MVS)

The metering valve sensor is a variable resistor that sends a metering valve position signal to the ECM. The resistance is lowest at wide open throttle. Voltage output will be about 5 volts. As the metering valve closes, the resistance will increase, changing the voltage signal back to the ECM.

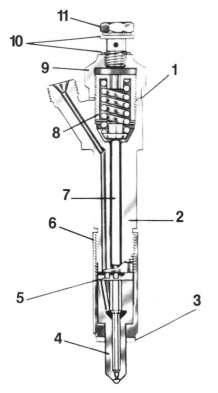

1. Protection Cap Gasket
2. Nozzle Holder Body
3. Nozzle Gasket
4. ADB Nozzle Assembly
5. Dowel Pin
6. Nozzle Cap Nut
7. Spindle Assembly
8. Spring Adjusting Retaining Cap Nut
9. Protection Cap
10. Retaining Screw Gasket
11. Retaining Screw

FIGURE 16–40. Fuel from the fuel injection pump is sent into the injector. As the pressure increases, the needle valve is lifted against the spring pressure. Fuel under high pressure now enters the combustion chamber. *(Courtesy of United Technologies, formerly American Bosch)*

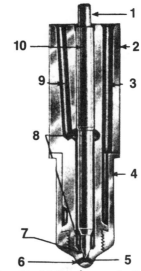

1. Stem (Needle Valve)
2. Body
3. Coolant Duct
4. Coolant Sleeve
5. Spray Hole Chamber
6. Spray Hole
7. Valve Seat
8. Pressure Chamber
9. Fuel Duct
10. Valve

FIGURE 16–41. As the pressure from the fuel pump increases, the needle valve will eventually open and allow fuel to pass into the combustion chamber. *(Courtesy of United Technologies, formerly American Bosch)*

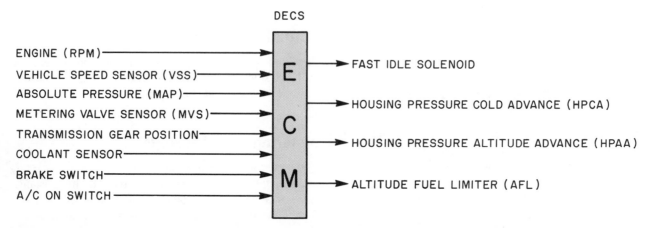

FIGURE 16–42. The diesel electronic control system (DECS) is used to control the fuel injection system. Both input and output systems are shown.

DRAG RACING CARS

This vehicle, designed and built by Jim Rennich, is a 1966 Chevy II. It is typical of the many low-cost drag cars built for Bracket Nationals. Bracket Nationals means racing within an estimated time (ET) that has been previously determined. (Courtesy of Jim Rennich, Bloomington, MN)

NOTE: *Many computer controls are used on fuel injection systems.* Each year the manufacturers improve the systems. In addition, each car manufacturer uses a different set of procedures for troubleshooting and for diagnosing problems in the computer fuel injection system. The following diagnosis and service hints are only general in nature, but they give an idea of how to diagnose and troubleshoot this area.

1. The ECM (electronic control module) used on each vehicle has a PROM. The word *PROM* stands for *Programmable Read Only Memory*. The PROM is a programmed set of data pertaining to the vehicle's weight, engine size, transmission, axle ratio, and so on. If any diagnostic procedure calls for the ECM to be replaced, the calibration unit (PROM) should be checked first to see if it is the proper PROM and if it is installed properly. If it is, the PROM should be removed from the faulty ECM and intalled in the new-service ECM. Manufacturers have designed a series of codes that are displayed on the dashboard whenever trouble occurs in the computer system. For example, a code 51 indicates an error in the PROM.

2. Make certain that all wiring harness connectors are properly connected on any computer system engine.

3. The electronic computers on the automobile continually monitor the operating conditions of the fuel and other systems. When a problem is detected by the computer, a two digit numerical "trouble code" will be displayed on the dashboard. These codes are shown as "Service Now" or "Service Soon" warnings on the dashboard. The computer will store the trouble code associated with the detected failure until the diagnostic system is cleared or until the ignition switch has gone through a certain number of on-off cycles. Depending upon the vehicle and the year, twenty or more cycles may be needed.

4. Whenever checking CCC (computer control combustion) or DECS (diesel electronic control system), determine if the system is in an open-loop or closed-loop cycle.
   a. *Closed loop* means that the computer continually changes the ignition timing and air-fuel ratio as it receives different inputs. The engine normally runs in a closed-loop mode.
   b. *Open loop* means the computer is providing a constant air-fuel ratio. This is usually done during starting, warm-up, and hard acceleration. During open-loop operation, there is usually a rich air-fuel mixture. After the engine warms up, the computer switches to a closed-loop mode.

5. On certain computer systems, a "Check Engine" self-diagnostic light system is used for diagnosis. By grounding a "test" terminal, called the ALCL (assembly line communication link), that is located under the instrument panel, the "Check Engine" light will start to flash. The flashes indicate a code. These codes lead the technician to different charts for quick and accurate diagnosis. The trouble code "test" lead is located in a 12-volt terminal con-

nector under the dash. To ground the terminal, connect a jumper wire between the "test" lead terminal and the adjacent ground terminal. *Figure 16–43* shows an example of the many codes that are used to diagnose computer systems. Note that each manufacturer has a different method of displaying codes. The codes and the circuits affected are also different with each manufacturer.

6. Use diagnostic sheet "trees" to determine the sequence of checks on any computer-controlled system. *Figure 16–44* shows how to use the diagnostic chart to check a typical code.

7. While checking computer systems, use a digital-readout voltmeter to check voltages at various points, based on the manufacturer's specifications.

8. Certain manufacturers use a *pinpoint test* along with a set of codes. Voltmeter readings are taken at different points and compared to the manufacturer's specifications. These readings can be found in many of the automotive service and repair manuals. Since there are so many variations, it is important to follow the exact procedure in the manual for the particular year and make of the vehicle.

9. Inspect all *fuel filters* for dirt or any damage.

10. Because of the complexity of fuel injection systems, follow the manufacturer's recommendations pertaining to assembly and disassembly of fuel injection components. These are listed in the service manuals. Fuel injectors for gasoline engines can usually be tested for flow rate and spray pattern. Refer to the manufacturer's service manual for instructions on cleaning, testing, and servicing injectors.

11. On high-pressure diesel fuel injection pumps, inspect the rollers and cam for signs of wear. Replace, if necessary.

12. Certain companies manufacture special diagnostic instruments for checking CCC systems. *Figure 16–45* shows such an instrument. These test instruments can be connected directly into the computer system (ALCL) to make rapid and accurate diagnostic checks. These hand-held testers are also known as scanners. They are manufactured by several companies.

13. Inspect all intake manifold vacuum *hose connections* for leaks. The service and repair manuals have drawings of each vehicle, the model year, and the hose connections. *Figure 16–46* shows an example of the vacuum hose routing for a specific type of vehicle.

CODE	CIRCUIT AFFECTED
■■ 12	NO DISTRIBUTOR (TACH) SIGNAL
□ 13	O$_2$ SENSOR NOT READY
□ 14	SHORTED COOLANT SENSOR CIRCUIT
□ 15	OPEN COOLANT SENSOR CIRCUIT
■■ 16	GENERATOR VOLTAGE OUT OF RANGE
□ 18	OPEN CRANK SIGNAL CIRCUIT
□ 19	SHORTED FUEL PUMP CIRCUIT
■■ 20	OPEN FUEL PUMP CIRCUIT
□ 21	SHORTED THROTTLE POSITION SENSOR CIRCUIT
□ 22	OPEN THROTTLE POSITION SENSOR CIRCUIT
□ 23	EST/BYPASS CIRCUIT PROBLEM
□ 24	SPEED SENSOR CIRCUIT PROBLEM
□ 26	SHORTED THROTTLE SWITCH CIRCUIT
□ 27	OPEN THROTTLE SWITCH CIRCUIT
□ 28	OPEN FOURTH GEAR CIRCUIT
□ 29	SHORTED FOURTH GEAR CIRCUIT
□ 30	ISC CIRCUIT PROBLEM
■■ 31	SHORTED MAP SENSOR CIRCUIT
■■ 32	OPEN MAP SENSOR CIRCUIT
■■ 33	MAP/BARO SENSOR CORRELATION
■■ 34	MAP SIGNAL TOO HIGH
□ 35	SHORTED BARO SENSOR CIRCUIT
□ 36	OPEN BARO SENSOR CIRCUIT
□ 37	SHORTED MAT SENSOR CIRCUIT
□ 38	OPEN MAT SENSOR CIRCUIT
□ 39	TCC ENGAGEMENT PROBLEM
■■ 44	LEAN EXHAUST SIGNAL
■■ 45	RICH EXHAUST SIGNAL
■■ 51	PROM ERROR INDICATOR
▼ 52	ECM MEMORY RESET INDICATOR
▼ 53	DISTRIBUTOR SIGNAL INTERRUPT
▼ 60	TRANSMISSION NOT IN DRIVE
▼ 63	CAR AND SET SPEED TOLERANCE EXCEEDED
▼ 64	CAR ACCELERATION EXCEEDS MAX. LIMIT
▼ 65	COOLANT TEMPERATURE EXCEEDS MAX. LIMIT
▼ 66	ENGINE RPM EXCEEDS MAXIMUM LIMIT
▼ 67	SHORTED SET OR RESUME CIRCUIT
.7.0	SYSTEM READY FOR FURTHER TESTS
.7.1	CRUISE CONTROL BRAKE CIRCUIT TEST
.7.2	THROTTLE SWITCH CIRCUIT TEST
.7.3	DRIVE (ADL) CIRCUIT TEST
.7.4	REVERSE CIRCUIT TEST
.7.5	CRUISE ON/OFF CIRCUIT TEST
.7.6	"SET/COAST" CIRCUIT TEST
.7.7	"RESUME/ACCELERATION" CIRCUIT TEST
.7.8	"INSTANT/AVERAGE" CIRCUIT TEST
.7.9	"RESET" CIRCUIT TEST
.8.0	A/C CLUTCH CIRCUIT TEST
-1.8.8	DISPLAY CHECK
.9.0	SYSTEM READY TO DISPLAY ENGINE DATA
.9.5	SYSTEM READY FOR OUTPUT CYCLING OR IN FIXED SPARK MODE
.9.6	OUTPUT CYCLING
.0.0	ALL DIANOSTICS COMPLETE
■■	TURNS ON "SERVICE NOW" LIGHT
□	TURNS ON "SERVICE SOON" LIGHT
▼	DOES NOT TURN ON ANY TELLTALE LIGHT

NOTE: CRUISE IS DISENGAGED WITH ANY "SERVICE NOW" LIGHT OR WITH CODES 60-67.

FIGURE 16–43. Diagnostic codes are used on computer-controlled systems to aid in diagnosing problems. Each manufacturer has a different set of codes. For example, if code 13 is displayed, there may be trouble in the oxygen sensor. Code 32 shows that there is an open in the MAP sensor circuit. *(Courtesy of Motor Publications, Auto Engine Tune-Up and Electronics Manual, 1986)*

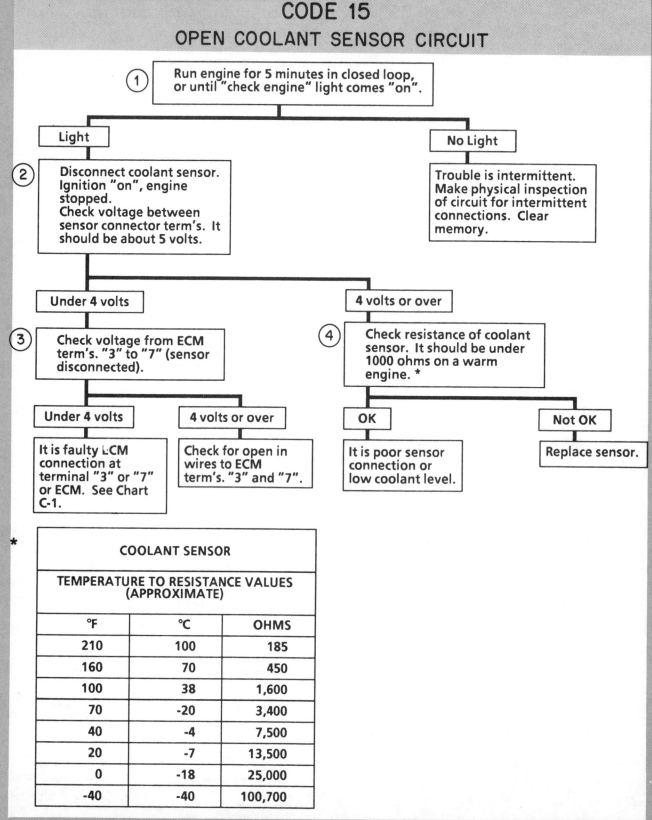

FIGURE 16–44. Codes are used to diagnose different ECM systems. Here Code 15 indicates trouble in the coolant sensor system. Diagnostic charts are then used to locate the exact problem. *(Courtesy of General Motors Product Service Training)*

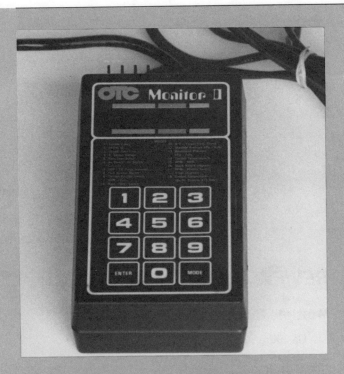

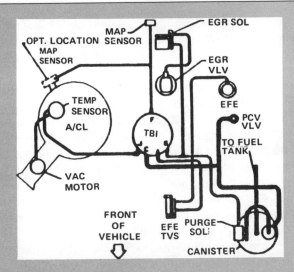

FIGURE 16–46. Always refer to the specification manuals for routing of vacuum hoses. Check vacuum circuits for leaks and for damage or broken parts. *(Courtesy of Motor Publications, Auto Engine Tune-Up & Electronics Manual, 1986)*

FIGURE 16–45. Various manufacturers sell test instruments used to diagnose computer-controlled systems. These instruments can save the technician valuable time and money. *(Courtesy of Owatonna Tool Company)*

SUMMARY

Fuel injection is the process of injecting fuel into the flow of air going into the engine. There are several types of fuel injection, including throttle body, port, and high-pressure diesel fuel injection. In addition, fuel injection systems are classified as either direct or indirect, timed or continuous, and port or throttle body injection.

As the demand for cleaner engines and better fuel economy has increased, automotive manufacturers have converted many fuel systems to fuel injection. The main goal of using fuel injection is to maintain a stoichiometric ratio of 14.7–1. However, there are also nonstoichiometric conditions such as when the engine is cold and during low manifold vacuum. Fuel injection is extremely precise compared to the standard carburetion systems.

Throttle body injection is designed so that an injector injects fuel directly into the air as it comes through the throttle body. It uses a throttle body injector that is controlled by a computer. The injector is able to produce a conical spray pattern when a solenoid lifts a metering valve. Fuel pressure is regulated by a fuel pressure regulator.

The throttle body injection system uses several other components for correct operation. A vacuum assist is used to monitor the vacuum. A signal is sent from the vacuum assist to the electronic control module to meter the right amount of fuel. A throttle position sensor is used to determine the exact position of the throttle. The idle air control is used to control engine rpm at idle. The manifold absolute pressure sensor signals the amount of manifold vacuum (absolute pressure) to the ECM. All of these components work in conjunction with the ECM to control the throttle body injector for all operations.

A second type of fuel injection is called port fuel injection. Here a small fuel injector is placed just before each intake valve. More accurate fuel control is obtained with this system than with the throttle body injection system. There are several types of port injection, including double-fire injection, sequential fuel injection, and tuned port injection.

The port fuel injector is a solenoid-operated valve with a pintle. Fuel is sent to the fuel injector through a fuel rail. The amount of air going into the intake manifold is controlled by the throttle body.

Several additional sensors are used in port fuel injection. The manifold air temperature sensor tells the ECM the temperature of the incoming air. Idle air control is used to control the amount of air bypassing the throttle blade during idle. The cold start injector is used to increase the amount of fuel during cold starting conditions. A thermal-time switch is used to tell the cold start injector when to operate. The mass air flow sensor tells the ECM the amount of air entering the engine. All of these components are integrated into the ECM to control the exact amount of fuel being delivered to the injectors.

High-pressure, direct, timed fuel injection is used on diesel engines. High pressure (4,000–10,000 psi) is needed to inject fuel into the cylinder during the compression stroke. High-pressure injection systems are designed to time, atomize, meter, and pressurize the fuel. The system consists of a drive shaft, housing, metering valve, transfer pump, distributor rotor, internal cam ring, automatic advance, pumping plungers, and a governor.

The transfer pump is used to draw the fuel from the fuel tank to the injection pump. Regulator valves are used to control the amount of pressure. The fuel is charged inside a rotor. As the rotor turns, a set of rollers and an internal cam ring are used to create the high pressure. The high pressure is sent to each piston through fuel lines. A delivery valve is used to assure that the fuel shuts off rapidly.

The governor is used to maintain correct speed by using a centrifugal weight force against a spring pressure. The balance of these two forces controls the position of the metering valve. An automatic advance is used on some injectors to advance the fuel injection time as rpm increases.

Fuel injector nozzles are used to inject the high-pressure fuel into the cylinder. A nozzle consists of high-pressure spring designed to hold a needle valve in the closed position. The high pressure from the fuel pump lifts the needle valve off its seat to inject the fuel.

Diesel engines are also incorporating computers for fuel control. The system is called the diesel electronic control system (DECS). Several sensors are used to control the fuel. A fast idle solenoid controls the amount of fuel at idle by controlling the governor spring tension with a solenoid. Advance timing is accomplished by rotating the cam ring. This is done by using two hydraulic pressures working against each other. Both altitude and cold starting conditions are improved by advance timing the fuel injection. This is done by reducing the housing pressure.

A series of codes is used to diagnose computer-controlled systems. The computer has built-in diagnostic checks that can also be made. When identifying a code, refer to a diagnostic chart. It is important to follow the chart to check voltages and other readings. These readings are then compared to the manufacturer's specifications.

TERMS TO KNOW

Can you explain each of the following terms? Review the chapter until you can use each term correctly.

Throttle body
Emission standards
Stoichiometric ratio
Intake manifold vacuum
Conical
Pulse width

Metered
Pintle
Tuned ports
Needle valve
Governor

REVIEW QUESTIONS

Multiple Choice

1. Gasoline fuel injection is considered:
   a. Indirect
   b. High pressure
   c. Direct
   d. Nonmetered
   e. Governed

2. Which type of fuel injection has one fuel injector feeding all pistons?
   a. Port fuel injection
   b. Direct fuel injection
   c. Throttle body injection
   d. All of the above
   e. None of the above

3. The correct stoichiometric ratio is:
   a. 12.2 to 1
   b. 14.7 to 1
   c. 16.7 to 1
   d. 18.1 to 1
   e. 19.2 to 1

4. Metering of throttle body fuel injection is controlled by the:
   a. Operator only
   b. Fuel pump
   c. Computer
   d. Cooling system
   e. Lubrication system

5. The fuel injector in the throttle body fuel injection system is opened and closed by using:
   a. Battery voltage
   b. A solenoid
   c. The operator's foot
   d. Mechanical linkage from the carburetor
   e. Lubricating oil pressure

6. The amount of fuel injection from a throttle body injector is measured by the:
   a. Pulse width
   b. Pulse frequency
   c. Size of the battery
   d. Temperature of the fuel
   e. Speed of the tires

7. The fuel pressure regulator in throttle body fuel injection is located on:
   a. The cooling thermostat
   b. The fuel pump
   c. The fuel rails
   d. The exhaust manifold
   e. The carburetor

8. The throttle position sensor is used as a/an _____ the ECM.
   a. Input to
   b. Output from
   c. Frequency sensor from
   d. Voltage from
   e. Speed sensor from

9. Which of the following units bypasses air around the throttle plate on the TBI system?
   a. Throttle position sensor
   b. Gauge pressure unit
   c. Manifold absolute pressure
   d. Idle air control
   e. Absolute pressure unit

10. Which of the following measures absolute pressure inside the manifold?
    a. MAF
    b. IAC
    c. MAP
    d. TPS
    e. None of the above

11. Which pressure scale starts at zero, when the atmospheric pressure is at 14.7 psi?
    a. Psig
    b. Psia
    c. CCC
    d. Port pressure
    e. None of the above

12. Which fuel system would eliminate fuel droplets forming inside the intake manifold due to the fuel being heavier than air?
    a. Throttle body
    b. Port
    c. Carburetors
    d. Central fuel system
    e. Controlled-combustion fuel system

13. Manifold air temperature is used as an _____.
    a. Input to the ECM
    b. Output from the ECM
    c. Input to the MAP
    d. Output from the MAP
    e. None of the above

14. The port fuel injector opens and closes because of the operation of a _____.
    a. Plunger
    b. Set of rollers
    c. Regulator
    d. Solenoid
    e. High-pressure diaphragm

15. The cold start injector on a port fuel injection system operates:
    a. Only during cold starting
    b. To add extra fuel during starting
    c. Based on the temperature of the coolant
    d. All of the above
    e. None of the above

16. The thermal-time switch operates along with the:
    a. Manifold air temperature
    b. Cold start fuel injector
    c. Mass air flow
    d. Carburetor
    e. High-pressure fuel injector

17. Which system has a sensing element that must be kept 75 degrees C above the temperature of the incoming air?
    a. Mass air temperature
    b. Manifold absolute pressure
    c. Idle air control
    d. Mass air flow
    e. None of the above

18. Diesel engines use injection systems that are classified as:
    a. Direct
    b. High-pressure
    c. Timed
    d. All of the above
    e. None of the above

19. Which of the following is not done on a diesel fuel injection system?
    a. Timed injection
    b. Atomized injection
    c. Pressurized injection
    d. Metered injection
    e. Produce continuous injection

20. High pressure is produced on a diesel fuel injection system by using a set of:
    a. Rollers inside a cam ring
    b. Bearings on a cam
    c. Plungers pushing outward
    d. Eccentric pumps
    e. None of the above

21. The purpose of the transfer pump on a diesel fuel injection system is to:
    a. Produce the high pressure
    b. Produce the governor action
    c. Produce the regulation action
    d. All of the above
    e. None of the above

22. Which item is used to measure and control the right amount of fuel in a diesel fuel injection system?
    a. Transfer pump
    b. Cam ring
    c. Plunger and rollers
    d. Metering valve
    e. Delivery valve

23. The delivery valve is used to control:
    a. The start of injection
    b. The middle of injection
    c. The end of injection
    d. The total pulse width
    e. Governor pressure

24. A governor is a device that is able to sense a change in:
    a. Load
    b. Speed
    c. Metering valve setting
    d. Spring forces
    e. Vehicle speed

25. When spring force is greater than centrifugal weight force on a governor, the fuel setting will:
    a. Increase
    b. Decrease
    c. Remain about the same
    d. Decrease and then immediately increase
    e. Increase and then immediately decrease

26. When the diesel fuel injection system is advanced, what object is moved or rotated?
    a. The transfer pump
    b. The delivery valve
    c. The cam ring
    d. The nozzle
    e. None of the above

27. The fuel nozzle is used to:
    a. Inject fuel into the combustion chamber
    b. Atomize the fuel in the combustion chamber
    c. Start and stop injection rapidly
    d. All of the above
    e. None of the above

28. Which of the following are used on the DECS?
    a. Altitude fuel limiter
    b. Housing pressure cold advance
    c. Housing pressure altitude advance
    d. All of the above
    e. None of the above

29. When diagnostic work is done on a computer-controlled system, a series of _____ are used.
    a. Codes
    b. Subsystems
    c. Voltage readings that are displayed on the dashboard
    d. Special instructions that are displayed on the dashboard
    e. None of the above

The following questions are similar in format to ASE (Automotive Service Excellence) test questions.

30. Technician A says when the air-fuel ratio is constant during starting, the computer system is in a closed-loop mode. Technician B says when the air-fuel ratio is constant during starting, the computer system is in an open-loop mode. Who is right?
   a. A only
   b. B only
   c. Both A and B
   d. Neither A nor B

31. Technician A says the PROM is a set of data that tells the fuel injector when to shut off. Technician B says the PROM is a speed sensor. Who is right?
   a. A only
   b. B only
   c. Both A and B
   d. Neither A nor B

32. The air-fuel ratio is being held constant. Technician A says the system is in a closed-loop mode. Technician B says the system is in an open-loop mode. Who is right?
   a. A only
   b. B only
   c. Both A and B
   d. Neither A nor B

33. Technician A says that the PROM is a set of programmable data for each specific car. Technician B says that the PROM is the same for all cars being manufactured. Who is right?
   a. A only
   b. B only
   c. Both A and B
   d. Neither A nor B

Essay

34. What is the difference between direct and indirect fuel injection?

35. What is the difference between port and throttle body injection?

36. Describe the difference between continuous and timed injection.

37. Describe the purpose and operation of the idle air control.

38. What is the purpose of the fuel rails?

39. Describe the purpose and operation of the mass air flow sensor.

40. What is the difference between diesel and gasoline fuel injection?

CHAPTER 17

Air Intake and Exhaust Systems

INTRODUCTION

For an engine to operate correctly, air must flow into and out of the engine without restriction. Air intake and exhaust systems are designed to clean the air coming in and reduce the noise coming out. This chapter deals with the components and operation of the air intake and exhaust systems.

OBJECTIVES

After reading this chapter, you will be able to:
- Analyze the use and operation of air filter systems.
- Define the use and operation of intake manifolds.
- State the purpose and operation of using turbochargers on gasoline and diesel engines.
- Define the operation of exhaust systems.
- Describe various diagnosis and service hints on the intake and exhaust systems.

CHAPTER HIGHLIGHTS

17.1 AIR INTAKE SYSTEMS
    A. Purpose of Air Filters
    B. Oil Bath Air Filters
    C. Dry-type Air Filters
    D. Intake Ducting
    E. Resonator
    F. Intake Manifolds
    G. Wet and Dry Manifolds
    H. Exhaust Crossover Intake Manifold

17.2 TURBOCHARGING
    A. Supercharging Defined
    B. Blower
    C. Turbocharging Principles
    D. Turbocharger Engine Changes
    E. Turbocharger Ducting

    F. Turbocharger Lag
    G. Turbocharger Wastegate
    H. Wastegate Control
    I. Turbocharger Construction

17.3 EXHAUST SYSTEMS
    A. Exhaust Manifold
    B. Heat Riser
    C. Vacuum-controlled Heat Valve
    D. Exhaust Piping
    E. Exhaust Muffler
    F. Resonator
    G. Tail Pipe

DIAGNOSIS AND SERVICE

SUMMARY

17.1 AIR INTAKE SYSTEMS

Purpose of Air Filters

The average gasoline engine brings in and exhausts approximately 10,000 gallons of air for every gallon of fuel consumed. The intake air must be clean. *Airborne* contaminants can shorten engine life or even cause premature failure. For example, the dirt that gets into the engine causes excessive wear on the rings, pistons, bearings, and valves. Depending upon the amount of dirt getting into the engine, its life can be shortened from 1/3 to 1/2.

As was indicated when studying fuel, the correct amount of air is very important for correct engine operation. Engines must breathe freely to provide maximum power. Air filters are used on every engine to trap contaminants, yet provide a free flow of air into the engine. Air filters that are dirty and not replaced can cause large restrictions to the air. This condition will cause the engine to run excessively rich. Fuel mileage can be substantially reduced by a dirty air cleaner. Exhaust emissions will also be increased.

The type of dirt and contaminants that enter the engine is determined by how and where the engine is used. The most common contaminants are leaves, insects, exhaust soot, dust, and road dirt. The geographic location of the vehicle also affects the amount of airborne contaminants. In dusty conditions, there may be more dirt than at high altitude conditions. All of these variables determine the exact type of filter needed and the degree of filtration.

Oil Bath Air Filters

The oil bath air cleaner has been used for many years on older automobiles. The oil bath air cleaner is shown in *Figure 17–1*. Here dirty air enters the air cleaner on the side. The air is drawn down into the cleaner and makes a 180 degree turn upward. At this point, the air must pass over a pool of oil. During this turn, some of the heavier dirt particles are thrown off into the oil. These dirt particles settle to the bottom of the air cleaner housing. As the air continues upward, it passes through an oil-wetted mesh. The remaining contaminants are trapped here and held inside the filter.

Oil bath cleaners depend upon high-velocity air moving through the oil. They are most efficient near the highest power requirements. This efficiency is about 95%. As the engine speed and load decrease, the air flow through the filter also decreases. The efficiency of the air cleaner may drop as much as 25% during lower speed operation.

The main advantage of this filter is that it can be cleaned without replacing any parts. In very dirty environments, this filter may be less expensive to operate and clean. The filter is cleaned with soap and water and new oil is added.

Dry-type Air Filters

The dry-type air filter is made of a paper element. *Figure 17–2* shows an example of a dry-type air filter. This filter permits air to flow into the engine with little resistance.

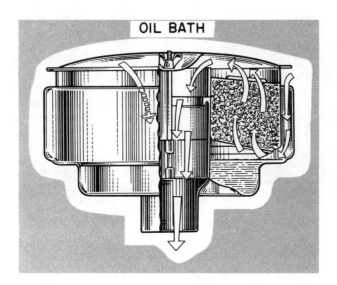

FIGURE 17–1. Oil bath air filters remove the dirt particles by forcing the air past an oil reservoir. *(Courtesy of Dana Corporation)*

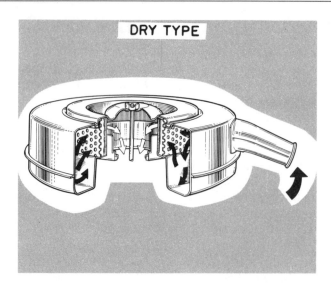

FIGURE 17-2. The dry-type air filter permits air flow into the engine with little restriction. (Courtesy of Dana Corporation)

However, it traps and holds contaminants inside the paper. When the dry-type air filter becomes plugged with dirt, it is replaced with another filter.

There are several types of dry-type air filters. The light-duty *paper-type* air cleaner is shown in **Figure 17-3**. These filters are generally used on passenger vehicles and small pickup trucks. The filter element is made of paper. These filters are usually small in size due to space restrictions under the hood. The efficiency of this type of filter is near 98% for most driving conditions. Because of the high efficiency, the dry air filter has replaced the oil bath air

cleaner. The filter can be designed in many shapes to fit different types of air cleaners. The design depends upon the type of ducting and housing used on the engine.

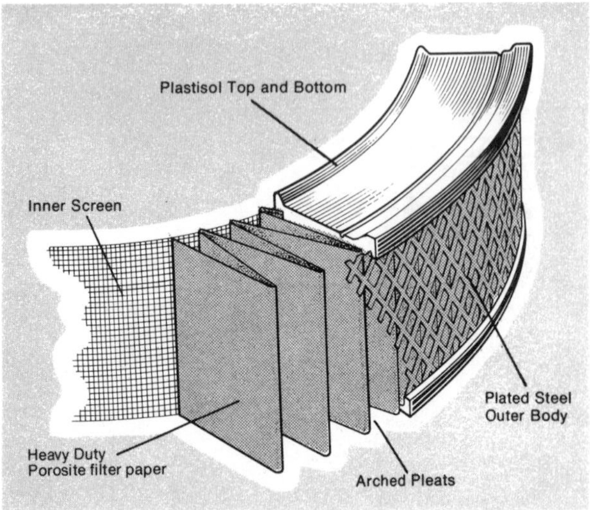

A second type of light-duty air cleaner is called the *polyurethane* filter. This filter is sometimes called the foam filter, **Figure 17-4**. It consists of a polyurethane wrapper stretched over a metal support. The material has thousands of pores and interconnecting strands that create a maze-like contaminant trap. It may be used dry or with a thin coat of oil. In both cases, this filter has about the same efficiency as a paper-type element. The advantage to this filter is that it can be removed, cleaned, and reused. Many polyurethane filters are used as *aftermarket* equipment.

There are also a series of heavy-duty filters used on certain equipment. Heavy-duty air cleaners are used in very dirty and contaminated areas. However, because the auto-

FIGURE 17-3. A paper-type filter is made of pleated paper, steel mesh, and seals on the top and bottom. (Courtesy of Dana Corporation)

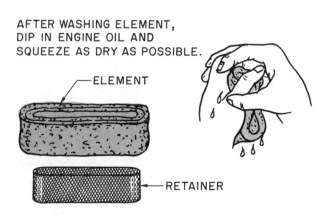

FIGURE 17-4. The polyurethane-wrapped filter stretches over a metal support.

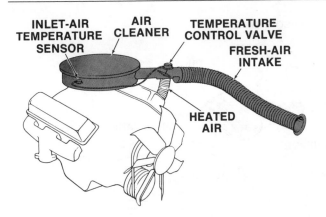

FIGURE 17–5. Intake ducting can take many forms. A simple type is shown with the air cleaner mounted directly on top of the carburetor. *(Courtesy of Chrysler Corporation)*

mobile is normally driven on paved roads, there is less call for heavy-duty air cleaners in cars.

Intake Ducting

On vehicles using carburetors, air cleaners are typically placed directly on top of the carburetor. *Figure 17–5* shows such an example. The air is drawn into the fresh-air intake, through a temperature-controlled valve, and into the air cleaner. From the air cleaner, the air is drawn directly into the venturi of the carburetor. (The temperature control valve is discussed in the chapter entitled "Pollution Control.")

When manufacturers started using fuel injection and other configurations on the automobile, the intake ducting was also changed. Intake ducting is also different on engines that are *turbocharged*. *Figure 17–6* shows two examples of different ducting. The air cleaners are light-duty paper elements. They are located for easy maintenance and service. The intake ducting has several clamps and rubber ducting to get the air into the engine.

Resonator

Certain engine applications use a *resonator* on the intake ducting. *Figure 17–7* shows a resonator used in a diesel engine application. The purpose of the resonator is to reduce induction noise produced on the intake system.

Intake Manifolds

The intake manifold is used to transfer or carry the air and/or fuel from the air cleaner to the intake valve. *Figure 17–8* shows a typical intake manifold for a V-8 cylinder diesel engine. On 4- and 6-cylinder engines, the intake and exhaust manifolds form an assembly.

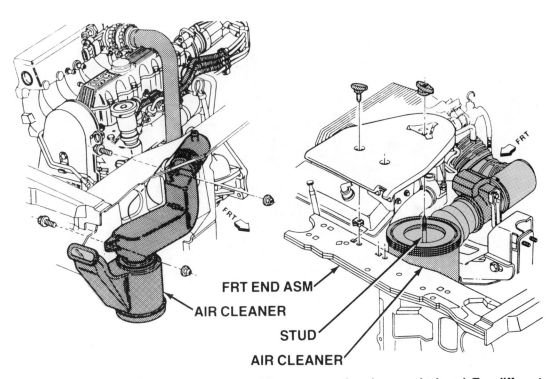

FIGURE 17–6. Intake ducting will change as different types of engines are designed. Two different ducting arrangements are shown. *(Courtesy of General Motors Product Service Training)*

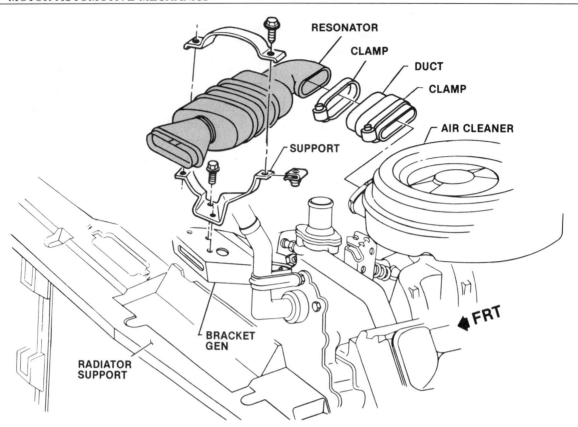

FIGURE 17–7. A resonator is used to reduce the intake noise. *(Courtesy of General Motors Product Service Training)*

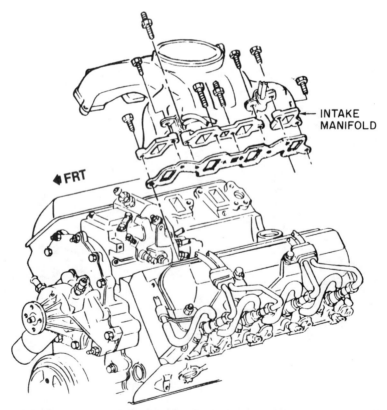

FIGURE 17–8. A typical V-8 manifold for a diesel engine. *(Courtesy of General Motors Corporation)*

The intake manifold is designed to deliver the right amount of air and fuel to each cylinder under all driving conditions. It would be best if all intake ports were the same length (tuned ports). However, on many engines with carburetors this design is compromised to reduce the cost. The most common manifolds are made of a one-piece casting of either cast iron or aluminum.

On four-cylinder engines, the intake manifold has either four *runners* or two runners that break into four near the intake manifold, *Figure 17–9*. On in-line six-cylinder engines, there are either six runners or three that branch off into six near the intake manifold. On V-configuration engines (V-6 and V-8), both open and closed intake manifolds are made. Open intake manifolds have an open space between the bottom of the manifold and the valve lifter valley. Closed intake manifolds act as the cover to the intake lifter valley.

Wet and Dry Manifolds

Manifolds can also be either wet or dry. Wet manifolds have coolant passages cast directly into the manifold. Dry manifolds do not have cooling passages.

Exhaust Crossover Intake Manifold

On some manifolds, there is an exhaust crossover passage. This passage allows the exhaust from one side of the engine to cross over through the intake manifold to the other side to be exhausted. The exhaust crossover provides heat to the base of the carburetor to improve the vaporization of the fuel while the engine is warming up. See *Figure 17–10*. The crossover also reduces carburetor icing.

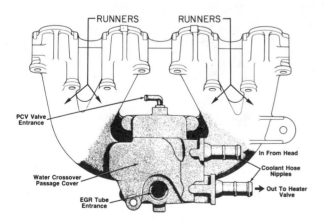

FIGURE 17–9. The intake manifold on an in-line four-cylinder engine can have two or four runners. Two runners are shown here. *(Courtesy of Chevrolet Division, General Motors Corporation)*

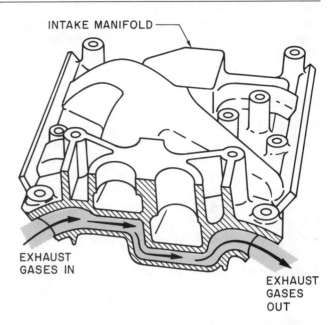

FIGURE 17–10. A crossover manifold is used to transfer the exhaust from one side of the engine to the other to be exhausted. The exhaust heat causes better vaporization of the fuel in the carburetor.

CAR CLINIC

PROBLEM: AIR INTAKE RESTRICTION

Customers often complain that the engine isn't running correctly. Technicians usually say that the problem may be in the air intake. What are the symptoms of an air intake restriction?

SOLUTION:

One of the most common problems today in all automobiles is air intake restrictions. If the owner does not change the air filter at regular intervals, the dirty air filter restricts the air as it tries to enter the engine. Common symptoms of an air intake restriction are:

1. The engine is hard to start.
2. The engine has a loss of power.
3. The coolant temperature decreases.
4. The exhaust smoke increases in density and is black in color.
5. The oil consumption and fuel usage increase significantly.

17.2 TURBOCHARGING

With the increased use of fuel injection and computer-controlled combustion systems, turbochargers have become common components on gasoline and diesel engines. In the past, only large engines had turbochargers. Today, with the precise control afforded by computers, turbochargers are making smaller engines more efficient and capable of producing more power. In order to study turbochargers, supercharging must first be defined.

Supercharging Defined

When the piston moves downward on the intake stroke, a vacuum is created. This vacuum causes air and fuel to be drawn into the engine. This process is called a *naturally aspirated* or normally aspirated engine. The amount of air entering the engine is based on atmospheric pressure. Most engines are considered to be naturally aspirated. However, with the increased use of smaller engines, there may be a lack of power for certain driving conditions.

To overcome this lack of power, an engine can be supercharged. *Supercharging* an engine means to deliver a greater volume of air to the cylinders than that delivered from the suction of the pistons alone. The engine is not naturally aspirated; it is supercharged. When more air is forced into the cylinders, there must be a corresponding increase in fuel. Fuel is needed to maintain a 14.7 to 1 air-fuel ratio. If these conditions occur, then a great increase in power will result. In some cases, up to 50% more power can be obtained by supercharging an engine.

Blower

Either a blower or turbocharger can be used to supercharge an engine. A blower is a mechanical air pump that forces air into the engine. It is driven by a set of gears or belts from the crankshaft. It produces a substantial frictional loss on the engine because it requires horsepower from the engine to operate. Blowers are used on certain heavy-duty diesel engines and on high-performance racing engines.

Turbocharging Principles

A turbocharger is a device that uses the exhaust gases, rather than engine power to turn an air pump or compressor. The air pump then forces an increased amount of air into the cylinders. Both diesel and gasoline engines in the automotive market use turbochargers. *Figure 17–11* shows a typical schematic of air and exhaust in a turbo-

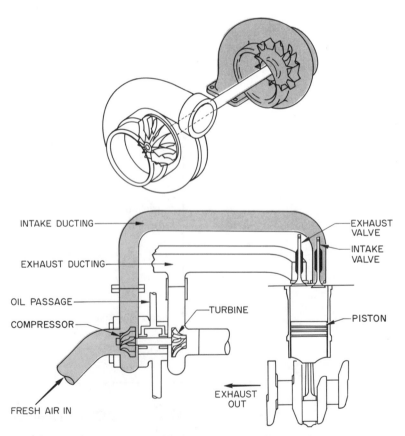

FIGURE 17–11. A turbocharger uses the exhaust gases to turn an air pump or compressor. The compressor turbine forces extra air into the engine. (Courtesy of Peugeot Motors of America, Inc.)

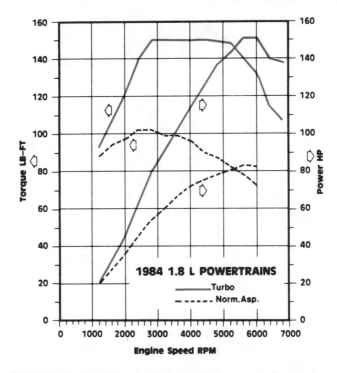

FIGURE 17–12. This chart shows the increase in power and torque when an engine is turbocharged. *(Courtesy of General Motors Product Service Training)*

charged engine. High-velocity exhaust gases pass out of the exhaust ports. From there they pass through a *turbine-*driven pump. Here the exhaust gases cause the exhaust turbine to turn very rapidly. The exhaust turbine causes the intake *compressor* to turn very rapidly also. As the compressor turbine turns, it draws in a large amount of fresh air. The intake air is pressurized and forced into the intake port. The increase in pressure in the intake manifold is called *boost*. Boost may produce pressure in the intake manifold of about 6–10 psi or more, depending upon the manufacturer.

If a corresponding amount of fuel is added, a large increase in power will result. ***Figure 17–12*** shows a chart that compares the difference between a turbocharged and normally aspirated engine. Note that both the torque and horsepower have been increased at all rpm. For example, at 5,000 rpm the normally aspirated engine produces about 80 hp. At this rpm, the turbocharged engine can produce about 140 hp. ***Figure 17–13*** shows a turbocharger used on an automobile.

Turbocharged Engine Changes

A few internal changes are necessary on turbocharged engines. These include strengthening the pistons, using different piston rings, and making sure the bearings can withstand the extra load. However, turbocharged and non-

turbocharged engines share the same compression ratio and emission control devices.

Turbocharger Ducting

The inlet ducting is changed when a turbocharger is used. See ***Figure 17–14***. The exhaust gases pass through the

FIGURE 17–13. A complete turbocharger unit for an automobile.

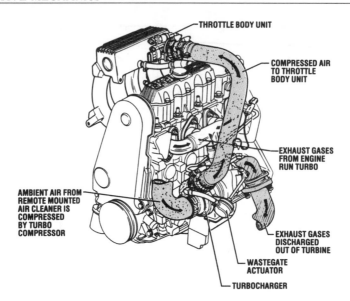

FIGURE 17–14. When using a turbocharger, the intake ducting will be different. Here the ducting is designed so the turbocharger can be located on the lower side of the engine. *(Courtesy of General Motors Product Service Training)*

exhaust manifold and into the exhaust turbine on the turbocharger. From the exhaust turbine, the exhaust gases are sent through the exhaust system into the environment. As the compressor turbine turns, air is drawn through the air cleaner into the intake manifold. The pressurized air is sent through ducting to the throttle body unit. Here the air and fuel are mixed in the correct proportion.

Turbocharger Lag

One problem associated with a turbocharged engine is called *lag*. Lag is defined as the time it takes for the turbocharger to increase the power. It is the delay between a rapid throttle opening and the delivery of increased boost. There is a lag between the time the operator calls for the extra power and the actual power produced. This is because it takes time for the turbine speed to increase and produce the necessary power. Turbochargers on automobiles operate at about 10,000 rpm at idle. They run most efficiently at about 100,000 to 150,000 rpm under maximum boost. It takes time for the turbocharger to increase to this speed for best efficiency.

Turbocharger Wastegate

A wastegate is connected to the turbocharger. The wastegate is used to bypass the exhaust gases when the turbocharger boost is too high. Too much pressure or too high a boost may

cause excessive detonation, engine damage, and/or destruction. The wastegate causes the exhaust gases to bypass the exhaust turbine. When this happens, there is less power turning the compressor turbine; thus, the turbocharger action is reduced. *Figure 17–15* shows a typical wastegate flow diagram. When the wastegate is closed, exhaust gases pass through the exhaust turbine. The engine is now being turbocharged. When the wastegate is opened, the exhaust gases bypass the turbine.

Wastegate Control

The wastegate is normally closed. It opens to bypass exhaust gases to prevent an overboost condition. The wastegate opens when pressure is applied to the *actuator*. The actuator is controlled by a wastegate control valve which is pulsed on and off by the ECM. See *Figure 17–16*. Under normal driving conditions, the control solenoid is energized 100% of the time. This means the exhaust gases pass through the exhaust turbine. During rapid acceleration, there may be an increase in boost pressure. As the boost increases, it is sensed by the MAP (manifold absolute pressure) sensor. The ECM now pulses the wastegate control valve above a boost of 15 psi. With the wastegate pulsing on and off, the manifold pressure decreases. If an overboost condition does occur, the ECM will also reduce the fuel delivery. *Figure 17–17* shows the wastegate actuator attached to the turbocharger.

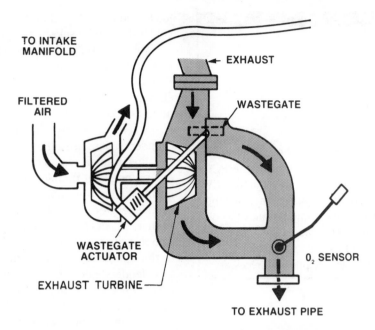

FIGURE 17–15. A wastegate is used to bypass the exhaust gases during times of high boost. *(Courtesy of General Motors Product Service Training)*

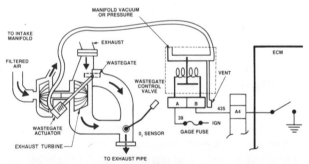

FIGURE 17–16. The wastegate is controlled by a wastegate actuator and manifold vacuum along with the electronic control module (ECM). *(Courtesy of General Motors Product Service Training)*

Turbocharger Construction

Figure 17–18 shows a complete turbocharger. The shafts on the intake and exhaust turbines are connected together. The intake turbine is designed to act as a centrifugal compressor. The exhaust turbine acts as a fan, causing the shaft to turn. A housing surrounds both turbines. Flanges are attached to the housing for mounting.

FIGURE 17–17. The wastegate is controlled by the wastegate actuator which is located on the turbocharger.

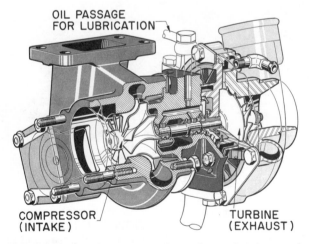

FIGURE 17–18. A cutaway view of a turbocharger is shown. Note the location of the oil passages for lubrication. *(Courtesy of Peugeot Motors of America, Inc.)*

CAR CLINIC ▰▰▰▰▰

PROBLEM: OILY AIR CLEANER

A customer says that the air cleaner is getting oil in it. The filter must be replaced about every 1,500 miles. What could be the problem?

SOLUTION:

The oil gets into the air cleaner via the PCV system. Crankcase vapors are normally sent through the PCV valve to the carburetor to be burned. However, if the PCV valve is plugged or blocked, the engine cannot breathe correctly. Crankcase vapors now come through the oil filter cap into the air cleaner. This causes the filter to be soaked with oil. Check the PCV valve by shaking it. It should rattle if it is not plugged. Replace, if necessary.

17.3 EXHAUST SYSTEMS

The exhaust system collects the high-temperature gases from each combustion chamber and sends them to the rear of the vehicle to be *dispersed*. An exhaust manifold, heat riser, mufflers, and pipes are used to accomplish this. A catalytic converter is also used. Catalytic converters are discussed in the chapter on pollution control.

Exhaust Manifold

The exhaust manifold is connected to the cylinder head of the engine. The exhaust gases from the exhaust valves pass directly into the exhaust manifold. The exhaust manifold is made of cast iron that can withstand rapid increases in temperature and expansion. See *Figure 17–19*. Under full-

load conditions the exhaust manifold may be red hot, yet cold water can be splashed on the manifold while driving.

Volumetric efficiency was defined in an earlier chapter as the efficiency of air moving in and out of the engine. Exhaust manifolds can be designed to improve volumetric efficiency. This can be done by designing the manifolds with more or less restriction.

Several types of exhaust manifolds are used. Four-cylinder engines use either three- or four-runner manifolds. On three-runner manifolds, the two center cylinders feed one runner. Four-runner manifolds have a runner for each cylinder. This will improve volumetric efficiency. On six-cylinder engines, the exhaust manifold is either a four or six runner. Again, the four-runner manifold has two of the center cylinders feeding one runner. On V-6 and V-8 engines, there is an exhaust manifold on each side.

Headers may be used on certain racing engines. Headers are welded steel tubing used for exhaust. They are designed to allow a smooth, even flow of exhaust gases out of the engine. This assures that each cylinder has equal exhaust back pressure. It also assures that each cylinder is completely cleaned of exhaust (scavenged). It has been found that headers improve only high-speed and load performance. They have a small effect on normal driving performance.

Heat Riser

Most engines manufactured have a type of heat riser attached to the exhaust manifold. The heat riser is a valve. Its purpose is to restrict the exhaust gases during starting and warm-up periods. This restriction tends to increase the engine to operating temperature quicker by aiding in vaporization of fuel. On in-line engines, the heat riser also helps to vaporize the fuel during cold starting.

FIGURE 17–19. The exhaust manifold is used to collect the exhaust gases and send them to the exhaust pipe.

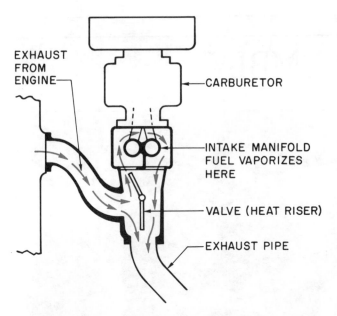

FIGURE 17–20. A heat riser is used to increase vaporization of fuel during starting. During cold starting, exhaust gases are routed internally to increase the temperature at the base of the carburetor.

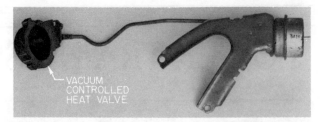

FIGURE 17–21. A vacuum-controlled heat valve keeps the riser closed until the temperature of the engine coolant is high. The vacuum is then removed and the valve opens.

The heat riser is controlled by a flat spring. When the engine is cold, the spring and a counterweight cause a valve in the exhaust manifold to close. As the spring heats up, it relaxes. This causes the counterweight to open the valve and allow normal exhaust. *Figure 17–20* shows how the exhaust gases are routed to improve vaporization.

Vacuum-controlled Heat Valve

On V-8 engines, the manifold heat control valve is located on one bank of the exhaust manifold. When the engine is cold, the heat control valve closes and directs some of the hot gases up through the intake manifold to the other side of the engine (crossover intake manifold). When the engine is warm, the valve opens and exhaust gases pass out each side normally. The valve is actuated by a vacuum-controlled motor. As the temperature of the engine increases, vacuum is removed from the motor and the heat valve opens. See *Figure 17–21*.

Exhaust Piping

The exhaust pipe is the connecting pipe between the exhaust manifold and the muffler or catalytic converter.

Many types of exhaust piping are used on vehicles. The shape depends upon the configuration of the engine, size of the engine, and undercarriage of the car. Exhaust piping can also be single or dual design.

Exhaust Muffler

The muffler is used to dampen the exhaust sound of the engine. Two types are primarily used. One uses a series of baffled chambers to reduce the sound. The other uses a perforated straight pipe enclosed in fiberglass and a shell. The straight pipe, which is also called a "glass pack," reduces exhaust back pressure, but it does not reduce the sound as much as the baffle type. Glass packs are illegal in most states because they alter emissions considerably.

Resonator

A resonator is another type of muffler. Most of the noise from an exhaust system is sound vibration. These vibrations cause louder noise. Resonators provide additional sound protection at critical points in the exhaust flow. They are used to absorb excessive sound vibration.

Tail Pipe

The tail pipe is a tube that is used to carry the exhaust gases from the muffler or resonator to the rear of the automobile. Many shapes and sizes are used, depending upon the vehicle. The tail pipe is supported by a series of hangers that allow the exhaust system to flex and move during driving. Rubber connectors help isolate the vibration from the rest of the vehicle.

POWER TRAIN ASSEMBLY

The assembly of an automobile is a highly complex and sophisticated process. This photo shows the entire power train assembly being installed on the underside of the vehicle. (Courtesy of Pontiac Motor Division, General Motors Corporation)

AUTOMATED MANUFACTURING

This photo shows the milling and drilling machine used on the Pontiac Fiero. This manufacturing process automatically completes the sequence of operations shown. Because of manufacturing processes such as these, cars can be manufactured with more efficiency and less cost. (Courtesy of Pontiac Motor Division, General Motors Corporation)

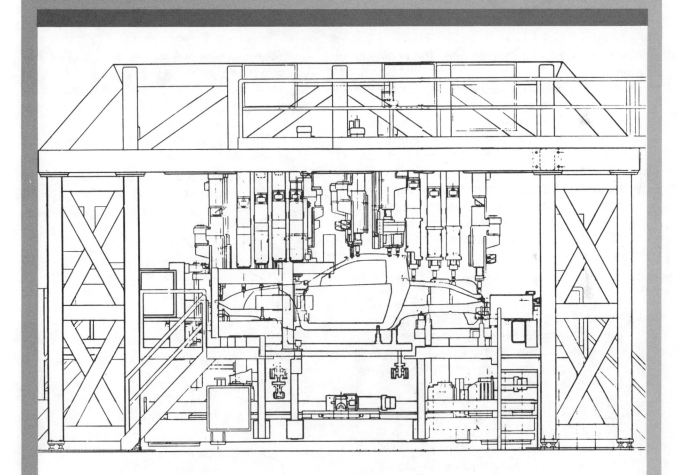

Sequence of Operations

1 Automatic on-line load
2 Clamp body at eight locations
3 Position body per LVDT readings
4 Set back-ups at front, rear, top
5 Verify position for drilling
6 Pierce four slots

7 Drill 39 vertical holes
8 Machine 39 horizontal surfaces
9 Verify operations completed
10 Unclamp and release back-ups
11 Automatic unload to conveyor line

DIAGNOSIS and SERVICE

1. *Replace air cleaners* every 10,000 miles or at tune-up time. More frequent replacement may be necessary in dusty conditions.
   a. Remove the air cleaner cover by removing the wing nut or other housing.
   b. Remove the air cleaner element.
   c. Inspect the housing and parts for oil and dirt, and clean if necessary.
   d. Replace with a new air cleaner element.
2. Clean the air cleaner housing when replacing the filter. All dirt and grease should be removed from the housing using a solvent that is not flammable.
3. Use soap and water to clean an oil bath air cleaner.
   a. Remove the air cleaner element from the housing.
   b. Using a clean solution of soap and water, clean the dirt from the element.
   c. Dry the air cleaner element and replace it on the engine.
4. Excessive black smoke may indicate a dirty air cleaner. A dirty air cleaner restricts the amount of air that can enter the engine. Less air causes a very rich air-fuel ratio to occur. The rich mixture produces excessive black smoke.

5. Tighten all wing nuts securely when replacing air cleaners. Be careful not to tighten the wing nut too much, as the carburetor may be damaged.
6. Use a light coat of oil on the polyurethane filter. The light coat of oil will help to capture more of the dirt and dust particles that flow through the air cleaner element.
7. Inspect all *intake ducting* for *cracks* or leaks, and replace where necessary. Any leak that occurs in the air inlet may also bring in dirty, unfiltered air. This dirty air can damage the engine.
8. The fresh air duct shown in **Figure 17–22** is designed to deliver cool air to the engine when the temperature of the air under the hood is hot. Bringing in hot air to the engine may cause starting or drivability problems such as vapor lock or hesitation. Check the duct by unclamping and removing it. Bend the duct slightly to inspect between the bellows for tears or rips.

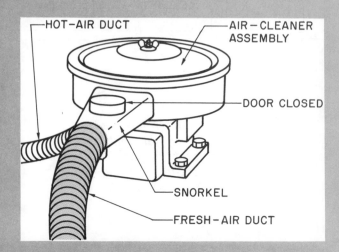

FIGURE 17–22. The fresh air duct should be checked for cracks. This is done by removing the duct and slightly bending it to observe cracks and damaged sections.

9. Always use the filter recommended by the manufacturer.

10. Whenever an intake manifold is removed, always replace the gaskets.

11. Engine hesitation and poor fuel economy are two symptoms of a plugged intake-manifold exhaust crossover passage. (The exhaust heat aids in fuel vaporization.) To check for blockage:
   a. Start the engine cold.
   b. Let the engine idle for several minutes.
   c. Carefully touch the manifold crossover area with your finger.
   d. If it is warm or cool, the passage is plugged. If it is not plugged, the crossover area should be very hot. See *Figure 17–23*.

12. *CAUTION*: *Whenever working on turbochargers, care must be taken to avoid contact with the hot exhaust pipes*. On turbocharged engines, these exhaust pipes may be positioned high in the engine compartment.

13. When *troubleshooting a turbocharged engine*, the following suggestions should be considered:
   a. Be aware of different sounds made by the turbocharger. Different noise levels during operation may signal air restrictions or dirt built up in the compressor housing.

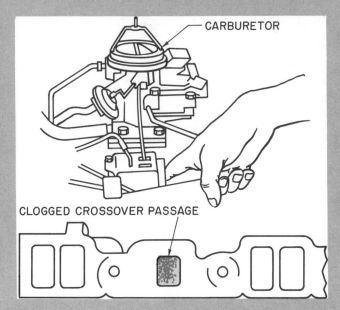

FIGURE 17–23. The exhaust crossover passage may become plugged. To check it, touch your finger to the intake manifold where the crossover passage is located. If it is cold (when the engine is hot), the passage is plugged.

b. A lack of power in the engine could be caused by a restriction, air leak, restricted exhaust, dirt in the compressor housing, or incorrect timing.
   c. Black exhaust smoke can be caused by a restricted air inlet, dirt in the compressor housing, or an exhaust gas leak to the input of the turbocharger.
   d. Other problems that are also associated with turbochargers include excessive oil consumption, blue exhaust smoke, noisy turbocharger, and turbocharger seals leaking.

14. Never shut a turbocharged engine off immediately after running it at high load and speed. Let the engine idle for a short period of time to cool the turbocharger.

15. On a turbocharged system, both overboost and underboost conditions can occur. *Overboost* can be caused by:
   a. A sticking wastegate or wastegate actuator.
   b. A control valve stuck in the closed position.
   c. A cut or pinched vacuum hose.
   d. A fault within the ECM.

   Underboost can be caused by:
   a. The wastegate sticking open.
   b. The control valve sticking open.
   c. A faulty ECM unit.

16. The wastegate control valve can be checked by using a vacuum tester. For example, the actuator should begin to move at 4 psi and obtain full travel at 15 psi. Readings will be determined by manufacturer recommendations.

17. Always follow the manufacturer's diagnostic procedures when checking the electrical ECM with the turbocharger. These procedures are listed in service and repair manuals.

18. A common failure for turbochargers is a lack of lubrication and/or oil. This occurs when oil pressure is insufficient to lubricate the internal bearings before the turbocharger reaches high speeds. The turbocharger bearings need more oil as the speed of the engine and the load increase. Lack of oil will usually cause the turbocharger to fail.

19. Check the exhaust system for any leaks or dents in the system that may restrict the flow of gases. Never run the engine in an unventilated area.

20. Exhaust manifolds can crack. Check the exhaust manifold for cracks around each runner. If a crack is found, install a new exhaust manifold.

21. Check the *heat riser* for movement. The heat valve should be able to move freely. On some engines, the heat riser has rusted closed and may cause poor volumetric efficiency, longer warm-ups, and hesitation when cold.

22. Replace any exhaust or tail pipes when broken or when leaks occur.

23. Many exhaust systems are supported by free-hanging rubber mountings. This permits some movement of the exhaust system, but it does not permit transfer of noise into the passenger compartment. Annoying rattles and noise vibrations in the exhaust system are usually caused by misalignment of parts. Loosen all bolts and nuts. Realign the exhaust system parts. Working from the front to the rear of the car, tighten each part.

24. The catalyst in catalytic converters is usually not serviceable. The bottom cover on some catalytic converters can be replaced, primarily to replace the insulation. Follow the manufacturer's procedure.

25. Exhaust restriction can be checked by checking the exhaust back pressure. Each manufacturer has a different procedure for checking exhaust back pressure. Typically, if the exhaust back pressure is above 2 psi, an exhaust restriction is indicated. Of course, the back pressure specification will change with different vehicles and manufacturers.

SUMMARY

This chapter dealt with the design and components of the air intake and exhaust systems. It included study in air filters, intake ducting, turbochargers, wastegates, exhaust manifolds, and heat risers.

Air filters are designed to keep dirt and other contaminants out of the engine. To do this, an oil bath or dry-type air filter can be used. The dry type is the most common. Dry-type filters are highly efficient, but they cannot be cleaned. They must be replaced. There are several types of dry filters, including the paper type and the polyurethane type.

The intake ducting is used to transfer the air from outside the engine to the filter and finally to the engine. Many designs are used, depending upon the configuration of the engine and air cleaner. Some intake ducting also includes a resonator to reduce the intake noise.

The intake manifold is used to transfer the air and/or fuel from the air cleaner to the intake valve. Several designs are used. The runner on the intake manifold is the port that carries the flow of air and/or fuel. Manifolds can also be wet or dry. Wet manifolds have coolant passing through them.

Turbochargers are being used more on smaller engines. In order to understand turbocharging, supercharging must first be defined. Supercharging forces air into the cylinder. The engine is not naturally aspirated. A turbocharger supercharges an engine by using hot exhaust gases to turn a compressor turbine. The compressor turbine compresses the air to get a boost in intake pressure.

Turbocharged engines must use stronger pistons and different rings, and make sure bearings can withstand the extra load. Ducting on turbocharged engines will also change.

The turbocharger wastegate is used to open a bypass tube. This tube is used so that if the boost pressure gets too high, the exhaust gases will bypass the turbocharger turbine. This gives the turbocharger protection under rapid acceleration. The wastegate is controlled in conjunction with the ECM.

The exhaust system is used to collect and carry the exhaust gases to the rear of the vehicle to be dispersed. The exhaust manifold collects the exhaust from each cylinder. The exhaust pipe transfers the gases to the catalytic converter and muffler. The tail pipe transfers the exhaust gases from the muffler to the atmosphere near the rear of the vehicle.

TERMS TO KNOW

Can you explain each of the following terms? Review the chapter until you can use each term correctly.

Airborne	Supercharging
Aftermarket	Turbine
Turbocharged	Compressor
Resonator	Actuator
Runner	Disperse
Naturally aspirated	Header

REVIEW QUESTIONS

Multiple Choice

1. If dirt gets into the engine, the engine life may be:
   a. Increased by 1/2
   b. Decreased by 1/3
   c. Affected very little — nothing to worry about
   d. Increased by 1/3
   e. None of the above

2. A dirty air cleaner will produce:
   a. White exhaust smoke
   b. More power in the engine
   c. Black smoke
   d. A rich mixture
   e. C and d

3. What is the efficiency of an oil bath air cleaner at maximum speed?
   a. 25%
   b. 50%
   c. 75%
   d. 80%
   e. 95%

4. One of the disadvantages of a paper-type air cleaner is that:
   a. It must be cleaned with oil
   b. It must be replaced and cannot be cleaned
   c. It usually doesn't fit correctly
   d. It must be replaced every 5,000 miles
   e. It cannot be removed

5. Which of the following is a type of dry air cleaner?
   a. Paper
   b. Polyurethane
   c. Runner
   d. Timed
   e. A and b

6. Which type of air cleaner can be cleaned?
   a. Paper
   b. Polyurethane
   c. Runner
   d. Timed
   e. A and b

7. Which of the following is used primarily to reduce noise on the intake system?
   a. Intake manifold
   b. Intake ducting
   c. Intake resonator
   d. Intake filter
   e. Catalytic converter

8. How many runners may an intake manifold for a four-cylinder engine have?
   a. One
   b. Two
   c. Three
   d. Four
   e. B and d

9. An engine that uses atmospheric pressure to force the air into the engine is called a _____ engine.
   a. Supercharged
   b. Turbocharged
   c. Naturally aspirated
   d. Blown
   e. None of the above

10. Which type of component uses exhaust gases to turn a turbine that forces air into the cylinder?
    a. Blower
    b. Wastegate
    c. Turbocharger
    d. Intake manifold
    e. Runner

11. The wastegate bypasses exhaust gases when:
    a. Turbine speed is low
    b. Turbine speed is high
    c. Too much fuel is added
    d. Too much boost is sensed by the MAP
    e. The load is removed

12. On a turbocharged engine:
    a. The pistons must be strengthened
    b. The rings must be strengthened
    c. The bearings must be strengthened
    d. All of the above
    e. None of the above

13. On a turbocharged engine, the delay between a rapid throttle opening and the delivery of boost is called:
    a. Turbocharger efficiency
    b. Volumetric efficiency
    c. Turbocharger lag
    d. Wastegate control
    e. Speed control

14. Steel tubes of the same length that are welded into an exhaust manifold are called:
    a. Turbocharging
    b. Headers
    c. Exhaust resonators
    d. All of the above
    e. None of the above

15. Which device is used to block off the flow of exhaust gases to improve vaporization during cold starting?
    a. Wastegate
    b. Blower
    c. Turbocharger
    d. Intake actuator
    e. Heat riser

16. After approximately how many miles should the air cleaner be replaced?
    a. 4,000
    b. 10,000
    c. 50,000
    d. 80,000
    e. 100,000

The following questions are similar in format to ASE (Automotive Service Excellence) test questions.

17. Technician A says that an exhaust restriction can be checked by testing for exhaust back pressure. Technician B says that an exhaust restriction can be checked by observing the timing. Who is right?
    a. A only
    b. B only
    c. Both A and B
    d. Neither A nor B

18. Technician A says that all air cleaners can be cleaned with gas and replaced. Technician B says that there are no air cleaners on older or newer vehicles that can be cleaned. Who is right?
    a. A only
    b. B only
    c. Both A and B
    d. Neither A nor B

19. A turbocharged engine is exhausting black smoke. Technician A says it could be caused by an air restriction. Technician B says it could be caused by a bad ECM (electronic control module). Who is right?
    a. A only
    b. B only
    c. Both A and B
    d. Neither A nor B

20. An engine is lacking power. Technician A says it could be caused by a dirty air cleaner. Technician B says it could be caused by a restriction in the air intake system. Who is right?
    a. A only
    b. B only
    c. Both A and B
    d. Neither A nor B

Essay

21. Describe the purpose and operation of a turbocharger.

22. What is the difference between a wet- and dry-type manifold?

23. What is the definition of supercharging?

24. What is the purpose and operation of a wastegate on a turbocharger?

25. What is the purpose of the vacuum-controlled heat valve?

CHAPTER 18

Basic Electricity Principles

INTRODUCTION

Electricity and electronics are being used more and more to control the automobile. The electronic control module (ECM), which was discussed earlier, is an example of the use of electronics in the automobile. Storage batteries, ignition systems, charging systems, and starting systems use electricity to operate. Therefore, the study of automotive technology must include a discussion of the principles of electricity. This chapter will help you to understand electricity and how it is applied to automotive systems.

OBJECTIVES

After reading this chapter, you will be able to:
- Define electricity in terms of voltage, amperage, and resistance.
- Calculate both Ohm's and Watt's Laws.
- Define voltage drop, simple circuits, and symbols.
- Analyze series, parallel, and series-parallel circuits.
- Apply magnetism principles to electromagnetic induction.
- Explain the principles of a simple generator.
- Identify the fundamentals of basic electronics.
- Describe the uses of diodes, transistors, integrated circuits, and microprocessors in the automobile.

CHAPTER HIGHLIGHTS

18.1 INTRODUCTION TO ELECTRICITY
  A. Atomic Structure
  B. Valence Ring
  C. Electricity Defined
  D. Electron Theory
  E. Conventional Theory
  F. Amperage Defined
  G. Voltage Defined
  H. Resistance Defined
  I. Ohm's Law
  J. Watt's Law

18.2 BASIC CIRCUITS
  A. Simple Circuit
  B. Voltage Drop
  C. Open, Shorts, and Grounds
  D. Series Circuit
  E. Parallel Circuit
  F. Series-parallel Circuit
  G. Ground Symbol
  H. Electrical Symbols

18.1 INTRODUCTION TO ELECTRICITY

Atomic Structure

The heart of all information concerning electricity is in the study of *atoms* and atomic structure. Everything — water, trees, buildings, and so on — is made up of atoms. They are very small, about a millionth of an inch across. There are millions of atoms in a single breath of air.

The structure of atoms can be illustrated as shown in *Figure 18–1*. Each atom has at its center a nucleus that contains both *protons* and neutrons. The nucleus is the major part of the atom. Protons are said to carry a positive charge (+). Neutrons carry no charge and are not considered in the study of electricity. Also present in the atom are *electrons* which orbit around the nucleus. Electrons are very light in comparison to the nucleus. They carry a negative charge (−).

There is an attraction between the negative electrons and the positive protons. The attractive force and the centrifugal forces cause the electron to orbit the nucleus or protons in the center, *Figure 18–2*.

The number of electrons in all orbits and the number of protons in the nucleus will try to remain equal. If they are equal, the atom is said to be balanced or neutral. *Figure 18–3* shows several atoms for comparison.

Valence Ring

In the study of electricity, we are only concerned with the electrons in the outer orbit. The outer orbit of the atom is

SYMBOL FOR AN ATOM WITH ELECTRONS

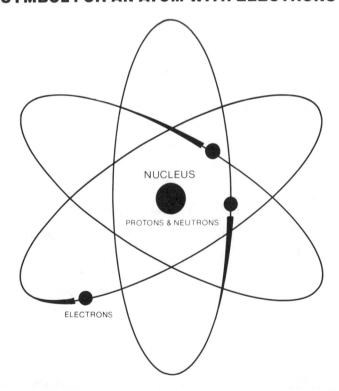

NUCLEUS

PROTONS & NEUTRONS

ELECTRONS

FIGURE 18–1. An atom with protons, neutrons, and electrons. *(Courtesy of Davis Publications, Inc.)*

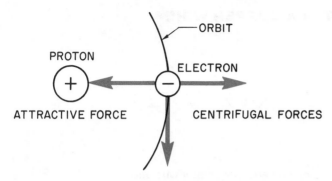

FIGURE 18–2. When an attractive force equals the centrifugal force, the electron will orbit around the proton.

called the *valence ring*. It holds the outermost electrons. Actually, electrons in the valence ring can easily be added or removed. Generally, an atom with several electrons missing will try to gain or capture other electrons. This will make the atom balanced. Also, if an atom has an excess amount of electrons in the valence ring, it may try to get rid of these electrons. This will also help balance the atom.

Certain materials can lose or gain electrons rather easily. This depends upon the number of electrons needed in the valence ring to balance the atom. If an atom loses electrons easily, the material is called a good *conductor*. If the atom cannot lose electrons easily, the material is called a good *insulator*. Insulators and conductors can be defined as:

COMPARISON OF DIFFERENT ATOMS

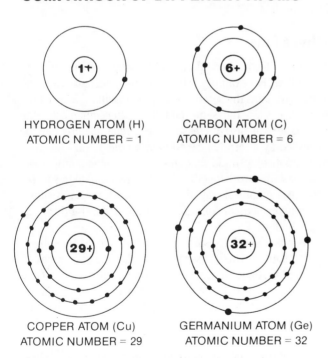

HYDROGEN ATOM (H)
ATOMIC NUMBER = 1

CARBON ATOM (C)
ATOMIC NUMBER = 6

COPPER ATOM (Cu)
ATOMIC NUMBER = 29

GERMANIUM ATOM (Ge)
ATOMIC NUMBER = 32

FIGURE 18–3. Different atoms have different numbers of protons and electrons. Hydrogen is the simplest atom while germanium is much more complex. (*Courtesy of Davis Publications, Inc.*)

a. three or less electrons — conductor
b. five or more electrons — insulator
c. four electrons — semiconductor

Those materials that have exactly four electrons in the outer orbit can be considered either a conductor or an insulator. These materials are called semiconductors. They are used in solid-state components, which will be discussed later in this chapter.

Electricity Defined

Electricity can be defined as the movement of electrons from atom to atom. This can only happen in a conductor material. An example is shown in ***Figure 18–4***. Copper atoms are shown with only the valence ring. Copper is a good conductor. If an excess amount of positive charges or protons is placed on the left, one of the positive atoms will try to pull the outer electron away from the copper atom on the far left.

This action will make the far left copper atom slightly positively charged. There is one more proton than electron in the total atom. This positively charged atom will then pull an electron from the one on its right side. This atom also becomes positively charged. In total, when this action continues to happen, electrons will be flowing from the right to the left. Remember, there must be an abundant amount of protons on the left and electrons on the right.

Electron Theory

In Figure 18–4 the electrons were flowing from a negative point to a positive point. When electricity is defined this way, it is called the *electron theory*. This means that electricity will flow from a negative point to a more positive point. This is one method of defining the direction of electrical flow.

Conventional Theory

Electricity can be defined another way. Electricity can be defined as flow from a positive point to a more negative point. This is called the *conventional theory*. For example, referring back to Figure 18–4, while negative charges are flowing from right to left, positive charges are flowing from left to right. This means that electrical charges could also flow from positive to negative. In the automotive field, this method has been used to define the direction of electrical flow. It is only important to be consistent with the method you choose. If electron theory is used, stay with electron theory. If conventional theory is used, stay with conventional theory.

ELECTRON MOVEMENT IN A COPPER WIRE

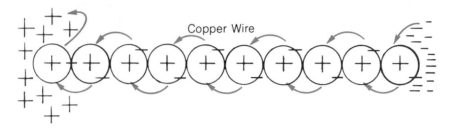

FIGURE 18-4. When (+) charges are placed on one end of a copper wire and (−) charges are placed on the other end, electrons will move through the wire. This is called the flow of electrons or electricity. *(Courtesy of Davis Publications, Inc.)*

Amperage Defined

The measurement of the amount of electrons flowing from a negative point to a positive point in a given time period is called *amperage*. This term is also called current. This term is linked to the example of water flowing through a pipe (electrons flowing through a wire). Amperage is defined as the amount of electrons passing any given point in the circuit in one second. The letter used to identify amperage or current is (I). This stands for *I*ntensity of current flow.

Voltage Defined

Voltage is defined as the push or force used to move the electrons. Referring back to Figure 18-4, the difference between the positive and negative charges is called voltage. This difference in charges has the ability to move electrons through the wire. *Figure 18-5* illustrates the definition of voltage.

Other terms are used to describe voltage. They include potential difference, electromotive force (emf), and pressure. When the term *voltage* is linked to a water system, water pressure is used to push water through a pipe. (Voltage is used to push electrons through a wire.) Voltage is represented by the letter (E), which stands for *E*lectromotive force.

Resistance Defined

The third component in electricity is called *resistance*. Resistance is defined as opposition to current flow. Because of their atomic structure, certain materials offer poor conductivity. This will slow down the electrons. Actually, as the electrons move through a wire, they bump into other atoms in the conductor. As this occurs, the material heats up and causes even more resistance.

Various types and values of resistors are designed today to control the flow of electrons. This depends upon how much current is needed to flow through a circuit. Resistance is identified by the letter (R), which stands for *R*esistance to electron flow. When linking resistance to a water circuit, flow valves and faucets control or restrict water flow. (Resistance in an electrical circuit controls or restricts electron flow.)

Ohm's Law

The three electrical components just described interact with each other. For example, if the resistance decreases and the voltage remains the same, the amperage will increase. If the resistance stays the same and the voltage increases, the amperage will also increase. These relationships can be identified by a formula called *Ohm's Law*. Ohm's Law is a mathematical formula that shows how voltage, amperage, and resistance work together. The triangle shown in *Figure 18-6* is a graphical way of showing this formula. It shows that if two electrical components are known, the third can easily be found.

For example, if resistance (R) and voltage (E) are known, cover the unknown (I) to see the formula. The amperage can be found by dividing the resistance into the voltage. If the amperage and resistance are known, voltage can be found by multiplying the amperage by the resistance. In actual practice, many electrical circuits are designed so that control of the current or amperage can be obtained by changing the voltage or resistance. In addition, when using the ECM (electronic control module), as studied in several earlier chapters, a certain resistance will set up a voltage signal to be sent to the ECM.

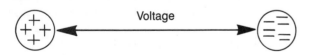

FIGURE 18-5. When there is a difference in charges from one side of a conductor to another, the difference is called voltage. *(Courtesy of Davis Publications, Inc.)*

OHM'S LAW
Voltage (E) = Amperage (I) × Resistance (R)

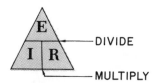

FIGURE 18–6. Ohm's Law states that Voltage (E) = Amperage (I) × Resistance (R). *(Courtesy of Davis Publications, Inc.)*

Watt's Law

Wattage is another term used to help analyze electrical circuits. It is measured using *Watt's Law*. Wattage is a measure of the power (P) used in the circuit. Wattage is a measure of the total electrical work being done per unit of time. When voltage (E) is multiplied by amperage (I), the result is wattage (P). Wattage, which is a measure of electrical power, may also be referred to as kilowatts (kw). 1,000 watts equals 1 kilowatt. *Figure 18–7* illustrates the relationship between voltage, amperage, and wattage. If the amperage and wattage are known, cover the voltage to see the formula. If the voltage and wattage are known, amperage can be calculated.

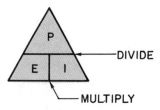

FIGURE 18–7. Wattage (P) = Voltage (E) × Amperage (I).

18.2 BASIC CIRCUITS

When studying electrical systems in the automobile, circuits are used to show the operation of electrical components. A circuit normally consists of several components. First, a power source is needed to provide the necessary voltage. Second, wire is needed to provide a path for the flow of electrons. Third, a load, which can be any resistance, is needed. Lights, radios, starter motors, spark plugs, CCC sensors, batteries, and wiper motors are all examples of loads. The load provides the resistance in the circuit.

Simple Circuit

The simplest circuit consists of a power source (a battery), a single unit or load to be operated (a light), and the connecting wires. *Figure 18–8* shows a simple circuit. Note that the wires must be connected to complete the circuit. In this case, electricity flows from the positive terminal on the battery, through the wire to the light, and back to the negative terminal of the battery.

Voltage Drop

When testing and troubleshooting an electrical circuit, *voltage drop* is usually measured. Voltage must be present

CAR CLINIC ▐████

PROBLEM: ELECTRICAL SYSTEM TOTALLY FAILS

The entire electrical system on a new vehicle with 500 miles on it began to loose voltage. While driving on the highway, it was noticed at night that the lights seemed to become dimmer. The radio and other electrical components also seemed to lose power. What could be the problem?

SOLUTION:

Since the vehicle is new, one of the electrical connections on the alternator may have come loose. The first and most obvious connection to check is the main wire between the alternator and the battery (red). If this connection is broken or faulty, any electrical current produced by the alternator will not get to the battery or the other electrical systems. When the battery is drained of all its electrical energy, the electrical circuits also lose power. Check the "BAT" wire to make sure it has a solid electrical connection.

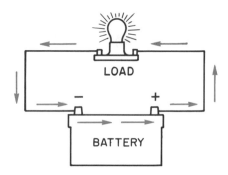

FIGURE 18–8. A simple circuit shows amperage flowing through the light to make it operate.

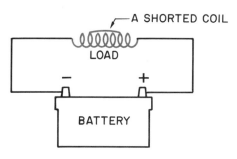

FIGURE 18–10. A shorted circuit bypasses part of the load.

for amperage to flow through a resistor. Voltage is dropped across each resistor that it pushes amperage through. To determine the voltage drop across any resistor, simply use Ohm's Law. In this case, however, use only the voltage, amperage, and resistance at that particular resistor. Voltage drop at any resistor is shown as:

$$\text{Voltage Drop} = \text{Resistance} \times \text{Amperage (I} \times \text{R)}$$
$$\text{(at any one resistor)}$$

Voltage drop can be measured by using a voltmeter. Usually, the voltmeter leads are placed across the component to be checked or across the component to ground.

Opens, Shorts, and Grounds

Electrical systems may develop an *open*, *shorted*, or *grounded circuit*. Each of these conditions will render the circuit ineffective.

An open circuit is one that has a break in the wire. Refer to *Figure 18–9*. This is called a break in continuity. If the circuit is open, there is not a complete path for the current to flow through. An open circuit acts the same as if the circuit had a switch in the open position. Voltage drop across an open circuit is always the same as the source, or maximum voltage.

A shorted circuit is one that allows electricity to flow past part of the normal load. An example of this is a shorted coil. See *Figure 18–10*. The internal windings are usually insulated from each other. However, if the insulation breaks and allows the windings to touch each other, part of the coil will be bypassed. Any load can be partially or fully bypassed by having a shorted circuit. If a load is fully bypassed, the voltage dropped across the load will be zero.

A grounded circuit is a condition that allows current to return to the battery before it has reached its intended destination. An example is a grounded tail light. See *Figure 18–11*. If a wire leading to the light were broken and touching the frame, the electricity would be grounded back to the battery. Grounded circuits can cause excessive current to be drained from the battery.

Series Circuit

A *series circuit* consists of two or more resistors connected to a voltage source with only one path for the electrons to follow. An example is shown in *Figure 18–12*. The series circuit has two resistors placed in the path of the electrons. The resistors are shown as R_1 and R_2. (The jagged line is a symbol used to represent a resistor.) All of the amperage that comes out of the positive side of the battery must go through each resistor, then back to the negative side of the battery.

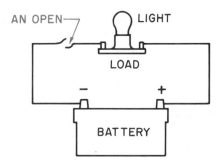

FIGURE 18–9. An open circuit will stop all current from flowing through the circuit.

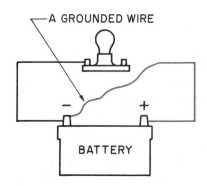

FIGURE 18–11. A grounded wire can cause excessive drain on the battery.

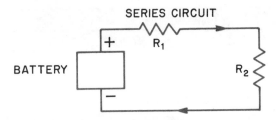

FIGURE 18–12. A series circuit with two resistors for current to flow through.

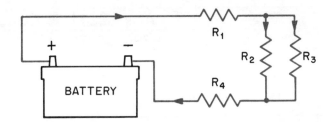

FIGURE 18–14. A series-parallel circuit. Resistors R_2 and R_3 are in parallel. The total resistance of R_2 and R_3 in parallel is added to R_1 and R_4 which are in series with R_2 and R_3.

In a series circuit, the resistors are added together to get the total resistance (R total).

Series circuits are characterized by the following facts:

1. The resistance is always additive. R total is equal to $R_1 + R_2 + R_3$, and so on.
2. The amperage through each resistor is the same. Amperage is the same throughout the circuit.
3. The voltage drop across each resistor will be different if the resistance values are different.
4. The sum of the voltage drops of each resistor equals the source voltage.

Parallel Circuit

Parallel circuits provide two or more paths for the current to flow through. Each path has separate resistors and operates independently from the other parallel paths. In a parallel circuit, amperage can flow through more than one resistor at a time. An example of a parallel circuit is shown in *Figure 18–13*. Note that if one branch of the circuit breaks or has an open circuit, the remaining resistors can still operate. The resistance in a parallel circuit is calculated by the formula:

$$R\ total = \frac{1}{1/R_1 + 1/R_2 + 1/R_3 + 1/R_4,\ etc.}$$

Total resistance in a parallel circuit will always be less than the smallest resistor. Most circuits on the automobile are parallel circuits. If more resistors are added to the circuit,

the total resistance will decrease. A parallel circuit can be characterized by the following:

1. The total resistance is less than the lowest resistor.
2. The amperage flowing through each resistor is different if the resistance values are different.
3. The voltage drop across each resistor is the same. This is also the source voltage.
4. The sum of the separate amperages in each branch equals the total amperage in the circuit.

Series-parallel Circuit

A series-parallel circuit is designed so that both series and parallel combinations exist within the same circuit. *Figure 18–14* shows four resistors connected in a series-parallel circuit. To calculate total resistance in this circuit, first calculate the parallel portions, then add the result to the series portions of the circuit. Total current flow will be determined by the total resistance and the total source voltage.

Ground Symbol

A ground symbol is sometimes used when analyzing a circuit. This means that the circuit is connected to the steel structure of the vehicle. The ground symbol is shown in *Figure 18–15*. The symbol indicates that the electricity is

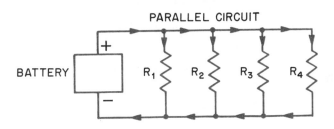

FIGURE 18–13. A parallel circuit showing there is more than one path for the current to follow.

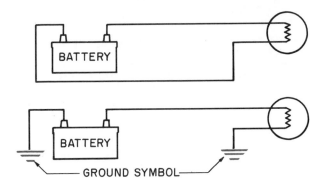

FIGURE 18–15. Ground symbols are used to make a circuit drawing easier to read.

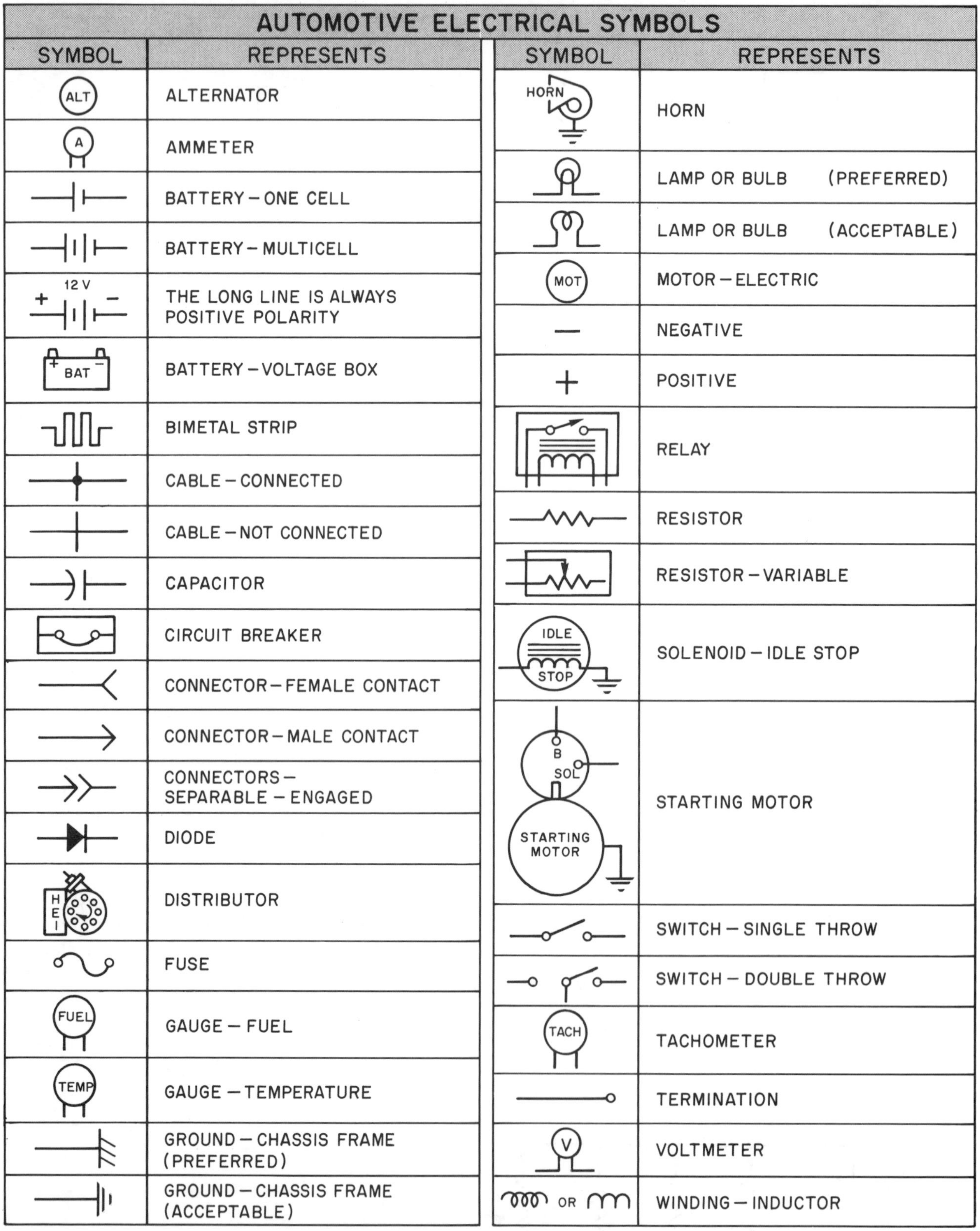

AUTOMOTIVE ELECTRICAL SYMBOLS

SYMBOL	REPRESENTS	SYMBOL	REPRESENTS
(ALT)	ALTERNATOR	HORN	HORN
(A)	AMMETER		LAMP OR BULB (PREFERRED)
—┤├—	BATTERY – ONE CELL		LAMP OR BULB (ACCEPTABLE)
—┤│├—	BATTERY – MULTICELL	(MOT)	MOTOR – ELECTRIC
+ 12 V – —┤│├—	THE LONG LINE IS ALWAYS POSITIVE POLARITY	—	NEGATIVE
BAT	BATTERY – VOLTAGE BOX	+	POSITIVE
⊓⊔⊓	BIMETAL STRIP		RELAY
—•—	CABLE – CONNECTED	—ᴧᴧᴧ—	RESISTOR
—┼—	CABLE – NOT CONNECTED		RESISTOR – VARIABLE
—)├—	CAPACITOR	IDLE STOP	SOLENOID – IDLE STOP
	CIRCUIT BREAKER	B SOL STARTING MOTOR	STARTING MOTOR
—<	CONNECTOR – FEMALE CONTACT		
—>	CONNECTOR – MALE CONTACT		
—»—	CONNECTORS – SEPARABLE – ENGAGED		
—▶├—	DIODE	—o͜ o—	SWITCH – SINGLE THROW
H E I DISTRIBUTOR	DISTRIBUTOR	—o͜ ͜o—	SWITCH – DOUBLE THROW
ᴖ⌣ᴖ	FUSE	(TACH)	TACHOMETER
(FUEL)	GAUGE – FUEL	—o	TERMINATION
(TEMP)	GAUGE – TEMPERATURE	(V)	VOLTMETER
⏚	GROUND – CHASSIS FRAME (PREFERRED)	ᴔᴔᴔ OR ᴖᴖ	WINDING – INDUCTOR
—╫	GROUND – CHASSIS FRAME (ACCEPTABLE)		

FIGURE 18–16. Various symbols used in electrical circuits.

returning to the battery via the frame of the vehicle. Any steel structure of the vehicle can actually act as the wire that returns electricity to the battery. This could include the body sheet metal, frame, engine block, or transmission case.

Electrical Symbols

A more complete listing of symbols used in automotive circuits can be found in *Figure 18–16*. Many symbols are used in the wiring diagrams in this textbook and in the automotive repair manuals. It is important to become familiar with these symbols in order to analyze more involved circuits.

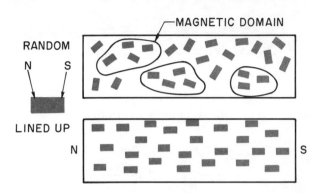

FIGURE 18–17. There is no magnetism in a metal bar when the domains are not lined up. When they are lined up, magnetism is produced.

CAR CLINIC ▮▮▮▮▮▮▮▮

PROBLEM: ENGINE RUNS ON

A customer has a vehicle with a carburetor. When the engine is hot and shut off, it continues to run. The points have been replaced. The carburetor has been cleaned. What could be another cause of dieseling or run on?

SOLUTION:

Quite often the throttle solenoid on the carburetor is either not working or is not adjusted correctly. The throttle solenoid is an electrical device that helps set the position of the throttle during engine operation. When the engine ignition is shut off, the throttle solenoid is released and the throttle valve is completely closed. Vehicles without a throttle solenoid are shut off with the throttle position at idle. If the pistons continue to draw in fuel, and if there is a hot spot in the cylinder, the engine may run on. Shutting off the throttle valve completely with the throttle solenoid valve reduces this possibility. Check to see that the throttle solenoid valve operates and is adjusted correctly.

18.3 MAGNETISM

One area of study in automotive systems is that of magnetism. The principles of magnetism are integrated into motors, generators, solenoids, and other electrical systems in the automobile. Magnetism can be best understood by observing some of its effects.

The effects of magnetism were first observed when fragments of iron ore, referred to as lodestones, were attracted to pieces of iron. It was further discovered that a long piece of iron would align itself so that one end always pointed toward the Earth's north pole. This end of the bar was called the north (N) pole, and the other end was called the south (S) pole. The bar was called a bar magnet.

Domains

Inside the bar magnet are many small *domains*. Domains are minute sections in the bar where the atoms line up to produce a magnetic field. Most of the domains must be lined up in the same direction in the bar magnet to form a magnetic field. *Figure 18–17* shows a bar of metal with the domains located randomly and with the domains lined up.

Lines of Force

Magnets can be further defined by the *lines of force* being produced. Referring to *Figure 18–18*, the magnetic field is

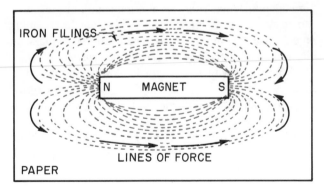

FIGURE 18–18. Bar magnet with iron filings show the invisible lines of force around a magnet. Magnetic lines of force always flow from the north pole to the south pole.

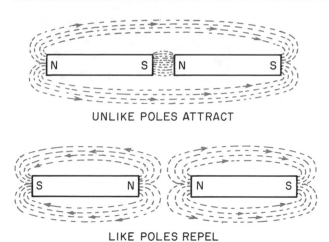

FIGURE 18-19. Like poles will repel each other. Unlike poles will attract each other.

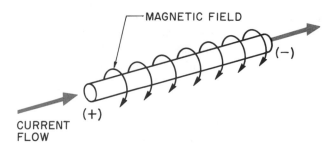

FIGURE 18-20. Any wire that has electricity flowing in it will produce a small magnetic field around the wire.

defined as invisible forces that come out of the north pole and enter the south pole. The shape of the magnetic lines of force can be illustrated by sprinkling iron filings on a piece of paper on top of the bar magnet. When the paper is tapped, the iron filings align to form a clear pattern around the bar magnet. Note that the lines of force will never touch each other. Also note that the lines of force are more concentrated at the ends of the magnet.

Repulsion and Attraction

If two bar magnets are placed together at unlike poles, they will snap together. If the ends have the same poles, they will repel each other. This is shown in *Figure 18–19*.

Electromagnetism

The bar magnet that was mentioned previously is called a permanent magnet. There are also temporary magnets that can be made from electricity. This can be done by wrapping an electrical wire around an unmagnetized bar to make an electromagnet. This is called *electromagnetism*.

When any wire has electricity flowing through it, a magnetic field develops around the wire. See *Figure 18–20*. If the wire is then placed in the shape of a coil, as shown in *Figure 18–21*, the magnetic field in the center of the coil will be additive. If a nonmagnetized bar is placed in the center of the coil, the bar will also be magnetized, *Figure 18–22*.

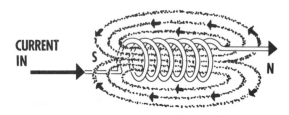

FIGURE 18-21. When electrical current passes through a conductor, a magnetic field is developed around the conductor. If the conductor is formed in the shape of a coil, the magnetic field is additive in the center. *(Courtesy of Delco Remy Division of General Motors Corporation)*

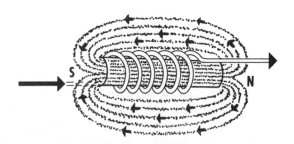

FIGURE 18-22. When a coil of wire has current passing through it, north and south poles are created. The strength of the magnetic field can be increased by inserting an iron bar. *(Courtesy of Delco Remy Division of General Motors Corporation)*

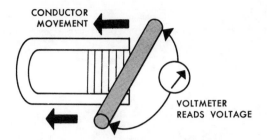

FIGURE 18-23. As a conductor moves or cuts through a magnetic field, a voltage is produced. If the voltage is applied to a circuit with a load or resistance, a complete circuit can be produced. *(Courtesy of Delco Remy Division of General Motors Corporation)*

Electromagnetic Induction

Through experimentation it was discovered that a conductor moving across or cutting a magnetic field would produce a voltage. This is called *electromagnetic induction*. Actually, an electromotive force (emf) will be induced or generated within the wire. Internally, a generator is converting mechanical energy to electrical energy. In the simplest form, *Figure 18-23* shows how voltage is produced with a magnet.

The direction of current flow produced by the voltage can be reversed if the movement is also reversed. In fact, if the wire were moved back and forth rapidly, the result would be alternating current. Alternating current will be discussed further in this chapter and in the chapter on charging systems.

Figure 18-24 shows a complete simple generator used to produce voltage in a wire. Note that rather than using permanent magnets, electromagnets are used. Also, twelve volts have been applied to produce the electromagnets. This voltage is called the field voltage. If the center wire moves up, a voltage will be produced on the voltmeter. If the wire moves down, a reverse voltage will be produced. This means the positive and negative points will have been reversed.

Note that three things are necessary to induce a voltage in a generator. These include:

1. a magnetic field producing lines of force,
2. conductors that can be moved, and
3. movement between the conductors and the magnetic field so that the lines of force are cut.

If any one of the preceding factors increases or decreases, the induced voltage will also increase or decrease. Note also that a conductor can be stationary while the lines of force are moved. This will still induce a voltage because the lines of force will be cut.

SIMPLE GENERATOR

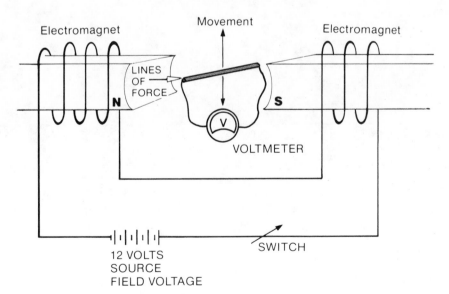

FIGURE 18-24. If a conductor is moved perpendicular to the lines of force from the magnet, a voltage will be induced in the conductor. *(Courtesy of Davis Publications, Inc.)*

COMPUTER-AIDED GRAPHICS IN VEHICLE DESIGN

Computer-aided graphics and engineering are used today to help plan and design many parts of the vehicle. Pictured here is a typical workstation that utilizes a software program called Product Design Graphics System (PDGS). (Courtesy of Ford Motor Company)

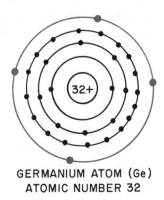

GERMANIUM ATOM (Ge)
ATOMIC NUMBER 32

FIGURE 18–25. Semiconductors have only four electrons in their outer orbit. *(Courtesy of Davis Publications, Inc.)*

18.4 SOLID-STATE COMPONENTS

One area of electrical study that has grown recently is that of electronics. Electronics is the study of solid-state devices such as diodes, transistors, and integrated circuits. Solid-state devices are those that have no moving parts except internal electrons.

To begin, let's review the study of conductors and insulators presented earlier in this chapter. It was mentioned that any material that has four electrons in its outer orbit is called a semiconductor. See *Figure 18–25*. This means that it could be either a good conductor or a good insulator.

Semiconductors are also called solid-state devices. Because of their characteristics, semiconductors are often used as switches. Circuits can be turned off and on by semiconductors with no moving parts.

Diodes

The *diode* is a semiconductor that permits current to flow through a circuit in one direction but not in the other. *Figure 18–26* shows a circuit with a diode. The alternator produces a current that flows back and forth 60 times per minute. This is called ac voltage. In the battery, however, direct current (dc) voltage is used to charge the battery. The battery is considered the load. The diode can be used to convert the ac to dc. Referring to Figure 18–26, electricity is able to flow in the direction of the arrow. If electricity tries to flow in the opposite direction, it will be stopped. There will be no current flow in the reverse direction.

Transistors

The *transistor* is also a type of semiconductor. In this case, the transistor has some semiconductor material added to it. A circuit using a transistor is shown in *Figure 18–27*. In operation, the circuit to be turned off and on is identified as circuit A. Circuit B is the controlling circuit. The transistor has three wires. These are called the base, emitter, and collector.

If a small amount of current flows (when the switch is closed) in the emitter-to-base circuit, the resistance between the emitter and collector circuits will be zero. Circuit A is then turned on to operate the coil. (This could be a primary coil in an ignition system.) When the current stops flowing in circuit B (switch open), the resistance between the emitter and collector is very high. This resistance shuts off circuit B.

Transistors can also be used to amplify an on-off sequence of a signal. A small amount of current flowing on

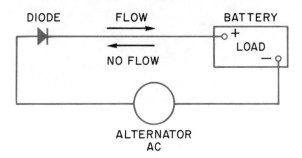

FIGURE 18–26. A diode in a circuit allows electricity to flow in only one direction.

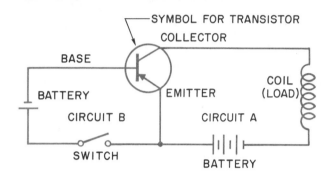

FIGURE 18–27. A transistor can control the on-off sequence of the coil. Circuit B controls circuit A.

and off in circuit B can control a large amount of current flowing in circuit A. The on-off sequence will be amplified from circuit B to circuit A.

Integrated Circuits

Over the past several years, engineers have found ways of making diodes and transistors extremely small. Because they are smaller, many diodes, transistors, and other semiconductors can be placed on a board called an inte-

grated circuit. See *Figure 18–28*. These circuits may contain many semiconductors. More recently, the chip has been designed to incorporate even smaller components on the integrated board. Some applications that use *integrated circuits* and chips in the automobile include solid-state ignition, electronic fuel injection, electronic engine systems, computer-controlled combustion, and speed and cruise controls. Other circuits are constantly being developed to make the automobile even more electronic. Many of these will be discussed in other chapters within this textbook.

Microprocessors

With the addition of integrated circuits and chips, the automobile is using more and more *microprocessors*. Microprocessors are small computers that can be used for a variety of tasks. They contain logic and control circuits. Various sensors relay input information to the microprocessor. Engine speed, temperature, outside weather conditions (barometric pressure), load and weight distribution of the vehicle, vehicle speed, throttle position, and so on can be fed into the microprocessor. The microprocessor uses this input information to control the vehicle to its optimum performance.

FIGURE 18–28. Integrated circuits and chips are used in many applications in the automobile today.

SCANNING DRAWINGS
TO PUT IN MEMORY

One of the more recent developments in automotive design is the process of scanning a life-size tape drawing of a new vehicle. The scanning device then records the shape and stores the information in a computer. The information is then available later for such functions as the automatic milling of three-eighths scale or full-scale clay models of cars. (Courtesy of Ford Motor Company)

SUMMARY

This chapter investigated the basic principles of electricity and how they are related to the automobile. Electricity is defined as the movement of electrons. Voltage, amperage, and resistance are components used to control electricity. Both Ohm's and Watt's Laws were defined to help determine the relationships between voltage, amperage, resistance, and power.

Several circuits, including the series, parallel, and series-parallel circuits, were studied. Each type of circuit has specific characteristics. Types of circuits, voltage drop, opens, shorts, and grounds were also defined.

As part of the study of electricity, magnetism was defined. Domains, lines of force, electromagnetism, and simple generators were discussed. The principles concerning magnetism are directly related to automobile systems such as starting and charging, ignition, and electronic controls.

Semiconductors are becoming a significant part of the automobile. Diodes, transistors, integrated circuits, and microprocessors are being used in all systems studied on the automobile.

TERMS TO KNOW

Can you explain each of the following terms? Review the chapter until you can use each term correctly.

Atom	Series circuit
Electron	Parallel circuit
Proton	Voltage drop
Valence ring	Open circuit
Conductor	Shorted circuit
Insulator	Grounded circuit
Semiconductor	Domain
Electricity	Lines of force
Voltage	Electromagnetism
Amperage	Electromagnetic induction
Resistance	Diode
Electron theory	Transistor
Conventional theory	Integrated circuit
Ohm's Law	Microprocessor
Watt's Law	

REVIEW QUESTIONS

Multiple Choice

1. Which of the following components has a negative charge in the atom?
   a. Protons
   b. Electrons
   c. Neutrons
   d. Watts
   e. Volts

2. Which of the following materials has been identified as being a good conductor?
   a. Those with four electrons in the valence ring
   b. Those with three electrons in the valence ring
   c. Those with five or more electrons in the valence ring
   d. Glass
   e. Wood

3. Which theory says that electricity flows from a positive point to a negative point?
   a. Conventional theory
   b. Magnetism theory
   c. Diode theory
   d. Electron theory
   e. Valence ring theory

4. Which theory says that electricity flows from a negative point to a positive point?
   a. Conventional theory
   b. Magnetism theory
   c. Diode theory
   d. Electron theory
   e. Valence ring theory

5. Pressure or push on the electrons is defined as:
   a. Amperage
   b. Voltage
   c. Wattage
   d. Resistance
   e. Magnetism

6. Wattage is found by multiplying the voltage by the:
   a. Resistance
   b. Power
   c. Amperage
   d. Protons
   e. Neutrons

7. Current is also defined as:
   a. Amperage
   b. Voltage
   c. Wattage
   d. Resistance
   e. Magnetism

8. Which circuit has only one path for the electricity to flow in?
   a. Series-parallel circuit
   b. Parallel circuit
   c. Series circuit
   d. Open circuit
   e. None of the above

9. A break in a wire is referred to as:
   a. A short
   b. An open
   c. A ground
   d. A series circuit
   e. A parallel circuit

10. What is necessary to produce an induced voltage?
    a. A magnetic field
    b. A conductor
    c. Movement between a conductor and magnetic field
    d. All of the above
    e. None of the above

11. A wire with electricity flowing through it:
    a. Has no magnetic field around it
    b. Has a magnetic field around it
    c. Is usually considered an open circuit
    d. Is usually considered ground
    e. Has maximum wattage

12. Which of the following devices can amplify an off-on signal?
    a. Diode
    b. Resistor
    c. Transistor
    d. Voltmeter
    e. Resistance meter

13. Which of the following devices causes electricity to flow in only one direction?
    a. Diode
    b. Resistor
    c. Transistor
    d. Voltmeter
    e. Resistance meter

14. In a parallel circuit, the voltage drop is _____ at each resistor.
    a. The same
    b. Different
    c. 0
    d. The same as the amperage
    e. None of the above

15. Which circuit on a transistor is considered to be the control part?
    a. Base
    b. Emitter
    c. Collector
    d. Negative terminal
    e. Positive terminal

The following questions are similar in format to ASE (Automotive Service Excellence) test questions.

16. Technician A says that inputs such as vehicle speed and engine temperature are fed to the microprocessor. Technician B says that inputs such as engine speed and barometric pressure are fed to the microprocessor. Who is right?
    a. A only
    b. B only
    c. Both A and B
    d. Neither A nor B

17. Technician A says that a transistor is the same as a diode. Technician B says that a transistor is the same as a resistor. Who is right?
    a. A only
    b. B only
    c. Both A and B
    d. Neither A nor B

18. Technician A says that voltage drop is the same as resistance. Technician B says that voltage drop is the amount of voltage dropped across each resistor. Who is right?
    a. A only
    b. B only
    c. Both A and B
    d. Neither A nor B

19. There is a break in an electrical wire. Technician A says this is called a short. Technician B says this is called a ground. Who is right?
    a. A only
    b. B only
    c. Both A and B
    d. Neither A nor B

Essay

20. Define Ohm's Law.

21. Define and state the difference between an open, short, and ground.

22. Define voltage drop.

23. What are three characteristics of a series circuit?

24. What are three characteristics of a parallel circuit?

25. Define the purpose and operation of a diode.

26. What is the difference between a diode and a transistor?

CHAPTER 19

Automotive Batteries

INTRODUCTION

The battery is a very important part of the automobile. It is used to supply electricity for many systems and components within the vehicle. The purpose of this chapter is to study the operation and maintenance of the automotive storage battery.

OBJECTIVES

After reading this chapter, you will be able to:
■ Identify the purpose of the automotive battery.
■ Analyze the internal parts, construction, and operation of the battery, including chemical action and specific gravity.
■ Determine the methods used to rate batteries.
■ Identify the methods used to test and maintain the battery.
■ Analyze the various diagnosis and service tips used on batteries.

CHAPTER HIGHLIGHTS

19.1 BATTERY DESIGN AND OPERATION

Purpose of the Battery

The purpose of the battery is to act as a reservoir for storing electricity. The battery receives, stores, and makes electrical energy available to the automobile. It is called a storage battery because it stores or holds electricity. The battery is considered an *electro-chemical* device. This means that it uses chemicals to produce and store electricity (electro-chemical). Energy in the battery is stored in a chemical form. The chemical energy is released as electrical energy for use in the automobile. The purpose of the battery is to provide sufficient electrical energy to crank the starter, and operate the ignition system, computers, solenoids, lights, and other electrical components.

Types of Batteries

Batteries can be subdivided into two major groups: primary and secondary batteries. *Primary batteries* are those that are nonrechargeable. Examples of primary batteries include those used in flashlights, calculators, smoke alarms, and radios. Various metals and chemicals are used to manufacture primary batteries.

Secondary batteries are those that can be discharged and recharged repeatedly. This can be done by reversing the normal current flow through the battery. The automobile battery is called a secondary battery because it can be charged and discharged many times.

Battery Cells

Batteries are made by putting together a number of cells. A *battery cell* is defined as that part of a battery that stores chemical energy for later use. In its simplest form, a battery cell consists of three components. These include a positive plate (one type of metal), a negative plate (another type of metal), and an *electrolyte* solution. When these three components, two dissimilar metals and an electrolyte solution, are placed together, electricity can be produced.

All batteries, both primary and secondary, have cells. The difference between battery cells is the type of metals and electrolyte chemicals used in the cells. For example, a two-volt primary battery cell uses different metals and chemicals than a nine-volt primary battery cell.

Lead-acid Battery

Figure 19–1 shows a simple cell. The positive plate is made of a metal called lead peroxide ($Pb\ O_2$). The negative plate is made of a metal called sponge lead (Pb). The electrolyte solution is made of a mixture of sulfuric acid and water ($H_2\ SO_4$). This is called a lead-acid battery cell. The metals are made of different types of lead and the electrolyte is made of acid. When these three components are arranged

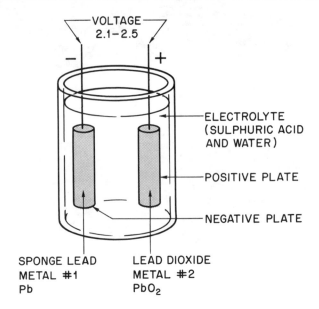

FIGURE 19–1. Battery cells consist of three components. These include two different metals and an electrolyte solution. When these are placed together, a voltage can be produced across the two metals.

as shown, approximately 2.1 to 2.5 volts are produced across the positive and negative plates. If an electrical circuit were placed across this voltage source, electrons would flow from the negative plate, through the circuit, back to the positive plate.

Battery Cell Symbol

The electrical symbol of a battery is shown in *Figure 19–2*. The longer line represents the positive plate. The shorter line represents the negative plate.

Combining Battery Cells

The automobile requires 12 volts to operate. Battery cells can be connected to produce different voltages. When battery cells are connected in series, the voltage is additive

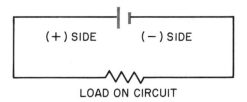

FIGURE 19–2. The symbol for a battery cell is shown. The long line represents the (+) side of the battery. The short line represents the (−) side.

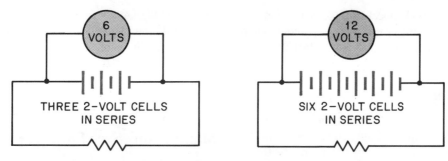

WHEN CELLS ARE IN SERIES, VOLTAGE IS ADDITIVE

FIGURE 19–3. When three 2-volt cells are placed in series, the total voltage is 6. When six 2-volt cells are placed in series, the voltage is 12.

for each cell. For example, in order to produce a 6-volt lead acid battery, three 2-volt cells are connected in series. A 12-volt battery has six 2-volt cells connected in series. See *Figure 19–3*.

Whenever battery cells are placed in parallel, the voltage remains the same. However, the amperage that the battery can produce is increased. This is called *amperage capacity*. This means that the length of time the battery can produce a certain amperage is increased as compared to the length of time a single cell could produce that amperage. See *Figure 19–4*.

Chemical Action in a Battery

Figure 19–5 shows how the chemical reactions occur inside the battery cell. During *discharging* or "load," the lead peroxide ($Pb\,O_2$) is separated. The sulfuric acid and water ($H_2\,SO_4$) are separated. Some of the sulfate (SO_4)

combines with the negative-plate sponge lead (Pb). Some of the sulfate (SO_4) combines with the lead (Pb) part of the positive plate. The remaining oxygen (O_2) combines with the hydrogen (H_2) left from the electrolyte. If the battery continues to be discharged, the two metals will eventually change to lead sulfate (Pb SO_4) and the electrolyte will become water (H_2O).

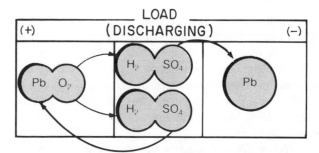

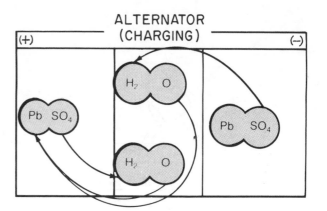

FIGURE 19–5. This diagram shows the chemical action taking place in a battery during discharging and charging. *(Courtesy of Davis Publications, Inc.)*

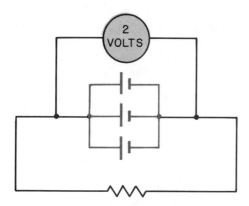

WHEN CELLS ARE IN PARALLEL, VOLTAGE REMAINS THE SAME, BUT AMPERAGE CAPACITY INCREASES.

FIGURE 19–4. When two-volt cells are connected in parallel, the voltage remains the same, but the amperage capacity increases.

During *charging*, the chemical and electrical actions are just the reverse. This is done by using the generator or alternator on the vehicle. At the end of a charge, the battery cell is the same chemical as it was before discharge.

Explosive Gases

Hydrogen and oxygen gases are released during the process of charging and discharging a battery. The gases escape the battery through the vent caps. The more rapidly a battery is charged or discharged, the greater the possibility of producing these gases. These gases can be very explosive. It is always important to keep any sparks or open flame away from a battery that is being charged or discharged. Always charge batteries in a well-vented room.

Electrolyte

The electrolyte in a battery is a combination of water and sulfuric acid. Normally, the ratio of acid to water is 40% acid to 60% distilled water. This combination of water and acid is the best electrolyte solution for the battery. During normal operation, the water will evaporate. The acid does not usually evaporate. Distilled water is the only ingredient that should be added to a battery. The right amount of acid will always be in the battery unless it is tipped over and spilled.

Specific Gravity

The condition of the battery electrolyte can be checked by measuring its *specific gravity*. Specific gravity is defined as the weight of a solution as compared to the weight of water. The specific gravity of water is rated as 1.000. Any solution heavier than water is expressed in terms of the ratio of its density to the density of water. Sulfuric acid is heavier than water. Adding sulfuric acid to water causes the density to exceed water or 1.000.

Figure 19–6 shows how the specific gravity of water and acid change when they are mixed. The specific gravity

of water (1.000) mixed with acid (1.835) is equal to an electrolyte solution of 1.270.

The condition or state of charge can be determined by measuring the specific gravity. *Figure 19–7* shows different specific gravity readings for different battery conditions.

Maintenance-free Batteries

In a standard battery, antimony (a metallic substance) is added to the lead plates to strengthen them. This also aids the battery in producing certain gases during charging and discharging. Calcium, rather than antimony, is used in maintenance-free batteries. A trace of tin is also included. Lead-calcium batteries normally do not discharge gases unless they are grossly overcharged. They do not have filler caps and never need water.

The purity of the lead-calcium plates reduces self-discharging and makes the battery harder to overcharge. Overcharging is a major cause of water loss in the battery. Because of these characteristics, the battery can be almost totally sealed. Since vent holes are eliminated, breathing takes place through a series of baffles inside the battery. These baffles collect acid vapor and condense it back into the solution.

Corrosion outside the battery is also reduced. Acid vapors that help cause outside corrosion are now kept inside the battery. While the battery appears sealed, it can breathe through small openings along the edge or on top of the battery casing.

Future Battery Developments

Newer chemicals are being tested for use in batteries. Some manufacturers use an electrolyte in the form of a gel to reduce further gassing. In fact, manufacturers are experimenting with different materials inside the battery to

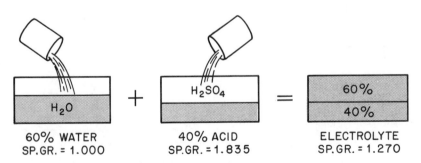

FIGURE 19–6. The ratio of water to acid is about 60% to 40%. Note how the specific gravity of each causes the electrolyte solution to be about 1.270.

SPECIFIC GRAVITY READINGS AT 80°F	
1.260–1.280	Fully charged
1.235–1.260	¾ charged
1.205–1.235	½ charged
1.170–1.205	¼ charged
1.140–1.170	Poorly charged
1.110–1.140	Dead

FIGURE 19–7. The specfiic gravity reading can determine the condition of charge on a battery.

increase the *cell density*. Cell density is a measure of how many watts can be discharged per hour, per pound of battery. For example, the lead-antimony battery has a cell density of 3–4 watt-hours/pound. The lead-calcium battery has a cell density of 6–8. A zinc-chloride (experimental) battery has a cell density of 10–15. Future batteries may have cell densities of 45–50 watt-hours/pound.

Battery Construction

Figure 19–8 shows the internal parts and construction of a typical battery. Following the diagram, first a grid is manufactured as the base of the positive and negative plates. Both lead peroxide and sponge lead, called the active materials, are manufactured and spread on the grid. The result is a completed plate.

Connectors are used to assemble the plates into a group of cells. Note that the plates are placed in parallel to increase the amperage capacity. Both positive and negative groups are manufactured. Separators are then placed inside the groups to keep the plates from touching each other. One cell or element is now complete. This cell is capable of producing about 2.1 to 2.5 volts. Amperage capacity can be increased by adding more plates to each cell.

Once the elements or cells are completed, they are connected in series and placed in a plastic container. A one piece cover is placed on top of the container and sealed. Here the battery posts are on top. Many batteries have the battery posts on the side as shown in *Figure 19–9*. The electrolyte is added and the vent caps are installed. The battery is now complete.

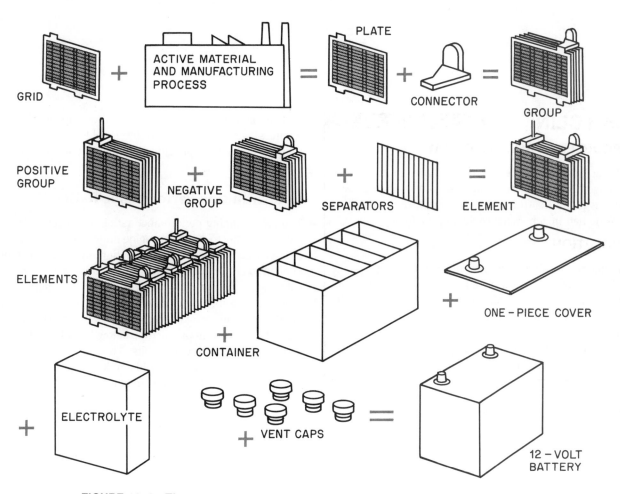

FIGURE 19–8. The components and basic steps in the construction of a battery.

FIGURE 19-9. Battery posts can be either on top of the battery or on the side. Here, side-mounted battery connectors are used in place of posts.

CAR CLINIC

PROBLEM: FROZEN BATTERY

A car has been stored over the winter in an outside garage. During the winter an inspection was made and it was found that the battery had cracked and the fluid inside had frozen. What could cause a frozen battery?

SOLUTION:

Batteries are able to freeze only if the battery is discharged severely. Electrolyte has a lower freezing point than water. The electrolyte changes to water when the battery is discharged. Then the water is able to freeze. When water freezes, it expands, cracking the case. Always store the battery inside a building to prevent this problem and keep the battery charged.

19.2 BATTERY RATINGS

Several ratings are used to identify the size and amperage capacity of the battery. Certain battery applications require cranking ability, while others may require long periods of high-amperage output. Because of these application differences, batteries are rated by 1) cold-cranking performance, 2) reserve capacity, 3) ampere-hour ratings and 4) watt-hour ratings.

Cold-cranking Performance Rating

The cold-cranking performance rating is a measure of the battery's ability to crank an engine under cold weather conditions. During cold weather operation, the engine is more difficult to crank. It takes much more amperage to crank a cold engine than to crank a warm engine. This rating indicates the number of cranking amperes the battery can deliver at 0 degrees F for a period of 30 seconds. During this time, the cell voltage cannot drop below 1.2 volts or the total battery voltage below 7.2 volts. This rating is listed, for example, as 500 CCA or cold-cranking amperes. When buying a battery, this rating should be compared to other batteries in the same range.

Reserve Capacity Rating

The reserve capacity rating is expressed in minutes. It measures a battery's ability to provide emergency power for ignition, lights, and so on if the charging system is not working. This rating involves a constant discharge at normal temperatures. The reserve capacity rating is defined as the number of minutes a fully charged battery at 80 degrees F can be discharged at 25 amperes and still maintain a minimum voltage of 1.75 volts per cell or 10.5 volts for the total battery. The higher the reserve capacity rating, the longer the battery can provide the emergency power for lights and accessories. A typical reserve capacity rating might be 125 minutes.

Ampere-hour Rating

The ampere-hour rating is another measure of the battery's capacity. It is obtained by multiplying a certain flow of amperes by the time in hours during which current will flow. This is usually expressed as a "20-hour rating." Again, the test is run at 80 degrees F, and until the cell voltage drops below 1.75 or the total battery voltage drops below 10.5.

The amount of amperage used in this test is one-twentieth of the published 20-hour capacity in ampere-hours. For example, if a battery is rated at 105 ampere-hours, its discharge amperage is 5.25 (1/20th or 0.05 of 105 equals 5.25). To pass the test, this battery would be discharged at 5.25 amps for 20 hours and still have a total voltage above 10.5.

Watt-hour Rating

The watt-hour rating is measured by multiplying the ampere-hour rating by the voltage of the battery. It is much the same as the ampere-hour rating, except watts are measured.

CAR CLINIC ▐████

PROBLEM: OVERCHARGED BATTERY

The battery on a customer's car (1982 Ford) seems to always overcharge. The battery seems very hot after operation and the customer has replaced the battery once. What could be the problem?

SOLUTION:

The problem is most likely tied to the voltage regulator. The voltage regulator is used to monitor and control the amount of current and voltage going back into the battery during charging. On older cars, the voltage regulator can be adjusted. If the voltage regulator is overcharging on a newer car, it should be replaced. This should solve the problem.

19.3 BATTERY MAINTENANCE AND TESTING

Hydrometer Testing

A *hydrometer* is used to measure the specific gravity of batteries that have filter caps. *Figure 19–10* shows a typical hydrometer. The hydrometer is made of a weighted float inside a glass tube. A bulb syringe is used to draw the electrolyte into the glass tube. Some hydrometers also have a temperature indicator or thermometer inside the float.

The object of the hydrometer is to determine the density of the electrolyte. The heavier or more dense the electrolyte, the greater the percentage of acid. The lighter or less dense the electrolyte, the less the percentage of acid. If the electrolyte is more dense, the float inside the hydrometer will not sink as deep into the solution. If the electrolyte is less dense, the float will sink deeper into the solution. The amount the float sinks into the solution is a measure of the specific gravity.

Figure 19–11 shows an example of how the hydrometer is used. The electrolyte solution is drawn into the glass tube. The depth the float sinks into the solution is read by reading the specific gravity stated at the top of the fluid. The inside of the float has a specific gravity scale to show the exact reading. Never take specific gravity readings immediately after the battery has been filled with water. The water and acid will not be completely mixed and the reading will be incorrect.

Temperature and Specific Gravity

The temperature of the electrolyte will have an affect on the exact specific gravity. Specific gravity readings must be changed to compensate for temperature. For each 10 degrees above 80 degrees F, 0.004 points must be added to the specific gravity reading. For every 10 degrees below 80 degrees F, 0.004 points must be subtracted from the measured specific gravity reading. For example, *Figure 19–12* shows a hydrometer correction scale. If the electrolyte solution is at 100 degrees F and the specific gravity is measured at 1.218, the corrected specific gravity will be 1.226. (For each 10 degrees above 80 degrees F, 0.004 points are added. 1.218 + 0.008 points = 1.226 corrected specific gravity reading.)

Battery Freezing

The freezing point of the electrolyte depends upon its specific gravity. A fully charged battery will never freeze. As the battery discharges and the specific gravity gets close to 1.000 or pure water, the chance of having the electrolyte freeze increases. If the electrolyte freezes, the battery case may crack and damage the battery. The table shown in

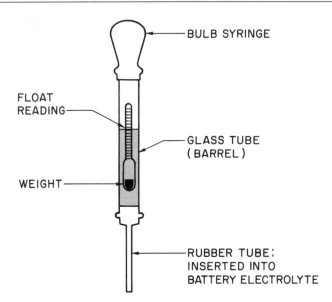

FIGURE 19–10. A hydrometer is used to measure the specific gravity of the battery electrolyte. Fluid is drawn into the glass tube and the float reading tells the specific gravity of the electrolyte.

FIGURE 19–11. When reading a hydrometer, the specific gravity reading is measured by lining up the level of the fluid with the floating bulb. The specific gravity reading is read on the floating bulb.

Figure 19–13 indicates the freezing temperatures of electrolytes at various specific gravities.

Specific Gravity on Maintenance-free Batteries

Since there are no filler vents on a maintenance-free battery, another method is used to measure specific gravity. Refer to *Figure 19–14*. On certain batteries, specific gravity is determined by a float device built into the battery. A small colored ball floats when the specific gravity is correct or in the right range. If the specific gravity is below the required level, the ball sinks. If the green dot is seen, the specific gravity is correct. If the indicator is black, the specific gravity is low. The battery should be charged. If the indicator shows light yellow, the battery electrolyte is too low. The battery may be damaged and should be replaced.

Other maintenance-free batteries have a test indicator that is also located on the top of the battery. When the test indicator shows "OK," the battery is satisfactory. This is shown by a blue or other color on the test indicator. If the test indicator is colorless, charging is necessary.

Load Tests

Several types of load tests can be performed on a battery. A light load test places a small 10–15 ampere load on the battery for a specific amount of time. The voltage of the cells is then checked against the manufacturer's specification. The specifications are listed in the service and repair

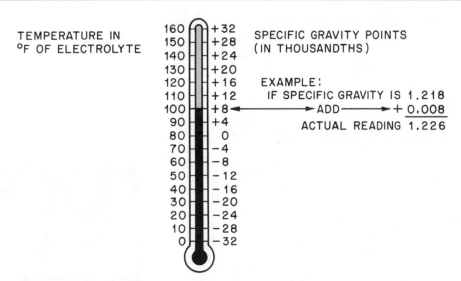

TEMPERATURE IN
°F OF ELECTROLYTE

SPECIFIC GRAVITY POINTS
(IN THOUSANDTHS)

EXAMPLE:
  IF SPECIFIC GRAVITY IS 1.218
           ADD   + 0.008
  ACTUAL READING 1.226

FIGURE 19–12. Specific gravity readings must be corrected for temperature according to this scale.

VALUE OF SPECIFIC GRAVITY	FREEZING TEMP. DEG. F.	VALUE OF SPECIFIC GRAVITY	FREEZING TEMP. DEG. F.
1.100	18	1.220	−31
1.120	13	1.240	−50
1.140	8	1.260	−75
1.160	1	1.280	−92
1.180	−6	1.300	−95
1.200	−17		

FIGURE 19–13. The freezing temperature of the electrolyte will change with different specific gravity readings.

manuals. The high-discharge and cold-cranking tests place a larger load, 300–500 amperes, on the battery for a short period of time, usually 15 seconds. The amount of amperes placed on the battery depends upon its size. Voltage readings are again taken to determine the voltage of each cell or the total battery. These readings are compared to specifications to determine the condition of the battery. The reserve-capacity test places the same load on the battery as the ampere-hour rating test. This rating is a time or minutes measurement.

Slow Charging Batteries

Batteries can be brought back to a fully charged state by two methods: the slow and fast charges. The *slow charge* is defined as charging the battery at a low ampere rating over a long period of time. Slow charging a battery is much more effective than fast charging. Slow charging causes the lead sulfate of the discharged battery plates to convert to lead peroxide and sponge lead throughout the thickness of the plate. See *Figure 19–15*.

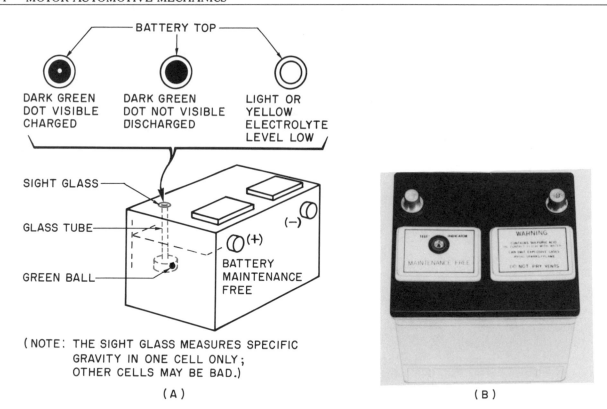

FIGURE 19-14. Specific gravity of some maintenance-free batteries is determined by an indicator. If the green dot can be seen, the specific gravity is correct. If there is a black spot, the specific gravity is too low, and the battery needs a charge. If the indicator is yellow, the battery possibly needs replacement. *(Courtesy of Chrysler Corporation, Sales and Marketing)*

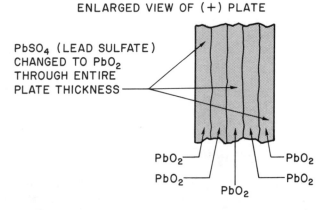

FIGURE 19-15. A slow charge will change the sulfated lead back to lead peroxide throughout the entire thickness of the plate.

The slow charge ampere rate is one ampere per positive plate in one cell. For example, if the cell has eleven plates, there are six negative and five positive plates. The slow charge ampere rate would then be 5 amps until the battery is fully charged. This may be in excess of 24 hours.

Fast Charging Batteries

Batteries can also be charged using a higher rate of amperes. Fast charging a battery only changes the lead sulfate on the outside of the battery plates. The internal parts of the plate are still lead sulfate. See *Figure 19-16*. Fast charges are used to "boost" the battery for immediate cranking power. *Figure 19-17* shows a table of time and charge rates for different specific gravity readings. Use this table only as a guideline.

FAST CHARGE

ENLARGED VIEW OF (+) PLATE

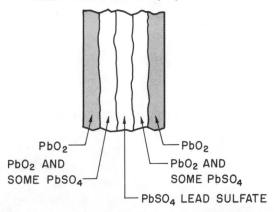

PbO$_2$

PbO$_2$ AND
SOME PbSO$_4$

PbO$_2$

PbO$_2$ AND
SOME PbSO$_4$

PbSO$_4$ LEAD SULFATE

FIGURE 19–16. When a battery is fast charged, only the outer surface of the plate changes to lead peroxide. The inner part of the plate is still lead sulfate.

BATTERY HIGH RATE CHARGE TIME SCHEDULE

SPECIFIC GRAVITY READING	CHARGE RATE AMPERES	BATTERY CAPACITY – AMPERE-HOURS			
		45	55	70	85
Above 1.225	5	★	★	★	★
1.200–1.225	35	30 min.	35 min.	45 min.	55 min.
1.175–1.200	35	40 min.	50 min.	60 min.	75 min.
1.150–1.175	35	50 min.	65 min.	80 min.	105 min.
1.125–1.150	35	65 min.	80 min.	100 min.	125 min.

★ Charge at 5-ampere rate until specific gravity reaches 1.250 @ 80°F.

FIGURE 19–17. This table shows the charging rates for different specific gravity ratings and battery capacity ratings. Use this chart as a guideline when fast charging a battery. Charge at 5-ampere rate only until the specific gravity reaches 1.250 at 80 degrees F. At no time during charging should the electrolyte temperature be higher than 125 degrees F.

IMPACT TESTING

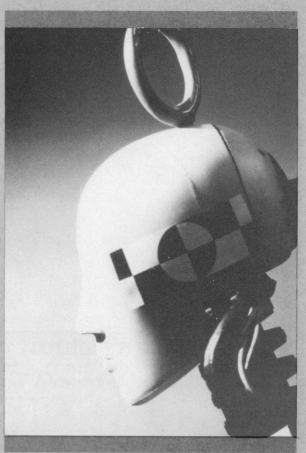

The purpose of impact testing is to find out precisely what happens in an impact-type accident. The data collected will provide engineers with information needed to design interiors, structural components, and passenger-restraint systems that help reduce occupant injuries.

These third generation dummies are the best developed to date. They approximate the human body in the way they suffer damage in accidents. They are available as males, females, children, and infants of different sizes. (Courtesy of General Motors Proving Grounds)

DIAGNOSIS and SERVICE

SAFETY TIPS

1. *The battery acid used in batteries is very toxic and corrosive. Never let the battery acid get on your clothes or skin. Battery acid can be neutralized by mixing baking soda with the acid. Always have a box of baking soda nearby when working on batteries. If battery acid gets into the eyes, flush with water immediately.*

2. *Always wear safety glasses when working around batteries.*

3. *Never place tools on the top of the battery, especially on older batteries that have the terminals on top. If the terminals are shorted by a tool, sparks will be produced and excessive current will heat the tool to its melting point, possibly causing injury to a person.*

4. *When batteries are discharged and charged, hydrogen gas, which can be highly explosive, is developed. Always make a point to keep sparks from the electrical system away from the battery.*

5. *Do not use an excessive charge rate or charge batteries with cells that are low on electrolyte.*

3. Always check batteries for cracks in the case and on the cover. Acid may leak out and damage other parts as well.

4. Never add sulfuric acid to a battery to increase its *specific gravity* readings. Only add water.
   a. Check the level of water in the battery by removing the vent caps (if not a maintenance-free battery).
   b. Using distilled water, add water to each cell until the water is level with the lower inside ring of the vent hole.
   c. Make sure the level of water is the same in each cell.
   d. Replace the vent caps and dry off any water spilled on the top of the battery case.

5. If a battery continually needs more water, this may be a sign of overcharging by the alternator. Check the regulator to see if it is operating correctly.

6. Keep the outer case of the battery clean from dirt and grime. A certain amount of battery leakage can occur through dirt and grime. Clean with a solution of water and baking soda.

CAUTION: *When working around batteries always wear safety glasses and protective clothing*.

1. *Clean* the *battery terminal* connections with a wire brush and baking soda. The battery terminals can build up with corrosion between the terminal and the battery cable connector, preventing the engine from starting. This is shown in *Figure 19–18*. This corrosion can also build up on batteries that have side terminals.

2. Batteries will *self-discharge* as shown in *Figure 19–19*. It is best to store a battery not in use, in a cool location. The cool temperature retards the chemical reaction and stops self-discharging.

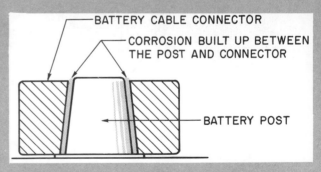

FIGURE 19–18. Corrosion can develop between the battery post and the cable connector. Use a wire brush to clean off the corrosion on the post and the cable connector.

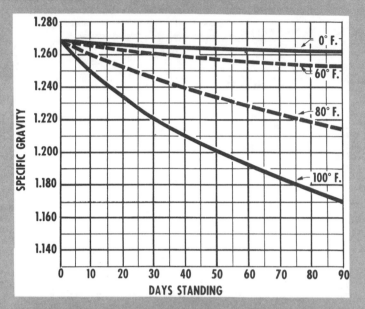

FIGURE 19–19. Batteries will self-discharge depending upon the temperature around the battery. Colder conditions slow down any chemical action that would take place during self-discharging. *(Courtesy of Delco-Remy Division of General Motors Corporation)*

7. If battery acid gets into your eyes, immediately flush the eyes with cool water and neutralize the acid with baking soda. Seek medical attention immediately.

8. If the battery terminal post on the top cannot be identified as positive or negative, the larger diameter top terminal post is usually the positive side. It may also have a (+) near it or be painted red.

9. Always connect booster or *jumper cables* as shown in *Figure 19–20*.

10. Be careful not to spill any electrolyte when using the hydrometer tester.

11. Use the proper tools, *Figure 19–21*, when servicing a battery.

12. When charging a battery, loosen or remove the vent cap plugs to allow gases to escape.

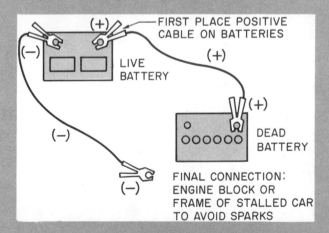

FIGURE 19–20. When boosting or jump starting a car, always connect the jumper cables as shown. Make the final connection between the engine block or frame to avoid sparks around or near the battery.

FIGURE 19–21. Always use the proper tools when servicing a battery. *(Courtesy of Snap-On Tools Corporation)*

13. Never lay tools across the battery top while working on the vehicle.

14. Always observe the manufacturer's recommendations when servicing, testing, or maintaining a battery.

15. Acid, dirt, corrosion, or cracks on the top of the battery can cause a slow discharge or leakage of the energy stored within the battery. If enough of this leakage occurs, there is a good chance that the battery will be too weak to start the engine. Use the following procedure to *check leakage*.
    a. Make sure the engine is off.
    b. Using a voltmeter, connect the black clip to the negative side of the battery.
    c. Lightly touch the red clip to various parts of the battery top and sides (not the battery post).
    d. Carefully watch the meter for any volt readings. There should be no volt readings.
    e. A volt reading, regardless of how small, indicates leakage. To correct it, thoroughly clean the battery with a water and baking soda solution, dry thoroughly, and re-test the battery for leakage.

16. Battery terminal clamps that are corroded beyond repair can be replaced. Remove the clamps from the battery post. Cut off the old clamp and strip about 3/4 inch of insulation from the cable. Clean the exposed copper wire strands and install a top- or side-mount replacement clamp. See *Figure 19–22*.

17. At no time during *charging a battery* should the electrolyte temperature exceed 125 degrees F. (The battery will feel very hot to the touch.) Depending upon the charger (see *Figure 19–23*), follow these general steps to charge a battery.
    a. Make sure the top of the battery is clean.
    b. Determine if the battery will be slow or fast charged. This will determine the amperage rating needed.
    c. With the battery charger disconnected from the power (110 volts), attach the positive lead of the charger to the positive terminal of the battery.
    d. Attach the negative lead of the battery charger to the negative terminal on the battery.
    e. At this point, plug in the charger to 110 volts and set the amperage reading.

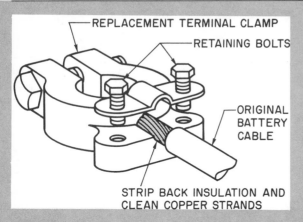

FIGURE 19-22. The cable connectors on the ends of battery cables can be replaced. Remove the old connector, cut back the insulation, and replace the connector with a new one.

FIGURE 19-23. Each battery charger may have a slightly different procedure to charge the battery. Always follow the battery charger manufacturer's recommended procedure for charging batteries. *(Courtesy of Snap-On Tools Corporation)*

f. Some battery chargers may also have a timer that should be set.

g. After the battery is charged, shut the charger off.

h. Remove the power cord (110 volts).

i. Now remove the positive and negative terminal leads between the battery and the battery charger. ***CAUTION****: The purpose of having the charger unplugged from the power when attaching or removing the terminal leads is to prevent an explosion. Any small spark could ignite explosive gases around the battery.*

18. Batteries can be tested for capacity. The capacity of a battery is its ability to furnish current and still maintain no less than a specified voltage. The testing procedure varies, but a *battery load tester* is needed. The general procedure is as follows:

a. Apply a load (current draw) that is approximately three times the ampere-hour rating of the battery to the battery.

b. Maintain this load for approximately 15 seconds while observing the voltage of the tester.

c. Check the voltage. It should be not less than 9.6 volts for a 12-volt battery.

d. If it is less than 9.6 volts, the battery may be discharged or defective.

e. If it is more than 9.6 volts, the condition of the battery is satisfactory.

SUMMARY

This chapter analyzed the storage battery and its design. The purpose of the storage battery is to store energy chemically so it can be used later as electrical energy. Two types of batteries are manufactured: the primary battery and secondary battery. The secondary battery can be recharged and is the type used in the automobile.

Batteries are made by combining a series of cells. A cell is made by using two metals and an acid. The positive plate metal is lead peroxide. The negative plate metal is sponge lead. The acid is a mixture of water and sulfuric acid. When these three components are placed together, approximately two volts are produced. This is called a lead-acid battery. Cells are added in series and parallel to get 12 volts and a larger amount of amperage capacity.

The combination water and acid is called an electrolyte. During discharging and charging, the electrolyte gives off explosive gases. Also, as the battery discharges, the two metals turn into lead sulfate and the electrolyte turns chemically into water.

The amount of acid in the electrolyte is measured by the specific gravity. Specific gravity is a measure of the weight of the electrolyte as compared to the weight of water. Water has a specific gravity of 1.000. A fully charged battery has a specific gravity of 1.270. As the battery is discharged, the specific gravity decreases.

The maintenance-free battery is used to reduce the gassing and maintenance. There are no caps on the battery for adding water. The chemicals inside are designed to reduce gassing and evaporation of water.

There are several types of battery ratings. The ratings usually tell the technician the amperage capacity or the ability of the battery to produce amperage. The cold-cranking performance rating is used to measure the cranking ability of the battery when the engine is cold. The reserve capacity rating measures how long the battery can supply electricity for emergency power when the engine is shut down. The ampere-hour rating measures the discharge of the battery at a specific amperage for a 20-hour period. The watt-hour rating is measured by multiplying the ampere-hour rating by the voltage of the battery.

There are several tests and maintenance procedures that can be performed on the battery. The hydrometer test is used to measure the specific gravity of the battery. When taking this test, the readings must be corrected for temperature. Also, note that if the battery's specific gravity gets too low, it may freeze and crack the battery.

Load tests are used to measure the battery's ability to handle an electrical load. The light load test places a small amperage on the battery to discharge it. The battery voltage is measured to determine its state or condition. The high-discharge or cold-cranking test places a larger amperage load on the battery, about 300–500 amps. Voltage is then measured to determine the battery's condition. The reserve-capacity test places a continuous load on the battery. The battery is discharged until the battery voltage drops below a certain point. The measurement is made in minutes.

Batteries can either be slow or fast charged. Slow charged batteries get a better and deeper charge. Fast charge batteries use more amperage initially to charge the battery, but the time of charge is relatively short.

There are many service tips that should be observed when working on batteries. Important tips include never work around a battery without safety glasses, use baking soda and water to flush battery acid from the eyes, use the proper tools when working on a battery, never smoke around a charging battery, and never spill acid on other parts or clothes.

TERMS TO KNOW

Can you explain each of the following terms? Review the chapter until you can use each term correctly.

Electro-chemical   Discharge
Primary battery   Charge
Secondary battery   Specific gravity
Battery cell   Cell density
Electrolyte   Hydrometer
Amperage capacity

REVIEW QUESTIONS

Multiple Choice

1. The positive plate of the lead-acid battery is made of:
   a. Lead oxide
   b. Lead peroxide
   c. Sponge lead
   d. Sulfuric acid
   e. None of the above

2. The negative plate of the lead-acid battery is made of:
   a. Lead oxide
   b. Lead peroxide
   c. Sponge lead
   d. Sulfuric acid
   e. None of the above

3. The purpose of the battery is to store _____ energy for use later as electrical energy.
   a. Radiant
   b. Chemical
   c. Thermal
   d. Nuclear
   e. Mechanical

4. Which type of battery can be recharged?
   a. Primary battery
   b. Antimony battery
   c. Secondary battery
   d. Sulfur battery
   e. None of the above

5. As a lead-acid battery is discharged, the negative plate is converted to:
   a. Lead sulfate
   b. Sponge lead
   c. Water
   d. Sulfuric acid
   e. None of the above

6. As a lead-acid battery is discharged, the positive plate is converted to:
   a. Lead sulfate
   b. Sponge lead
   c. Water
   d. Sulfuric acid
   e. None of the above

7. When battery cells are placed in series, the voltage of the battery will:
   a. Be reduced to zero
   b. Be additive
   c. Remain the same
   d. Increase by one half of each cell voltage
   e. None of the above

8. When battery cells are placed in parallel, the amperage capacity of the battery will:
   a. Be reduced to zero
   b. Be increased
   c. Remain the same
   d. Decrease
   e. None of the above

9. What is the ratio of water and sulfuric acid in a lead-acid battery?
   a. 10% water, 90% acid
   b. 20% water, 80% acid
   c. 30% water, 70% acid
   d. 50% water, 50% acid
   e. None of the above

10. What is the specific gravity of a fully charged battery?
    a. 1.000–1.002
    b. 1.128–1.130
    c. 1.270–1.295
    d. 1.940–2.300
    e. 1.750–2.100

11. The number of watts that can be discharged per hour/pound of battery is called:
    a. Electrolyte
    b. Amperage capacity
    c. Cell density
    d. Active materials
    e. Hydrometer testing

12. Which battery rating is used to measure the battery's ability to provide electricity when the engine is shut off?
    a. Cold-cranking performance
    b. Reserve capacity rating
    c. Ampere-hour rating
    d. Watt-hour rating
    e. None of the above

13. The object of a hydrometer test is to determine the _____ of the battery.
    a. Voltage
    b. Amperage
    c. Wattage
    d. Specific gravity
    e. Amount of lead sulfate

14. What correction is needed to compensate for the temperature of the electrolyte when taking specific gravity readings?
    a. Increase 0.004 for each 10 degrees above 80 degrees F.
    b. Increase 0.008 for each 10 degrees above 80 degrees F.
    c. Decrease 0.004 for each 10 degrees above 80 degrees F.
    d. Decrease 0.004 for each 1 degree above 80 degrees F.
    e. None of the above.

15. Which type of charge will give the battery the deepest and most effective charge?
    a. Fast charge
    b. Boost charge
    c. Intermediate charge
    d. Slow charge
    e. No charge

16. What should be used when battery acid gets into the eyes?
    a. Water only
    b. Lead sulfate
    c. Sulfuric acid
    d. Baking soda and water
    e. Sponge lead

17. When a battery is stored or not used, it is better to store the battery in a _____ location.
    a. Warm
    b. 80 degrees or above
    c. Cool
    d. Damp, moist
    e. None of the above

The following questions are similar in format to ASE (Automotive Service Excellence) test questions.

18. Technician A says that maintenance-free batteries cannot be charged. Technician B says that maintenance-free batteries cannot be checked for specific gravity using a hydrometer. Who is right?
    a. A only
    b. B only
    c. Both A and B
    d. Neither A nor B

19. Battery acid has just been spilled on a person's clothing. Technician A says to wash the acid off with baking soda and water. Technician B says to wash the acid off with vinegar. Who is right?
    a. A only
    b. B only
    c. Both A and B
    d. Neither A nor B

20. While being charged, a battery becomes very hot. Technician A says the heat is normal and there is no need for concern. Technician B says that the battery could be overcharging and the amount of charge should be reduced. Who is right?
    a. A only
    b. B only
    c. Both A and B
    d. Neither A nor B

21. Technician A says the larger battery terminal is positive and the smaller terminal is negative. Technician B says the smaller battery terminal is positive and the larger terminal is negative. Who is right?
    a. A only
    b. B only
    c. Both A and B
    d. Neither A nor B

Essay

22. What is the difference between primary batteries and secondary batteries?

23. What is a definition of the specific gravity of a battery?

24. Define cell density of a battery.

25. What is the definition of the cold-cranking performance rating of a battery?

26. What is the purpose of a hydrometer when used with a battery?

27. What happens to the specific gravity as temperature increases?

VISUALIZATION OF WAKE FLOW

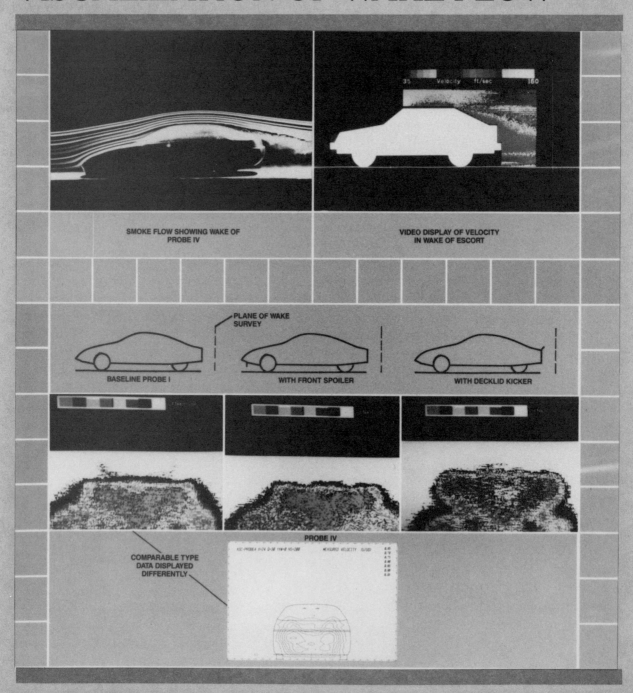

SMOKE FLOW SHOWING WAKE OF
PROBE IV

VIDEO DISPLAY OF VELOCITY
IN WAKE OF ESCORT

PLANE OF WAKE
SURVEY

BASELINE PROBE I

WITH FRONT SPOILER

WITH DECKLID KICKER

PROBE IV

COMPARABLE TYPE
DATA DISPLAYED
DIFFERENTLY

The amount of wake that is produced at the rear of the vehicle can have significant effects on the drag of the vehicle. Computerized color video displays are currently being used to aid in visualization. Note the differences in the wakes of the
three different vehicle designs shown. (Courtesy of Ford Motor Company)

CHAPTER 20

Ignition Systems

INTRODUCTION

In order for proper combustion to take place, the air and fuel mixture must be ignited. The ignition system is designed to produce the spark within the combustion chamber at a precise moment of time. This chapter discusses the components that make up the ignition system. These include the coil, points, spark plugs, advance mechanisms, and electronic and computer systems for ignition.

OBJECTIVES

After reading this chapter, you will be able to:
- Identify the parts and operation of the conventional ignition system that uses contact points.
- Define the operation of the primary and secondary circuits.
- Examine the operation and purpose of advance mechanisms.
- Identify spark plug design and operation.
- Analyze the electronic spark control systems used on vehicles today.
- State common diagnosis and service procedures used on the ignition system.

CHAPTER HIGHLIGHTS

20.1 CONTACT POINT (CONVENTIONAL) IGNITION SYSTEMS
    A. Purpose of the Ignition System
    B. Conventional Ignition System Components and Operation
    C. Ignition Switch
    D. Ballast Resistor
    E. Ignition Coil
    F. Breaker Points
    G. Dwell
    H. Contact Points' Gap Adjustment
    I. Effect of Pitting on Contact Points
    J. Distributor
    K. Condenser
    L. Distributor Rotor and Cap

20.2 ADVANCE MECHANISMS
    A. Purpose of Advancing the Spark
    B. Initial Timing
    C. Centrifugal Advance
    D. Vacuum Advance
    E. Advance and Emissions

20.3 SPARK PLUGS
    A. Spark Plug Design
    B. Spark Plug Heat Range
    C. Factors Affecting Spark Plug Temperatures
    D. Factors Affecting Spark Plug Voltages
    E. Variations in Spark Plug Design

CHAPTER HIGHLIGHTS (CONTINUED)

20.1 CONTACT POINT (CONVENTIONAL) IGNITION SYSTEMS

Purpose of the Ignition System

The air and fuel in the combustion chamber must be ignited at a precise point in time during the four-stroke cycle. This is done by causing an electrical spark to jump across a gap on a spark plug. About 5,000 to 50,000 volts are needed to force the electrical current to jump across the spark plug gap. However, there is only a 12-volt battery within the automobile. The ignition system is used to increase the voltage to the necessary amount at the right time for the spark to occur. In addition, the time of spark must also be altered as speed and load increase or decrease. Advance and retard mechanisms are used to accomplish this.

The ignition system has been developed and changed as the automobile has changed. The conventional ignition system using a mechanical set of points and condensers has been updated to an electronic ignition system using semiconductors and transistors. From this development, electronic ignition systems now work directly with computer-controlled systems. In order to analyze these ignition systems, the conventional ignition system will be studied and compared to the newer computer-controlled systems.

Conventional Ignition System Components and Operation

There are two separate circuits in the ignition system: the *primary circuit* and the *secondary circuit*. The primary circuit is considered the low-voltage circuit. Low voltage is battery voltage, or about 12 volts. The secondary circuit is called the high-voltage circuit. Components in this circuit operate at voltages between 5,000 and 50,000 volts, depending upon the type of system being used.

The primary circuit operates as shown in *Figure 20–1*.

1. *The battery* is used to provide a source of electrical energy needed to operate the system. The negative side is grounded to the frame and the positive side is fed directly to the ignition switch.

2. *The ignition switch* connects or disconnects the flow of electricity to the ignition system. The ignition switch will direct the current through a bypass route during cranking and through the *ballast resistor* during normal operation. The ballast resistor can also be a resistive-type wire.

3. *The ballast resistor* controls the current flow to the coil during normal operation. The resistor reduces the voltage available to the coil at low engine speeds. It increases the voltage at higher rpm when the voltage requirement increases.

4. *The resistor bypass circuit* is used only when the engine is being cranked. During this time, more voltage is needed at the coil to produce spark. When the operator stops cranking the engine, the electrical current flows through the ballast resistor.

5. *The primary coil windings* are used to convert the electrical energy into a magnetic field. When electricity passes through the primary windings, a strong magnetic field is produced. This magnetic field also surrounds the secondary windings.

6. *Breaker points* are used to close and open the primary circuit. As the distributor shaft rotates, it also causes a distributor cam to rotate. The *distributor cam* causes a small set of points to open and close at each cam lobe position. When the points are closed, current flow in the primary circuit causes the magnetic field to build up in the coil. When the points are opened, the current flow stops. This causes the magnetic field in the coil to collapse. The sudden collapse of the primary magnetic field produces a strong induced voltage in the secondary windings.

7. *The condenser* is used to reduce the amount of arcing when the points open the primary circuit. Whenever an electrical circuit is broken, it produces arcing. This arcing can cause the points to be pitted or corroded. The condenser helps to reduce corrosion on the points.

The secondary circuit is called the high-voltage circuit. Depending upon the system, the voltage in the secondary circuit may be as high as 50,000 volts. The components in the secondary circuit operate as shown in Figure 20–1.

8. *The secondary coil windings* are used to capture the voltage produced by the collapsing primary

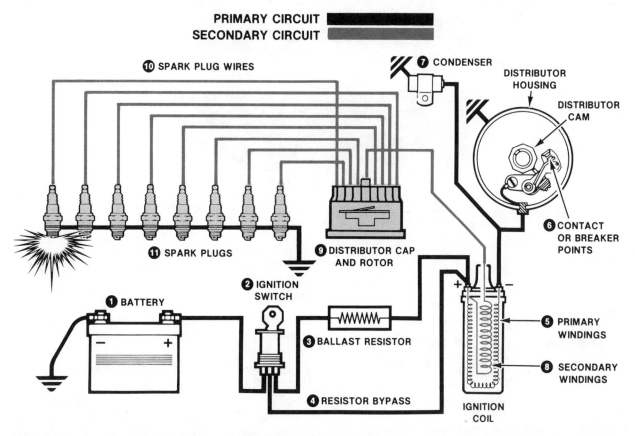

FIGURE 20-1. Both primary and secondary circuits are used in the ignition system. The primary circuit operates on 12 volts and the secondary circuit operates on 5,000 to 50,000 volts. *(Courtesy of Sun Electric Corporation)*

magnetic field. There is approximately a 12–20,000 volt (1–1,666) ratio of windings between the primary and the secondary circuits. This means that for each volt in the primary windings, there are about 1,666 volts in the secondary windings. This will produce 20,000 volts on the secondary circuit. This high voltage available at the secondary windings is fed to the coil tower when the primary magnetic field collapses.

9. *The distributor cap and rotor* are used to distribute the surges of high voltage available at the coil tower. The high-voltage surge is sent from the coil to the center of the distributor by the coil wire. The surges are then directed, one at a time, to each outer terminal of the distributor cap by the rotor. The rotor is turned by the distributor shaft.

10. *The spark plug wires* are used to connect the high-voltage surge to each spark plug. They are arranged in the firing order of the engine.

11. The spark plugs provide a predetermined gap within the combustion chamber so that each time

a high-voltage surge is delivered, a quality spark will occur in the combustion chamber.

Ignition Switch

The ignition switch shown in *Figure 20–2* uses a key and is attached to the dashboard of the engine. There are several connections made by the ignition switch. When the key is in the off position, the ignition switch is off and no current flows in the circuit. When the ignition switch is turned to crank or start the engine, current flows from the battery directly to the ignition coil. When the switch is on run, the current flows through the ballast resistor to the coil. Current also flows to other circuits. *Figure 20–3* shows the common positions of the ignition switch. Note that the reason for having an accessory position is to operate the radio and lights when the engine is shut off. If the ignition switch is placed in the run or on position, the accessories will work. However, if the breaker points are closed, the ignition coil will be energized. This may cause the battery to discharge, the coil to overheat, or the ignition points to burn out.

FIGURE 20–2. The ignition switch turns the ignition circuit on and off.

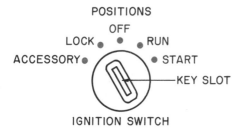

FIGURE 20–3. The ignition switch can be placed in several positions for controlling different circuits.

Ballast Resistor

The ballast resistor is used to reduce the voltage to the coil during low-speed operation. During this time there is too much voltage. The points may become hot and start to pit. The resistor is used to drop the voltage to about 7–10 volts, depending upon the engine, before going to the coil. For example, when the engine is at low speed, the contact points are opening and closing slower for a specific period of time. There is more time for the current to flow in the primary circuit. This heats up the wire in the resistor and causes a larger voltage drop. As the engine speeds up, the opening and closing of the points occurs faster. The current in the primary circuit is less for an equal period of time. Now the resistive wire cools and there is less resistance. This means the voltage available at the coil increases. Some vehicles use a resistive wire that is built into the wiring harness to change the voltage drop as the speed changes. In other vehicles, the resistor is placed inside the coil itself. In older vehicles, the resistor was made of a resistive wire built into a ceramic block.

Ignition Coil

The ignition coil is considered a *transformer*. **Figure 20–4** shows a typical 12-volt coil. Its purpose is to change 12 volts at the battery to 20,000–30,000 volts for the spark plug. This is done by using primary and secondary windings located inside the coil. The primary coil has several hundred turns of thick wire, while the secondary coil has thousands of turns of very thin wire. The ratio of turns between the primary and secondary windings is determined by the ratio of voltage from the primary winding to the secondary winding.

In operation, as the contact points close, current is sent through the primary coil. During this time, a magnetic field is built up around the primary winding. When the points open, primary current stops and the magnetic field col-

FIGURE 20–4. The ignition coil, a transformer, converts 12 volts into 20,000 volts. It uses a set of primary and secondary coil windings.

lapses. At this point, three components are needed to produce a voltage in the secondary winding. These include wire, a magnetic field, and movement between the two. The movement between the two is the collapsing magnetic field. The faster the magnetic field collapses, the greater the voltage produced in the secondary winding. The slower the magnetic field collapses, the less voltage produced in the secondary winding.

Breaker Points

The purpose of the breaker points is to open the primary circuit so there can be a collapse of the primary magnetic field. This is done by using a set of contact points as shown in *Figure 20–5*. Contact points are considered switches. The cam located on the distributor shaft is used to open and close the points. There is one cam lobe for each cylinder of the engine. There is a rubbing block on the points that rides against the distributor cam. The point gap is adjusted by repositioning the stationary contact. A spring is used to keep the contacts closed.

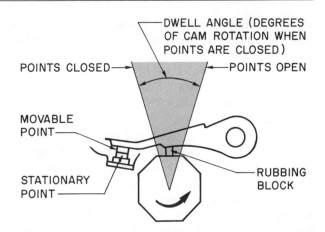

FIGURE 20–6. Dwell is defined as the length of time the points are closed in degrees of rotation.

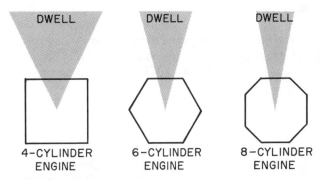

FIGURE 20–7. Four-cylinder engines have a greater dwell time than 8-cylinder engines.

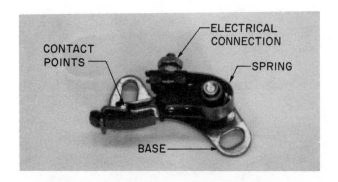

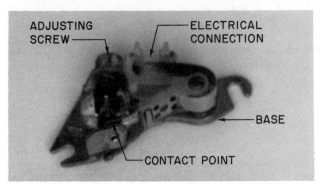

FIGURE 20–5. The ignition contact points are used to open and close the primary circuit. They are operated by a cam on the distributor shaft.

Dwell

Dwell is defined as the length of time the points remain closed. *Figure 20–6* shows the dwell on a typical engine. As the distributor cam rotates, the points are opened and closed. Dwell is important because during dwell the magnetic field is increasing. As the dwell time increases, the magnetic field buildup increases, producing a greater secondary voltage. *Figure 20–7* shows the different dwells for different cylinder engines. Four-cylinder engines have a greater amount of time that the points remain closed (greater dwell). This causes a greater magnetic field buildup around the primary coil, which produces a greater secondary voltage.

Contact Points' Gap Adjustment

It is critical that the contact or breaker points be adjusted to a precise clearance. This adjustment determines exactly when the points are opened and closed by the distributor cam. This affects the time of spark. Points are adjusted by setting the points in the most open position on a cam lobe. The clearance is then changed by adjusting the stationary contact. The clearance will also change as the rubbing block on the point assembly wears down. A common adjustment for a vehicle is 0.017 inch.

The point gap affects timing and dwell on the ignition system. If the points are adjusted too small or the gap decreases, the timing of the spark will be retarded. If the points are adjusted too large or the gap increases, the timing of the spark will be advanced. Also, as the point gap decreases, the dwell is increased slightly and vice versa. See *Figure 20–8*. Normal wear causes the point gap to decrease and ignition timing to be retarded.

Effect of Pitting on Contact Points

It is critical to the ignition operation that the points stop the primary current from flowing as quickly as possible. The quicker the current stops, the faster the collapse of the magnetic field, and the greater the secondary voltage produced. As points operate over a period of time, they become pitted and corroded. *Figure 20–9* shows what happens if

the points are corroded. There is little arcing on a good set of points when they are open. The current in the primary coil stops rapidly. When the points are corroded, there is greater arcing and the points stop the primary current much slower. This will produce a lower secondary voltage and cause the engine to operate incorrectly or run poorly.

Distributor

The distributor is used to hold many of the ignition components. It has a center shaft that is driven by the main camshaft of the engine. This is done by using a small helical gear that meshes with a similar gear on the camshaft. On many engines, there is also a slot on the bottom of the distributor shaft. This slot is used to turn the oil pump. It must be aligned during installation. In addition to a helical gear, certain manufacturers use a simple slot or offset slot to turn the distributor shaft.

The distributor is used to turn the distributor cam. In addition, the distributor holds the contact points, condenser, and advance mechanisms, and supports the rotor and distributor cap. *Figure 20–10* shows a typical distributor for a conventional ignition system.

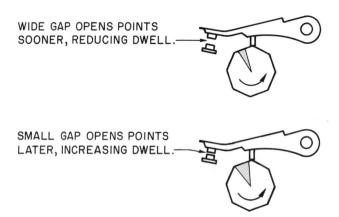

FIGURE 20–8. As point gap decreases, the dwell is increased. As point gap increases, the dwell is decreased.

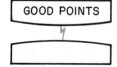 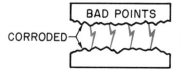

FIGURE 20–9. If the points are corroded, there will be arcing. The arcing makes the primary current stop at a much slower rate than it would with a good set of points.

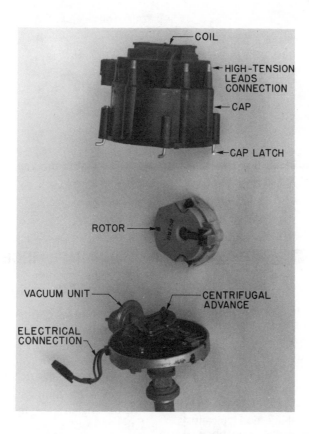

FIGURE 20–10. The distributor holds the primary ignition parts, the rotor, and the distributor cap.

Condenser

The purpose of the *condenser* in the ignition system is to protect the ignition points from pitting and corrosion. The condenser also increases the spark to the plugs. It does this in several ways. When the points open, a small amount of arcing is produced. This arcing reduces the speed at which the magnetic field collapses. The condenser is used to absorb the current that normally would arc across the contact points. This causes the current to stop in the primary much faster, increasing the spark plug voltage.

The condenser, also called a *capacitor*, is made of two metal plates that accept an electrical charge. These two plates are rolled together to increase their capacity. No current actually runs through the condenser. It stores the excess electrons that would have arced across the points.

The condenser also increases the voltage produced at the spark plug in another way. When the points open, the condenser will fully charge. The instant the spark occurs on the spark plug, the condenser tries to discharge itself back through the coil in a reverse direction. This action causes the magnetic field in the coil to collapse even faster, producing a greater sustained spark.

Distributor Rotor and Cap

The distributor rotor and cap are used to distribute the secondary voltage to each spark plug. *Figure 20–11* shows a rotor and cap used on an 8-cylinder engine. They are made from insulating materials such as bakelite, plastic, or epoxy that are easily shaped. Conductors are placed inside the material to allow high-voltage electricity to pass through.

The high voltage produced each time the points open is sent from the high voltage terminal on the coil to the center of the distributor cap, which is called the coil tower. The high voltage is then sent inside the cap to the rotor, which is being turned by the distributor shaft. As the rotor revolves inside the cap, it distributes the high voltage to each cylinder according to the firing order of the engine. There are many shapes and sizes of rotors used on older engines.

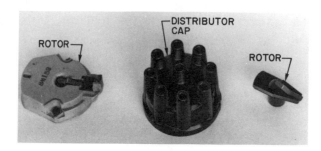

FIGURE 20-11. Examples of rotors and a distributor cap used on an engine.

CAR CLINIC

PROBLEM: TOO HIGH IDLE

A car that has a 2.5 liter engine with fuel injection has just been purchased. The vehicle should normally get about 35 mpg, it is only getting about 25 mpg. The idle is very high and the car is very hard to slow down. What can be done?

SOLUTION:

GM cars that use a computer and fuel injection typically have few controls for idle. On some vehicles, the PROM may be programmed for different altitudes. The engine is getting too much fuel and is running very rich. In this case, the PROM should be replaced. Go to the manufacturer to get the PROM. The manufacturer is the only source for the correct PROM.

20.2 ADVANCE MECHANISMS

Purpose of Advancing the Spark

Ignition timing is defined as the time in degrees of crankshaft rotation that the spark occurs during idle. This is called the initial timing. The initial timing is adjusted when the engine is at idle. It is usually adjusted several degrees before top dead center (BTDC). As the engine speed and load increase, the timing must also increase.

As the engine speed increases, the piston moves faster and the time of spark must also be advanced. This is because the crankshaft will move farther during the time the combustion occurs. There must be enough time for the combustion to be complete. Referring to *Figure 20-12*, when the engine is at 1,200 rpm, the spark occurs about 6 degrees BTDC. At 23 degrees after top dead center (ATDC), the combustion ends. This will produce an even power pulse to the

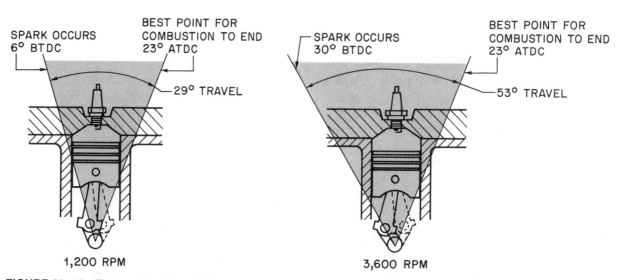

FIGURE 20-12. The speed of the engine determines how much spark advance is needed. As the engine speeds up, more advance is required to complete the combustion on time.

piston on the power stroke. As the engine is run faster, say at 3,600 rpm, the timing must be increased to as far as 30 degrees BTDC so the combustion can end at 23 degrees ATDC. Of course, these figures will vary with different engines. The principle, however, is the same.

Load also affects when the timing should occur. If the load is increased, different amounts of air and fuel are needed in the combustion chamber. It takes longer for the combustion to occur and the timing must be adjusted.

Initial Timing

The initial timing is set by adjusting the distributor. When the distributor is placed in the engine, the distributor shaft gear meshes with the engine camshaft. The distributor housing can, however, be rotated. When the housing is rotated, the timing of the engine will be either advanced or retarded. Timing is set by using a timing light attached to the number one cylinder. When the number one cylinder fires, the timing light flashes. If this light is pointed toward the timing marks on the crankshaft of the engine, the exact timing of the engine can be determined. See *Figure 20–13*.

Centrifugal Advance

The centrifugal advance increases or decreases the timing based on the speed of the engine. There are two weights on the distributor shaft, *Figure 20–14*. Springs are used to hold them inward. As these weights turn, their centrifugal force causes them to spin outward. The outward movement of the weights causes the time in which the points open and close to be different. On some engines, weight movement causes the breaker plate (the plate the points are attached to) to move so the points open earlier. On other engines, the movement of the weights causes the position of the distributor cam to change. This causes the cam to open the points at a different time. *Figure 20–15* shows the effect of engine rpm and the increase in timing advance.

Vacuum Advance

The vacuum advance is used to increase the timing of the engine as the load increases. As load is applied to the engine, the intake manifold vacuum is reduced. The change in the intake manifold vacuum is used to advance the timing.

A diaphragm is attached to the side of the distributor. One side of the diaphragm is mechanically attached to the

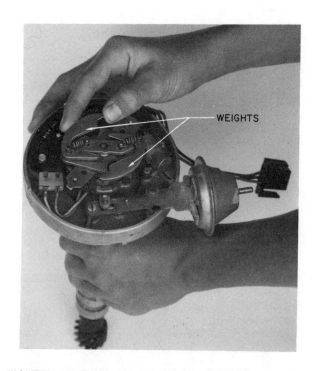

FIGURE 20–14. Centrifugal weights are used to advance the timing of the spark based on rpm.

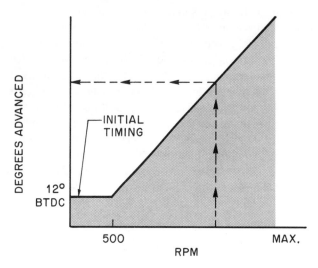

FIGURE 20–15. This chart shows the effect of engine rpm and the increased timing of the spark. The initial timing is set at idle.

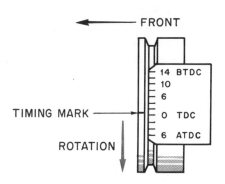

FIGURE 20–13. Timing marks are used to help determine the spark advance of the engine.

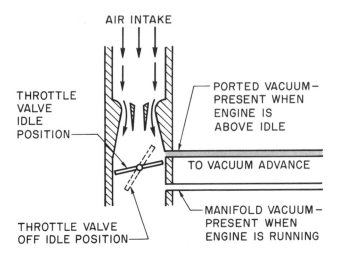

FIGURE 20-16. Ported vacuum is taken from slightly above the throttle plate.

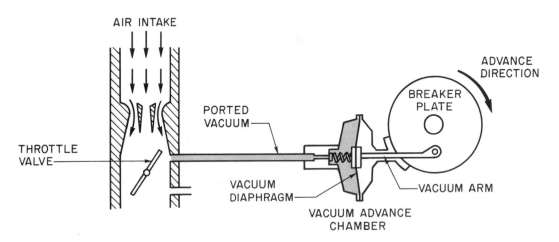

FIGURE 20-17. Vacuum is used to rotate the breaker plate so the timing can be retarded or advanced based on load.

breaker plate by a small rod. The other end of the diaphragm is attached to the intake manifold vacuum. The intake manifold vacuum used is considered *ported vacuum*. Ported vacuum is present only when the engine is above idle. It is taken from slightly above the throttle plate when the throttle plate is in the closed position. See *Figure 20-16*. A spring is used to force the diaphragm and breaker plate to a retarded position. As the engine throttle is opened for increased load, the vacuum pulls back on the diaphragm against spring pressure. This rotates the breaker plate and advances the engine spark. See *Figure 20-17*. *Figure 20-18* shows how the vacuum advance is added to the centrifugal advance to obtain the total advance of the engine.

Advance and Emissions

Since the introduction of emission control devices, many systems have been added to control the spark more effectively. The goal has been to retard the spark during idle, or

to delay the advance. By controlling the advance more accurately, combustion chamber temperatures have been raised. This reduces the amount of hydrocarbon emissions.

Many systems have been used. The following systems are the most common:

1. *Spark delay* — a valve placed in the vacuum line from the distributor vacuum advance to the carburetor. The spark delay is used to delay the spark during acceleration.

2. *Temperature sensing vacuum control valve* — used to increase the advance when the engine coolant temperatures are too high. Advancing the engine causes the rpm to increase, which helps to cool the engine.

3. *Dual diaphragm vacuum advance* — used to apply both the ported and manifold vacuum to the advance. Its purpose is to improve starting and to retard initial timing to reduce exhaust emissions.

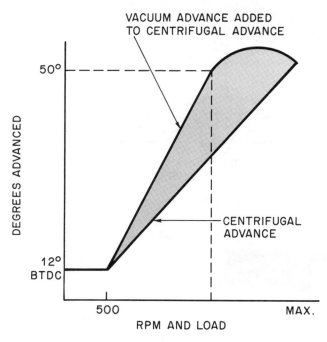

FIGURE 20–18. Vacuum advance is added to centrifugal advance during engine operation.

4. *Solenoid retard-advance* — used to control the timing by using a solenoid to operate the advance for starting.

CAR CLINIC

PROBLEM: SPARK KNOCK

A 1983 GM engine is constantly knocking when the vehicle goes up hill. This started after about 50,000 miles. Using a higher octane fuel helps a little, but the engine manufacturer says a lower octane fuel is OK. Many ignition parts have already been replaced with no improvement.

SOLUTION:

On GM engines, the ignition is controlled by the PROM in the vehicle computer. The original PROM may have allowed too much advance on the ignition under engine load and 1983 octane conditions. The new PROM advances timing less at low intake manifold pressures, when knock is hard to control. Check with the manufacturer and replace the PROM.

20.3 SPARK PLUGS

Spark Plug Design

The purpose of the spark plug is to provide a place for a spark that is strong enough to ignite the air/fuel mixture to occur inside the combustion chamber. This is done by

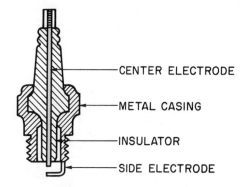

FIGURE 20–19. A spark plug is made of a center electrode, an insulator, a metal casing, and the side electrode. *(Courtesy of Champion Spark Plug Company)*

causing a high voltage to arc across a gap on the spark plug. Spark plugs are designed as shown in *Figure 20–19*. The center electrode is a thick metal wire that runs through the plug. Its purpose is to conduct electricity from the high-voltage wire to the combustion chamber area. The insulator is a porcelain-like casing that surrounds the center electrode. The upper and lower portions of the center electrode are exposed. The metal casing is a threaded casing used for installing the spark plug into the cylinder head. It has threads and is hex-shaped to fit into a spark plug socket. The side electrode is a short, thick wire made of nickel alloy. It extends about 0.020–0.080 inch away from the center electrode. Its position creates the gap for the spark to jump across.

Spark Plug Heat Range

The heat range of a spark plug refers to its thermal characteristics. The thermal characteristics of a plug are a measure of how fast the plug can transfer combustion heat away from its firing end to the cylinder head of the engine, *Figure 20–20*. Plugs are considered to be cold or hot. There is a certain amount of thermal temperature at the time of the spark. If the plug tip temperature is too cold, the plug may foul out with carbon, oil, and other combustion deposits. If the plug tip temperature is too hot, preignition occurs and the plug and piston may be damaged. *Figure 20–21* shows the effect on the engine when plugs are too hot or cold. The tip temperature is shown as the vertical axis. The load on the engine is shown as the horizontal axis. When too cold a plug is used, carbon and oil fouling occur at idle conditions. When the plug is too hot, electrode burning and preignition occur at full load.

The spark plug heat range is changed by changing the length of the insulator nose. Hot plugs have relatively long insulator noses with a long heat flow path to the cylinder head. Cold plugs have a short insulator nose with a short heat flow path to the cylinder head. *Figure 20–22* shows the difference in design of hot and cold plugs.

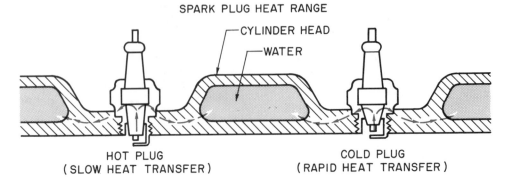

FIGURE 20–20. Plug heat ranges are based on the length of time it takes to remove the heat from the tip of the spark plug.

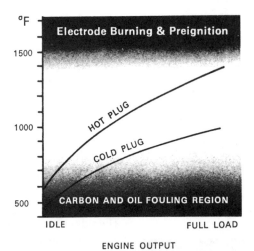

FIGURE 20–21. When plugs are too cold, they may foul out. If plugs are too hot, they may cause preignition. *(Courtesy of Champion Spark Plug Company)*

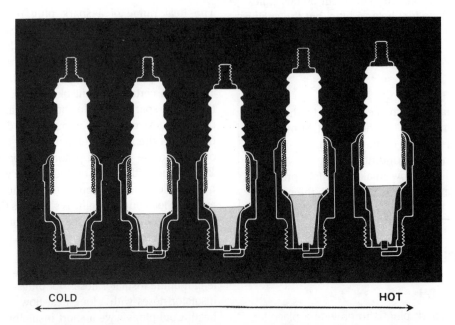

FIGURE 20–22. Spark plug heat ranges are changed by changing the length of the insulator nose inside the spark plug. *(Courtesy of Champion Spark Plug Company)*

DRAG COEFFICIENT

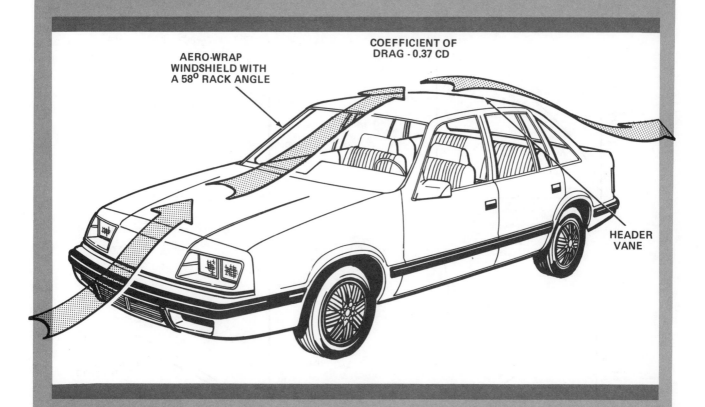

AERO-WRAP
WINDSHIELD WITH
A 58° RACK ANGLE

COEFFICIENT OF
DRAG - 0.37 CD

HEADER
VANE

The aerodynamic design of many new vehicles helps to reduce the drag coefficient. The external shape of the vehicle (windshield angle, header, and so on) helps to reduce the drag coefficient. This vehicle has a drag coefficient of 0.37. (Courtesy of Chrysler Corporation)

Factors Affecting Spark Plug Temperatures

Many operational factors also affect the temperature of spark plugs. *Figure 20–23* shows some of the factors. Insulator tip temperature is shown on the left axis of each chart. Different factors are listed on the bottom axis of each chart.

Each chart shows the type of relationship. Ignition timing causes spark plug temperature to change. Figure 20–23 (A) shows that as the engine is over-advanced, tip temperature increases. Figure 20–23 (B) shows that as coolant temperature increases, the tip temperature also increases. Figure 20–23 (C) shows that as more detonation occurs from lower octane fuel, tip temperature also increases. Figure

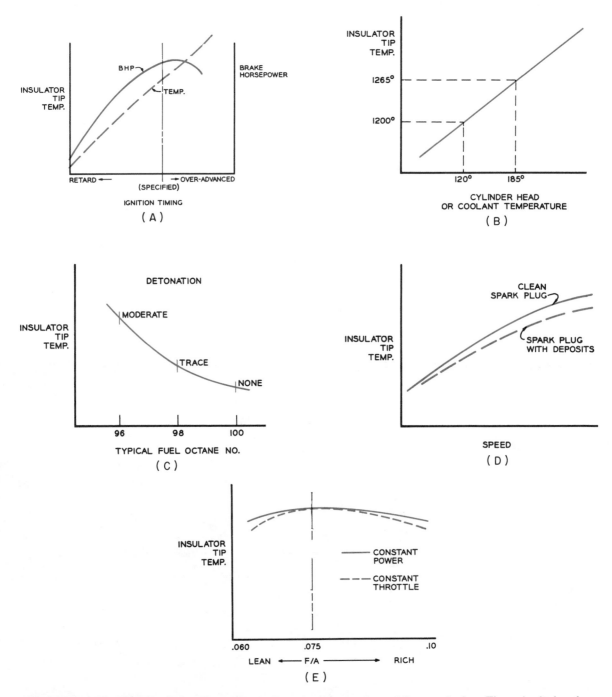

FIGURE 20–23. There are many factors that affect the temperature of the spark plug. These include advance (A), coolant temperature (B), detonation (C), condition of spark plugs (D), and air-fuel ratio (E). *(Courtesy of Champion Spark Plug Company)*

20–23 (D) shows that clean spark plugs have higher tip temperatures than spark plugs with deposits. Figure 20–23 (E) shows that as the air-fuel ratio changes from rich to lean, insulator tip temperatures also change.

Factors Affecting Spark Plug Voltages

The voltage required by spark plugs also changes with various factors. *Figure 20–24* shows how new and old plugs require a different voltage from the ignition system. On the chart, line A is the voltage available from the ignition system. Line B is the kilovolts required to fire the spark plug. When old plugs are used, the voltage required to spark is increased. Older plugs may misfire at certain times.

Other factors also affect the voltage requirements. *Figure 20–25* shows some of these factors. Figure 20–25 (A) shows that as compression pressures increase, voltage requirements also increase. Figure 20–25 (B) shows that as plug gap is widened, voltage requirements increase. Figure 20–25 (C) shows that as the throttle or load is increased, the voltage requirements also increase. Figure 20–25 (D) shows that as the engine timing is advanced, voltage requirements decrease.

Variations in Spark Plug Design

Automotive manufacturers today require many variations in the spark plug. These may include:

1. The number of threads on the spark plug.
2. The design of the gasket for sealing the spark plug to the cylinder, or the use of a tapered seat rather than a gasket.
3. The type of resistor element inside the plug to eliminate interference in radios and TVs.

4. The type of electrode used to establish the spark plug gap.

20.4 ELECTRONIC IGNITION SYSTEMS

Purpose of Electronic Ignition Systems

One of the biggest problems with the conventional ignition system is the wear on the points and the speed at which the primary current is stopped. Conventional ignition systems use a set of contact points to do this. With the introduction of solid-state components and transistors in the 1970s, however, many manufacturers converted the conventional ignition system to solid state. These ignition systems were developed and improved over a period of time. Systems were called capacitive discharge ignition systems, breaker-less ignition systems, solid-state ignition systems, electronic ignition systems, and high-energy ignition systems, but they all had one thing in common. They used semiconductors (transistors) in circuits to open the primary circuit faster. The secondary circuit remained much the same, except that higher voltages were produced.

Transistors Used in Ignition Systems

Transistors are used to open and close the primary side of the ignition system. It was found that a transistor can open and close a circuit much faster than a set of breaker points. The magnetic field will then collapse much faster, producing a higher voltage in the secondary circuit. The secondary voltage may be as high as 60,000 volts with a transistor ignition system. This will improve the combustion and emission characteristics of the engine.

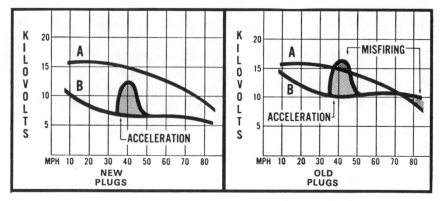

FIGURE 20–24. When old plugs are used, voltage requirements are higher. There may be times in which older plugs may misfire. (Courtesy of Champion Spark Plug Company)

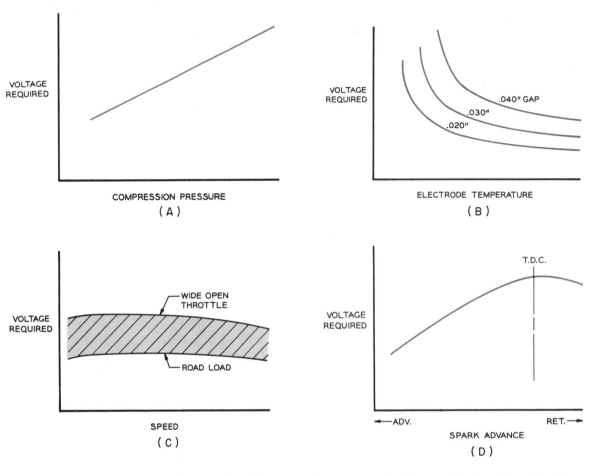

FIGURE 20-25. There are many factors that affect the voltage required of the spark plug. These include compression pressures (A), plug gap (B), speed/load (C), and timing advance (D). *(Courtesy of Champion Spark Plug Company)*

Triggering the Primary

Figure 20-26 shows how a simple transistor is used to trigger the primary side of the ignition coil. A small signal voltage is used to trigger the emitter-base circuit. This voltage signal is produced by a pickup coil and *reluctor* that act much like a small generator. The pickup voltage is a precisely timed signal. It triggers the electronic circuitry and transistors in the control unit. This interrupts the current flowing in the primary circuit, causing the ignition coil's magnetic field to collapse. The difference is that the emitter-collector circuit can be stopped very rapidly. This causes the magnetic field to collapse rapidly, producing a higher secondary voltage. The electronic circuit is much more complex than just one transistor. Dwell and timing advance can also be designed electronically into the system. *Figure*

20-27 shows some of the more common types of reluctors used to trigger the transistor.

Figure 20-28 shows an example of the control unit. There is a complex electronic circuit inside that turns the primary ignition coil on and off. The input includes the pickup coil signal and the output is the primary side of the ignition coil. A vacuum advance and a centrifugal advance are still used on this system.

Computer-controlled Spark

The next logical step in the development of electronic ignition systems is to incorporate the computer to control the spark. An example of this system is the computer command control (CCC) system. The CCC system, which was dis-

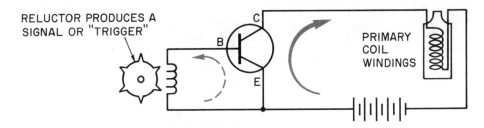

FIGURE 20–26. The transistor in a solid-state ignition system is used to turn off and on the primary side of the ignition coil. A small signal voltage triggers the transistor.

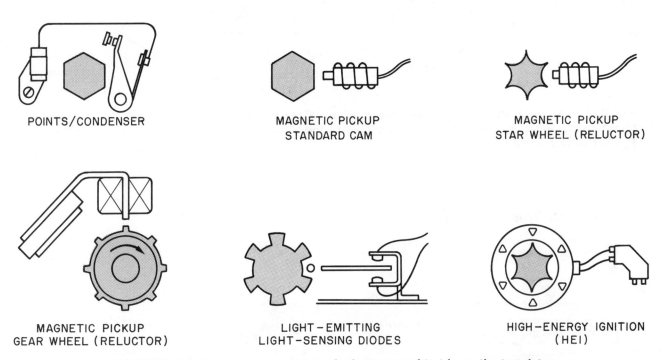

FIGURE 20–27. There are many types of reluctors used to trigger the transistor.

cussed in earlier chapters, uses several inputs from different sensors. These inputs are combined and decisions are made on the amount of fuel to be added to the engine and, in this case, the amount of spark advance to be allowed. The centrifugal weights and the vacuum advance are controlled by the computer. *Figure 20–29* shows a similar system in block diagram. The ECCS stands for Electronic Constant engine Control System. There are several inputs to the computer. These include engine speed, amount of air, temperature of the engine, throttle position, vehicle speed, start signal, engine knocking, and battery voltage. Based on these inputs, the power transistor in the electronic ignition system is operated to give the correct spark timing.

Electronic Spark Timing (EST)

Electronic spark timing uses many signals sent into the computer to electronically control the spark. This electronic advance is much more exact and reliable than conventional advance mechanisms. *Figure 20–30* shows how the timing of the spark occurs in a high-energy ignition system. To help understand the operation, a relay with a double set of contact points is shown in the HEI module. Solid-state circuitry is used in the module, but adding the relay makes it easier to visualize how EST functions.

During cranking, the relay is in the de-energized position. The pickup coil is connected to the base of the tran-

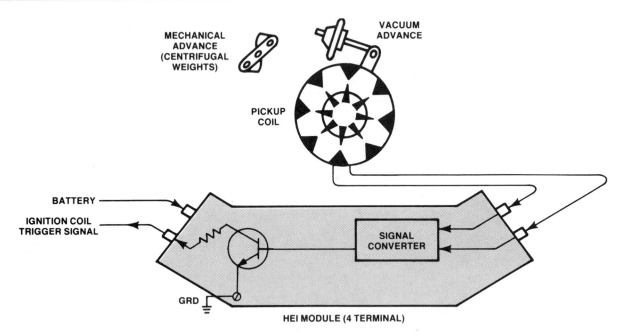

FIGURE 20–28. The control unit in a high-energy ignition (HEI) system uses the signal produced from the reluctor to turn the transistor on and off. The transistor controls the primary coil current. *(Courtesy of General Motors Product Service Training)*

IGNITION TIMING CONTROL

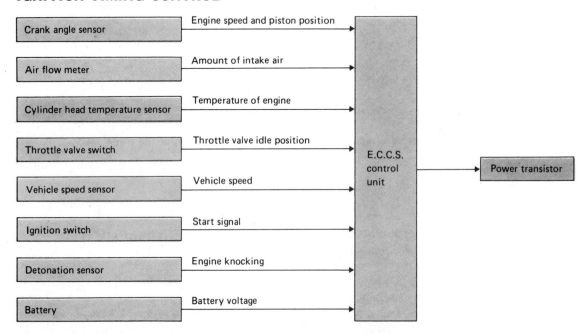

FIGURE 20–29. A computer-controlled spark system uses many inputs to electronically control the exact timing of the spark. *(Courtesy of Nissan Motor Corporation in USA)*

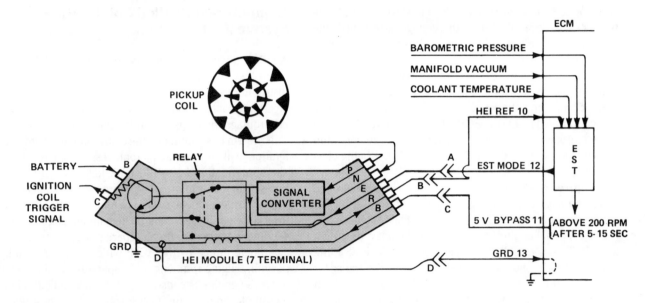

FIGURE 20–30. This circuit shows how timing of the spark is produced during cranking of the engine. *(Courtesy of General Motors Product Service Training)*

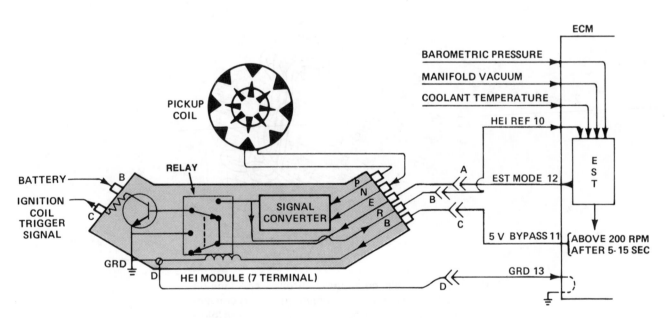

FIGURE 20–31. This circuit shows how timing of the spark is produced while the engine is running. *(Courtesy of General Motors Product Service Training)*

sistor. When the pickup coil applies a positive voltage to the transistor, it turns on. When the voltage is removed, the transistor turns off. It then accomplishes what the contact points did in the old ignition systems. When the transistor turns on, current flows through the primary winding of the ignition coil. When the transistor turns off, the primary current stops and the spark will be developed. The EST circuit is located inside the ECM (electronic control module). Several inputs are also shown for reference. The condition shown is for starting the engine. Timing is not being electronically controlled at this point.

Figure 20–31 shows how the timing is controlled when the engine is running. At about 200 rpm, the ECM applies about 5 volts to the bypass line. This voltage enters the HEI module at pin B and energizes the relay, causing it to shift. This is actually done electronically. The EST line from the ECM is now connected directly to the transistor base. The HEI system is now controlled by the signal from the ECM. The time at which the spark occurs is now determined by a circuit in the ECM based on the many inputs to the ECM. Timing is now controlled electronically.

Computer-controlled Coil Ignition System (C³I)

Another advance in ignition systems is called the computer-controlled coil ignition system. It is considered to be a distributorless ignition system. This system consists of an electronic control module (ECM), ignition (coil) module, and electromagnetic camshaft and crankshaft position sensors. This system has eliminated the distributor and conventional ignition coil. It uses a microprocessor that receives and alters information from the crankshaft and camshaft position sensors. This information is processed to determine the proper firing sequence. The system then triggers each of three interconnected coils on 6-cylinder engines (two on 4-cylinder engines) to fire the spark plugs. Ignition timing is again determined by the ECM, which monitors crankshaft position, engine rpm, engine temperature, and the amount of air the engine is consuming. It then signals the ignition module to produce the necessary spark at the right time. *Figure 20–32* shows a typical C³I ignition module installation.

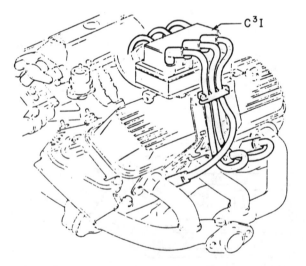

FIGURE 20–32. This computer-controlled coil ignition system (C³I) has eliminated the distributor and conventional ignition coil. It uses several sensors that feed information to the ignition module which, in turn, triggers the coils to produce the high voltage. *(Courtesy of Motor Publications, Auto Engine Tune-up & Electronics Manual, 1986)*

DIAGNOSIS and SERVICE

SAFETY TIPS

1. *When replacing parts or doing other work on the ignition system, always make sure the ignition system is off. If it is left on, a spark may be produced, causing gasoline fumes to ignite or damage the parts.*
2. *Always wear safety glasses when diagnosing the ignition system.*
3. *If it is necessary to crank the engine when working on the distributor, disconnect and ground the coil wire to make sure you will not get a shock from the high-voltage wire.*
4. *Never use only your hands to pull a high-tension wire off of the spark plug when the engine is running. Use the correct pliers to remove the plug wire.*

1. If the ballast resistor or resistive wire is shorted out, the points may burn prematurely due to the increased amount of current in the primary circuit. Replace the resistor or resistive wire.

2. If the ballast resistor or resistive wire is burned out or opened, the engine will start in the crank position on the ignition key, but it will shut off when the key is released.

3. A common problem is to have the ignition coil primary develop an open circuit due to high current. This will cause the total ignition system to shut down. This can be checked by using an ohmmeter to check the resistance of the ignition coil primary. An open circuit will show infinite resistance. Check the resistance between the positive side of the primary circuit and the negative side of the primary circuit.

4. As the contact points become pitted, the engine may misfire erratically. Use the following general procedure to *replace the points.*

a. Disconnect the negative side of the battery.
b. Remove the distributor cap.
c. Disconnect all wires attached to the contact points.
d. Using a screwdriver, remove the screws holding the contact points to the distributor.
e. Replace the old points with a new set of contact points by reversing the disassembly procedure.
f. Always use the small capsule of grease that comes with the points to prevent premature wear on the rubbing block. Any wear will change the contact points adjustment.
g. Set the contact points by rotating the distributor cam until the points are fully open.
h. Using a feeler gauge, set the points to the manufacturer's specifications. This adjustment will put the points close enough to run the engine.
i. Use a dwell meter to more accurately adjust the points. Hook the dwell meter to the coil and ground. Follow the directions for the specific dwell meter. Now run the engine. Measure the dwell and compare it to the manufacturer's specifications. Further adjustment may be necessary to get the points adjusted accurately.

5. If the *condenser is bad*, the points may corrode faster. Always change the condenser when replacing the points.

6. Make sure the positive side of the coil primary is connected to the switch. The coil will then operate with the correct polarity. If it is connected backwards, the coil output will be reduced by approximately one half. Whatever the vehicle ground polarity, that side of the coil goes toward the points.

7. Adjust the contact points before setting the timing of the engine because the point adjustment will affect the timing.

8. Check the distributor cams for excessive wear. If these cams are worn, the points will open and close at the incorrect time.

9. The distributor bearings may be worn. This will affect the mechanical opening and closing of the points.

10. Moisture condensing inside the distributor may cause the engine to misfire or not start at all.

11. Place the rotor on the distributor shaft correctly. It should only fit in one position.

12. A cracked distributor cap can cause the engine to misfire or not start at all. A cracked distributor cap can be found by removing the distributor cap and carefully observing the inside of the cap for small cracks.

13. If sparks are continually jumping from one spark plug wire to another, the insulation could be bad. The wire will have to be replaced. A new set of spark plug wires can be purchased. The *spark plug wires* often must be cut to the correct length. Only replace one spark plug wire at a time so that they don't get mixed up. Use the following procedure if the plug wires have been mixed up and cannot be replaced in the correct order. ***CAUTION:*** *Carelessly pulling spark plug wires off the spark plug, bending them sharply, or stretching them will damage the new wires and cause the engine to misfire.*

   a. With all plug wires removed, remove the number one spark plug. The number one spark plug can be determined by looking in the maintenance manual to determine its location.

   b. Slowly crank the engine until the number one spark plug is at top dead center on the compression stroke. Top dead center can be determined by observing the timing marks and lining them up correctly. When turning the engine to get to top dead center on the compression stroke, a slight pressure should be felt if a finger is placed over the spark plug hole.

   c. Remove the distributor cap and notice the location of the rotor. At this point (top dead center for the number one spark plug), the rotor should be pointing at the number one spark plug wire on the distributor cap.

   d. Place a spark plug wire on the correct hole on the distributor and on the number one spark plug.

   e. With the distributor cap removed, observe the direction of rotation of the rotor as the engine is cranked.

   f. Place the distributor cap back on the distributor.

   g. Using the firing order listed on the intake manifold or in the maintenance manual, continue placing the spark plug wires on the

distributor and the corresponding spark plug. For example, if the firing order is 1, 5, 3, 6, 2, 4, then after number one fires, the next cylinder to fire is number five. Place a spark plug wire from the next hole in the distributor cap to spark plug number five. Continue this procedure until all spark plug wires have been replaced.

14. If the springs on the centrifugal advance break, the engine will be advanced at the wrong time. Always check the *spring weights* for correct tension, and lubricate moving parts.

15. Check the *vacuum diaphragm* for leakage. A leaking vacuum diaphragm on the distributor will prevent the vacuum advance from operating. This will cause lack of power, poor fuel economy, and increased exhaust emission. Use a vacuum tester to check the diaphragm.

16. ***Figure 20–33*** shows different *spark plug conditions*. Match the spark plugs with the correct characteristics. Replace the spark plugs, if necessary.

17. Check the correct spark plug heat range with the manufacturer's specifications. When purchasing new spark plugs, always make sure the number on the spark plug is that which is recommended by the manufacturer.

18. Always make certain the plug gaps are adjusted to the manufacturer's specifications. These can be found in service repair manuals.

19. Note that electronic ignition systems use wider spark plug gaps.

20. Always follow the recommended diagnosis and service procedures listed in service and repair manuals to troubleshoot and diagnose electronic and computer-controlled spark systems.

21. Engines today use *electronic ignition systems*. Diagnosis procedures are different for each of these systems. In addition, there is a constant flow of service bulletins that help the service technician troubleshoot and diagnose the ignition systems. Today's engines are often checked on the electronic oscilloscope. Several specific procedures can, however, be followed.

   a. Using an ohmmeter, check the spark plug wire resistance through the distributor cap. The resistance should be 5,000 ohms per inch or less. If the resistance of the wire is greater, replace the wire.

   b. Spark plug wires should be checked for road salt deposits, dirt, damaged boots, and cuts and punctures. Replace the wires, where required. Clean the wires using a mild soap solution.

   c. Various tests can be made to check voltage to

Match Plug End Condition Description At Right with Correct Illustration on Left

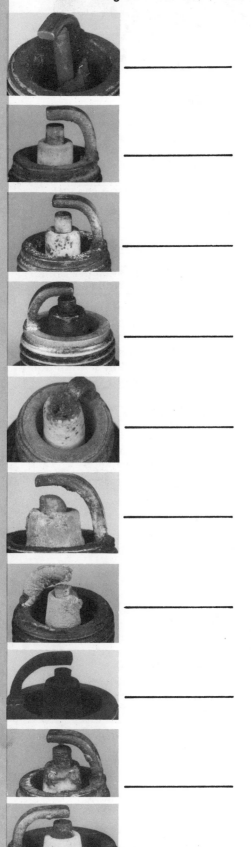

1 CRACKED INSULATOR

Small portion broken off or vertical crack in the insulator.

Check ignition timing, replace plug.

2 NORMAL

Brown to greyish tan color and slight electrode wear.

Service plug or replace with same range.

3 SPLASHED DEPOSITS

Leopard-like deposits. Occurs shortly after long delayed tune-up.

Service and reinstall plugs.

4 OIL DEPOSITS

White or yellowish deposits on one side of plug.

Check engine condition. Use non turbo action plug.

5 PREIGNITION

Melted electrodes. Center electrode generally melts first and ground electrode follows.

Check engine conditions and timing, use colder plug.

6 WORN OUT

Electrodes eroded, heavy deposits, pitted insulater.

7 MODIFIER DEPOSITS

Powdery white or yellow deposits that build up on shell, insulator and electrodes.

Service plug or replace with same range.

8 CARBON DEPOSITS

Dry soot, or oily deposits.

Check engine condition or use hotter plug.

9 HIGH SPEED GLAZING

Insulator usually has yellowish varnish-like color.

Service plug or replace with colder type.

10 TOO HOT

Blistered, lily white insulator, eroded electrodes and absence of deposits.

Reset ignition timing or use colder plug.

FIGURE 20–33. The condition of the spark plug can tell many things about the engine. The condition of the spark plug should always be checked to see if problems are developing in the engine. *(Courtesy of Champion Spark Plug Company)*

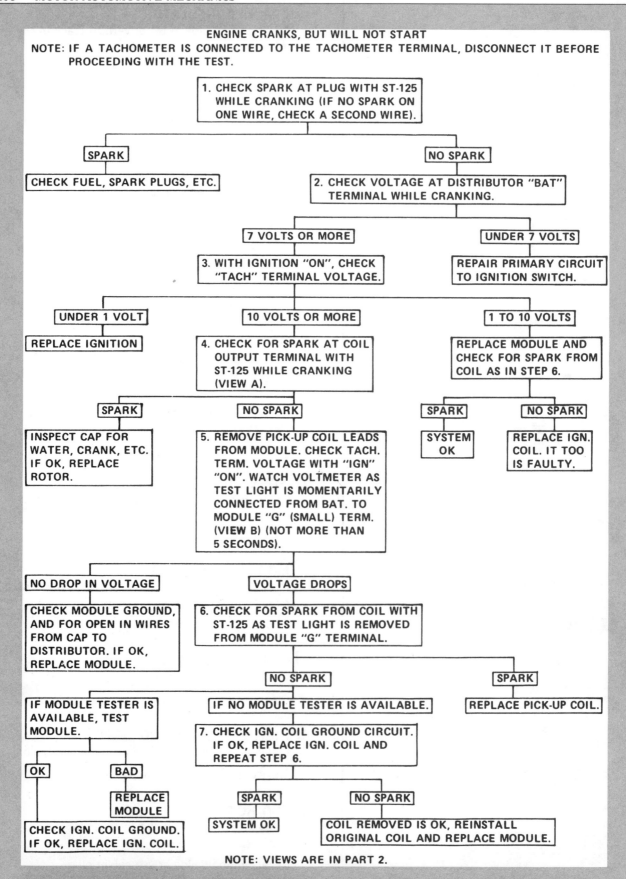

FIGURE 20–34. Because of the complexity of many of the ignition systems, various diagnosis charts are available in the service and repair manuals. Follow the correct diagnosis chart to help troubleshoot the electronic ignition system. *(Courtesy of Motor Publications, Auto Engine Tune-up & Electronics Manual, 1986)*

the spark plug. Because several methods can be used, follow the specific manufacturer's recommendation to determine if spark exists.

22. When troubleshooting and diagnosing an electronic ignition system, refer to the diagnosis charts available in the maintenance and service manuals. *Figure 20–34* shows an example of one of many diagnosis charts. As each step is performed, the problem will eventually be found by taking voltage readings and testing for spark.

23. In many maintenance and service manuals, specific connections are shown for ohmmeter tests. These tests should be done as part of the diagnosis and troubleshooting procedures on electronic ignition systems.

24. Use a *timing light*, **Figure 20–35**, to check the timing of an engine. Attach the clip-on leads to the battery. Place the inductive lead over the number one spark plug wire. Every time the number one spark plug fires, the timing light will flash. When the engine is running, the flashes look like a strobe light. When shown on the timing marks on the

front of the engine during idle, the amount of advance or retard can be observed. With the distributor base loosened, as the distributor is turned, advancing or retarding can be observed by the timing light flashes.

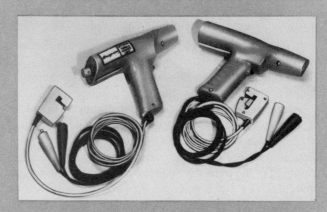

FIGURE 20–35. A timing light is used to adjust the time of the spark. The light is attached to the battery and to the number one spark plug wire. When the plug fires, a light flashes to show the timing marks. *(Courtesy of Snap-On Tools Corporation)*

SUMMARY

The many types of ignition systems used on vehicles today were studied in this chapter. The purpose of the ignition system is to convert the 12 volts produced by the battery to 20,000 to 50,000 volts for use at the spark plug. To do this, primary and secondary systems are used in the ignition system. The primary circuit operates on low voltage and includes the ignition switch, primary coil windings, ballast resistor or resistive wire, condenser, and contact points. The secondary circuit operates on high voltages of 20,000 to 30,000 volts and includes the high-voltage windings in the coil, the rotor, distributor cap, spark plug wires, and spark plugs.

The ignition switch is used to turn the ignition system on and off. The ballast resistor or resistive wire is used to drop the primary voltage to 7–10 volts to protect the ignition coil and points. The ignition coil is used to change the 12 volts to 5,000–50,000 volts. The breaker points act as a primary circuit switch to turn the circuit on and off. When the points are closed, a magnetic field is built up around the coil. When the points open, the magnetic field collapses and produces a high secondary voltage. The length of time in distributor degrees of rotation that the points remain closed is defined as dwell.

The points must also be adjusted. Breaker points are adjusted when the points are fully open. If the points are corroded or burned, they should be replaced. Pitted points slow

down the collapse of the magnetic field, producing less secondary voltage. The points mechanism is held in place on the distributor and is opened and closed by the distributor cam.

The secondary components in the ignition system operate on higher voltages. The rotor and distributor cap are used to distribute the high-voltage spark. The high voltage then passes through the spark plug wires to the spark plugs.

Spark timing must be changed when the speed or load is increased. This is done by using advance mechanisms. Two types are used. The centrifugal advance increases the advance of the spark as the engine speed increases. The vacuum advance increases the spark advance as the load is increased. It is connected to the engine intake manifold which is sensitive to engine load. Several systems are incorporated to help retard the spark during certain conditions. These include spark delay systems, temperature sensing vacuum control valves, dual diaphragm vacuum advance, and solenoid retard-advance systems.

One critical component in the ignition system is the spark plug. Spark plugs have heat ranges. These are determined by how long it takes to transfer the heat at the tip of the plug to the cylinder head. The longer it takes, the hotter the heat range.

Spark plugs are affected by many conditions. Spark plug temperature is affected by the amount of advance, coolant temperature, detonation, plug condition, and air-fuel ratio. Spark plug voltage is affected by plug condition, compression pressures, plug gap, load, and engine timing.

Because of the wear and maintenance on conventional ignition systems, manufacturers have developed electronic and computer-controlled ignition systems. Electronic ignition systems use transistors to open and close the primary circuit much faster. More voltage, about 30,000 to 60,000 volts, is produced. The only additional component is a reluctor which is used to signal or trigger the transistor on and off at the right time. These systems require minimal maintenance, mostly checking parts for abnormal wear.

Once electronic systems had been improved, the next logical step was to control the spark with computers. Spark advance is controlled by the CCC, computer control combustion system. The computer is used to signal the exact triggering of the transistor so that correct spark advance is produced. The signal is based on several inputs to the computer, including engine speed, amount of air, throttle position, engine knocking, start signal, battery voltage, engine coolant temperature, vehicle road speed, and air intake temperature.

There are many service tips and suggestions for working on the ignition system. Conventional ignition systems typically require more service than electronic or computer-controlled systems. Dwell must be adjusted, and the breaker points must be changed periodically. The condenser may also fail. Electronic ignition systems require little maintenance. Dwell is built directly into the circuits. Because of the complexity of computer-controlled systems, it is important to refer to the specific type of ignition system and the appropriate service and repair manual for diagnosis.

TERMS TO KNOW

Can you explain each of the following terms? Review the chapter until you can use each term correctly.

Primary circuit	Dwell
Secondary circuit	Condenser
Ballast resistor	Capacitor
Breaker point	Ported vacuum
Distributor cam	Reluctor
Transformer	

REVIEW QUESTIONS

Multiple Choice

1. The primary circuit on the ignition system operates on _____ volts.
   a. 5 volts
   b. 7 volts
   c. 12 volts
   d. 20,000 volts
   e. 50,000 volts

2. The secondary circuit on the ignition system operates on _____ volts.
   a. 5 volts
   b. 7 volts
   c. 12 volts
   d. 20,000 volts
   e. 70,000 volts

3. The ballast resistor is placed into the ignition circuit when:
   a. The engine is cranking
   b. The engine is stopped
   c. The engine is running
   d. The ignition switch is turned off
   e. The ignition switch is turned to start

4. The purpose of the ballast resistor is to:
   a. Increase the voltage to the ignition coil
   b. Decrease the voltage to the ignition coil
   c. Protect the points from changing the dwell
   d. Increase the voltage to the condenser
   e. Protect the secondary circuit

5. The spark is produced on the secondary windings when the:
   a. Points close
   b. Points open
   c. Switch is turned on
   d. Magnetic field increases
   e. Condenser charges

6. Which of the following components is not part of the primary circuit?
   a. Condenser
   b. Rotor
   c. Points
   d. Ballast resistor
   e. Large size windings in the coil

7. Which of the following components is not part of the secondary circuit?
   a. Points
   b. Condenser
   c. Ballast resistor
   d. All of the above
   e. None of the above

8. Dwell is defined as:
   a. The length of time in degrees the points are closed.
   b. The length of time in degrees the points are open.
   c. Point gap in thousandths of an inch.
   d. The voltage at the secondary circuit.
   e. The voltage at the primary circuit.

9. If the points are adjusted too large, the spark will be:
   a. Retarded
   b. Advanced
   c. Kept the same
   d. Stronger
   e. Greater in voltage

10. If the magnetic field collapses faster, the voltage at the secondary will:
    a. Be increased
    b. Be decreased
    c. Remain about the same
    d. Totally stop
    e. Recharge the primary

11. The current produced by the arcing when the points are opened is stored at the:
    a. Ignition switch
    b. Ballast resistor
    c. Secondary coil windings
    d. Condenser
    e. Battery

12. How may cams are placed on the distributor cam for a 6-cylinder engine?
    a. 2
    b. 4
    c. 6
    d. 8
    e. 12

13. Which type of advance is used to increase the timing as speed increases?
    a. Contact point advance
    b. Centrifugal advance
    c. Vacuum advance
    d. Ported advance
    e. Computer advance

14. Which type of advance is used to increase the timing as load increases?
    a. Contact point advance
    b. Centrifugal advance
    c. Vacuum advance
    d. Ported advance
    e. Computer advance

15. A plug that is fouling out with carbon and soot may be:
    a. Too hot a plug
    b. Too high a voltage on the plug
    c. Too cool a plug
    d. Too long a plug
    e. Too short a plug

16. The tip temperature of the plug will increase with:
    a. An over-advanced engine
    b. An increase in coolant temperature
    c. Clean spark plugs
    d. All of the above
    e. None of the above

17. The spark plug voltage required will increase with:
    a. Decreases in compression
    b. Decreases in spark plug gap
    c. Decreases in load on the engine
    d. All of the above
    e. None of the above

18. What component is used to open and close the ignition primary on an electronic ignition system?
    a. A condenser
    b. A diode
    c. A computer
    d. A transistor
    e. A set of contact points

19. What is an average voltage on the secondary for an electronic ignition system?
    a. 10,000 volts
    b. 20,000 volts
    c. 12 volts
    d. 50,000 volts
    e. 6 volts

20. Which of the following is used to trigger the electronic circuit to open and close the primary windings in the coil?
    a. Conductor
    b. Reluctor
    c. Condenser
    d. Transistor
    e. Secondary coil windings

21. The spark on a computer-controlled ignition system is timed by:
    a. The centrifugal weights
    b. The vacuum advance
    c. The computer or electronic control module
    d. The springs on the weights
    e. None of the above

22. As the breaker points become corroded, the speed at which the magnetic field collapses:
    a. Increases
    b. Remains about the same
    c. Decreases
    d. Is proportional to the voltage
    e. None of the above

23. When diagnosing the secondary circuit, which of the following should be checked?
    a. Moisture in the distributor cap
    b. Cracks in the distributor cap
    c. Condition of the spark plug wires
    d. All of the above
    e. None of the above

The following questions are similar in format to ASE (Automotive Service Excellence) test questions.

24. Technician A says that as the contact points become pitted, the gasoline engine mileage goes down. Technician B says that as the contact points become pitted, the voltage to the spark plug is decreased. Who is right?
    a. A only
    b. B only
    c. Both A and B
    d. Neither A nor B

25. Technician A says that moisture inside a distributor cap will have no effect on the ignition system operation. Technician B says that moisture inside a distributor cap will cause the spark plug voltage to increase and improve ignition system operation. Who is right?
    a. A only
    b. B only
    c. Both A and B
    d. Neither A nor B

26. Technician A says that the resistance of the spark plug wire can be checked with an ohmmeter. Technician B says that the resistance of the spark plug wire cannot be checked at all. Who is right?
    a. A only
    b. B only
    c. Both A and B
    d. Neither A nor B

27. An engine seems to start during cranking but shuts off when the key is released from cranking. Technician A says the problem is a burned out ballast resistor. Technician B says the problem is bad spark plugs. Who is right?
    a. A only
    b. B only
    c. Both A and B
    d. Neither A nor B

28. Sparks are noticed jumping from spark plug wire to wire. Technician A says the problem is too high a voltage to the spark plugs. Technician B says the problem is the ballast resistor. Who is right?
    a. A only
    b. B only
    c. Both A and B
    d. Neither A nor B

Essay

29. What is the difference between the primary and secondary ignition circuits?

30. Describe the purpose and operation of the ballast resistor.

31. Define the term dwell.

32. State two types of advance and identify how they operate.

33. Describe spark plug heat ranges.

34. Identify two things that affect spark plug temperature.

35. What is a reluctor and why is it used?

36. State the purpose of electronic spark timing.

CHAPTER 21

Charging Systems

INTRODUCTION

The purpose of this chapter is to study the charging system. This includes the study of generators, alternators, and regulation of the charging system.

OBJECTIVES

After reading this chapter, you will be able to:
- Identify the purpose of the charging system.
- Analyze the principles of converting the mechanical energy of the engine to electrical energy for charging.
- Define the parts and operation of dc generators.
- State the operation of dc regulation on a generator.
- Define the parts and operation of the alternator.
- Identify how three-phase voltages from the alternator are rectified.
- State the operation of solid-state electronic regulation using diodes.
- Identify basic diagnosis and service procedures on the charging system.

CHAPTER HIGHLIGHTS

413

21.1 GENERATOR PRINCIPLES

Purpose of the Charging System

A charging system is used on all automobiles to convert the mechanical energy of the engine to electrical energy. The electrical energy is used to operate the vehicle during normal driving conditions and to charge the battery. Each time electrical energy is removed from the battery, the battery must be recharged. If it is not recharged, the battery will eventually become discharged. The charging system is used to accomplish this. In addition to charging, it is important to be able to regulate the amount of charge. If too little charge is produced, the battery will have a low charge. If the charging system puts too much back into the battery, the battery may be overcharged and damaged. The regulation of the charge ensures that the exact amount of electricity needed is being generated.

Producing Electricity

A generator uses three things to change mechanical energy to electrical energy. These include a magnetic field, a conductor or wire, and movement between the two. This principle is called electromagnetic induction. ***Figure 21–1*** shows a simple generator used to produce electricity. When wire is wound around a metal core, a magnetic field is produced. The magnetic lines of force are moving from the north pole to the south pole. When a wire or conductor is moved to cut the magnetic field, a voltage is produced within the wire. This voltage is used to push electrons back to the battery for charging. It is also used to supply current to the rest of the vehicle during normal operation.

Generated voltage (and, therefore, current available) can be increased by increasing any one of the three components in the generator. If the magnetic field is increased, the voltage will increase. If the amount of copper wire is increased, the voltage will increase. If the speed of motion is increased, the voltage will increase.

Difference Between Generators and Alternators

For years the automotive charging system used a generator to charge the battery. Today an alternator is used. The difference between a generator and an alternator is the method of physical construction. A generator has stationary poles, and the wire moves across the field. An alternator has moving poles (magnetic field) and a stationary wire. Although both produce electricity, the alternator is much more efficient.

There are also several minor differences that will be discussed in this chapter. Generators use brushes and a *commutator* to remove the produced voltage from the rotating windings. The alternator uses diodes to *rectify* the voltage to dc.

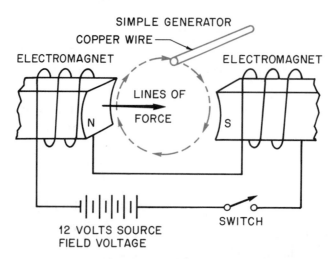

FIGURE 21–1. A simple generator is made of a set of poles, copper wire, and movement between the two in which the wire cuts magnetic lines of force. All three are needed to produce a voltage.

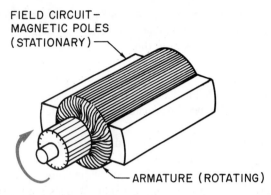

FIGURE 21-3. A dc generator has several parts, including the poles (which are stationary) and the armature (copper wire) which rotates inside the poles.

21.2 DC GENERATORS

Although all vehicles being manufactured today use an alternator, there are many vehicles that still have generators. It is important to study these systems so the newer electronic systems can be understood. *Figure 21-2* shows a typical dc generator. It is driven by a belt from the crankshaft.

Dc Generator Parts

The dc generator is made of several parts. These are shown in *Figure 21-3*. The poles are stationary and produce the magnetic field needed to generate electricity. This circuit is called the *field circuit*. Rotating inside the field poles is an *armature*. Voltage is produced in the armature as it rotates within the magnetic field.

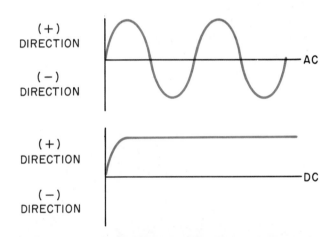

FIGURE 21-4. Ac voltages alternate from a positive direction to a negative direction and back. Dc voltages always have current flow in only one direction. These graphs illustrate the differences between ac and dc voltages.

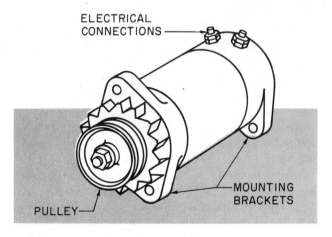

FIGURE 21-2. A generator is used on older vehicles to produce a voltage for charging the battery.

Ac and Dc Voltages

The voltage produced within the armature is ac or alternating current. As studied in an earlier chapter, ac voltage is continually changing back and forth as shown in *Figure 21-4*. The current flows first in a positive direction, then in a negative direction. Dc voltage is continuous, or flows in one direction. Dc voltage is always flowing in a positive or negative direction.

Rectifying Ac Voltages

The ac voltage within the armature must be converted to dc voltage for automotive circuits. A commutator and a set of brushes are used to accomplish this. The commutator is also called a split-ring commutator. Ac voltage is changed to dc voltage when it goes from the commutator to the brushes. The brushes are made of carbon and rub against the commutator. The voltage at the brushes is dc voltage that can be used to charge the battery or operate the vehicle circuits. *Figure 21–5* shows the armature, the commutator, and the brushes.

A graph of the voltage rectified by the split-ring commutator is shown in *Figure 21–6*. Note that although the voltage is pulsating from zero to a maximum point, it continually goes in the same direction. It does not reverse direction as ac voltages do.

Purpose of Regulating Voltage

The output of the dc generator must be regulated and controlled. Regulation is needed to protect the generator, the battery, and the electrical circuits in the automobile from too much voltage or current. It is very important to keep the voltage output of the generator slightly above 12 (about 13.5) volts. If the voltage is too high, the light bulbs may

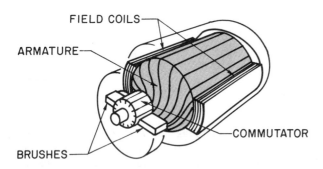

FIGURE 21–5. The generator is made of an armature, field coils, the commutator, and the brushes. *(Courtesy of Delco-Remy Division of General Motors Corporation)*

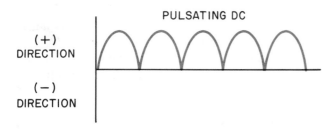

FIGURE 21–6. The dc voltage that is changed by the split-ring commutator is pulsating dc and is illustrated as shown in the graph.

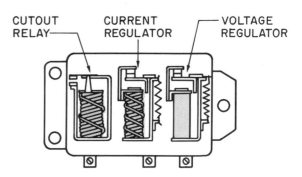

FIGURE 21–7. A three-unit voltage regulator has a cutout relay, a current regulator, and a voltage regulator. These units were used on older vehicles before electronic regulators were developed. *(Courtesy of Delco-Remy Division of General Motors Corporation)*

burn out and the battery may overheat. Too low a voltage will cause poor charging of the battery and incorrect operation of the electrical circuits. The voltage regulator is used for this control.

Voltage Regulator

Figure 21–7 shows a three-unit, or coil, regulator. A coil with a set of points or contacts is also called a *relay*. It has three coils inside. They are called the cutout relay, the current regulator, and the voltage regulator. When these coils are energized, the contact points close. As semiconductors were used more, automotive manufacturers improved these circuits. Depending upon the vehicle make and year, three, two, or one coil could be used. Today, all regulators are solid state. Some are built into the alternators and no relay coils are used.

Cutout Relay

The cutout relay is used to protect the generator when it is not charging as much voltage as the battery is producing. If the battery voltage is greater than the generator voltage, the cutout relay will open. This action stops any current from going backwards into the generator. This could cause the battery to discharge unnecessarily. When the generator increases its voltage above the battery voltage, the coil is energized and closes the set of contacts on top of the cutout relay. Now the generated voltage forces electrons from the generator to the battery and/or into the vehicle circuits. See *Figure 21–8*. If the battery is fully charged, the generated voltage supplies current only to the circuits in the vehicle. If the generator stops producing voltage, the cutout relay will de-energize and the contact points will open. This will protect the generator from receiving any unnecessary battery voltage.

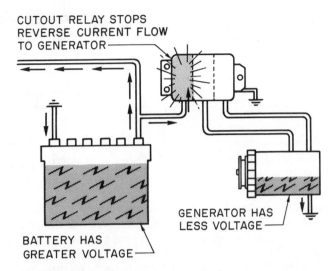

FIGURE 21–8. The cutout relay is used to stop battery voltage from entering the generator aramature. This would happen only if the battery voltage were greater than the generated voltage. *(Courtesy of Delco-Remy Division of General Motors Corporation)*

Voltage Relay

The voltage relay controls the generator's field current so the voltage doesn't get too high. The field current controls the strength of the magnetic field. If the field current increases, its magnetic field will also increase. Voltage output will now increase. If the field current decreases, the magnetic field will also decrease. Voltage output will now decrease.

A voltage-sensitive coil is used to control field current. When voltage is too high, say 15 volts, the relay coil is energized. This opens the voltage relay contact points. This action limits the amount of current that can flow into the field windings. The voltage output will now decrease. See *Figure 21–9*. When the generator output decreases to 12 volts or lower, the voltage relay closes again so full current can flow to the field windings.

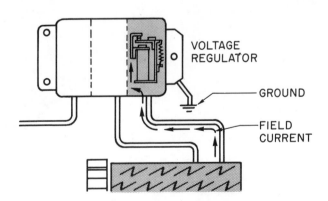

FIGURE 21–9. The voltage relay is used to control output voltage to slightly higher than battery voltage. The field current is grounded when the voltage output is too high. *(Courtesy of Delco-Remy Division of General Motors Corporation)*

Current Relay

There may be times when the electrical system will demand more current than the generator can produce. This could happen if too much load were placed on the vehicle's electrical circuits. The current relay controls the amount of generator current output. Here a current-sensitive coil is used, *Figure 21–10*. During normal operation, current output from the generator passes into the current regulator, through the contact points, and back to the field windings. When too much current is demanded by external circuits, the coil is energized. This opens the points and again limits the field current. Many regulator designs have been used on different vehicles. All circuits, however, must control three things: the reverse current protection (cutout), voltage control, and current control.

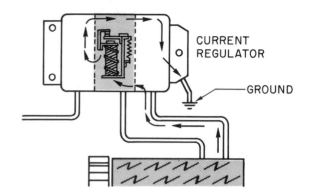

FIGURE 21–10. The current relay is used to control the output current of the generator to protect the generator. The field current is grounded when the current output gets too high. *(Courtesy of Delco-Remy Division of General Motors Corporation)*

CAR CLINIC ▮▮▮▮▮▮▮

PROBLEM: PINGING PROBLEM

A 4-cylinder engine has a problem with pinging at high speeds and hot temperatures. The timing has been checked and is right on specifications. All of the advance mechanisms, including the weights and the vacuum systems, have been checked. What other problem might cause pinging?

SOLUTION:

The only items not checked were the spark plugs. Check the spark plugs and make sure they are the correct ones for the engine. If a spark plug range is too hot, the tip of the plug might not be cooling enough. This could cause pinging at high speeds and temperatures. The only other solution might be a large amount of carbon buildup on the top of the piston. This could cause the compression ratio to increase, which could cause compression temperatures to increase and cause preignition.

21.3 ALTERNATORS

An alternator, rather than a generator, is used on today's vehicles to charge the battery and operate the electrical circuits. This is because the alternator is much more efficient than a generator. Alternators are much smaller, lighter in weight, and produce more current than generators. The alternator has a set of rotating poles and a stationary set of windings. In addition, there is no split-ring commutator. Solid-state diodes are used to convert ac to dc voltages. The alternator is made of a stator, rotor, and slip-ring and brush assembly. Many modern (late model) alternators have the regulator built into the housings as a complete unit.

Stator

The *stator*, *Figure 21–11*, is made of a circular, *laminated* iron core. There are three separate windings wound on the core. The windings are arranged so that a separate ac voltage *waveform* is induced in each winding as the rotating magnetic field cuts across the wires.

Rotor

The *rotor*, *Figure 21–12*, is made of a coil of wire wound around an iron core on a shaft. When current is passed through the windings, the assembly becomes an electromagnet. One side is a north pole and the other side is a south pole. Iron claws are placed on both ends. Each projec-

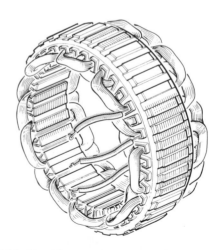

FIGURE 21–11. Alternators use a stator or stationary set of windings to generate voltage. *(Courtesy of Lucas Industries, Inc.)*

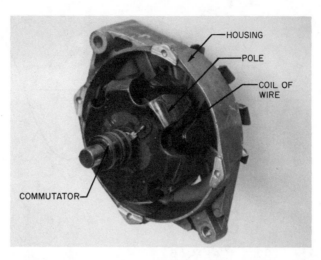

FIGURE 21–12. The rotor is made of a coil of wire wound around an iron core.

tion has the same polarity as the ends of the coil. When the claws are meshed together from each side, pairs of north and south poles are formed around the rotor circumference, *Figure 21–13*. The number of north and south poles is determined by the manufacturer. Four, six, and seven sets of poles are common in alternators today. The poles are designed to rotate inside the stator, producing the voltage needed to charge the battery.

Slip-ring and Brush Assembly

The ends of the rotor coil are connected to *slip-rings* that are mounted on the shaft. Current is supplied from the battery

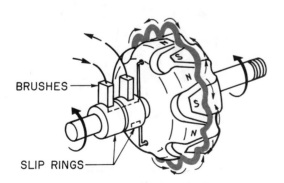

FIGURE 21–13. The coil of wire on the rotor has two end plates meshed together. Seven sets of north and south poles produce the magnetic fields.

through the brushes and slip-rings to energize the rotor field windings. This produces the magnetic field needed for making the north and south poles in the alternator.

Three-phase Voltage

As the rotor revolves inside the stator windings, voltage is produced in the stator. Three separate windings are spaced evenly inside the stator. The voltage produced is called *three-phase* voltage. Each winding produces a separate voltage. This is shown in *Figure 21–14*. The voltages, just like the windings, are 120 degrees apart from each other.

The three-phase stator windings can be connected in one of two ways. The star connection is also called the wye connection. This is the most common hookup in an alternator. The delta connection is used for higher output alternators. Each winding produces a separate voltage. The dark solid voltage line (Figure 21–14) is produced from coil (A). The lighter solid voltage line is produced from coil (B). The dotted voltage line is produced from coil (C). Each voltage is separate and independent. However, it is still ac voltage and must be converted or rectified to dc for use in the automobile.

Rectifying Three-phase Ac Voltage

The three-phase voltage in the alternator stator windings must be converted or rectified to dc. Six diodes are used to do this. Three are considered positive diodes and three are considered negative diodes. Diodes allow current to flow only in one direction. (See Chapter 18.) *Figure 21–15* shows a diode circuit (A), the electrical symbol for a diode (B), and a sectioned view of a diode (C).

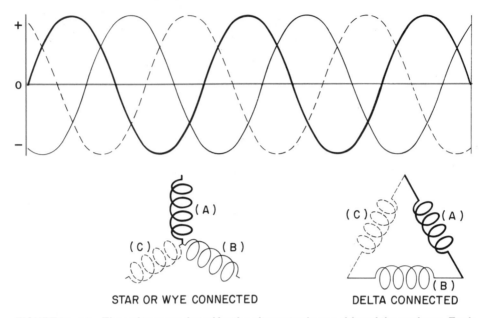

STAR OR WYE CONNECTED   DELTA CONNECTED

FIGURE 21-14.   The voltage produced by the alternator is considered three phase. Each phase is produced by a separate set of stator windings. This graph shows the three-phase voltage produced. Also, the stator can be connected as a "star" or Y connection or as a "delta" connection. (Courtesy of Lucas Industries, Inc.)

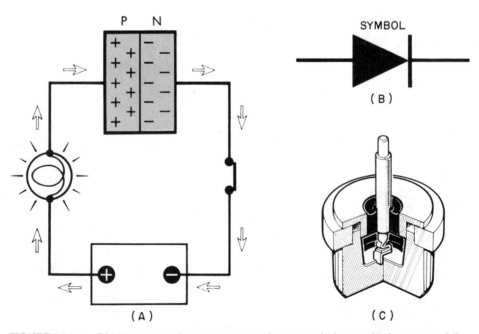

FIGURE 21-15.   Diodes are used to convert ac voltages to dc for use in the automobile. Diodes allow current to flow only in one direction. A diode circuit is shown in (a), a diode symbol (b), and a sectioned view of a diode (c). (Courtesy of Lucas Industries, Inc.)

Figure ***Figure 21–16*** shows a circuit that uses six diodes to rectify three-phase ac voltage to dc. Coils (A), (B), and (C) produce an ac voltage. Each voltage is sent to the rectifier. No matter where the current enters the rectifier, it is sent out on top to the positive side of the battery. The current at the battery has been rectified to dc.

Current Flow in a Three-phase Rectifier

Referring to Figure 21–16, the first half cycle produced from coil (A) passes current from point 1 to diode (C) to the positive side of the battery. The current returns to coil (A) through diode (D) and coil (B). On the negative half cycle, coil (A) passes current from point 2, through coil (B), through diode (A), to the positive side of the battery. The

current returns from the battery through diode (F) to point 1 on coil (A). The voltage produced by coil (A) is now rectified to dc. It goes through the battery in the same direction.

Coil (B) works the same way. The first half cycle produced by coil (B) passes current from point 4 to diode (A) to the positive side of the battery. The current returns to coil (B) through diode (E) and coil (C). On the negative half cycle, coil (B) passes current from point 3, through coil (C), to diode (B), to the positive side of the battery. The current returns from the battery through diode (D) to point 4 on coil (B). The voltage produced by coil (B) is now rectified to dc.

The flow through the rectifier can be followed the same way for coil (C). It is also rectified to dc. ***Figure 21–17*** shows the resultant dc voltage wave that is produced by rectifying three-phase ac voltage. It is called pulsating dc and is considered *full-wave rectification*.

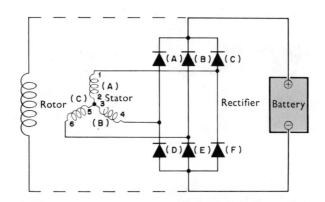

FIGURE 21–16. Six diodes are used to convert three-phase ac to pulsating dc. *(Courtesy of Lucas Industries, Inc.)*

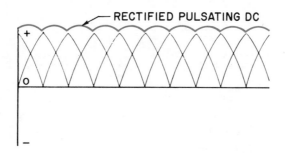

FIGURE 21–17. The voltage that is rectified on an alternator is called pulsating dc. It is made by rectifying all three phases of ac to dc in the rectifier. *(Courtesy of Lucas Industries, Inc.)*

COMPUTER-GENERATED SURFACE DRAWINGS

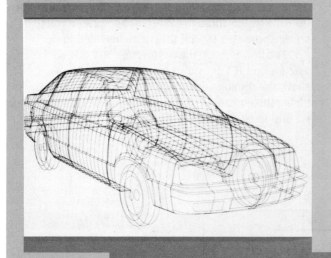
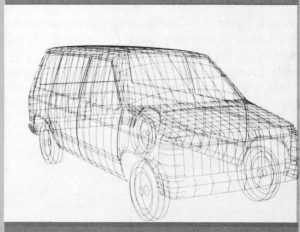

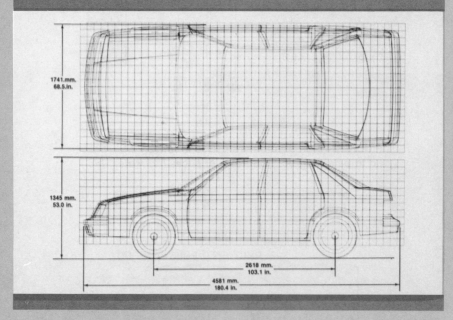

Computer-aided design (CAD) systems are used to examine, test, and modify body design in three-dimensional form. This type of design will help to insure the most efficient design at relatively low cost. (Courtesy of Chrysler Corporation)

21.4 ALTERNATOR REGULATION

Many types of regulators have been used on automotive charging systems. As more solid-state circuitry was developed, the regulation systems slowly changed from relays and coils to electronics. Today regulators are all electronic and require very little maintenance.

All regulators are designed to control the amount of current sent into the field windings. *Figure 21–18* shows the relationship between rotor speed, field current, and the regulated voltage. The solid line represents the regulated voltage from the alternator. The dotted line represents the field current. As the rotor speed increases, the field current

is reduced to keep the regulated voltage controlled.

In dc generators, a relay was used to control the amount of current to the field windings. In electronic regulators, the switching of the field current is controlled by transistors turning off and on. No cutout relay is needed because diodes in the rectifying circuit prevent current from flowing in the reverse direction.

Two-unit Regulators

When alternators were first introduced in the automobile, the regulator was designed using relays and diodes. *Figure 21–19* shows an example of a typical alternator wiring

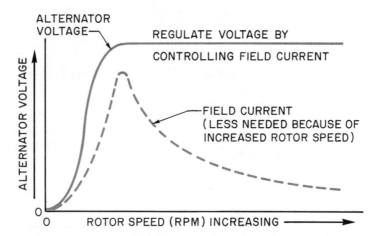

FIGURE 21–18. This chart shows the relationship between rotor speed, current in the field windings, and output voltage. As the rotor speed increases, field current is reduced to keep the regulated voltage controlled.

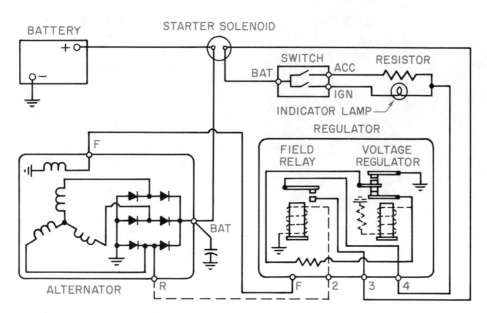

FIGURE 21–19. When alternators were first introduced, a two-unit regulator was used. Diodes replaced the cutout relay, but a field relay and voltage relay were still used. *(Courtesy of Delco-Remy Division of General Motors Corporation)*

diagram. The cutout relay has been removed. When diodes are used, the battery voltage cannot force current back into the alternator. The diodes prevent current flow in this direction.

There are still two relays. One relay is called the field relay. The second relay is called the voltage relay. Referring to Figure 21–19, when the ignition key is turned on, the battery is connected to the No. 4 terminal of the regulator. Current now flows through the indicator lamp on the dashboard. This lights the lamp to show no charging. Current also flows from the No. 4 terminal to the voltage regulator through its contacts, and to terminal F over to the field windings. The field windings are now energized to produce a strong north and south pole.

When the engine starts, the alternator starts to produce voltage. This voltage is used to charge the battery or operate other electrical circuits. One phase of the voltage is also sent to terminal R on the alternator. This voltage sends a current to terminal 2 on the regulator and to the field relay, closing its set of points. When the points close in the field relay, this action causes the battery voltage to be connected directly to the field windings. This is done by connecting terminal 3, through the field relay points, to the points in the voltage relay, and finally to terminal F to the field windings.

When the points close, battery voltage is impressed directly on terminal 4. This action stops the current flowing in the switch circuit, turning off the indicator lamp. This indicates that there is a voltage being produced by the alternator.

As the voltage from the alternator increases because of high speed or too much current sent to the field windings, the bottom set of contacts on the voltage regulator opens. This action causes the field current to be sent through the resistor. The resistor reduces the field current, causing the generated voltage to be reduced. In operation, the voltage regulator action is very rapid when maintaining an accurate voltage.

When the alternator speed increases rapidly and there is a chance of excessive voltage, the upper contact on the voltage regulator is forced down. Now all current to the field windings is grounded through the set of contact points. This action stops current flow to the field windings, reducing the alternator output.

Electronic Regulators

As solid-state circuitry was developed, voltage regulators eventually became totally electronic. *Figure 21–20* shows an alternator with the voltage regulator built directly into the alternator. *Figure 21–21* shows a complete charging

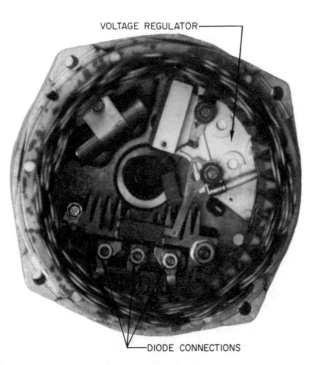

FIGURE 21–20. Voltage regulators are totally electronic today. Transistors turn the field windings off and on for control.

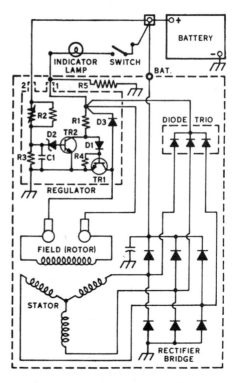

FIGURE 21–21. This is a complete circuit of an alternator, field windings, electronic voltage regulator, and diodes for rectification. *(Courtesy of Motor Publications, Auto Repair Manual, 1981–1987)*

circuit for the alternator. The action of the regulator is similar to relay types, except transistors are used to turn the field windings off and on.

A *zener diode* is used to tell the circuit exactly when too much voltage is being produced by the alternator. The zener diode is identified as D2. When the alternator voltage is too high, the zener diode turns on. When the zener diode turns on, transistor TR1 turns off, shutting off the current to the field windings. The reverse happens when the voltage drops to an acceptable level. Although the action is more complicated, the concept of controlling the field current is the same.

Alternator Construction

Alternators are designed today as complete integral units. *Figure 21–22* shows a dismantled alternator. Note the position of each part previously discussed in this chapter. Housings are used on both ends for support. They also include the necessary bearings for supporting the rotor. The rotor has six fingers on the north pole and six fingers on the south pole. The stator is connected to the slip-ring end bracket. This bracket is used to house the brush set. The rectifier and regulator are located in the slip-ring end cover. The suppression capacitor is used to eliminate radio static produced by the alternator.

Rectifier Assembly

Figure 21–23 shows several rectifier assemblies. Rectifiers with diodes are constructed on plates of metal. One plate holds one polarity of diodes, and another plate holds the opposite polarity of diodes. Depending upon the manufacturer and the circuit, different diode combinations are used.

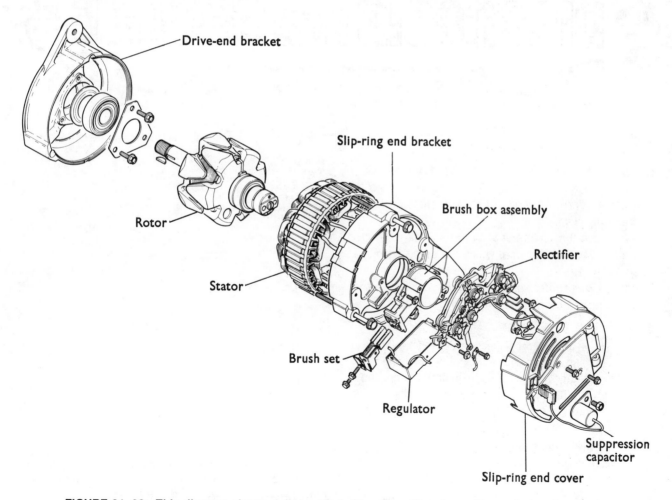

FIGURE 21–22. **This diagram shows a dismantled alternator.** *(Courtesy of Lucas Industries, Inc.)*

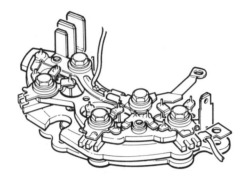
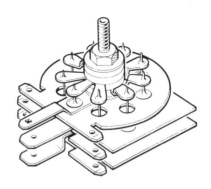

FIGURE 21-23. Rectifier packs are designed in many styles. Usually the diodes are grouped together on a plate for correct operation and polarity. *(Courtesy of Lucas Industries, Inc.)*

DIAGNOSIS and SERVICE

SAFETY TIPS

1. *Never work on the alternator when the engine is running.*
2. *When checking belt tension on the alternator, always make sure the battery is disconnected to eliminate the possibility of the engine accidently cranking over.*
3. *When using a voltmeter with the engine running, make sure the electrical wires are away from the spinning fan.*
4. *Always wear safety glasses when working on the alternator or other charging components.*

3. Make sure that all electrical leads are in the correct position and all alternator connections are clean and tight.

1. There are typically four problems found in the charging circuit. These include a low battery, an overcharged battery, a faulty indicator lamp, and a noisy alternator. Follow the manufacturer's diagnosis and service procedures when determining the cause of the problem.

2. Always check the drive belt tension on the generator or alternator and compare it to the manufacturer's specifications. Refer to *Figure 21-24*.

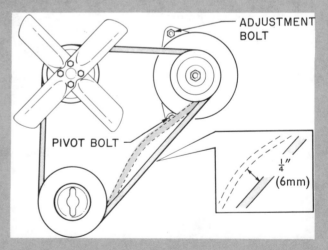

FIGURE 21-24. When checking the charging circuit always make sure the belt tension on the alternator is correct. *(Courtesy of Lucas Industries, Inc.)*

4. Noise from an alternator may be caused by worn or dirty bearings, loose mounting bolts, a loose drive pulley, a defective diode, or a defective stator.

5. A *rotor* can be checked for *shorts* or *opens* with an ohmmeter as shown in **Figure 21–25**. An ohmmeter is used to check for shorts between the rotor shaft, frame poles, and windings.

6. A *stator* can be checked for *shorts* or *opens* as shown in **Figure 21–26**. An ohmmeter is used to check each winding.

7. An ohmmeter can be used to check for *defective diodes*. When placed across the diode, high resistances should be noted in one direction and low resistances should be noted in the reverse direction. If both readings are very low or both readings are very high, the diode is defective. Defective diodes can cause a loud vibration to occur in the alternator during operation and reduced alternator current output. This is because during part of each revolution of the rotor, the alternator is charging. During part of each revolution (because of the bad diode), there is no charging. This on and off charging during one revolution of the rotor causes a loud vibration.

8. Inspect the brushes for excessive wear, damage, or corrosion. If there is any doubt about their condition, replace the brushes. Always install *new brushes* whenever rebuilding an alternator.

9. *Voltage output* on an alternator can be checked by placing a voltmeter from the BAT (battery) connection to ground. To check the overall condition of the charging system, use the following general procedure.
   a. Run the engine at idle for about 15 minutes to get the engine warm.
   b. Stop the engine and place the tester (a voltmeter) on the alternator. Clip the black lead to the negative side of the battery. Attach the positive side of the meter to the positive post on the battery.
   c. **CAUTION**: *Keep wires and body parts away from any part of the engine during running*.
   d. Start the engine and gradually increase the engine speed to about 1,800 rpm. Note the voltage reading on the meter scale.

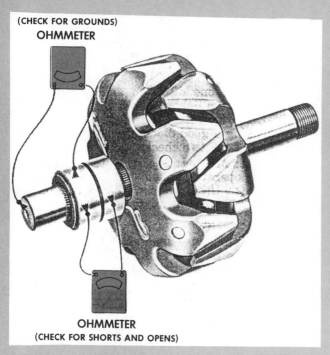

FIGURE 21–25. The rotor can be checked by using an ohmmeter. Grounds, shorts, and opens can be checked. *(Courtesy of Delco-Remy Division of General Motors Corporation)*

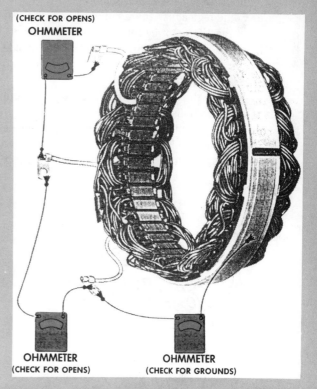

FIGURE 21–26. The stator can be checked using an ohmmeter. All three windings should be checked for grounds and opens. *(Courtesy of Delco-Remy Division of General Motors Corporation)*

e. If the meter pointer climbs steadily and comes to an abrupt halt within a range of 13.4 to 15.5 volts for a 12-volt system, the entire charging system is operating correctly.

f. If the meter climbs but stops below the correct range of 13.4 to 15.5 volts or above it, or if the meter fluctuates and will not stop, the regulator is probably defective.

g. If the meter does not climb to the correct range, the alternator is probably defective.

10. Voltage regulators that use a magnetic relay can be checked for correct voltage and amperage settings. Refer to the manufacturer's specifications for the correct procedure. Current solid-state regulators cannot be adjusted or repaired. When they are defective, they must be replaced. Follow the following general procedure to *replace* the *solid-state regulator.*

a. Loosen and remove the alternator belts from the front of the engine.

b. Disconnect and mark all of the electrical connections attached to the alternator.

c. Carefully remove the bolts that hold the alternator to the brackets on the engine.

d. Scribe marks on the alternator front and rear housings for use in reassembly.

e. Some regulators can be removed from the rear of the alternator by removing several screws.

f. Other alternator regulators are located inside the alternator, and the alternator must be disassembled.

g. Remove the through-screws that hold the front and rear housings together.

h. Pry the assembly apart, making sure to keep the stator assembly with the rear half, and remove the center rotor assembly from the end housing.

i. Remove the regulator and other parts to the assembly as necessary.

j. Reverse the procedure for assembly. However, the brushes must be retracted to allow the rotor and the slip-rings to be installed into the end frame.

k. After the rotor has been replaced in the end frame, align the scribe marks and reassemble the through-bolts.

11. Check cable continuity by disconnecting the wires and connecting a voltmeter between it and ground. The voltmeter should indicate battery voltage. Any zero reading could indicate an open-circuit lead.

12. Never ground any terminal on an alternator system.

13. Certain types of generators should be polarized. Never, however, polarize an alternator. Follow the procedure recommended in the service manual to polarize a generator.

14. Never allow an alternator to operate on an open circuit. Open circuits may result in high voltage and possibly damaging the alternator. Never remove a battery cable or disconnect the battery wire from the alternator when the engine is running.

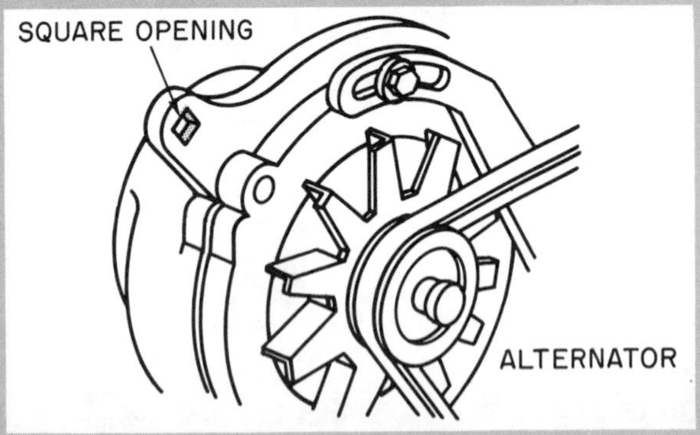

FIGURE 21-28. Charging systems can be checked using this tester. Both voltage and amperage can be checked to determine the charging system condition. *(Courtesy of Snap-On Tools Corporation)*

FIGURE 21-27. The alternator can be moved for tightening the belt tension by placing a ratchet in the square opening on the alternator housing.

15. Check the generator or alternator for bad bearings. *Bad bearings* can be determined by rotating the armature and feeling for rough turning. Bearings should be replaced if there is any evidence of wear.

16. On certain late model vehicles, there is a half-inch square opening in the belt tension bracket. This square opening, as shown in ***Figure 21–27***, allows the use of either a torque wrench or drive socket. Using this approach will eliminate stress on the components. Remember also that when adjusting the alternator to tighten the tension on the belt, always loosen the pivot and mounting bolts.

17. The output of any generator or alternator can be checked using the tester shown in ***Figure 21–28***. Both voltage output and amp or current *output test* can be performed. Each manufacturer's system may be different. Follow the manufacturer's specifications and directions when checking alternator or generator output. The output amperage is typically measured between the alternator output terminal and the battery with an ammeter. The charging system tester usually contains a voltmeter and an ammeter. These testers also have a specific set of instructions for use.

SUMMARY

The charging system was covered in this chapter. Generator principles, the dc generator, alternators, and voltage regulators were covered.

Three things are needed to produce electricity. These include a magnetic field, electrical wire, and having the wire cut across the magnetic lines of force. Both dc generators and alternators are designed to accomplish this. As any one of the three factors increase, the output voltage will also increase. If any one of the three decrease, the voltage output will also decrease.

One method used to convert mechanical energy into electrical energy involves using a dc generator. Here the magnetic field is stationary. The wire rotates within the magnetic field. A split-ring commutator is used to remove the voltage produced in the rotating part. The split-ring commutator is also used to convert the ac in the rotor to dc for use in the automotive electrical system.

The charging system must have a regulator. A regulator is used to control the voltage and amperage. The regulator uses a cutout relay to stop the battery voltage from forcing current back into the generator when it is at a low charge. In operation, the regulator controls the amount of current being sent to the field windings. If more current goes to the field windings, generator output increases. If less current goes to the field windings, generator output decreases.

The regulator used for dc charging systems has three relays. The cutout relay is used to stop the flow of reverse current from the battery back to the generator. The voltage relay controls the exact amount of voltage output from the generator. The amperage relay controls the maximum amperage being produced by the generator.

Alternators are much like generators, but they are more efficient. In an alternator, the field windings rotate, and the wire where the voltage is produced is stationary. Several sets of magnetic poles are usually used. Seven sets of north and south poles are very common. The commutator used on an alternator is called a slip-ring type. Its purpose is to connect electricity to the rotating fields.

Alternators also produce three-phase voltages, rather than single-phase voltages as in the generator. These three phases are rectified to pulsating dc by using six diodes. This is called three-phase full-wave rectification.

Alternator output must also be regulated. Voltage and current regulation are built into the circuits. There is no need for a cutout relay because the diodes prevent any reverse current flow from the battery back to the alternator. Depending upon the year and manufacturer, voltage regulators may have electrical relays or, as in the present system, be fully solid state. Zener diodes and transistors are used to regulate the alternator output.

When servicing alternators and generators, four major symptoms can be observed. These include: 1) the battery may be getting a low charge, 2) the battery may be getting too high a charge, 3) the indicator lamp may be faulty, 4) the alternator may be noisy. These faults can be caused by bad diodes; grounded, open, or shorted circuits; bad bearings; or a faulty regulator. Always follow the manufacturer's recommended procedures and specifications when troubleshooting the charging circuit.

TERMS TO KNOW

Can you explain each of the following terms? Review the chapter until you can use each term correctly.

Commutator	Laminated
Rectify	Rotor
Field circuit	Slip-ring
Armature	Three phase
Relay	Full-wave rectification
Stator	Zener diode

REVIEW QUESTIONS

Multiple Choice

1. The charging system is used to convert _____ energy to electrical energy.
   a. Radiant
   b. Mechanical
   c. Chemical
   d. Nuclear
   e. Thermal

2. What is necessary to produce a voltage in an alternator or generator?
   a. Conductors
   b. Magnetic field
   c. The conductors cutting the magnetic field
   d. All of the above
   e. None of the above

3. In order to increase the output of a generator or alternator:
   a. Increase the magnetic field strength
   b. Increase the amount of wire or conductors
   c. Increase the speed of movement between the magnetic field and the wire
   d. All of the above
   e. None of the above

4. The field circuit in a generator is also called:
   a. The magnetic core
   b. Field poles
   c. Slip-ring commutator
   d. Diode
   e. Voltage circuit

5. Which component is used to extract the voltage produced in the armature on a dc generator?
   a. Slip-ring commutator
   b. Diode
   c. Split-ring commutator
   d. Laminations
   e. Stator

6. Voltage produced inside the dc generator or alternator is always:
   a. Dc, and must be converted to ac
   b. Ac, and must not be converted to dc
   c. Dc, and must not be converted to ac
   d. Ac, and must be converted to dc
   e. Greater than the battery voltage

7. A voltage regulator on a charging system is always controlling the:
   a. Diodes
   b. Ac voltage
   c. Strength of the magnetic field
   d. Speed of operation
   e. Zener diode

8. Which of the following relays is used on a dc generator system to stop the reverse flow of electricity when the generator is at a low charge?
   a. Voltage relay
   b. Current relay
   c. Field relay
   d. Cutout relay
   e. All of the above

9. Which of the following components are rotating on an alternator?
   a. Magnetic north and south poles
   b. Slip-rings
   c. Magnetic pole fingers
   d. All of the above
   e. None of the above

10. The alternator produces:
    a. Single-phase voltages
    b. Double-phase voltages
    c. Three-phase voltages
    d. Dc without diodes
    e. Ac without rectification being needed

11. How many diodes are used on an alternator?
    a. 1
    b. 3
    c. 4
    d. 5
    e. 6

12. Regulators used in current charging systems today:
    a. Use two relays
    b. Are all solid state
    c. Use one relay
    d. Have no voltage regulator
    e. Use three relays

13. A zener diode is used to:
    a. Close off the field current
    b. Tell the transistors when there is too much voltage
    c. Generate voltage
    d. Rectify three-phase voltages
    e. Rectify single-phase voltages

14. Which of the following is not a problem in a charging system?
    a. Diodes shorting out
    b. Field windings grounding
    c. Bad bearings
    d. Too high rpm
    e. Brushes wearing down

The following questions are similar in format to ASE (Automotive Service Excellence) test questions.

15. Technician A says that if a constant 11 volts is being produced on the alternator, the system is alright. Technician B says that if a constant 14 volts is being produced on the alternator, the system is alright. Who is right?
    a. A only
    b. B only
    c. Both A and B
    d. Neither A nor B

16. A noise seems to be coming from an alternator, like a bearing is damaged. Technician A says that the bearings should be replaced. Technician B says that the diodes should be checked. Who is right?
    a. A only
    b. B only
    c. Both A and B
    d. Neither A nor B

17. An alternator does not have enough current output. Technician A says the problem is bad brushes. Technician B says the stator is reversed. Who is right?
    a. A only
    b. B only
    c. Both A and B
    d. Neither A nor B

18. An alternator is not charging correctly. Technician A says the problem is in the brushes and they may be worn. Technician B says the problem is that the alternator frame has too much magnetism. Who is right?
    a. A only
    b. B only
    c. Both A and B
    d. Neither A nor B

Essay

19. What three things are necessary to produce a voltage?

20. What is the difference between a dc generator and an alternator?

21. Describe the definition of rectifying.

22. What is the purpose of a stator?

23. What is a zener diode and how is it used in an electronic regulator?

24. Describe how to check the output of an alternator.

CHAPTER 22

Starting Systems

INTRODUCTION

The starter system is a type of electrical circuit that converts electrical energy into mechanical energy. The electrical energy contained in the battery is used to turn a starter motor. As this motor turns, the engine is cranked for starting. This chapter is concerned with the starting system, its components, operation, and service.

OBJECTIVES

After reading this chapter, you will be able to:
- Identify the operating principles of starter motors.
- List the parts of the starter motor and state their purpose.
- Compare the operation and parts of different starter drive and clutch mechanisms.
- Describe the purpose and operation of the solenoid and related circuits.
- Describe various diagnostic and service and repair procedures for the starting system.

CHAPTER HIGHLIGHTS

22.1 STARTER SYSTEM OPERATING PRINCIPLES
   A. Purpose of the Starter System
   B. Magnetic Field Around a Conductor
   C. Producing Motion from Electricity
   D. The Principle of Commutation
   E. Producing the North and South Poles

22.2 STARTER MOTORS
   A. Parts of the Starter
   B. Series Wound Field Circuits
   C. Series-parallel Wound Field Circuits
   D. Armatures
   E. Brushes

22.3 DRIVE AND CLUTCH MECHANISMS
   A. Purpose of Drives and Clutches
   B. Inertia Drives
   C. Overrunning Clutch Drives

22.4 SOLENOIDS
   A. Definition of a Solenoid
   B. Solenoid Operation
   C. Closing and Hold-on Coils
   D. Solenoid Drive Linkage

22.5 STARTER CIRCUIT OPERATION

DIAGNOSIS AND SERVICE

SUMMARY

22.1 STARTER SYSTEM OPERATING PRINCIPLES

Purpose of the Starter System

Although energy cannot be created, it can be changed from one form to another. In the starter system, the starter motor is used to convert electrical energy from the battery to mechanical energy to crank the engine. Several parts are needed to do this. These include the starter motor, a drive and clutch mechanism, and a solenoid that is used to switch on the heavy current in the circuit.

A considerable amount of mechanical power is necessary to crank and start a car engine. About two horsepower, or approximately 250 to 500 amps of electricity, is normally needed. The amperage must be higher on a diesel engine, but the basic principles are the same. Because of the high amounts of current, heavy cables must be used to carry this current. A switch on the dashboard is used to energize the heavy-current circuits. *Figure 22–1* shows a complete electrical circuit of the starting system. It includes the starter to cranking motor, solenoid, drive and clutch mechanisms, electrical wire, and ignition switch. Each component will be studied in more detail in this chapter.

Magnetic Field Around a Conductor

When current is passed through a conductor, a magnetic field is built up around the conductor. This is shown in *Figure 22–2*. When current passes from the back of the wire to the front, a magnetic field is created. It circles around the conductor in a counterclockwise direction. The lines of magnetic force are formed in a definite pattern. If the current is reversed, the magnetic lines of force will flow in a reverse or clockwise direction.

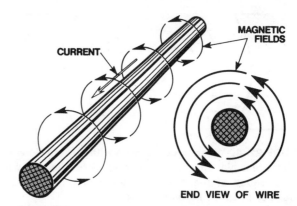

FIGURE 22–2. When current passes through a conductor, a magnetic field is built up around the conductor. The magnetic field surrounds the conductor. *(Courtesy of Lucas Industries, Inc.)*

Producing Motion from Electricity

A starter motor has a north pole and a south pole. This is shown in the top of *Figure 22–3*. The north and south poles produce a magnetic field. The lines of force between the two poles flow from the north pole to the south pole. A wire carrying electricity is placed within the magnetic field. When this is done, both magnetic fields below the wire travel in the same direction. Above the wire, the lines of force travel in opposite directions. The result is a stronger magnetic field on the bottom of the wire and a weaker magnetic field on the top. This difference in magnetic fields causes the wire to move upward.

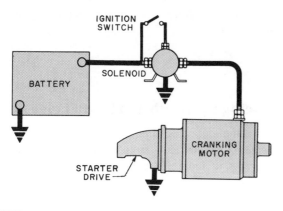

FIGURE 22–1. The starter system is made of the starter or cranking motor, solenoid, ignition switch, and starter drive mechanism. *(Courtesy of Delco-Remy Division of General Motors Corporation)*

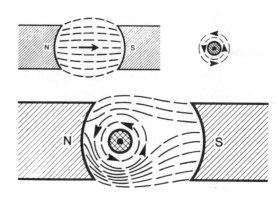

FIGURE 22–3. A north pole and a south pole are needed to create the action in a motor. If the current-carrying conductor is placed within the magnetic field, the wire tends to move. This movement is caused by differences in the magnetism on top of and below the wire. *(Courtesy of Lucas Industries, Inc.)*

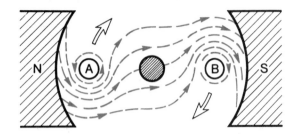

FIGURE 22-4. When a loop of wire with current flowing in the wire is placed within a magnetic field, the loop will turn because of the direction of the magnetic fields. *(Courtesy of Lucas Industries, Inc.)*

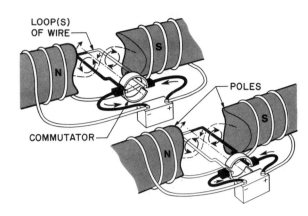

FIGURE 22-5. A simple starter motor consists of a set of poles, a loop of wire, and a commutator. *(Courtesy of Lucas Industries, Inc.)*

There is another way to define this movement. Magnetic lines of force traveling in the same direction tend to repel each other. Magnetic lines of force are often compared to a stretched elastic band. If there is any distortion, the bands (magnetic lines of force) will try to realign themselves. This causes the wire to move away from the area where the lines of force are concentrated.

In a starter, the wire is wound into loops. *Figure 22-4* shows how a loop of wire would be turned if it were placed within a magnetic field. The electricity flows into point B and loops around to come out at point A. The wire is placed on the center axis. The direction of the magnetic field around the left and right conductors is opposite. When this condition occurs, the magnetic lines of force strengthen below the left side and above the right side. The concentration of magnetic lines of force causes the wire loop to turn in a clockwise direction on the center axis. When electricity is passed through wire, electrical energy is converted to mechanical energy.

The Principle of Commutation

Referring to Figure 22-4, when the loop has rotated one-half revolution, point A is positioned at the south pole and point B is positioned at the north pole. Because the magnetic field around point B is clockwise, magnetic distortion occurs above "B" and below "A." This means that the wire loop will reverse direction to return to the original position.

To keep the loop turning in the same direction, the electrical current in the loop must be reversed precisely at the right time. This is done by using a commutator. Commutation, or current reversal, is achieved by joining the ends of the loop to two metal segments. The contact surface between the battery and the segments is formed by the brushes.

Figure 22-5 shows a simple starter with a *commutator* and brushes. The loop of wire is shown as having a black side and a white side. Both ends of the loop of wire are connected to a segment of metal. On the top diagram the white half of the loop (on the right) is connected to the positive side of the battery. The current will flow from the positive terminal on the battery, through the loop, and return to the negative side of the battery. This will produce a magnetic field around the wire in the loop. In this position, the direction of the magnetic field on the black half is counterclockwise and of the white half is clockwise. This causes the loop to turn.

In the lower drawing, the loop has moved through a half turn. The white section is now on the left and the black section is on the right. Because of the position of the metal segments under the brushes, the current is now reversed in the loop. This causes the magnetic field around both sides to reverse as compared to the top drawing. The magnetic field around the conductor will now keep the loop turning in the same direction.

Producing the North and South Poles

The north and south poles are created by using soft iron that can be easily magnetized. This is done by wrapping a coil of wire around the soft iron core and passing current through the wire. These coils are called field coils or poles. The soft iron core is specially shaped to concentrate the magnetic lines of force in the space provided. *Figure 22-6* shows the main magnetic fields. In this case, the north pole is formed on the left, and the south pole is formed on the right. The lines of magnetic force always flow from the north pole to the south pole.

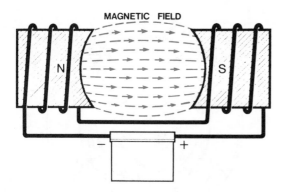

FIGURE 22-6. All starter motors have a set of north and south poles that are produced by winding wire around a soft iron core. *(Courtesy of Lucas Industries, Inc.)*

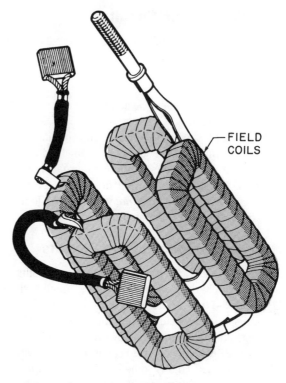

FIGURE 22-7. Four field coils are used on a starter motor. *(Courtesy of Lucas Industries, Inc.)*

CAR CLINIC

PROBLEM: BAD STARTER BEARINGS

A vehicle has 60,000 miles on it. No work has been done on the starter system. There have been times recently when the engine would not even crank over. It sounds like the solenoid is trying to engage the starter pinion, but starter just won't turn over.

SOLUTION:

Quite often the starter bushings will wear. When the bushings wear, the armature sits lower in the starter field windings, often scraping on them. Remove the starter and check the no-load speed test. This should show that the starter bushings are bad and should be replaced.

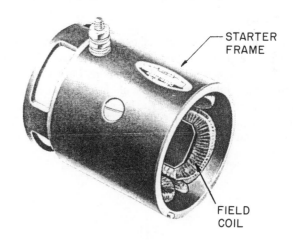

22.2 STARTER MOTORS

Parts of the Starter

The parts of the starter motor will now be considered in more detail. All starters have a set of field coils. Coils are made of copper or aluminum strips wound in the correct direction to produce the poles. Typically, four field coils are used on a starter. *Figure 22-7* shows an example of the field coils. These coils are assembled over the soft iron core. The iron core or shoe is attached to the inside of a heavy iron frame on the motor. See *Figure 22-8*. The iron frame and *pole shoes* provide a place for the field coils. They also provide a low-resistance path for the magnetic lines of force.

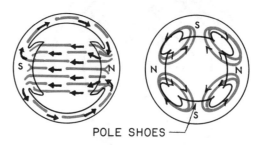

FIGURE 22-8. The four field coils are attached to the starter frame. The shoes provide a place for the field coils. *(Courtesy of Delco-Remy Division of General Motors Corporation)*

Series Wound Field Circuits

The four field coils in the starter can be connected in several ways. Two common connections are the series and the series-parallel wound field. The one selected is determined by the application, engine speed, torque requirements on the starter, cable size, battery capacity, and current-carrying capacity of the brushes.

Figure 22–9 shows the series wound circuit. Here the field coils are connected in series. The four field coils are wound in different directions to produce alternate north and south poles. The electrical current flows from the battery into each coil, then to the *armature* windings. All four field coils are in series with the armature windings.

Series-parallel Wound Field Circuits

Another type of electrical circuit used in starters is called the series-parallel field connection. This arrangement is shown in *Figure 22–10*. The electrical current comes into the starter and splits. The field circuit consists of two paths for the current. The field windings, however, are still in series with the armature, while one field coil is in parallel with the armature windings. This is sometimes called a *shunt* coil connection. It is used to control the maximum speed of the starter during operation.

Armatures

The armature is the rotational part of the starter. Earlier in this chapter, the armature was called the loop of wire. There are many loops of wire in the armature. Each wire loop is connected to the copper segments arranged in a "barrel" shape. See *Figure 22–11*. The segments are insulated from each other, and from the armature shaft. These segments of copper form the commutator and provide the running surface for the brushes.

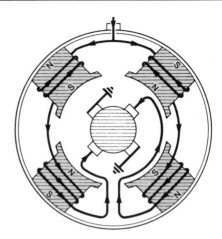

FIGURE 22–10. A series-parallel wound circuit in the starter has two paths for the current to flow through the poles. *(Courtesy of Lucas Industries, Inc.)*

The armature assembly consists of a stack of iron *laminations* located on a steel shaft. The steel shaft is supported by bearings located on the end plates of the starter. The iron laminations are used to concentrate the magnetic field. Laminations are also used to reduce the heating effect of *eddy currents*. Eddy currents are small electrical currents that are produced inside the iron core. The windings are made of heavy copper ribbons that are inserted into the slots in the iron laminations. *Figure 22–12* shows a photo of an actual armature used in a starter motor.

Brushes

Brushes are used to make electrical contact between the armature, which rotates, and the battery, which is stationary. Brushes are made from a high percentage of copper and carbon. This material minimizes electrical losses due to overheating. *Figure 22–13* shows an example of the brushes held in place on the starter end plate. Each brush uses a spring metal to force the brush against the commutator during operation. There are typically four brushes. Two brushes are used to feed electricity into the armature. Two brushes are grounded and used to return the electricity to ground, which is the negative side of the battery.

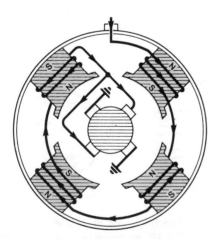

FIGURE 22–9. A series wound circuit in the starter has all four coils in series with the armature. *(Courtesy of Lucas Industries, Inc.)*

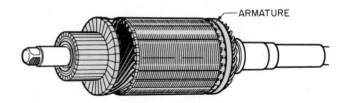

FIGURE 22–11. The armature in a starter is the rotational part that has many loops of wire with a commutator. *(Courtesy of Lucas Industries, Inc.)*

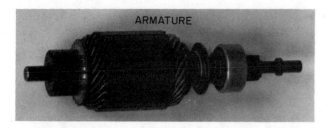

FIGURE 22–12. An armature used in a starter motor.

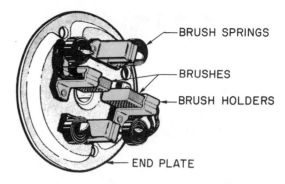

- BRUSH SPRINGS
- BRUSHES
- BRUSH HOLDERS
- END PLATE

FIGURE 22–13. Brushes are held in place on the end plate of the starter. *(Courtesy of Lucas Industries, Inc.)*

CAR CLINIC

PROBLEM: STARTER DRAG

A vehicle with about 30,000 miles on it starts OK the first time. After the engine is hot, however, it usually takes several times to crank the engine. The starter seems to be very sluggish and seems to have a drag.

SOLUTION:

Check the condition of the starter bearings. If they are in good condition, check the current on the starter when it is hot. Suspect high current draw. If the starter and solenoid have no electrical faults, exhaust gases may be heating up the starter, causing it to drag and be sluggish. A heat baffle may be needed between the starter and the exhaust pipe. Also, make sure that as the car is moving forward, there is sufficient air flowing over the starter to help cool it.

22.3 DRIVE AND CLUTCH MECHANISMS

Purpose of Drives and Clutches

The starting system uses several types of drives and clutches. The purpose of the drive is to engage and disengage the *pinion gear* from the flywheel. The pinion gear is the small gear located on the armature shaft. When the engine is running, the pinion gear cannot be in contact with the flywheel. When the starter is cranked, however, the pinion gear must slide on the shaft and engage the flywheel, only at that time.

When the engine starts to turn on its own power, the flywheel turns faster. Now the pinion gear must be removed rapidly. There are several types of drives and clutches used for this purpose. Two popular types include the *inertia drive* and the *overrunning clutch drive*. Most manufacturers use a design similar to one of these two drives.

Inertia Drives

There are several types of *inertia* drives. Inertia drives are also called Bendix Drives. Although they may differ considerably in appearance, each drive operates on the principle of inertia. Inertia is a physical property of an object. An object at rest tends to stay at rest. An object in motion tends to remain in motion. Inertia is used to move the pinion gear to engage the engine ring gear when the starter motor is energized.

Figure 22–14 shows an inertia drive system. There is a pinion gear, a screw sleeve, the armature shaft, and springs. All of the parts of the drive mechanism are assembled onto the armature drive shaft. The drive shaft has a spline that meshes with the internal spline on the screw sleeve. Thus, the screw sleeve can move along the axis of the armature shaft.

When the armature turns during starting, the armature shaft and the screw sleeve rotate. Due to the inertia of the pinion (remaining at rest), the screw sleeve turns inside the pinion. This causes the pinion to move out and engage with the flywheel ring gear. As soon as the engine fires and runs under its own power, the flywheel is driven faster by the engine. As the pinion gear is forced faster, it is moved back along the screw sleeve and out of engagement with the flywheel. At this point, the starter switch is released. The spring is used to absorb excessive twisting of the shaft. On certain drives, a compression spring is used to reduce the shock at the moment of engagement.

Overrunning Clutch Drives

A second type of drive is called the overrunning clutch drive. This type of drive is also referred to as a roll-type drive or *sprag* drive. The sprag drive uses a series of 30 sprags, rather than rollers. It is commonly used on diesel engines and on some gasoline engines. The roller clutch drive is shown in *Figure 22–15*. It consists of a drive and driven

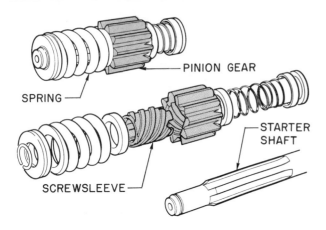

FIGURE 22–14. An inertia drive system is used on some starter systems to engage the pinion gear with the ring gear. *(Courtesy of Lucas Industries, Inc.)*

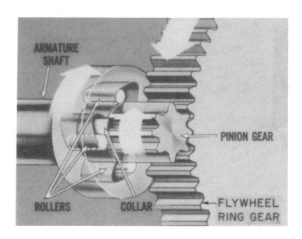

FIGURE 22–16. When the armature shaft rotates during cranking, small rollers become wedged, causing the pinion gear to lock up and rotate the ring gear. *(Courtesy of Delco-Remy Division of General Motors Corporation)*

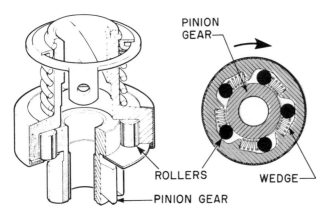

FIGURE 22–15. A roller clutch, also called the overrunning clutch, is used on some starter systems to engage the pinion gear with the ring gear. *(Courtesy of Lucas Industries, Inc.)*

When the engine starts, *Figure 22–17*, the flywheel spins the pinion faster than the armature. This action releases the rollers, unlocking the pinion gear from the armature shaft. The pinion then "overruns" safely and freely until being pulled out of mesh. Note that the overrunning clutch is moved in and out of mesh by linkage operated by the solenoid.

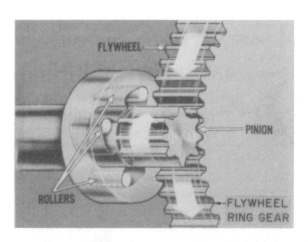

FIGURE 22–17. When the engine starts, the flywheel spins the pinion gear faster, which releases the rollers in the wedge. *(Courtesy of Delco-Remy Division of General Motors Corporation)*

member and a series of cylindrical rollers. These rollers are placed in wedge-shaped tracks in the clutch housing. Referring to *Figure 22–16*, when the armature shaft rotates during cranking, the small rollers become wedged against the collar attached to the pinion gear. This wedging action locks the pinion gear with the armature shaft. The pinion gear now rotates with the shaft, cranking the engine.

TIRE AND WHEEL PERFORMANCE

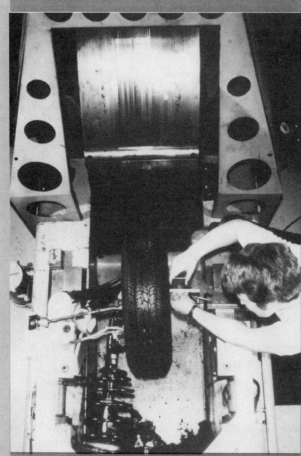

Tires must be tested for many characteristics. At the GM Proving Grounds in Milford, Michigan, tires are tested for traction, cornering, structural durability, uniformity, and road hazard resistance. Other tests include dynamometer testing of wheels under overload conditions, and corrosion resistance testing of wheel attachments. (Courtesy of General Motors Proving Grounds)

22.4 SOLENOIDS

Definition of a Solenoid

A *solenoid* is an electromechanical device that switches electrical circuits on and off. *Figure 22–18* shows the operation of a solenoid. When current flows through the electrical coil, a magnetic field is created inside the coil. If a soft iron core of metal is placed near the center of the coil, the metal core tends to center itself inside the coil. If a spring is used to hold the metal core outside the center, it again tries to center itself. Thus, when current passes through the coil, the metal core will move. This process, then, converts electrical energy into mechanical energy. The movement of the metal core is used to open and close electrical circuits. Therefore, a solenoid is considered an electromechanical switch.

Solenoid Operation

Figure 22–19 shows the internal operation of a solenoid used on certain vehicles. The solenoid is used to start and stop the heavy current that flows to the starter motor during cranking. The electrical winding is placed around the iron core. This core is called the armature or plunger. On the bottom of the armature is a set of moving contacts. On the bottom of the solenoid is a set of fixed contacts connected to the heavy-current terminals. When the solenoid windings are energized, the center armature is forced downward magnetically. This causes an electrical connection between the two fixed contact points. When the current is stopped in the windings, the armature returns to its original position. Some older armatures can also be pushed manually to accomplish the same result.

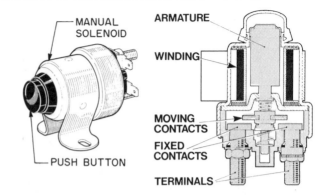

FIGURE 22–19. When the solenoid is operated, a plunger or armature moves in and makes contact with two terminals used to switch the motor on. *(Courtesy of Lucas Industries, Inc.)*

Closing and Hold-on Coils

Many solenoids contain two coils. One is called the closing or pull-in coil. One is called the hold-on coil. This is shown in the electrical circuit in *Figure 22–20*. The heavy gauge winding is called the closing coil or pull-in winding. It has low resistance. The finer or thinner coil is the hold-on winding. It has higher resistance.

The closing coil has less resistance. This means that there will be more current and more force in the coil. This force is needed along with the hold-on coil to move the center plunger. Once the plunger movement has been completed, much less magnetism is needed to hold the plunger in. When the contact disk touches the terminals, the pull-in winding is shorted out, thus no current flows through the winding. This reduces current draw on the battery during cranking.

Solenoid Drive Linkage

On many vehicles, there is a linkage attached to the opposite end of the plunger on the solenoid. This linkage is used to engage the pinion with the flywheel gear. *Figure 22–21* shows an example of this linkage. Any movement of the plunger results in a similar but opposite movement of the pinion gear. The movement of the linkage is designed so the pinion gear meshes completely before the solenoid makes contact with the terminals. This assures that the gears are meshed before the starter turns.

On occasion, the pinion gear and flywheel gear will not mesh. This is caused by the teeth butting up against each other. A spring on the overrunning clutch assembly will overcome this problem. When the teeth butt up against each other, the plunger and operating lever move their full

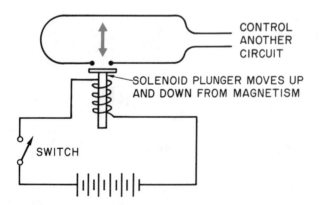

FIGURE 22–18. A solenoid is a device that converts electrical energy to mechanical energy by the use of magnetism.

distance. However, the pinion gear is still not meshed. When the starter begins to turn, the teeth immediately mesh and cause the engine to crank.

22.5 STARTER CIRCUIT OPERATION

Now that all of the components have been studied on the starting system, the complete circuit can be analyzed. *Figure 22–22* shows the complete operation of the starting circuit. Electrical energy is made available by the battery. A large, heavy cable of wire is connected from the battery to the starter solenoid. The ignition switch is also connected to the positive side of the battery.

A neutral safety switch is also included in the circuit. This switch remains open except when the transmission is

in park or neutral. When the transmission is in any gear, the switch is open and the engine cannot be cranked. When the transmission is in park, the switch is closed and the engine can be cranked.

If the transmission is in park and the start switch is closed, electric current passes to the solenoid. The solenoid is now energized and the plunger moves. This causes the linkage to engage the pinion gear and crank the engine.

When the ignition switch is closed, current causes the solenoid to engage the starter. At the same time, the mechanical linkage engages the pinion gear with the ring gear. When the engine starts, the pinion gear is pushed away from the ring gear. When the ignition switch is released, the solenoid also releases. The starter circuit is now ready to operate again when the switch is turned to the crank position.

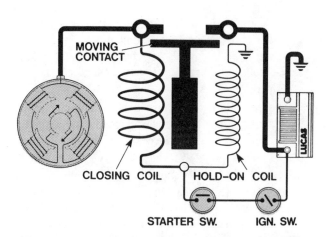

FIGURE 22–20. The two windings used in a solenoid close the plunger and hold the plunger in. *(Courtesy of Lucas Industries, Inc.)*

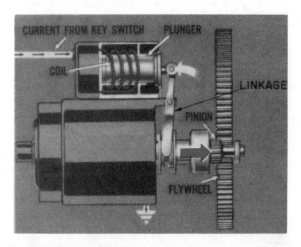

FIGURE 22–21. Linkage is used to engage the pinion gear with the flywheel gear. *(Courtesy of Delco-Remy Division of General Motors Corporation)*

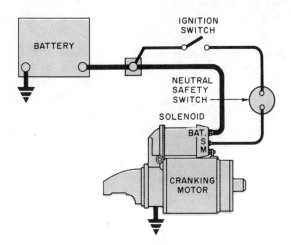

FIGURE 22–22. The complete electrical circuit of the starter system. *(Courtesy of Delco-Remy Division of General Motors Corporation)*

DIAGNOSIS and SERVICE

1. An ohmmeter can be used to *check high resistance* between any two points on the starter circuit. **Figure 22–23** shows several examples of where to hook up the meter.

2. The *voltage drop* in the starter can be checked during cranking with a voltmeter. Excessive voltage drop could mean the starter circuit has too high resistance. There are several checks that can be made. **Figure 22–24** shows these checks. The voltage can be checked between the vehicle frame and the grounded battery terminal post (A), between the vehicle frame and the starter field frame (B), or between the ungrounded battery terminal post and the battery terminal on the solenoid (C). If any of these readings show more than one-tenth (0.1) volt drop when the starting motor is cranking, the electrical connections will have to be cleaned. A light coating of grease on the battery cables and terminal clamps will retard further corrosion.

3. *Bushings* in the end plates wear and need to be replaced in the starter. If the bushings wear excessively, the starter armature may rub on the pole shoes, causing damage to the starter. If the bushings are bad, the starter will require very high current draw during cranking. A growling or scraping sound may also be heard during cranking. Replace the bushings if wear is suspected.

4. If the solenoid engages but the starter does not turn, the solenoid moving contact may be corroded. On certain starters, this moving contact can be cleaned and replaced in the solenoid. This is done by taking the solenoid apart and using emery cloth to clean the solenoid moving contact.

5. If the starter appears inoperative, the field coils can be checked again for shorts, opens, or grounds. Use an appropriate ohmmeter to measure the *resistance of the field coils*. The starter must be disassembled first.

6. When the starter is disassembled, check the *roller clutch drive* for smoothness, instantaneous lockup, and free movement along the shaft. If any of these conditions are missing, replace the roller clutch.

7. Check the armature *end play* and compare the readings to the manufacturer's specifications. Shims may be needed to correct excessive end play. Depending upon the manufacturer, the clearance is between 0.010 and 0.140 inch. See **Figure 22–25**. In addition, make sure that during assembly of the starter the small friction disk on the end of the pinion gear is replaced in the same position.

8. Check the *starter brushes* for excessive wear. Check the manufacturer's specifications for the amount of wear allowed on the brushes. Replace the brushes if necessary. Also check the brushes to be sure that the brush springs are providing the correct tension. Depending upon the manufacturer, the tension may be from 30–100 or more

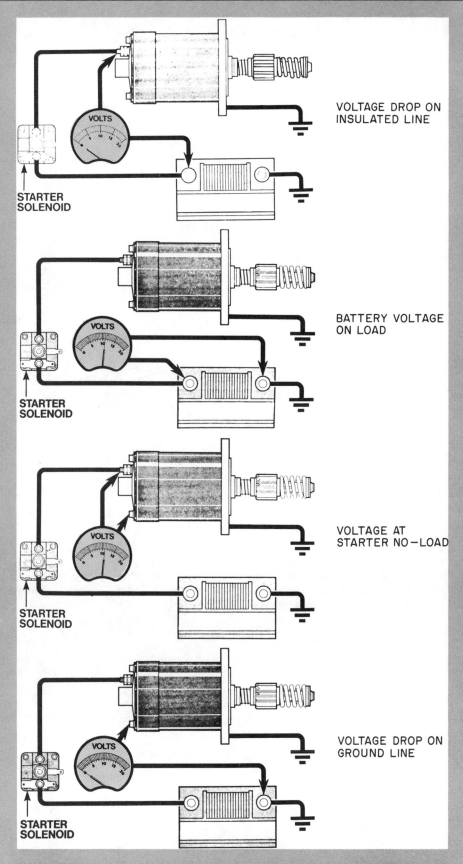

FIGURE 22–23. Several voltage checks can be made to test the voltage drop on the starter circuits. *(Courtesy of Lucas Industries, Inc.)*

VOLTAGE DROP CHECKS

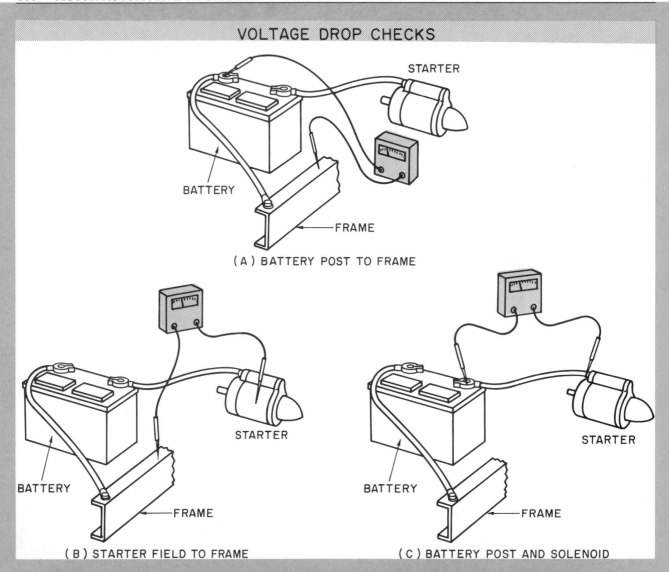

(A) BATTERY POST TO FRAME

(B) STARTER FIELD TO FRAME

(C) BATTERY POST AND SOLENOID

FIGURE 22-24. Several voltage drops can be checked to determine the condition of the engine. Voltage can be checked between the frame and the negative battery terminal, between the frame and the starter field frame, and between the positive side of the battery and the solenoid terminal.

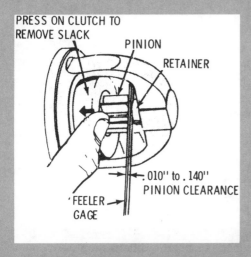

FIGURE 22-25. The starter can be checked for end play between the pinion gear and the retainer. *(Courtesy of Motor Publications, Auto Repair Manual, 1981-1987)*

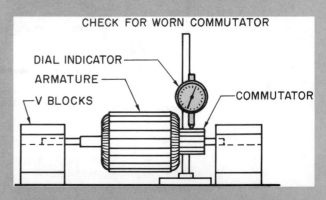

FIGURE 22-26. If the commutator is burned or worn, it can be turned on a lathe and cleaned up.

ounces. Check the manufacturer's specifications for the exact amount required for each starter. If the tension is not within the specifications, replace the springs.

9. Check the *commutator* segments for evidence of burning. This can be identified by observing the commutator and looking for burned copper on the commutator. If there is evidence of burning, the commutator can be cut down on a lathe. The normal procedure is to cut down or turn the commutator with light cuts until the worn or bad spots are removed. This can be observed on a dial indicator. See *Figure 22–26*. Then remove the burrs with No. 00 sandpaper.

10. When servicing a starter, always clean the parts before reassembling them.

11. The starter armature can be checked for short circuits by using a *growler*, *Figure 22–27*. A growler is a tool that creates a magnetic field around the armature. A hacksaw blade is placed on top of the armature. When it vibrates, it indicates a short in the armature.

12. The *free speed test* can also be performed on starters. Using the circuit shown in *Figure 22–28*, a tachometer can be used to measure the speed of

the starter per minute. Failure of the motor to perform to specifications may be due to bad bearings or high-resistance connections. During this test, check the amperage, volts, and rpm and compare them to manufacturer's specifications.

13. Both the hold-in and pull-in windings on the solenoid can be checked for resistance and voltage drops. Refer to the manufacturer's specifications for the correct readings.

14. Lubricate all bushings in the starter with high-temperature grease before assembly.

15. Always check each electrical connection on the starter system for solid and good electrical contact. Clean and replace connections where necessary. Poor connections can cause high-voltage drop across that particular connection, which, in turn, will cause the starter to perform poorly.

16. Check for broken or distorted springs on the starter linkage mechanism. Replace springs where necessary.

17. There are several other checks that can be done on the starter circuits and components. These depend, however, upon the exact car manufacturer and the style of starter.

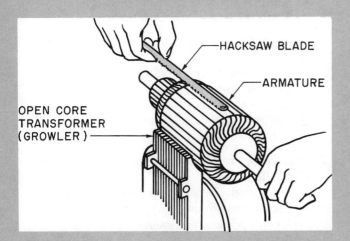

FIGURE 22–27. A growler is used to check for shorts in the armature.

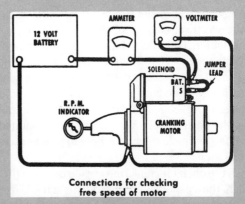

FIGURE 22–28. The free speed test can be performed on a starter when it is out of the vehicle. Voltage, amperage, and rpm are compared to specifications given by the manufacturer. *(Courtesy of Motor Publications, Auto Repair Manual, 1981–1987)*

SUMMARY

The purpose of the starting system is to convert electrical energy in the battery to mechanical energy to crank the engine. This is done by using a starter motor, ignition switch, solenoid, and starter drive mechanism.

In order to understand the starting system, several

principles are important. Whenever a wire carries electrical current, a magnetic field is built up around the conductor. If this conductor is placed within a strong magnetic field made from a north and south pole, the wire will move. If the wire is placed in the shape of a loop on an axis, the loop will revolve. A commutator is used to provide the correct path for electricity into the loop of wire. The north and south poles are made by using electromagnets.

A starter typically has four windings that make up the field coils. These coils are wound on pole shoes and attached to the starter housing. The rotating part of the starter is called the armature. It is made of many turns of wire. The connections between the armature and the poles can either be series wound or series-parallel wound. The type of circuit will depend upon the starter speed, load, size of the engine, cable size, and so on. The armature core is made of iron laminations to reduce eddy currents. Brushes are used to rub on the commutator for continuous connections to the armature.

The starter uses several types of drive mechanisms. Their purpose is to engage the pinion gear to the flywheel gear for cranking the engine. The inertia drive system uses a pinion gear on a screw sleeve to engage the two gears. When the starter begins to turn, the pinion gear slides outward and meshes with the ring gear.

The overrunning clutch drive uses a set of rollers that become wedged in the drive housing when the starter begins to turn. When the engine fires and turns faster than the starter, the pinion gear unlocks from the wedges and spins freely. This protects the starter from being turned at high rpm by the engine, if the operator continues to crank the engine. If the engine does turn the starter, it will be damaged.

The solenoid is used for several purposes. The solenoid makes the electrical contact for the high current used to crank the starter. The solenoid also moves the pinion gear in mesh with the ring gear. There are typically two windings in the solenoid. One is called the pull-in or closing winding. The other is called the hold-in winding. When the solenoid closes the electrical contacts for the starter, the pull-in winding pulls the solenoid plunger in. Then it is shorted out when contact is made. Only the hold-in winding is used during cranking.

When the operator cranks the engine by using the ignition switch, the solenoid is energized. This action closes the moving contacts in the solenoid and makes the electrical connection to the starter. As the starter cranks, the pinion gear is in mesh with the ring gear, thus cranking the engine. When the ignition switch is released, the starter motor stops turning and drive linkage is released, removing the pinion gear from the ring gear.

Several checks can be used to determine the condition of the starter circuit. A voltmeter is used to check voltage drops. The free speed test tells the current, voltage, and rpm when the starter is removed from the vehicle. Other checks include observing the spring tension, brush wear, bushing wear, and solenoid operation.

TERMS TO KNOW

Can you explain each of the following terms? Review the chapter until you can use each term correctly.

Commutator	Inertia
Pole shoe	Inertia drive
Armature	Overrunning clutch
Shunt	drive
Lamination	Solenoid
Eddy current	Sprag
Pinion gear	

REVIEW QUESTIONS

Multiple Choice

1. The starter system converts:
   a. Mechanical energy to chemical energy
   b. Thermal energy to electrical energy
   c. Mechanical energy to electrical energy
   d. Electrical energy to mechanical energy
   e. Thermal energy to mechanical energy

2. Any wire that has electrical current passing through it:
   a. Moves toward a north pole
   b. Moves toward a south pole
   c. Has a magnetic field around it
   d. Has no magnetic field around it
   e. Reduces its internal resistance

3. What causes the movement to occur on the armature of a motor?
   a. Two magnetic fields opposing and aiding each other
   b. Thermal energy forces
   c. A north pole pushing against a south pole
   d. Gravity
   e. Eddy currents

4. The _____ keeps the current flowing in the correct direction in the armature.
   a. Eddy currents
   b. Stator
   c. Commutator
   d. Brushes
   e. Solenoids

5. The north and south poles in a starter are:
   a. Made of permanent magnets
   b. Made from electromagnets
   c. Repelling each other
   d. Turned off during starting
   e. Operated in pulses

6. The _____ holds the field windings.
   a. Solenoid
   b. Brushes
   c. Hold-on coil
   d. Pole shoes
   e. Sprags

7. Eddy currents can be reduced in the armature by using a core made of:
   a. Plastic
   b. Silicon
   c. Copper
   d. Carbon
   e. Laminations

8. Which type of starter drive mechanism uses the principle that an object tends to remain at rest or stay in motion?
   a. Overrunning clutch
   b. Roller clutch
   c. Inertia drive
   d. Sprag drive
   e. All of the above

9. Which type of starter drive mechanism uses a series of sprags?
   a. Overrunning clutch
   b. Inertia drive
   c. Bendix drive
   d. Roller clutch
   e. None of the above

10. On the overrunning clutch, when the engine starts and turns faster than the starter:
    a. The rollers are not being wedged in
    b. The pinion gear moves out of mesh with the ring gear
    c. The starter will not be forced to turn faster
    d. All of the above
    e. None of the above

11. A solenoid is used to convert:
    a. Mechanical energy to thermal energy
    b. Electrical energy to chemical energy
    c. Electrical energy to mechanical energy
    d. All of the above
    e. None of the above

12. When the plunger is energized and the contact disk makes electrical contact inside the solenoid:
    a. The hold-in winding releases
    b. The starter motor begins to turn
    c. The pinion gear is disengaged
    d. The ring gear stops turning
    e. Brushes release contact on the armature

13. Which coil in the solenoid is shorted out when the starter motor begins to turn?
    a. The field coils
    b. The hold-in coil
    c. The pull-in coil
    d. All of the above
    e. None of the above

14. Which of the following is *not* considered a service check on the starter system?
    a. Checking the voltage drop of the starter
    b. Checking for opens on the field poles
    c. Checking the amount of wear on the brushes
    d. Checking the operation of the diodes in the circuit
    e. Checking the condition of the commutator

15. Which of the following is *not* considered a service check on the starter system? Checking the
    a. Bushings for wear
    b. Circuit for grounds and opens
    c. Drive mechanism for voltage drop
    d. Commutator for signs of wear
    e. Drive mechanism for damaged springs

The following questions are similar in format to ASE (Automotive Service Excellence) test questions.

16. Technician A says that if the commutator is slightly burned, it must be replaced. Technician B says that if the commutator is slightly burned, it may be machined on a lathe to remove the burn spots. Who is right?
    a. A only
    b. B only
    c. Both A and B
    d. Neither A nor B

17. When tested on a growler, the hacksaw blade vibrates at a specific spot. Technician A says that the armature has a short in it. Technician B says that the field poles have a short in them. Who is right?
    a. A only
    b. B only
    c. Both A and B
    d. Neither A nor B

18. The spring tension on the brushes on the starter is weak and below specifications. Technician A says that the springs should be bent to obtain more tension. Technician B says that the weaker spring tension will not cause any damage and that they need not be replaced. Who is right?
    a. A only
    b. B only
    c. Both A and B
    d. Neither A nor B

19. The armature of the starter has been rubbing on the field coils. Technician A says the problem is the spring tension of the brushes. Technician B says the problem is misalignment of the solenoid. Who is right?
    a. A only
    b. B only
    c. Both A and B
    d. Neither A nor B

Essay

20. Describe how magnetism is produced in a coil of wire.

21. What is the definition of commutation?

22. What is the purpose of pole shoes on a starter?

23. What is the purpose of using laminations of steel on the armature?

24. Describe the operation of the inertia drive clutch.

25. What is the purpose of the solenoid?

26. What is the purpose of the hold-in and pull-in coil on the solenoid?

CHAPTER 23

Characteristics of Air Pollution

INTRODUCTION

Before the 1960s very little emphasis was placed on *pollution* in our society. All industries, including the automotive manufacturers, were producing goods without concern for how the waste products affected the environment. Any product that is manufactured always produces waste. By the mid 1960s, there was serious concern about how these pollutants could be reduced. In the early 1960s, government emission standards were set. Because of these standards, automobile *emission controls* were developed to reduce pollution. This chapter is about the problem of air pollution, the types of pollutants being produced, and how our society is improving the air quality in the environment.

OBJECTIVES

After reading this chapter, you will be able to:
- State why there is a problem with pollution in our society.
- Compare the definitions and effects of the major types of air pollution produced by the automobile.
- Identify how pollution is being controlled through certification of new cars, regulation agencies, emissions requirements, and fuel economy standards.

CHAPTER HIGHLIGHTS

23.1 POLLUTION IN OUR SOCIETY
    A. Defining the Air Pollution Problem
    B. Smog
    C. Photochemical Smog
    D. Temperature Inversion
    E. Units Used to Measure Pollution

23.2 TYPES OF POLLUTANTS
    A. Carbon Monoxide
    B. Hydrocarbons
    C. Nitrogen Oxides
    D. Sulfur Dioxide
    E. Particulate Matter
    F. Additional Diesel Pollutants

23.3 CONTROLLING POLLUTION
    A. New Car Certification
    B. Progress in Reducing Emissions
    C. Regulating Agencies
    D. Emission Requirements
    E. Fuel Economy Standards

SUMMARY

23.1 POLLUTION IN OUR SOCIETY

Defining the Air Pollution Problem

Pollution is defined as the contamination of the environment by harmful products. If these *contaminants* are in large enough numbers, they can harm plants, animals, and humans. There are many sources of pollution. Industries, power plants, home heating, gasoline and diesel vehicles, and refuse disposal plants all produce pollution. There are many forms of pollution such as chemical, thermal, radiation, air, water, and noise. Pollution is also produced by natural causes such as volcanos and brush fires.

One of the greatest sources of pollution in our society in the past has been the automobile, with both gasoline and diesel engines. Today, however, automobile pollution has been reduced drastically. This chapter is primarily concerned with air pollution.

Air pollution is a by-product of any combustion process. Air pollution is produced when coal, oil, or natural gas is burned in the combustion process. Since the automobile uses both diesel fuel and gasoline, it is a source of air pollution.

Smog

Smog is the result of having too much pollution suspended in the air. Smog is defined as a fog made heavier and darker by the addition of smoke and chemicals to the air. Both the smoke and chemicals are floating in the air.

Photochemical Smog

There are known types of smog. One type is called *photochemical smog*. The word *photochemical* means the mixing of sunlight with the chemicals in the air. See *Figure 23–1*. This combination causes photochemical smog. Photochemical smog is more dangerous to plants, animals, and humans than plain smog. For example, when photochemical smog is breathed into the body, these chemicals and acids cause coughing, and irritate the nose, throat, and lungs. This pollution has been linked to such health problems as asthma, skin damage, cancer of internal organs, emphysema, and heart and circulatory problems.

Temperature Inversion

Under normal weather patterns, the warm air near the ground tends to rise and cool itself. When this happens, the smog and pollution in the air *dissipate* and are reduced. During a temperature inversion, the warm air becomes trapped and cannot rise. A temperature inversion happens when a layer of warm air acts as a cap or lid to the air near the ground or surface of a city. In this case, the warm air prevents the smog and air pollutants from rising and dissipating into the atmosphere. The inversion layer is usually within 1,000 feet of the surface of the ground. Temperature inversions are common in cities that are located in large valleys. See *Figure 23–2*. Los Angeles and Denver are examples of places where temperature inversions occur.

Units Used to Measure Pollution

In order to control pollution on automobiles, measurements are taken to determine how much pollution actually exists. There are several units that have been used to measure pollution. One common unit is a measurement in grams of

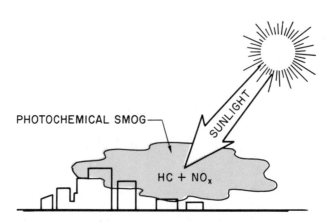

FIGURE 23–1. When the sunlight mixes with HC and NO$_x$, photochemical smog is produced.

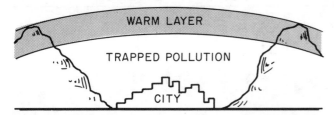

FIGURE 23-2. When a warm layer of air traps the air around a city, a temperature inversion results. The temperature inversion causes the pollutants to stay trapped in the air around the city.

pollution per mile of operation. This unit is abbreviated as g/mi. For example, a vehicle driving down the road may produce 1.5 grams of a certain type of pollution per mile of operation.

Another unit used to measure the amount of pollution is parts per million. It is abbreviated as ppm. This unit represents the number of parts of pollution per million parts of air being put into the atmosphere. This unit is also referred to as micrograms per cubic meter (μg/m^3). Both of these units are used as guidelines to control the exact amount of pollution that a vehicle is allowed to emit into the air.

A third method of referring to the amount of pollution is to use percentages. For example, the exhaust of a gasoline engine has about 2–3% carbon monoxide. This means that about 2–3% of the exhaust volume is carbon monoxide.

Other units also appear in many readings. These may include grams per hour, grams per ton mile, and grams per horsepower. The reason there are so many units is that many manufacturers like to make comparisons between different engines and different applications. These units allow easier comparisons to be made.

CAR CLINIC ▐▬▬▬

PROBLEM: MULTIPLE ENGINE PROBLEMS

An engine has a computer-controlled system for emissions, fuel, and ignition adjustments. The engine seems to stall at times when the throttle is reduced. Also, at 40–45 mph the vehicle seems to have surges. There also seems to be pinging in up-hill conditions.

SOLUTION:

The computer and associated parts are out of adjustment. Special diagnostic equipment and testing instruments are needed. Take the car to an authorized shop to have the computer system checked out thoroughly.

23.2 TYPES OF POLLUTANTS

Carbon Monoxide

Carbon monoxide is considered a deadly poison gas that is colorless and odorless. When people inhale carbon monoxide in small quantities, it causes headaches and vision difficulties. In larger quantities, it causes sleepiness and, in many cases, death.

Carbon monoxide (CO) is a by-product of combustion. Carbon monoxide emissions are increased as the combustion process becomes less efficient. During perfect combustion, the by-products are water (H_2O) and carbon dioxide (CO_2). Carbon monoxide forms in the engine exhaust when there is insufficient oxygen to form the carbon dioxide. Thus, whenever the engine operates in a rich air-fuel mixture, increased CO is the result. The best way to reduce CO is to increase the amount of oxygen in the combustion chamber. Carbon monoxide is not one of the chemicals that produces photochemical smog.

In general, carbon monoxide concentration is much lower in diesel exhaust than in gasoline exhaust. *Figure 23–3* shows the different carbon monoxide percentages in gasoline and diesel exhaust as compared to air-fuel ratios. Note that CO is considerably lower in diesel engines than in gasoline engines. This is because diesel engines are typically operated with a leaner air-fuel ratio than gasoline engines.

Hydrocarbons

Hydrocarbons (HC) are another type of pollution produced by the automobile combustion process. Hydrocarbons are

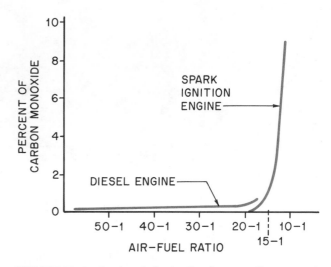

FIGURE 23-3. As the air-fuel ratio changes, the percentage of carbon monoxide also changes. (Courtesy of General Motors Corporation)

AERODYNAMIC DESIGN

Automobile manufacturers are trying to make the vehicle cut through the air with less resistance. Notice the smooth lines to these vehicles. Most parts are housed so as not to restrict the air flow over the vehicle. (Courtesy of Pontiac Motor Division, General Motors Corporation)

also called *organic materials*. Fossil fuels are made of various hydrogen and carbon molecules. Unburned hydrocarbons emitted by the automobile are largely unburned portions of fuel. Any fuel that is partially burned contains hydrocarbons. For example, gasoline in the combustion chamber burns extremely rapidly. Some of the hydrogen and carbon molecules near the sides of the combustion chamber may get partially burned. This action produces HC in the exhaust. All petroleum products produce small traces of HC during the combustion process. For example, gasoline in the fuel tank and carburetors also gives off traces of HC.

Most hydrocarbons are poisonous at concentrations above several hundred parts per million. Although they are not as dangerous by themselves as CO, hydrocarbons are the main ingredient in the production of photochemical smog.

Nitrogen Oxides

Nitrogen oxides (NO_x) are formed by a chemical union of nitrogen molecules with one or more oxygen molecules. Nitrogen oxides form more freely under extreme heat conditions. As the combustion process becomes leaner, combustion temperatures typically increase. Higher temperatures cause nitrogen oxides to be produced. When the combustion temperatures reach 2,200–2,500 degrees F, the nitrogen and oxygen in the air-fuel mixture combine to form large quantities of nitrogen oxide, ***Figure 23–4***.

Nitrogen oxide by itself does not appear to have any important harmful effects. However, nitrogen oxide reacts

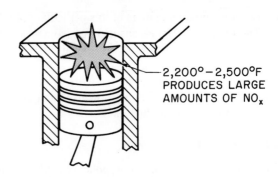

2,200°–2,500°F PRODUCES LARGE AMOUNTS OF NO_x

FIGURE 23–4. Whenever combustion temperatures increase into the 2,200–2,500 degree F range, large quantities of NO_x are produced.

with hydrocarbons to form harmful irritating *oxides*, and gives photochemical smog its characteristic light brown color.

Gasoline engines that are running very lean with higher temperatures produce nitrogen oxides. Diesel engines generally produce higher concentrations of nitrogen oxides because their internal temperatures are usually higher than gasoline engines. The best method of reducing the emission of nitrogen oxides is to reduce the temperature of the combustion process. Doing this, however, will result in less efficient burning and increases in HC and CO emissions.

To illustrate this, ***Figure 23–5*** shows how the three emissions discussed are affected by the air-fuel ratio and conversion efficiency. For example, as the air-fuel ratio becomes leaner, nitrogen oxide conversion efficiency

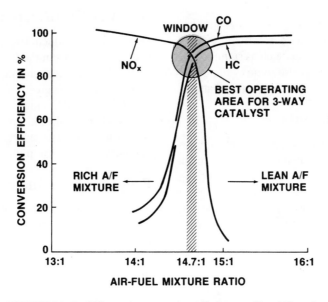

FIGURE 23–5. Whenever steps are taken to reduce HC and CO emissions, NO_x emissions increase. It is very important to keep the air-fuel ratio close to 14.7 to 1 for more effective emission control. (*Courtesy of General Motors Product Service Training*)

decreases, thus producing more of this particular pollution form. The CO and HC emissions increase (conversion efficiency decreases on the chart) as the air-fuel mixture becomes richer. By observing this chart, it can be seen that the best air-fuel ratio for any engine is in the "window" area, or close to 14.7–1. Manufacturers now use computer controls to obtain accurate air-fuel ratio.

Sulfur Dioxide

Sulfur dioxide is another form of air pollution. Sulfur is present in many of the hydrocarbon fuels that are used today in engines and power plants. Diesel fuel has slightly more sulfur than gasoline; therefore, sulfur emissions are greater with diesel engines than with gasoline engines. When sulfur gets into the atmosphere, it breaks down and combines with water in the air to produce sulfuric acid. This acid is very corrosive and produces the commonly known acid rain form of pollution. NO_x also contributes significantly to the acid rain problem.

Particulate Matter

Particulates are defined as a form of solid air pollution such as microscopic solid particles of dust, soot, and ash. They can be solid or liquid matter that floats in the atmosphere. Examples of particulate substances are lead and carbon produced by burning leaded gasoline. These particulates are absorbed directly into the body and can cause severe health hazards.

Engines that use unleaded fuel produce very little particulate matter. Engines that have the carburetor set very rich, however, may also produce carbon particulates that could be a health hazard.

Particulate matter is damaging because it reduces air visibility and allows less sunlight to reach the ground. Particulates carry damaging materials such as sulfuric acids to the surfaces they strike. There is also some evidence that particulate matter is having an effect on weather patterns.

Additional Diesel Pollutants

The diesel engine uses different fuels, has different compression ratios, and uses a different ignition system than the gasoline engine. Because of these characteristics, diesel engines present two other emission concerns. Diesel engines in general have an exhaust odor that can be considered a nuisance. These exhaust odors have recently been measured so that proper technological steps can be taken to reduce them. In addition, smoke is evident on many diesel engines. The blue-white smoke is caused by liquid droplets of lubricating or fuel oil. This indicates that maintenance is required. The black-gray smoke often seen on diesel vehicles is caused by unburned carbon particles. This is caused largely by inaccurate air-fuel ratios. Black smoke will

increase as the load is increased. In some cases, smoke charts have been developed to measure the amount of smoke being emitted by diesel engines.

CAR CLINIC ▌

PROBLEM: COLD STARTING PROBLEMS

An engine started fine all summer. As the temperature outside got colder, the engine began to have problems starting. When the temperature is very cold outside, the operator has to continually pump the accelerator to keep the engine running. This must be done until the engine is at operating temperature, at which time the engine runs fine.

SOLUTION:

The most common problem with starting in cold weather is the setting of the choke. It is obvious that the engine is not getting enough fuel when the temperature is cold. Check the choke setting when the engine is cold. The choke plate should be closed during this condition. When the engine is warm, the choke plate should be fully open. Adjust the choke plate accordingly. It may also be necessary to replace the bimetal strip on the choke.

23.3 CONTROLLING POLLUTION

New Car Certification

One method of reducing pollution from automobiles is to set certain emission standards for new vehicles. Standards are established by the Environmental Protection Agency (*EPA*). Families of cars are given emissions tests. They are tested for HC, CO, and NO_x emissions. Tests are performed under cold start, normal, and hot-start conditions. If the vehicle passes the emission standards set by the EPA, it can be offered for sale to the public.

Progress in Reducing Emissions

Much has been done recently to the automobile to reduce emissions. These emission controls have drastically reduced the amount of CO, particulate matter, SO_x, and NO_x in the atmosphere. *Figure 23–6* shows the drop in carbon monoxide and particulate matter over a ten-year period from 1970–80. This chart shows an approximately 37% drop for CO and a 50% drop for particulate matter for the ten-year period. HC emissions dropped by 11% during this same ten-year period.

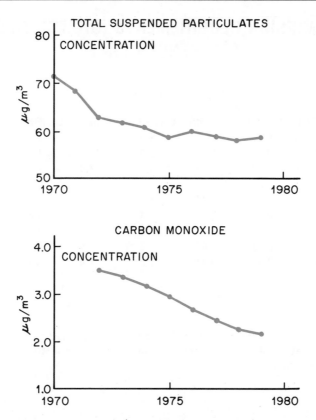

FIGURE 23–6. These charts show the drop in different forms of pollution over a ten-year period.

Although these emissions were reduced, NO_x increased during this ten-year period. When pollution standards were started in 1968, the automotive manufacturers tried to reduce HC and CO. As these were reduced, however, the engine combustion process became leaner, increasing the production of NO_x. Only in the past 5–8 years have efforts been made to reduce NO_x in vehicles. See **Figure 23–7**. Various pollution control devices are used today to effectively reduce NO_x.

Regulating Agencies

Several regulating agencies were established in the 1960s to help identify standards and regulations for automobiles. California became the first state to use air pollution standards. California also established the Air Resources Board (ARB) which helps to enforce the emission standards. California standards are typically much more strict than those of other states.

In 1968 the Clean Air Act was passed by the U. S. Congress. This act set allowable levels of HC, CO, and NO_x. The EPA is the agency responsible for enforcing the Clean Air Act. Several other states also passed their own laws.

These laws must be approved by the EPA. These laws were based on different altitude operation such as in mountainous regions. As a result, there were several emission standards. These included 1) California vehicles, 2) federal 49 state vehicles, 3) high-altitude vehicles.

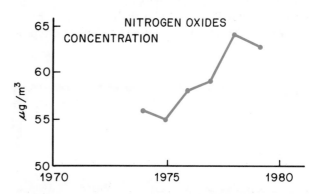

FIGURE 23–7. During the initial years of pollution control on vehicles, NO_x increased. Now it is being reduced substantially by several pollution controls.

EMISSION REQUIREMENTS (GRAMS/MILE)
(PASSENGER CARS)

MODEL YEAR	HYDROCARBON (HC)		CARBON MONOXIDE (CO)		OXIDES OF NITROGEN (NOx)	
	CALIFORNIA	FEDERAL	CALIFORNIA	FEDERAL	CALIFORNIA	FEDERAL
1978	0.41	1.5	9.0	15.0	1.5	2.0
1979	0.41	0.41	9.0	15.0	1.5	2.0
1980	0.39	0.41	9.0	7.0	1.0	2.0
1981	0.39	0.41	7.0	3.4	0.7	1.0
1982	0.39	0.41	7.0	3.4	0.4	1.0
1983	0.39	0.41	7.0	3.4	0.4	1.0
1984	0.39	0.41	7.0	3.4	0.4	1.0
1985	0.39	0.41	7.0	3.4	0.7	1.0
1986	0.39	0.41	7.0	3.4	0.7	1.0
1960 (No Control)		10.6		84		4.1

FIGURE 23–8. Emission standards have been established by the EPA and must be met by all manufacturers. (Courtesy of General Motors Product Service Training)

The manufacturers must be able to meet the standards set by the EPA or other agency. It is up to the manufacturer, however, to determine the type of technology used to reduce emissions.

Emission Requirements

Emission requirements were established by the EPA for each year vehicle. *Figure 23–8* shows an example of the emission standards set by the EPA for various years. Newer vehicles (last five years) have stricter standards compared to earlier years to help reduce emissions even further. Emission control standards are often placed on the vehicle's hood, *Figure 23–9*. These are used to identify the type of

vehicle and to help service technicians identify the amount of emissions allowed for each vehicle.

Fuel Economy Standards

As the emission requirements became stricter, the fuel economy of vehicles was also reduced. In 1973 the United States faced an "energy crisis." There were severe supply and demand problems, especially in obtaining petroleum for gasoline. The supply of gasoline was much less than the demand. This resulted in federally mandated economy standards.

Congress enacted the Energy Policy Conservation Act. This act set standards for the manufacturers to follow. These standards are referred to as *CAFE* or Corporate Average Fuel Economy. *Figure 23–10* shows the corporate average fuel economy for cars through several years. While certain vehicle lines may obtain less than the standard, others in the fleet will obtain higher than the standard.

In order for the manufacturers to meet both the fuel economy and emission level standards, it is necessary to control HC, CO, and NO_x emissions. This must be done while maintaining or increasing fuel economy standards and performance. This is accomplished with the use of electronic fuel-air control, the catalytic converter, and the addition of port fuel injection on engines.

FIGURE 23–9. Certain pollution control information can be found under the hood of the vehicle.

FUEL ECONOMY STANDARDS

MODEL YEAR	MPG	Cumulative Improvement Over MY 1974
1978	18.0	50%
1979	19.0	58%
1980	20.0	67%
1981	22.0	83%
1982	24.0	100%
1983	26.0	116%
1984	27.0	125%
1985	27.5	129%
1986	27.5	129%

FIGURE 23–10. The automotive manufacturers must meet CAFE (Corporate Average Fuel Economy) standards. (Courtesy of General Motors Product Service Training)

SUMMARY

Pollution is a very harmful contamination within the environment. It is produced by many sources, with combustion of hydrocarbon fuels being a major source. Air pollution is of greatest concern when discussing the automobile.

Pollution produces photochemical smog. Photochemical smog results when NO_x, HC, and sunlight are combined, producing irritation to the nose, throat, and lungs. One reason for increases in the smog is temperature inversions. A temperature inversion occurs when a warm layer of air traps pollutants near the ground. Several units can be used to measure the amount of pollution. These include grams of pollution per mile, parts per million, and percentages.

There are several major types of pollution being produced by automobiles. Carbon monoxide is a colorless, odorless gas that is produced by a lack of oxygen. It is deadly in large quantities. Hydrocarbons are also produced when incomplete combustion occurs. Nitrogen oxides are produced when there are high temperatures within the combustion process. Usually temperatures of 2,200–2,500 degrees F produce large quantities of NO_x. Sulfur dioxides are produced because the crude oil carries sulfur before refining. Sulfur and nitrogen oxides also contribute to the acid rain problem. Particulates are solid forms of air pollution, either water or air that floats in the atmosphere. Diesel engines also produce odors and smoke that are considered pollution forms.

Many methods have been used to reduce pollution from the automobile. New cars must now be certified by the EPA for control of HC, CO, and NO_x. Considerable progress has been made in reducing these pollutants by using many pollution devices on the automobile engine. Many regulating agencies and acts have been established by Congress to help enforce the regulations in different states. Emission requirements were established by the EPA. Each manufacturer must design and manufacture vehicles that pass these standards. In addition, fuel economy standards were implemented to improve the fuel mileage of vehicles. The CAFE (Corporate Average Fuel Economy) standards must be met by each manufacturer.

TERMS TO KNOW

Can you explain each of the following terms? Review the chapter until you can use each term correctly.

Pollution

Emission control

Contaminant

Photochemical smog

Dissipate

Carbon monoxide

Hydrocarbon

Organic material

Nitrogen oxide

Oxide

Particulates

Environmental Protection Agency (EPA)

Corporate Average Fuel Economy (CAFE)

REVIEW QUESTIONS

Multiple Choice

1. Photochemical smog is created when _____ mixes with NO_x and HC emissions.
   a. CO
   b. Particulates
   c. Sunlight
   d. Smog
   e. None of the above

2. Which of the following are considered acceptable units for measuring emissions from the automobile?
   a. Parts per million
   b. Grams per mile
   c. Percentages
   d. All of the above
   e. None of the above

3. Which of the following emissions from the automobile is a deadly, odorless, tasteless gas that can cause death?
   a. CO
   b. NO_x
   c. HC
   d. SO_x
   e. None of the above

4. Which of the following emissions from the automobile is a contributor to acid rain?
   a. CO
   b. NO_x
   c. HC
   d. All of the above
   e. None of the above

5. Which of the following emissions from the automobile increases with combustion temperatures above 2,200 degrees F?
   a. CO
   b. NO_x
   c. HC
   d. SO_x
   e. Particulates

6. As combustion efficiency increases to reduce HC and CO emissions, what happens to NO_x?
   a. NO_x increases
   b. NO_x decreases
   c. There is no effect on NO_x
   d. NO_x is completely eliminated
   e. None of the above

7. Another term for HC is _____.
   a. CO
   b. Organic materials
   c. NO_x
   d. Particulates
   e. SO_x

8. Lead in gasoline is considered what type of pollution?
   a. NO_x
   b. CO
   c. HC
   d. Particulates
   e. Photochemical smog

9. What is the CAFE for the newest vehicles being manufactured?
   a. 22.5 mpg
   b. 27.5 mpg
   c. 31.5 mpg
   d. 38.5 mpg
   e. CAFE has nothing to do with miles per gallon.

10. What agency will help enforce emission standards on automobiles?
    a. U.S. Congress
    b. U.S. CAFE
    c. EPA
    d. ARB
    e. None of the above

The following questions are similar in format to ASE (Automotive Service Excellence) test questions.

11. Technician A says that as the engine air-fuel ratio becomes leaner, nitrogen oxides increase. Technician B says that as the engine air-fuel ratio becomes richer, hydrocarbons increase. Who is right?
    a. A only
    b. B only
    c. Both A and B
    d. Neither A nor B

12. Technician A says that photochemical smog is caused by sulfur mixing with sunlight. Technician B says that photochemical smog is caused by lead compounds mixing with sunlight. Who is right?
    a. A only
    b. B only
    c. Both A and B
    d. Neither A nor B

13. Technician A says that nitrogen oxide, carbon monoxide, and sulfur emissions are all checked before the manufacturer sells the vehicle. Technician B says that nitrogen oxide, carbon monoxide, and hydrocarbon emissions are all checked before the manufacturer sells the vehicle. Who is right?
    a. A only
    b. B only
    c. Both A and B
    d. Neither A nor B

14. Technician A says that amounts of air pollution can be measured in parts per million. Technician B says that amounts of air pollution can be measured in grams per mile. Who is right?
    a. A only
    b. B only
    c. Both A and B
    d. Neither A nor B

Essay

15. Define carbon monoxide.

16. How is photochemical smog produced?

17. What three forms of air pollution are commonly checked on a car?

18. What does CAFE stand for?

CHAPTER 24

Pollution Control

INTRODUCTION

During the middle 1960s, our society was very concerned about the amount of pollution produced by both industries and automobiles. Because of this interest, it became mandatory that emission controls be installed on all automobiles. The automobile has been analyzed very carefully to determine the types of pollution that it produces. *Figure 24–1* shows where most of the pollution comes from in an automobile. Usually the fuel tank, exhaust gases, crankcase area, and carburetor produce the majority of pollution in the automobile. Automobile emissions can be controlled with various devices. This chapter examines pollution control systems in detail.

OBJECTIVES

After reading this chapter, you will be able to:
■ Explain how the PCV system works.
■ Describe how evaporative emission controls operate.
■ Identify the purpose of the carbon canister.
■ Examine how the intake and exhaust emissions are controlled.
■ Define several devices that are used to control combustion efficiency.
■ Examine electronic control of emissions.
■ Identify common diagnosis and service procedures used on pollution control devices.

CHAPTER HIGHLIGHTS

24.1 POSITIVE CRANKCASE VENTILATION (PCV) SYSTEM
    A. Purpose of Crankcase Ventilation
    B. Positive Crankcase Ventilation System Operation
    C. PCV Valve Operation

24.2 EVAPORATIVE EMISSION CONTROLS
    A. Carbon Canister
    B. Vapor-liquid Separator
    C. Rollover Check Valve
    D. Fuel Tank
    E. Vapor Control From Carburetor

24.3 INTAKE AND EXHAUST EMISSION CONTROLS
    A. Heated Air Intake
    B. Exhaust Gas Recirculation (EGR)
    C. Air Injection System
    D. Air Pump
    E. Check Valve
    F. Diverter Valve
    G. Aspirator Valve
    H. Catalytic Converters
    I. Types of Catalytic Converters
    J. Two-way Catalytic Converter
    K. Three-way Catalytic Converter

CHAPTER HIGHLIGHTS (CONTINUED)

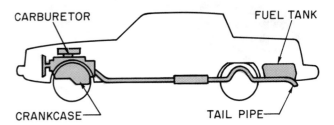

FIGURE 24-1. Sources of pollution on the automobile include the carburetor, crankcase, tail pipe, and fuel tank.

24.1 POSITIVE CRANKCASE VENTILATION (PCV) SYSTEM

Purpose of Crankcase Ventilation

During normal engine operation, a considerable amount of dirty air passes through the engine crankcase. This air is a result of a process called *blow-by*. Referring to **Figure 24-2**, blow-by is a product of the combustion process.

Every time combustion occurs, a certain amount of blow-by (from combustion) escapes past the piston rings. This blow-by produces a small *crankcase pressure*. The gases from blow-by are very acidic in nature. If they are allowed to stay in the crankcase area, the acids will attack the oil and metal within the engine. To help prevent this on older engines, air was drawn into the engine through the oil filler cap. The air flowed through the crankcase area, picking up the acidic gases. It was then directed out through a tube into the atmosphere. This tube was referred to as the "draft" or "breather" tube. Obviously, these gases are a great source of air pollution from the automobile.

Positive Crankcase Ventilation System Operation

Crankcase emissions are easily controlled by the positive crankcase ventilation (*PCV*) system. Refer to **Figure 24-3**. In this system, any crankcase vapors produced are directed back into the base of the carburetor to be reburned.

With this system, which is called a closed system, air is drawn through the carburetor air cleaner assembly, into the engine valve compartment and crankcase. These vapors are

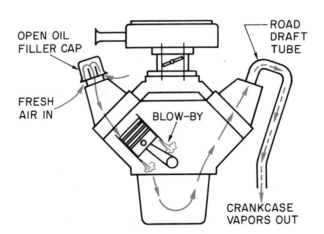

FIGURE 24-2. Crankcase ventilation was accomplished in older vehicles by passing the vapors in the crankcase out of the engine through a draft tube.

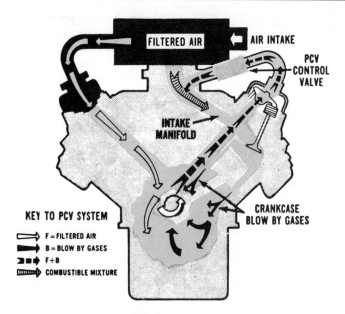

FIGURE 24–3. A closed PCV system brings air through the air cleaner, through the closed oil filler cap, and into the crankcase. The vapors are then sent through a PCV valve and back into the base of the carburetor to be reburned. *(Courtesy of Motor Publications, Emission Control Manual, 1985–86)*

then drawn up through a vacuum-and-spring-controlled ventilating valve (PCV valve) and into the intake manifold. The vapors are then mixed with the air-fuel mixture and burned in the combustion process.

Not all PCV systems were designed like this. Earlier systems, which were called open systems, did not bring the fresh air through the carburetor air cleaner assembly. The air came through the open oil filler cap. This means that any air that entered the open system contained dirt and other materials. The open system was replaced by the closed system shown in Figure 24–3.

PCV Valve Operation

If crankcase vapors are allowed to flow into the carburetor during all loads and engine rpm, the crankcase gases will upset the basic carburetor air-fuel ratios. During idle and low speeds, less vapor must be burned. Therefore, a PCV valve is placed in the flow just before the carburetor. A PCV valve is shown in *Figure 24–4*. *Figure 24–5* shows the internal operation of the PCV valve. There are two forces that are working against each other. Vacuum from the intake manifold is working against spring pressure inside the valve. When the engine is stopped, no intake manifold vacuum exists. During this time, the PCV valve is held closed by the force of the internal spring. The tapered valve is moved fully to the right, closing off the entrance to the valve.

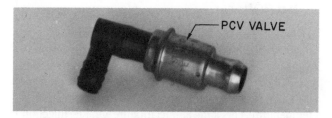

FIGURE 24–4. A PCV valve is located in the positive crankcase ventilation system to control the flow of vapors back to the carburetor.

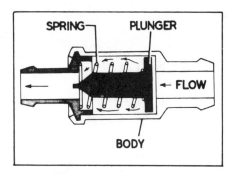

FIGURE 24–5. A typical PCV valve is shown. Spring pressure pushes the valve to the right while intake manifold vacuum pulls the plunger to the left. *(Courtesy of Motor Publications, Emission Control Manual, 1985–86)*

When the engine is at idle or deceleration, vacuum in the intake manifold is very high. The tapered valve plunger is drawn to the left. This action closes off the metered opening. Little or no crankcase vapor is allowed to enter the carburetor.

When the engine operates at normal loads and speeds, the vacuum in the intake manifold drops. This allows the spring to push the plunger to the right. This action causes the metered opening to increase in size. The amount of crankcase vapor sent back to the intake manifold is now increased.

During acceleration or heavy loading, the intake manifold vacuum is very low. The spring moves the tapered valve fully to the right. Now a maximum amount of crankcase vapor can enter the carburetor. Therefore, when the engine is at low speeds, little crankcase vapor is sent to the carburetor. As the engine speed and load increase, more and more of these vapors are allowed to enter the carburetor.

CAR CLINIC

PROBLEM: SMELLY EXHAUST

The exhaust from a vehicle smells much like rotten eggs. This smell usually occurs during acceleration. What could be the cause of the smell?

SOLUTION:

The most common cause of the rotten egg smell is a rich mixture of fuel. The three-way catalytic converter is probably not up to temperature yet. Whenever the air-fuel mixture is rich, the rotten egg smell is produced. If the smell is also there after the engine is at operating temperature, check the air-fuel mixture. On fuel injection vehicles, the PROM may have to be replaced.

24.2 EVAPORATIVE EMISSION CONTROLS

Any gasoline released from the fuel tank or carburetor in a liquid or vapor form is also pollution. These liquids and vapors are made of hydrocarbons. A variety of devices are being used to help reduce these emissions. These systems are called evaporative emission controls. *Figure 24–6* shows the typical parts to the evaporative emission controls.

Carbon Canister

The *carbon canister* is used to store any vapors from the fuel tank or carburetor. Refer to *Figure 24–7*. The carbon canister is made of a bed of activated charcoal (carbon). Warm weather can cause the fuel to expand in the fuel tank. This causes the vapors to be forced out of the tank and into the carbon canister. The carbon absorbs or stores gasoline vapors from the fuel tank and carburetor when the engine is not running.

When the engine is running, the clean vapors in the carbon canister are sent back to the carburetor to be burned. As shown in *Figure 24–8*, when the engine is operating, vacuum inside of the *purge* line draws air through the air cleaner. Air then passes through the filter at the bottom of the canister and into the activated charcoal. This action evaporates the gasoline vapors trapped within the charcoal. The remaining fumes are carried to the intake manifold and burned during combustion.

Vapor-liquid Separator

It is important that only vapors, not liquid fuel, enter the carbon canister. Liquid fuel may leak out when the engine is stopped, or excess fuel may enter the carburetor during running conditions. Certain conditions may cause liquid gasoline to enter the carbon canister. For example, a full fuel tank that expands from heat, or a vehicle that accidentally

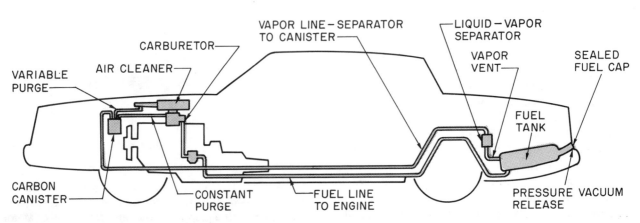

FIGURE 24–6. Typical evaporative emission control systems. Vapors and liquids from the fuel tank and carburetor are being controlled.

rolls over may cause fuel to enter the carbon canister. To prevent this, a vapor-liquid separator is used.

One type of vapor-liquid separator passes the vapors through open-cell foam. Liquid gasoline cannot pass through the foam. Referring to **Figure 24–9**, the vapors will pass through the open-cell filter material and be sent to

the carbon canister. The separator is usually located directly on or near the fuel tank. Figure 24–6 shows the location of the liquid-vapor separator near the fuel tank.

A second type of vapor-liquid separator is called the float type. **Figure 24–10** shows an example of this type of separator. Vapors come into the separator from the fuel tank and cause the float to rise. This forces the needle into its seat, closing off the vent line until the liquid fuel drains back into the tank.

Rollover Check Valve

Another method of controlling the liquid gasoline is by using a rollover check valve. This valve is located in series with the main vapor line from the fuel tank to the canister.

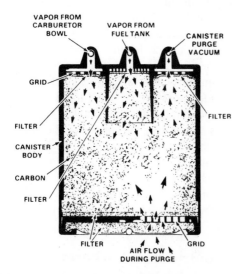

FIGURE 24–7. A schematic of a charcoal (carbon) canister used to store fuel vapors. (Courtesy of Motor Publications, Emission Control Manual, 1985–86)

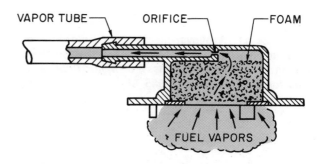

FIGURE 24–9. A vapor-liquid separator is shown. This system uses a foam to separate the liquid and vapors. (Courtesy of Motor Publications, Emission Control Manual, 1985–86)

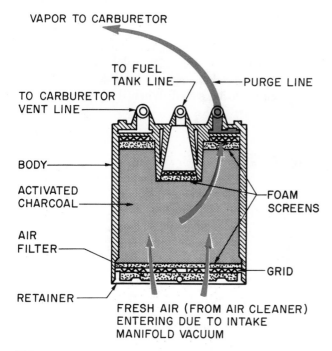

FIGURE 24–8. Arrows show the flow of air and vapors during canister purging or when the engine is running. (Courtesy of Motor Publications, Emission Control Manual, 1985–86)

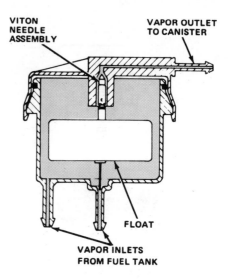

FIGURE 24–10. A float type of vapor-liquid separator is shown. Fuel entering the valve raises the float. This forces the needle valve to seal off the vent line to the carbon canister. (Courtesy of Motor Publications, Emission Control Manual, 1985–86)

If the vehicle rolls over in an accident, a small stainless steel ball will force the plunger to close off the vent line. See *Figure 24–11*.

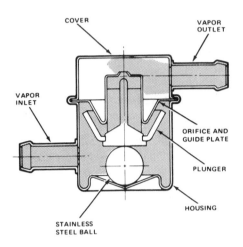

FIGURE 24–11. A rollover check valve will stop any liquid fuel from entering the carbon canister if the vehicle rolls over in an accident. *(Courtesy of Motor Publications, Emission Control Manual, 1985–86)*

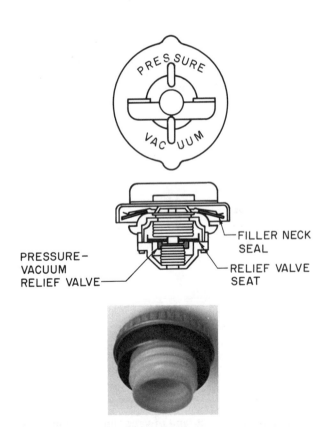

FIGURE 24–12. The pressure-vacuum relief cap opens with either high pressure or vacuum in the fuel tank. This action will prevent the tank or other components in the fuel system from being damaged.

Fuel Tank

Most fuel tanks today are sealed with a special pressure-vacuum relief filler cap. This is shown in *Figure 24–12*. The relief valves in the cap are used to regulate pressure or vacuum in the fuel tank. Excess pressure or vacuum could be caused by some malfunction in the evaporative emission control system or by excess heating or cooling. The fuel tank is allowed to "breathe" through the filler cap when gasoline expands or contracts from heating or cooling.

When pressure in the fuel tank rises above 0.8 psi, the pressure cap opens to let the pressure in the fuel tank escape. If the vacuum in the fuel tank is greater than 0.1 Hg (inches of mercury), the valve is also activated.

Some fuel tanks are designed to allow for expansion when the tank has been filled. The tank shown in *Figure 24–13* has a built-in expansion tank. The orifices leading to the expansion tank are so small that it takes 10–15 minutes for gasoline from the main tank to enter the expansion tank. The expansion tank area will accept some fuel from the main fuel tank after being filled at the fuel pump. This also provides room for fuel expansion and vapors in the main tank.

Vapor Control from Carburetor

The carburetor float bowl is also vented into the carbon canister. Fuel vapors are produced in the float bowl, especially when the engine is shut off. During this time, heat is transferred from the engine to the carburetor. This heat causes excess vapor to be produced inside of the carburetor bowl. *Figure 24–14* shows the carburetor bowl area connected with a line to the carbon canister.

When the engine is shut off, vapors from the float bowl are sent to the bowl vent valve. This valve is open when the engine is off. The carburetor vapors now go directly to the carbon canister to be stored. When the engine starts, a

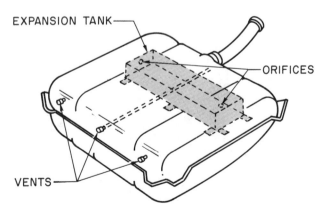

FIGURE 24–13. An expansion tank is used on certain vehicles to allow for fuel expansion when the tank is filled with gasoline.

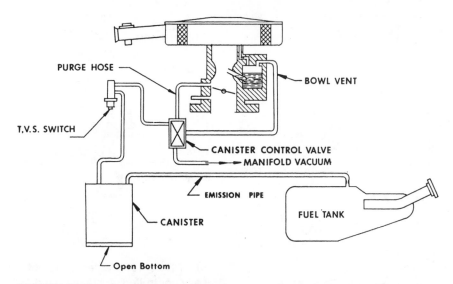

FIGURE 24–14. A complete evaporative control system is shown. This type of evaporative control system uses a bowl vent valve. *(Courtesy of Motor Publications, Emission Control Manual, 1985–86)*

vacuum line from the carburetor closes off the bowl vent valve. The valve stops the vapors coming from the carburetor to the carbon canister.

When the engine is running, the idle purge line draws stored vapors out of the carbon canister to the carburetor. These vapors are burned during combustion.

Some carburetors use an insulator, *Figure 24–15*, to reduce heat transfer from the engine intake manifold to the carburetor and float bowl. The insulator is normally placed directly below the base of the carburetor on the intake manifold.

CAR CLINIC

PROBLEM: ERRATIC ENGINE OPERATION

An engine has developed several problems. The engine has 22,000 miles on it. The system uses computer-controlled EFI (Electronic Fuel Injection). The complaints are that the engine stops after cold starting, stops at idle after deceleration, surges during cruise, and has a rough idle. What would be a good starting point for troubleshooting?

SOLUTION:

Erratic engine operation such as this indicates a bad or inoperative EGR valve. The symptoms suggest the EGR valve is providing too much flow of exhaust back into the intake. Too little EGR flow would produce problems with spark knock, engine overheating, and emission test failure. Make sure the vacuum system on the EGR valve

operates. Use a vacuum pump to check the diaphragm. Also make sure the EGR valve passages are clean from carbon buildup. Repair or replace the EGR valve to eliminate the erratic engine operation.

24.3 INTAKE AND EXHAUST EMISSION CONTROLS

In addition to the pollution controls already mentioned, certain changes can be made to the intake and exhaust systems. These particular systems alter or change the intake temperatures, the exhaust hydrocarbons, the nitrogen oxides, and the carbon monoxide.

Heated Air Intake

Automotive manufacturers have made the air-fuel mixture leaner over the past few years. Leaner air-fuel mixtures have lower hydrocarbon and carbon monoxide emissions. This is true especially during idle conditions. Faster warm-up and quicker choke action have also helped reduce emissions. During starting, the carburetor usually produces a richer mixture than what is actually needed. If the amount of time the choke is used could be reduced, less emissions would result. It is also known that a lean air-fuel mixture will ignite easier when the air is warm.

To accomplish this, air going into the carburetor can be preheated. *Figure 24–16* shows how a heated air intake system operates. This system is also referred to as a thermostatically controlled air cleaner. When heated air is required, the air control motor attached to the air cleaner assembly

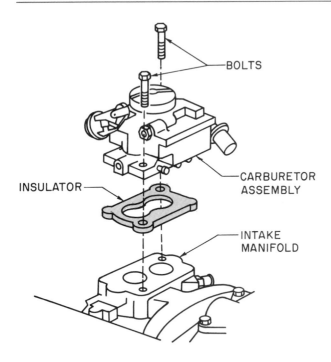

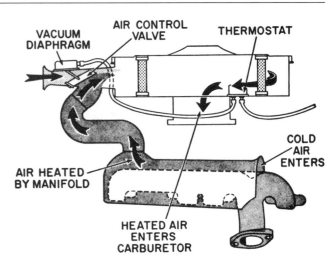

FIGURE 24-16. Heated air is obtained on a thermostatically controlled air cleaner by drawing the air going into the air cleaner across the exhaust manifold. *(Courtesy of Motor Publications, Emission Control Manual, 1985–86)*

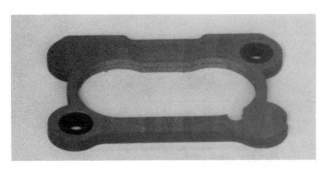

FIGURE 24-15. Some carburetors use an insulator to reduce heat transfer from the intake manifold to the carburetor. This reduces the formation of vapors in the carburetor float bowl.

closes off cold air coming into the engine. All air coming into the engine now passes over the exhaust manifold and through the hot air pipe. The area where the air is heated is called the *heat stove*.

The motor is operated by the vacuum of the engine and a bimetal strip. When the engine is off, no vacuum is produced in the vacuum chamber. This is shown in *Figure 24–17*. In this position, the control damper assembly is closing off the hot air pipe. When the engine starts, vacuum is produced. The vacuum lifts the diaphragm and closes off the air inlet. Now only warm air drawn over the exhaust manifold can enter the carburetor.

Temperature can also be used to control the direction of air flow. This is done by a temperature sensor located on the base of the air cleaner. When the engine is operating, and the under-hood temperature is below 85 degrees F, the

temperature sensing spring closes the air bleed valve. This admits full engine vacuum to the vacuum chamber. Now all outside air below 85 degrees F (cold air) is shut off. Only warm air can enter the carburetor.

When the under-hood temperature reaches about 135 degrees F, the bimetal strip spring in the temperature sensor opens the air bleed valve. This reduces any vacuum in the vacuum chamber. The diaphragm spring now pushes the control damper assembly in a position to close off the hot air pipe.

When the air temperature under the hood is between 85 and 135 degrees F, the control damper assembly is between a fully open and fully closed position. The actual opening and closing temperatures can vary somewhat, depending upon the manufacturer. Also, when the engine is accelerated heavily, vacuum to the chamber will drop off instantly. This causes the spring to snap the damper downward, closing off the hot air tube. This action will allow maximum air flow through the *snorkel* tube. Under heavy acceleration, engines need all the intake air they can get for good performance.

Exhaust Gas Recirculation (EGR)

When combustion temperatures are in the range of 2,200 to 2,500 degrees F, nitrogen mixes with oxygen and produces oxides of nitrogen (NO_x). This type of emission has a detrimental effect on the environment. The method used to reduce oxides of nitrogen is to cool down the combustion process. This is done by using an *EGR* valve.

In operation, part of the exhaust gas (usually less than

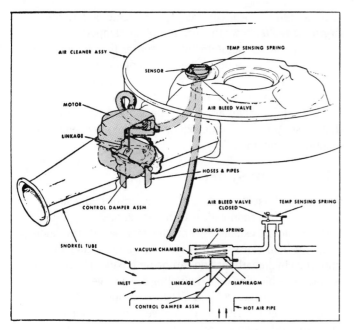
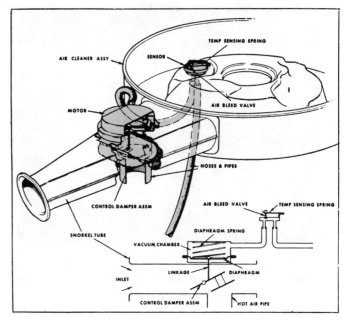
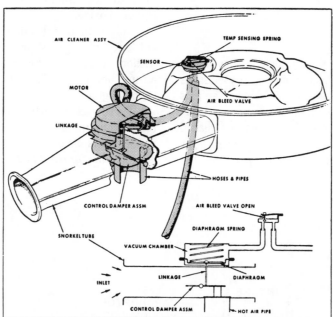

FIGURE 24–17. The control damper and the temperature sensor and spring control when and how much warm air will enter the air cleaner. *(Courtesy of Motor Publications, Emission Control Manual, 1985–86)*

10%) is sent back through the intake manifold. The exhaust gases, which are considerably cooler than the combustion temperature, cool down the process of combustion. **Figure 24–18** shows a picture of the EGR valve. This valve is located near the back of the carburetor on the intake manifold. The exhaust is picked up on the crossover passageways in the intake manifold. On in-line engines, an external line must be connected from the exhaust side to the intake side.

The EGR valve is controlled by engine vacuum.

Referring to **Figure 24–19**, exhaust gases are present at the base of the EGR valve. The *pintle* valve opens or closes to regulate the exhaust gases flowing to the intake manifold.

The EGR valve also has a diaphragm which is controlled by spring pressure against a vacuum from the carburetor. When the throttle valve is at a specific position, it uncovers the carburetor vacuum port. This action pulls the diaphragm upward against spring pressure. The EGR valve is now opened, allowing exhaust gases to enter the intake manifold.

FIGURE 24-18. The EGR valve located on the intake manifold recirculates exhaust gases back into the intake for cooler combustion and less NO$_x$ emissions.

When the engine is at idle, the throttle valve has not yet uncovered the port. During this condition, no vacuum is available to operate the EGR valve. The EGR valve remains closed during idle or heavy deceleration.

A temperature-sensitive control is sometimes placed between the EGR valve and the carburetor. Exhaust gas recirculation is not needed below certain temperatures. If it

is allowed to occur, it will cause poor engine performance. *Figure 24-20* shows the location of the temperature control. The EGR temperature control senses the temperature of the engine coolant. When the engine temperature reaches a certain point, the EGR temperature control opens and allows the vacuum to operate the EGR valve.

Some EGR valves also have a back-pressure *transducer* that senses pressure in the exhaust manifold to change the amount of opening.

Air Injection System

One method used to reduce the amount of hydrocarbons (HC) and carbon monoxide (CO) in the exhaust is by forcing fresh air into the exhaust system after combustion. This additional fresh air causes further oxidation and burning of the unburned hydrocarbons and carbon monoxide. The process is much like blowing on a dwindling fire. Oxygen in the air combines with the HC and CO to continue the burning, which reduces the HC and CO concentrations. This allows them to oxidize and produce harmless water vapor and carbon dioxide.

This method of cleaning the exhaust does not affect the efficiency of combustion. However, a small amount of frictional horsepower (up to 3 hp) is needed to operate the air injection pump. This acts as a form of horsepower loss to the engine.

These systems, which are referred to as air injection reactors, thermactors, air guards, and so on, use an air pump, air manifolds, and valves to operate. A typical system is shown in *Figure 24-21*.

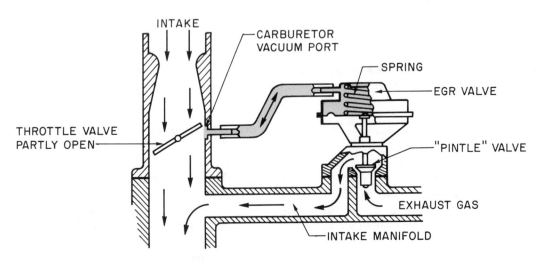

FIGURE 24-19. The EGR valve controls the amount of exhaust flowing back into the intake.

EGR TEMPERATURE VALVE

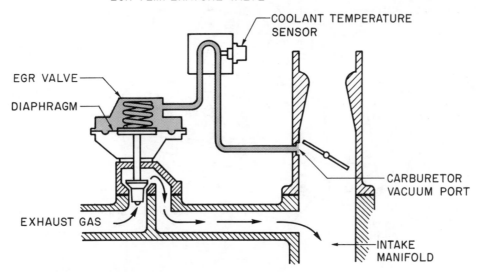

FIGURE 24-20. The EGR valve can also be controlled by a temperature valve. The EGR valve is designed to be closed below a certain temperature in the engine.

The air pump is driven by a belt from the crankshaft. The air pump produces pressurized air that is sent through the exhaust manifold to the injection tubes located at each cylinder. The air is then injected directly into the exhaust flow. *Figure 24-22* shows where the air is injected into the exhaust flow. This is only one setup. The air can be injected at the base of the exhaust manifold as shown, or it can be injected directly through the head at the exhaust port.

Air Pump

The air pump is a rotating vane (also called an eccentric-type) *positive displacement* pump. See *Figure 24-23*. A relief valve is placed on the pump to control the amount of pressure the pump can develop. When pressure inside the pump exceeds a predetermined level, the relief valve opens and allows the excess pressure to escape.

As each rotating vane passes the intake chamber, a vacuum is produced. This vacuum draws in a fresh charge of air. The vane carries the air charge around to the compression area. Because of the shape of the internal housing, the air charge is compressed and sent to the exhaust chamber. The pressurized air is then sent to the manifold and finally to the exhaust manifolds.

Check Valve

A check valve is also placed in the air injection system. This check valve is used to prevent back flow of exhaust gases. Backfire may occur whenever the exhaust pressure is greater than the air pump pressure. A check valve is shown in *Figure 24-24*.

Diverter Valve

The diverter valve is used to bypass the air from the air injection pump when the throttle is closed quickly. It is also called the deceleration or gulp valve. During this condition, there is a high intake manifold vacuum. The high vacuum produces a rich air-fuel mixture from the carburetor. The mixture is too rich to be ignited in the combustion chamber. It will, therefore, go out the exhaust valve and into the exhaust manifold. If excess air were pumped into this rich mixture, violent burning action might be produced. This could cause a backfire and possibly damage the exhaust system.

The diverter valve operates as shown in *Figure 24-25*. During normal operation, pressurized air enters the inlet and goes directly to the exhaust manifold through the outlet. When deceleration occurs, the high vacuum is felt at the vacuum signal line. This vacuum causes the metering valve to rise, closing off the line to the outlet and the exhaust manifold. Pressurized air is then directed internally to the diverted air outlet.

Aspirator Valve

The *aspirator valve* is a system that permits air to be injected into the exhaust without the use of an air pump. This unit consists of a steel tube containing a one-way aspirator valve that is attached to the exhaust manifold. *Figure 24-26* shows the external connections on the aspirator valve.

This unit uses exhaust pressure pulsations to draw fresh air into the exhaust system. This reduces carbon monoxide and, to a lesser degree, hydrocarbon emissions.

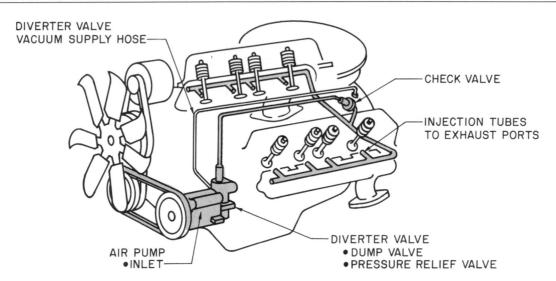

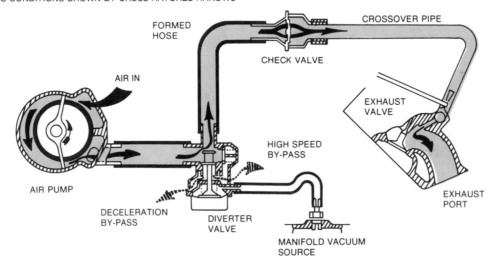

FIGURE 24–21. A typical air injection system injects air (oxygen) directly into the exhaust flow. This will assist the burning of any unburned hydrocarbons and carbon monoxide in the exhaust system. (Courtesy of Motor Publications, Emission Control Manual, 1985–86)

The aspirator valve is shown in *Figure 24–27*. During the exhaust pulses, when the pressure in the exhaust system is positive (view B), the aspirator valve is closed. Air is not sent into the exhaust to reduce emissions. When the exhaust has a negative pressure during the pulses (view A), the aspirator valve opens and allows a small amount of air to enter the exhaust. This small amount of fresh air is first brought through the carburetor. It has a similar effect as injecting air into the exhaust. This action reduces emissions.

Catalytic Converters

Catalytic converters provide another method of treating exhaust gases. Catalytic converters are located in the exhaust system, between the engine and the muffler. They are used to convert harmful pollutants such as HC, CO, and NO_x into harmless gases. In operation, the exhaust gases pass over a large surface area that is coated with some form of *catalyst*. A catalyst is a material that causes a chemical

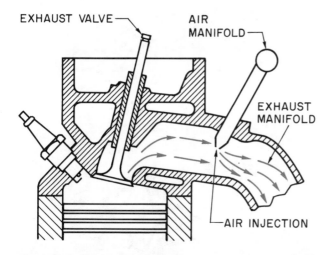

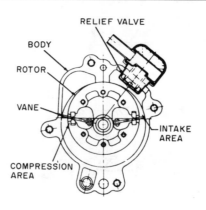

FIGURE 24–22. Air is injected directly into the exhaust gases in the air injection system. This extra fresh air will cause the unburned hydrocarbons and carbon monoxide to burn more completely, reducing emissions.

FIGURE 24–23. A cutaway view of an air injection system pump is shown. This is a positive displacement pump using vanes to produce the air pressure. *(Courtesy of Motor Publications, Emission Control Manual, 1985–86)*

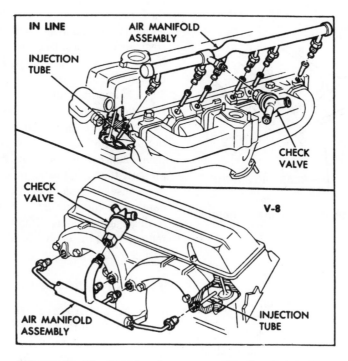

FIGURE 24–24. Check valves are used in the air injection system to control any reverse flow of exhaust gases. The check valve allows air flow in one direction but not in the other. *(Courtesy of Motor Publications, Emission Control Manual, 1985–86)*

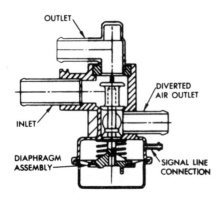

VALVE IN OPEN POSITION

FIGURE 24-25. A diverter valve or deceleration valve is used to stop the flow of pumped air to the exhaust during deceleration periods. This action eliminates the possibility of backfire damage. *(Courtesy of Motor Publications, Emission Control Manual, 1985-86)*

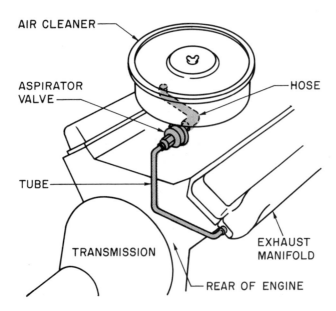

FIGURE 24-26. External connections for the aspirator air injection system are shown.

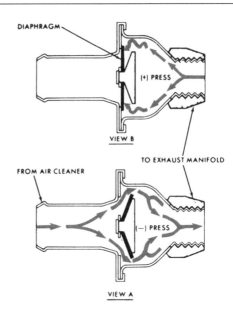

FIGURE 24-27. The aspirator valve works in response to the positive and negative pulses in the exhaust. *(Courtesy of Motor Publications, Emission Control Manual, 1985-86)*

reaction without becoming part of the reaction process. The catalyst's chemicals are not changed in the process.

The catalyst used on a catalytic converter depends upon the exact type of pollutant being removed. When exhaust gases are passed through a bed of platinum- or palladium-coated pellets or through a coated honeycomb core, the HC and CO react with the oxygen in the air. The result is the formation of water and carbon dioxide. When the metal rhodium is used as the catalyst, the NO_x in the exhaust gases is reduced to harmless nitrogen and oxygen. In this case, rhodium is known as a "reducing" catalyst.

The reaction within the catalyst produces additional heat in the exhaust system. Temperatures up to 1,600 degrees F are normal. This additional heat is necessary for the catalyst to operate correctly. Because of these high temperatures, catalytic converters are made of stainless steel. Special heat shields are used to protect the underbody from excessive heat. Each car manufacturer has its own unique heat shielding.

It is important that only unleaded fuel be used with a catalytic converter. Leaded gasoline will destroy the effectiveness of the catalyst as an emission control device. Under normal conditions, the catalytic converter will not require maintenance. It is important, however, to keep the engine properly tuned. If it is not properly tuned, engine misfiring may cause overheating of the catalyst. This heat may cause damage to the converter. This situation can also occur during engine testing. If any spark plug wires have been removed and the engine is allowed to idle for a prolonged period of time, damage may occur.

Types of Catalytic Converters

There are two types of catalytic converters: the two-way converter and the three-way converter. Both types can employ either a *monolith* or pellet design. The monolith and pellet designs are shown in *Figures 24-28* and *24-29*, respectively.

Two-way Catalytic Converter

The two-way catalytic converter reduces the carbon monoxide and hydrocarbon particles. It does not reduce any

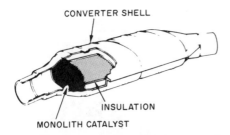

FIGURE 24–28. The two-way monolith catalytic converter uses a catalyst made like a honeycomb construction to allow gases to pass through. *(Courtesy of Motor Publications, Emission Control Manual, 1985–86)*

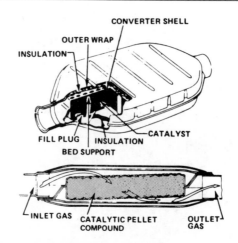

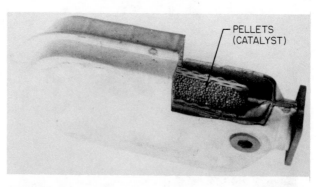

FIGURE 24–29. The two-way pellet-type catalytic converter uses pellets as the catalyst. Exhaust gases are forced through the catalyst pellets. *(Courtesy of Motor Publications, Emission Control Manual, 1985–86)*

nitrogen oxide emissions. Figures 24–28 and 24–29 are examples of the two-way catalytic converter. Here, only platinum and palladium are used as a catalyst to reduce hydrocarbons and carbon monoxide.

Three-way Catalytic Converter

The three-way catalytic converter is designed to reduce nitrogen oxide emissions. An additional catalyst bed coated with platinum and rhodium is used. The bed not only helps reduce hydrocarbons and carbon monoxide but also lowers the levels of nitrogen oxide emissions.

Figure 24–30 shows a three-way monolith catalytic converter. The front bed or inlet is treated with platinum and rhodium and is termed a "reducing" catalyst. The rear bed is coated with palladium and platinum and is referred to as the "oxidizing" catalyst.

Referring to Figure 24–30, exhaust gases first pass through the reducing catalyst. This causes the levels of NO_x to be reduced. Pressurized air from the air injection system is forced into the space between the catalyst beds. Extra air supplies additional oxygen. The extra air causes more oxidation of the gases.

As the treated exhaust gases from the first bed continue, they eventually pass through the conventional oxidation catalyst made of palladium and platinum. Here, hydrocarbons and carbon monoxide emissions are reduced.

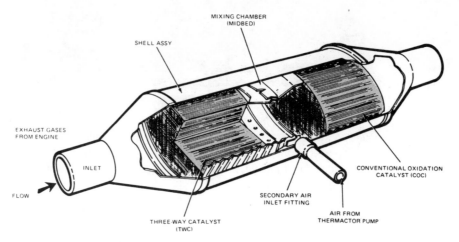

FIGURE 24–30. A three-way, dual-bed, monolith catalytic converter is shown. Secondary air from the air injection system is used to cause greater oxidation of the emissions. *(Courtesy of Motor Publications, Emission Control Manual, 1985–86)*

SERVICE CENTERS

Service centers are the link between the customer and the manufacturer of the automobile and its parts. Service centers are designed to make the customer comfortable and to develop attractive product display areas. (Courtesy of The Firestone Tire and Rubber Company)

TECHNICAL TRAINING CENTERS

The automotive technician must have a great deal of information. To help improve the technician's level of technical competence, many training centers are available to learn about automobiles. (Courtesy of The Firestone Tire and Rubber Company)

24.4 CONTROLLING COMBUSTION AND AIR-FUEL MIXTURES

Many other types of emission control systems are used on the automobile today. These systems are designed to change or control the air-fuel ratio and/or the combustion process. A knowledge of the fuel and ignition system is necessary to understand this information.

Idle Limiters

Certain types of carburetors have an idle limiter screw. Its purpose is to assure that the carburetor will deliver a leaner air-fuel mixture, especially during idle. The idle mixture adjustment screw is adjusted leaner by the manufacturer. Then, the idle limiter cap is installed. The cap permits only a small idle mixture adjustment. The cap should only be removed during carburetor repair. *Figure 24–31* shows an example of the idle limiter screw in one model of carburetor. On late-model carburetors, the idle mixture screw has been completely sealed. No adjustment is possible on newer carburetors except by authorized technicians.

Combustion Chamber

Combustion chambers have also been changed to produce fewer emissions. During combustion, the layer of air and fuel mixture next to the cooler cylinder head and piston head typically does not burn. The metal surfaces chill this area below the combustion temperature. A certain amount of unburned fuel is swept out of the cylinder on the exhaust stroke.

These emissions can be controlled by reducing the S/V ratio of the combustion chamber. The S/V ratio is the ratio between the surface (S) area and the volume (V) of the combustion chamber. A wedge-design combustion chamber has the highest S/V ratio. The hemispheric combustion chamber has a lower S/V ratio. *Figure 24–32* shows the wedge and the hemispheric combustion chambers.

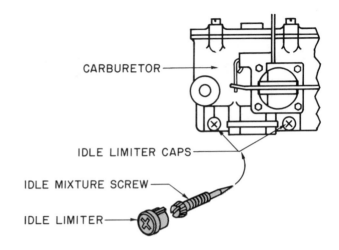

FIGURE 24–31. The idle limiter is located at the base of the carburetor. It assures a leaner air-fuel mixture during idle conditions. This leaner mixture reduces emissions. On newer carburetors, the idle mixture screw has been sealed to eliminate adjustments.

Combustion chambers have also been changed to reduce close clearances that tend to quench the flame before all of the air-fuel mixture is burned. Quench heights have been increased. This permits more complete burning of the air-fuel mixture in these areas. On late-model engines, manufacturers have gone to a more "open" style combustion chamber to reduce exhaust emissions.

Lower Compression Ratios

During the 1960s, the compression ratio of engines being built was about 9.5 to 1. In fact, compression ratios were being increased each year to produce more power. Higher compression ratios also produced higher combustion temperatures. These higher temperatures then increased the amount of nitrogen oxide being produced. Because of this problem, automobile manufacturers have lowered the compression ratios to about 8 to 1. Reducing the compres-

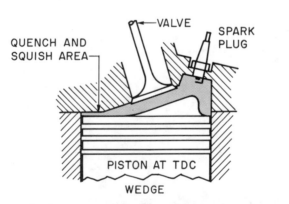
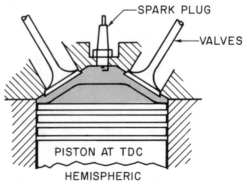

FIGURE 24–32. Both the wedge and hemispheric combustion chambers are shown. The hemispheric combustion chamber has a lower S/V (surface to volume) ratio, which produces less pollution in the cylinder.

sion ratio reduces peak combustion temperatures, which, in turn, reduces the amount of nitrogen oxide produced. However, reducing compression ratios also reduces engine performance and efficiency to some extent.

Intake Manifolds

Intake manifolds have been modified to assure more rapid vaporization of fuel during warm-up periods. The exhaust crossover flow area of the intake manifold between the inlet and exhaust gases has been made thinner. The time required to get the heat from the exhaust gases into the inlet gases has been reduced.

Ignition Timing

Ignition timing of the automobile engine is very important to the reduction of emissions. For example, during part throttle operation, the distributor vacuum advance normally advances the ignition timing. This provides more time for the leaner air-fuel mixture to burn. The added time, however, also allows more NO_x to develop. Several systems have been incorporated to control vacuum advance under certain driving conditions such as reverse, neutral, or low forward speeds. These include the TRS (transmission-regulated spark) and CCC (computer command control). For example, the CCC system does not use mechanical or vacuum advance for distributor spark control or timing. It is all done by the computer.

Control of Vacuum Advance: TRS System

A transmission-regulated spark (*TRS*) system is one method used to regulate the vacuum at low speeds. Referring to *Figure 24–33*, the system works so that the solenoid valve is normally open. This allows the distributor to have full vacuum advance when the transmission is in high gear only. In the lower gears, the transmission switch is closed. This closes the solenoid valve. When the transmission switch is closed, vacuum to the distributor vacuum advance is shut off. Now there is no vacuum advance applied to the distributor during lower speeds.

Electronic Spark Control

Automotive manufacturers have recently designed and incorporated electronic control of spark and timing. These systems do not use the conventional mechanical and vacuum advance. A number of sensors are used in their place. Basic timing is provided by a magnetic pickup coil sensor in the distributor. Referring to *Figure 24–34*, signals such as barometric pressure, manifold vacuum, coolant temperature, and rpm are fed into a solid-state computer. Based on these input signals, the computer determines and provides the optimum spark advance within milliseconds.

Computer Command Control

The use of computers to control spark advance has led to the development of computer command control systems, EFI, and ECCS (Electronic Concentrated engine Control Systems) used on automobiles today. These systems provide very sensitive and precise control of air-fuel ratios and ignition system timing advance.

A computer system is shown in *Figure 24–35*. This system is capable of sensing up to eleven variables and controlling six functions with the information. Other systems, including injection timing, injection of air into the exhaust manifold, injection of air into the converter, and idle speed, can also be controlled. Precise control of these systems makes possible a sizable reduction in emissions.

The computer used on automobiles is also tied into some of the existing pollution control equipment. These include carbon canister purge control, EGR control, and *EFE* control.

   a. The carbon canister purge is controlled by a computer. A solenoid is placed in the manifold vacuum purge line. The computer controls when the purging takes place.

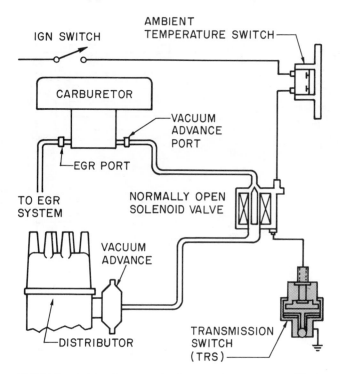

FIGURE 24–33. A transmission-regulated spark (TRS) system is shown. A signal from the transmission operates the normally open solenoid valve to stop vacuum advance during low speeds.

b. The EGR valve can also be controlled by the computer. Based on the coolant temperature, mass air flow, engine rpm, and throttle position, the computer will cycle the EGR valve open and closed.

c. The EFE (early fuel evaporation) is used to cause exhaust gases on one side of a V engine to be directed over to the other side. The excess heat causes the fuel to evaporate faster, reducing emissions. The EFE system is also controlled by the computer.

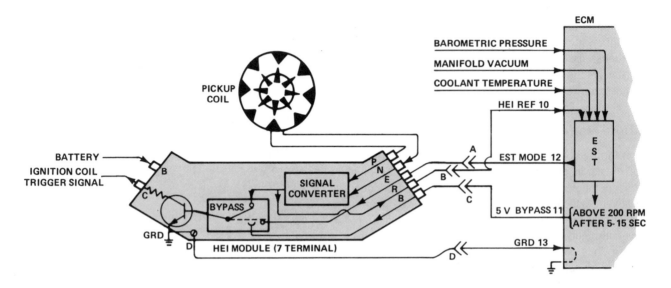

FIGURE 24–34. On the electronic spark timing (EST) system, various sensors feed information into the computer to control the spark advance for optimum emission control. *(Courtesy of General Motors Product Service Training)*

E.C.C.S. CHART

Inputs	E.C.C.S. control unit	Controls	Outputs
Crank angle sensor		Fuel injection & mixture ratio control	Injectors
Air flow meter			
Water temperature sensor		Ignition timing control	Power transistor
Exhaust gas sensor			
Ignition switch		Idle speed control	Idle-up solenoid valve
Throttle valve switch			
Neutral switch		Automatic transaxle control	Automatic transaxle lock-up solenoid valve
Air conditioner switch			
Heater fan switch / Lightening switch / Power steering oil pressure switch		Fuel pump control	Fuel pump
Detonation sensor			
Battery voltage		Exhaust gas sensor monitor & self diagnosis	Inspection lamps

FIGURE 24–35. A schematic of a closed loop electronic control system which monitors and controls a wide range of units and functions to help reduce emissions. This system is called the electronic concentrated engine control system. *(Courtesy of Nissan Motor Corporation in USA)*

DIAGNOSIS and SERVICE

SAFETY TIPS

1. *After running the engine, the EGR valve may be very hot. Let the engine cool down so that your hands will not be seriously burned.*

2. *Never weld anything on or near the gasoline tank, even if the fuel has been removed. There may still be gasoline fumes inside the tank.*

3. *When checking the heated intake air system, be careful not to burn your hands. The tube is connected to the exhaust manifold and can be very hot.*

4. *Whenever checking or diagnosing pollution control components with the engine running, make sure there is adequate ventilation for the exhaust fumes.*

5. *The catalytic converter normally operates at a higher temperature than the exhaust. Always be extra careful when working around the catalytic converter so as to eliminate the possibility of severe burns to the skin.*

1. Always check the condition of emission control hoses and tubes for leaks or small cracks. Any leak could cause pressure or vacuum to leak. This, in turn, may cause incorrect operation of the emission control devices.

2. Hoses, vacuum diaphragms, and vacuum tubes can be checked with a *vacuum tester* as shown in ***Figure 24–36***. To check a component, remove the hose attached to the device to be checked. Using the vacuum tester, apply vacuum to the component by pumping the vacuum tester. The device should be able to hold a vacuum without leaking. Leakage can be observed on the vacuum scale.

3. Always check the specification manual for the correct hook-up of vacuum hoses. This is referred to as a vacuum hose routing diagram. ***Figure 24–37*** shows a typical vacuum hose routing.

4. Do not remove any pollution control devices on any engine and run the vehicle. It is against the law to do so.

5. Check the PCV valve for correct operation by removing the valve, running the engine, and observing a small vacuum. A hissing sound should also be heard. If there is no hissing sound, vacuum,

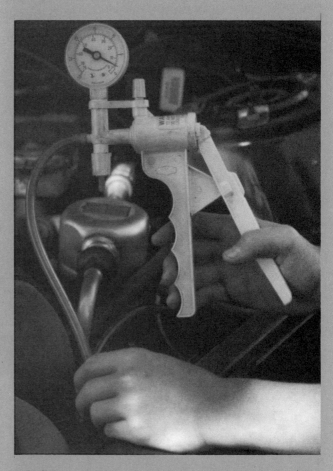

FIGURE 24–36.   **A vacuum tester can be used to check for vacuum leaks and the condition of diaphragms and other pollution control equipment.**

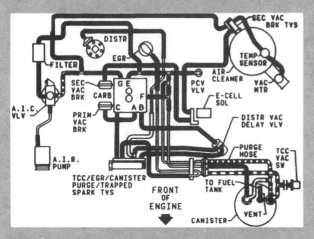

FIGURE 24–37. Vacuum hose routing diagrams are available in many of the maintenance and service manuals. Refer to these diagrams when checking for correct vacuum connections. *(Courtesy of Motor Publications, Emission Control Manual, 1985–86)*

or if there is a slight pressure, the PCV valve may be damaged or plugged. If this is the case, remove the valve and replace it with a new one.

6. When it is shaken, a PCV valve should rattle, indicating that the valve is free to move. If there is no rattle, this indicates that the valve may be plugged or inoperative. Replace the valve, if necessary.

7. A collapsed fuel tank may indicate that the fuel tank cap is defective. This is because as the fuel is removed during normal driving, no air can get into the tank to replace the volume of fuel. If this is the case, replace the cap.

8. Check for the correct tension on the *air injection reactor* pump belt. This can be done using the belt tension checker discussed in the cooling system chapter.

9. There should be no air leaks in the AIR system components. Look for air leaks at the hoses and the air injection tubes on the exhaust manifold. If there are any leaks, replace the component as necessary.

10. Replace the filter on the *carbon canister* according to the manufacturer's recommended time intervals.

11. If the *diverter valve* is operating incorrectly, the engine may backfire. Use the following procedure to test the diverter valve.

   a. Start the engine and run it until it reaches normal operating temperature.

   b. Disconnect the vacuum hose from the diverter valve to ensure the presence of a vacuum at the hose.

   c. Reconnect the hose to the valve. Air should be vented through the diverter valve for at least one second.

   d. Momentarily accelerate the engine to full throttle while observing the diverter valve. Each time the engine is accelerated, air should be discharged from the diverter valve muffler for about one second. If the diverter valve fails to perform as outlined, the valve is defective.

12. Check the *heated intake air* system for damaged tubes and parts. When this system is defective, the engine usually runs rough, hesitates, or stalls during cold starting conditions. The vacuum motor can be checked with the vacuum tester. First, disconnect the vacuum hoses that lead to the vacuum motor. When a vacuum from the vacuum tester is applied to the motor, the linkage and control damper should move freely and the damper should remain in position. If the damper assembly returns slowly, there is a leak in the motor and it should be replaced.

13. Replace the air filter and breather element at correct intervals to maintain correct air-fuel ratios and engine operation. To remove the air filter, remove the wing nut or other locking device and the air cleaner housing top.

14. Since there are several types of EGR systems, refer to the manufacturer's specifications for correct testing and service of each. Common testing procedures include several checks.

   a. Place a finger under the EGR valve and lift up on the diaphragm plate. *CAUTION: Be careful not to burn your fingers or hands.* The plate should move up and down freely without binding or sticking. If it doesn't move freely, replace the valve.

   b. Check the vacuum by connecting a vacuum gauge to the EGR valve signal line using a T-fitting. Start the engine and run it at part throttle. The vacuum gauge should read at least 5 inches Hg (the reading will be different for each vehicle) at part throttle and at normal engine operating temperature. If the gauge reads less, check the hoses and vacuum source, and repair as needed.

   c. With the engine running as in step (b), disconnect the vacuum hose to the EGR valve. The diaphragm plate should move and the engine speed should increase. If the plate does not move, replace the EGR valve.

   d. Reconnect the vacuum hose. The EGR valve diaphragm plate should move and the engine speed should decrease.

e. If the diaphragm moves but there is no change in the engine speed, check the manifold passages for blockage. If the passages are clear, the EGR valve is defective.

f. The intake manifold can be cleaned with the EGR valve removed. Use a suitable screwdriver and wire brush to remove deposits.

15. A rotten egg smell is produced by vehicles that have *catalytic converters* when there is a rich mixture in the carburetor. Check the carburetor adjustments or try using another type of fuel to eliminate this odor.

16. If pellets are coming out of the exhaust, this indicates that the catalytic converter is damaged internally. Replace if necessary.

17. Never use leaded gasoline in a vehicle that has a catalytic converter. Lead particles plug up the catalyst and restrict the exhaust flow.

18. Rough idle is often caused by leaks in the vacuum hoses. Check all vacuum hoses with the vacuum tester.

19. Check the carbon canister for cracks or fuel leaking from the fittings. Repair and replace where necessary.

20. A plugged PCV valve can cause rough idle, stalling at low speeds, oil leaks, oil in the air cleaner, or sludge in the engine. Check the condition of the PCV valve as mentioned earlier in this section.

21. On vehicles with CCC systems, codes are used for troubleshooting. These codes are shown on the instrument panel when a fault is detected. By grounding a trouble code "TEST" lead terminal under the dash, the "CHECK ENGINE" light will flash a trouble code indicating the problem area. Refer to the manufacturer's manual for the correct identification of codes for each vehicle.

22. On certain vehicles, there is a replaceable filter on the EGR valve. Replace this filter at the manufacturer's recommended interval.

23. Check all wires and connectors for solid electrical connection on CCC systems.

24. Follow the manufacturer's service procedure for repairing the *air injection reactor* pump. When servicing the AIR system:
a. Inspect the check valves for air flow in only one direction.
b. Check for excessive belt noise.
c. Check for a seized air pump.
d. Check the pump for noises such as knocking, chirping, or rumbling.
e. Check for loose air pump hoses.
f. Check the mounting bolts for looseness, and if the pump is aligned correctly.
g. Check for a defective pressure relief valve.

SUMMARY

This chapter covered the operation of the crankcase ventilation system, evaporative systems, and intake and exhaust controls. In addition, several devices designed to control combustion more effectively to reduce emissions were also studied.

The positive crankcase ventilation system is used to prevent crankcase vapors from entering the atmosphere. This is done by using a PCV valve. Blow-by gases are produced by the combustion process. These blow-by gases are then sent through a PCV valve and back to the intake manifold of the engine for reburning. This prevents any crankcase emissions from entering the atmosphere.

Another area of emissions from the automobile is that produced from fuel. This emission is made of hydrocarbons. Several devices are used to prevent any hydrocarbon vapors from being released into the atmosphere. The carbon canister (also called the charcoal canister) is a small canister used to hold carbon. The fuel vapors from the fuel tank and carburetor are sent to the carbon canister to be absorbed. During engine operation, some of these vapors are sent directly into the intake to be burned in the combustion process.

Other components are also used to control evaporative emissions. A vapor-liquid separator is used to stop any raw fuel from entering the canister. The rollover check valve is used to stop liquid fuel from going into the canister if the vehicle rolls over in an accident. Many fuel tanks also have an expansion tank to allow for fuel expansion.

Another set of pollution controls is designed to control the intake and exhaust of the engine. The heated air intake system is used to heat the intake air for the most efficient operation during cold starting. Air is drawn over the exhaust manifold to heat the air. The EGR (exhaust gas recirculator) is used to reduce NO_x. It allows cooler exhaust gases to flow back into the intake to reduce combustion temperatures. The AIR (air injection reactor) system forces excess air into the exhaust to continue burning emissions that have been partially burned within the combustion chamber. This system uses an air pump driven by the crankshaft and a set of tubes to each exhaust port. Several other devices are used as part of an AIR system. These include check valves, diverter valves, and aspirator valves.

A catalytic converter is also used on vehicles today. The catalytic converter uses a catalyst to change the chemistry of the exhaust. All three emissions, including HC, CO, and NO_x, can be controlled by the converter. Two types of catalytic converters are used: the two-way and three-way converters. Pellet and monolith designs are used for the converters.

Emissions can also be controlled by changing the combustion at various operational speeds and loads. Idle limiters are used on certain carburetors to prevent tampering with precise carburetor air-fuel mixture adjustments which affect emissions at idle. Combustion chambers are being reshaped to reduce emissions. Lower compression ratios allow for nonleaded fuels. Intake manifolds are being designed to enhance vaporization of fuel quicker. Timing is also being electronically adjusted and controlled to help reduce emissions.

In addition to the changes mentioned, computer controls are now being used to control emissions even more accurately. Computers that monitor various inputs such as engine and vehicle speed, coolant temperature, position of the crankshaft, load, air flow into the engine, intake air temperature, and throttle position, are being added to vehicles. These inputs help the computer to precisely control the fuel and timing advance for optimum performance and reduced emissions.

There are many service procedures used on pollution control systems. Each manufacturer has its own procedures and set guidelines to follow. On computer-controlled systems, codes are used to identify problems in the system.

TERMS TO KNOW

Can you explain each of the following terms? Review the chapter until you can use each term correctly.

Blow-by	Transducer
Crankcase pressure	Positive displacement pump
Positive crankcase ventilation (PCV)	Aspirator valve
Carbon canister	Catalytic converter
Purge	Catalyst
Heat stove	Monolith
Snorkel tube	Transmission regulated spark (TRS)
Exhaust gas recirculation (EGR)	Early fuel evaporation (EFE)
Pintle	

REVIEW QUESTIONS

Multiple Choice

1. Crankcase pressures are produced by:
   a. The cooling system
   b. The ignition system
   c. Power output
   d. Blow-by
   e. The starter

2. The PCV valve _____ the amount of vapor being burned in the combustion chamber during low speeds.
   a. Increases
   b. Decreases
   c. Doubles
   d. Has no effect on
   e. None of the above

3. The carbon canister stores:
   a. Fuel vapor from the fuel tank
   b. Fuel vapor from the carburetor
   c. Blow-by vapors
   d. All of the above
   e. A and b

4. The carbon canister:
   a. Stores fuel vapors
   b. Has a filter
   c. Has a purge line back to the carburetor
   d. All of the above
   e. None of the above

5. Which of the following stops raw fuel from entering the carbon canister?
   a. Check valve
   b. Vapor-liquid separator
   c. Air pump
   d. PCV valve
   e. Idle limiter screws

6. Which device is used to allow a place for expanding fuel in a full tank of gasoline in warm weather?
   a. Check valve
   b. Rollover valve
   c. Expansion tank
   d. PCV valve
   e. AIR system

7. Warming up the intake air during cold starting will:
   a. Reduce emissions
   b. Increase emissions
   c. Reduce fuel consumption
   d. Cause rough idling
   e. Increase power

8. The "heat stove" is used on what pollution control device?
   a. AIR
   b. PCV
   c. Thermostatically Controlled Air Cleaner
   d. Transmission Controlled Spark
   e. CCC

9. Which of the following pollution control devices is designed to reduce the NO_x emissions?
   a. AIR
   b. PCV
   c. EGR
   d. TCS
   e. None of the above

10. On the EGR system _____ is/are sent back to the intake.
    a. Exhaust gases
    b. Blow-by
    c. Extra spark
    d. Liquid coolant
    e. Catalyst

11. Which pollution control system forces air into the exhaust stream to help burn emissions?
    a. AIR
    b. PVC
    c. EGR
    d. TCS
    e. CCC

12. Which of the following will have little or no effect on the combustion process?
    a. PCV
    b. Carbon canister
    c. AIR
    d. Evaporative emission systems
    e. Heated air intake systems

13. What type of system allows air to be injected into the exhaust without the use of a positive displacement pump?
    a. CCC
    b. Aspirator system
    c. Carbon canister
    d. All of the above
    e. None of the above

14. Which of the following systems change(s) the exhaust system to reduce emissions?
    a. Two-way catalytic converter
    b. Three-way catalytic converter
    c. Diverter valve
    d. AIR system
    e. A, b, and d

15. Which of the following can be found in a catalytic converter?
    a. A monolith catalyst
    b. A pellet catalyst
    c. An air connection from the AIR system
    d. A platinum catalyst
    e. All of the above

16. Which catalytic converter system uses air from the AIR system?
    a. Two-way converter
    b. Three-way converter
    c. Four-way converter
    d. All of the above
    e. None of the above

17. Carburetors today:
    a. All have idle limiters
    b. Have no pollution controls on them
    c. All are designed to use lead
    d. All of the above
    e. None of the above

18. Which compression ratio is for engines that are designed to reduce emissions?
    a. 12–1
    b. 11–1
    c. 10–1
    d. 8–1
    e. 4–1

19. Which of the following has/have been changed to enhance vaporization of the fuel?
    a. Combustion chambers
    b. Intake manifolds
    c. Exhaust flow
    d. Catalytic converters
    e. PCV systems

20. Which of the following is/are part of the computer-controlled systems to reduce pollution?
    a. EGR valves
    b. Carbon canister
    c. EFE (early fuel evaporation)
    d. All of the above
    e. None of the above

The following questions are similar in format to ASE (Automotive Service Excellence) test questions.

21. Technician A says that the PCV valve should not rattle when it is shaken. Technician B says that the PCV valve is in good shape when the valve rattles when it is shaken. Who is right?
    a. A only
    b. B only
    c. Both A and B
    d. Neither A nor B

22. Technician A says that the EGR valve can be tested by pushing up inside of the valve and seeing if the diaphragm moves freely. Technician B says that the EGR valve can be checked using the vacuum tester. Who is right?
    a. A only
    b. B only
    c. Both A and B
    d. Neither A nor B

23. Technician A says that the charcoal canister has no filter to be changed. Technician B says that the charcoal canister has a filter that should be changed at regular recommended intervals. Who is right?
    a. A only
    b. B only
    c. Both A and B
    d. Neither A nor B

24. Technician A says that a collapsed fuel tank is caused by too much vacuum in the tank. Technician B says a collapsed fuel tank is caused by a bad fuel tank cap. Who is right?
    a. A only
    b. B only
    c. Both A and B
    d. Neither A nor B

Essay

25. Describe the purpose and operation of the PCV system.

26. Describe the purpose and operation of the EGR system.

27. What is the carbon canister used for?

28. What is a heat stove used for on a pollution control system?

29. Describe the operation of a catalytic converter.

30. Define the S/V ratio when dealing with combustion chambers.

31. List several ways in which pollution has been reduced by changing the combustion chamber and/or air-fuel mixtures.

CHAPTER 25

Manual Transmissions

INTRODUCTION

Torque that is produced at the end of the crankshaft by the engine must be transmitted to the driving wheels. To accomplish this, torque must first pass through the clutch and transmission. This chapter is about these two components.

OBJECTIVES

After reading this chapter, you will be able to:
- Identify the purpose and operation of the clutch.
- Define the purpose of the standard or manual transmission.
- Analyze the purpose of different gear ratios.
- Describe the operation and gear selection of the manual transmission.
- State the purpose and operation of synchronizers.
- Compare different types of transmissions.
- Identify the operation of linkages and accessories used on manual transmissions.
- State common diagnosis and service suggestions pertaining to manual transmissions.

CHAPTER HIGHLIGHTS

25.1 CLUTCH SYSTEMS

Purpose of Clutches

All standard or manual transmissions have a *clutch* to engage or disengage the transmission. While the engine is running, there are times when the driving wheels must not turn. The clutch is used as a mechanism to engage or disengage the transmission and driving wheels. If clutches were not used, every time the vehicle came to a stop, the engine would stop. Since this is not practical, the operator can engage or disengage the clutch when needed.

The clutch is designed to engage the transmission gradually. This will eliminate jumping abruptly from no connection at all to a direct solid connection to the engine. This is done by allowing a certain amount of slippage between the input and the output shafts on the clutch. Several components are needed to do this. These include the pressure plate, driven plate or friction disc, flywheel, clutch release bearing, clutch fork, and clutch housing. These are shown in *Figure 25–1*.

Figure 25–2 shows the basic principle of engaging a clutch. The left side is considered the drive member or input to the clutch. It is composed of the flywheel and *pressure plate*. The output is the center driven member or *friction disc*. The output of this shaft drives the transmission. When the pressure plate is withdrawn, the engine can revolve freely and is disconnected from the driven member and the transmission. When the pressure plate moves in the direction of the arrows, however, the friction disc is forced to turn at the same speed of the input or driving member. *Figure 25–3* shows the actual components including the flywheel, pressure plate, and friction disc.

Pressure Plate Assembly

The pressure plate is designed to squeeze or clamp the friction disc between itself and the flywheel. The pressure plate has several components, *Figure 25–4*. They include the cover, a series of springs, release fingers, and the pressure plate. The pressure plate and the flywheel have smooth

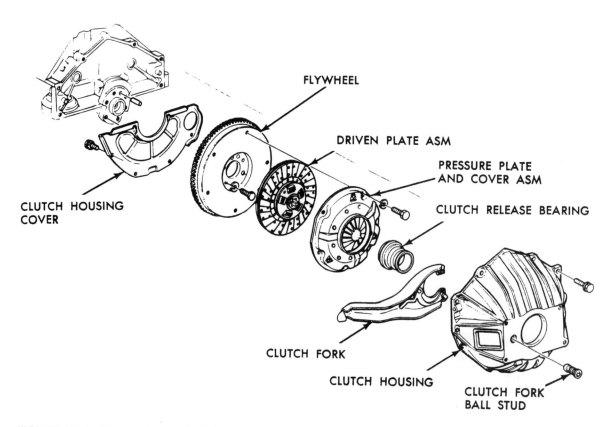

FLYWHEEL

DRIVEN PLATE ASM

PRESSURE PLATE
AND COVER ASM

CLUTCH RELEASE BEARING

CLUTCH HOUSING
COVER

CLUTCH FORK

CLUTCH HOUSING

CLUTCH FORK
BALL STUD

FIGURE 25–1. The main parts of the clutch. (Courtesy of Chevrolet Motor Division, General Motors Corporation)

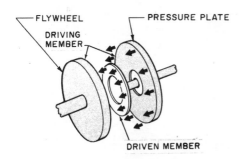

FIGURE 25-2. When the clutch is engaged, the driven member is squeezed between the two driving members. The transmission is attached to the driven member. (Courtesy of General Motors Corporation)

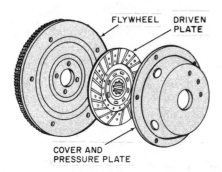

FIGURE 25-3. The actual components of a typical clutch system are shown. (Courtesy of General Motors Corporation)

machined surfaces to clamp the clutch disc. The pressure plate cover is bolted to the flywheel and turns at exactly the same speed as the flywheel. The springs are used to squeeze or clamp the friction disc between the two plates. Several types of springs are used. The *diaphragm spring* shown in *Figure 25-5* is the type currently used. It consists of a series of spring plates used to force or clamp the friction disc.

Pressure plates also have some form of linkage used to engage or disengage the friction disc, *Figure 25-6*. When the release levers are pushed in at the center, the pressure plate surface is moved to the right. This action disengages the friction disc. When there is no force on the release levers, the pressure plate springs force the pressure plate to tighten down and squeeze the friction disc to start it rotating.

Friction Disc

The friction disc is the output of the clutch system. When the clutch assembly spins, it drives the manual transmission. The clutch is made of several parts. These are shown in *Figure 25-7*. The center of the clutch has a spline hole for the transmission input shaft connection. The grooves on both sides of the clutch disc lining prevent sticking of the plate to the flywheel and pressure plate. Frictional facings are attached to each side of the clutch disc. These are made of cotton and *asbestos* fibers woven or molded together. On some clutches copper wires are also woven into the material for additional strength.

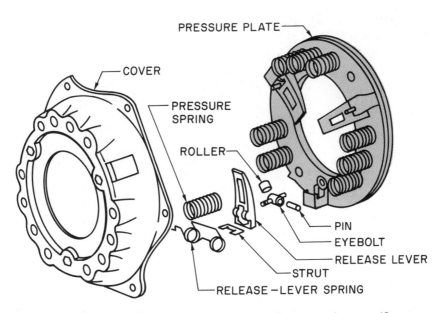

FIGURE 25-4. The pressure plate is constructed of several parts. (Courtesy of Chevrolet Division, General Motors Corporation)

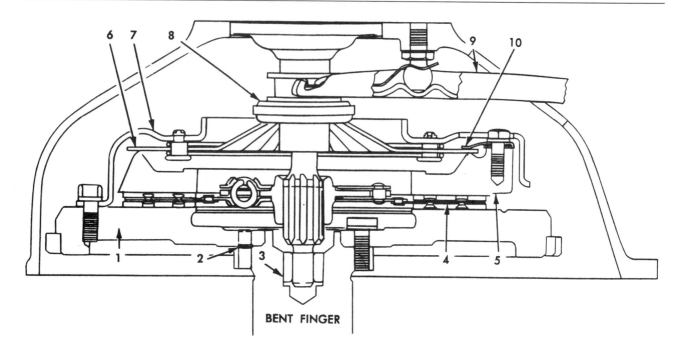

BENT FINGER

1. Flywheel   3. Pilot Bushing   5. Pressure Plate   7. Cover   9. Fork
2. Dowel-hole   4. Driven Disc   6. Diaphragm Spring   8. Clutch Release Bearing   10. Retracting Spring

FIGURE 25-5. The diaphragm spring is also used to help force the pressure plate against the clutch driven disk. *(Courtesy of Chevrolet Division, General Motors Corporation)*

The clutch has a flexible center to absorb the *torsional vibration* of the crankshaft. Steel compression springs permit the disc to rotate slightly in relation to the pressure plate.

The cushion springs are raised to eliminate chatter when the clutch is engaged. These cause the contact pressure on the facings to rise gradually as the springs flatten out when the clutch is engaged.

Clutch Release Bearing (Throwout Bearing)

The clutch release bearing, also called a throwout bearing, is a ball-thrust bearing held within the clutch housing. The bearing is moved by the clutch pedal and linkage to engage or disengage the pressure plate. When the clutch pedal is

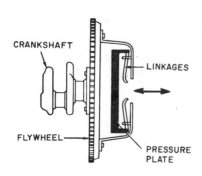

FIGURE 25-6. All clutches must have some type of linkage to engage and disengage the clutch. *(Courtesy of General Motors Corporation)*

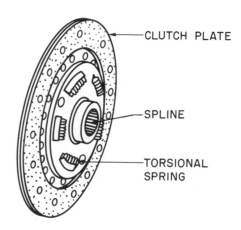

FIGURE 25-7. The clutch disk has torsional springs to eliminate jerky starting.

pressed down, the clutch release bearing pushes against the revolving pressure plate release levers. This action disengages the clutch. When the clutch pedal is released, the clutch release bearing moves back. The springs in the pressure plate now engage the clutch disc, *Figure 25–8*.

Clutch Linkage

The purpose of the clutch linkage is to engage and disengage the clutch. As the operator moves the clutch pedal down, the clutch release bearing pushes against the release levers to disengage the clutch. Depending upon the design of the vehicle, several clutch linkages are used. These may include the rod and lever, the cable type, and the hydraulic linkage. The rod and lever system is shown in *Figure 25–9*. As the clutch is pushed in, the clutch fork moves to the left. Note the pivot points. This causes the clutch release bearing to move to the right. This action disengages the clutch from the pressure plate.

The cable type of linkage uses a flexible cable connected from the clutch pedal to the clutch fork. As the clutch pedal is pressed down, the flexible cable forces the clutch fork to move. *Figure 25–10* shows an example of how the cable is attached to the clutch fork.

The hydraulic system consists of a master cylinder and a *slave cylinder*. When pressure is applied to the clutch pedal, hydraulic pressure is built up in the master cylinder. The pressure is sent through hydraulic tubing to the slave cylinder. Here the pressure is used to move the clutch fork to engage or disengage the clutch disc, *Figure 25–11*.

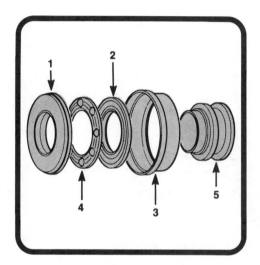

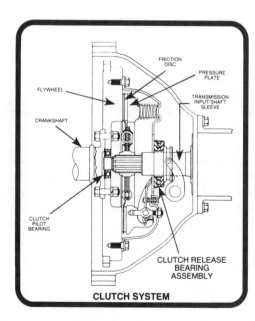

FIGURE 25–8. The main elements in the clutch release bearing are the two washer-shaped rings (1, 2) in which the raceways for the balls are machined. A steel housing (3) holds the rings and balls (4) together. The bearing is pressed onto the carrier (5) to complete the assembly. The clutch bearing is located on the transmission input shaft sleeve. *(Courtesy of Federal-Mogul Corporation)*

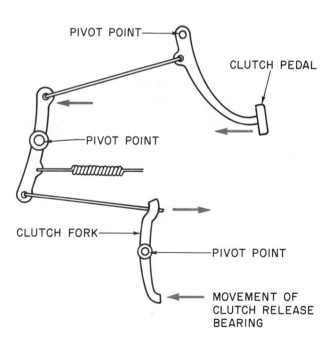

FIGURE 25–9. This rod and lever system is used to engage and disengage the clutch.

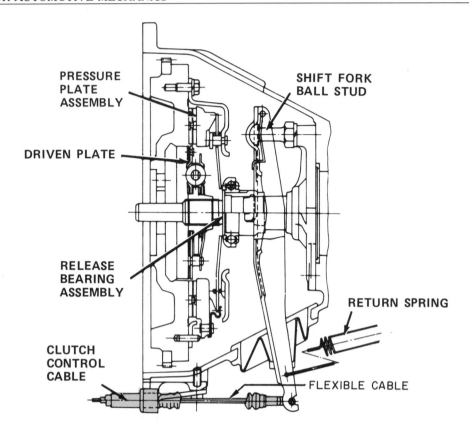

PRESSURE PLATE ASSEMBLY

SHIFT FORK BALL STUD

DRIVEN PLATE

RELEASE BEARING ASSEMBLY

RETURN SPRING

CLUTCH CONTROL CABLE

FLEXIBLE CABLE

FIGURE 25–10. A flexible cable can be used to engage the clutch. *(Courtesy of American Isuzu Motors, Inc.)*

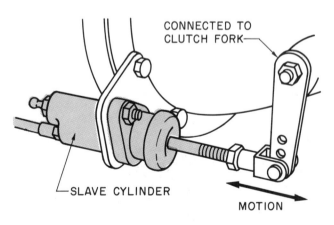

CONNECTED TO CLUTCH FORK

SLAVE CYLINDER

MOTION

FIGURE 25–11. A slave clutch can be used to engage and disengage the clutch fork.

CAR CLINIC

PROBLEM: POOR SHIFTING

A customer indicates that the five-speed manual transmission has been having trouble downshifting into first gear and is getting worse. The vehicle is used mostly in the city. Downshifting into first gear makes driving much easier in traffic. It is becoming more difficult to make the downshift, and gears are clashing.

SOLUTION:

Manual transmissions have synchronizers to enable ease of downshifting. The synchronizers are used to get both gears at the same speed before engaging them. Replace the first gear synchronizer and the problem should be eliminated.

25.2 DESIGN OF TRANSMISSIONS

Purpose of the Transmission

The purpose of a transmission is to apply different torque forces to the driving wheels. Vehicles are required to perform under many types of loads. Stopping and starting, heavy loads, high speeds and small loads are examples of the different demands placed on the vehicle. The transmission is designed to change the torque applied to the driving wheels for different applications. In addition, the transmission is used to reverse the vehicle direction for parking, and to provide neutral (no power) to the wheels. Normally, less torque is needed with higher speeds and smaller loads. When slower speeds and higher loads are used, more torque is needed.

Levers and Forces

In order to understand how gears in a transmission work, levers and forces must first be defined. A *lever* is a device that changes forces and distances. For example, in *Figure 25–12 (A)*, a lever has an input, a *fulcrum*, and an output. When a force is applied to one side over a certain distance, a resulting force is produced with a certain distance. Force times distance on the input will always equal force times distance on the output. Therefore, work input always equals work output.

If the fulcrum is positioned further to the right as shown in Figure 25–12 (B) several things occur. First, the work input will again equal the work output. However, the output force will be much greater while the input distance will be much greater. By changing the position of the fulcrum, different forces can be obtained while giving up distance.

Torque and Torque Multiplication

The principle of levers can also be applied to gears. A gear can be considered a set of spinning levers. A set of gears can increase or decrease torque the same way that levers increase or decrease force. This is shown in *Figure 25–13 (A)*. One lever is pushing a second lever. The length of each lever determines how much force will be created on them. For example, if the short lever pushes against the longer lever, the longer lever will move less distance but will have more force. If a series of levers is attached as shown in Figure 25–13 (B), then a continuous torque is developed. When applied to gears as shown in Figure 25–13 (C), a small gear can increase torque in a larger gear. However, as with levers, the larger gear moves less distance than the input gear.

Gear Ratios

The amount of torque increased from the input to the output depends upon the relative size of the gears. The difference in size between the input and output gears is called the gear

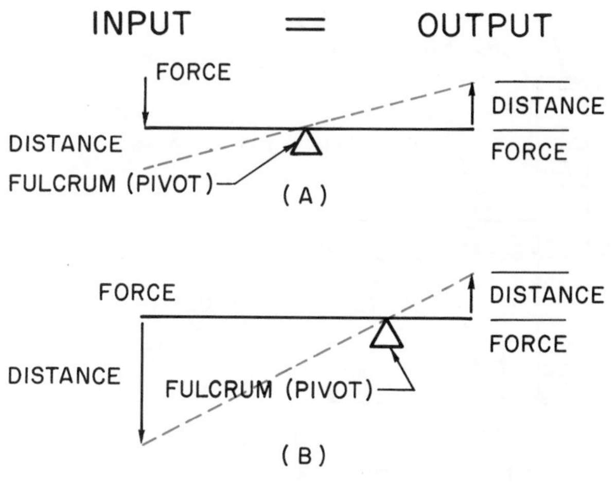

FIGURE 25–12. When the pivot point is moved to the right as in (B), increased forces can be obtained. However, the distance traveled is less. Force times distance input will always equal force times distance output.

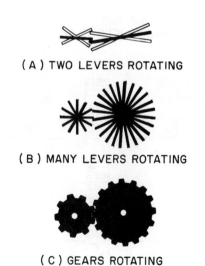

(A) TWO LEVERS ROTATING

(B) MANY LEVERS ROTATING

(C) GEARS ROTATING

FIGURE 25–13. Gears can also be used to increase torque just like levers are used to increase force. *(Courtesy of General Motors Corporation)*

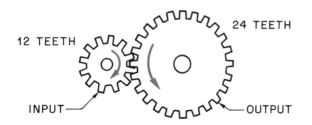

GEAR RATIO = 2 – 1
(2 REVOLUTIONS INPUT TO
1 REVOLUTION OUTPUT)

12 TEETH   24 TEETH

INPUT   OUTPUT

FIGURE 25–14. Gear ratios are determined by the number of teeth on the output compared to the input. This gear ratio is 24 to 12 or 2 to 1.

ratio. The best way to show a gear ratio is by counting the number of teeth on each gear. *Figure 25–14* shows a set of gears. The input gear has 12 teeth and the output gear has 24 teeth. It will take two revolutions of the input gear to get the output gear to turn once. The speed of the output gear will be half the speed of the input gear. However, the torque will be doubled on the output. The gear ratio is stated as 24 to 12 or 2 to 1. This means that the input gear will turn two times while the output gear will turn one time.

Manual Transmission Parts

The transmission is a case of gears located behind the clutch. The output of the clutch drives the set of gears. The case is attached to the clutch housing. There are several shafts with different sized gears inside the case. As the gears are shifted to different ratios, different torques can be selected for different operational conditions. *Figure 25–15* shows a diagram of the internal parts of a manual transmission. Although there are many parts in the transmission, the main ones include the drive or clutch gear, the countershaft gear (also called the cluster gear) with four gears on it, the main shaft with two gears on it, and the reverse idler gear. The countershaft gears all turn at the same speed. The low and reverse gear and the second and high-speed gear are able to slide on the main shaft spline.

Gear Selection

First gear is selected by connecting the gears so as to produce the greatest torque and the lowest output speed. This condition is used to start the vehicle moving or to go up a very steep hill. This is done as shown in *Figure 25–16*. When the shift lever is moved to low or first gear, the low and reverse gear on the main shaft meshes with the low gear on the countershaft. Power is now transmitted from the clutch shaft to the countershaft. From the countershaft, the power is transmitted to the main shaft. Because of the gear

NEUTRAL

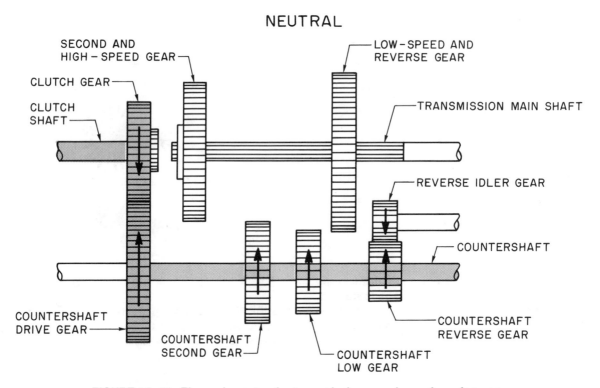

SECOND AND
HIGH – SPEED GEAR

CLUTCH GEAR

CLUTCH
SHAFT

LOW – SPEED AND
REVERSE GEAR

TRANSMISSION MAIN SHAFT

REVERSE IDLER GEAR

COUNTERSHAFT

COUNTERSHAFT
DRIVE GEAR

COUNTERSHAFT
SECOND GEAR

COUNTERSHAFT
LOW GEAR

COUNTERSHAFT
REVERSE GEAR

FIGURE 25–15. The main parts of a transmission are shown for reference.

FIRST

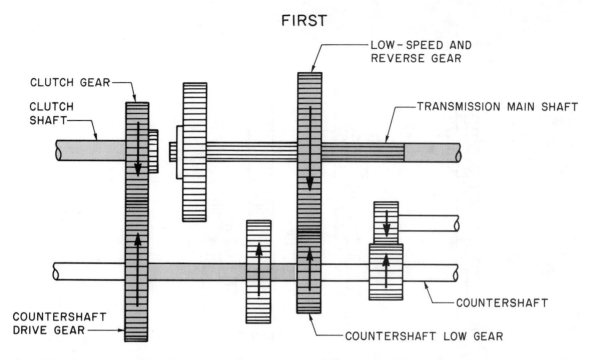

FIGURE 25-16. Low or first gear is selected by moving the low and reverse gear in mesh with the countershaft low gear.

ratios, the clutch shaft speed is about three times as fast as the speed of the output main shaft. However, the output shaft has about three times as much torque as the clutch shaft.

Second gear is selected by connecting the second and high-speed gear on the main shaft in mesh with the second gear on the countershaft. See *Figure 25-17*. Note that the gears used for low or first gear have been disengaged and are not in mesh. In this case, the ratio of input shaft to output shaft is about 2 to 1. The speed has been increased slightly, while torque has been decreased.

Third gear is selected by connecting the second and high-speed gear directly to the clutch shaft. This causes the input and output to rotate at the same speed, *Figure 25-18*. The ratio is now 1 to 1. In this condition, maximum speed with minimum torque is produced.

Reverse gear is selected by connecting the low-speed and reverse gear with the reverse *idler gear*. When the extra idler gear is added to the gear train, the direction of the output speed reverses. Reverse is considered a very low-speed, high-torque gear arrangement.

Types of Gears

Several types of gears are used in manual transmissions. Spur gears are found in older transmissions. These gears are typically very noisy. In order to reduce noise, helical gears are now used. Helical gears have the teeth at an angle. This produces a smoother meshing of teeth between the two gears. It also causes more surface contact between the teeth which makes them much stronger. Most manual transmissions use helical gears. A third type of gear is called the internal gear. The gear teeth are machined inside a ring. A fourth type of gear is called the compound gear. A compound gear is composed of several gears of different sizes placed on a shaft. Each gear spins at the same speed. The countershaft gear in a manual transmission is called a compound gear. *Figure 25-19* shows examples of different types of gears.

Purpose of Synchronizers

During normal shifting patterns, gears must be engaged and disengaged. Most gears, however, are not turning at the same speed before being engaged. To eliminate gears clashing with each other during normal downshifting or upshifting, *synchronizers* are used. Synchronizers are used in all forward speeds in most of today's manual transmissions. There are several types of synchronizers, but they all work on a similar principle. If the gear speeds can be synchronized or put at the same speed, they will mesh easily and there will be little clash or clatter.

SECOND

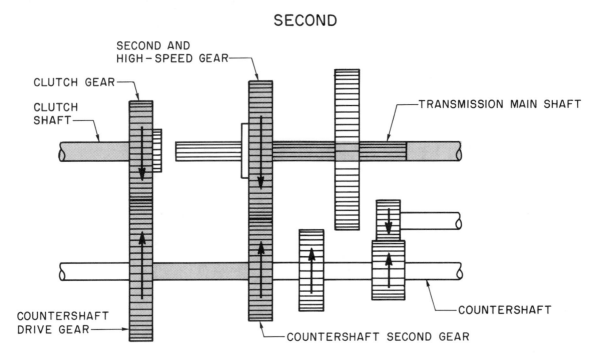

FIGURE 25-17. Second gear is selected by meshing the second and high-speed gear with the counter-shaft second gear.

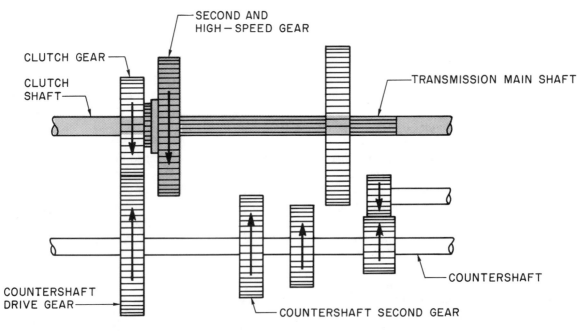

FIGURE 25-18. Third gear is selected by connecting the second and high-speed gear with the clutch gear.

Synchronizer Operation

Figure 25-20 shows how a synchronizer operates. The assembly is made of an input gear, several cone surfaces, a ring gear sliding sleeve, the hub, several spring-loaded steel balls, and the internal gear. The goal is to mesh the input gear with the internal gear. If the internal gear turns, then the output shaft will also turn.

For ease of understanding, view the input gear as spinning and the output shaft, hub, and internal gear as not spinning. As the shift fork is moved to the left, it moves the ring

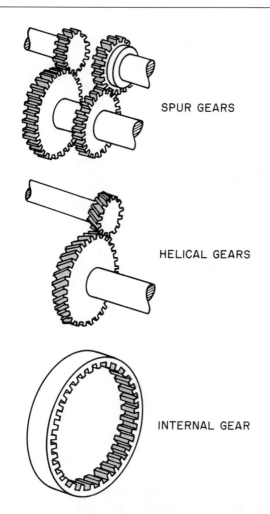

FIGURE 25–19. Several types of gears are used in manual transmissions. These may include the spur, helical, and internal gears.

gear sliding sleeve and the hub to the left. The hub and the ring gear are splined together. As the movement to the left continues, the two cone surfaces eventually begin to rub together. This action brings both objects to the same speed. The internal gear is still not in mesh with the input gear at this point. As the shift fork is moved further to the left, the ring gear sliding sleeve continues to move while the hub remains in the same position. As the ring gear continues to move left, it eventually meshes easily with the input gear because both are spinning at the same speed.

In most transmissions, the synchronizers are moved rather than the gears. This type of transmission is called a constant mesh type. *Figure 25–21* shows this principle. The gears are meshed at the start, but they are spinning on bearings on the output shaft. When the first gear synchronizer is moved to the left, it brings the output shaft to the same speed as the first gear. When the second gear synchronizer is moved to the right, it brings the second output shaft to the same speed as second gear. When the third gear syn-

chronizer is moved to the left, the output shaft is brought to the same speed as the input gear. For simplicity, the reverse gear is not shown here.

Although many synchronziers may differ in appearance, the basic design is the same. Most synchronizers use a synchronizer ring (called a blocker ring) to replace the internal cone. Keys, rather than the small spring-loaded balls are used on some synchronizers. *Figure 25–22* shows a disassembled synchronizer.

Overdrive Systems

The purpose of an *overdrive* system in a transmission is to decrease the speed of the engine during high vehicle speed operation. If this is done, the engine fuel mileage will improve. However, the torque at high speeds will be reduced. On transmissions without overdrive, the highest gear ratio is usually 1 to 1. On overdrive transmissions, the ratio may be 0.8 to 1 as shown in *Figure 25–23*. This means the engine crankshaft has slowed down slightly. This is typically done by adding another gear in the gear train. The size of the gear and the direction of power through the transmission cause the engine to slow slightly at the higher vehicle speeds. On older transmissions, the overdrive gear was engaged by mechanical means. On modern transmissions, a synchronizer clutch or electrical solenoid is used to engage the overdrive gear. If the transmission has an overdrive gear, the gear shift knob often has the letters *OD* stamped on it.

Transmission Case

The transmission case is used to hold the gears for proper shifting. The case is bolted to the clutch housing with several bolts. It is made of aluminum or cast iron for strength and support. The transmission case also includes an extension housing that contains the transmission output shaft. The housing is bolted to the rear of the transmission case. *Figure 25–24* shows the complete transmission with the case and the extension housing. The housing is used as an engine mount base and holds a rear seal to the transmission. The extension housing also includes a hole for the speedometer cable. The speedometer cable is driven from the output shaft by the speedometer gear.

Transmission Lubrication

Manual transmissions use gear oil for lubrication. Gear oil is a type of lubrication used to keep gears well lubricated. Each manufacturer typically recommends its own type of gear oil. In addition, gear oils have different SAE viscosities. Typical viscosity ranges recommended by manufacturers are from 75 to 90 or more, depending upon the outside temperature. The viscosity recommendation increases as the outside temperature increases.

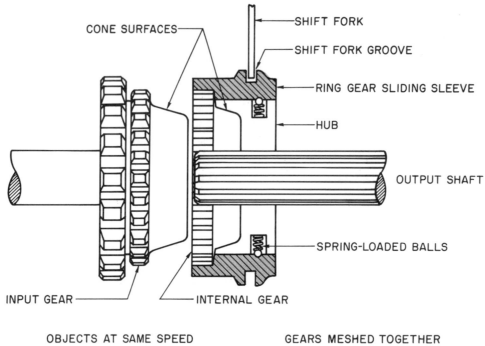

FIGURE 25–20. On a synchronizer, as the hub and internal gear assembly is moved to the left, the cone surfaces begin to touch. This brings both shafts to the same speed. As the shift fork continues to move, it eventually slides the ring gear sliding sleeve in mesh with the input gear.

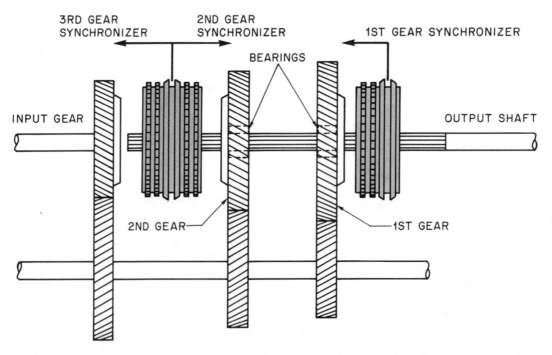

FIGURE 25–21. In most manual transmissions, the gears are always meshed. However, the synchronizers actually engage and disengage the gears on the main output shaft for correct operation.

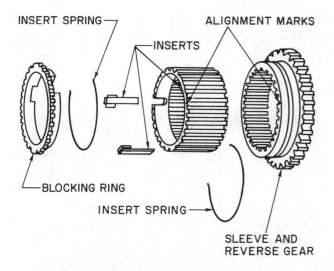

FIGURE 25–22. A disassembled synchronizer.

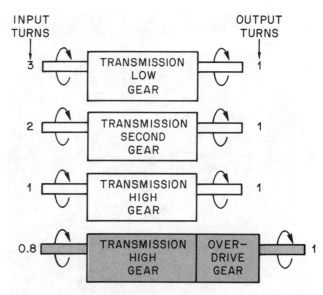

FIGURE 25–23. An overdrive gear has a ratio of approximately 0.8 to 1. This means the input shaft is going slower than the output shaft. This reduces engine speed and improves fuel mileage. However, torque at high speeds is reduced.

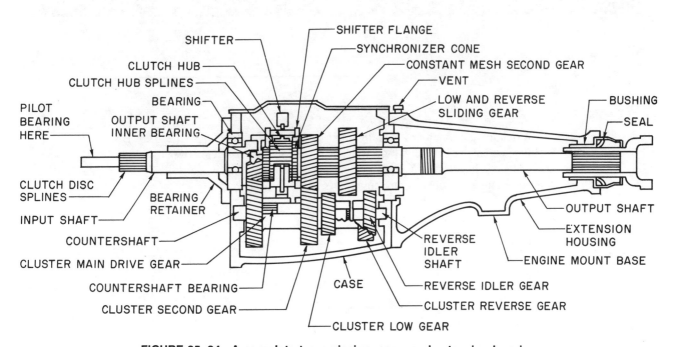

FIGURE 25–24. A complete transmission, case, and extension housing.

COMPUTER TREAD DESIGN

A drafting programmer uses a light pen to create a tire tread design as shown on the screen of this computer terminal. The special computerized instrument helps the company to create a variety of new tire designs rapidly. (Courtesy of The Firestone Tire and Rubber Company)

COMPUTER DESIGN

Tires can be tested through several computer models that tell how a specific tread will work on the automobile. The tread and profile of a tire can be tested to determine driveability, comfort of ride, wear characteristics, and tire life. (Courtesy of The Firestone Tire and Rubber Company)

25.3 TYPES OF TRANSMISSIONS

Purposes of Different Transmissions

Now that basic transmission components have been studied, other styles and types can be discussed. Vehicle designs have been changing over the past several years. Front wheel drive vehicles are now very popular. In addition, many manufacturers are using smaller engines. Because of these changes, several types of transmissions have been developed. These include four-speed, five-speed, and transaxle transmissions.

Four-speed Transmissions

The four-speed manual transmission has an additional gear set to produce four forward gears rather than three. With four gears, a wider range of torque characteristics is available. Increased torque range is usually needed for smaller engines. *Figure 25–25* shows an example of a four-speed transmission. The major difference is that the countershaft has four forward gears rather than three. Synchronizers are also used to perform the shifting.

Five-speed Transmissions

Five-speed transmissions are used for small engine displacements. With five gears, the range of torque is increased so that a smaller engine can be used. *Figure 25–26* shows an example of a five-speed transmission. In order to get five speeds, an additional set of gears are added to the transmission. Synchronizers are again used to engage and disengage the gears. In each gear shown, the transmission of mechanical energy is from the input shaft, to the countershaft, to the specific gear on the main or upper shaft, through the synchronizer, and to the output shaft. In this particular case, the fifth gear is considered an overdrive gear. Fourth gear is direct drive.

Transaxle System

The *transaxle* system is used on front wheel drive vehicles that have the transmission and final drive gearing placed together. It is mounted on the rear of the engine in such a way that the output of the transaxle feeds directly to the front wheels. There are many designs used for transaxles. The principles are much the same. *Figure 25–27* shows an example of a typical four-speed transaxle. The countershaft, called the input cluster gear, is attached directly to the clutch system. The gears are in constant mesh with synchronizers which are used to engage and disengage each gear. The output of the main shaft is used to drive the differential. The differential is discussed in a later chapter.

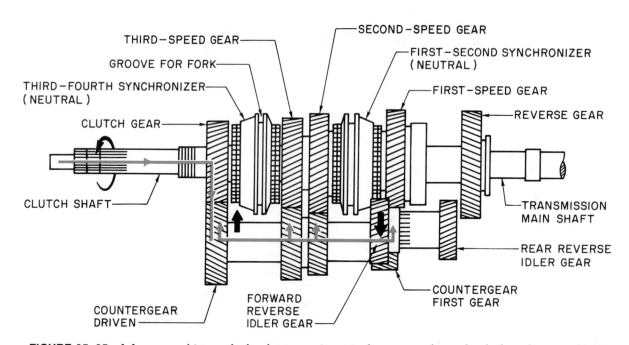

FIGURE 25–25. A four-speed transmission has an extra set of gears on the main shaft and countershaft.

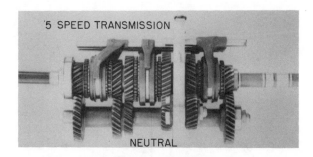

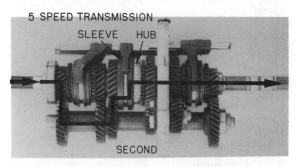

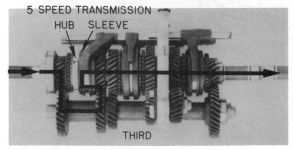

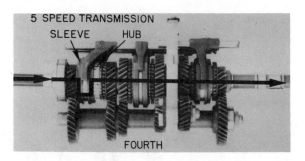

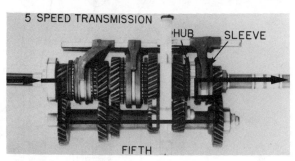

FIGURE 25–26. A five-speed transmission is shown with the power flow for each of the five gears. *(Courtesy of American Isuzu Motors, Inc.)*

CAR CLINIC

PROBLEM: CLUTCH CHATTER

An older vehicle with about 70,000 miles on it has a standard or manual transmission. When the clutch pedal is released, a noticeable chatter is felt in the vehicle. What could be the problem?

SOLUTION:

Clutch chatter is most often caused by oil getting on to the clutch and pressure plate surface. The friction of the clutch against the flywheel and pressure plate causes the oil to become very sticky. This causes the chatter when the clutch pedal is released and the clutch tries to engage. Most often the oil comes from a leaky rear main seal on the engine. Replace the rear main seal on the engine to eliminate the problem.

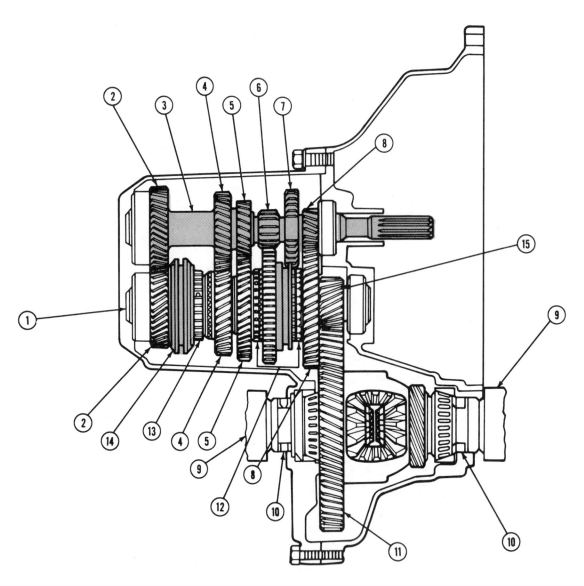

LEGEND:

1. MAINSHAFT
2. 4TH SPEED GEARS
3. INPUT CLUSTER
4. 3RD SPEED GEARS
5. 2ND SPEED GEARS
6. REVERSE GEAR
7. REVERSE IDLER GEAR
8. 1ST SPEED GEARS

9. HALF SHAFTS
10. DIFFERENTIAL OIL SEALS
11. FINAL DRIVE RING GEAR
12. 1ST/2ND SPEED SYNCHRONIZER
    BLOCKER RINGS
13. 3RD/4TH SPEED SYNCHRONIZER HUB
14. 3RD/4TH SPEED SYNCHRONIZER SLEEVE
15. PINION GEAR — PART OF MAINSHAFT

FIGURE 25–27. A four-speed transaxle has the differential attached to the manual transmission. Most front wheel drive vehicles have similar designs. (Courtesy of Ford Motor Company, Parts and Service Division)

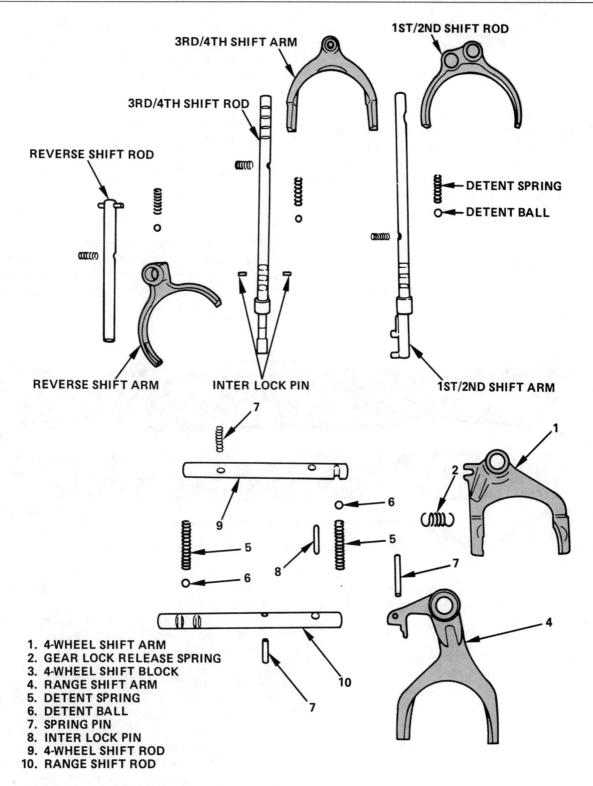

3RD/4TH SHIFT ARM

1ST/2ND SHIFT ROD

3RD/4TH SHIFT ROD

REVERSE SHIFT ROD

DETENT SPRING

DETENT BALL

REVERSE SHIFT ARM

INTER LOCK PIN

1ST/2ND SHIFT ARM

7

1

2

6

9

5

8

5

6

7

4

1. 4-WHEEL SHIFT ARM
2. GEAR LOCK RELEASE SPRING
3. 4-WHEEL SHIFT BLOCK
4. RANGE SHIFT ARM
5. DETENT SPRING
6. DETENT BALL
7. SPRING PIN
8. INTER LOCK PIN
9. 4-WHEEL SHIFT ROD
10. RANGE SHIFT ROD

10

7

FIGURE 25–28. Clutch forks are used to move the synchronizers. *(Courtesy of American Isuzu Motors, Inc.)*

25.4 LEVERS, LINKAGES, AND ACCESSORIES

Shifting Forks

Shifting forks are used to move the synchronizers back and forth to engage the gears. There are several types of shifting forks. *Figure 25–28* shows an example of a typical shifting fork. It is positioned in such a manner that it fits over the synchronizer assembly. As the operator moves the shift linkage, the fork moves the synchronizer for a specific gear. Detent springs and balls are used to hold the shifting fork in the correct gear position to ensure full engagement.

Shift Mechanisms

There are many designs of shifting mechanisms. The style and design depend upon the position of the shift lever, the type of vehicle style, the year of the vehicle, the manufacturer, the number of gears, and the type of transmission. Certain vehicles have the shift lever on the steering wheel column, while others have a floor shift system. Both mechanical linkage and cables can be used to transfer the shift lever movement to the shifting forks. Refer to the specific maintenance manual specifications for the type and correct adjustment of the shift mechanisms.

Other Accessories

The speedometer in most vehicles is driven from the output shaft on the manual transmission. A small drive gear is attached to the output shaft. The speedometer gear is then inserted into a hole in the transmission extension housing. The speedometer gear meshes with the gear attached to the transmission output shaft. On most vehicles, there is also an electrical hookup for the backup light switch and overdrive solenoid.

SAFETY TIPS

1. *Always use the proper tools when working on manual transmissions. Incorrect tools may damage the parts and may cause you to be injured.*
2. *Always wear safety glasses when working on manual transmissions.*
3. *When removing the transmission, always support the vehicle safely on the hoist or on the jack stands.*
4. *The transmission is very heavy. When removing manual transmissions, use a transmission jack stand to aid in this process.*
5. *When working under the vehicle on the transmission, make sure that no oil or antifreeze has been spilled. You may slip and cause serious injury to yourself.*
6. *Always disconnect the battery cables from the battery before working on any part of the manual transmission. This will eliminate any possibility of the engine accidentally being cranked during maintenance.*
7. *The manual transmission is often used to support the rear of the engine. Always support the bottom of the engine before removing the transmission.*

1. Gears that clash during shifting usually indicate a bad or defective synchronizer. To *remove and replace the synchronizer*, the transmission must be disassembled. Use the following general procedure to remove and disassemble the transmission.
   a. Using the proper safety precautions, position the car so that the transmission can be removed.
   b. Remove the drive shaft(s).
   c. Check to see if the rear of the engine is supported by the transmission. If it is, the engine must be supported while the transmission is being worked on.

d. Drain the gear oil from the transmission.

e. Disconnect the linkage used to shift the transmission.

f. Remove the speedometer cable from the transmission.

g. Remove any electrical wires attached to the transmission and mark for reassembly.

h. Support the transmission with a transmission jack.

i. Remove the bolts holding the transmission to the clutch housing. The transmission should now be easily removed and set on the workbench.

j. Remove the bolts holding the side covers on the transmission.

k. Remove the main drive gear. See **Figure 25–29**. Snap rings may have to be removed first.

l. Remove the countershaft. This may require removing the woodruff key from the rear of the case.

m. Remove all other gears inside the case.

n. Follow the manufacturer's recommended procedure to remove the synchronizer assembly for replacement.

2. A *clutch release bearing* can be damaged by constantly applying a slight pressure to the clutch pedal. This causes the clutch release bearing to spin constantly.

3. Use the following general procedure to *remove the clutch* assembly.

a. Remove the transmission gear case from the clutch housing.

b. Disconnect the clutch linkage to the clutch fork.

c. At this point, the clutch release bearing should be in such a position to remove it easily.

d. Remove the bolts holding the clutch housing to the engine block.

e. At this point, the clutch pressure plate assembly should be observable.

f. Using the correct wrench size, remove the clutch pressure plate. The clutch disc should now be loose and easily removed. Be careful not to breathe any of the asbestos particles from the clutch disc.

4. If the rear main oil seal leaks oil, the oil may get on the friction disc. This condition will cause a jerking motion when the clutch is engaged. To repair, the rear main seal on the engine must be replaced.

5. Always adjust the *clutch linkage* according to the manufacturer's specifications. If the clutch linkage is adjusted incorrectly, the clutch may not engage completely, causing it to slip or to be hard to shift. Each manufacturer has a different procedure for adjusting the shift linkage. A typical procedure is 1) place the transmission in neutral, 2) place a gage pin in the shift levers for alignment, 3) tighten the linkage to a certain specification.

6. Check the shifting linkage if there is a hard shifting pattern. The linkage may be damaged or bent. Repair and readjust the linkage if necessary.

7. If the transmission slips out of a gear, the shifting linkage may be adjusted incorrectly. Again, follow the manufacturer's recommended procedure for adjusting the shift linkage.

8. When removing the transmission, always use a transmission jack for proper support. The engine

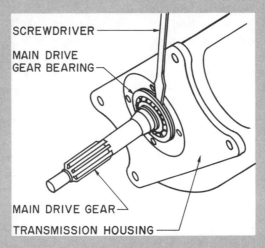

FIGURE 25–29. When disassembling the transmission, the main drive gear must be removed first.

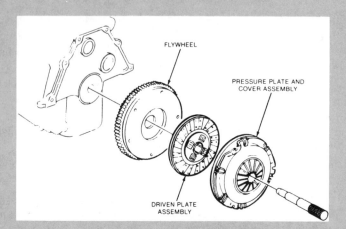

FIGURE 25-30. Before tightening the pressure plate, insert the correct alignment tool. This will help align the pressure plate, friction disc, and flywheel correctly so that the transmission can be inserted easily. *(Courtesy of Motor Publications, Auto Repair Manual, 1981-1987)*

may also have to be supported if the rear engine mount is on the transmission.

9. Before the transmission is replaced, the friction disc and pressure plate must be correctly aligned. Use the following general procedure to *align the disc* correctly.

    a. Place the friction disc in the correct position on the flywheel.

    b. Using the correct bolts, attach the pressure plate. However, do *not* tighten the bolts at this point. The clutch disc must still be loose enough to be moved slightly.

    c. Using the correct alignment tool (same as the transmission main shaft), align the pressure plate assembly and clutch disc to the flywheel. The alignment tool positions the clutch with the flywheel so that the transmission main shaft will be inserted easily into the clutch disc and flywheel. See *Figure 25-30*.

    d. Using the correct torque specifications, tighten the pressure plate bolts.

    e. Remove the alignment tool and insert the transmission. If it is aligned correctly, the transmission will slip into place easily.

10. Use the correct manufacturer's recommendation for the type and viscosity of gear oil when replacing the transmission fluid.

11. Check the transmission shift effort as outlined in the specific maintenance manual.

12. When troubleshooting the transmission, always refer to the specific manual for the "condition,"

"probable cause," and "correction." A sample diagnosis sheet is shown in *Figure 25-31*.

13. When the transmission is disassembled, check for wear on the gear teeth, clutch release bearing, pressure plate, synchronizers, and internal bearings. Replace, where necessary.

14. A small presence of metal in the transmission fluid may indicate gear wear. With the transmission disassembled, check for gear wear on all internal gears in the transmission.

15. Inspect the reverse gear bushing for wear, and replace if necessary.

16. Check the synchronizer sleeves to see that they slide freely on their hubs. Replace where necessary.

17. If the friction disc is worn too thin, it may score and scratch the pressure plate and flywheel surfaces, causing grooves to be cut into the pressure plate. If they are scored or scratched, the pressure plate and flywheel will have to be machined or replaced.

18. Check for excessive wear or bent diaphragm spring fingers when the clutch is disassembled. If the spring fingers are worn, the pressure plate must be replaced.

19. Always use the correct tools when working on manual transmissions. Common tools other than the standard wrenches include a snap ring pliers, gear pullers, drift punch, and hammer.

20. Torque all bolts to their correct specification as indicated by the manufacturer's maintenance manual.

DIAGNOSIS

TROUBLE DIAGNOSIS

Condition	Possible Cause	Correction
Hard Shifting	1. Clutch 2. Synchronizers worn or broken. 3. Shift shafts or forks worn.	1. Adjustment 2. Replace 3. Replace
Slips out of Gear	1. Shift shafts worn. 2. Bearings worn. 3. Drive gear retainer broken or loose. 4. Excessive play in synchronizers.	1. Replace 2. Replace as necessary. 3. Tighten or replace retainer. 4. Replace
Noisy in All Gears	1. Insufficient lubricant. 2. Worn countergear bearings. 3. Worn or damaged drive gear and countergear. 4. Damaged drive gear or main shaft 5. Worn or damaged countergear.	1. Fill to correct level. 2. Replace countergear bearings and shaft. 3. Replace worn or damaged gears. 4. Replace damaged bearings or drive gear. 5. Replace countergear.
Noisy in Neutral	1. Damaged drive gear bearing. 2. Damaged or loose pilot bearing. 3. Worn or damaged countergear. 4. Worn countergear bearings.	1. Replace damaged bearing. 2. Replace pilot bearing. 3. Replace countergear. 4. Replace countergear bearings and shaft.
Noisy in Reverse	1. Worn or damaged reverse idler gear or idler bushing. 2. Worn or damaged reverse gear. 3. Damaged or worn countergear.	1. Replace reverse idler gear assembly. 2. Replace reverse gear. 3. Replace countergear assembly.
Leaks lubricant	1. Excessive amount of lubricant in transmission. 2. Loose or broken drive gear bearing retainer. 3. Drive gear bearing retainer gasket damaged. 4. Center support gaskets either side. 5. Rear extension seal. 6. Speedo driven gear	1. Drain to correct level. 2. Tighten or replace retainer. 3. Replace gasket. 4. Replace gaskets. 5. Replace. 6. Replace O ring seal

FIGURE 25–31. When troubleshooting a manual transmission, always refer to the manufacturer's diagnosis information as shown here. *(Courtesy of American Isuzu Motors, Inc.)*

SUMMARY

The purpose of this chapter was to learn about the manual or standard transmission. This chapter studied the principles and operation of the clutch system, the basic transmission, including synchronizers and overdrives, and the levers and linkages used on the transmission.

The clutch is used to engage and disengage the transmission. It is composed of the friction disc, the pressure plate, the clutch linkage, and the clutch release bearing. In operation, when the operator pushes the clutch pedal down, the linkage and clutch release bearing push against springs so the pressure plate is brought away from the friction disc. Under these conditions, the engine is not connected to the transmission. When the operator releases the clutch, the mechanism causes the pressure plate to squeeze the friction disc between the flywheel and the pressure plate. Because the friction disc is squeezed between the two, it causes the disc to rotate. The output of the friction disc then drives the transmission.

On some vehicles, rather than using mechanical linkages to move the pressure plate, a slave cylinder is used. A hydraulic cylinder operated by the clutch pedal moves the clutch fork in and out to engage or disengage the friction disc.

Transmissions are used to produce different torque capabilities at the driving wheels. When the vehicle is starting, or going up a steep hill, high torque and low speed are needed. When the vehicle is moving fast along a straight highway, low torque and high speed are required. A transmission is designed to give these different torque requirements.

Torque can be changed by using different gears and gear ratios. Spur, helical, and internal gears are used in a transmission. Typically, when a smaller gear drives a bigger gear, torque is multiplied. First, second, third, fourth, or fifth gear is determined by moving different size gears in mesh with the input gear on the transmission. A three-speed transmission will typically have several gear ratios. First gear is normally 3 to 1. Second gear is 2 to 1, and third gear is 1 to 1. Reverse is also a very low gear ratio. When four- and five-speed transmissions are used, more gear ratios are available.

To eliminate the clashing and grinding of gears, transmissions today have synchronizers. Synchronizers are used to get both gears spinning at the same speed before meshing. In late model transmissions, all gears are in mesh. The synchronizers engage and disengage to produce the power flow through the transmission.

On some transmissions, an overdrive gear set is used. Overdrive reduces the speed of the engine somewhat to produce better fuel mileage. A 0.8 to 1 gear ratio is typically used when in the highest gear.

Transaxles are used on front wheel drive vehicles. When the engine is mounted in the front, a transaxle is used to transmit the output of the transmission to the front wheels for driving. The differential is usually included inside the case. The output of the differential is used to drive the front wheels.

All transmissions have some form of linkage from the foot pedal to the clutch release bearing. Most vehicles have mechanical linkages. Cable linkage is also used to move the shift forks that are placed around the synchronizers. Several other components are also placed on the transmission. These include the speedometer cable, backup light switch, and solenoids to control overdrive.

There are many diagnosis and service suggestions that are important for manual transmissions. Possible areas of service include the synchronizers, bearings, clutch linkage adjustment, and gears. Service may also be necessary on the clutch disc, pressure plate, and throwout bearing.

TERMS TO KNOW

Can you explain each of the following terms? Review the chapter until you can use each term correctly.

Clutch	Lever
Pressure plate	Fulcrum
Friction disc	Idler gear
Diaphragm spring	Synchronizer
Asbestos	Overdrive
Torsional vibration	Transaxle
Slave cylinder	

REVIEW QUESTIONS

Multiple Choice

1. The purpose of the clutch system is to:
   a. Start the engine
   b. Engage the crankshaft to the oil pump
   c. Engage the engine to the transmission
   d. Reduce the speed of the overdrive
   e. Increase the speed of the synchronizers

2. What device squeezes the friction disc against the flywheel?
   a. Torsional springs
   b. Clutch linkage
   c. Clutch release bearing
   d. Pressure plate
   e. Synchronizers

3. The clutch release bearing pushes against the
_____ to engage the friction disc.
   a. Pressure plate
   b. Pressure plate springs and linkage
   c. Flywheel
   d. Transmission case
   e. Torsional springs

4. _____ are placed on the friction disc to eliminate a jerky or chatter motion when the clutch is
engaged.
   a. Asbestos springs
   b. Torsional springs
   c. Splines
   d. Gears
   e. Pressure plates

5. Which device is/are used to push the clutch release
bearing in and out of engagement?
   a. Clutch disc
   b. Torsional springs
   c. Clutch fork
   d. Synchronizers
   e. Pressure plate

6. The purpose of the standard or manual transmission
is to:
   a. Increase torque output at the right time
   b. Change the speed of the output shaft
   c. Decrease the input shaft at high speeds in overdrive
   d. All of the above
   e. None of the above

7. On a set of gears, if the input gear has 12 teeth and
the output gear has 24 teeth, what is the gear ratio?
   a. 3 to 1
   b. 2 to 1
   c. 1 to 1
   d. 1 to 3
   e. 1 to 0.2

8. If a smaller gear is used to drive a larger gear, the:
   a. Torque output will be decreased
   b. Torque output will be increased
   c. Speed output will be increased
   d. Input speed will be decreased
   e. None of the above

9. When reverse gear is selected in a transmission:
   a. Two extra gears are added to the gear train
   b. Three extra gears are added to the gear train
   c. One extra gear is added to the gear train
   d. The overdrive gear is always used
   e. The clutch is always disengaged

10. What is the gear ratio for a three-speed transmission
in first gear?
   a. About 1 to 4
   b. About 1 to 5
   c. About 1 to 3
   d. About 3 to 1
   e. About 2 to 1

11. In high or third gear on a three-speed transmission,
the countershaft:
   a. Is driving all gears
   b. Is freewheeling and not driving any gears
   c. Is driving the second and first gears
   d. All of the above
   e. None of the above

12. Which type of gear has the teeth at an angle to the
gear axis, rather than parallel to the gear axis?
   a. Spur gear
   b. Helical gear
   c. Spur gear that is internal
   d. Synchronizer gear
   e. Overdrive gear

13. In current transmissions that use synchronizers, the
countershaft gears in the transmission:
   a. Are not in mesh during shifting
   b. Are always in mesh during shifting
   c. Are not spinning
   d. Must first engage the reverse gear
   e. Mesh only after the synchronizer engages

14. The center of a synchronizer assembly that slides on
the main shaft is called the:
   a. Cone
   b. Synchronizer ring
   c. Synchronizer sleeve
   d. Hub
   e. Key

15. What type of gear actually engages inside the synchronizer assembly?
   a. External clutch gear
   b. External spur gear
   c. Internal gear
   d. All of the above
   e. None of the above

16. The shift forks on a synchronized transmission control or move the:
   a. Hub
   b. Synchronizer sleeve
   c. Synchronizer keys
   d. Synchronizer cones
   e. Splines

17. What would be a common gear ratio in highest gear with a transmission using an overdrive system?
    a. 1 to 1
    b. 1 to 1.8
    c. 1 to 2
    d. 2 to 1
    e. 0.8 to 1

18. Which type of transmission would be used on a front wheel drive vehicle?
    a. 2 speed
    b. 6 speed
    c. Transaxle
    d. All of the above
    e. None of the above

19. The shift lever for the operator of the vehicle:
    a. Moves the shifting forks
    b. Has several types of linkages used to shift the transmission
    c. Can be on the column of the steering or on the floor
    d. All of the above
    e. None of the above

20. Which component is not attached to the transmission?
    a. Speedometer cable drive
    b. Backup lights
    c. Oil seals
    d. Oil pan
    e. Extension housing

21. Gear clashing during shifting may be caused by:
    a. A bad speedometer cable
    b. A misaligned clutch
    c. Damaged or worn synchronizers
    d. A worn clutch release bearing
    e. A leaky rear seal

22. If the clutch chatters or causes a jerky motion when engaged, the problem could be caused by:
    a. A leaky rear transmission seal
    b. Bad synchronizers
    c. Oil leaking from the engine to the friction disc
    d. A broken extension housing
    e. A bad clutch release bearing

The following questions are similar in format to ASE (Automotive Service Excellence) test questions.

23. Technician A says that if oil leaks on the clutch disc, the clutch will slip during engagement. Technician B says that if oil leaks on the clutch disc, the clutch will chatter during engagement. Who is right?
    a. A only
    b. B only
    c. Both A and B
    d. Neither A nor B

24. Technician A says that gear clashing is caused by misalignment of the shift linkage. Technician B says that gear clashing is caused by a bad synchronizer. Who is right?
    a. A only
    b. B only
    c. Both A and B
    d. Neither A nor B

25. During assembly of the transmission to the clutch assembly, the transmission cannot be inserted into the clutch assembly. Technician A says that the transmission main shaft is damaged. Technician B says that the clutch disc is not properly aligned. Who is right?
    a. A only
    b. B only
    c. Both A and B
    d. Neither A nor B

26. A small amount of ground metal filings are found in a transmission. Technician A says that they are clutch particles, not filings. Technician B says that these filings are metal and indicate possible gear wear. Who is right?
    a. A only
    b. B only
    c. Both A and B
    d. Neither A nor B

Essay

27. What is the purpose of a pressure plate?

28. Define torsional vibration.

29. What is the purpose of the clutch release bearing?

30. Describe the gear ratios for first, second, third, and fourth gears.

31. Describe the purpose and basic operation of a synchronizer.

32. Why are overdrive systems used on transmissions?

33. What is a transaxle?

DIAGNOSTIC CENTERS

Advanced diagnostic service centers are able to check more than 200 elements of engine systems and electrical performance. The diagnostic center is then able to print out the type of service needed. (Courtesy of The Firestone Tire and Rubber Company)

CHAPTER 26

Automatic Transmissions

INTRODUCTION

The automatic transmission is designed to shift gears automatically from low to high without the driver's control. Several major components are used in the automatic transmission. These include the torque converter, planetary gear system, hydraulic components, different types of clutches and bands, and various controlling parts. This chapter explains how these components are used in the automatic transmission.

OBJECTIVES

After reading this chapter, you will be able to:
- Identify the purpose and operation of the torque converter.
- Explain the purpose and operation of the planetary gear system used on automatic transmissions.
- Analyze the different types of clutches and bands used on automatic transmissions.
- State the purpose and basic operation of the hydraulic systems used on automatic transmissions.
- Define the purpose and operation of various control devices used on automatic transmissions.
- Identify common diagnosis and service suggestions for automatic transmissions.

CHAPTER HIGHLIGHTS

26.1 BASIC DESIGN AND REQUIREMENTS

Purpose of the Automatic Transmission

The purpose of the automatic transmission is to connect the rotational forces of the engine to the drive wheels and to provide correct torque multiplication. This must be done at varying loads, speeds, and driving conditions. For example, to get the car started, low speed and high torque are required. As the speed increases, the torque requirements decrease. The automatic transmission is designed to produce the correct torque needed for these varying driving conditions.

Major Parts of the Automatic Transmission

The torque converter is used for the same reason that a clutch is used in a standard transmission. The torque converter connects the engine to the transmission gearing. At times, the engine and gears must be directly connected. At other times (at a stop light), the two must be disconnected. In addition, the torque converter multiplies the torque sent to the transmission for varying loads.

The planetary gear system is used to produce the correct gear ratio for different torque and speed conditions. Low speeds, drive, and reverse gear ratios can be achieved by using the planetary gear system.

Various hydraulic controls are used to lock up parts of the gear system in an automatic transmission. Clutches, bands, and other components help to control which gears are used in the planetary gear system.

The automatic transmission uses several hydraulic control valves to operate the clutches and bands for correct operation. Complex hydraulic circuits have been incorporated into the automatic transmission to accomplish this goal.

26.2 TORQUE CONVERTER

In a standard transmission, a clutch is used to engage and disengage the engine from the transmission. This is called a friction drive. In the automatic transmission, a *fluid coupling* is used to engage and disengage the engine from the transmission. This coupling is called the *torque converter*. It uses a hydraulic or fluid coupling.

Fluid Coupling

A common way of describing a fluid coupling is by using two electric fans. Referring to *Figure 26–1*, one fan produces a pressure and blows the air against the other fan. The air pressure produces enough energy to rotate the second fan. This action couples the input and output. The first fan is called the pump and the second fan is called the *turbine*. The faster the pump turns, the better the fluid lockup between the input and output. In the actual torque converter, a pump and turbine are used, and transmission oil, rather than air, is used.

Parts of the Torque Converter

Several parts make up a torque converter to produce the fluid or, in this case, hydraulic coupling. *Figure 26–2* shows the major parts of the torque converter. These include the pump (also called impeller), the turbine, and the guide wheel or *stator*. Note that each part is made of a series of vanes to direct oil through the torque converter. These parts are contained in a sealed housing which is completely filled with oil. Motion and power are transferred by the pressure of the mass of the flowing oil. There is no direct mechanical contact between the input and the output drives, only a fluid connection. Because of this type of connection, torque converters operate essentially wear-free.

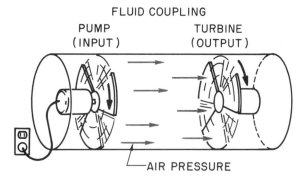

FLUID COUPLING

FIGURE 26–1. A torque converter is similar to the operation of two fans to produce a fluid coupling. The pressure from the pump fan causes the blades on the second fan or turbine to turn. The input and output are now connected.

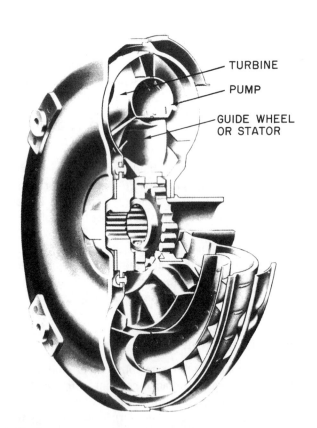

FIGURE 26–2. The major parts of the torque converter are the pump, turbine, and guide wheel or stator. *(Courtesy of Sachs Industries, Inc.)*

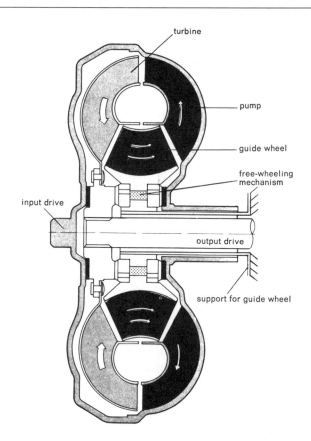

FIGURE 26–3. As the pump rotates, oil is sent to the turbine, through the stator or guide wheel, and back to the pump. This action produces a hydraulic coupling. *(Courtesy of Sachs Industries, Inc.)*

Operation of the Torque Converter

Figure 26–3 shows the operation of the torque converter. The entire torque converter is bolted to the engine crankshaft through a flexplate. See *Figure 26–4*. As the engine turns, the entire outside housing of the torque converter turns. The pump vanes are attached directly to the inside of the housing of the torque converter. This causes the pump vanes to turn. As the pump rotates, the oil in the pump vanes is forced to the outer *periphery* by centrifugal force. The oil now reaches the turbine vanes with high velocity (just like the air hitting the fan as shown in Figure 26–1). The high-velocity oil causes the turbine vanes to rotate and lock up hydraulically with the pump.

Within the turbine vanes, the oil flow is changed to mechanical rotation by the sharply curved vanes of the turbine. The turbine is connected to the output drive of the torque converter. The stator or guide wheel is used to direct the oil flow back to the pump in the most efficient manner. In effect, the oil flows in a circular fashion throughout the torque converter as shown in *Figure 26–5* (much like a spring bent in a circle).

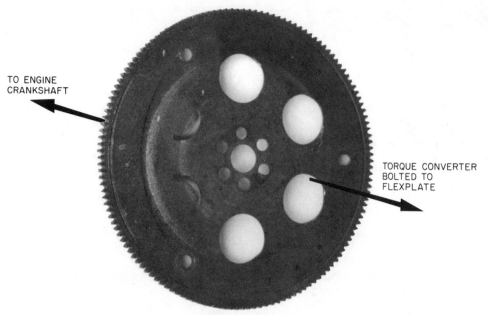

FIGURE 26-4. The flexplate is used to connect the engine crankshaft to the torque converter.

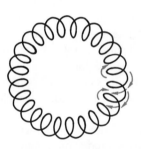

OIL FLOWS INSIDE THE TORQUE CONVERTER FROM THE PUMP, TO THE TURBINE, TO THE STATOR AND BACK TO THE PUMP.

FIGURE 26-5. Oil flows in a circular fashion throughout the torque converter.

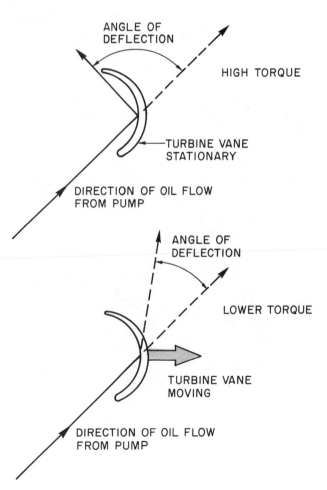

FIGURE 26-6. Torque is multiplied inside the torque converter. The greater the angle of deflection of oil, the greater the torque. As the speed of the turbine increases, the torque decreases.

Torque Multiplication

The torque converter is capable of producing different torque ratios. When starting, the torque converter multiplies torque. The torque may be increased 2–3.5 times the engine torque. As the speed of the turbine blades increases, the torque is continuously reduced until the torque ratio is 1 to 1. At this point, the pump and turbine are spinning at the same speed.

Torque on the turbine wheel is a direct result of how the oil is deflected from the blades. Referring to *Figure 26–6*, when oil hits the turbine vane, it is forced to deflect. It has a greater *deflection angle* when the turbine vane is station-

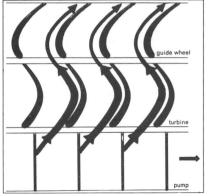
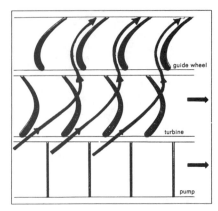
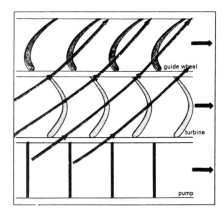

(A) HIGH TORQUE;
GREAT DEFLECTION ANGLE;
LOW OUTPUT SPEED.

(B) MEDIUM TORQUE;
MEDIUM DEFLECTION ANGLE;
MEDIUM OUTPUT SPEED.

(C) LOW TORQUE
LOW DEFLECTION ANGLE;
HIGH OUTPUT SPEED.

FIGURE 26–7. Figure (A) shows high torque multiplication. Figure (B) shows medium torque multiplication. Figure (C) shows low torque multiplication. *(Courtesy of Sachs Industries, Inc.)*

ary. The greater the deflection angle, the greater the torque on the turbine wheel. When the turbine wheel is stopped, there is a maximum deflection angle. As the turbine blade increases in speed, the oil is not deflected as much. This is because the turbine blade is now moving to the right. Since the turbine blade is also moving, there is less deflection. Less deflection means less torque multiplication.

Figure 26–7 shows the oil flow through all three components inside the torque converter. Figure 26–7 (A) shows the pump rotating with the turbine and guide wheel (stator) stopped. Notice that there is maximum angle of deflection of oil, causing greater torque multiplication.

The oil flows from the torque converter to the stator. Here the oil flow direction wants the stator to turn to the left, but the stator cannot turn that way. The stator is designed so that it can only rotate to the right. It is supported by a one-way, freewheeling sprag or roller clutch drive system. See *Figure 26–8*. The sprag clutch drive system allows rotation in one direction but locks up in the opposite direction. The oil then is redirected to the correct angle to reenter the pump.

As the speed of the turbine increases, the angle of deflection decreases, producing less torque multiplication. This is shown in Figure 26–7 (B). The stator is still not mov-

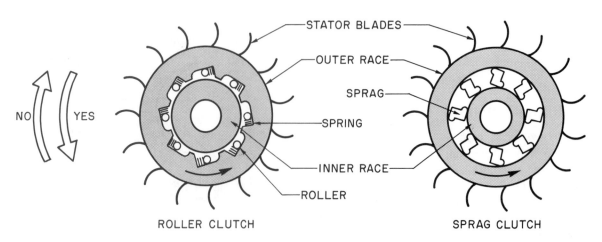

FIGURE 26–8. The stator must turn in one direction only. This is done by using either a roller or sprag clutch drive system. When the inner hub is held stationary, the outer stator vanes can only rotate in a counterclockwise direction. When forced in the opposite direction, the stator vanes cannot turn.

ing because there is still pressure to move it to the left. Notice that the pattern of oil flow is becoming straighter.

Figure 26–7 (C) shows the pump and turbine at approximately the same speed. During this condition, the oil flow has no deflection. The speed of the oil and the turbine blade, moving to the right, are the same. There is no torque multiplication at this point. However, the oil flow is now hitting the back of the stator or guide wheel blades. If left this way, the oil will not flow back to the pump correctly. To eliminate this, the stator starts to turn to the right. The freewheeling action of the stator is in effect.

Torque Converter Lockup

Torque converters never operate without any loss. There is usually a certain slip between the turbine speed and the pump speed. The pump speed may be 2–8% faster than the turbine speed. In most transmissions, this loss is not accounted for, but it does reduce gasoline mileage. In some newer transmissions, there is a lockup system. This system locks the pump and the turbine only at highway or cruising speed. It is unlocked again when high torque multiplication is required. A disk-type "piston" with a friction material locks and unlocks the pump and turbine. Oil pressure to one side of the piston forces the two together.

Torque Converter Location

On rear wheel drive vehicles, the torque converter is located directly behind the engine. The transmission is located directly behind the torque converter. The torque converter and transmission are one combined unit.

On certain front wheel drive cars, a different arrangement is used. The torque converter is located along side the transmission. A large heavy chain is used to connect the torque converter to the transmission. See *Figure 26–9*.

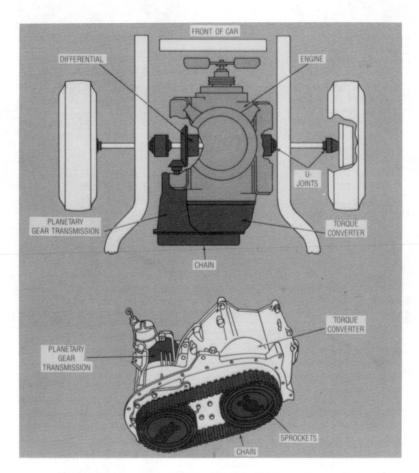

FIGURE 26–9. On front wheel drive vehicles, a heavy chain is used on some transmissions to connect the torque converter to the transmission case. (Courtesy of General Motors Corporation)

ELECTRONIC ENGINE ANALYZER

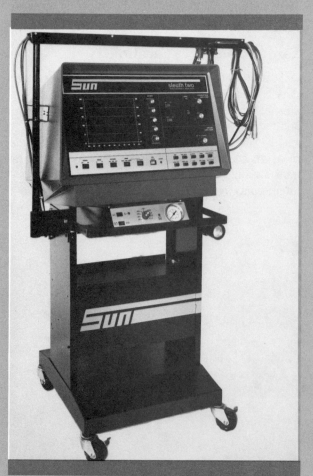

With the increased technology used on engines today, electronic engine analyzers are used to help the technician diagnose and repair the engine. These analyzers are able to check emissions, timing, ignition components, and computer systems used on vehicles today. (Courtesy of Sun Electric Corporation)

26.3 PLANETARY GEAR SYSTEM

Purpose of the Planetary Gear System

The *planetary gear system* is used to produce different gear ratios within the automatic transmission. The output of the torque converter (turbine) is considered the input to the planetary set of gears. Power flow comes from the torque converter, through a hydraulic clutch, to the planetary gear system. Based upon the operation of the planetary gear system, either low speeds, drive, or reverse can be obtained. In addition, certain automatic transmissions may have D1, D2, D3, and overdrive. Planetary gear systems are designed to produce various gear ratios. More than one planetary gear system can also be used on transmissions to produce more gear ratios.

Parts of the Basic Planetary Gear System

All planetary gear systems have several major parts. They include the:

    a. sun gear
    b. ring gear
    c. planet gears (also called pinion gears)
    d. planet gear carrier

Figure 26–10 shows an example of a simple planetary gear system. The sun gear is the center gear. The planet (pinion) gears are held in position by the *planet carrier*. The ring gear is an internal gear that surrounds all of the planet gears. All gears are helical in design and are in constant mesh.

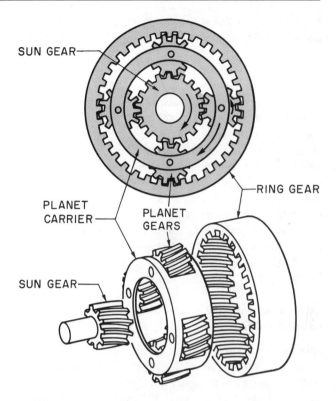

FIGURE 26–10. A planetary gear system is made of a sun gear, ring gear, several planet gears, and a planet carrier.

Operation of the Planetary Gear System

To select different gear ratios, one of three gears, the sun, ring, or planets, is held stationary. The input and output then occurs on the remaining gears. For example, if the ring gear is held stationary, and the input is the sun gear, the carrier is the output with a lower speed. The most common

GEAR COMBINATIONS								
GEAR	1	2	3	4	5	6	7	8
RING	Output	Hold	Input	Hold	Input	Output	HOLD ANY TWO GEARS	FREE ALL GEARS
CARRIER	Input	Input	Output	Output	Hold	Hold		
SUN	Hold	Output	Hold	Input	Output	Input		
SPEED CHANGE (Input to Output)	Increase	Increase	Lower	Lower	Increase Reverse	Lower Reverse	Direct Drive	Neutral

FIGURE 26–11. There are eight combinations of gear ratios that can be produced in a planetary gear system. By holding one gear stationary, and changing the input and output, various gear ratios can be achieved.

arrangement for low gear is to have the sun gear held stationary, the input on the ring gear, and the output on the carrier.

Eight combinations can be produced in a planetary gear system. These are shown in **Figure 26–11**. In addition, direct drive occurs when any two gears of the planetary gear system are turned at the same speed. The third gear must then rotate at the same speed. In this condition, the planet gears do not rotate on their shafts. The entire unit is locked together to form one rotating part. Neutral occurs when all members of the planetary gear system are free. When the transmission is placed in park, a small pawl is engaged with the teeth on the planet carrier.

Overdrive conditions can also be obtained by using planetary gears. Overdrive can be used when the sun gear or the ring gear is held stationary. Input would be as shown in Figure 26–11, condition 1 or 2.

Variations in the Planetary Gear System

In order to meet the many gear ratio demands of vehicles today, there are several variations to the basic planetary gear system. One type of planetary gear system is called the simple type. It has two sets of planet gears. Another type of planetary gear system uses two sets of planet gears, two sun gears, and a planet carrier. **Figure 26–12** shows an example of this system. This is called the compound planetary gear system. On certain transmissions, two complete sets of planetary gears are used.

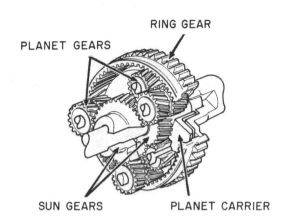

FIGURE 26–12. A compound planetary gear system has two sets of planet gears and two sun gears. *(Courtesy of Chrysler Corporation)*

26.4 CLUTCHES, BANDS, AND SERVO PISTONS

Purpose of Clutches and Bands

The planetary gear system provides the gear ratios, but the clutches and bands control which gears are held or released. The clutches and bands are used to lock up the correct gear to get the right gear ratio. Depending upon which clutch or band is activated, one member of the planetary gear system is held, while the other is driven.

Multiple-Disc Clutch

One common clutch used in automatic transmissions is called the *multiple-disc clutch*. It is made of a series of friction discs placed between steel discs or plates. The exact

---

CAR CLINIC ▆▆▆▆▆▆

PROBLEM: TOWING PROBLEM

A person would like to use a vehicle with a 6-cylinder engine to tow a trailer and boat. What problems might occur using the standard radiator and transmission?

SOLUTION:

The most common problem when towing a boat or trailer is engine overheating. The engine must do more work to pull a boat or trailer. This means the engine will run hotter and the transmission will be under more stress. The transmission is cooled by the engine coolant. If the engine is running hotter, the transmission fluid is also hotter. Damage to the transmission could result.

The best protection is to have the manufacturer install a towing package. The most common package includes placing an extra transmission cooler on the front of the radiator. Some manufacturers also recommend changing the rear end differential to a different ratio. The cost is much higher, but the engine will not be loaded as much.

FIGURE 26–13. Multiple-disc clutches use both a friction disc and a steel disc. When pressure is applied, the two lock up and transmit the necessary torque.

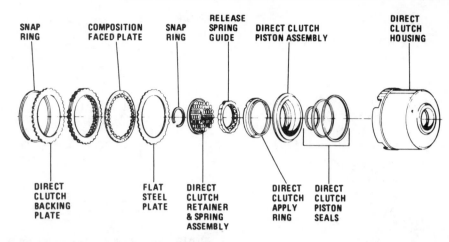

FIGURE 26–14. An exploded view of the multiple-disc clutch with the piston and other operational parts. *(Courtesy of Motor Publications, Automatic Transmission Manual)*

number of discs depends upon the vehicle manufacturer. See *Figure 26–13*. The friction discs or composition-faced plates have rough gripping surfaces. The steel discs have smooth metal surfaces. These two components make up the input and output of the clutch. The clutch pack also has a piston and return springs.

When fluid pressure is applied to the clutch, the piston moves and compresses the clutch pack together. The action locks up the input and output of the clutch. When the pressure is released, the springs help to remove the pressure on the discs. This action unlocks the clutch. *Figure 26–14* shows an exploded view of the multiple-disc clutch. This type of clutch could be used to connect two shafts together. For example, the torque converter output must be connected and disconnected to the planetary gear system. This is called the forward clutch and it uses a multiple-disc clutch.

Transmission Band

Another type of clutch is called the *transmission band*. It is made of a flexible piece of steel wrapped around a clutch housing or drum. The inside of the band has a friction surface to help grip the clutch housing. The band is tightened or loosened to hold or free the clutch housing or drum. If the clutch housing is attached to the sun gear, control of the sun gear is made by this transmission band. *Figure 26–15* shows an example of a band type of clutch used on a transmission.

Servo Piston

The *servo* piston is used to control the transmission band operation. In *Figure 26–16*, the band is controlled by a servo piston. The servo piston is made of the case, piston, stem, spring, and cover. When oil pressure from the

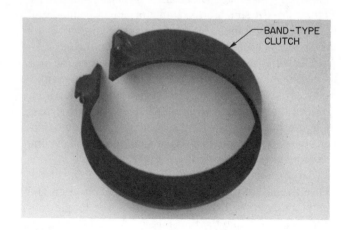

FIGURE 26–15. A transmission band is used to lock up a clutch housing.

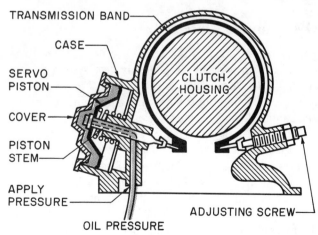

FIGURE 26–16. A servo piston is used to operate the transmission band.

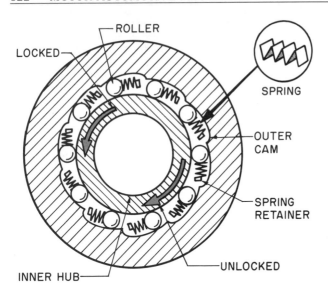

FIGURE 26-17. The overrunning clutch assembly is used to provide a smooth shift from low to drive.

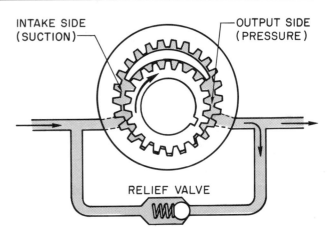

FIGURE 26-18. A gear-type, eccentric pump is used to produce the needed oil pressure in the automatic transmission.

hydraulic system is applied, it pushes the servo piston to the right. This action causes the stem to move and tighten the band. When oil pressure is removed, the spring pushes the servo piston back and releases the band. The band clearance can be changed by turning the adjusting screw.

Accumulator

An *accumulator* is a device that cushions the motion of the clutch and servo actions. This is typically done by using a smaller piston inside a servo piston. The smaller piston makes contact with the linkage quicker, thus acting as a cushion.

Overrunning Clutch

The overrunning clutch is used to prevent backward rotation of certain parts of the planetary gear system during shifting. The overrunning clutch shown in *Figure 26-17* is made of the inner hub, outer cam, and a series of rollers and springs. It operates the same as a sprag drive, allowing rotation only in one direction.

It is difficult to get a smooth upshift from low to drive on most transmissions. If one band releases slightly before the second band engages, the engine may have a rapid increase in rpm. The overrunning clutch provides smooth engagement and disengagement without any delay. This improves the shift quality and timing from low to drive gear.

26.5 HYDRAULIC SYSTEM

Purpose of the Hydraulic System

The automatic transmission is controlled by the hydraulic system. When the operator shifts into drive, low, reverse, or any other gear, hydraulic pressure is used to lock up different clutches and bands on the planetary gear system. The torque converter also uses transmission fluid.

Oil Pump

All pressurized oil in the automatic transmission is produced by the transmission oil pump. It creates the pressure used to lock up the clutches or operate the servo to lock up the transmission bands. It also sends oil to the torque converter.

Several types of oil pumps are used on automatic transmissions. These may include the gear, vane type, or rotor type. These are all positive displacement pumps. *Figure 26-18* shows an example of how a common oil pump operates. It uses a gear-type, eccentric pump. As the inside gear is turned, a suction is created on the left side. Automatic transmission fluid is drawn in between the gear teeth. The oil is then carried to the other side of the pump. Pressure is produced here by forcing the oil out from between the gear teeth. The relief valve is used to bypass high-pressure oil back to the suction side. This only happens when higher rpm is producing oil pressure above the recommended psi.

The oil pump usually is built into the front of the transmission case, directly behind the torque converter. The pump is driven by the torque converter. The torque converter is inserted into the transmission body. Two *tangs* or

FIGURE 26–19. The oil pump on most transmissions is located directly in front of the transmission and is driven by the torque converter.

feet on the torque converter index with the center gear of the oil pump. *Figure 26–19* shows an example of the oil pump and body near the front of the transmission.

Automatic Transmission Fluid

Within the transmission, there are high shearing and pressure conditions. The automatic transmission fluid must be designed to withstand extreme pressure, friction, and shearing stresses. In addition, both cold and hot conditions, which might change viscosity, may exist. To do this, the automatic transmission fluid has the following characteristics:

a. corrosion inhibitors
b. detergents
c. pour-point depressants
d. friction modifiers
e. antifoam agents
f. viscosity-index improvers

Because of these special characteristics and requirements, only the recommended fluid should be used. The fluid has a reddish color. It is identified as *ATF* (automatic transmission fluid) and is referred to as DEXRON II, Type F or other designation. Follow the manufacturer's recommendation when selecting the proper type of ATF.

Valve Body

The oil from the oil pump is sent to the valve body. The valve body is usually located on the underside of the

FIGURE 26–20. The valve body is used to house the transmission controls. It is also used to direct the oil to the proper part for correct operation.

transmission case and is covered by the transmission oil pan. The oil is sent through many passageways inside the valve body to the various parts of the transmission. The oil is then used to lock up or release the clutches. There are a series of valves, controls, and springs inside the valve body that are used to control the shifting of the automatic transmission. *Figure 26–20* shows an example of a typical valve body.

Oil Cooler

A great deal of friction is produced on most transmissions. Friction tends to increase the temperature of the transmission fluid. If the temperature of the transmission fluid is allowed to get too high, the lubricating properties of the ATF may deteriorate. This may cause the transmission to fail. To overcome this potential problem, the transmission oil is cooled by an oil cooler. Oil is typically sent out of the transmission to a heat exchanger placed inside the radiator. Transmission fluid operates about 40–50 degrees Fahrenheit hotter than engine coolant. Most heat is produced by the friction in the clutch and the torque converter. Engine coolant absorbs thermal energy from the transmission to cool the fluid. Vehicles that have excess loads placed on them may also have an additional external cooler on the front of the radiator.

CAR CLINIC

PROBLEM: THROTTLE STICKING

A customer complains that while passing a vehicle on a highway, the throttle was pushed completely to the floor. The vehicle then continued to accelerate even with the foot removed from the pedal. The operator shut off the ignition and coasted to a stop. What could be the problem?

SOLUTION:

This problem sounds like the throttle linkage between the foot pedal and the carburetor stuck in the open position. Most likely causes include possible interference of the linkage, a kinked or bent throttle cable, or a sticky throttle plate in the carburetor. With the engine shut off, push the accelerator completely to the floor and try to observe where the linkage is sticking. If the sticky throttle still occurs after disconnecting the throttle from the carburetor, the problem is in the throttle linkage, not in the carburetor.

26.6 TRANSMISSION CONTROLS

Purpose of Transmission Controls

Several controls are used to lock up or release the clutches and transmission bands to control the shifting. The type of valves used are determined by the year and manufacturer of the transmission. Pressure regulator valves, kickdown valves, shift valves, governors, throttle valves, vacuum modulators, and manual valves are used. A working knowledge of these valves will help you to visualize the internal operation of different transmissions. Certain valves are also used to increase the smoothness and quality of shifting.

Pressure Regulator Valve

The *pressure regulator valve* is used to regulate the correct amount of oil pressure. Various oil pressures are required within the transmission. Certain pressures are needed to control locking of clutches, and other valves. The pressure regulator valve uses a *spool valve* to control these pressures. The pressure from the transmission oil pump is sent to the regulator valve so that correct pressures can be maintained throughout the transmission.

Manual Shift Valve

Manual shifting is done by placing the vehicle shift lever in the proper position inside the car. Depending upon the vehicle, park, reverse, neutral, drive, and low are common positions to place the shift lever. On many cars, there is also a D1 and D2. When the operator shifts this lever, a set of shift linkages causes a *manual shift valve* to move inside the valve body. *Figure 26–21* shows an example of the valve that is moved. It is very similar to a spool valve.

Throttle Valve

Under heavy acceleration, it is necessary to increase the force on the transmission bands and clutches to reduce slippage. This can be done by increasing the oil pressure to the servo that controls the transmission band. The *throttle valve* (TV) is used to increase pressure. As the throttle is increased, the throttle linkage is used to change the position of the valve.

Different positions produce different pressures. This valve is used in conjunction with the shift valve.

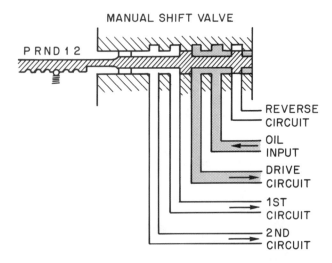

FIGURE 26–21. When the gear shift lever is moved, a manual shift valve is operated inside the transmission valve body. The movement of this valve causes different hydraulic circuits to be put into operation.

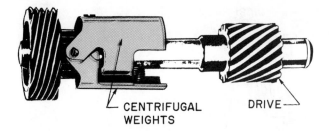

FIGURE 26–22. The governor is used to sense road/engine speed and shift the transmission accordingly. *(Courtesy of Oldsmobile Division, General Motors Corporation)*

Governor

The governor is used to help shift the transmission based upon vehicle speed. For example, when the vehicle is in low or first gear, it must automatically shift to second or drive at a certain speed. Also, as the vehicle slows to a stop, the transmission must downshift. This is done by using a governor assembly as shown in *Figure 26–22*. The speed of the vehicle is sensed by the governor from the output shaft on the transmission. A set of small weights moves out and in from centrifugal force. Centrifugal force is working against a spring force. These two forces acting against each other cause the valve to open and close a set of hydraulic ports. The output oil pressure controlled by the governor is sent to the *shift valve* to make the necessary shift.

Shift Valve

The throttle valve and the governor do not actually send oil to the clutches. The oil is first sent to the shift valve. Here, governor oil pressure works against throttle valve oil pressure to position the shift valve. *Figure 26–23* shows a simplified example of this. Governor oil pressure enters at the bottom. Throttle valve pressure enters at the top. These two pressures balance the valve. Any time this balance is disturbed, the shift valve moves upward or downward. Line pressure (pressure used to control the clutches) is also directed to the center of the shift valve. This pressure does not enter into shift valve movement. This pressure is simply ready to be directed to the correct clutch in the planetary gear system to lock up or release the correct clutch. Note that the shift valve is actually more complicated than this and has several outputs to different clutches.

Vacuum Modulator

Another way in which engine load can be determined to help shifting is by using a *vacuum modulator* valve instead of the throttle valve. Both valves are still used in vehicles today.

As the throttle opening is increased, the intake manifold vacuum decreases. The intake manifold vacuum is sent to the transmission and is hooked to a vacuum diaphragm. Spring pressure acting against the vacuum causes a rod to be moved. The rod controls the position of the throttle valve. *Figure 26–24* shows an example of the vacuum modulator.

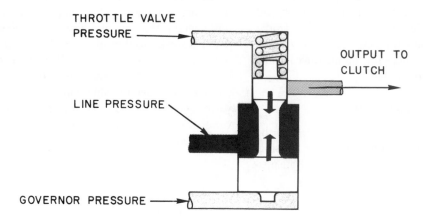

SIMPLIFIED SHIFT VALVE

FIGURE 26–23. A shift valve is controlled by the throttle valve oil pressure working against the governor oil pressure. Its position tells exactly when and to which clutch to send the oil pressure. *(Courtesy of Chrysler Corporation)*

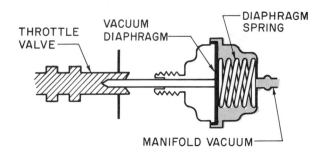

FIGURE 26–24. The vacuum modulator is used to sense engine load based on intake manifold vacuum.

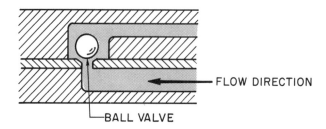

FIGURE 26–25. Ball check valves are used to control the direction of oil flow through the passageways.

Kickdown Valve

A kickdown valve is used on some transmissions. This valve, which is also called a detent valve, is part of the throttle valve assembly. It operates by using a variable-shaped cam that controls the position of the throttle valve. Under full load, or when the throttle is pushed all the way to the floor, the kickdown valve positions the throttle valve in such a position as to shift to a lower gear.

Converter Clutch Control Valve

As indicated earlier, certain torque converters lock up the turbine wheel directly with the torque converter housing. A converter clutch control valve is used on many transmissions to send oil pressure to the lock-up assembly.

Ball Check Valves

Ball check valves are used to control the flow of oil in the transmission. Check valves prevent flow until a certain pressure is reached. They also close passages to prevent back flow of transmission fluid. *Figure 26–25* shows the action of a ball check valve.

Oil Flow Circuits

Many oil flow circuits are outlined in the service manuals. *Figure 26–26* shows a typical oil flow circuit. It is not necessary to understand the complete flow and operation of this diagram. More importantly, observe how all of the components work together to control the oil in the automatic transmission. Note that most of the major components have been discussed previously. These components include the oil pump, manual and shift valves, accumulator, regulator valve, servo, governor, oil cooler, clutches, torque converter, pressure relief valve, ball check valves, and converter clutch control valve.

Neutral Start Switch

On most vehicles, there is an electrical neutral start switch. This switch protects the vehicle from starting when the shift linkage is in any position except neutral or park. If the vehicle were able to start in a forward gear, a possible safety hazard could exist.

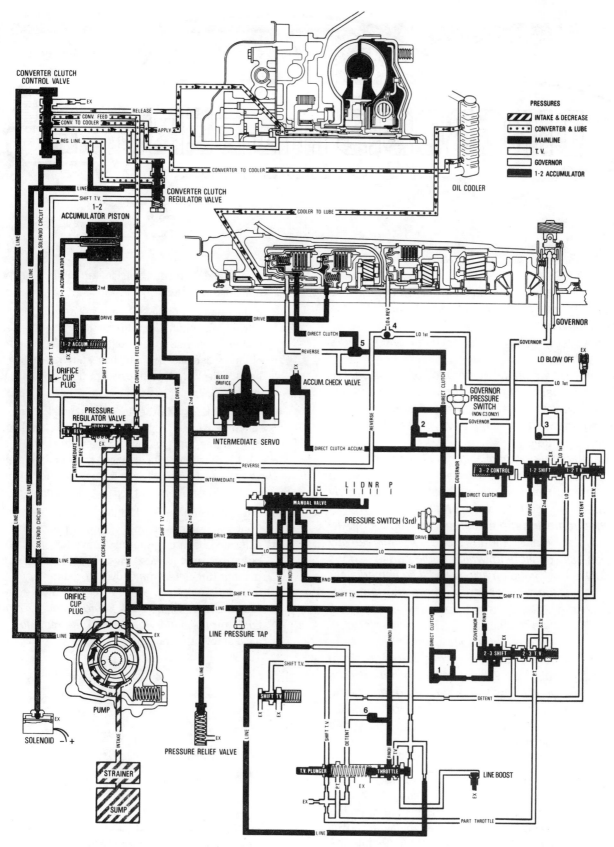

FIGURE 26–26. Oil flow circuits are used to help understand the complete oil flow within an automatic transmission. *(Courtesy of Oldsmobile Division, General Motors Corporation)*

DIAGNOSIS and SERVICE

SAFETY TIPS

1. *The automatic transmission is very heavy. Always use a transmission stand to aid in removing the transmission.*
2. *Always wear safety glasses when working under the vehicle.*
3. *When working on the automatic transmission, the vehicle must be supported on a hoist or hydraulic stand. Always follow the correct procedure when lifting the vehicle, and always use extra jack stands for safety.*
4. *The automatic transmission is often used as the rear support for the engine. Never remove the automatic transmission without first adequately supporting the rear of the engine.*
5. *When removing the automatic transmission, ATF fluid often leaks onto the floor. To eliminate any possibility of slipping, which could cause serious injury, always wipe up any spilled fluid immediately.*
6. *Transmission fluid operates above the temperature of the radiator. Be very careful when checking the fluid when the vehicle has just been stopped. A serious burn could result.*
7. *When disconnecting the torque converter from the flywheel, the flywheel must be turned. The ring gear on the flywheel is very sharp and could accidentally scrape your knuckles and hands. Be very careful and use the correct tools when removing the torque converter.*

1. Always check the *transmission fluid* for the correct level. The fluid level should be checked when the transmission is at operating temperature (about 200 degrees F) and in park or neutral while the engine is running. Refer to ***Figure 26–27***. *CAUTION: When adding fluid do not over fill, as foaming and loss of fluid may occur through the vent.* Also, if the fluid level is too low, complete loss of drive may occur, especially when the transmis-

sion is cold. This condition can further damage the transmission internally.

2. If the engine cooling system is not working properly (overheating), the transmission my operate at higher temperatures as well. Higher temperature oil may damage the clutches and transmission band. Check the cooling system if overheating is suspected.
3. Check ATF for a burned odor, darkened color, or metal particles. A burned odor may indicate overheating, and possibly damaged clutches. Metal particles may indicate damaged gears or extremely worn clutches. Refer to the service manual for the proper disassembly and repair procedure.
4. If transmission fluid is leaking from the front, the front seal is damaged or not sealing. The front seal is located behind the torque converter on the oil pump housing. *Replace the seal* according to the manufacturer's recommended procedure. Use the following general procedure to remove the transmission. ***CAUTION:*** *Remove the ground cable on the battery before starting removal.*
   a. Remove the transmission dipstick.

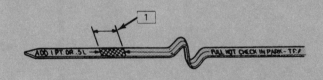

1 LEVEL TO BE IN CROSS-HATCHED AREA
ON FLUID LEVEL INDICATOR BLADE.
CHECK AT OPERATING TEMPERATURE.

FIGURE 26–27. Follow the manufacturer's recommendation when checking the fluid level. (Courtesy of Oldsmobile Division, General Motors Corporation)

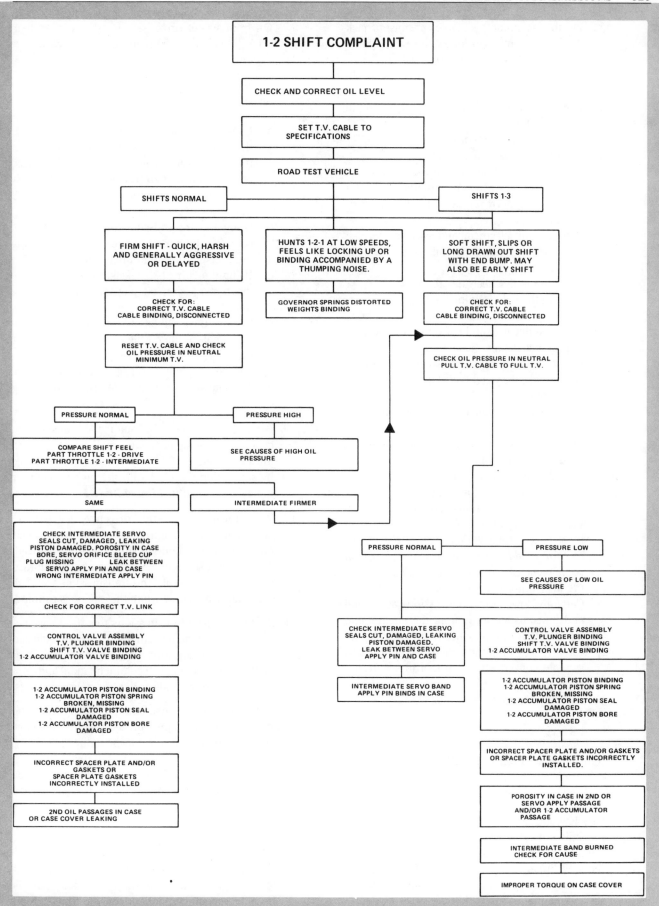

FIGURE 26–28. Flowcharts are used to help diagnose problems in automatic transmissions. *(Courtesy of Oldsmobile Division, General Motors Corporation)*

b. Lift the car up on a suitable car hoist or jack. Always remember to follow the safety guidelines when operating the car hoist.

c. Remove the drive shaft from between the transmission and the differential.

d. Remove the shift linkage connected to the transmission.

e. Remove the transmission cooler lines.

f. Remove the speedometer cable.

g. Drain the fluid from the transmission into a suitable container by removing the drain plug (if equipped) or by removing the transmission oil pan.

h. Remove any electrical and vacuum connections to the transmission and mark for reassembly.

i. Remove the flywheel undercover.

j. Remove the bolts holding the torque converter to the flywheel. There are usually three bolts. Mark the flywheel and converter for reference during installation.

k. Using a suitable jack, lift the rear of the engine so as to take the weight off of the rear transmission mount.

l. Remove the rear transmission mount from the transmission.

m. Place a transmission jack stand under the transmission to support it after it is disconnected from the engine block.

n. Slowly and carefully lower the engine until the bolts between the transmission torque converter housing and the block of the engine can be removed. *CAUTION: Lowering the engine too far could press the distributor cap on some vehicles against the fire wall and cause the distributor cap to crack.*

o. Carefully lower the transmission jack stand and transmission away from the engine, and remove them from beneath the vehicle. Be careful not to drop the torque converter. Use a suitable converter holding tool to secure the converter.

p. Reverse the procedure to install.

5. *Diagnosis flowcharts* can be used to diagnose transmissions. *Figure 26–28* is an example of one such chart. Charts are usually available for each make and year of transmission.

6. Very few adjustments are necessary on automatic transmissions. However, there is usually an adjustment for the manual valve, the throttle valve, and the shift linkage. Each vehicle is different. To get the exact procedure and specification, refer to the service manual.

7. On front wheel drive vehicles that use a chain drive, check the chain for excessive wear. See *Figure 26–29*.

8. *Replace the oil filter* and ATF at recommended intervals. To change the filter, use the following general procedure.

a. Using a suitable jack, support the vehicle so that work can easily be done on the bottom of the transmission. The vehicle is usually placed on a car hoist. *CAUTION: Follow all safety procedures when lifting a vehicle up on a hoist.*

b. Drain the transmission fluid into a suitable container by loosening the oil pan and letting the transmission fluid drain out.

c. Using the correct size socket wrench, remove the transmission oil pan. At this point, the transmission filter can be observed.

d. Remove the oil filter and replace it with a new one. Replace the oil filter and gaskets in their proper positions. Failure to do so may damage the transmission. See *Figure 26–30*.

e. To reassemble, reverse the preceding procedure.

9. Periodically inspect the automatic transmission cooler lines for signs of damage. Worn or crimped lines may cause ATF loss. Prevent future damage by supporting the lines with the proper clamp or hanger as shown in *Figure 26–31*.

10. When towing a rear-wheel driven vehicle with an automatic transmission and with the rear wheels on the ground, the drive shaft must be removed. This is because there may not be enough lubrication on the rear seal. The transmission seals are being lubricated only when the engine is running.

11. Some automatic transmissions require occasional transmission band adjustment. Bands are adjusted by rotating the adjusting screw on the bands. Refer to the manufacturer's recommendation and procedure to correctly adjust the transmission bands.

12. Bubbles in the transmission fluid indicate the presence of a high-pressure leak in the transmission. A leak such as this may be in the valve body, valve, or hydraulic passageways. If this is the case, the transmission hydraulic system will need to be checked and repaired. Normally, the valve body can be replaced as a unit.

13. If varnish or gum is showing on the dipstick, change the filter.

14. A poorly tuned engine that has low or insufficient vacuum may cause shifting problems. This is because several components on the transmission

operate off of a vacuum. Before testing any part of the automatic transmission, make sure that the engine vacuum is correct.

15. Inspect all linkages to the transmission for binding and damage. Incorrect linkage adjustment may cause bad shifting. Follow the service manual for the correct adjustment on the linkage for each particular manufacturer. The adjustment typically calls for the shift linkage to be placed in the park or neutral position. After the shift levers have been loosened, the levers are placed in a certain detent position. With the levers in this position, the linkage is adjusted so that the levers and the shifter assembly are in the same position.

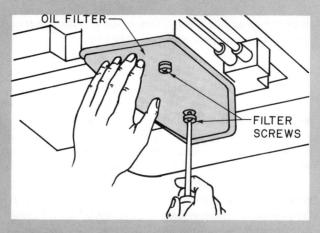

FIGURE 26–30. Replace the filter and oil at the recommended intervals. Always replace the filter and gasket in the correct and proper position.

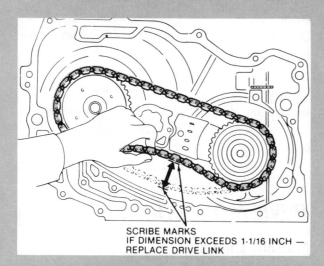

FIGURE 26–29. Check front wheel drive transmissions for chain wear, and replace if necessary. (Courtesy of Oldsmobile Division, General Motors Corporation)

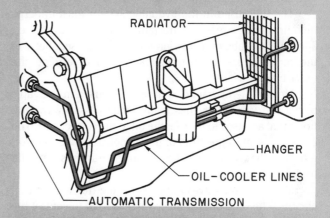

FIGURE 26–31. Inspect the transmission cooler lines for damage or leakage. Support the lines with the correct type of hanger.

SUMMARY

This chapter discussed the purpose, parts, and operation of the automatic transmission. The purpose of the automatic transmission is to connect the engine to the drive wheels. It is made of several major parts and systems. These include the torque converter, planetary gear system, hydraulic system, and controls.

The torque converter connects the engine to the transmission and multiplies the torque at lower speeds. It is considered a fluid coupling. The torque converter is made of several parts, including the pump, turbine, and stator. As the engine turns, the pump forces oil centrifugally into the turbine blades. The turbine blades then turn the transmission gears. The stator is used to correctly redirect the flow of oil back to the pump.

When the turbine blade is stopped under load, the oil deflection is greatest. The angle of deflection produces the amount of torque produced on the turbine blades. As the

speed of the turbine increases, the angle of deflection decreases. Torque is also decreased as speed increases.

The torque converter is placed in front of the transmission case on rear wheel drive vehicles. On front wheel drive vehicles, the torque converter is located to the side of the transmission case. A large chain from the torque converter to the transmission is used to transfer the torque.

Different gear ratios are produced in the transmission by using a planetary gear system. The planetary gear system is made of several parts. These include the sun gear, ring gear, planet gears, and planet gear carrier. If any one of these are held stationary, different gear ratios can be produced.

Several types of clutches and bands are used to lock up or hold the correct gear in the planetary gear system. The multiple-disc clutch uses a set of friction discs working against a set of metal discs. Oil pushes a piston against the discs to lock up the system. One of the planetary gears is connected to the discs which then lock up.

A transmission band is used to lock up a clutch housing or ring gear. The band is wrapped around the housing. A servo is operated hydraulically to engage and disengage the transmission band. Some transmissions also use an accumulator to cushion the action of the servo.

To aid in smoothness of shifting, an overrunning clutch is used on transmissions. This clutch is used to provide smooth engagement and release of the other clutches.

The heart of the automatic transmission is the hydraulic system. Oil is used to operate and control the transmission. Oil pressure is produced by the oil pump, which is located in the front of the transmission case. There are several styles of oil pumps, including the gear, rotor, and vane-type pumps. All use an eccentric design to produce suction and pressure for pumping oil. The oil pump is driven by the torque converter with two tangs.

The fluid used in the hydraulic system must undergo high pressures, stresses, temperatures, and shearing forces. Special automatic transmission fluid (ATF) is used. Refer to the manufacturer's recommendations to determine the exact type to use.

Most of the hydraulic controls are housed in the valve body. The valve body has many small passageways and holes that are used to transfer the oil to the correct part. The oil is also sent to the radiator to be cooled. At the radiator, a heat exchanger extracts heat from the transmission fluid to the radiator coolant.

Many control valves are used on the transmission oil circuits. The pressure regulator valve is used to control the exact pressure for the different components. The manual shift valve is used to tell the transmission what gear is to be used. Its operation is controlled by the operator moving the gear shift lever. The throttle valve is used to help shift gears, based on engine load and amount of throttle. The governor is used to help shift gears, based on vehicle speed. Based upon the speed and load of the vehicle, the shift valve operates the clutches to lock and unlock the exact gear in the planetary gear system.

Several other valves are also used. The vacuum modulator tells the transmission the amount of load, based on intake manifold vacuum. The kickdown valve is used to downshift when the operator pushes the throttle fully to the floor. Various ball check valves are used to control the direction of oil flow and prevent back flow.

There are several major considerations when servicing the automatic transmission. Filters and ATF should be checked and changed at the correct intervals. When the transmission oil smells burned or contains fine metal particles, the clutches might be damaged. The engine temperature also affects the operation and temperature of the transmission fluid. Before checking any part of the transmission, always be sure the engine is running correctly. Various other diagnosis and service tips can be found in the appropriate service manuals.

TERMS TO KNOW

Can you explain each of the following terms? Review the chapter until you can use each term correctly.

Periphery
Fluid coupling
Torque converter
Turbine
Stator
Deflection angle
Planetary gear system
Planet carrier
Overdrive
Multiple-disc clutch
Transmission band
Servo

Accumulator
Tang
Automatic transmission fluid (ATF)
Valve body
Pressure regulator valve
Manual shift valve
Throttle valve
Shift valve
Vacuum modulator
Spool valve

REVIEW QUESTIONS

Multiple Choice

1. Which of the following is not considered a major part in an automatic transmission?
   a. Torque converter
   b. Planetary gear system
   c. Oil pump
   d. Power steering gear
   e. Hydraulic valves

2. Which of the following parts is/are used as the input inside the torque converter?
   a. Turbine
   b. Pump
   c. Stator
   d. Seal
   e. All of the above

3. The output of the torque converter is taken off of the:
   a. Turbine
   b. Stator
   c. Pump
   d. Seal
   e. All of the above

4. The coupling between the engine and transmission on the automatic transmission is considered:
   a. An air coupling
   b. A fluid coupling
   c. An electrical coupling
   d. A mechanical coupling
   e. All of the above

5. As the turbine speed decreases, the angle of deflection:
   a. Increases
   b. Decreases
   c. Remains the same
   d. Causes the stator speed to change
   e. Causes the pump speed to increase

6. Which of the following is a part of a planetary gear system?
   a. Sun gear
   b. Planet carrier
   c. Planet gears
   d. Ring gear
   e. All of the above

7. The planetary gear system is able to produce a/an _____ in output speed.
   a. Increase
   b. Decrease
   c. Reverse direction
   d. All of the above
   e. None of the above

8. Which of the following uses a set of friction discs and steel discs pressed together to engage a clutch?
   a. Transmission band
   b. Overrunning clutch
   c. Multiple disc
   d. Servo
   e. Accumulator

9. What component aids in the operation of the transmission band?
   a. Servo piston
   b. Accumulator
   c. Vacuum modulator
   d. Multiple-disc
   e. Torque converter

10. What component is used to help smooth the shifting from one gear to another?
    a. Overrunning clutch
    b. Servo
    c. Sun gear
    d. Ring gear
    e. Torque converter

11. The oil pump on the automatic transmission uses the _____ type of design.
    a. Eccentric
    b. Vane
    c. Gear
    d. All of the above
    e. None of the above

12. The oil pump on the transmission housing is located:
    a. Inside the planetary gear system
    b. On the hydraulic circuit body
    c. On the front of the transmission case
    d. On the drive shaft
    e. None of the above

13. Which component is used to turn the oil pump on an automatic transmission?
    a. Planetary gear system
    b. Accumulator
    c. Servo
    d. Crankshaft
    e. Torque converter

14. Which component in the automatic transmission houses the valves and contains the many oil passageways?
    a. The accumulator
    b. The modulator valve
    c. The valve body
    d. The planetary gear system
    e. None of the above

15. Which of the following components helps to cool the transmission fluid?
    a. Radiator
    b. Torque converter
    c. Accumulator
    d. Planetary gear system
    e. Valve body

16. Which of the following transmission control valves is used to measure vehicle speed and, in turn, direct oil flow to shift gears?
    a. Manual shift valve
    b. Shift valve
    c. Throttle valve
    d. Governor
    e. Modulator valve

17. Which of the following transmission control valves uses intake manifold vacuum to measure vehicle load?
    a. Shift valve
    b. Modulator valve
    c. Shift valve
    d. Throttle valve
    e. Governor

18. Which type of valve is used to control or direct oil flow and prevent back flow of oil?
    a. Ball check valve
    b. Governor
    c. Throttle valve
    d. Shift valve
    e. Modulator valve

19. If the transmission fluid has a burned odor, the trouble may be in the _____ on the automatic transmission.
    a. Torque converter
    b. Clutches
    c. Seals
    d. Governor
    e. Shift valve

20. What parts may need service on the automatic transmission?
    a. Filter may be dirty
    b. Clutches may be worn
    c. Seals may leak
    d. All of the above
    e. None of the above

The following questions are similar in format to ASE (Automotive Service Excellence) test questions.

21. Technician A says that if the engine coolant is too hot, the transmission may be overheated and damaged. Technician B says that the cooling system has no effect on the operation or temperature of the transmission. Who is right?
    a. A only
    b. B only
    c. Both A and B
    d. Neither A nor B

22. Bubbles are noticed in the transmission fluid. Technician A says that there is too much transmission fluid, and foam is being produced. Technician B says that there is an internal leak in the hydraulic system. Who is right?
    a. A only
    b. B only
    c. Both A and B
    d. Neither A nor B

23. Transmission fluid is leaking from the front of the transmission housing. Technician A says to remove and replace the transmission filter. Technician B says to replace the oil pump. Who is right?
    a. A only
    b. B only
    c. Both A and B
    d. Neither A nor B

24. Technician A says transmission fluid level should be checked when the transmission is at operating temperature. Technician B says it should be checked when the transmission is cold. Who is right?
    a. A only
    b. B only
    c. Both A and B
    d. Neither A nor B

Essay

25. What is the purpose of the torque converter?

26. What does the "deflection angle" refer to on automatic transmissions?

27. Describe the purpose and operation of a planetary gear system.

28. What are the purposes of clutches and bands in an automatic transmission?

29. What is the purpose of the valve body in an automatic transmission?

30. What is the purpose of using a governor in an automatic transmission?

31. Describe the purpose and operation of a neutral start switch.

CHAPTER 27

Drive Lines, Differentials, and Axles

INTRODUCTION

To get the vehicle moving, power or torque must be transferred from the transmission to the drive wheels. This is done by using drive lines, differentials, and axles. The purpose of this chapter is to define the parts and operation of these components.

OBJECTIVES

After reading this chapter, you will be able to:
- State the parts and operation of rear wheel drive shafts and universal joints.
- State the parts and operation of front wheel drive shafts and constant velocity universal joints.
- Identify the parts and operation of differentials, including the nonslip (limited slip) differentials.
- Describe common diagnosis and service tips used with drive lines, differentials, and axles.

CHAPTER HIGHLIGHTS

535

27.1 REAR DRIVE SHAFTS AND OPERATION

Purpose of the Drive Shaft

The torque that is produced from the engine and transmission must be transferred to the rear wheels to push the vehicle forward and reverse. The drive shaft must provide a smooth, uninterrupted flow of power to the axles. The *drive shaft* and differential are used to transfer this torque. *Figure 27–1* shows the location of the drive shaft and differential.

There are several functions of the drive shaft. First, it must transmit torque from the transmission to the axle. During operation, it is necessary to transmit maximum low-gear torque developed by the engine. The drive shaft must also be capable of rotating at the very fast speeds required by the vehicle.

The drive shaft must also operate through constantly changing angles between the transmission and the differential and axles. The rear axle is not attached directly to the frame of the vehicle. It rides suspended by the springs and travels in an irregular floating motion. As the rear wheels roll over bumps in the road, the differential and axles move up and down. This changes the angle between the transmission and the differential.

The length of the drive shaft must also be capable of changing while transmitting torque. Length changes are caused by axle movement due to torque reaction, road deflections, braking loads, and so on. *Figure 27–2* shows

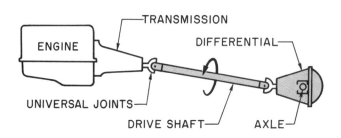

FIGURE 27–1. The drive shaft is used to transmit the power from the engine and transmission to the differential and axles on rear wheel drive vehicles.

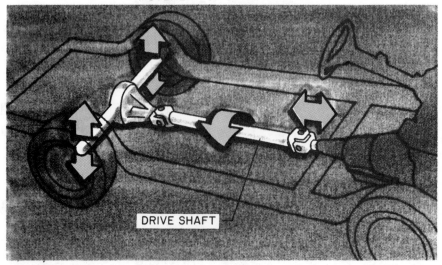

FIGURE 27–2. The drive shaft must be designed to allow for both up and down motion on the differential, and shortening and lengthening between the differential and transmission. *(Courtesy of Dana Corporation)*

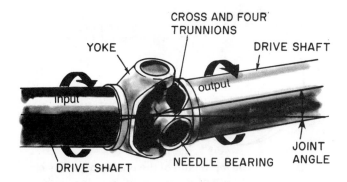

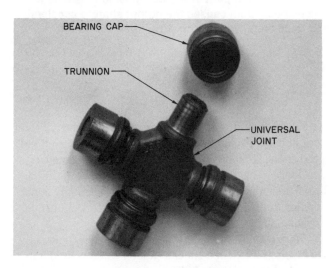

FIGURE 27–3. Universal joints allow the drive shaft to turn through different angles. (Courtesy of Drivetrain Service Division, Dana Corporation)

the movements that the drive shaft may undergo during normal driving operation.

Purpose of Universal Joints

Universal joints (U-joints) are used to permit the drive shaft to operate at different angles. *Figure 27–3* shows a drawing and a photograph of a simple universal joint. Note that the rotation can be transmitted when two shafts are at different angles. This type of universal joint is called the cross and bearing type *Cardan* universal joint or four-point joint. It consists of two *yokes* and a journal assembly with four *trunnions*. A yoke is a Y-shaped assembly that is used to connect the U-joint together. The trunnion is a protrusion on the journal assembly. The journal assembly is also called a *cross* and bearing assembly or spider.

Universal Joint Parts

The universal joint is made of several parts. The center of the universal joint is called the cross and bearing assembly.

Its purpose is to connect the two yokes. The yokes are attached directly to the drive shafts. Four *bearing caps* are placed on each trunnion. Each bearing cap is a needle-type bearing that allows free movement between the trunnion and yoke. The needle bearing caps are attached to the yokes by several methods. They can be pressed into the yokes, bolted to the yokes, or held in place with bolts, nuts, U-bolts, or metal straps. Snap rings are also used to hold bearing caps in place. See *Figure 27–4*. On most replacement universal joints, there is a lubrication fitting to put grease into the needle bearings.

Slip Joints

The drive shaft must also be able to lengthen and shorten during operation with irregular road conditions. A *slip joint* is used to compensate for this motion. The slip joint is usually made of an internal and external spline. It is located on the front end of the drive shaft and is connected to the transmission. See *Figure 27–5*. The slip joint can also be placed in the center of the drive shaft.

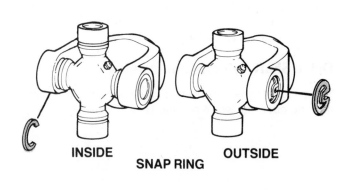

INSIDE   OUTSIDE
SNAP RING

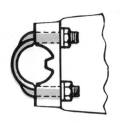 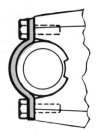

U-BOLT TYPE   STRAP TYPE

FIGURE 27–4. Universal joints are held in the yoke in several ways. Snap rings, U-bolts, and straps are used. (Courtesy of Dana Corporation)

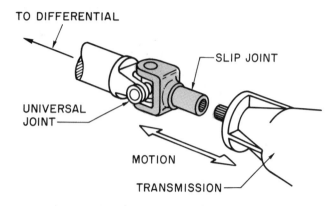

FIGURE 27–5. The slip joint is designed to allow the drive shaft to shorten and lengthen when the differential goes over an irregular road surface.

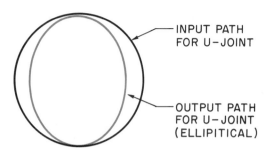

FIGURE 27–7. The input path of the universal joint is circular. The output path of the universal joint is elliptical. The elliptical path causes the output shaft to accelerate and decelerate within each drive shaft revolution.

Drive Shaft Vibration

Vibration is the most common drive shaft problem. It can be either transverse or torsional. As shown in *Figure 27–6*, transverse vibration is the result of an unbalanced condition acting on the shaft. This is usually caused by dirt or foreign material on the shaft, and it can cause a rather noticeable vibration in the vehicle.

Torsional vibration occurs from the power impulses of the engine or from improper universal joint angles. It causes a noticeable sound disturbance and can cause a mechanical shaking. In excess, both types of vibration can cause damage to the universal joints and bearings.

Drive Shaft Velocity and Angles

The popular cross and bearing assembly universal joint has one disadvantage. When the U-joint transmits torque

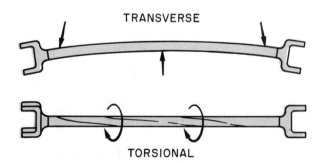

FIGURE 27–6. Vibration in the drive shaft can be either transverse or torsional. Both types of vibration can cause damage to the universal joints.

through an angle, the output shaft increases and slows down twice in each revolution of the shaft. The rate at which the speed changes depends upon the steepness of the universal joint angle. The speed changes are not normally visible during rotation. However, they may be felt as torsional vibration due to improper installation, steep and/or unequal operating angles, and high speeds.

Figure 27–7 shows how this happens. The input path and bearings rotate in a circular motion when viewed from the end of the drive shaft. The output path and bearings rotate in an elliptical motion or path. The output path looks like an ellipse because it is viewed at an angle instead of straight on. This is because the output shaft is at an angle from the input shaft.

In operation, the input shaft speed has a constant velocity. The output shaft speed accelerates and decelerates (catching up and falling behind in rotation) during one complete revolution. Referring to *Figure 27–8*, from 0–90 degrees rotation, the output shaft accelerates. From 90–180 degrees rotation, the output shaft decelerates. From 180–270 degrees rotation, the output shaft accelerates again. From 270–360 degrees rotation, the output shaft decelerates. The input and output shafts complete one rotation at exactly the same time. The greater the output angle, the greater the change in velocity of the output shaft per revolution.

The torsional vibrations mentioned before travel down the drive shaft to the next universal joint. At the second universal joint angle, similar acceleration and deceleration occur. However, these take place at equal and reverse angles to the first joint. See *Figure 27–9*. Now the speed changes cancel each other when both operating angles are equal. This is why drive shafts must have at least two universal joints, and operating angles must be small and equal. Any variations from this may cause excessive needle bearing and trunnion wear.

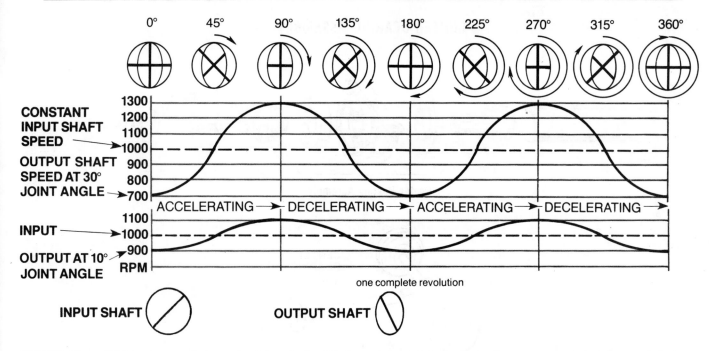

FIGURE 27–8. Within one revolution of a drive shaft with the universal joint at an angle, the output speed accelerates and decelerates two times. The greater the universal joint angle, the greater the change in speed. *(Courtesy of Drivetrain Service Division, Dana Corporation)*

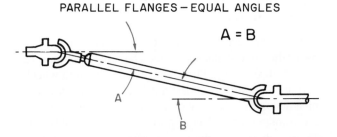

FIGURE 27–9. To offset the velocity changes in a drive shaft, a second universal joint is used at an equal and opposite angle.

Drive Shaft Construction

There are several designs for drive shafts. The type depends upon several factors. These include the length of the vehicle, the amount of power the engine must transmit, the rpm of the drive shaft, and the year of the vehicle. *Figure 27–10* shows a typical drive shaft. In most cases, a universal joint is placed on both sides. The most common location for the slip joint is on the output of the transmissions. On certain vehicles, the slip joint is located in the center of the drive shaft.

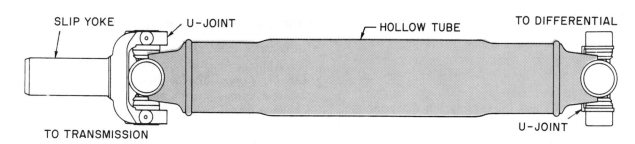

FIGURE 27–10. The parts of a standard drive shaft. *(Courtesy of Chevrolet Division, General Motors Corporation)*

CENTER BEARING ASSEMBLY

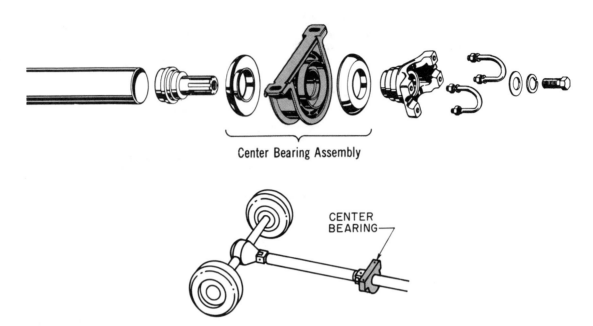

Center Bearing Assembly

CENTER BEARING

FIGURE 27–11. The center bearing is used to support longer drive shaft systems. The center bearing assembly is attached directly to the frame. *(Courtesy of Universal Joint Division, Dana Corporation)*

Center Bearing

On many large vehicles, it may be necessary to use a center bearing for support on the drive shaft. *Figure 27–11* shows a typical center bearing assembly. When a center bearing is used, there will be two drive shafts and usually more than two universal joints. The bearing assembly supports the end of the first drive shaft by being bolted directly to the frame of the vehicle.

CAR CLINIC ▊▊▊

PROBLEM: NOISE IN DRIVE SHAFT

A car has developed a squeaky noise in the drive shaft. The noise is not heard all of the time. However, when the car is shifted into reverse and then into forward again, a slight clunk is heard and the sound returns.

SOLUTION:

The most common cause of noise in the drive shaft is a bad universal joint. When it is worn, the U-joint makes a squeaky sound. The squeak comes from the lack of lubrication in the U-joint needle bearings. Remove the drive shaft and check the U-joints. When moving the U-joints to check them, check for any irregular bumps or motion to its movement. Replace the U-joints if necessary.

27.2 FRONT WHEEL DRIVE SYSTEMS

Requirements of Front Wheel Drive Systems

A front wheel drive vehicle presents several unique problems for the drive shafts and joints. The drive shaft must be able to do three things. First, it must allow the front wheels to turn for steering. Second, it must be able to telescope, a similar action to a slip joint on rear drive vehicles. Third, it must transmit torque continuously without vibration. Because of these characteristics, a cross and bearing universal joint would not work well, especially during sharp turning. A constant velocity (CV) joint is used instead. See *Figure 27–12*.

Constant Velocity Joints

The constant velocity drive is designed much the same as a set of bevel gears. Balls and grooves are used, rather than gears. See *Figure 27–13*. Balls and grooves, rather than gear teeth, connect the input and output shafts. If the balls are placed in elongated grooves, the result is called a *constant velocity (CV) joint*. With this type of joint, there will be no speed or velocity changes on the output shaft as with cross and bearing type universal joints.

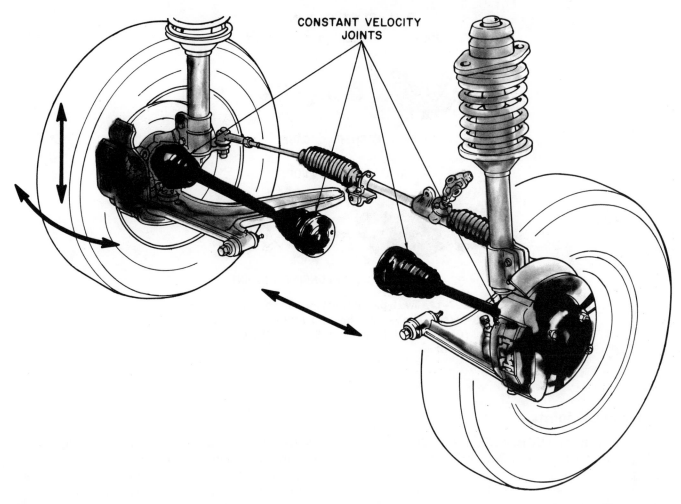

FIGURE 27-12. Front wheel drive systems use constant velocity joints on both axles to transmit power to the wheels. *(Courtesy of Drivetrain Service Division, Dana Corporation)*

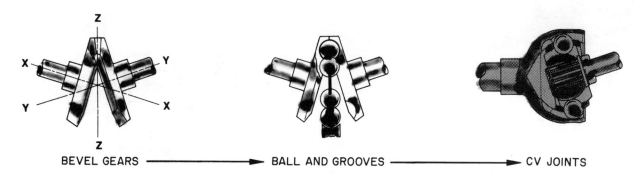

BEVEL GEARS ⟶ BALL AND GROOVES ⟶ CV JOINTS

FIGURE 27-13. Constant velocity joints act much the same as bevel gears or ball and groove connections. *(Courtesy of Drivetrain Service Division, Dana Corporation)*

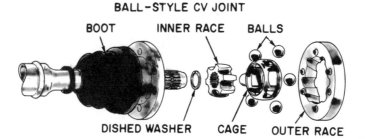

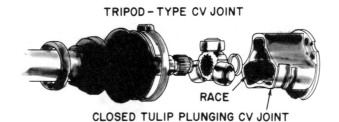

FIGURE 27–14. Two types of constant velocity joints are commonly used. These are the ball style and the tripod type. *(Courtesy of Drivetrain Service Division, Dana Corporation)*

Types of CV Joints

There are two types of CV joints. *Figure 27–14* shows the ball-style and the *tripod*-type CV joints. The ball style uses a series of balls, a cage, and inner and outer races. The tripod uses a tulip assembly and three rollers. Both perform well in front wheel drive cars.

The typical front wheel drive car uses two drive shaft assemblies. One assembly drives each wheel. Each assembly has a CV joint at the wheel end. This CV joint is called the fixed joint or outboard joint. A second joint on each shaft is located at the transaxle end. This CV joint is called the inboard or plunging joint. This may either be a ball or tripod CV joint. It allows the slip motion that is required when the drive shaft shortens or lengthens because of irregular surfaces. See *Figure 27–15*.

Other Applications for Constant Velocity

On certain makes of cars, the engine is located in the rear. When this is the case, constant velocity joints may also be used. *Figure 27–16* shows such a system. Two drive shafts

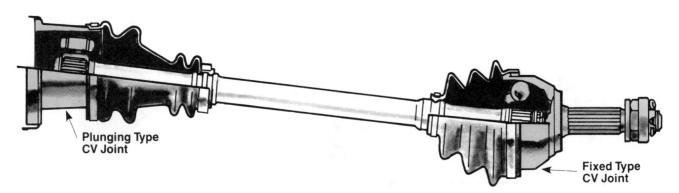

FIGURE 27–15. The plunging CV joint is used on the transaxle end. The fixed-type CV joint is used on the wheel end. *(Courtesy of Drivetrain Service Division, Dana Corporation)*

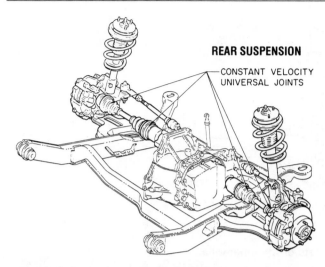

FIGURE 27–16. Constant velocity joints are also used on certain types of rear engine vehicles. *(Courtesy of Pontiac Division, General Motors Corporation)*

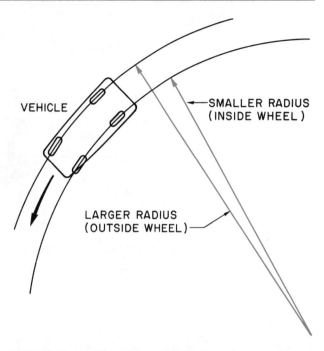

FIGURE 27–17. The differential is needed because the rear wheels turn at different speeds around a corner.

are used on this system. In addition, two constant velocity joints are used on each drive shaft.

CAR CLINIC ▬▬▬

PROBLEM: REAR END WHINE

A car has a noticeable whine in the rear end or differential at 30–45 mph. When the engine is decelerated, the whine stops. What could be the problem?

SOLUTION:

The whine is most often coming from the ring and pinion gear in the differential. However, the rear wheel bearings may also be bad. Check the rear wheel bearings first. If they are OK, check the differential ring and pinion gear. Check for wear and backlash specifications. The ring and pinion gear will probably have to be replaced.

27.3 DIFFERENTIALS AND AXLES

Purpose of the Differential

The purpose of the *differential* is to transmit the torque from the drive shaft to the axles and drive wheels of the vehicle. On front wheel drive vehicles, the differential is located inside the transaxle and is a part of the total assembly. Torque is transmitted from the engine, through the transmission, and through the drive shaft. The differential then splits the torque and sends it to the drive wheels.

In addition, the differential allows the rear wheels to turn at different speeds during cornering. *Figure 27–17* shows why the wheels turn at different speeds on corners. The inside rear wheel turns at a smaller radius than the outside rear wheel. The differential is designed to keep the power transmitting equally to both wheels while they are traveling at different speeds.

Differential Main Parts

Figure 27–18 shows the parts of the differential. These include:

1. *Drive pinion* — The drive pinion is the main input shaft to the differential. It is driven from the drive shaft.

2. *Ring gear* — The ring gear is driven from the drive pinion. Its purpose is to drive the remaining parts of the differential.

3. *Differential case* — The differential case holds several bevel gears. The entire differential case is driven from the ring gear which is bolted to it.

4. *Two differential side gears and two pinion gears* — These four gears are placed inside the differential case. All four gears are meshed together. The pinion gears have a shaft running through their center. The shaft is secured to the differential case. Thus, as the differential case turns, the shaft rotates (end to end) at the same speed as the differential case. Since

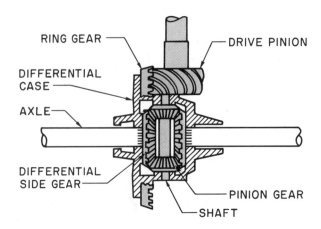

FIGURE 27-18. The main parts of the differential.

all four gears are meshed, the differential side gears also rotate at the same speed.

5. *Axle* — The axles are attached to the differential side gear by a spline on the axle and inside the differential side gear. As the differential side gear rotates, the axles also rotate, causing the vehicle to move.

Differential Operation

When the vehicle is moving straight down the road, the transmission of power comes from the drive pinion to the ring gear and differential case. As the differential case turns, the four bevel gears inside the case also move with the case. However, although the four gears are meshed during this condition, they are not rotating among themselves. In fact, the pinion gears are not spinning on their shafts at this point.

As the car goes around a left corner, the left axle slows down and the right axle speeds up. During this slowing down and speeding up, the four bevel gears inside the differential case begin to rotate among themselves. The pinion gears are now rotating or turning on the shaft. The pinion gear is said to walk around the differential side gears. When the vehicle returns to a straight line, both axles are spinning at the same speed, but the four gears are no longer turning among themselves. As the car goes around a right corner, the right axle slows down and the left axle speeds up. Again the four gears inside the differential case begin turning among themselves. (The pinion gear walks around the differential side gears.) This is necessary so that the axles will spin at different speeds while still transmitting power.

Complete Differential

Figure 27–19 shows an example of a complete differential. In addition to the parts already mentioned, several other parts are included. These are:

1. *Pinion gears* — This differential shows the addition of two extra pinion gears for more torque carrying capabilities.
2. *Adjusters* — As they are turned in and out, adjusters move the entire differential case (gear case) from side to side. This adjustment is used to change the gear backlash between the drive pinion and the ring gear, and to preload bearings.
3. *Bearings* — Various bearings are used to support each shaft.
4. *Thrust washers* — Thrust washers are used to absorb any side thrust produced by the gears in the differential.
5. *Spacer* — A spacer is used on the drive pinion shaft to adjust the position of the drive pinion in relation to the ring gear.
6. *Carrier* — The carrier housing is used to hold all of the differential parts together.

Limited Slip Differential

On a standard differential, equal torque is transmitted to the rear wheels if the load on each wheel is the same. However, if one wheel hits a patch of ice or other slippery surface, the torque will be different between the two wheels. In this case, one wheel may spin freely while the other wheel produces torque.

To eliminate this condition, *limited slip differentials* are used. Limited slip differentials use a set of clutches or cones to lock up both wheels. The clutches or cones apply a pressure to the side gears. This additional pressure prevents one wheel from spinning more rapidly than the other. The clutch assembly or cone assembly is located inside the differential case, usually between the side gears and the carrier housing or differential case.

Figure 27–20 shows a limited slip differential system. Three additional parts are shown. These include the:

1. *Clutch plate* — used to produce the necessary friction to lock up both wheels.

2. *Preload spring* — used to produce the pressure against the clutches.

3. *Pressure ring* — a surface against which the clutches rub for the side gear.

Although there are many designs, the principles of operation remain the same. Springs or some other force must push clutches or cones against a surface to lock the side gears to the differential case.

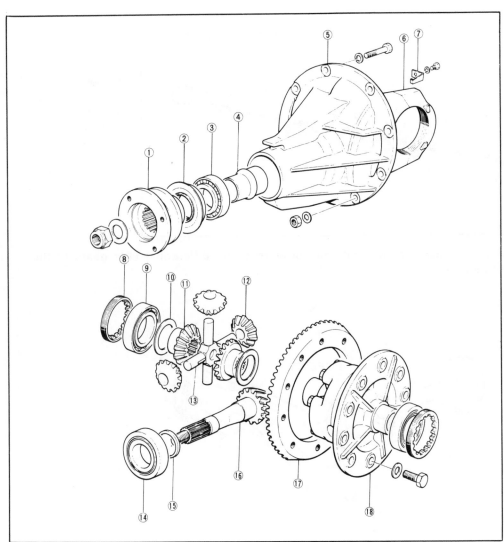

Rear axle components
1. Companion flange
2. Oil seal
3. Front bearing
4. Collapsible spacer
5. Carrier
6. Bearing cap
7. Adjuster lock plate
8. Adjuster
9. Differential bearing
10. Thrust washer
11. Side gear
12. Pinion gear
13. Spider
14. Rear bearing
15. Spacer
16. Drive pinion
17. Ring gear
18. Gear case

FIGURE 27–19. The differential assembly can also use four pinion gears. (Courtesy of Mazda Motor Corporation)

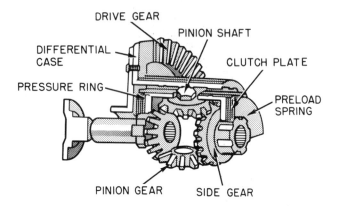

FIGURE 27–20. A limited slip differential uses a set consisting of a clutch plate, springs, and a pressure ring to lock up the two side gears. *(Courtesy of Chrysler Corporation)*

Rear Axle

The rear axle is used to connect the differential to the rear wheels. The rear axle shown in ***Figure 27–21*** is attached to the side gears inside the differential case. The two are connected by an internal and external spline. As the side gears are turned inside the differential case, the axles are also turned.

The axle is supported in the axle tube shown in ***Figure 27–22***. Bearings, gaskets, shims, oil seals, bearing retainers, bearing collars, and spacers are used to support the shaft and keep differential lubricant from leaking out of the assembly.

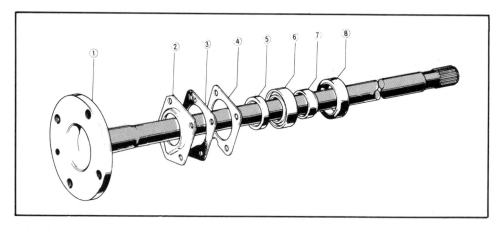

Rear axle shaft

1. Rear axle shaft
2. Bearing retainer
3. Gasket
4. Shim
5. Spacer
6. Bearing
7. Bearing collar
8. Oil seal

FIGURE 27–21. The axle is used to transmit the power from the differential side gears to the wheels. *(Courtesy of Mazda Motor Corporation)*

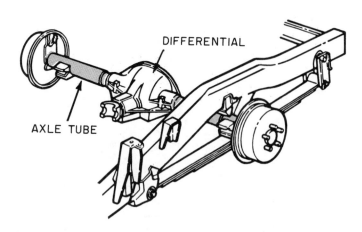

FIGURE 27–22. The axle fits inside the axle tube. *(Courtesy of Chevrolet Division, General Motors Corporation)*

DIAGNOSIS and SERVICE

SAFETY TIPS

1. *Always disconnect the battery before working on any part of the drive lines and differential. This will eliminate any possibility of the engine accidentally cranking over.*

2. *When working on the drive line, differential, and axle the vehicle must be lifted. Use the correct procedure when lifting the vehicle on a hoist and always use an extra jack when jacking up a vehicle with a hydraulic or air jack. Also, always block the front wheels to eliminate the possibility of the vehicle moving.*

3. *Always wear safety glasses and correct clothing when working on drive lines, differentials, and axles.*

4. *The axle and differential are very heavy components. Never try to lift these items without the use of proper lifting equipment.*

5. *When working on the differential, oil may be spilled on the floor. Always wipe up the oil immediately to eliminate the possibility of slipping and causing serious injury.*

6. *When replacing universal joints, high pressures are needed to remove the components. Make sure that all components are securely fastened to the tool to eliminate the possibility of injury to hands and face.*

1. Check the universal joints for wear by checking for brinelling. *Brinelling* is the process of producing grooves in the trunnion from the needle bearings. Refer to *Figure 27–23*. If brinelling has occurred, *replace the universal joints*. Use the following general procedure to disassemble and reassemble.

   a. *CAUTION: The vehicle must be positioned so that the drive shaft can be removed.* Follow all safety procedures when putting the vehicle on the hoist.

   b. Remove the drive shaft by removing the two U-bolts which attach the rear universal joint to the differential yoke.

   c. Pull the drive shaft away from the yoke and pull backward to remove the assembly.

   d. Remove the snap rings and retainer plates that hold the bearings in the yoke and drive shaft.

   e. Select a wrench or socket with an outside diameter slightly larger than the U-joint bearing diameter. Select another socket with an inside diameter slightly larger than the U-joint bearing diameter.

   f. To remove the U-joint from the drive shaft, place the sockets on opposite ends of the bearings on the U-joint. The smaller socket will become the bearing pusher and the larger socket will become the bearing receiver.

   g. Close the vise with the assembly between the jaws.

   h. As the vise is closed, one bearing will be pushed out of the drive shaft and into the larger socket. The other bearing will be pushed to the center.

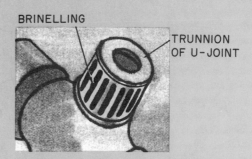

FIGURE 27–23. Brinelling, grooves from needles marking and burning into the trunnion, is caused by improper angles, lack of lubrication, or too much load. (Courtesy of Dana Corporation)

i. To remove the opposite bearing, place it in the vise with the pusher socket on the exposed cross journal of the U-joint. Then tighten the vise jaws, pressing the bearing back through the drive shaft into the receiving socket.

j. The spider or cross of the U-joint should now be easily removed.

Use the following procedure to assemble a universal joint to the drive shaft.

a. Make sure the bearings are packed with grease. Note that too much grease may damage the seals on the U-joint.

b. Using the pusher socket, press one bearing part way into the drive shaft. Position the spider into the partially installed bearings. *CAUTION: Make sure that no needle bearings have fallen to the center of the bearing.* Place the second bearing into the drive shaft.

c. Place the assembly in the vise so the bearings can be pressed in by the jaws of the vise.

d. As the vise is tightened, press the bearings into the drive shaft. Make sure the spider is held in such a position as to avoid binding or damage to the needle bearings.

e. The U-joints are removed and replaced the same way for the yoke.

2. Lubricate the spider or cross and bearing caps when replacing the universal joints. See *Figure 27–24*.

3. When replacing universal joints, obtain the correct replacement kit from the parts center. Since there are many types and sizes of U-joints, it is possible to accidentally obtain the wrong replacement kit.

4. When replacing the universal joint bearings that are held in place with the U-bolts, be very careful to avoid having any of the needle bearings fall into the center of the cap. See *Figure 27–25*.

5. Replace constant velocity joints according to the manufacturer's suggested procedure.

6. Always check universal joints for end *galling* (wear on the end of the trunnion and inside bearing caps).

7. A humming noise during driving could indicate the early stages of insufficient or incorrect lubricant on CV drive shafts. Replace when necessary.

8. A popping or clicking sound on sharp turns indicates possible CV joint wear in the outer, or wheel-end joint. Replace when necessary.

9. When installing a universal joint kit, be sure the snap rings are installed correctly and are completely touching the trunnion. Note that some late model cars use an injected nylon retainer on the universal joint bearing. When service is necessary, pressing the bearing out as described previously will shear the nylon retainer. Replacement U-joints must be the steel snap-ring type. See *Figure 27–26* which shows the retaining rings being installed.

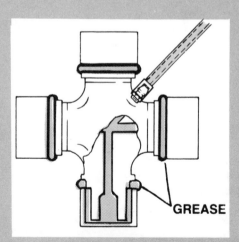

FIGURE 27–24. The universal joints should be lubricated before installation. Grease must flow from all four bearing seals. *(Courtesy of Dana Corporation)*

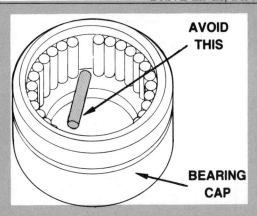

FIGURE 27–25. Make sure that all needle bearings are lined up inside the cap. *(Courtesy of Dana Corporation)*

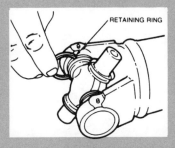

FIGURE 27–26. Retaining rings should be installed on the universal joints to hold the spider in place. *(Courtesy of Motor Publications, Auto Repair Manual, 1981–1987)*

10. When the differential is suspected of being noisy or damaged, first make sure the noise is *not* any of the following:
    a. road noise
    b. tire noise
    c. front bearing noise
    d. transmission and engine noise
    e. drive shaft and U-joint noise

    If tests show that the differential or rear axle is noisy, it could be coming from one of several parts. These include the rear wheel bearings, differential side gear and pinion, side bearing, and ring and pinion gear. Follow the manufacturer's recommended disassembly and reassembly procedures to repair the differential assembly.

11. *Figure 27–27* shows the *wear* characteristics on the *ring gear*. Based upon these wear patterns, several checks and adjustments can be made on the differential.

a. *Backlash adjustment* is made by mounting a dial indicator on the ring gear as shown in *Figure 27–28*. The pointer is set to zero. The backlash is checked by moving the ring gear back and forth, and noting the amount of space (backlash) indicated on the dial indicator. If the adjustment is greater than that allowed by the manufacturer, loosen the right-hand nut one notch and tighten the left-hand nut one notch. If the backlash is less than the allowable minimum, loosen the left-hand nut one notch and tighten the right-hand nut one notch. The adjustment nuts are located on the sides of the bearing caps.

b. The position of the drive pinion can also be adjusted. A shim pack is placed between the pinion head and the inner race of the rear bearing. Increasing the shim pack will move the pinion close to the ring gear. Decreasing the shim pack will move the pinion away from the ring gear. See *Figure 27–29*.

12. Never use different size tires on the rear wheels. Excessive wear may develop on the differential because the differential will act as if the vehicle is cornering. Also, limited slip differentials will have excessive wear on the clutches or cones when different sized tires are used.

13. A humming noise in the differential is an indication that the clearance between the ring gear and the pinion gear is incorrect or that the differential ring gear has been damaged or worn.

14. Make sure the correct type of lubricant is used on limited slip differentials. Refer to the manufacturer's recommendation.

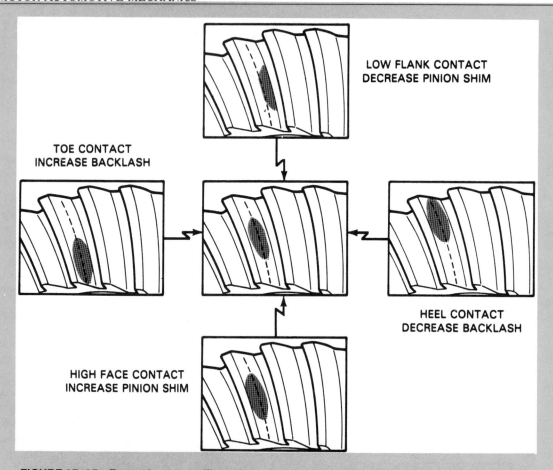

LOW FLANK CONTACT
DECREASE PINION SHIM

TOE CONTACT
INCREASE BACKLASH

HEEL CONTACT
DECREASE BACKLASH

HIGH FACE CONTACT
INCREASE PINION SHIM

FIGURE 27–27. Excessive wear will result on the ring gear if the differential backlash is incorrect. Adjust the backlash by using shims and the adjusters. *(Courtesy of Motor Publications, Auto Repair Manual, 1981–1987)*

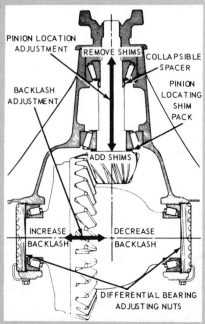

PINION LOCATION
ADJUSTMENT   REMOVE SHIMS   COLLAPSIBLE SPACER

BACKLASH
ADJUSTMENT   PINION LOCATING SHIM PACK

ADD SHIMS

INCREASE BACKLASH   DECREASE BACKLASH

DIFFERENTIAL BEARING
ADJUSTING NUTS

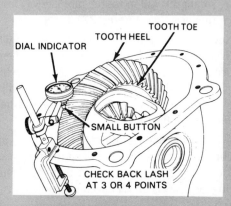

DIAL INDICATOR   TOOTH HEEL   TOOTH TOE

SMALL BUTTON

CHECK BACK LASH
AT 3 OR 4 POINTS

FIGURE 27–28. A dial indicator should be used to check the backlash between the ring gear and the pinion. *(Courtesy of Motor Publications, Auto Repair Manual, 1981–1987)*

FIGURE 27–29. This drawing shows the direction of movement when adjusting the ring gear and pinion gear. Shims are used to position the pinion gear and bearing adjusting nuts are used to position the ring gear. *(Courtesy of Motor Publications, Auto Repair Manual, 1979–1985)*

WHEEL ALIGNMENT

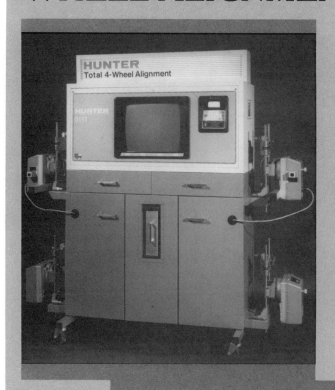

Wheel alignment has become very sophisticated in recent years. Wheel alignment equipment is now computerized to help maintain accuracy. (Courtesy of Hunter Engineering Company)

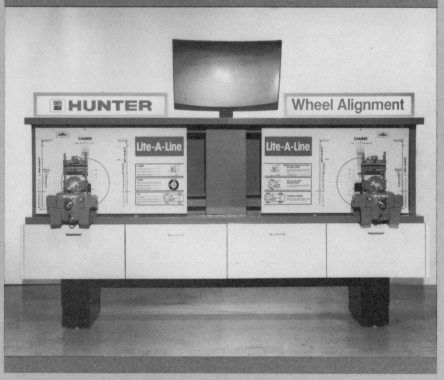

COMPUTERIZED ALIGNMENT

The procedure and specifications for correct wheel alignment are now computerized. Specifications are stored within the computer's memory so that the technician does not have to refer to a specifications manual. Visual procedures and instructions are also given on the computer screen. (Courtesy of Hunter Engineering Company)

SUMMARY

The purpose of this chapter was to explain how torque is transmitted through the drive train to the wheels. In order to do this, power must be sent through the drive shaft, differential, and axles.

Rear wheel drive vehicles use a drive shaft with universal joints to transmit the power to the differential. During normal operation, the drive shaft must transmit torque through a variety of angles. Universal joints are used to allow this to happen. Universal joints are made with two yokes and a cross assembly with trunnions on the end. Needle bearings are placed inside bearing caps and placed on the trunnions. The universal joint is held in place by bolts or straps, or by being pressed into the yoke.

The drive shaft must also have a slip joint. Irregular road conditions cause the distance from the transmission to the differential to change. A slip joint compensates for this lengthening and shortening.

Vibration is the most common drive shaft problem. Both torsional and transverse vibration are common. Transverse vibration is a result of an unbalanced condition. Torsional vibration is a result of the power pulses from the engine. Vibration can cause excessive wear on the universal joints. Torsional vibration can also be caused by too great an angle or unequal angles on the universal joints.

Because of their length, some drive shafts use a center bearing. Usually two drive shafts are used with the center bearing supporting the end of the first drive shaft. The center bearing is bolted to the frame for support.

Front wheel drive systems require constant velocity joints to be used. Here the speed of the drive shaft will be constant throughout its complete revolution. Two types of constant velocity joints are used: the ball style and the tripod. Slip motion is also necessary and is designed into the CV joints.

The differential is used to transmit the torque from the drive shaft to the drive wheels. The torque must be distributed to both wheels. The differential includes a ring gear, pinion gear, differential case, bevel gears, and side gears. The torque is transmitted to each part before going to the axle. When the vehicle is going around a corner, the differential side gears and bevel gears work together to keep torque equal on both the inner and outer wheel.

The limited slip differential is designed to provide equal torque to the rear wheels when one wheel is slipping on ice or other slippery surfaces. Clutches or cones are used to lock up the two side gears connected to the axles. With these gears locked up, equal torque will be applied to both wheels during these abnormal conditions.

The rear axle is connected to the side gears inside the differential. As the differential side gears are turned, the axles also turn. The axle and side gear are splined together. The axle is then held in place in the axle tube by bearings.

In addition, seals and gaskets are used to keep the differential lubricant inside the axle tube and differential.

TERMS TO KNOW

Can you explain each of the following terms? Review the chapter until you can use each term correctly.

Cardan universal joint	Drive shaft
Constant velocity joint (CV joint)	Slip joint
	Galling
Bearing cap	Tripod
Trunnion	Differential
Yoke	Limited slip differential
Cross	

REVIEW QUESTIONS

Multiple Choice

1. Which of the following is used to transmit the torque to the differential?
   a. Axles
   b. Universal joints
   c. Drive shaft
   d. Axle bearings
   e. B and c

2. Which of the following is/are used to allow the drive shaft to rotate through changing angles?
   a. Differential side gears
   b. Differential ring gear
   c. Universal joints
   d. Slip joint
   e. None of the above

3. Which of the following are used to allow the length of the drive shaft to change (lengthen and shorten) during operation?
   a. Differential side gears
   b. Bearings
   c. Universal joints
   d. Differential ring gears
   e. Slip joints

4. Which of the following is part of the universal joint?
   a. Yoke
   b. Trunnion
   c. Needle bearings
   d. All of the above
   e. None of the above

5. _____ vibration is caused by the power pulses of the engine.
   a. Transverse
   b. Elongated
   c. Trunnion
   d. Torsional
   e. Velocity

6. Cardan universal joints have one major disadvantage. That is that:
   a. The input shaft continually slows down during operation.
   b. The output shaft continually slows down during operation.
   c. There is always excessive wear on the tripods.
   d. The output shaft changes velocity in each revolution due to the angle of operation.
   e. The input shaft changes velocity in each revolution due to the angle of operation.

7. A _____ is used on vehicles that require a long drive shaft distance.
   a. Center bearing
   b. Single universal joint
   c. Double slip joint
   d. Constant velocity joint
   e. All of the above

8. Front wheel drive systems:
   a. Use constant velocity joints
   b. Use two drive shafts
   c. Must have a slip joint on each wheel
   d. All of the above
   e. None of the above

9. Which type of joint on a drive system does not have a velocity change within one revolution?
   a. Cross/bearing type
   b. Constant velocity
   c. Cardan type
   d. All of the above
   e. None of the above

10. The fixed-type constant velocity joint is used:
    a. On the inboard side of the drive shaft
    b. On the outboard side of the drive shaft
    c. On rear wheel drive systems
    d. On the inside of the differential
    e. On rear axles

11. Which of the following is/are *not* a part of the differential?
    a. Pinion gear
    b. Ring gear
    c. Side and bevel gears
    d. Universal joints
    e. Differential case

12. The differential is used on the drive train of a vehicle:
    a. To transmit equal power during cornering
    b. To allow for changing angles of the drive train
    c. To reduce the velocity changes during rotation
    d. To support the axles only
    e. None of the above

13. When the vehicle is moving in a straight line and there is equal resistance at each wheel:
    a. The differential case is turning
    b. The side gears and bevel gears are in mesh but are not rotating with each other
    c. The ring gear is turning
    d. All of the above
    e. None of the above

14. The adjusters on the differential are used to change the backlash between the pinion gear and the _____.
    a. Side gears
    b. Axles
    c. Differential case
    d. Bevel gears
    e. Ring gear

15. To eliminate one wheel spinning on ice or other slippery surface, the _____ is/are used.
    a. Double ring gear
    b. Universal joints
    c. Limited slip differential
    d. Spacer
    e. Bearing

16. The axle and differential side gears are attached by:
    a. Bolts
    b. A strong welded section
    c. An internal and external spline
    d. Grease seals
    e. Gaskets

17. The process of needle bearings producing grooves in the trunnion is called:
    a. Metal wear
    b. Brinelling
    c. End galling
    d. Yoke
    e. Torsional vibration

18. Which of the following is not a service concern on drive shafts and differentials?
    a. Correct lubricant in U-joints
    b. Adjustment between ring gear and pinion gear
    c. Backlash adjustment
    d. End galling
    e. Piston scraping

The following questions are similar in format to ASE (Automotive Service Excellence) test questions.

19. Technician A says that backlash can be checked on the differential by using a vernier caliper to measure the correct clearance. Technician B says that backlash can be checked by using a dial indicator. Who is right?
    a. A only
    b. B only
    c. Both A and B
    d. Neither A nor B

20. Technician A says that too much lubrication can damage the seals in the U-joint. Technician B says that too much lubrication can damage the spider in the U-joint. Who is right?
    a. A only
    b. B only
    c. Both A and B
    d. Neither A nor B

21. A humming noise is heard in the rear end of the vehicle. Technician A says the differential backlash may be out of adjustment. Technician B says the differential ring gear may be damaged or worn. Who is right?
    a. A only
    b. B only
    c. Both A and B
    d. Neither A nor B

22. Technician A says that the U-joints can be replaced by using common tools in the automotive shop. Technician B says that the universal joints are replaced by using special presses and hydraulic jacks. Who is right?
    a. A only
    b. B only
    c. Both A and B
    d. Neither A nor B

Essay

23. Define the purpose of using universal joints.

24. What is the purpose of a slip joint?

25. What is the difference between constant velocity joints and Cardan universal joints?

26. State the purpose of a differential.

27. What is a limited slip differential?

28. What is galling on universal joints?

CHAPTER 28

Braking Systems

INTRODUCTION

The automobile braking system is used to control the speed of the vehicle. The system must be designed to enable the vehicle to stop or slow down at the driver's command. The brake system is composed of many parts, including friction pads on each wheel, master cylinder, wheel cylinders, and a hydraulic control system. This chapter discusses the total braking system used on the automboile today.

OBJECTIVES

After reading this chapter, you will be able to:
- Identify the principles of friction, hydraulic circuits, and basic braking system operation.
- State the name and operation of all braking system components.
- Analyze the purpose and operation of power brake systems.
- Analyze various diagnosis and service tips and procedures used on braking systems.

CHAPTER HIGHLIGHTS

28.1 BRAKING SYSTEM PRINCIPLES
  A. Friction
  B. Heat Energy
  C. Friction and Braking Systems
  D. Basic Operation of Drum Brakes
  E. Shoe Energization
  F. Servo Type Brakes
  G. Primary and Secondary Shoe Operation
  H. Basic Operation of Disc Brakes
  I. Caliper Operation
  J. Fixed Caliper Design
  K. Floating Caliper Design
  L. Fluids
  M. Hydraulic Principles
  N. Force and Pressure

28.2 BRAKING SYSTEM COMPONENTS AND OPERATION
  A. Total System Operation
  B. Master Cylinder Operation
  C. Forward Stroke
  D. Return Stroke
  E. Compensating Port
  F. Residual Pressure Check Valve
  G. Reservoir Diaphragm Gasket
  H. Dual Master Cylinder
  I. Diagonal Brake System
  J. Additional Brake Fluid Components
  K. Brake Lines
  L. Brake Linings
  M. Drum Brake Wheel Cylinder

28.1 BRAKING SYSTEM PRINCIPLES

Friction

Friction is defined as a resistance to motion between two objects. When two surfaces rub against each other, there is friction. See **Figure 28–1**. The amount of friction depends on two things: the roughness of the surfaces and the amount of pressure between the two surfaces.

Heat Energy

From an energy viewpoint, when there is friction, *kinetic energy* (energy in motion) is converted to thermal (heat) energy. The larger the amount of kinetic energy that must be brought to rest, the greater the amount of heat produced.

The energy of motion or kinetic energy depends upon the weight of the vehicle and the speed of the vehicle. Brakes must also then be able to remove the heat that is produced.

Friction and Braking Systems

In any braking system, the amount of friction is controlled by the operator. By varying friction, the vehicle can be stopped, and its speed can be modified on curves, grades, and in different driving conditions. The control of friction is obtained by forcing a stationary brake shoe or pad against a rotating drum or disc. As the driver presses harder on the brake pedal, friction increases.

When the wheel is being slowed down by the brake friction, the tire is also slowed down. However, friction is also produced between the tire and the road. The friction on the brakes must be matched by the friction of the tires and the road. If the wheels cannot produce the friction, the tire will lock up and skid. A car stops better when the wheels are not locked. Locked wheels can produce dangerous results, especially since there is no driver control of the friction between the tires and the road. *Computer-controlled brakes* are now being used to control the slowing down of each wheel without skidding.

Basic Operation of Drum Brakes

A drum brake assembly consists of a cast drum that is bolted to and rotates with the wheel. Inside the drum, there is a *backing plate* that has a set of *brake shoes* attached to it. Other components are also attached to the backing plate, including a hydraulic cylinder and several springs and linkages. The brake shoes are lined with a frictional material. The frictional material contacts the inside of the drum when the brakes are applied. See **Figure 28–2**. When the

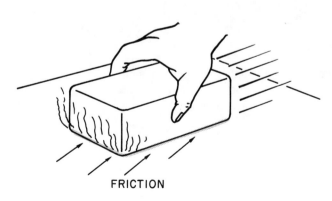

FIGURE 28–1. When two surfaces rub together, friction is produced. Brakes produce friction to stop or slow down the vehicle.

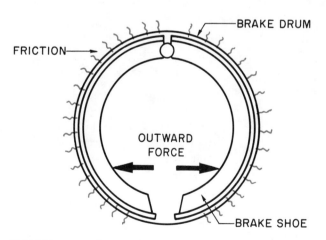

FIGURE 28–2. On a drum brake system, the shoes are forced outward against a brake drum to produce the necessary friction.

FIVE-CYLINDER ENGINE

This 2.2 liter, five-cylinder engine produces approximately ten percent more horsepower and nine percent more torque than the engine it replaced. It has a sophisticated electronic fuel injection system and electronic ignition. (Courtesy of Volkswagen United States, Inc.)

brakes are applied, the brake shoes are forced out and produce friction against the inside of the drum.

Shoe Energization

When the brakes are applied, it is important for the shoe to be *self-energizing*. When the brake shoe is engaged, the frictional drag acting around the shoe tends to rotate the shoe about its hinged point as shown in *Figure 28–3*. When the drum rotates in the same direction, the frictional drag between the two will cause the shoe to become tighter against the inside of the drum. This action is called self-energizing.

Servo Type Brakes

A *servo* is a device that converts a relatively small force into a larger motion. In most vehicles today, *servo brakes* are used to cause the brake shoes to move outward from a hydraulic pressure inside a cylinder. The pressure is produced by the operator's foot. The motion is the outward push of the brake shoes against the drum.

Primary and Secondary Shoe Operation

In a drum brake system, there is a primary and secondary shoe as shown in Figure 28–3. When the brakes are applied, the *primary shoe* reacts first. It has a weaker return spring.

The shoe lifts off the anchor and contacts the drum surface. The anchor normally acts as a stop. As the shoe begins to contact the drum, the shoe is energized, forcing it to rotate deeper into the drum.

During this time, there is also action on the *secondary shoe*. As the primary shoe moves, it tends to push or move the secondary shoe at the bottom. This motion forces the secondary shoe in the same direction as the drum. Note that the secondary shoe cannot move upward because it is forced against the anchor. This causes the secondary shoe to also be energized. The servo brake system then acts or behaves as if it were one continuous shoe.

The actuating force from the hydraulic cylinder only pushes on one shoe. A small amount of pressure is usually produced against the secondary shoe by the hydraulic cylinder. This force is not, however, used to energize the brakes, except in a reverse direction.

Basic Operation of Disc Brakes

Many vehicles today use *disc brakes* along with drum brakes. On many vehicles, disc brakes are used on the front of the vehicle, while drum brakes are used on the rear wheels. Disc brakes resemble the brakes used on a ten-speed bicycle. The friction is produced by using pads. This is shown in *Figure 28–4*. These pads are squeezed or

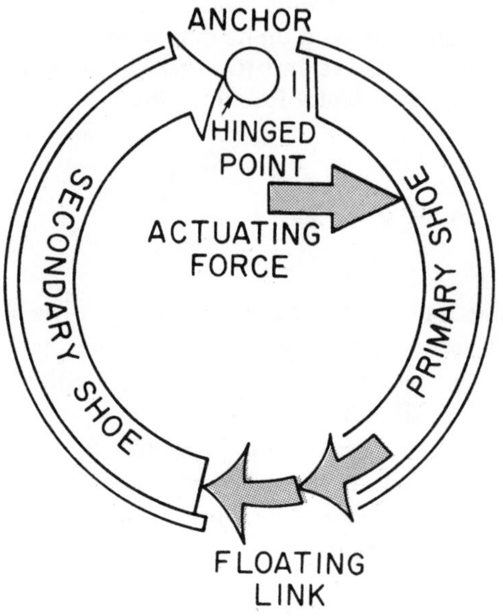

FIGURE 28–3. When the brakes are applied, the primary shoe reacts first. Then the primary shoe pushes against the secondary shoe to produce additional friction. *(Courtesy of EIS Brake Parts, Division Standard Motor Products, Inc.)*

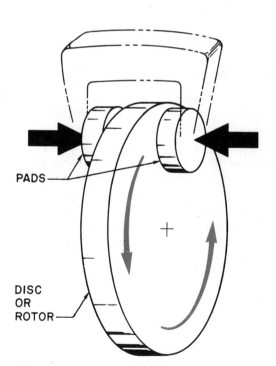

FIGURE 28–4. A disc brake system uses two pads reacting against a rotor to produce the friction and stop the vehicle. *(Courtesy of EIS Brake Parts, Division Standard Motor Products, Inc.)*

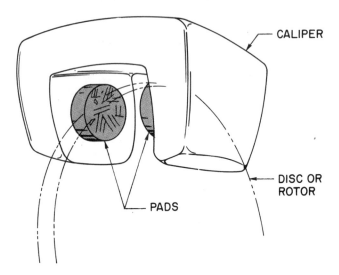

FIGURE 28–5. On a disc brake system, the pads work perpendicular to the rotor. *(Courtesy of EIS Brake Parts, Division Standard Motor Products, Inc.)*

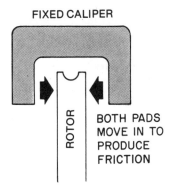

FIGURE 28–6. The fixed caliper remains stationary and pads move in and out to produce friction. There are two pistons on this system. *(Courtesy of EIS Brake Parts, Division Standard Motor Products, Inc.)*

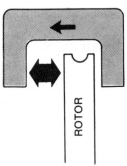

FIGURE 28–7. The floating caliper is permitted to slide in and out a small amount on its mountings. There is only one piston on this system. *(Courtesy of EIS Brake Parts, Division Standard Motor Products, Inc.)*

clamped against a rotating wheel. The wheel, also called the rotor, is attached to the rim and tire. The rotor is made of cast iron that is machined on both sides. The pads are attached to metal plates which are actuated by pistons from the hydraulic system.

Caliper Operation

The pistons in a disc brake system are contained or held in place by the *caliper*. The caliper does not rotate because it is attached to the vehicle's steering mechanism. The caliper is a housing that contains hydraulic pistons and cylinders. It also contains seals, springs, and fluid passages that are used to produce the movement of the piston and pads.

The pads act perpendicular to the rotation of the rotor. See *Figure 28–5*. This is different from the drum brake system. Disc brakes are said to be "non-self-energized." This means that they require more force to achieve the same braking effort. For this reason, disc brakes are usually used with power brakes.

Fixed Caliper Design

There are two types of caliper designs: the fixed caliper and the floating caliper. The *fixed caliper* design has the caliper assembly attached directly to the frame or steering components. Each pad is actuated by a piston. *Figure 28–6* shows the fixed caliper deisgn.

Floating Caliper Design

The *floating caliper* design has the main housing of the caliper able to slide in and out a small amount on the

mountings. There is a piston on only one side. The other has only a friction pad. When the brakes are applied, the hydraulic pressure within the cylinder pushes the piston in one direction. The entire caliper housing is free to slide in the opposite direction. As the pads contact the rotor, the force of the piston pad is matched by an equal force from the pad on the other side of the caliper. *Figure 28–7* shows a floating caliper design.

Fluids

Fluids play an important part in braking systems. Brake fluid is used to transfer the motion of the operator's foot to the cylinders and pistons at each brake. Fluids cannot be compressed, while gases are compressible as shown in *Figure 28–8*. Any air in the brake hydraulic system will compress as the pressure increases. This action will reduce the amount of force that can be transmitted. This is why it

FIGURE 28–8. Hydraulic fluid is used in brake systems because it is noncompressible. Air, however, is compressible and must be removed from the hydraulic brake system. *(Courtesy of Delco Moraine Division, General Motors Corporation)*

is very important to keep all air out of the hydraulic system. To do this, air must be bled from brakes. This is called *bleeding* the brake system.

Hydraulic Principles

The automotive braking system uses hydraulic pressure to transfer the force of the operator's foot to press the friction surfaces together. Referring to *Figure 28–9*, when the foot pedal is pressed down, a pressure is built up in the master cylinder. This pressure is then transferred throughout the hydraulic lines to each wheel cylinder. Note that the pressure at each point in the system is the same.

Force and Pressure

There is a specific relationship between the force of the pedal and the piston area in a closed *hydraulic system*. If a force of 100 pounds were applied to a piston with an area of 1 square inch, a pressure of 100 pounds per square inch would be produced. Also, as shown in *Figure 28–10*, if there are other pistons in the hydraulic system, they may

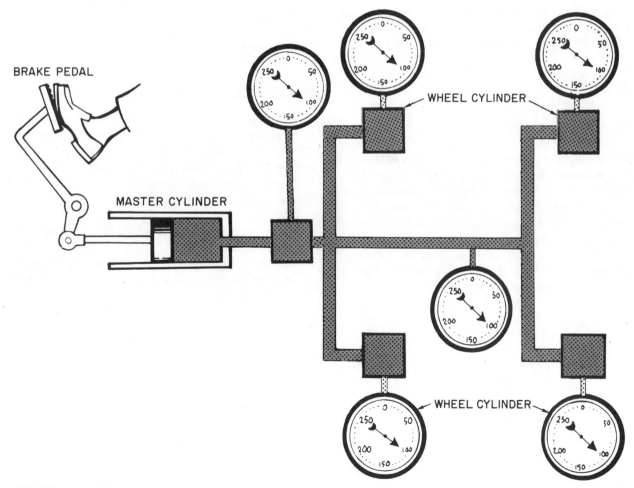

FIGURE 28–9. When the operator pushes the brake pedal, hydraulic pressure increases and is sent to each wheel cylinder to operate the brake mechanism. *(Courtesy of Delco Moraine Division, General Motors Corporation)*

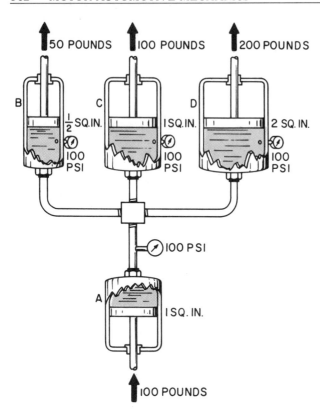

FIGURE 28-10. Pressures in the hydraulic system can be changed by changing the size of the piston. *(Courtesy of EIS Brake Parts, Division Standard Motor Products, Inc.)*

produce different pressures because of their size. A 1/2 square inch piston produces 50 pounds of force. The 1 square inch piston produces 100 pounds of force. The 2 square inch piston produces 200 pounds of force. This example shows that a certain force applied to a hydraulic system can produce different forces, depending upon the piston size.

Very little movement of fluid occurs in the hydraulic system. It is the pressure and the forces that do the job needed in the braking system. In actual practice, the fluid is only used to transfer the force and pressure of the operator's foot to the piston and friction pads.

CAR CLINIC ▮▮▮▮▮▮

PROBLEM: REAR BRAKE NOISE

A vehicle with 80,000 miles on it is producing a scraping noise on the rear right brake when it is raining or very damp outside. What could be the problem?

SOLUTION:

Brake linings can become glazed and covered with road dirt and grease under wet and damp conditions. Glazed linings are very hard, which can produce noises. In addition, some of the brake retainer parts and mechanisms may be worn or locked up. Remove the rear brake and check all parts, including the drums, for wear and correct operation. Also, consider replacing the brake linings.

28.2 BRAKING SYSTEM COMPONENTS AND OPERATION

Total System Operation

A common brake system is shown in *Figure 28-11*. In this system, drum brakes are used on all wheels. The system starts at the brake pedal which is attached to the *master cylinder*. The master cylinder is used to produce the necessary pressure in the hydraulic system. Hydraulic lines are connected from the master cylinder, through the distribution block and warning light switch, to the individual wheels. Here the hydraulic pressure is sent to each *wheel cylinder* which finally moves the drum brake mechanism. In addition to the hydraulic system, there is also a parking brake system that uses mechanical linkage to the rear wheels. Either the pedal mechanism or the hand parking brake can be used.

Master Cylinder Operation

The purpose of the master cylinder is to convert the mechanical force of the operator's foot to hydraulic pressure. The main components of the master cylinder are shown in *Figure 28-12*. Although many designs are used for master cylinders, the principles remain the same. The important parts include:

a. the push rod which is moved by the movement of the operator's foot,
b. the piston which is used to produce the pressure,
c. the primary and secondary sections which are used to help produce the pressure,
d. the return spring which is used to return the pedal after braking, and
e. the compensating and intake ports which are used to enhance the speed of the fluid flow.

Forward Stroke

When the operator pushes the brake pedal, the piston (push rod) moves to the left as shown in *Figure 28-13*. This action causes the pressure on one side of the piston to be increased. Pressure is first produced when the primary cup

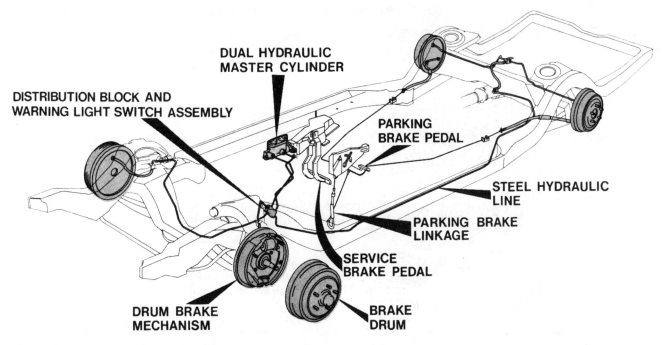

FIGURE 28–11. A common brake system and the major parts are shown on the vehicle. *(Courtesy of Delco Moraine Division, General Motors Corporation)*

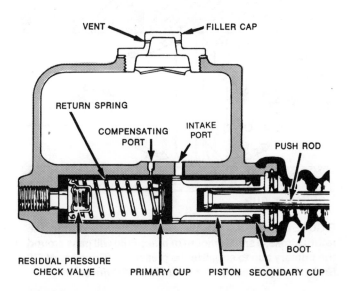

FIGURE 28–12. The main components of a master cylinder. *(Courtesy of EIS Brake Parts, Division Standard Motor Products, Inc.)*

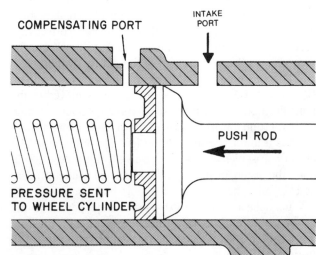

FIGURE 28–13. Once the push rod moves past the compensating port, pressure starts to build up. This pressure is then sent to the wheel cylinders. *(Courtesy of EIS Brake Parts, Division Standard Motor Products, Inc.)*

passes the *compensating port*. The vacuum produced on the backside is relieved because brake fluid is drawn in from the reservoir above the piston through the intake port. The reservoir is also needed because as the brake pads and linings wear, there will be more fluid displacement during braking. The reservoir holds this extra amount of fluid when needed.

Return Stroke

During the return stroke, the brake pedal is pushed back to its original position. The spring is used to move the piston back. During this action, a low pressure is created on the left

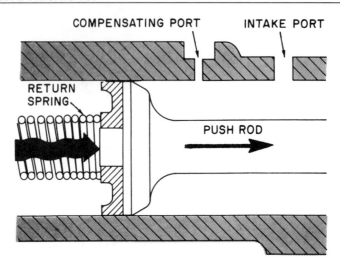

FIGURE 28–14. The return spring is used to push the master cylinder piston back to its original position. *(Courtesy of EIS Brake Parts, Division Standard Motor Products, Inc.)*

side of the piston. Refer to **Figure 28–14**. The piston moves back faster than the fluid coming from the brake lines. If the operator immediately reapplied the brakes, there would not be enough fluid for them to operate correctly. In order to remedy this, fluid must be able to flow from the secondary to the primary port of the master cylinder.

The shape of the primary cup allows this to happen. **Figure 28–15** shows that as the piston is moved to the right during the return stroke, a certain amount of fluid passes around the primary cup. This can only happen when there is a lower pressure on the left side of the cup. The piston cup serves as a one-way valve. Under these conditions, the piston, push rod, and brake pedal will return very quickly, allowing for successive rapid brake strokes.

To make up for the extra brake fluid passed to the left of the piston, a certain amount of fluid is drawn from the reservoir into the secondary area of the piston. See **Figure 28–16**.

Compensating Port

Now that the piston has been fully returned, the fluid from the brake lines will continue to enter the area left of the piston. This area is now, however, full of brake fluid from the return stroke. If there were no place for the fluid from the lines to go, the brakes would not release as necessary. Another passage, called the compensating port, is used. The compensating port allows the excess fluid to return to the reservoir when the pedal is released. The compensating port is only uncovered when the piston is fully returned. See **Figure 28–17**.

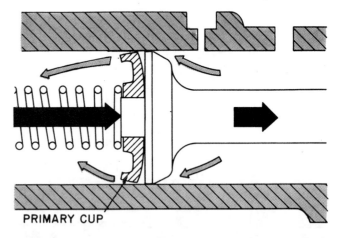

FIGURE 28–15. As the piston moves back during the return stroke, a small amount of brake fluid will pass around the primary cup to equalize the pressure. *(Courtesy of EIS Brake Parts, Division Standard Motor Products, Inc.)*

Residual Pressure Check Valve

In the past, a *residual pressure* check valve was used on drum brake systems. In theory, it was felt that a small amount of pressure on the brake lines would be beneficial. This could reduce the possibility of air getting into the system. Also, a slight pressure on the system would take up any slack in the linkages in the brake mechanism at each wheel. A small pressure check valve was placed inside the primary area of the master cylinder. Since disc brakes don't

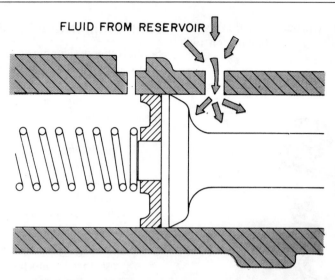

FLUID FROM RESERVOIR

FIGURE 28-16. Any extra brake fluid needed on the right side of the primary cup will be admitted through the intake port. *(Courtesy of EIS Brake Parts, Division Standard Motor Products, Inc.)*

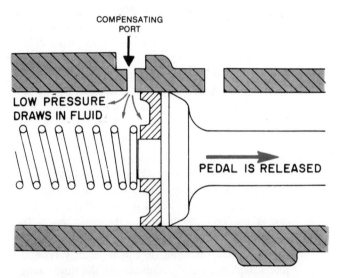

COMPENSATING PORT

LOW PRESSURE DRAWS IN FLUID

PEDAL IS RELEASED

FIGURE 28-17. The compensating port allows excess fluid to return to the reservoir when the pedal is released. *(Courtesy of EIS Brake Parts, Division Standard Motor Products, Inc.)*

employ shoe return springs, these valves are omitted. Also, in some drum brake systems, the check valve has been eliminated by improved design of the wheel cylinders.

Reservoir Diaphragm Gasket

A flexible rubber gasket is used between the reservoir and the master cylinder cap. It is used to stop moisture and dirt from getting into the brake fluid reservoir. The reservoir must be vented to the atmosphere because of the rising and falling of the brake fluid level during brake operation. The diaphragm gasket separates the brake fluid from the air above it, while remaining free to move up and down with fluid level changes. *Figure 28-18* shows a typical reservoir diaphragm gasket.

Dual Master Cylinder

The dual master cylinder provides two separate and distinct pressure chambers in one bore. This design was required by federal law in the late 1960s. Should a failure occur in one master cylinder piston, the second piston will still work. *Figure 28-19* shows a typical dual master cylinder in the applied position. One chamber is used for the front brakes and the other chamber is used for the rear brakes. Note that on this system the by-pass holes act as the compensating port. The operation of the dual master cylinder is the same as that of the single cylinder just described, except that two cylinder pressures are being developed. One cylinder is actuated by the push rod. The second cylinder is operated by a spring and the "plug" of fluid between the two. This system is sometimes called a *tandem* master cylinder.

Diagonal Brake System

Another variation in brake systems is the diagonal brake system. In the diagonal system, the right-front and left-rear brakes are connected to one chamber of the master cylinder. The left-front and right-rear brakes are connected to the other master cylinder chamber. The purpose of this system is again to make sure that there is still braking on two wheels if the master cylinder fails to work.

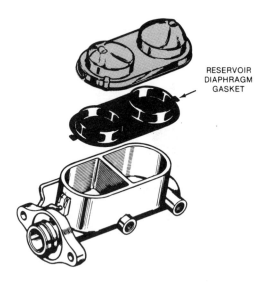

FIGURE 28–18. The reservoir diaphragm gasket keeps dirt out of the reservoir and allows the brake fluid to rise and fall during normal braking operation. It is also vented to the atmosphere. *(Courtesy of EIS Brake Parts, Division Standard Motor Products, Inc.)*

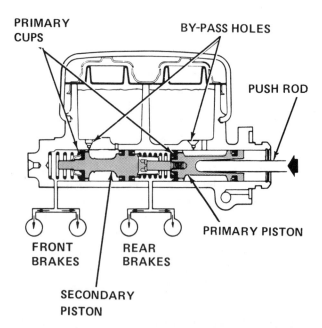

FIGURE 28–19. The dual master cylinder has both a primary and secondary piston to produce pressure. If one master cylinder fails, the other will stop the vehicle. *(Courtesy of Delco Moraine Division, General Motors Corporation)*

Additional Brake Fluid Components

Several other designs are used on brake fluid systems. These all depend upon the type of vehicle and the year the vehicle was manufactured. They include the quick-take-up master cylinder, warning light switches, and proportioning and metering valves.

1. *Quick-take-up master cylinder* — This system has a master cylinder with two different bore sizes. Its purpose is to displace a larger amount of brake fluid during the initial stages of brake pedal movement. This helps to take up shoe return spring linkage more quickly.

2. *Warning light switches* — A warning light switch is cast directly into the master cylinder. This switch senses the pressure in the master cylinder. If one half of the system has failed, the pressure in the other half will be greatly increased. This pressure trips an electric switch and lights up a warning light.

3. *Proportioning valve* — The *proportioning valve* is used to proportion the pressure to the rear brakes and the front brakes. It is located in the brake line after the master cylinder. The harder the brakes are applied, the greater the shift to the front of the vehicle. If the pressure is equal to all wheels, the back wheels may lock up, causing loss of vehicle control. As the force on the brake pedal increases, the proportioning valve causes the pressure to the rear brakes to be less than to the front brakes. This action reduces the possibility of rear wheel skidding.

4. *Metering valve* — The metering valve is used on disc/drum brake systems. The metering valve keeps the front discs from operating until the rear drums have started to work. This is needed because the disc system operates faster than the drum brakes. In operation, the fluid to the front disc brakes must go through the metering valve. The metering valve acts like a regulator valve. It holds back the fluid to the front brakes until a certain amount of pressure has been developed. When this pressure is reached, the metering valve opens and the system operates normally.

Brake Lines

Brake lines are used to carry the brake fluid and pressure from the master cylinder to the individual cylinders. The brake lines are made of double-walled, rust-resisting steel except where they have to flex. Flexing usually occurs between the chassis and the front wheels. Here flexible high-pressure hoses are used. All brake lines are designed for high pressure by using double-flared ends and connectors.

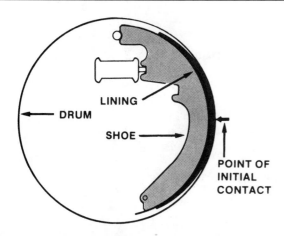

FIGURE 28-20. When a brake shoe is applied on a drum brake system, the initial point of contact is near the center. *(Courtesy of Allied Aftermarket Division, Allied Automotive)*

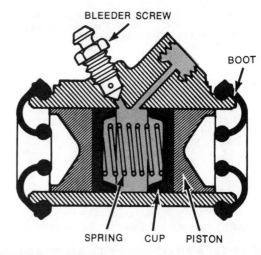

FIGURE 28-21. The standard wheel cylinder is shown with each part identified. *(Courtesy of EIS Brake Parts, Division Standard Motor Products, Inc.)*

Brake Linings

Brake linings provide the friction against the drum to stop the car. There are many kinds of linings. The lining is attached to the shoe either by riveting or by bonding. The primary shoe has the shorter length lining. The secondary shoe has a full-length lining because it carries a bigger load. In addition, most brake linings used today are ground so that they are slightly thicker at the center. This design improves the ease with which the lining comes in contact with the drum. This is shown in *Figure 28-20*. When the shoe pressure is increased, the lining and shoe flex slightly to produce full contact.

Drum Brake Wheel Cylinder

The purpose of the wheel cylinder is to convert hydraulic pressure to mechanical force. *Figure 28-21* shows a typical wheel cylinder for a drum brake system. The assembly includes two pistons, two cups, two boots, a bleeder screw, and an internal spring. When two pistons are used, it is called a duo servo system. When the brakes are applied, hydraulic pressure inside the wheel cylinder forces both pistons outward, causing the brakes to be applied.

Duo Servo Drum Brake Assembly

There are many variations on the drum brake assembly. One common type is shown in *Figure 28-22*. The parts include:

a. Primary shoe — used to produce friction for stopping (the forward shoe).
b. Secondary shoe — used to produce friction for stopping (the rear shoe).

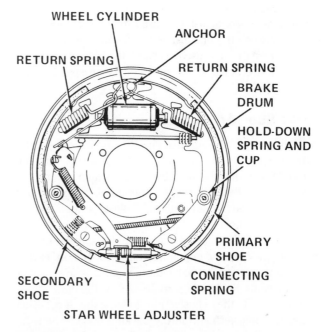

FIGURE 28-22. A duo servo drum brake and associated parts. *(Courtesy of Delco Moraine Division, General Motors Corporation)*

c. Return springs — used to pull the shoes back away from the drum after the brakes have been released.
d. Wheel cylinder — used to produce the mechanical motion to move the brake shoes.
e. Hold-down spring and cup — used to hold the brake shoes against the backing plate.
f. Anchor — used for self-energization and as a stop for the brake shoes.

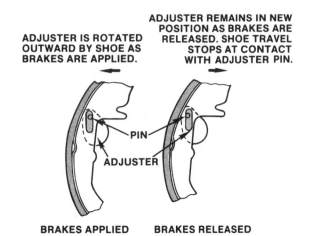

FIGURE 28-23. A set of eccentric cams is used on some brake systems to adjust the brakes automatically. *(Courtesy of Allied Aftermarket Division, Allied Automotive)*

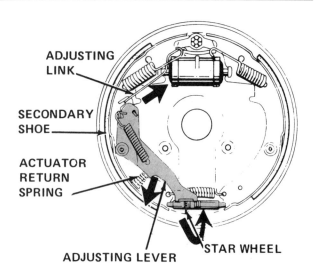

FIGURE 28-24. Certain types of brake systems use an adjusting lever that works against the star wheel to adjust the brakes. *(Courtesy of EIS Brake Parts, Division Standard Motor Products, Inc.)*

g. Connecting spring — used to hold the brake shoes together on the bottom.

h. Star wheel adjuster — used to adjust the distance between the brake shoe linings and the drum.

i. Brake drum — used to absorb the friction produced by the brake shoes and reduce the speed of the wheels.

Brake Shoe Adjustment

Drum brakes may be adjusted either manually or by automatic adjusters to compensate for lining wear. Manually adjusted brakes have an adjusting screw, which is normally called a *star wheel adjuster*, for this purpose. As the star wheel is turned by an external adjusting tool, excess clearance is removed.

Automatically adjusted brakes are designed so that as the shoes move in and out during normal operation, excess lining clearance is removed. One method is to use a set of eccentric cams. Refer to *Figure 28-23*. As the brake shoes travel outward, the adjuster pin follows the shoe. This rotates the adjuster cam on the backing plate. When the brake is released, the adjuster will remain in the new position.

A second method of automatically adjusting brakes is to use the ratchet adjuster. *Figure 28-24* shows this system. The adjusting lever acts like a ratchet on the star wheel. Each time the brake shoe moves outward, the ratchet mechanism tries to advance the star wheel to make the adjustment. This happens whenever the brake is applied when the vehicle is moving in reverse.

Parking Brakes

The parking brake is a hand- or foot-operated mechanical brake designed to hold the vehicle while it is parked. A sim-

ple parking brake system is shown in *Figure 28-25*. The system uses a series of mechanical cables that are operated by the parking brake pedal. When the parking brake is applied, the parking brake cables and equalizer apply a balanced pull on the parking brake levers in the rear wheels. The levers and the parking brake strut move the brake shoes outward against the brake drum. This position is held until the parking brake pedal is released.

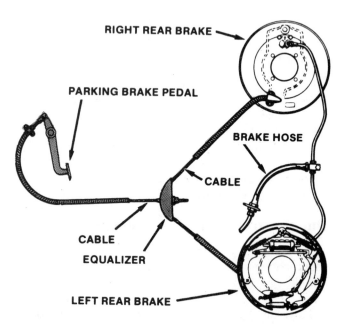

FIGURE 28-25. A typical integral parking brake system. *(Courtesy of Allied Aftermarket Division, Allied Automotive)*

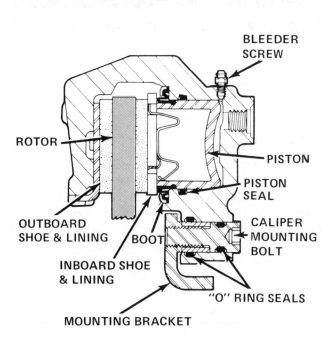

FIGURE 28–26. A floating-caliper disc brake assembly is shown with all parts identified. *(Courtesy of Delco Moraine Division, General Motors Corporation)*

FIGURE 28–27. A complete assembly of a disc brake system.

Disc Brake Assembly

As with the drum brake assembly, there are many arrangements for disc brake assemblies. *Figure 28–26* shows a floating-caliper type of disc brake assembly. The parts include:

a. Inboard and outboard shoe and lining — used to produce the friction against the rotor.
b. Rotor — attached to the wheel and used to absorb the friction to slow down the wheel.
c. Piston — used to produce the pressure from the hydraulic system to force the shoes against the rotor.
d. Piston seal — used to seal the brake fluid inside the piston bore.
e. Boot — used to keep dust and dirt out of the piston bore.
f. Mounting bracket — used to hold the assembly on the vehicle.
g. Bleeder screw — used to remove air from the hydraulic fluid.

Although the parts on other vehicles may look different, the principles of operation are still much the same. *Figure 28–27* shows a complete assembly of a disc brake system.

Brake Fluid

A wide variety of materials is used within a standard braking system. Several types of metal, rubber, and plastics that come in contact with the brake fluid are used. Brake fluid must be compatible with all materials in the brake system and maintain stability under varying conditions, both in temperature and pressures. Because of these conditions, brake fluid must possess the following characteristics:

1. *Viscosity* — must be free flowing at all temperatures.
2. *High boiling point* — must remain liquid at high operating temperatures without vaporization.
3. *Noncorrosive* — must not attack metal, plastic, or rubber parts.
4. *Water tolerance* — must be able to absorb and retain moisture that collects in the system. The characteristic is called hygroscopic. Water causes pitting in the brake system.
5. *Lubricating ability* — must lubricate pistons and cups to reduce wear and internal friction.
6. *Low freezing point* — must meet a certain freezing point as established by Federal Motor Vehicle Safety standards.

It is best to refer to the vehicle manufacturer's recommendations to determine the exact type of brake fluid to use. The Department of Transportation (DOT) specifies brake fluid for vehicles. Manufacturers recommend a specific DOT specification.

Brakelight Switches

The brakelight switch is a spring-loaded electrical switch that comes on when the brake pedal is depressed. There are generally two types of switches. These include:

1. *Mechanically operated switches* — used on most recent model vehicles.
2. *Hydraulically operated switches* — used on older vehicles.

The mechanically operated brakelight switch is operated by contact with the brake pedal. It is usually attached to a bracket on the brake pedal. The hydraulic switch is operated by hydraulic pressure developed in the master cylinder. In both types, there is no electrical current through the switch when the brakes are not being applied. When the brakes are applied, the circuit through the switch closes and causes the brakelight to come on. *Figure 28–28* shows examples of various mechanically and hydraulically operated switches.

CAR CLINIC

PROBLEM: IS A WATER FILTER NEEDED?

A gasoline engine has a very dirty cooling system. The coolant looks brown and seems to have a lot of rust in it. No antifreeze has been mixed with the water for about three years. There is an overflow tank on the system. Would a water filter help the situation?

SOLUTION:

Rust in the coolant is usually caused by oxygen in the cooling system or the lack of a rust inhibitor in the coolant. The overflow tank will help eliminate some of the oxygen. The key here is that no antifreeze has been added to the water in the coolant. Antifreeze has a rust inhibitor in it, so rust should be even further reduced. Running an engine without antifreeze will certainly produce rust in the coolant. Normally, a water or coolant filter isn't needed on a gasoline engine. Antifreeze today has most of the chemicals needed for correct operation.

28.3 POWER BRAKES

Power brakes are used today on many passenger cars. Power brakes are designed to have an extra pressure called a

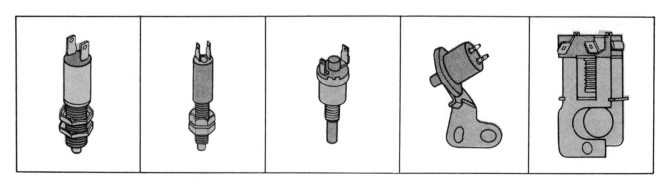

Mechanically Operated Switches

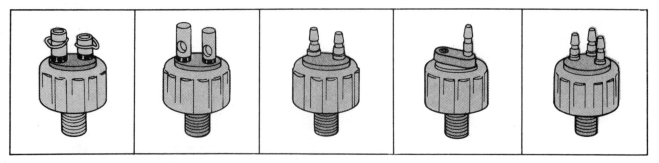

Hydraulically Operated Switches

FIGURE 28–28. There are many brakelight switches used. Both the mechanically operated and hydraulically operated switches are common. (Courtesy of Allied Aftermarket Division, Allied Automotive)

booster. The boost is produced either by a vacuum or by hydraulic fluid acting as an extra force for the brake pedal. When the brake pedal is applied, the booster unit multiplies the pedal force for the master cylinder. This means the operator puts less force on the brake pedal, making it easier to stop the car. The booster unit is placed between the brake pedal and the master cylinder. The master cylinder and the rest of the brake system parts are all identical to a regular brake system.

Vacuum-assisted Brakes

Intake manifold vacuum can be used as the booster in a power brake system. Referring to *Figure 28–29*, power brakes use a diaphragm with a vacuum placed on each side. The center shaft of the diaphragm is connected to the master cylinder. If a vacuum is placed on both sides of the center diaphragm, the diaphragm will not move. However, if the vacuum is removed and atmospheric pressure is admitted to the right side of the diaphragm, the center shaft will be forced to the left. This motion can then be used to operate the master cylinder. When vacuum is returned to the right side of the diaphragm, the brakes are released. The brake pedal is used to open, hold, or close two internal valves to allow atmospheric pressure or vacuum to enter the right side of the diaphragm. An example of a booster unit attached to the master cylinder is shown in *Figure 28–30*.

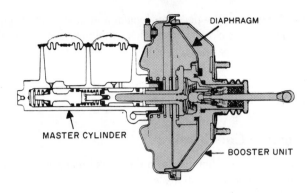

FIGURE 28–30. The booster unit for power brakes is attached directly to the master cylinder. *(Courtesy of EIS Brake Parts, Division Standard Motor Products, Inc.)*

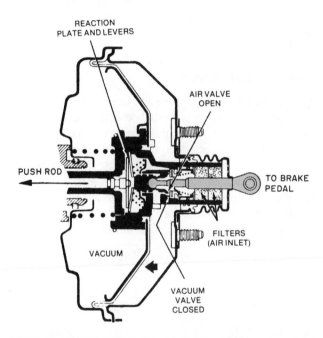

FIGURE 28–31. When the operator pushes the brake pedal, vacuum on the right is released which pushes the push rod to the left. *(Courtesy of EIS Brake Parts, Division Standard Motor Products, Inc.)*

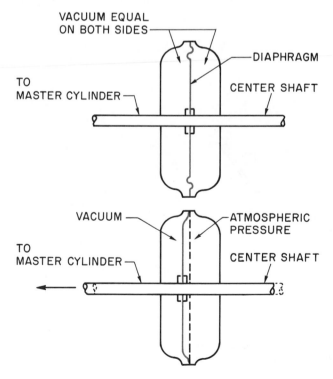

FIGURE 28–29. Power brakes use vacuum to produce the push against the push rod on the master cylinder.

Referring to *Figure 28–31*, the power brake system operates in one of three modes. These include apply, hold, and release. Normally, the entire inside of the canister is under a vacuum. When the driver steps on the brake pedal, the mechanism closes the vacuum valve and opens the air valve. Air is allowed to come into the right side of the diaphragm through air filters. Now the diaphragm is forced to move to the left. As braking increases, the back pressure on the hydraulic system in the master cylinder is felt on the brake pedal. This gives the driver the feel of braking. When the driver stops the pedal movement and holds a position, both valves are closed. Any further pedal movement, either applying or releasing, causes the mechanism to unseat the appropriate valve. For example, if the driver releases the brake pedal, the vacuum valve will open, causing the diaphragm to return to its original position.

Vacuum Check Valve

The vacuum is obtained from the intake manifold on the engine. Since fluctuations occur in the intake manifold vacuum, there is a reservoir in the system to act as a storage for vacuum. The reservoir is the large canister surrounding the diaphragm. See *Figure 28–32*. Also, there is a check valve between the manifold and the reservoir. This check valve prevents the vacuum from escaping the reservoir during conditions of wide-open throttle. The check valve is also a safety device in the event of a leaking supply line or other vacuum failure.

Hydro-boost Brakes

Hydro-boost brakes are designed to use hydraulic pressure from the power steering pump to boost the pressure for the master cylinder. This system does the same job as a vacuum booster and is connected in the brake system much the same way. One reason the hydro-boost unit is used is because of federal regulations. These regulations require that the vehicle stop in fewer number of feet with less pressure on the brake pedal. This could be done with vacuum boost systems, but the vacuum diaphragm would

have to be increased in size. The trend today is to make parts smaller, not larger.

In certain applications, the hydro-boost has many advantages over vacuum boosters. Hydro-boost systems work well with diesel and turbocharged engines which have, at times, inadequate vacuum available. In addition, its compact size allows it to be installed where under-hood space is at a premium such as in vans or compact cars. Also, because its boost is much higher than that of vacuum units, it can be used where greater master cylinder pressures are required. Light-to-medium-duty trucks and cars equipped with four-wheel disc brakes are good examples.

Hydro-boost Operation

Figure 28–33 shows a hydro-boost system. Pressure from the steering pump and reservoir is sent to the hydro-boost. Note its small size as compared to the vacuum-boost system. Hydraulic fluid is used to multiply the pressure for the master cylinder.

The hydro-boost system works with the use of a spool valve that is built into the unit. The spool valve shown in *Figure 28–34* is operated or moved by the movement of the brake pedal. The position of the spool valve directs the high-pressure fluid either back to the steering system or to the power piston. This directs the high-pressure fluid to a cavity behind the power piston. The pressure forces the power piston forward, applying the pressure to the output push rod. The output push rod is used to operate the master cylinder.

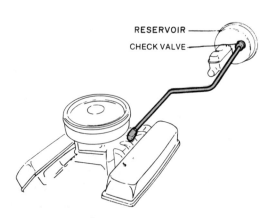

FIGURE 28–32. Vacuum for power brakes is taken directly from the intake manifold. (Courtesy of EIS Brake Parts, Division Standard Motor Products, Inc.)

Variations in Power Brake Systems

Although the principles remain the same, there are two variations for power-assisted brakes. These variations are determined by the type of manufacturer, the application, and the year of production. These include:

a. Tandem power head — a power brake booster with two diaphragms in tandem or series. This provides additional boost to the master cylinder.
b. Dual power brake system — a power brake system that uses both a vacuum-assisted and hydraulically assisted (hydro-boost) design. This system is used on heavy-duty applications such as buses and trucks.

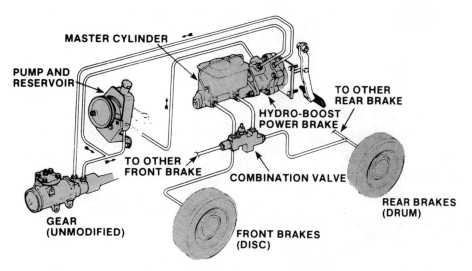

FIGURE 28–33. A hydro-boost system uses hydraulic pressure from the steering pump to produce the increased pressure working on the master cylinder. *(Courtesy of Allied Aftermarket Division, Allied Automotive)*

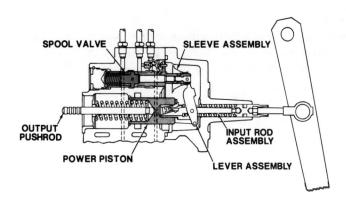

FIGURE 28–34. The hydro-boost unit uses a spool valve inside the master cylinder. *(Courtesy of General Motors Product Service Training)*

ELECTRONIC FUEL INJECTION

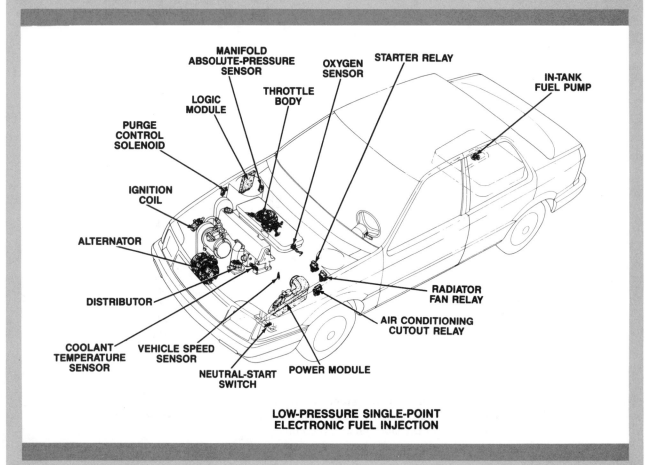

MANIFOLD ABSOLUTE-PRESSURE SENSOR

OXYGEN SENSOR

STARTER RELAY

IN-TANK FUEL PUMP

LOGIC MODULE

THROTTLE BODY

PURGE CONTROL SOLENOID

IGNITION COIL

ALTERNATOR

DISTRIBUTOR

RADIATOR FAN RELAY

AIR CONDITIONING CUTOUT RELAY

COOLANT TEMPERATURE SENSOR

VEHICLE SPEED SENSOR

NEUTRAL-START SWITCH

POWER MODULE

LOW-PRESSURE SINGLE-POINT ELECTRONIC FUEL INJECTION

Electronic fuel injection (EFI) helps to improve fuel economy and driveability. The components shown in the photo are all part of the EFI system. They are tied together and controlled by the on-board computer. (Courtesy of Chrysler Corporation)

DIAGNOSIS and SERVICE

SAFETY TIPS

1. *The springs that are used on drum brakes are under high pressure. Always use the correct tools and wear safety glasses when removing or installing these springs.*
2. *Always wipe up any brake fluid that has been spilled to eliminate the possibility of slipping and causing injury.*
3. *When working on the brake system, the vehicle must be jacked up and correctly supported with a hoist or hydraulic/air jack. Use the correct safety procedure when lifting the vehicle.*
4. *When bleeding the brakes, brake fluid must be forced out of the hydraulic system, along with the air. To eliminate the possibility of spilling the fluid, use a hose connected from the bleeding valve to a canister to catch the excess brake fluid.*
5. *Be careful not to breathe the dust particles left in the brake assembly when removing the brakes. The dust may contain asbestos and cause serious injury in your lungs.*

1. A leaking wheel cylinder will usually have brake fluid on the inside surface of the tire. Also, the leaky brake cylinder will cause the wheel to grab first when the brakes are applied. This means that if the left front wheel cylinder is leaking, when the car is braking, it will pull to the left. Use the following general procedure to *replace the wheel cylinder* on drum brakes..

   a. With the car's emergency brake on, jack up the vehicle following the safety guidelines.
   b. Remove the hubcap and remove the wheel.
   c. Remove the brake drum. If it is the front brake drum, the front wheel bearings and hub must also be removed.
   d. Check the brake drums for wear and distortion as shown in *Figure 28–35*. If the drums are out of specifications, have them machined to

correct specifications. Typically, the maximum amount to be taken off is about 0.060 inch.

   e. With the brakes exposed, remove the springs that hold the brake shoes in place. Also remove the two hold-down springs. Remove all other parts on the brake drum assembly. The exact parts will be determined by the manufacturer and the type of vehicle.
   f. Remove the link between the brake drums and the wheel cylinder.
   g. When replacing or rebuilding a wheel cylinder, always purchase a wheel-cylinder repair kit. Never use old parts of the wheel cylinder.
   h. Remove the two rubber boots on either side of the wheel cylinder.
   i. Pull out the pistons, cups, springs, and expanders.
   j. Before replacing the old wheel cylinder kit with a new one, the cylinder should be slightly honed. This will help the cups seal better within the cylinder. Follow the manufacturer's recommendation concerning honing.
   k. Replace the new cylinder kit and reassemble the drum brake assembly using the correct tools.
   l. Because air has entered the hydraulic system, the brakes must be bled. *Bleeding the brakes* gets the air out of the hydraulic system. Use the following general procedure to bleed the brake system.

      1. With the vehicle placed properly on the jack stand, have another person pump the brakes inside the vehicle, then press down on the brake pedal. The operator should feel a spongy brake pressure from the air being compressed.
      2. With pressure applied to the brake system and using the correct wrench, release the hydraulic pressure by opening the bleed valve on the back of the wheel cylinder. A rubber hose can be attached to the end of the

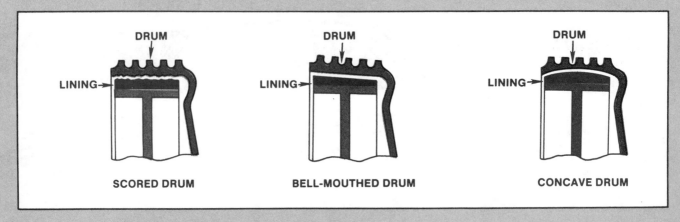

Drum Conditions

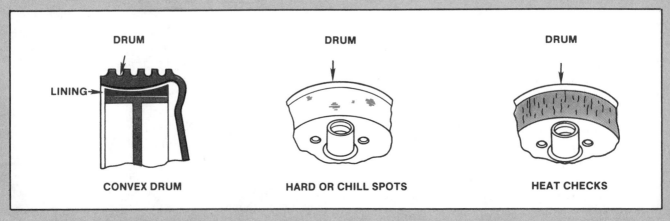

Drum Conditions

FIGURE 28–35. Always check the drum for excessive wear as shown. *(Courtesy of Allied Aftermarket Division, Allied Automotive)*

bleed valve so the hydraulic brake fluid can be directed into a can. ***CAUTION***: *Be careful not to get any brake fluid in your eyes*. Note that the brake fluid should have small bubbles of air mixed with it at this time.

3. When the valve is opened, the operator's foot should go to the floor. When the pedal is completely to the floor, tighten the bleed valve before the operator lets the pedal spring back. If the valve is not closed first, air will be drawn back into the hydraulic system at the wheel cylinder. Have the operator pump the brakes again. There should be less of a spongy feeling.

4. Continue bleeding air until there is a firm and solid brake pressure when the brake is applied. Note that it is important to continually check the brake fluid level in the master

cylinder so there is enough fluid in the system. Also, when completed there should be no air coming out with the brake fluid. Note that depending upon the service performed, one or all of the wheel cylinders may have to be bled. ***CAUTION***: *If more than one wheel cylinder is being bled, check the master cylinder to make sure there is always sufficient brake fluid available*.

2. Check the brake linings for excessive wear. New brake linings have approximately 0.250 inch of thickness to the width of the lining. When worn excessively, the lining has been worn down to the metal or rivet heads. Replace if necessary.

3. When rebuilding wheel cylinders, make sure there is no grease or oil in the system. This will cause damage to the rubber parts.

4. Flush and keep all wheel cylinder parts in clean brake fluid. Any dirt that gets into the hydraulic system may cause wear on the pistons and cups.

5. Inspect brake hoses for leaks, cuts, cracks, twists, and loose supports when servicing the brake system. Replace where necessary.

6. Always use the correct size and type of flare wrench when tightening or loosening hydraulic fittings on the hydraulic part of the brake system.

7. Follow the same rules when rebuilding a master cylinder as when rebuilding the wheel cylinders.

8. When servicing special valves on the brake system, always refer to the manufacturer's service manual for correct procedures.

9. When removing drums, back off the adjusting screw or release the shoe adjusting cams to provide ample lining-to-drum clearance. If this is not done, the brake drum may be very difficult to get off. *Figure 28–36* shows how the adjusting lever can be lifted to allow the star wheel to be turned.

10. Use the correct tools when working on braking systems. When working on brakes, there are several specially designed tools used to remove the return springs and hold-down springs.

11. Use wheel cylinder clamps, *Figure 28–37*, to hold the wheel cylinder pistons in place during disassembly and reassembly.

WHEEL CYLINDER CLAMP

FIGURE 28–37. Use a wheel cylinder clamp to hold the wheel cylinder in place during repair. (Courtesy of Allied Aftermarket Division, Allied Automotive)

12. Check all springs and other parts for loss of tension and damage. Replace weak springs and other badly damaged parts.

13. Follow the manufacturer's procedure for *adjusting brakes*. This typically involves tightening the star wheels until there is a very slight drag between the drum and the brake linings. Another method used is to measure the inside of the brake drum with a suitable brake drum to shoe gauge. With the brake drum off of the wheel, adjust the shoes to this dimension. See *Figure 28–38*.

14. Check and inspect disc brakes for cracks or chips on the pistons, amount of wear on the pads, even wear on the brake pads, damage to the rotor, damaged seals, and cracks in the caliper housing.

15. Follow the correct disassembly procedure stated by the manufacturer during disassembly and reassembly of the disc brake system.

16. Check the disc brake rotor for scoring, runout, parallelism, and thickness. Runout can be checked using a dial indicator as shown in *Figure 28–39*.

17. Do not interchange rotors or other parts on the disc from one side of the vehicle to the other.

18. Use the correct tools to remove the caliper assembly as determined by the manufacturer.

19. Always refer to the manufacturer's troubleshooting guides when checking disc and drum brakes and all master cylinders, including power brakes.

20. Some disc brakes use a sensor spring to determine when the brake pads are worn out. See *Figure 28–40*.

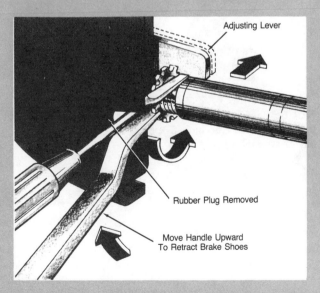

Adjusting Lever

Rubber Plug Removed

Move Handle Upward To Retract Brake Shoes

FIGURE 28–36. To back off the self-adjusters on most rear drum brakes, push in the adjusting lever and loosen the star wheel. (Courtesy of Motor Publications, Motor Magazine)

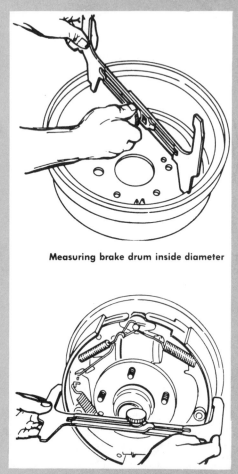

Measuring brake drum inside diameter

Adjusting brake shoes to brake drum
inside diameter

FIGURE 28–38. To adjust the brake shoes, measure the drum first, then set the brake shoes to this dimension. *(Courtesy of Motor Publications, Auto Repair Manual, 1981–1987)*

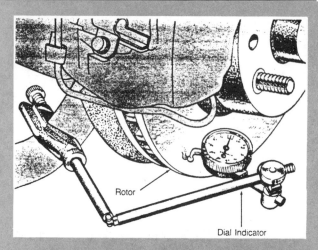

Rotor

Dial Indicator

FIGURE 28–39. Runout on the disc of a disc brake system can be checked using a dial indicator. *(Courtesy of Motor Publications, Motor Magazine)*

Sensor Spring

FIGURE 28–40. The sensor spring will squeak if the disc brake pads wear too far. *(Courtesy of Motor Publications, Motor Magazine)*

21. Certain discs may be out-of-round or may vary in disc thickness. Measure the thickness at six or more locations around the circumference of the lining contact surface. Use a suitable micrometer. If there is a difference greater than the manufacturer's specification, have the disc machined. See *Figure 28–41*.

22. Always inspect the wheel cylinder bore during disassembly for scoring, pitting, and corrosion. An approved cylinder hone, *Figure 28–42*, may be used to remove light roughness or deposits in the bore. Aluminum wheel cylinders and certain other wheel cylinders should not be honed. Check the manufacturer's service manual.

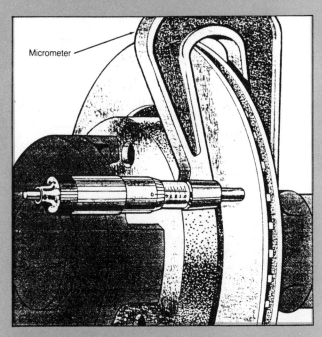

FIGURE 28–41. Use a suitable micrometer to check the thickness variation of the disc on a disc brake system. At least six or more measurements should be taken and compared to the manufacturer's specifications. *(Courtesy of Motor Publications, Motor Magazine)*

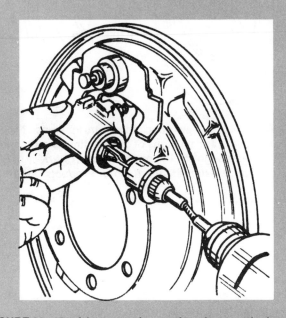

FIGURE 28–42. A hone can be used to clean up the internal surface of the wheel cylinder. Lubricate the surfaces with brake fluid while honing. *(Courtesy of EIS Brake Parts, Division Standard Motor Products, Inc.)*

23. Check to see if the piston moves freely within the caliper bore. Road dirt and rust can cause the piston to stick in the bore. When removing the seal, use a soft object such as a pencil, so as not to scratch the cylinder walls. See *Figure 28–43*.

24. Only use precision equipment to refinish or machine rotors or drums. Always follow the manufacturer's instructions for use of this equipment. See *Figure 28–44*.

25. *Figure 28–45* shows areas where the drum on drum brakes could crack. Always check drums carefully for cracks before reinstalling them in the vehicle.

26. Scoring and grooving will sometimes occur on the disc as shown in *Figure 28–46*. The rotors will require reconditioning for maximum performance to be developed.

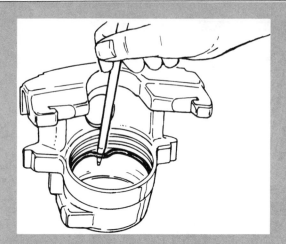

FIGURE 28–43. When removing the seal inside the cylinder on disc brakes, use a soft pointed tool so as not to scratch the cylinder surfaces. *(Courtesy of EIS Brake Parts, Division Standard Motor Products, Inc.)*

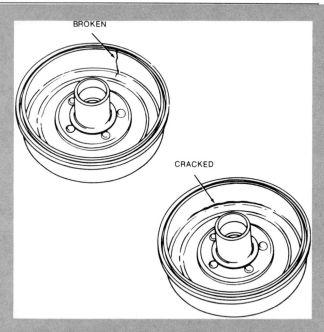

FIGURE 28–45. Always inspect the drums for cracks. *(Courtesy of EIS Brake Parts, Division Standard Motor Products, Inc.)*

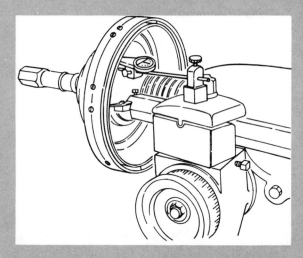

FIGURE 28–44. Use the correct procedure and equipment to recondition the internal surface of the drum brake. *(Courtesy of EIS Brake Parts, Division Standard Motor Products, Inc.)*

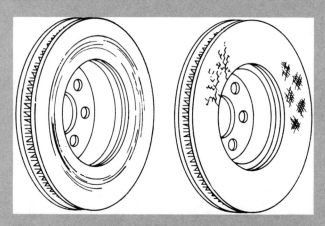

FIGURE 28–46. Always check the rotors for signs of scoring or grooving and for signs of overheating. Overheating also causes a slight bluing on the rotor. *(Courtesy of EIS Brake Parts, Division Standard Motor Products, Inc.)*

SUMMARY

This chapter covers the principles and operation of braking systems. Braking systems operate on the principle of friction. Friction is the resistance to motion between two objects. The brakes produce friction to slow or stop a vehicle. In any braking system, the amount of friction is con-

trolled by the operator. As the operator pushes harder on the brake pedal, friction is increased and the vehicle slows.

There are several types of brake systems used on vehicles today. The drum brake assembly has a cast drum bolted to the wheel. A set of brake shoes is inside of the drum. As the brakes are applied, the shoes move out against the inside of the drum, producing the friction needed to control the vehicle. Current drum brake systems are designed for self-energization. This means that as the brake is applied and the drum rotated, the shoe gets tighter against the inside of the drum. Both a primary and a secondary shoe action is designed into the brake system.

A second type of brake system is called the disc brake. In this system, pads are forced against a rotor to stop the vehicle. This is much like a bicycle brake system, only more complex. The caliper is used to hold the two disc pads in place. A fixed or a floating caliper may be used, depending upon the vehicle design and manufacturer.

Brake systems also use a hydraulic system to transfer the operator's foot pressure to the brakes. The hydraulic system uses servos and different sizes of pistons to produce the forces and pressures needed.

The braking system uses a master cylinder to convert the mechanical force of the operator's foot to hydraulic pressure. There are several principles of operation during the forward stroke and the return stroke of the piston in the master cylinder. The compensating port allows excess fluid to return to the reservoir. A residual pressure check valve is used to keep a small amount of pressure on the hydraulic system. The reservoir diaphragm gasket is used to keep moisture and dirt from getting into the system.

Other types of master cylinders are also used. The dual master cylinder provides two separate pressure chambers. It is required on certain vehicles by law. One chamber is used for the front brakes and one chamber is used for the rear brakes. In the diagonal system, the left-front and right-rear brakes are connected to one chamber. The other chamber is connected to the remaining wheels.

Other designs are also used on master cylinders. These include the quick-take-up master cylinder, warning light switches, proportioning valve, and metering valves. Depending upon the manufacturer and year, the vehicle may have any one of these designs.

Once the pressure has been developed in the master cylinder, brake lines are used to transfer it to the wheel cylinders. The brake linings then move outward to produce the necessary friction.

Drum brakes use a wheel cylinder to operate and move the brake linings. The wheel cylinder converts hydraulic pressure to mechanical force. The most common type of brake assembly is called the duo servo drum brake. It includes both the primary and secondary shoe, return springs, the wheel cylinder, hold-down springs and cup, anchor, connecting springs, star wheel adjuster, and brake drum.

Brake shoes must be adjusted either manually or by automatic adjusters. When adjusted correctly, the brakes move only a certain distance before touching the brake drum.

Disc brakes are used more extensively on the front wheels of most American vehicles. The common parts of a disc brake are the inboard and outboard shoe or lining, the rotor, pistons, piston seals, boots, mounting brackets, and bleeder screws.

Brake fluid is a very important part of braking systems. Brake fluid must have a certain viscosity, a certain boiling point, be noncorrosive, have water tolerance, be a good lubricant, and have a low freezing point. Always refer to the manufacturer's specifications when buying brake fluid.

Many vehicles with disc brakes use power brakes as well. Power brakes are designed to have an additional pressure exerted on the hydraulic system. This is done either by using vacuum or by using hydraulic pressure from the power steering system. Vacuum-assisted power brakes use vacuum from the intake manifold. Vacuum is fed to the diaphragm and used to increase the pressure in the master cylinder. Hydro-boost brakes use pressure from the power steering. The extra pressure also works against the master cylinder.

There are many service tips that are used on brake systems. Service procedures are available in the manufacturer's service manuals. Several important service tips include using correct tools and checking all parts, including the disc, rotors, drum and shoes, springs, brake lines, and so on for damage and excessive wear. In addition, brakes should be bled and adjusted according to the manufacturer's specifications. At times, both the master cylinder and wheel cylinder may also be rebuilt.

TERMS TO KNOW

Can you explain each of the following terms? Review the chapter until you can use each term correctly.

Friction	Fixed caliper
Kinetic energy	Floating caliper
Computer-controlled brakes	Bleeding
Backing plate	Hydraulic system
Brake shoes	Master cylinder
Self-energizing	Wheel cylinder
Servo brake	Compensating port
Primary shoe	Residual
Secondary shoe	Tandem
Disc brake	Proportioning valve
Caliper	Star wheel adjuster

REVIEW QUESTIONS

Multiple Choice

1. Friction on the automobile brakes depends upon which of the following?
   a. Power steering pressure
   b. Tire size
   c. Speed of vehicle
   d. Pressure between the drum and linings
   e. Amount of spring pressure on the parking brake

2. Brake shoes are used on _____.
   a. Front disc brakes
   b. Drum-type brakes
   c. Brakes with a caliper
   d. The back of the master cylinder
   e. Piston-type brakes

3. When a brake shoe becomes tighter against the drum from rotation, this is called:
   a. A master cylinder
   b. Wheel cylinder
   c. Hydraulic pressure
   d. Self-energization
   e. Compensating system

4. The _____ is located to the front of the wheel on vehicles with drum brakes.
   a. Secondary shoe
   b. Caliper
   c. Master cylinder
   d. Primary shoe
   e. Relief valve

5. Which of the following uses a caliper?
   a. Disc brakes
   b. Drum brakes
   c. Master cylinder
   d. Star adjuster
   e. Compensating port

6. The pads on a disc brake system are _____ the rotor.
   a. Perpendicular to
   b. Parallel to
   c. Never touching
   d. Attached to
   e. None of the above

7. The _____ is/are directly attached to the steering assembly and frame on a disc brake system.
   a. Floating caliper
   b. Linings
   c. Fixed caliper
   d. Pads
   e. Master cylinder

8. Hydraulic fluid used on a braking system:
   a. Transfers the pressure from the master cylinder to the wheel cylinder
   b. Cannot be compressed
   c. Must withstand very high pressures
   d. Is pressurized in the master cylinder
   e. All of the above

9. The purpose of the master cylinder is to:
   a. Reduce pressure on the brakes during braking
   b. Produce the pressure on the brakes during braking
   c. Produce the necessary friction during braking
   d. Be used as a parking brake
   e. Adjust the brakes

10. The _____ allows fluid to flow from the reservoir to the master cylinder.
    a. Forward stroke
    b. Compensating port
    c. Primary cup
    d. Wheel cylinder
    e. Brake linings

11. Brake fluid passes around the primary cup in the master cylinder:
    a. During the return stroke
    b. During the forward stroke
    c. During acceleration
    d. During parking brake action
    e. During rapid stopping

12. A small amount of pressure is kept on the hydraulic fluid on certain brake systems:
    a. To stop the car completely
    b. By the compensating port
    c. By the residual pressure check valve
    d. By the parking brake
    e. By the springs in the wheel cylinder

13. Two wheels braking from one chamber and two wheels braking from a second chamber:
    a. Use a dual master cylinder
    b. Use a diagonal brake system
    c. Protect the system if one master cylinder fails
    d. All of the above
    e. None of the above

14. A _____ is used to reduce the pressure to the rear wheels during braking.
    a. Proportioning valve
    b. Compensating valve
    c. Relief valve
    d. Metering valve
    e. None of the above

15. The primary shoe is _____.
    a. Shorter than the secondary shoe
    b. Longer than the secondary shoe
    c. The same length as the secondary shoe
    d. Placed to the rear on the wheel cylinder
    e. Placed with the disc against the drum

16. Brake shoes can be adjusted _____.
    a. Manually
    b. Automatically
    c. Using the star wheel
    d. All of the above
    e. None of the above

17. On a disc brake system, the pads are forced against the _____.
    a. Rotor
    b. Parking shoe
    c. Bleeder screw
    d. Piston seal
    e. Piston

18. Which of the following is *not* a good characteristic of brake fluid?
    a. Must have extremely low viscosity (thick)
    b. Must have a high boiling point
    c. Must be able to absorb or retain moisture
    d. Must be able to lubricate
    e. Must have a low freezing point

19. Power brakes use _____ for increasing the pressure.
    a. Vacuum
    b. Lubricating oil
    c. Cooling system fluid
    d. The rpm of the engine
    e. The operator's foot

20. Which of the following is a mode of operation on power braking systems?
    a. Apply mode
    b. Hold mode
    c. Release mode
    d. All of the above
    e. None of the above

21. Hydro-boost brake systems use _____ to increase the pressure for the master cylinder.
    a. Vacuum
    b. Power steering fluid
    c. Cooling system fluid
    d. The rpm of the engine
    e. The operator's foot

22. Which of the following is a good service tip when working on brakes?
    a. Bleed the system of air.
    b. Replace linings that show excessive wear.
    c. Keep all wheel cylinder parts clean and coated with brake fluid.
    d. All of the above
    e. None of the above

The following questions are similar in format to ASE (Automotive Service Excellence) test questions.

23. Technician A says that a spongy brake pedal is caused by air in the hydraulic system. Technician B says that a spongy brake pedal is caused by a weak return spring. Who is right?
    a. A only
    b. B only
    c. Both A and B
    d. Neither A nor B

24. When a car is being stopped, it pulls to the right. Technician A says that the left wheel cylinder is leaking. Technician B says that the right wheel cylinder is leaking. Who is right?
    a. A only
    b. B only
    c. Both A and B
    d. Neither A nor B

25. The left front wheel cylinder has just been replaced. Technician A says that the brake system should not be bled of air. Technician B says that the brakes must be bled to remove the air. Who is right?
    a. A only
    b. B only
    c. Both A and B
    d. Neither A nor B

26. A wheel cylinder is leaking slightly. Technician A says that the wheel cylinder should be honed and the old parts replaced. Technician B says a new wheel cylinder kit should be put in after the cylinder is honed slightly. Who is right?
    a. A only
    b. B only
    c. Both A and B
    d. Neither A nor B

Essay

27. Describe the process of bleeding the brakes.

28. What are the primary and the secondary shoes?

29. Describe the operation of a floating caliper.

30. What is a compensating port in the master cylinder?

31. Describe the purpose and operation of the proportioning valve.

32. Describe the operation of power brakes as compared to manual brakes.

CHAPTER 29

Suspension Systems

INTRODUCTION

The suspension system of a car is used to support its weight during varying road conditions. The suspension system is made of several parts and components. These include both the front and rear suspensions, *shock absorbers*, and the *MacPherson strut system*. The objective of this chapter is to analyze the parts and operation of different suspension systems.

OBJECTIVES

After reading this chapter, you will be able to:

■ Define the parts and operation of the front suspension system.
■ Define the parts and operation of the rear suspension system.
■ Analyze the purpose, parts, and operation of different types of shock absorbers.
■ Compare the MacPherson strut suspension to other suspension systems, including parts and operation.
■ Identify the purpose and operation of automatic level control and air suspension systems.
■ State common diagnosis and service suggestions concerning different types of suspension systems.

CHAPTER HIGHLIGHTS

29.1 FRONT SUSPENSION
    A. Purpose of the Front Suspension
    B. Parts of the Front Suspension System
    C. Ball Joints
    D. Control Arms
    E. Sway Bar and Link (Stabilizer Bar and Link)
    F. Strut Rods
    G. Coil Springs
    H. Torsion Bars
    I. Steering Knuckle and Spindle
    J. Front Wheel Bearings

29.2 REAR SUSPENSION
    A. Purpose of Rear Suspension
    B. Leaf Spring
    C. Coil Spring Rear Suspension
    D. Independent Rear Suspension

29.3 SHOCK ABSORBERS
    A. Purpose of Shock Absorbers
    B. Shock Absorber Operation
    C. Parts of a Shock Absorber
    D. Compression and Expansion Valve Operation
    E. Spiral-grooved Shock Absorbers
    F. Gas-filled Shock Absorbers

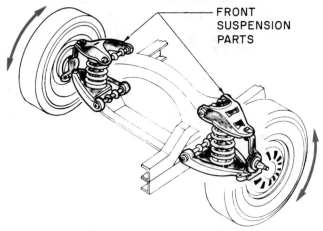

INDEPENDENT FRONT SUSPENSION

FIGURE 29–1. The front suspension is called independent front suspension. This means that each wheel acts independently when going over a bump. *(Courtesy of Dana Corporation)*

29.1 FRONT SUSPENSION

Purpose of the Front Suspension

The purpose of the front suspension is to support the weight of the vehicle. The suspension must also be designed to provide a smooth passenger ride over varying road conditions and speeds.

There are several types of front-end suspension systems. Automobiles commonly use the independent front suspension system. This means that each wheel is independent from the other. For example, if the left wheel goes over a bump in the road, only the left wheel will move up and down. See *Figure 29–1*. Certain types of trucks and other heavy-duty vehicles may use an I-beam suspension. This system has one main beam connecting each front wheel. This is not independent front suspension.

Parts of the Front Suspension System

Although there are different types of front suspension systems, many of the parts are the same. *Figure 29–2* shows a common type of front suspension system and the related parts. These parts include the:

a. *ball joints* (both upper and lower)
b. *control arms*, shaft bushings, and shims
c. *sway bar*, bushings
d. *strut rod*, bushings
e. coil springs

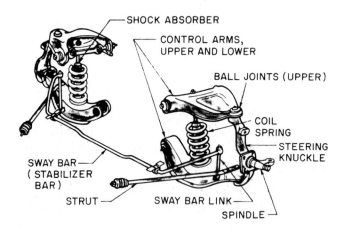

FIGURE 29–2. The parts of the front suspension. *(Courtesy of Dana Corporation)*

f. stabilizers
g. shock absorbers
h. *steering knuckle* and spindle

These parts are assembled to provide the entire front suspension. Each of these parts becomes a vital link in the front suspension operation. They must work properly to insure driving safety and comfort.

Ball Joints

The ball joints are used to connect the spindle and steering knuckle to the upper and lower control arms. They are

COMPUTER-CONTROLLED TEMPERATURE CONTROL

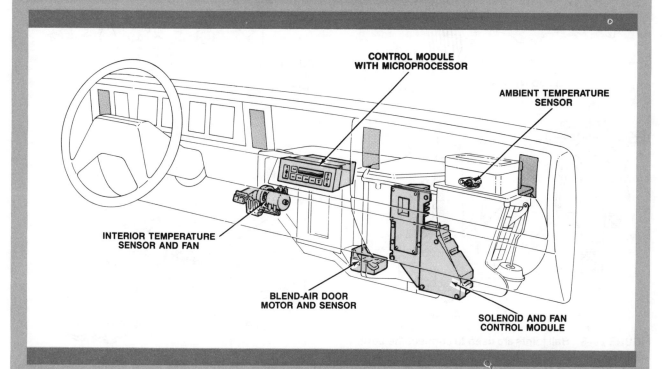

CONTROL MODULE
WITH MICROPROCESSOR

AMBIENT TEMPERATURE
SENSOR

INTERIOR TEMPERATURE
SENSOR AND FAN

BLEND-AIR DOOR
MOTOR AND SENSOR

SOLENOID AND FAN
CONTROL MODULE

Now that computers have been placed on most vehicles, a computerized automatic temperature control can be used. This system offers more choices of airflow and outlets than previous systems have. It operates with an 8-bit dedicated microprocessor with 4 K memory. The temperature of the incoming air can be regulated to the desired temperature selected by the driver. The microprocessor measures the interior temperature and makes adjustments every seven seconds. (Courtesy of Chrysler Corporation)

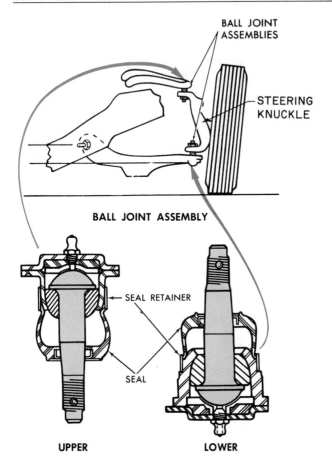

FIGURE 29-3. Ball joints are used to connect the control arms to the steering knuckle. *(Courtesy of Chevrolet Division, General Motors Corporation)*

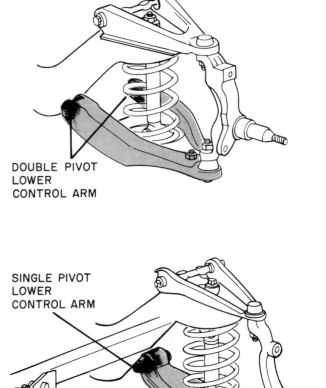

FIGURE 29-4. Both a single and a double lower control arm can be used. The single lower control arm also uses a strut for support. *(Courtesy of Dana Corporation)*

designed to do several things. The ball joints must carry the weight of the vehicle. They provide a pivot for the wheel to turn. They also allow for vertical movement of the control arms when the vehicle goes over irregularities in the road. *Figure 29-3* shows a typical set of ball joints for upper and lower control arms.

The frame of the upper ball joint is either riveted or bolted to the upper control arm. The steering knuckle is attached to the tapered stud and is held in place with a nut. The lower ball joint is usually bolted, riveted, or pressed into the lower control arm. The steering knuckle is placed on the tapered stud and held in place with a nut. A rubber boot is placed around the assembly to keep grease in and dirt out.

Control Arms

There are two control arms: an upper and a lower control arm. Several arrangements are used for the control arms. There are single pivot control arms, double pivot control

arms, and long and short control arms. *Figure 29-4* shows a comparison of control arms. The type of control arm depends upon the year and manufacturer of the vehicle.

The other end of the control arm is attached to the frame of the vehicle. *Figure 29-5* shows how upper control arms are attached to the frame. The upper control arm has a shaft that is bolted to the frame. The ends of the shaft carry bushings that are attached to the control arm. Shims are used to adjust the position of the shaft on the frame for alignment of the front suspension. The lower control arm is attached to the frame by bushings and bolts.

Sway Bar and Link (Stabilizer Bar and Link)

The sway bar and sway bar link are also called the stabilizing bar and link. The sway bar link connects the lower control arm to the sway bar. The sway bar twists like a *torsion*

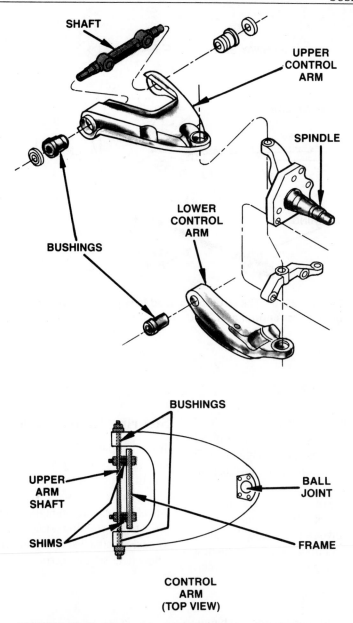

FIGURE 29–5. The parts of the upper control arm are shown. The upper control arm is attached to the frame by using bushings, shims, and the upper control arm shaft. *(Courtesy of Dana Corporation)*

bar during turns. It transmits cornering forces from one side of the vehicle to the other. This helps to equalize the wheel loads and prevent excessive leaning of the car on turns. The sway bar link and sway bar are attached to the frame with rubber bushings and bolts.

Strut Rods

The strut rod is used on vehicles that have single pivot lower control arms. They can be located either in front of or behind the control rod. They are designed to retain the lower control arms in their intended positions. They also provide a method of keeping the wheel in the right position for alignment. *Figure 29–6* shows the position of the strut rod. The sway bar and the sway bar link are also shown.

Coil Springs

The coil springs are used to support the car's weight, maintain the car's *stance* or height, and position all the other

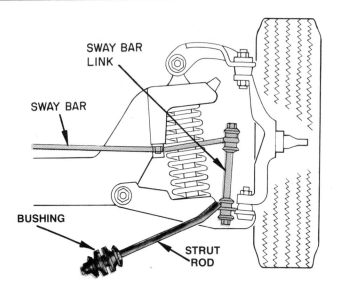

FIGURE 29-6. The strut rod is used to support the single pivot lower control arm. *(Courtesy of Dana Corporation)*

suspension parts correctly. Thus, if a spring sags a slight amount, the tires, shocks, ball joints, and control arms all work outside their normal positions. This condition may cause excessive or abnormal wear throughout the suspension systems.

Springs may be very flexible or very stiff. The purpose of the springs is to absorb road shock and then return to their original position. A stiff spring gives a rough ride. A flexible spring may cause the vehicle to bounce too much. The best combination is to use a softer spring with a shock absorber.

Torsion Bars

Another method used to provide the desired ride and handling characteristics is to use torsion bars rather than springs. Torsion bars are made so that as the car goes over bumps, the torsion bar will twist. The resistance to twisting produces an effect similar to that produced by springs. A torsion bar is attached to each side of the vehicle. One end of the torsion bar is attached to the frame. The other end of the torsion bar is attached to the lower control arm. As the lower control arm moves due to bumps in the road, the torsion bar twists and reduces the car's motion. *Figure 29-7* shows a typical torsion bar installation. Other arrangements have also been used. *Figure 29-8* shows a torsion bar connected between the back wheels. As the rear drive shaft moves up and down, it causes the torsion bar to twist. Certain torsion bars are also adjustable for setting the vehicle height.

Steering Knuckle and Spindle

Two other parts of the front suspension include the steering knuckle and the *wheel spindle*. The wheel spindle is the unit that carries the hub and bearing assembly with the help of the knuckle. In some vehicles, the steering knuckle and wheel spindle are one unit. *Figure 29-9* shows an example of the steering knuckle and spindle. The steering knuckle is attached to the two control arms with ball joints. The wheel spindle carries the entire wheel load. Bearings

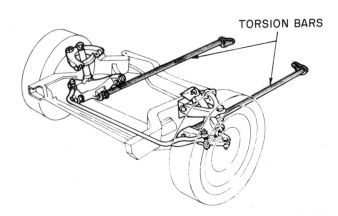

FIGURE 29-7. Torsion bars are used in place of springs on some vehicles. *(Courtesy of Dana Corporation)*

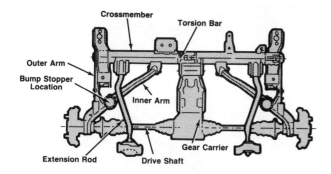

FIGURE 29–8. The torsion bar can also be located between the two rear wheels. In this case, the torsion bar is located between the outer arms on the rear suspension. *(Courtesy of Chrysler Corporation)*

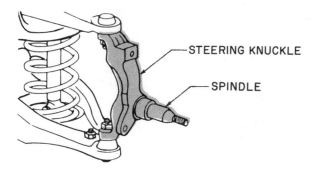

FIGURE 29–9. The spindle holds the wheel bearings and is attached to the steering knuckle. *(Courtesy of Dana Corporation)*

are used to reduce friction between the wheel and the spindle. The inner bearing on the spindle is usually larger than the outer bearing. It absorbs the greatest load because the wheel is placed as close to the knuckle as possible.

Front Wheel Bearings

The front wheel bearings are also considered part of the front suspension. There are two bearings on each front wheel spindle to support the wheel. Both bearings are called tapered roller bearings. On certain types of front end suspension systems, ball bearings are used as well. The inner race of the bearing rides on the spindle. The outer race is lightly pressed into the wheel hub.

CAR CLINIC

PROBLEM: SHIMMY PROBLEMS

A GM travel van with about 15,000 miles on it seems to have a severe shimmy when the vehicle hits a bump. This is especially noticeable on high-speed highway turns. The bias ply tires from the manufacturer are on the vehicle. The alignment has been checked and adjusted. The shimmy disappeared for about 500 miles, then returned.

SOLUTION:

The problem could be caused by two things. First, is the vehicle used on bumpy roads? Severe bumps may cause the alignment to go out of specifications even after 500 miles. The second cause could be the tires. The weight distribution of large vehicles may cause the problem. It is extremely important to have the tires grip the road under all conditions. The better the grip, the less chance of shimmy. With a lot of weight in the rear of a van, the road grip on ply tires may not be enough. Try replacing the tires with a good set of radial tires. This will most likely solve the problem.

29.2 REAR SUSPENSION

The rear suspension system is an integral part of the total suspension system. There are typically two types of rear suspension systems. These include the solid axle type and the *independent rear suspension* type.

Purpose of Rear Suspension

All rear suspension systems serve the same purpose. They are designed to keep the rear axle and wheels in their proper position under the car body. The rear wheels must always track exactly straight ahead. The rear suspension axle allows each of the rear wheels to move up and down somewhat independently from the frame. This helps to maintain alignment and good vehicle control and provides passenger comfort. The spring assembly must also absorb a large amount of rear end torque from acceleration (on rear drive vehicles), side thrust from turning, and road shock from bumps.

Leaf Spring

The most common type of spring used on rear suspensions is called the leaf spring. It consists of one or more leaves and usually has its ends formed into eyes for connection to the vehicle frame. A U-bolt is used to hold the rear axle to the

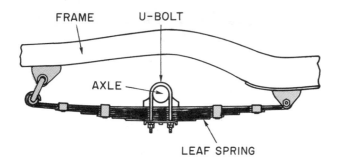

FIGURE 29-10. Both ends of the rear leaf spring are attached to the frame. The axle is attached to the rear spring by a U-bolt. *(Courtesy of Dana Corporation)*

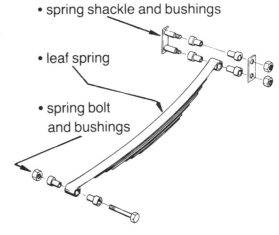

FIGURE 29-11. The parts of a leaf spring. *(Courtesy of Dana Corporation)*

spring. This type of spring is called the semi-elliptical spring. The ends are higher than the center arch as shown in *Figure 29–10*. One end of the spring is fixed to the frame. The other end of the spring is mounted to the frame by using a spring *shackle* and bushing. The bushings are used to dampen noise and vibration from the road to the frame of the car. The spring shackle allows the spring to change length slightly during driving. During normal operation, the spring also bends due to acceleration, braking, or road conditions. The leaf spring supports the car frame, but it allows independent movement of the rear wheels. *Figure 29–11* shows the individual parts of a leaf spring.

Coil Spring Rear Suspension

In a coil spring rear suspension, the spring is placed between a bracket mounted on the axle and the vehicle frame. The coil design is much the same as the front wheel coil spring. In addition to the coil springs, control arms and bushings are used. Control arms provide stability to the rear wheels during driving. The control arms are attached with bushings to the rear axle housing and the car frame. *Figure 29–12* shows a coil spring rear suspension system.

Independent Rear Suspension

Independent rear suspension is used on many front wheel drive cars. Independent rear suspension means that each rear wheel is independent in its movement. This is much the same as the front suspension system. Although there are many designs for independent rear suspension, most systems include coil springs, control arms, struts, and *stabilizer bars*.

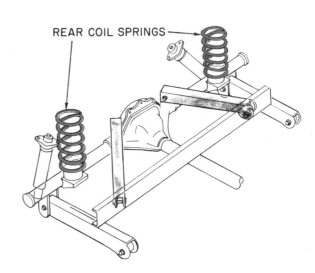

FIGURE 29-12. The rear suspension can also use coils rather than leaf springs. *(Courtesy of Dana Corporation)*

CAR CLINIC

PROBLEM: ENGINE MISSES AT HIGH SPEED

A customer complains that when running at fuel throttle up a slight grade, the vehicle begins to slow down, sputters, and misses. Several items have been repaired and replaced. These include the carburetor, timing, fuel pump, filters, and the EGR valve. What other areas should be checked?

SOLUTION:

The problem sounds like the fuel tank is developing a vacuum and the fuel is being restricted to the carburetor. Check the fuel tank cap. The caps are designed to vent the tank of fumes and to allow air to come into the tank as the fuel is removed. Also check the carbon canister for damage or lines that are crimped.

29.3 SHOCK ABSORBERS

Purpose of Shock Absorbers

Shock absorbers are hydraulic devices that help to control the up, down, and rolling motion of a car body. One shock absorber is used on each wheel. Each shock must control one wheel and axle motion. The car's springs support the body, but the shock absorbers work with the springs to control movements of the car body. A shock absorber can be considered a *damper* that controls energy stored in the springs under load. For this reason, shock absorbers are also called *oscillation* dampers.

The shock absorber is placed parallel to the upward and downward motion of the car. It has two tasks:

1. To prevent excessive rolling and bouncing of the car body.
2. To rapidly terminate the oscillation of the wheels and axle when they start moving up and down.

These two factors are of major importance for driving comfort and safety.

Figure 29–13 shows how a shock absorber works. When there is a rise or bump in the road surface, the car axle immediately rises. Now the spring is compressed and starts to push up the car body. The impact acting on the vehicle is absorbed by the spring. The spring prevents the axle from touching the car body. After the springs have been compressed, they try to expand. This helps separate the car body from the axle. This entire action causes an oscillation motion to develop.

The oscillation motions are also shown. The shock absorber is placed between the axle and the car body. It is designed to reduce the number of oscillations produced after hitting a bump. For comparison, the oscillations are also shown when a shock absorber is placed between the car body and the axle.

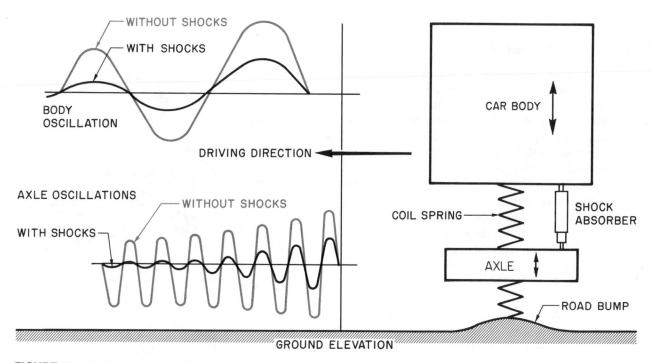

FIGURE 29–13. As the axle goes over a bump, the spring is compressed. The oscillations that are produced are shown with and without a shock absorber. *(Courtesy of Sachs Industries, Inc.)*

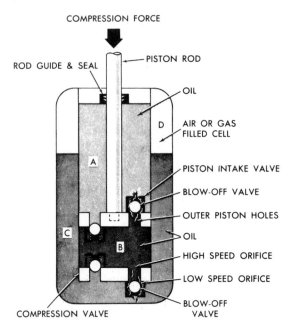

FIGURE 29–14. Under a compression force, the oil is forced through small valves into reservoirs C and A. (Courtesy of Chevrolet Division, General Motors Corporation)

Shock Absorber Operation

The shock absorbers are made to force a noncompressible liquid through small openings. *Figure 29–14* shows an example of how they work. Referring to this figure, when a compression force is produced from a road bump, the piston rod is forced down. A pressure is produced in the oil below the piston in chamber B. The oil pressure forces oil outward and through both the blow-off valve and the piston intake valve. Oil can only pass through these passageways at a certain speed. The pressure forces oil out into chamber C. During this time, air is being compressed in area D. The damping force originates from the resistance of oil flow at the narrow passages of the valve parts. In addition, oil passes into chamber A. Oil must flow into this chamber because it is getting larger as the piston is moved downward.

When the shock absorber rebounds, oil flows in the reverse direction. *Figure 29–15* shows the flow of oil when the shock absorber rebounds or returns to its original position. During rebounding, the piston rod is forced to extend out and upward. This action causes a vacuum to be produced inside chamber B. This vacuum draws oil from chamber A into chamber B through the rebound valve. Oil also flows from chamber C into chamber B. Air in chamber

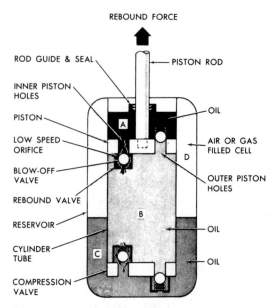

FIGURE 29–15. During rebound force, oil is forced from reservoirs C and A into B. (Courtesy of Chevrolet Division, General Motors Corporation)

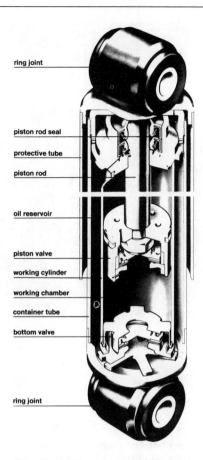

ring joint

piston rod seal

protective tube

piston rod

oil reservoir

piston valve

working cylinder

working chamber

container tube

bottom valve

ring joint

FIGURE 29-16.   The parts of a shock absorber. *(Courtesy of Sachs Industries, Inc.)*

D then expands to compensate for the loss of oil in chamber C. Since a part of the oscillating energy is converted to heat energy, shock absorbers that are working correctly get warm during operation.

Parts of a Shock Absorber

There are many types of shock absorbers, but the main principles of operation are the same. Although they may be different sizes, the parts are much the same. *Figure 29–16* shows a typical shock absorber. It is called the double-tube shock absorber. The more important parts include:

1. *Ring joint* — used to attach the shock to the axle and frame.
2. *Piston rod seal* — used to keep oil from leaking past the rod and into the atmosphere during pressure conditions.
3. *Piston valve* — used to control the flow of oil and produce the pressure and vacuum in the chamber above the piston during compression and rebound.
4. *Bottom valve* — used to control the flow of oil into the oil reservoir during compression and rebound.

5. *Protective tube* — used to keep dirt and road dust away from the seals and piston rod.
6. *Container tube* — used to house the internal parts of the shock absorber.
7. *Working chamber* — area where the pressure and vacuum are produced.

Compression and Expansion Valve Operation

The compression and expansion valves are shown in more detail in *Figure 29–17*. The shock absorber is compressed by the oscillation of the vehicle. The oil displaced by the downward-moving piston rod flows into the reserve chamber above the piston. The oil also flows through the bottoming valve at a certain flow rate, damping the motion.

The shock absorber is expanded by the oscillation of the vehicle. During this condition, the piston valve controls the damping. The piston valve resists the oil that is trying to flow from above the piston to the working chamber. The upward motion of the piston is retarded. The bottom valve allows the necessary oil to be sucked easily from the oil reservoir.

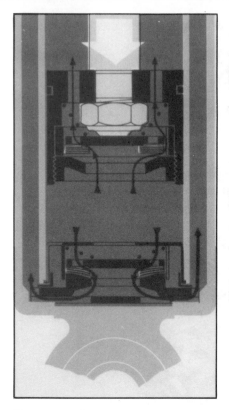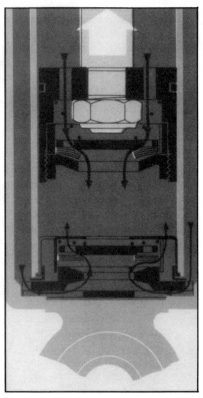

Compression   **Expansion**

FIGURE 29-17. The compression and expansion forces cause oil to flow through the piston valve. The resistance to this flow through the valve causes the motion to be dampened. *(Courtesy of Sachs Industries, Inc.)*

Spiral-grooved Shock Absorbers

When the oil is passed through the valves rapidly, some *aeration* or foaming is produced in the oil. Aeration is the mixing of air with the oils. When aeration occurs, the shock develops lag (piston moving through an air pocket that offers no resistance). This causes the shock absorber to work incorrectly and produces a poor ride.

One method used to reduce aeration is to use a spiral-grooved reservoir tube. The spiral grooves on the shock reservoir tend to break up the air bubbles. This action reduces lag.

Gas-filled Shock Absorbers

Gas-filled shocks are also used to reduce aeration. If a pressure gas replaces the air in the shock absorber, air cannot mix with the oil to produce aeration. *Figure 29–18* shows two types of gas-filled shock absorbers. Note that both compression and expansion valves are built into the piston. The deflection disc is used to help separate the oil and the gas. When oil is forced through the small holes in the valve, high-pressure jets of oil are produced. These jets of oil are deflected by the deflection disc before they get to the gas. This action reduces foaming and aeration. The separating disc, which is also shown, is arranged so that it completely separates the oil from the gas. It is movable to allow for differences in volume during compression and expansion.

Purpose of Air Shock Absorbers

When an increased amount of load is placed in the car, the springs may not be able to support the vehicle correctly. This is shown in *Figure 29–19*. This condition can cause several problems. These include:

1. Increased intensity of light beams to oncoming drivers even when the lights are on dim.

2. A change in steering geometry.

3. Worsening of comfort for the passengers.

4. Less steering control for the driver.

5. The possibility of bottoming out on bumps.

One way to overcome these problems caused by heavy weight is to use air shock absorbers.

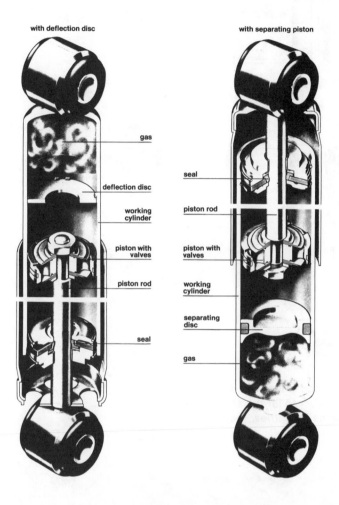

FIGURE 29–18. Gas-filled shock absorbers are used to reduce aeration. *(Courtesy of Sachs Industries, Inc.)*

Operation of Air Shock Absorbers

Figure 29–20 shows an example of an air shock absorber. The unit is made by including an air chamber in the shock. A *bellows* is used to keep the air chamber sealed from the outside while the shock absorber is in different positions. The pressure inside the air chamber determines the amount of load that the vehicle can carry. The entire unit also uses the typical shock absorber system discussed previously.

The air is admitted to the air chamber by use of a standard tire valve. Pressure is produced by a small air pump called an air compressor. A height-sensing control valve is also used. *Figure 29–21* shows a typical installation. Other common components include an air reserve tank, different types of air compressors, and the control valve.

Automatic Level Control

The automatic level control system is used to adjust the carrying load of the car when weight is added or removed from the vehicle. The system consists of several components. These include the air compressor, air dryer, manual switch, exhaust solenoid compressor relays, electronic height sensor, and the shock absorbers and air line fittings.

a. The air compressor is a positive-displacement, single-piston pump. It is powered by a 12-volt dc permanent-magnet motor. The casting contains intake and exhaust valves for correct operation.

b. The air dryer is used to dry the air by using a chemical. When air passes through this chemical, moisture is absorbed.

c. A manual switch is used to control the compressor on certain systems. When it is in the "off" position, the shock absorbers act like standard shocks. When it is in the "auto" position, the load leveling system is in operation.

d. The exhaust solenoid is used to exhaust air from the system and to control maximum output pressure from the air compressor.

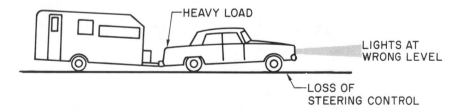

FIGURE 29–19. Heavy load in the rear of the vehicle can cause steering geometry to change, less steering control, and increased intensity of light beams to oncoming drivers. *(Courtesy of Sachs Industries, Inc.)*

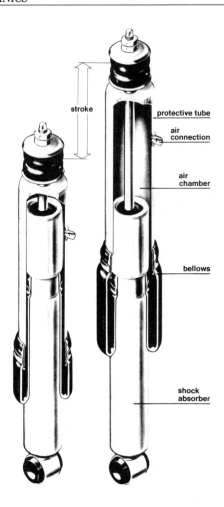

FIGURE 29–20. Air shocks are used to level the stance of the car. Air is forced into a chamber to lift the shock and level the car. *(Courtesy of Sachs Industries, Inc.)*

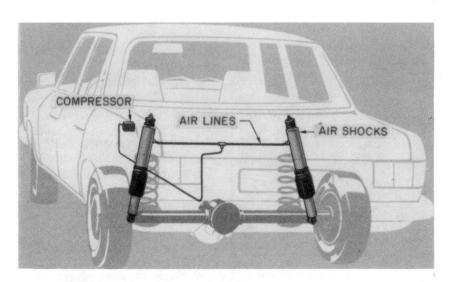

FIGURE 29–21. Air shocks use a small compressor to produce air for extra support. *(Courtesy of Sachs Industries, Inc.)*

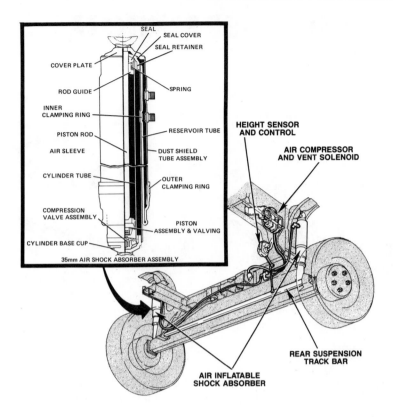

ELECTRONIC AUTOMATIC LOAD LEVELING

FIGURE 29–22. Height sensor and control units are used to measure the amount of load placed in the vehicle. The sensor tells the air compressor how much air should be used in the air shocks. *(Courtesy of Chrysler Corporation)*

e. The compressor relays are used to control the different functions of the system.

f. The electronic height sensors are used to measure the amount of drop and rise when weight is changed in the vehicle. This signal is then sent to the compressor relays to change the amount of air sent to the system.

g. The shocks are the same as previously mentioned. The air lines connect the air compressor to the shocks.

A similar system is shown in *Figure 29–22*. A height sensor linked to the rear suspension track bar monitors load changes. Solid-state circuitry then either turns on the compressor to inflate the shock absorbers or exhausts the air to maintain the desired pressure. The parts of the air shock absorber assembly are also shown for reference.

Air Suspension

Another type of suspension used on vehicles is called the air suspension system, *Figure 29–23*. Although there are shock absorbers, the system also uses four air springs, one on each wheel. As the front and rear height sensors feed information to the control module, the correct amount of air is sent to each air spring.

System operation is maintained by the addition or removal of air in the air springs. There is a predetermined height for both the front and rear sections of the car. The height sensors will lengthen or shorten, depending upon the amount of suspension travel. As weight is added, the body settles. As weight is removed, the body rises. The height sensors signal the control module. The control module then activates the air compressor through relays to change the amount of air in the air springs.

AIR SUSPENSION

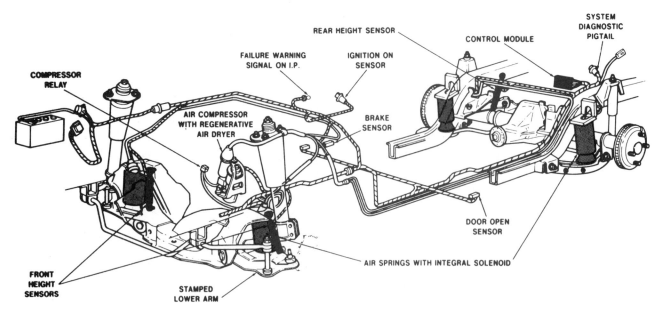

FIGURE 29–23. An air suspension system uses front height sensors, a rear height sensor, a control module, and air springs to keep the vehicle level under varying loads. *(Courtesy of Motor Publications, Auto Repair Manual, 1979–1985)*

29.4 MACPHERSON STRUT SUSPENSION

General Description of the MacPherson Strut Suspension

One other popular type of independent suspension system is called the MacPherson strut suspension. Many imported and domestic vehicles utilize this system on front wheel drive vehicles. Certain vehicles also use this system on the rear wheels. There is also a modified version of the MacPherson strut system. The MacPherson strut system is favored where space and weight savings are important. It is used by American, European, and Japanese auto manufacturers.

Parts of the MacPherson Strut Suspension

Figure 29–24 shows a complete MacPherson strut suspension system. It is very much like a regular shock absorber

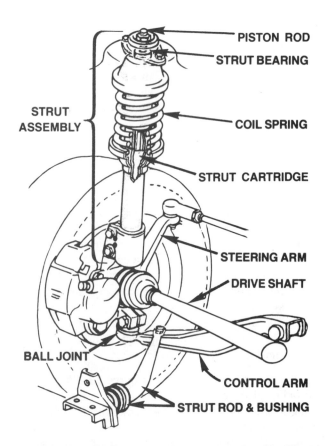

FIGURE 29–24. The basic parts of the complete MacPherson strut suspension system. *(Courtesy of Dana Corporation)*

LAVA MACHINES

These "lava machines" have a jet engine in the vehicle. There is no direct connection between the engine and the wheels. The forward thrust is produced the same as in a jet aircraft. (Courtesy of Jim Rennich, Bloomington)

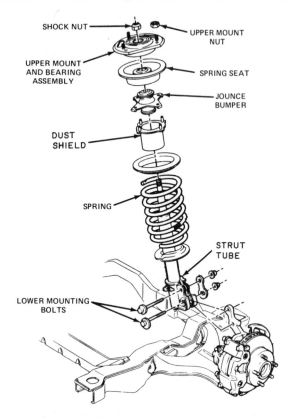

FIGURE 29–25. The detailed parts of the MacPherson strut suspension. *(Courtesy of General Motors Product Service Training)*

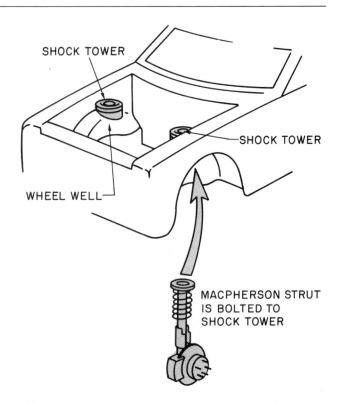

FIGURE 29–26. The MacPherson strut suspension is bolted to the shock tower. The shock tower is part of the wheel well.

and spring combined. The only difference is that the strut assembly is used as a structural part of the vehicle's suspension system. A more detailed drawing is shown in *Figure 29–25*. The system consists of the strut tube, suspension spring, dust shield, *jounce* bumper, upper spring seat, and upper mount and bearing assembly.

The MacPherson strut suspension has eliminated the need for several common suspension parts. There is no upper control arm and the upper ball joint is not needed. There is a lower ball joint, but the ball joint does not carry as much load as in other suspension systems. It is isolated from the vehicle weight. Vehicle weight is supported at the top of the strut assembly. The strut bearing is bolted directly to the shock tower. The shock tower is the part of the car body to which the MacPherson strut is attached. *Figure 29–26* shows an example of the shock tower built into the car body. The lower part of the strut assembly is attached by bolts to the steering knuckle. The steering knuckle is attached to the lower control arm through a ball joint.

The lower control arm is bolted to the frame with conventional rubber bushings. The lower control arm ball joint is riveted to the lower control arm.

Operation of the MacPherson Strut Suspension

During turning, the entire strut assembly is also turned. The strut assembly can turn because there is a bearing assembly located on top of the strut assembly and a ball joint on the bottom of the assembly. The upper bearing and mount assembly takes the place of the upper control arm. The steering arm and linkage, disc brake caliper, and lower control arm ball joint are all attached to the steering knuckle. The drive shaft is connected directly to the wheel spindle through the steering knuckle. The spring is used for the same purpose as on other suspension systems. It is used to support the vehicle weight and maintain the car stance and height. The shock absorber, which is built into the system, helps to smooth out the oscillations from the spring.

Advantages of Using MacPherson Strut Suspension

There are several advantages of using the MacPherson strut suspension systems. These include:

1. They are lighter in weight than the conventional two control arm system.

2. The system spreads the suspension load over a wider span of the car's chassis.

3. They take up less room in the engine compartment, which allows room for other components.

4. There are fewer moving parts than in the conventional two control arm system.

Modified MacPherson Strut Suspension

Another type of MacPherson strut suspension is called the modified system. *Figure 29–27* shows such a system. The system is basically the same, except the spring is placed between the frame and the lower control arm. With the spring located here, minor road vibrations are absorbed by the chassis rather than fed back to the driver through the steering system. A lower ball joint supports the vehicle weight. This system also eliminates the upper control arm, bushings, and upper ball joints used on the conventional suspension system.

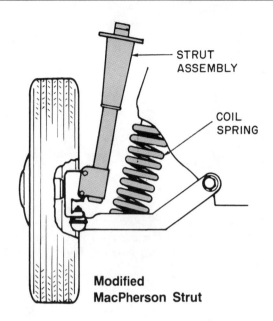

Modified MacPherson Strut

STRUT ASSEMBLY

COIL SPRING

FIGURE 29–27. The modified MacPherson strut suspension has the strut assembly separated from the coil assembly. *(Courtesy of Dana Corporation)*

SAFETY TIPS

1. *Use the correct procedure for lifting the vehicle when working on the ball joints, springs, and shock absorbers.*

2. *The suspension system has many parts that are under high pressure and tension. These include the shocks, springs, and torsion bars. When these parts are removed incorrectly, they may spring out violently, causing serious injury. Always make sure all tension has been removed from the components before disassembly.*

3. *Always wear safety glasses when working on any suspension parts.*

4. *Never use high-pressure air to dry off bearings after they have been cleaned. Never spin the bearings with a high-pressure air hose, as the balls could dislodge themselves and cause serious injury.*

5. *When removing MacPherson strut components, remember that some parts may have high tension on them. Always remove the tension or pressure before removing the components from the vehicle. This means that extra support will be needed.*

1. *Ball joints* can be checked for *wear* by inspecting the lower section of the ball joint. New ball joints will have a 0.050 clearance between the end of the grease zerk and the body of the ball joint. Refer to *Figure 29–28*. Replace the ball joint if the

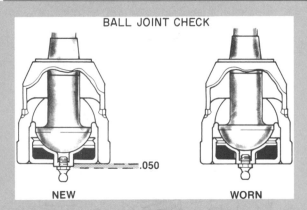

FIGURE 29–28. Ball joints can be checked for wear. There should be a 0.050 clearance between the grease zerk and the body of the ball joint. *(Courtesy of Dana Corporation)*

clearance is less than 0.050. Use the following general procedure to *replace a ball joint*.

a. Raise the vehicle and support it at the frame. Then remove the wheel and tire.

b. Position a suitable jack under the lower control arm spring seat and raise the jack to compress the coil spring. ***CAUTION:*** *The jack must remain in place when the ball joint is being replaced to hold the spring and control arm in position.*

c. Remove the cotter pins and nuts holding the ball joint stud to the steering knuckle. Now disconnect the joint from the knuckle using a pickle fork.

d. The ball joint must now be removed from the control arm. Remove the heads of the rivets that hold the ball joint to the control arms. Certain ball joints may have to be pressed out of the control arm. The ball joint should now be removable.

e. Place a new ball joint in the control arm. The new ball joint may have to be pressed or bolted in.

f. Install the ball joint stud into the steering knuckle and torque the nut to the manufacturer's specifications.

g. Install the cotter pin.

h. Replace the wheel and tire.

2. A bad shock absorber has the following characteristics:

a. Continuous bouncing of the body with every road bump.

b. Oscillation of the body with rough surface roads.

c. Lifting of the body when the car is accelerated.

Bad shock absorbers cannot be repaired. In all cases, the shocks are replaced with new ones. Use the following procedure to *replace the front shock absorbers*.

a. Raise and support the vehicle as needed.

b. Hold the shock absorber shaft with a suitable wrench, and remove the upper retaining nut.

c. Remove the lower bolts that hold the shock absorber pivot arm to the control arm. Pull the shock absorber from the coil spring. Replace the shock absorber with a new one by reversing the removal procedure.

The rear shocks are removed much the same way, except the vehicle is supported in the rear. On some vehicles, the upper retaining nut may be located in the trunk area. ***CAUTION:*** *On some vehicles, the rear springs may have to be supported by a jack to remove the shock absorber.*

3. ***Figure 29–29*** shows a troubleshooting chart for shocks. Refer to this chart when determining the trouble, possible cause, and what to do when servicing shock absorbers.

4. Front wheel shimmy can be caused by several conditions. These are shown in ***Figure 29–30***.

5. Noises in the front end can be caused by many malfunctions. ***Figure 29–31*** shows examples of some of the parts that may have malfunctioned.

6. Certain vehicles may have "dog" tracking. This means the rear wheels track to the right or left of the front wheels. Possible causes for this condition are shown in ***Figure 29–32***.

7. Ball joints can be damaged in several ways. The Ball Joint Diagnostic Procedure shown in ***Figure 29–33*** can be followed to determine the correct procedure when checking ball joints.

8. When testing for wheel bearing noise, follow these procedures to determine the *bad wheel bearing*:

a. Drive the car at low speed on a smooth road.

b. Turn the car to develop left and right motions, traffic permitting.

c. Noise should change due to cornering loads.

d. Jack up the wheels to verify roughness at the wheels.

e. Replace the worn wheel bearing.

Trouble Shooting Chart

Trouble	Possible Cause	What to do
1. Shock absorber breaks down	Vehicle spring suspension travel limit stop defective	Check rubber stop on the spring suspension travel, if necessary, replace it
	Shock absorber performs improperly	Replace shock absorber
2. Shock absorber noises (rattling, rumbling)	Shock absorber mounting loose	Fasten shock absorber properly
	Protective tube loose	Replace shock absorber
	Protective tube grazes on cylinder tube	Check offset between top and bottom shock absorber mountings
	Shock absorber worn	Exchange shock absorber
3. Shock absorber inefficient	Oil loss due to defective seals or worn valves	Exchange shock absorber
4. Shock absorber leaky	Defective piston rod seal. **Attention! There is a difference between an oil mist which is harmless and fresh oil – shock absorber offset**	Exchange shock absorber see § 2 also
5. Shock absorber works too hard	Wrong shock absorber type installed	Install correct type according to vehicle specification
	Valves not in order	Exchange shock absorber
6. Shock absorber works too smooth	Wrong shock absorber installed	Install correct type according to vehicle specification
	Shock absorber worn out	Install new shock absorber
7. Bad driving quality	damping efficiency fades	Install new shock absorber
8. Washing out (flattening) of tyre profile	Damping efficiency has vanished or ceased to exist	Install new shock absorber

FIGURE 29–29. This troubleshooting chart can be used to solve problems with the shock absorbers. *(Courtesy of Sachs Industries, Inc.)*

FRONT WHEEL SHIMMY

a. Tire and wheel out of balance	a. Balance tires
b. Worn or loose wheel bearings	b. Adjust wheel bearings
c. Worn tie rod ends	c. Replace tie rod end
d. Worn ball joints	d. Replace ball joints
e. Incorrect front wheel alignment	e. Check and align front suspension
f. Shock absorber inoperative	f. Replace shock absorber

FIGURE 29–30. Front wheel shimmy can be caused by several problems in the front suspension. (Courtesy of Chevrolet Division, General Motors Corporation)

NOISE IN FRONT END

a. Ball joints need lubrication	a. Lubricate ball joint
b. Shock absorber loose or bushings worn	b. Tighten bolts and/or replace bushings
c. Worn control arm bushings	c. Replace bushings
d. Worn tie rod ends	d. Replace tie rod ends
e. Worn or loose wheel bearings	e. Adjust or replace wheel bearings
f. Loose stabilizer bar	f. Tighten all stabilizer bar attachments
g. Loose wheel nuts	g. Tighten the wheel nuts to proper torque

FIGURE 29–31. Common noises in the front end can be caused by several malfunctions. (Courtesy of Chevrolet Division, General Motors Corporation)

"DOG" TRACKING

LEAF TYPE REAR SPRING

a. Rear leaf spring broken	a. Replace spring
b. Bent rear axle housing	b. Replace housing
c. Frame or underbody out of alignment	c. Align frame

COIL TYPE REAR SPRING

a. Damaged rear suspension arm and/or worn bushings	a. Replace suspension arm and/or bushings
b. Frame out of alignment	b. Align frame
c. Bent rear axle housing	c. Replace housing

FIGURE 29–32. Dog tracking (rear wheels not tracking the same as the front wheels) can be caused by several problems in the rear suspension. (Courtesy of Chevrolet Division, General Motors Corporation)

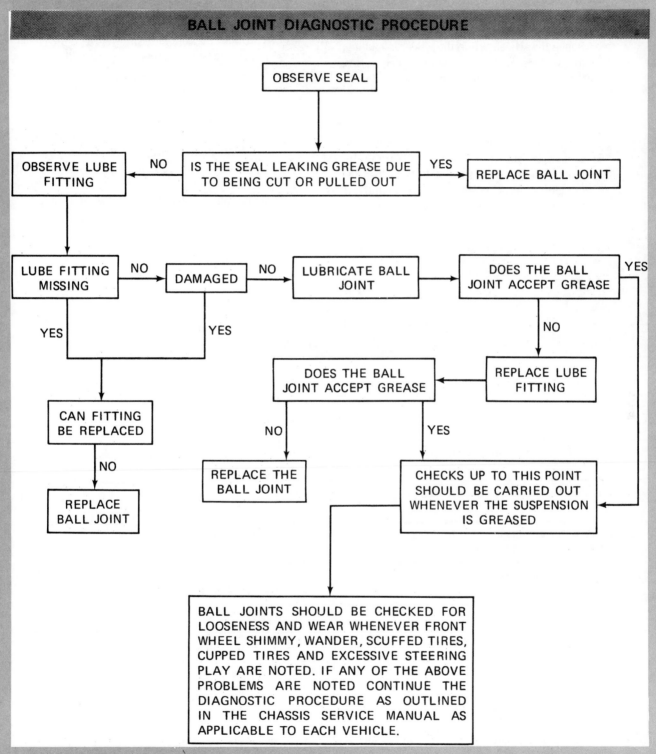

FIGURE 29–33. Follow this ball joint diagnostic procedure to determine problems in the ball joints. *(Courtesy of Chevrolet Division, General Motors Corporation)*

9. Front wheel bearings must be adjusted correctly so that the right amount of load is placed in the bearings. Use the following general procedure to *adjust the wheel bearings*.
   a. Support the front of the vehicle so that the wheels can be rotated freely.
   b. Rotate the wheel. While the wheel is rotating, torque the spindle nut to approximately 12 lbs. (Check the manufacturer's specifications for the correct torque.)
   c. Back off the nut until it is just loose, then retighten it by hand.
   d. Loosen the spindle nut until the cotter pin can be inserted. Do not, however, loosen the nut more than 1/2 flat on the nut. With the bearing properly adjusted, there should be about 0.001 to 0.005 inch end play.
   e. Check the front wheel rotation to see if it rotates noisily or roughly. Clean and inspect or replace the wheel bearing as necessary.
10. When handling bearings, always follow the recommended procedures shown in *Figure 29–34*.
11. The following conditions indicate a *damaged bearing*. Always check bearings carefully for each condition.
    a. Galling — metal smears on roller ends due to overheating, lubricant failure, or overload.
    b. Step wear — wear pattern on roller ends caused by fine abrasives.
    c. Indentations — surface depressions on race and rollers caused by hard particles of foreign materials.
    d. Etching — bearing surfaces appear gray or grayish black in color.

HANDLING BEARINGS:	
Things To Remember	**Things To Avoid**
1. Remove all outside dirt from housing before exposing bearing.	1. Working in dirty surroundings.
2. Treat a used bearing as carefully as you would a new one.	2. Using dirty, brittle or chipped tools.
3. Work with clean tools in clean surroundings.	3. Using wooden mallets or working on wooden bench tops.
4. Handle with clean, dry hands or, better, clean canvas gloves.	4. Handling with dirty, moist hands.
5. Use clean solvents and flushing oils.	5. Using gasolines containing tetraethyl lead, as they may be injurious to health.
6. Lay bearings out on clean paper.	6. Spinning uncleaned bearings.
7. Protect disassembled bearings from rust and dirt.	7. Spinning bearings with compressed air.
8. Use clean rags to wipe bearings.	8. Using cotton waste or dirty cloths to wipe bearings.
9. Keep bearings wrapped in oilproof paper when not in use.	9. Exposing bearings to rust or dirt at all times.
10. Clean inside of housing before replacing bearing.	10. Scratching or nicking of bearing surfaces.

FIGURE 29–34. Bearings should be handled with care. Always follow these guidelines. (Courtesy of Chevrolet Division, General Motors Corporation)

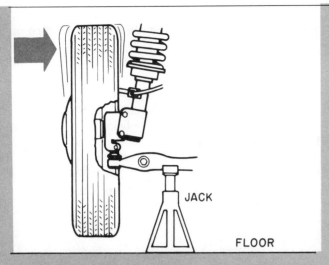

FIGURE 29–35. Check for loose or worn wheel bearings with the weight of the car off the wheel. *(Courtesy of Dana Corporation)*

 e. Heat discoloration — bearing surfaces appear faint yellow to dark blue, resulting from overload and lubricant breakdown.

 f. Brinelling — surface indentations in raceway caused by rollers either under impact loading or vibration while bearing is not rotating.

12. A weak strut assembly on the MacPherson strut suspension can be checked by pushing downward, then quickly releasing near the fender over each strut. Any tendency to bounce more than once means the shock may be in poor condition and should be replaced.

13. Check for loose or worn wheel bearings with the weight of the car off the wheel. This is shown in *Figure 29–35*.

14. Worn strut rod bushings may be checked by firmly grasping the strut rod and shaking it. Any noticeable play indicates excess wear, and replacement is needed. Use the following general procedure to *remove and replace the strut.*
 a. Raise and support the front of the vehicle.
 b. Remove the wheel and tire.
 c. Support the lower control arm with a suitable jack stand.
 d. Remove the brake hose bracket.
 e. Remove the strut-to-knuckle bolts.
 f. Remove the cover from the upper end of the strut at the shock tower area.
 g. Remove the nuts from the upper strut assembly.
 h. Remove the strut assembly from the vehicle.
 i. Reverse the procedure to install the strut.
 j. Torque all bolts and nuts to the correct manufacturer's specifications.

15. Visual inspection of the rear suspension system can reveal loose, worn, or broken parts. Leaf springs should bow upward at the ends. If the leaf springs are flat, they have either broken or have lost tension. Replace if necessary.

16. Check all coil springs for bright spots and cracks. Inspect the mounting plates for broken or missing pads. Make sure each coil is an equal distance from the coil above and below it. (Note that some springs are manufactured so that the spring coils are closer at the top.)

SUMMARY

This chapter dealt with suspension systems. The major areas that were studied included the front suspension, rear suspension, shock absorbers, MacPherson strut suspension, and service and troubleshooting information.

The front suspension system is used to support the front of the vehicle and provide a smooth ride for the passengers. The system is called an independent front suspension system (each wheel acts independently of the other). This is done by using several major components. Ball joints are used to connect the spindle and steering knuckle to the control arms. There are upper and lower control arms. Bushings are used to attach the control arms to the frame. With the use of control arms and ball joints, the wheel can turn, as well as move up and down during operation.

Several other components play an important part in the front suspension system. Sway bars and links are used to help transmit cornering forces from one side of the vehicle to the other. Strut rods are used help retain the lower control arm in its position. Coil springs are used to support the weight of the car and to help position the car correctly. Torsion bars are used on some vehicles in place of springs. A torsion bar is made of a long spring steel rod that is connected to the frame on one end and the lower control arm on the other. As the vehicle moves up and down, the torsion bar twists. The twisting produces much the same effect as that produced by springs. The spindle is used to carry the wheel bearings for the wheel. The spindle is connected to the control arms by the steering knuckle. These parts play a combined role as part of the front suspension system.

The rear suspension system is used to support the rear of the vehicle. It must also keep the rear wheels in line with the front wheels. The rear suspension also uses springs. Both coil and leaf springs are used. The leaf springs are made of spring steel. Each end is mounted to the frame of the vehicle. The axle is attached to the upper center of the leaf spring by a U-bolt. One end of the leaf spring uses a shackle. The shackle is a small link between the end of the leaf spring and the frame. It allows the spring to lengthen and shorten slightly when going over bumps.

All suspension systems, both front and rear, use shock absorbers. A shock absorber is a device that is placed on each wheel to dampen or slow down the bouncing of the car caused by bumps in the road. It is made by using a series of hydraulic cylinders, a piston, and valves. As the vehicle goes over bumps, the piston moves up and down. This action causes hydraulic oil to be forced through small valves. The valves only allow a certain amount of oil to pass through. The net result is that the number of oscillations after hitting a bump in the road is reduced. The shock absorber action happens both on upward and downward motions.

Several designs have been used on shock absorbers. Spiral grooves are placed on the cylinders to reduce aeration of the oil inside. Gas-filled shocks are also used to reduce aeration. Certain types of shock absorbers are called air shocks. Air shocks have small air lines attached to the cylinders. A compressor produces air pressure that is used to raise or lower the shock absorber. When there is extra load on the car, the air shocks can be pressurized to keep the vehicle level.

The automatic level control system can also be used to keep the vehicle level with changing loads. The system uses an air compressor, air dryer, electrical switches and relays, and electronic height sensor to operate. When an extra load is placed on the system, the right amount of air is admitted to each shock to keep the vehicle level.

The air suspension system is also used to maintain level conditions on the car with varying loads. This system uses air springs rather than air shocks. The remaining parts of the system, however, work in a manner similar to the automatic level control system.

MacPherson strut suspension systems are now being used on many smaller cars, both imported and domestic. The MacPherson strut suspension is favored where space and weight savings are important. This system is much like a shock absorber and spring built into one unit. The system has eliminated the need for upper control arms and upper ball joints. The entire unit is attached to the lower control arm on the bottom and to the shock tower (part of the body or fender well) on the top. The entire unit turns when the vehicle is turned. An upper bearing is needed to allow for the turning action. There are also certain modifications to the MacPherson strut suspension that are being used on vehicles today.

Service is an important part of the suspension system. Ball joints should always be checked for wear. Front wheel bearings should be checked for damage and handled according to the manufacturer's guidelines. Shocks should be checked for leakage, excessive bouncing, and broken parts. The struts and sway bars and links should be checked for damaged bushings or broken parts. Many of these parts can be checked by visual inspection. If any of these parts are damaged, it may cause uneven tire wear, poor handling, or uncomfortable rides.

TERMS TO KNOW

Can you explain each of the following terms? Review the chapter until you can use each term correctly.

Ball joint	Independent rear
Bellows	suspension
Control arm	Strut rod
Jounce	Sway bar
MacPherson strut system	Wheel spindle
Shock absorber	Stance
Stabilizer bar	Shackle
Steering knuckle	Oscillation
Torsion bar	Aeration

REVIEW QUESTIONS

Multiple Choice

1. The front suspension system on cars today is called:
   a. Independent front suspension
   b. Rigid suspension
   c. I beam suspension
   d. Leaf spring suspension
   e. Strut suspension

2. How many ball joints are used on each side of the standard front suspension system?
   a. 1
   b. 2
   c. 3
   d. 4
   e. 5

3. The ball joints are attached to the:
   a. Steering knuckle
   b. Upper control arms
   c. Lower control arms
   d. All of the above
   e. None of the above

4. Which of the following helps to transmit cornering loads to the opposite wheel?
   a. Strut rods
   b. Control arms
   c. Sway bars and links
   d. Ball joints
   e. Coil springs

5. To keep the single pivot lower control arm held in place, _____ are used.
   a. Coil springs
   b. Leaf springs
   c. Ball joints
   d. Strut rods
   e. Stabilizer bars

6. Which of the following are used instead of coil springs on the front suspension?
   a. Shock absorbers
   b. Ball joints
   c. Torsion bars
   d. Stabilizer bars
   e. Sway bar and link

7. Front wheel bearings are of the _____ type.
   a. Tapered-roller or ball-bearing
   b. Needle-bearing
   c. Bushing
   d. Triple-ball
   e. None of the above

8. The inner race of the wheel bearings rides on the:
   a. Steering knuckle
   b. Ball joint
   c. Stabilizer bar
   d. Control arm
   e. Spindle

9. The rear suspension systems used on cars:
   a. Never need service
   b. Use a solid axle
   c. Can be independent rear suspension
   d. All of the above
   e. B and c

10. One end of the leaf spring is attached to the frame. The end of the leaf spring is attached to:
    a. The shackle
    b. The frame
    c. The axle
    d. The wheel
    e. The bearing

11. Shock absorbers will help to:
    a. Reduce the number of oscillations of motion
    b. Provide a smoother ride
    c. Increase the stability of the car
    d. All of the above
    e. None of the above

12. Which of the following is/are not considered part of the shock absorber?
    a. Piston rod and piston
    b. Piston valves
    c. Strut
    d. Bottom valve
    e. Container tube

13. Which of the following helps to reduce or dampen the motion of a shock absorber?
    a. The position of the valves
    b. The amount of oil that can pass through the valve
    c. The size of the piston
    d. The type of material on the valves
    e. The addition of air or aeration of oil

14. Which of the following is considered a problem with shock absorbers?
    a. They can heat up too much
    b. They produce aeration inside, causing poor performance
    c. They don't support the car properly
    d. They leak transmission fluid
    e. They lock up, causing a rough ride

15. Gas-filled shock absorbers are designed:
    a. To reduce aeration
    b. To improve safety
    c. For automatic leveling
    d. All of the above
    e. None of the above

16. Which of the following systems uses an air pump and compressed air to level the car?
    a. Air shock absorbers
    b. Automatic leveling systems
    c. Leaf spring systems
    d. MacPherson strut systems
    e. Air suspension systems

17. The air suspension system uses which of the following?
    a. Shock absorbers
    b. Automatic height sensors
    c. Control module
    d. All of the above
    e. None of the above

18. A bellows is used on the _____ .
    a. Springs
    b. Air shock absorbers
    c. MacPherson strut suspension
    d. Standard shock absorbers
    e. Torsion bars

19. The air suspension system forces air into the:
    a. Shock absorbers
    b. Air springs
    c. Bellows
    d. Stance
    e. Coil springs

20. The upper portion of the MacPherson strut suspension:
    a. Is attached to the shock tower
    b. Uses a bearing
    c. Turns as the car turns
    d. All of the above
    e. None of the above

21. Which is an advantage of a MacPherson strut suspension system?
    a. It is lighter
    b. It takes up more room
    c. It is heavier
    d. It requires increased maintenance
    e. It has more moving parts

22. The MacPherson strut suspension system
    a. Uses a standard shock absorber
    b. Requires no maintenance
    c. Uses extra control arms
    d. Uses extra ball joints
    e. Uses no bearings

23. To check the condition of the ball joint, there should be a difference of _____ between the grease zerk and the body of the ball joint.
    a. 0.010
    b. 0.020
    c. 0.030
    d. 0.040
    e. 0.050

24. To check for loose wheel bearings, the car must be:
    a. Driven on a straight line at high speed
    b. Placed on a jack with the weight removed
    c. Placed on the road surface and shaken
    d. Turned in one direction only
    e. Loaded down with extra weight

25. A bad set of shocks has which of the following characteristics?
    a. Car lifts when accelerated
    b. Car has more oscillations after a bump
    c. Car has less stability on the road
    d. All of the above
    e. None of the above

26. Front wheel shimmy can be caused by:
    a. Tires being in balance
    b. Worn tie rods
    c. Worn ball joints
    d. All of the above
    e. B and c

The following questions are similar in format to ASE (Automotive Service Excellence) test questions.

27. Technician A says that the front wheel bearings should be adjusted as tightly as possible by tightening the spindle nut to 100 ft. lbs. Technician B says that there should be approximately 0.500 inch play in the front wheel bearings. Who is right?
    a. A only
    b. B only
    c. Both A and B
    d. Neither A nor B

28. Technician A says that there is no check for testing the conditions of ball joints. Technician B says that ball joints can be checked for a clearance between the end of the grease zerk and the body of the ball joint. Who is right?
    a. A only
    b. B only
    c. Both A and B
    d. Neither A nor B

29. There is a continuous bouncing of the vehicle body when the car goes over a bump. Technician A says the problem is bad shock absorbers. Technician B says the problem is a bad steering knuckle. Who is right?
    a. A only
    b. B only
    c. Both A and B
    d. Neither A nor B

30. Technician A says there is no way to check the condition of a ball joint. Technician B says that a dial indicator can be used to check the movement of a ball joint. Who is right?
    a. A only
    b. B only
    c. Both A and B
    d. Neither A nor B

Essay

31. What is the purpose of the sway bar?

32. Define the purpose and operation of a torsion bar.

33. Describe how to check the condition of shocks.

34. What is the purpose of using gas-filled shocks?

35. What are several advantages of using MacPherson strut suspension systems?

36. Describe the purpose of a wheel spindle.

37. What is the purpose of using ball joints?

CHAPTER 30

Steering Systems

INTRODUCTION

The steering system is used to control the direction of the vehicle. The steering system is designed to control the front wheels over all types of road conditions, through turns, and at different speeds. It is made of a linkage system that is attached to the front wheels, the steering wheel, and the steering gear. Manual and power steering units are used.

OBJECTIVES

After reading this chapter, you will be able to:
■ Define the parts and operation of the standard steering system.
■ Examine the operation of the steering gear.
■ Define front end geometry including caster, camber, toe, steering axis inclination, and turning radius.
■ Identify the operation of power steering units and pumps.
■ State common diagnosis and service procedures on the steering system.

CHAPTER HIGHLIGHTS

30.1 STEERING SYSTEM PARTS AND OPERATION
    A. Parts on a Steering System
    B. Steering Wheel and Column
    C. Manual Steering Gear
    D. Pitman Arm Steering Gear
    E. Rack and Pinion Steering Gear
    F. Steering Ratio
    G. Standard Steering Linkage
    H. Pitman Arm
    I. Center Link
    J. Idler Arm
    K. Tie Rods and Adjusting Sleeve

30.2 FRONT END GEOMETRY AND ALIGNMENT
    A. Purpose of Wheel Alignment
    B. Caster
    C. Camber
    D. Toe
    E. Steering Axis Inclination
    F. Included Angle
    G. Scrub Radius
    H. Turning Radius

30.3 POWER STEERING
    A. Purpose of Power Steering
    B. Major Parts on Power Steering Systems

30.1 STEERING SYSTEM PARTS AND OPERATION

Parts on a Steering System

The steering system is composed of three major subsystems, *Figure 30–1*. They include the steering column and wheel, the steering gear, and the steering linkage. As the steering wheel is turned by the operator, the steering gear transfers this motion to the steering linkage. The steering linkage turns the wheels to control the vehicle direction.

Although there are many variations to this system, these three major assemblies make up the steering system. Other variations may include power steering and *rack and pinion* steering.

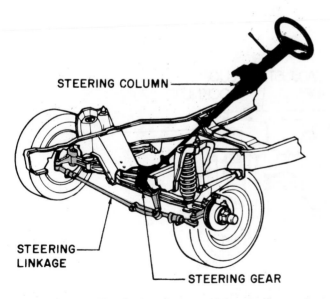

STEERING COLUMN

STEERING LINKAGE

STEERING GEAR

FIGURE 30–1. The three main parts of the steering system are the steering column, steering gear, and steering linkage. *(Courtesy of Dana Corporation)*

Steering Wheel and Column

The purpose of the steering wheel and column is to produce the necessary force to turn the steering gear. The steering column is made of many parts. The exact type of steering wheel and column depends upon the year and the car manufacturer. Major parts shown in *Figure 30–2* include:

    a. The steering wheel which is used to produce the turning effort.
    b. The lower and upper covers which conceal parts.
    c. The universal joints which rotate at angles.
    d. Support brackets which are used to hold the steering column in place.
    e. Assorted screws, nuts, bolts, pins, and seals, which are used to make the steering wheel and column perform correctly.

Other differences in the steering wheel and column include energy-absorbing or collapsible steering columns, tilt steering wheels, steering lock systems, and location of turn signals and flasher controls.

Manual Steering Gear

The purpose of the steering gear is to change the rotational motion of the steering wheel to reciprocating motion to move the steering linkage. There are two styles currently in use. These include the *pitman arm* or recirculating ball steering gear and the rack and pinion steering gear.

Pitman Arm Steering Gear

One of the most common types of manual steering gears is called the pitman arm steering gear. Many manufacturers call this the "recirculating ball and worm" system. *Figure 30–3* shows such a system. In operation, as the steering shaft is turned, the wormshaft also turns. The wormshaft has spiral grooves on the outside diameter. The ball nut, which has mating spiral grooves inside, is placed over the wormshaft. Small steel balls circulate in the mating grooves and ball guides. As the balls move through the grooves and out, they return to the other side through the guides. This provides a low-friction drive between the wormshaft and the ball nut.

Teeth on the ball nut mesh with the teeth on the sector shaft. The sector shaft is also known as the pitman shaft. As the wormshaft is rotated, the ball nut moves back and forth, to the left and right. As the ball nut moves back and forth, it causes the sector shaft or pitman shaft to rotate through a partial circle. The sector shaft is connected directly to the pitman arm which controls the steering linkage.

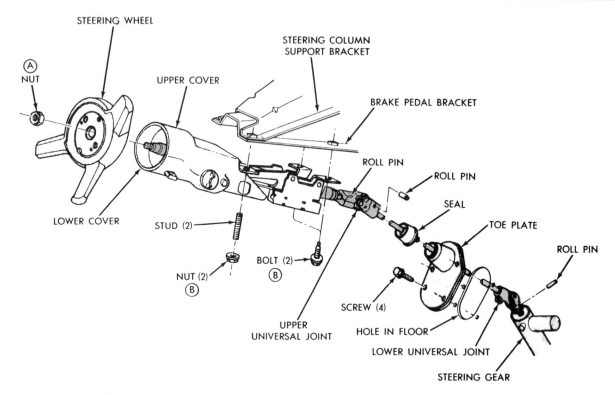

FIGURE 30–2. Steering columns have many parts. Steering columns vary in design, depending upon the manufacturer and the year of the vehicle. *(Courtesy of Motor Publications, Auto Repair Manual, 1981–1987)*

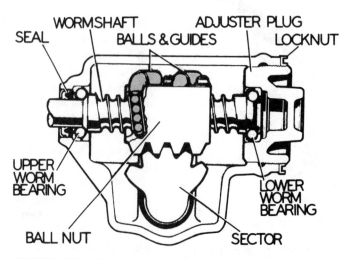

FIGURE 30–3. The manual steering gear uses a recirculating ball and worm system. *(Courtesy of Motor Publications, Auto Repair Manual, 1981–1987)*

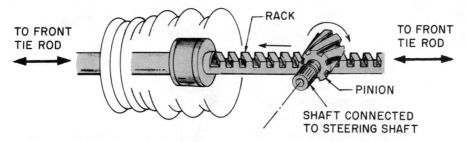

FIGURE 30–4. The rack and pinion uses a flat gear called the rack and a pinion gear attached to the steering column. *(Courtesy of Dana Corporation)*

Rack and Pinion Steering Gear

The rack and pinion system is fast becoming a standard system on most front wheel drive cars sold in the U.S. The rack and pinion system is used in conjunction with MacPherson struts and gives more engine compartment room for transverse-mounted engines.

Rack and pinion steering consists of a flat gear (the rack) and a mating gear called the pinion, *Figure 30–4*. When the steering wheel and shaft turn, the pinion meshes with the teeth on the rack. This causes the rack to move left or right in the housing. This motion moves the remaining steering linkage to turn the front wheels. This system is very practical for small cars that require lighter steering capacity. It is a direct steering unit which is more positive in motion (less lost motion) than the standard steering linkages. *Figure 30–5* shows the complete rack and pinion system with the housing and tie rods.

Key No.	Part Name
1 —	FLANGE ASSY, COUPLING & STRG.
2 —	BOLT, PINCH
3 —	HOUSING ASSY, RACK & PINION
4 —	BEARING ASSY, ROLLER
5 —	PINION ASSY, BEARING
6 —	RING, RETAINING
7 —	SEAL, STEERING PINION
8 —	RACK, STEERING

Key No.	Part Name
9 —	CLAMP, BOOT
10 —	BOOT
11 —	CLAMP, BOOT
12 —	ROD ASSY, INNER TIE
13 —	NUT, JAM
14 —	ROD ASSY, OUTER TIE
15 —	SEAL, TIE ROD
16 —	BEARING, RACK

Key No.	Part Name
17 —	SPRING, ADJUSTER
18 —	PLUG, ADJUSTER
19 —	NUT, ADJUSTER PLUG LOCK
20 —	GROMMET, GEAR MOUNTING
21 —	THIS NUMBER NOT USED
22 —	BUSHING, RACK
23 —	RING, RETAINING
24 —	FITTING, LUBRICATION
25 —	PIN, COTTER
26 —	NUT, HEXAGON SLOTTED

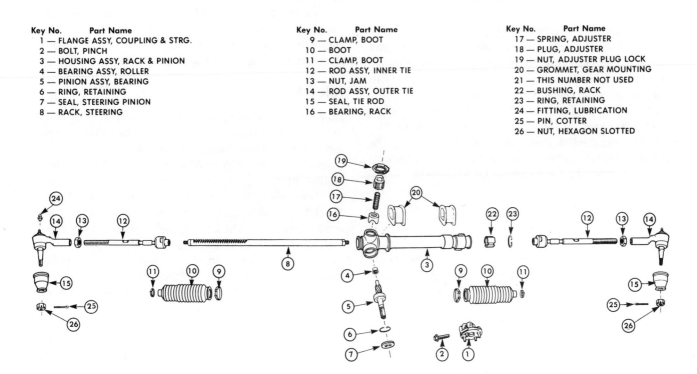

FIGURE 30–5. All parts of the rack and pinion system are shown and referenced with numbers and part names. *(Courtesy of General Motors Product Service Training)*

Steering Ratio

When the steering wheel is turned, a certain effort is needed. The amount of effort is determined by the mechanical advantage of the steering gear. *Steering ratio* is defined as the ratio between the degrees turned on the steering wheel and the degrees turned on the front wheels. The ratio is stated for exactly one degree of movement on the front wheels. For example, a 30 to 1 steering ratio means the steering wheel will turn 30 degrees for each degree of front wheel turn. The lower the ratio, the harder the steering. Lower steering ratios are called quick steering. The higher the ratio the easier the steering. When the steering ratio increases, however, the steering wheel must be turned farther to make a turn.

The steering ratio used on a car depends upon several factors and differences in the steering system. These include:

   a. manual or power steering
   b. weight and size of the vehicle
   c. type of steering gear
   d. size of the steering wheel

Standard Steering Linkage

The steering linkage is defined as the pivoting parts necessary to turn the front wheels. The linkage connects the motion produced by the pitman shaft to the front wheels on the vehicle.

The parts of a standard steering linkage are shown in *Figure 30–6*. The motion from the steering gear and sector shaft (pitman shaft) causes the pitman arm to rotate through a partial circle (reciprocating motion or back and forth). This motion causes the *center link* to move back and forth also. The idler arm is attached to the frame of the vehicle for support. Tie rods are connected to each side of the center link. As the center link moves, both tie rods also move. The tie rods are then attached directly to the wheel for turning. Sleeves are placed on each tie rod for adjustment.

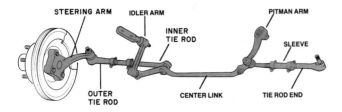

FIGURE 30–6. The steering linkage is made of tie rods, an idler arm, a center link sleeve, and the pitman arm. (Courtesy of Dana Corporation)

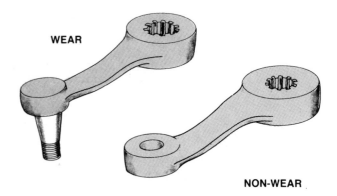

FIGURE 30–7. A vehicle can use one of two types of pitman arms. These include the wear and the non-wear types. (Courtesy of Dana Corporation)

Pitman Arm

Pitman arms can be of the wear or non-wear type. The wear-type pitman arm has a tapered ball stud that is connected to the center link. The other end is mounted on the steering gear sector shaft. The non-wear-style pitman arm has a tapered hole and seldom needs replacement. *Figure 30–7* shows the different types of pitman arms.

Center Link

The center (drag) link can be designed in several ways. *Figure 30–8* shows several styles. The major difference is the method in which the other linkage is connected to the center link. The point of connection can be the pivot point, stud end, bushing end, or open taper end (non-wear).

Idler Arm

Idler arms come in different designs as well. They differ mainly on the wear end of the arm. The different types include the bushing, taper, threaded, and constant torque types of idler arm. *Figure 30–9* shows the different styles of idler arms.

The constant torque type of idler arm shown in *Figure 30–10* is manufactured to precision tolerances and uses synthetic bearings. The bearings reduce friction and absorb road shock. The bearings are preloaded and preset at the factory. This type of idler arm has very low friction characteristics and is used on many new vehicles.

The taper type of idler arm is shown in *Figure 30–11*. It contains synthetic bearings, heat-treated tapered support brackets, and a compensating spring. The compensating spring takes up clearances produced by wear. The spring also maintains the steering resistance desired for good vehicle handling.

The bushing type of idler arm is shown in *Figure 30–12*. It uses a resilient, lubricated bushing. These

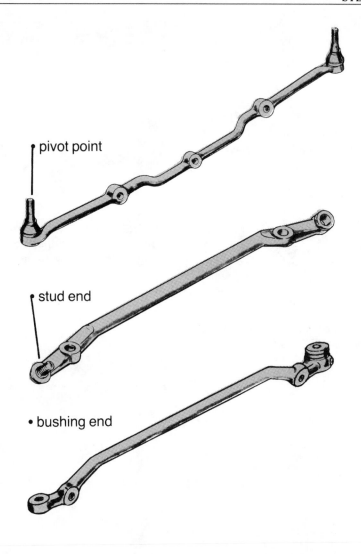

• pivot point

• stud end

• bushing end

• open taper end (non-wear)

FIGURE 30–8. Different types of center links are used by different manufacturers. (Courtesy of Dana Corporation)

bushings are designed to accept high shock loads and maintain good vehicle handling. The special waffle design bushing traps the lubricant. Seals are used on the end to make the unit self-contained.

Tie Rods and Adjusting Sleeve

Tie rod assemblies consist of an inner tie rod end, an outer tie rod end, and an *adjusting sleeve*. The adjusting sleeve looks like a piece of internally threaded pipe. See *Figure 30–13*. The unit has a slot that runs through the center. Adjusting sleeves also have two crimping or squeezing clamps. These are located at each end to lock the tie rod together after adjustment.

The adjusting sleeve has threads inside. One end of the thread has a left-hand thread and the other end has a right-hand thread. This allows adjustment without disassembling the tie rods.

The tie rod ends have a rounded ball stud to allow both lateral and vertical movement. *Figure 30–14* shows an example of the rounded ball stud. The tension spring inside the tie rod end is used to reduce road shock throughout the steering system. On some tie rod ends, there is also a grease *zerk*.

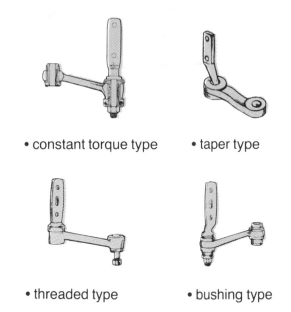

• constant torque type   • taper type

• threaded type   • bushing type

FIGURE 30-9. Different types of idler arms are used on the steering linkage. *(Courtesy of Dana Corporation)*

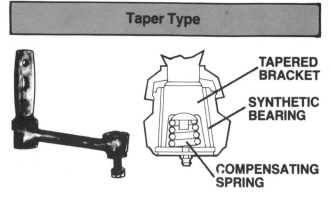

Taper Type

TAPERED BRACKET

SYNTHETIC BEARING

COMPENSATING SPRING

FIGURE 30-11. The taper type idler arm contains a tapered bracket, synthetic bearings, and a compensating spring. *(Courtesy of Dana Corporation)*

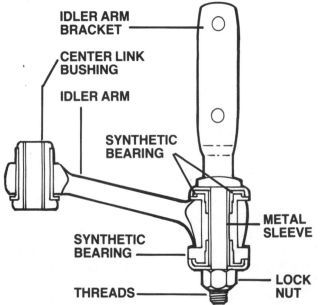

Constant Torque Type

IDLER ARM BRACKET

CENTER LINK BUSHING

IDLER ARM

SYNTHETIC BEARING

SYNTHETIC BEARING

METAL SLEEVE

LOCK NUT

THREADS

FIGURE 30-10. The constant torque type idler arms use precision tolerances and synthetic bearings. *(Courtesy of Dana Corporation)*

Bushing Type

LUBRICANT POCKETS

LIP SEALS

FIGURE 30-12. The bushing type idler gear has lubricant pockets for lubrication. *(Courtesy of Dana Corporation)*

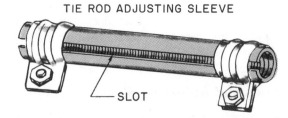

TIE ROD ADJUSTING SLEEVE

SLOT

FIGURE 30-13. This sleeve is attached to the tie rods to make adjustment for length. *(Courtesy of Dana Corporation)*

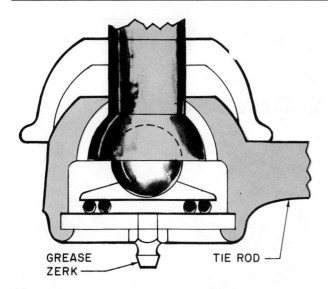

FIGURE 30–14. A rounded ball stud is used on the end of the tie rods. *(Courtesy of Dana Corporation)*

CAR CLINIC

PROBLEM: CAR PULLS TO ONE SIDE DURING ACCELERATION

The rear end of a vehicle with a limited slip differential seems to pull to one side during acceleration. Limited slip differentials are designed to provide equal push on both rear wheels. What could be the problem?

SOLUTION:

The first thing to check is the rear tires. Make sure that the tires are both the same diameter. The tires have most likely been mismatched in the past. Using a limited slip differential with two tire sizes will cause the car to pull to one side.

30.2 FRONT END GEOMETRY AND ALIGNMENT

Purpose of Wheel Alignment

Alignment is defined as the balancing of all forces created by friction, gravity, centrifugal force, and momentum while the vehicle is in motion. It is very important for the wheels of the vehicle to contact the road correctly. Wheel alignment is a check of how the wheels contact the pavement. The main purpose of wheel alignment is to allow the wheels to roll without scuffing, dragging, or slipping on the road. Good alignment results in:

  a. better fuel economy
  b. less strain on the front-end parts
  c. directional stability
  d. easier steering
  e. longer tire life
  f. increases safety

There are typically five angles that affect the steering alignment. These include caster, camber, toe, steering axis inclination, and turning radius.

Caster

Caster is defined as the backward or forward tilt at the top of the spindle support arm. Backward tilt is called positive caster. Forward tilt is called negative caster. Caster angle is the distance between the center line of the spindle support arm and the true vertical line. An example of caster is an ordinary furniture caster or a bicycle. See *Figure 30–15*.

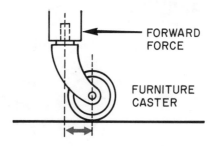

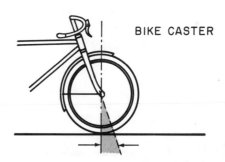

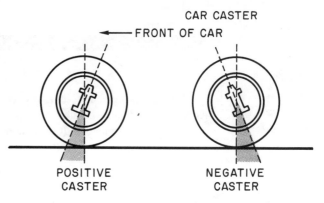

FIGURE 30–15. Caster is defined as the backward or forward tilt of the spindle support arm. It is much like a furniture caster.

CORROSION PROTECTION

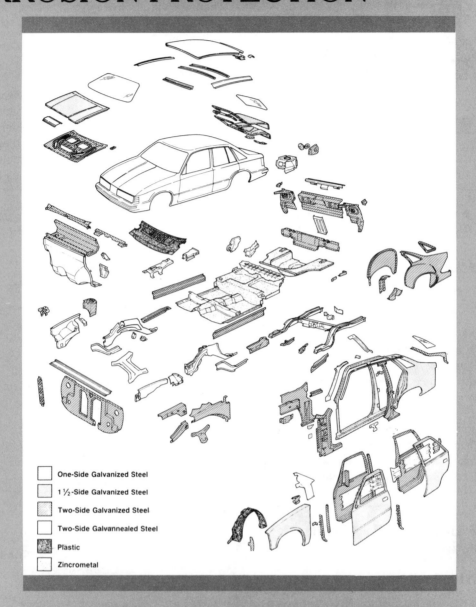

One-Side Galvanized Steel

1 ½-Side Galvanized Steel

Two-Side Galvanized Steel

Two-Side Galvannealed Steel

Plastic

Zincrometal

Corrosion protection in many vehicles includes a combination of various galvanized and galvannealed steel, plastic, and zincrometal. All of these materials offer extended service life. (Courtesy of Chrysler Corporation)

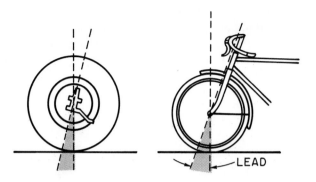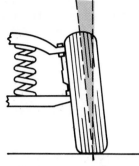

FIGURE 30–16. Lead is defined as the distance at ground level between the two center lines. A bicycle is shown as an example. *(Courtesy of Dana Corporation)*

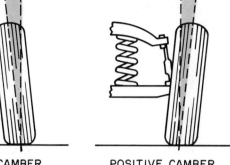

NEGATIVE CAMBER   POSITIVE CAMBER

FIGURE 30–17. Negative and positive camber. *(Courtesy of Dana Corporation)*

The furniture caster is an example of negative caster. The bicycle is an example of positive caster. Both examples tend to keep the rolling object going in a straight line. For example, a person can take his/her hands off of the handle bars and still go in a straight line. The *lead* is defined as the distance at ground level between the two center lines. ***Figure 30–16*** shows the lead.

Caster is designed into the front-end suspension of a car to do several things. Caster:

a. aids in the directional stability of the car by making the front wheels maintain a straight-ahead position.
b. aids in returning the front wheels to a straight-ahead position when coming out of a turn.
c. offsets the effect of road crown or curvature of the road.

Too much caster causes hard steering, excessive road shock, and wheel shimmy. Too little caster causes wander, weave, and instability at high speeds. Unequal caster causes pulling to the side of least caster.

While positive caster does aid in directional stability, it also increases steering effort by the driver. This can be compensated for by using power steering. Cars with manual steering usually require a caster setting of near zero or even a negative angle. Negative caster settings are required on some newer cars.

Camber

Camber is defined as the inward and outward tilt of the front wheels at the top. ***Figure 30–17*** shows an example of both negative and positive camber. Camber is measured as an angle in degrees from the center line of the wheel to a true vertical line. The purpose of checking camber is to make sure the tire is vertical to the road. This will make the tire tread uniform on both sides of the tire. This results in equal distribution of load and wear over the whole tire tread.

When the camber setting is correct:

a. there will be maximum amount of tire thread in contact with the road surface.
b. the road contact area of the tire will be directly under the point of load.
c. the result will be easier steering because the vehicle weight is placed upon the inner wheel bearing and spindle.

When the camber is incorrect:

a. there will be wear on the ball joint and wheel bearings.
b. the steering will pull to one side.
c. there will be excessive tire wear. This is shown in ***Figure 30–18***. Too much negative camber will cause wear on the inside of the tire. Too much positive camber will cause wear on the outside of the tire.

Wear on tires with incorrect camber is further explained in ***Figure 30–19***. Referring to this figure, notice the rolling radii at different parts of the wheel. At each point, the tire is rolling at different diameters. This causes the wheel to act as a cone. The cone has several rolling diameters and tends to want to roll in a circle. But since it is forced to roll in a straight line, the outer or smaller diameter tries to roll faster. This results in the outer parts of the tread being ground off by slipping and scuffing.

Toe

Toe is defined as the difference in the distance between the front and back of the front wheels. ***Figure 30–20*** shows toe. When dimension "B" is smaller than "A", it is called a toe-in condition. When dimension "B" is greater than "A", it is called a toe-out condition. Toe is measured in inches or parts of an inch.

When a vehicle is moving forward, certain forces are developed. Braking and the rolling resistance of the tires force the front wheel outward in front. Vehicles are generally set with just a small amount of toe-in to help overcome these forces. Once in motion, clearances in the steering linkage allow the front of the tires to swing out. At this point, there should be zero amount of toe-in. In front wheel drive vehicles, the tires may be purposely toed out to allow for other forces. Front wheel drive vehicles tend to return the wheels to their proper straight-ahead position. Incorrect toe adjustment will cause the tires to wear excessively and will cause harder steering.

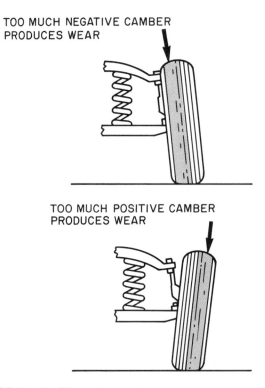

FIGURE 30–18. When the camber is incorrect, wear will increase on the sides of the tire. *(Courtesy of Dana Corporation)*

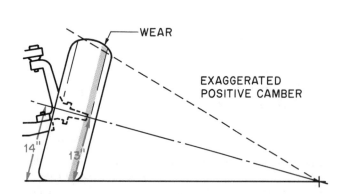

FIGURE 30–19. When camber is excessive, wear occurs because the tire is rolling at different radii. *(Courtesy of Hunter Engineering Company)*

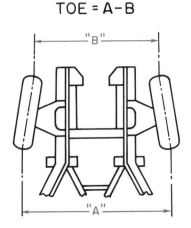

FIGURE 30–20. Toe is defined as the difference in distance between the front and back of the front tires. *(Courtesy of Dana Corporation)*

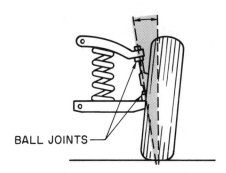

FIGURE 30–21. Steering axis inclination is the angle between the true vertical and the center line of the ball joints. *(Courtesy of Dana Corporation)*

Steering Axis Inclination

Steering axis inclination is defined as the inward tilt of the spindle support arm ball joints at the top. Steering axis inclination angle is the distance between the ball joint center line and true vertical. This angle is not adjustable. It is shown in **Figure 30–21**. The purpose of having a steering axis inclination angle is to:

a. reduce the need for excessive camber.
b. provide a pivot point about which the wheel will turn, producing easy steering.
c. aid steering stability.
d. lessen tire wear.
e. provide directional stability.
f. distribute the weight of the vehicle more nearly under the road contact area of the tire.

Included Angle

Certain manufacturers use the term *included angle* to illustrate information about steering axis inclination. The included angle is defined as the sum of the steering axis inclination angle and the camber. For example, **Figure 30–22** shows the included angle. Certain alignment charts specify the included angle instead of the steering axis inclination angle.

Scrub Radius

Scrub radius is the distance between the center line of the ball joints and the center line of the tire at the point where the tire contacts the road surface. The greater the scrub radius, the greater the effort required to steer, and the less the stability on the steering. During turning, when the ball joint center line is inside the tire contact point, the tire doesn't pivot where it touches the road. Instead, it has to move forward and backward to compensate as the driver turns the steering wheel. Steering effort is greatly increased because the tires scrub against the road during turns. **Figure 30–23** shows scrub radius. Note that both positive camber and steering axis inclination combine to reduce scrub radius to a minimum.

Turning Radius

Turning radius, also called "toe-out on turns," is defined as the amount one front wheel turns more sharply than the other. It is measured in degrees. The major purpose for having the correct turning radius is to make the front wheels pivot around a common center. This is shown in **Figure 30–24**. As the car turns around a corner, the outside wheel

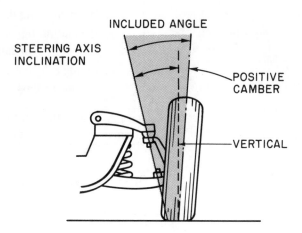

FIGURE 30–22. The included angle is defined as the camber plus the steering axis inclination angle. *(Courtesy of Hunter Engineering Company)*

turns a radius of 18 degrees. The inside wheel turns a radius of 20 degrees. If the turning radius is incorrect, the front wheels will scrub against the road surface on turns.

Turning radius is usually not adjustable. On certain vehicles, however, it can be checked. Turning radius is checked after all other alignment checks have been made. If turning radius is out of specifications, this usually indicates that some part of the steering linkage is bent or alignment is incorrect.

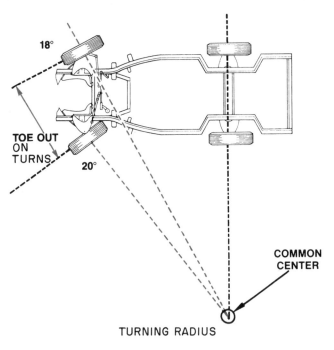

FIGURE 30-24. Turning radius is defined as the amount one wheel turns more sharply than the other. *(Courtesy of Dana Corporation)*

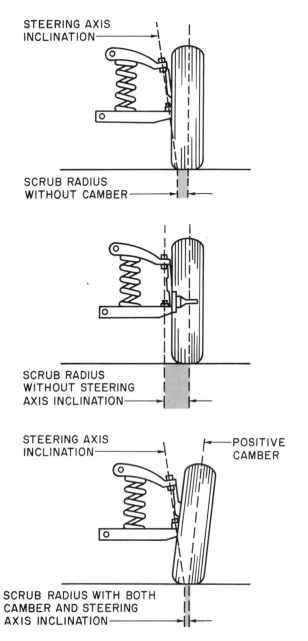

FIGURE 30-23. Scrub radius is the distance at the road surface between the center line of the tire and the center line of the ball joints. *(Courtesy of Dana Corporation)*

CAR CLINIC

PROBLEM: WARPED FRONT DISC BRAKE ROTORS

What is the most common reason the rotors on front disc brakes warps? A customer has noticed that there is a pulsing as the brakes are applied.

SOLUTION:

Several factors can cause the disc on front disc brakes to warp. These include:

1. Overtightening the front wheel lugs.
2. Having the rotors machined too far so that the heat of friction cannot be removed.
3. Hitting cold water in the road immediately after the brakes have been used excessively.

In most cases, the front rotor will have to be machined to make the rotor an even thickness.

30.3 POWER STEERING

Purpose of Power Steering

With manual steering, the driver is creating the forces needed to turn the steering gear. The only advantage that can be produced is by changing the steering ratio. Power steering is used on many vehicles today to make steering easier for the driver. This is especially true with heavier vehicles. The power steering system is designed to reduce the effort needed to turn the steering wheel. It reduces driver fatigue and increases safety during driving. Power steering systems are used both on the pitman arm steering gear system and the rack and pinion steering system.

Major Parts on Power Steering Systems

Although there are many designs of power steering systems, there are two major parts in all power steering system designs. These include a hydraulic pump and the steering unit. These two units are connected by high-pressure hoses. See *Figure 30–25*. The hydraulic pump is used to produce fluid pressure. The pump is driven by a belt running from the crankshaft. It supplies the hydraulic pressure needed to operate the steering gear.

The power steering unit is an integral part of either the steering gear or the rack and pinion arrangement. The hydraulic pressure is used to assist the motion of the steering gear or the rack and pinion gear. *Figure 30–26* shows a typical rack and pinion steering gear arrangement.

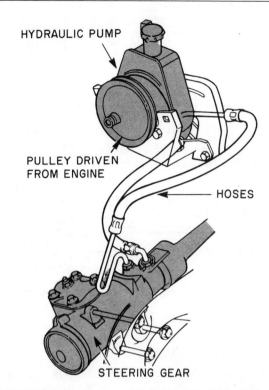

FIGURE 30–25. The power steering system uses a hydraulic pump and the steering unit to provide added pressure to turn the wheels. *(Courtesy of Chevrolet Division, General Motors Corporation)*

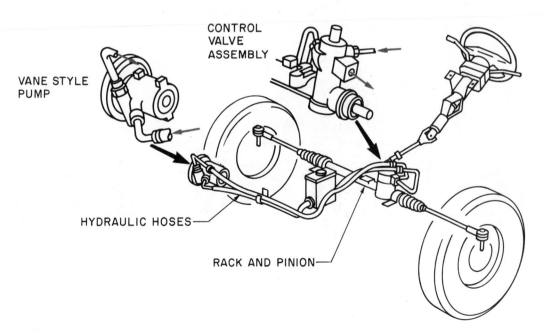

FIGURE 30–26. A power steering rack and pinion system. *(Courtesy of Volkswagen of America, Inc.)*

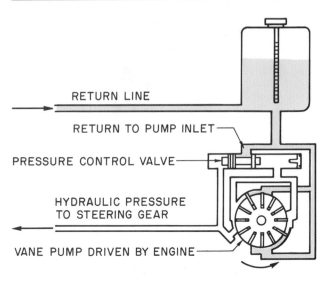

FIGURE 30-27. The parts of a power steering pump.

vent foreign matter from entering the system. A pressure relief valve is used to control excess pressure when the pump is increased in speed. *Figure 30-27* shows a power steering pump.

The pump can be of several designs. Three types of pumps are commonly used: the vane type, slipper type, and roller type. These are shown in *Figure 30-28*. All three types work on the same principle. The center of the pump turns within an *eccentric* area. A suction is produced on one side of the pump housing. A pressure is produced on the other side of the housing.

Integral Power Steering Gear

Referring to *Figure 30-29*, the sector shaft is turned by the piston and ball nut assembly. Normally, as the worm gear is turned by the steering wheel, the oil pressure is sent to the unit from the power steering pump. Oil is sent to both sides of the piston. This keeps the piston in a stable position. When the car is moving in a straight line, the pressures are equal on both sides of the piston. When the steering wheel is turned, higher oil pressure is directed to one side or the other to assist movement of the piston and ball nut assembly. Assisting the movement of this assembly makes it easier for the driver to turn the steering wheel.

Control Valves

Power steering control valves are built directly into the power steering gear assembly. The purpose of the control valve is to direct the oil pressure to one side or the other on the piston and ball nut assembly. When the steering wheel is turned, the control valve is positioned in such a way as to direct oil to the correct location. Two types of valves are commonly used: the rotary spool and sliding spool valves.

Power Steering Pump

All power steering pumps are called constant displacement or positive displacement pumps. They deliver different pressures, depending upon the type and make of the vehicle. They use special power steering fluid that is recommended by the manufacturer. Automatic transmission fluid should not be used in power steering systems except in small quantities and then only to bring the fluid level up to the fill mark. If more transmission fluid is used in an emergency situation, the system should be drained, flushed, and refilled with power steering fluid as soon as possible.

The fluid is stored in a reservoir that is attached to the pump. There is usually a filter in the reservoir to pre-

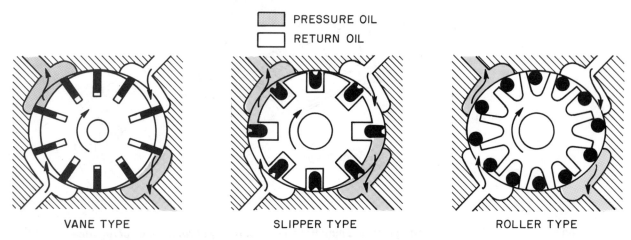

FIGURE 30-28. There are three types of power steering pumps. These include the vane type, slipper type, and roller type.

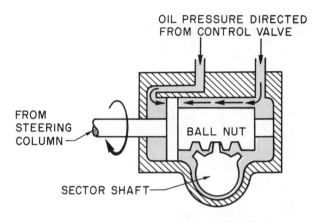

FROM STEERING COLUMN

OIL PRESSURE DIRECTED FROM CONTROL VALVE

BALL NUT

SECTOR SHAFT

FIGURE 30–29. The parts of an integral power steering gear are shown. Oil pressure pushes the ball nut to assist steering.

Sliding Spool Valve

The sliding spool valve is shown in **_Figure 30–30_**. As the steering wheel and the worm gear are turned, the sliding spool valve is moved slightly by linkage attached to the worm gear shaft. As this movement occurs, the internal spool opens a set of ports to allow high-pressure fluid to enter the correct side of the piston and ball nut assembly. Oil flows through internal passageways to get to the piston

assembly. When the steering wheel is turned the other way, the oil is sent to the other side of the piston.

Rotary Spool Valve

The rotary spool valve is also used on many vehicles to control the direction of oil through the steering gear. The rotary spool valve is shown in **_Figure 30–31_**. When the steering wheel is turned, a twisting effort is produced through a torsion bar to rotate the internal spool slightly. This is done on an internal spline. As the spool rotates slightly, a different set of ports is opened and closed to allow oil pressure to flow to the correct side of the piston assembly. If the steering wheel is turned in the opposite direction, the oil will flow to the opposite side of the piston assembly. **_Figure 30–32_** shows a cutaway view of a typical power steering gear assembly.

Rack and Pinion Power Steering

The rack and pinion power steering principles are much the same as the integral power steering principles. The major difference is that the pressure from the control valve operates the rack assembly. A power cylinder and piston assembly are placed on the rack. See **_Figure 30–33_**. Oil pressure from the control valve then pushes or assists the movement of the rack. The control valve is attached to and is operated from the pinion gear as in other power steering systems.

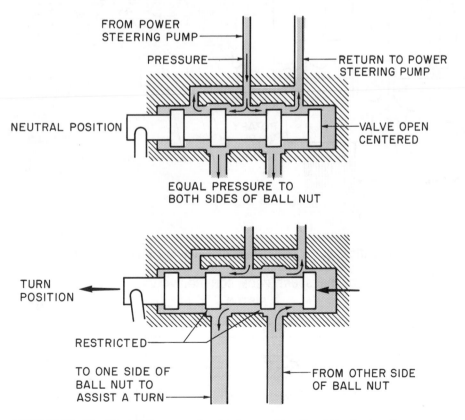

FROM POWER STEERING PUMP

PRESSURE

RETURN TO POWER STEERING PUMP

NEUTRAL POSITION

VALVE OPEN CENTERED

EQUAL PRESSURE TO BOTH SIDES OF BALL NUT

TURN POSITION

RESTRICTED

TO ONE SIDE OF BALL NUT TO ASSIST A TURN

FROM OTHER SIDE OF BALL NUT

FIGURE 30–30. The sliding spool valve is used to direct the hydraulic pressure to the correct side of the ball nut for a left or right turn.

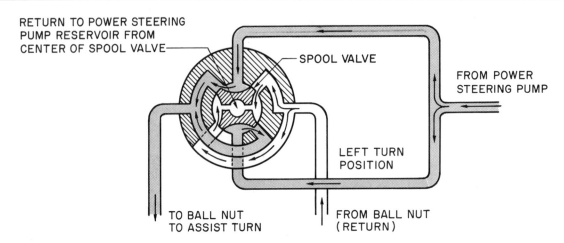

FIGURE 30-31. The rotary spool valve controls the direction of hydraulic pressure by twisting slightly to open or close different ports.

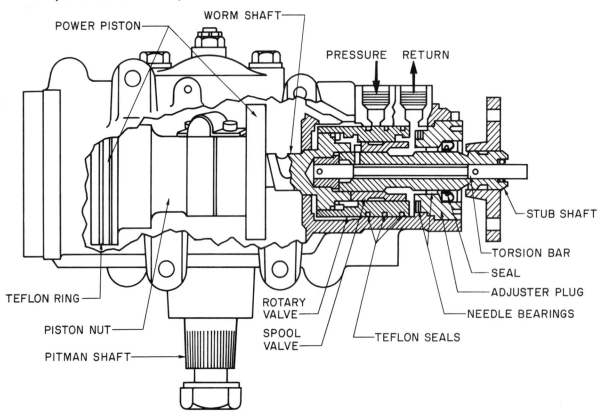

FIGURE 30-32. A complete power steering gear assembly.

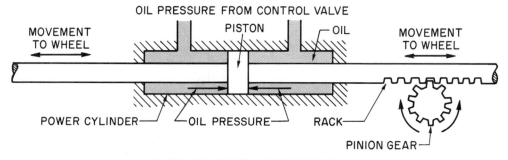

RACK AND PINION POWER STEERING

FIGURE 30-33. Power steering is accomplished on rack and pinion steering by using a piston and power cylinder attached to the rack.

AUTOMOTIVE ASSEMBLY LINE

The automotive assembly line is very complete and produces high-quality vehicles. Workers are positioned at various stations to complete the assembly of fenders and other external parts on the vehicle. (Courtesy of Pontiac Motor Division, General Motors Corporation)

DIAGNOSIS and SERVICE

1. Because there are so many types of steering columns, refer to the manufacturer's maintenance manuals for the correct disassembly and assembly procedures. The following service items can be done on *steering columns*:
   a. Use a wheel puller to remove the steering wheel to get at the horn and turn signal mechanisms.
   b. The universal joints can be replaced, but the procedure and parts replaced will depend upon the manufacturer.
   c. Electrical and mechanical problems can be serviced in the turn signal and flasher systems and the tilt mechanism.
2. *Figure 30–34* shows a list of problems and their possible causes that can be used to *diagnose the steering system*.
3. When working on the steering system, do the following preliminary checks.
   a. Check all tires for proper inflation pressures and approximately the same tread wear.
   b. Check front wheel bearings for proper adjustment. Correct if necessary.
   c. Check for loose ball joints, tie rod ends, and control arms.
   d. Check for runout of wheels and tires.

   e. Consider excess loads.
   f. Consider the condition and type of equipment being used to check alignment.
4. All wheel alignment angles are interrelated. The adjustment order should be caster, camber, and toe. *Figure 30–35* shows an example of the alignment specifications.
5. The alignment angles can be adjusted by several methods, depending upon the car. *Figure 30–36* shows that *shims*, eccentric bolts, cams, and *slotted frame* adjustments can be used to change the alignment.
6. When replacing a tie rod end, the retaining nut and cotter pin must be removed. Before removal, measure from the adjusting sleeve to the tie rod end center point. This allows a preliminary toe setting by placing the new tie rod end in approximately the same position. Refer to *Figure 30–37*.
7. *Figure 30–38* shows a diagnosis chart for alignment problems.
8. When removing rubber boots and retainers from rack and pinion steering systems, be careful not to puncture or damage the rubber boot.
9. Check rack and pinion steering systems for leaks at the end seals, rack seals, and pinion shaft seals. Replace where necessary by following the manufacturer's recommended procedure.
10. The list shown in *Figure 30–39* shows suggested causes for typical rack and pinion steering system problems.
11. To *remove a tie rod end*:
    a. Loosen the tie rod adjusting sleeve clamp nuts.
    b. Remove the tie rod end nut and cotter pin.
    c. Use a pickle fork to remove the tie rod from the steering knuckle. A pickle fork is shown in *Figure 30–40*.
12. Common power steering problems include:
    a. Steering column U-joint binding that needs lubrication.

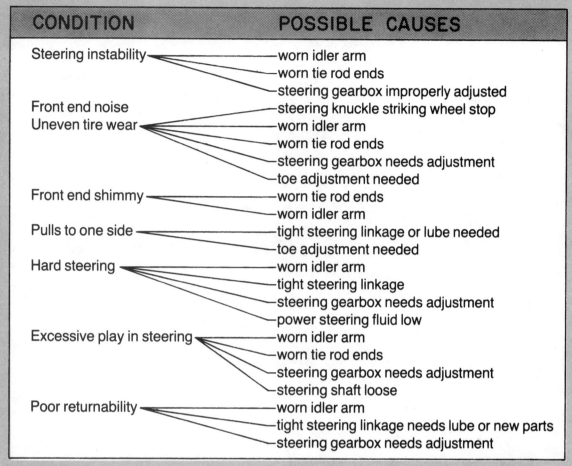

CONDITION	POSSIBLE CAUSES
Steering instability	worn idler arm worn tie rod ends steering gearbox improperly adjusted
Front end noise	steering knuckle striking wheel stop
Uneven tire wear	worn idler arm worn tie rod ends steering gearbox needs adjustment toe adjustment needed
Front end shimmy	worn tie rod ends worn idler arm
Pulls to one side	tight steering linkage or lube needed toe adjustment needed
Hard steering	worn idler arm tight steering linkage steering gearbox needs adjustment power steering fluid low
Excessive play in steering	worn idler arm worn tie rod ends steering gearbox needs adjustment steering shaft loose
Poor returnability	worn idler arm tight steering linkage needs lube or new parts steering gearbox needs adjustment

FIGURE 30–34. Each condition has several causes that need to be checked when troubleshooting the steering system. *(Courtesy of Dana Corporation)*

Wheel Alignment Specifications

Year	Model	Caster Angle, Degrees		Camber Angle, Degrees				Toe-In Inch	Toe Out on Turns, Deg.	
		Limits	Desired	Limits		Desired			Outer Wheel	Inner Wheel
				Left	Right	Left	Right			
1981	Century, Regal①	+2.5 to +3.5	+3	0 to +1	0 to +1	+.5	+.5	1/16 to 3/16	—	—
	Riviera	+2 to +3	+2.5	−.5 to +.5	−.5 to +.5	Zero	Zero	②	—	—
	Others	+2.5 to +3.5	+3	+.3 to +1.3	+.3 to +1.3	+.8	+.8	1/16 to 3/16	—	—
1982	Regal	+2.5 to +3.5	+3	0 to +1	0 to +1	+.5	+.5	1/16 to 3/16	—	—
	Riviera	+2 to +3	+2.5	−.5 to +.5	−.5 to +.5	Zero	Zero	②	—	—
	Others	+2.5 to +3.5	+3	+.3 to +1.3	+.3 to +1.3	+.8	+.8	1/16 to 3/16	—	—
1983–84	Regal	+2.5 to +3.5	+3	0 to +1	0 to +1	+.5	+.5	1/16 to 3/16	—	—
	Riviera	+2 to +3	+2.5	−.5 to +.5	−.5 to +.5	Zero	Zero	②	—	—
	Others	+2.5 to +3.5	+3	+.3 to +1.3	+.3 to +1.3	+.8	+.8	1/16 to 3/16	—	—
1985	Regal	+2 to +4	+3	−.3 to +1.3	−.3 to +1.3	+1/2	+1/2	1/16 to 1/4	—	—
	Riviera	+1.5 to +3.5	+2.5	−.8 to +.8	−.8 to +.8	0	0	−1/8 to +1/8	—	—
	Others	+2 to +4	+3	0 to +1.6	0 to +1.6	+.8	+.8	1/16 to 1/4	—	—
1986	Regal	+2 to +4	+3	−.3 to +1.3	−.3 to +1.3	+1/2	+1/2	1/16 to 1/4	—	—
	Estate Wagon	+2 to +4	+3	0 to +1.6	0 to +1.6	+.8	+.8	1/16 to 1/4	—	—

①—Power steering.   ②—1/16″ toe-in to 1/16″ toe-out.

FIGURE 30–35. This chart gives an example of different alignment specifications. *(Courtesy of Motor Publications, Auto Repair Manual, 1981–1987)*

b. Loose power steering belt which causes momentary steering difficulty. When it is loose, the belt will squeal when the engine is accelerated. This shows that the belt is slipping on the pulleys.

c. Loose mounting bolts in the steering gear which cause abnormal steering wheel kickback and poor control.

d. Leaky power steering lines. Check for loose fittings and damaged hoses.

e. A lack of hydraulic pressure which can be caused by internal leaks past piston rings, valve body, worn seals, or misaligned housing bore.

13. When power steering systems are serviced, there is a possibility of air getting into the hydraulic system.

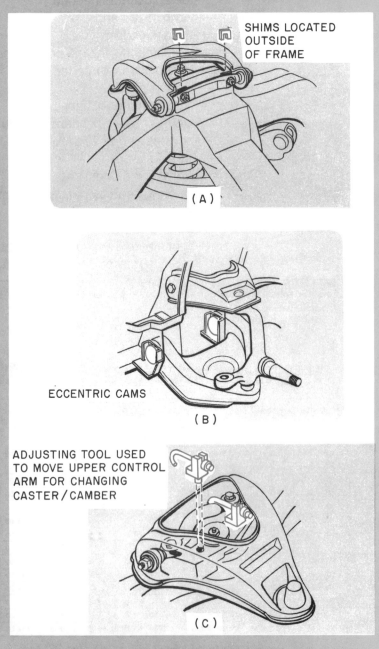

FIGURE 30-36. Shims, eccentric bolts, cams, and slotted frame bolts are used to adjust the steering angles. Other methods can also be used. *(Courtesy of Hunter Engineering Company)*

If this is the case, the power steering will make a buzzing sound. To bleed the system of air, start the engine and turn the steering wheel throughout its range. Keep checking the level of power steering fluid as needed. Turn the steering wheel throughout its range again, and add fluid. Continue this procedure until all of the air has been removed from the hydraulic system.

14. The steering gear can be checked and adjusted for correct clearance or lash. This adjustment is the clearance between the sector shaft and the ball nut. Refer to the correct maintenance manual for the procedure. *Figure 30–41* shows the position of the lash adjuster located on top of the steering gear. The general procedure for adjusting the clearance is as follows:

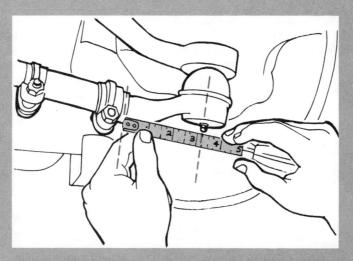

FIGURE 30–37. When removing a tie rod, check the distance from the adjusting sleeve to the ball nut grease zerk. This allows a preliminary toe setting when put back in the same position. *(Courtesy of Dana Corporation)*

•DIAGNOSIS CHART

SYMPTOM	PROBABLE CAUSE	SYMPTOM	PROBABLE CAUSE
• Excessive tire wear on outside shoulder.	• Excessive positive camber.	• Vehicle vibrates.	• Defective tires. One or more of all 4 tires out-of-round. One or more of all 4 tires out-of-balance. Drive shaft bent. Drive shaft sprayed with undercoating.
• Excessive tire wear on inside shoulder.	• Excessive negative camber.	• Car tends to wander either to the right or left.	• Improper toe setting. Looseness in steering system or ball-joints. Uneven caster. Tire pull.
• Excessive tire wear on both shoulders.	• Rounding curves at high speeds. Under-inflated tires.	• Vehicle swerves or pulls to side when applying brakes.	• Uneven caster. Brakes need adjustment. Out-of-round brake drum. Defective brakes. Under-inflated tire.
• Saw-tooth tire wear.	• Too much toe-in or toe-out.		
• One tire wears more than the other.	• Improper camber. Defective brakes. Defective shock absorber.	• Car tends to pull either to the right or left when taking hands off steering wheel.	• Improper camber. Unequal caster. Tires worn unevenly. Tire pressure unequal.
• Tire treads cupped or dished.	• Out-of-round tires. Out-of-balance condition. Defective shock absorber.	• Car is hard to steer.	• Tires under-inflated. Power steering defective. Too much positive caster. Steering system too tight or binding.
• Front wheels shimmy.	• Defective idler arm bushing. Out-of-round tires. Out-of-balance condition. Excessive positive caster. Uneven caster.	• Steering has excessive play or looseness.	• Loose wheel bearings. Loose ball-joints or kingpins. Loose bushings. Loose idler-arm. Loose steering gear assembly. Worn steering gear or steering gear bearings.

FIGURE 30–38. Diagnosis charts can help the service technician troubleshoot the steering system. *(Courtesy of Hunter Engineering Company)*

CONDITION	POSSIBLE CAUSE	CORRECTION

Hard steering — excessive effort required at steering wheel.	1. Low or uneven tire pressure.
	2. Tight outer tie rod end or ball joints.
	3. Incorrect front wheel alignment.
	4. Bind or catch in gear.
		1. Inflate to specified pressures.
		2. Lube or replace as required.
		3. Align to specifications.
		4. Remove gear, disassemble, and inspect.
		   a) Replace damaged or badly worn components (OPH)*.
		   b) If housing and tube assembly, rack, or pinion are damaged, replace with service assembly (TPH)**.

Poor returnability.	1. Tight ball joints or end housing pivots.
	2. Bent tie rod(s).
	3. Incorrect front wheel alignment.
	4. Bind or catch in gear.
		1. Lube or replace as required.
		2. Replace bent tie rod(s). Align front end.
		3. Align front end.
		4. Remove gear, disassemble, and inspect.
		   a) Replace damaged or badly worn components (OPH)*.
		   b) If housing and tube assembly, rack, or pinion are damaged, replace with service assembly (TPH)**.

Excessive play or looseness in steering system.	1. Front wheel bearings loosely adjusted. 2. Worn couplings or steering shaft U-joints. 3. Worn upper ball joints. 4. Loose steering wheel on shaft, tie rods, steering arms, or steering linkage ball studs. 5. Worn outer tie rod ends. 6. Loose frame to gear mounting bolts. 7. Deteriorated mounting grommets. 8. Excessive internal looseness in gear. 9. Worn rack bushing(s).	1. Adjust or replace as required. 2. Replace worn part(s). 3. Replace. 4. Tighten to specified torque. 5. Replace outer tie rod ends. 6. Tighten to specified torque. 7. Replace mounting grommets. 8. Readjust gear. If still loose, disassemble and inspect. 9. Replace rack bushing(s).

  *(OPH) One-piece housing only
** (TPH) Two-piece housing only

FIGURE 30–39. Use this diagnosis chart to help troubleshoot the power steering rack and pinion system.

a. If power steering is used, rotate the wormshaft through the complete range of travel. This is done to bleed air from the system. Then refill the reservoir to the top.

b. Place the steering gear in the center of its movement.

c. Loosen the lock nut on the adjusting screw.

d. Tighten the adjusting screw until all backlash is removed. Then tighten the lock nut.

e. Operate the unit through its range of operation.

f. Loosen the lock nut and the adjusting screw again.

g. Tighten the adjusting screw again until all backlash is again removed.

h. Now tighten the adjusting screw an additional 3/8 of a turn. Now tighten the lock nut.

15. Alignment is somewhat different on the MacPherson strut and other front wheel drive suspension systems. Depending upon the manufacturer, certain adjustments cannot be made on the front suspension. Camber, for example, is sometimes built into the suspension and cannot be changed. On

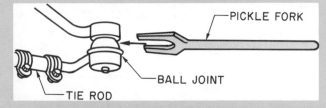

FIGURE 30–40. A pickle fork can be used to separate the tie rod ball joints correctly.

other vehicles, there is no caster adjustment. Refer to the vehicle manufacturer to determine exactly what adjustments can be done. *Figure 30–42* shows one method used to adjust camber on a knuckle-strut assembly. This method has an elongated bolt hole. A cam washer is placed on the end of the bolt to adjust for camber.

16. Tie rod end wear can be found by grasping the tie rod end firmly and forcing up and down or sideways to check for any lost motion. See *Figure 30–43*.

17. Specialized alignment machines and instruments should always be used to check and adjust for correct alignment. Computers are being used to help in measuring and recording steering angles and alignment. *Figure 30–44* shows an alignment machine.

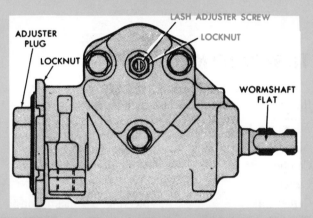

FIGURE 30–41. The lash adjuster screw is located on top of the steering gear assembly. Its purpose is to adjust the clearance between the sector shaft and the ball nut. *(Courtesy of Motor Publications, Auto Repair Manual, 1981–1987)*

FIGURE 30–43. The tie rod ends can be checked for wear by grasping the rod firmly and forcing it up and down. There should be no lost motion.

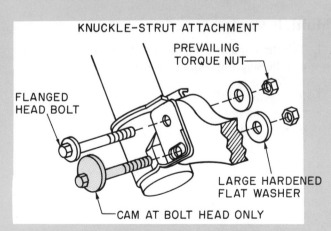

FIGURE 30–42. The cam washer is used to adjust the camber on the MacPherson strut suspension.

FIGURE 30–44. An alignment machine should always be used to check for correct alignment angles on the steering system. *(Courtesy of Hunter Engineering Company)*

SUMMARY

The steering system is made of three major components. These include the steering column, the steering gear, and the steering linkage. The steering column and wheel are used to produce the necessary force to turn the front wheel. The steering gear is used to change the motion of the steering wheel to reciprocating motion on the steering linkage. The steering linkage is used to connect the steering gear to the front wheels.

There are two types of steering gears currently in use. These include the pitman arm style and the rack and pinion style. The pitman arm style uses a wormshaft with a spiral groove in it. A ball nut is placed over the wormshaft. The ball nut also has an internal spiral groove. Small steel balls are placed inside the groove. This assembly provides a low-friction drive between the wormshaft and the ball nut.

As the ball nut is moved, it meshes with the sector shaft. The sector shaft causes the pitman arm to move. Thus as the wormshaft is turned by the driver, the end result is the moving of the pitman arm.

The rack and pinion steering system uses a flat gear called a rack. It is meshed with a mating gear called a pinion. The pinion is attached to the steering column. As the steering column is turned, the rack moves back and forth to force the wheel to turn.

Steering ratio is defined as the ratio between the degrees turned on the steering wheel compared to the degrees turned on the front wheels. The ratio is for exactly one degree of motion on the front wheels. The steering ratio can be changed by changing the gear ratio between the wormshaft and the ball nut, or between the pinion and the rack.

The steering linkage is used to connect the pitman arm to the front wheels. The components include the idler arm, center link, tie rods, and pitman arm. There are many styles and differences between each part. The style depends upon the manufacturer and the year of the vehicle.

Wheel alignment is defined as the balancing of all forces created by friction, gravity, centrifugal force, and momentum while the wheel is in motion. There are five angles that are related to wheel alignment. These include caster, camber, toe, steering axis inclination, and turning radius. Caster is the forward or backward tilt of the spindle support. Camber is the inward or outward tilt of the wheel. Toe is the inward or outward pointing of the front wheels. Steering axis inclination is the inward tilt of the spindle support arm ball joints at the top. Turning radius is the amount in degrees that one front wheel turns more sharply than the other front wheel while the vehicle is turning.

Power steering is used to assist the driver when turning the vehicle. Power steering is accomplished by using hydraulic pressure produced by a positive displacement pump. The pressure produced is used to help turn the steering gear or push the rack on a rack and pinion system. The power steering pump is run by a belt from the crankshaft.

Power steering fluid is used as the hydraulic fluid. Automatic transmission fluid should only be used in emergency situations. There are several designs of pumps including the vane, slipper, and roller types.

The hydraulic pressure in a power steering system is sent to the integral gear. Here the hydraulic pressure is used to help move the ball nut. A sliding spool or a rotary spool valve is used to direct the hydraulic pressure to the correct area for both a left and right turn. Rack and pinion power steering is accomplished by using hydraulic pressure as well. A piston is placed on the rack, inside a cylinder. Hydraulic pressure pushes the piston one way or the other to turn left or right.

TERMS TO KNOW

Can you explain each of the following terms? Review the chapter until you can use each term correctly.

Adjusting sleeve	Rack and pinion steering
Alignment	Scrub radius
Camber	Shim
Caster	Slotted frame
Center link	Steering axis inclination
Eccentric	Zerk
Eccentrics	Turning radius
Idler arm	Toe (in, out)
Pitman arm	

REVIEW QUESTIONS

Multiple Choice

1. The three major components of the steering system are the steering column, steering gear, and
   a. Pitman arm
   b. Steering toe
   c. Steering linkage
   d. Rack and pinion
   e. Steering wheel

2. Universal joints are used in the _____ on a typical steering system.
   a. Steering wheel
   b. Steering column
   c. Pitman arm
   d. Tie rods
   e. Steering linkage

3. The pitman arm steering gear uses:
   a. Recirculating balls
   b. A sector shaft
   c. A ball nut
   d. All of the above
   e. None of the above

4. Which of the following is/are *not* adjustable on a steering system?
   a. Tie rods
   b. Camber
   c. Steering gear
   d. Toe in
   e. Pitman arm

5. The purpose of the steering gear is to:
   a. Change rotary motion to rotary motion
   b. Change reciprocating motion to rotary motion
   c. Change rotary motion to reciprocating motion
   d. Adjust camber
   e. Adjust caster

6. Which type of steering system uses a flat gear?
   a. Tie rod system
   b. Rack and pinion system
   c. Integral gear system
   d. Universal system
   e. Scrub system

7. When the steering wheel turns 60 degrees and the front wheel turns 3 degrees, the steering ratio is:
   a. 15 to 1
   b. 20 to 1
   c. 25 to 1
   d. 30 to 1
   e. 60 to 1

8. Which of the following is/are part of the steering linkage?
   a. Idler arm
   b. Center link
   c. Tie rod
   d. All of the above
   e. None of the above

9. The adjusting sleeve is attached to the:
   a. Pitman arm
   b. Tie rods
   c. Center link
   d. Idler arm
   e. U-joint

10. Which of the following is defined as the backward and forward tilt of the spindle support arm?
    a. Caster
    b. Camber
    c. Toe
    d. Steering axis inclination
    e. Turning radius

11. Which of the following is defined as the inward and outward tilt of the front wheels at the top?
    a. Caster
    b. Camber
    c. Toe
    d. Steering axis inclination
    e. Turning radius

12. Which of the following is defined as the distance between the front and back of the front wheels?
    a. Caster
    b. Camber
    c. Toe
    d. Steering axis inclination
    e. Turning radius

13. Which of the following is defined as the toe-out on turns or the amount one wheel turns more sharply on turns?
    a. Caster
    b. Camber
    c. Toe
    d. Steering axis inclination
    e. Turning radius

14. The included angle is a combination of the steering axis inclination and the _____.
    a. Toe in
    b. Caster
    c. Camber
    d. Turning radius
    e. Toe out

15. Power steering can be used:
    a. Only on pitman arm steering systems
    b. Only on rack and pinion steering systems
    c. Both a and b
    d. Neither a nor b
    e. Only when the vehicle is extremely light

16. The power steering pump is driven from the:
    a. Steering linkage
    b. Crankshaft by a belt
    c. Differential
    d. Drive shaft
    e. Steering gear

17. The control valves used in power steering are placed:
    a. On the steering linkage
    b. In the integral gear and pitman arm assembly
    c. In the steering wheel
    d. On the rack and pinion
    e. All of the above

18. Which of the following spool valves change passageways by rotating?
   a. The sliding spool valve
   b. The rotary spool valve
   c. The positive displacement spool valve
   d. The linkage spool valve
   e. The vane spool valve

19. Power steering on a rack and pinion steering system pushes which component?
   a. The pitman arm
   b. The rack
   c. The pinion
   d. The U-joint
   e. The spool valve

20. Which of the following are methods used to adjust the alignment on a car?
   a. Shims
   b. Eccentric bolts
   c. Cams
   d. All of the above
   e. None of the above

21. Which of the following cannot be adjusted?
   a. Camber
   b. Steering axis inclination
   c. Steering gear lash
   d. Caster
   e. Toe

The following questions are similar in format to ASE (Automotive Service Excellence) test questions.

22. A buzzing sound is heard from the power steering unit. Technician A says that the fluid is low. Technician B says that there may be air in the hydraulic system. Who is right?
   a. A only
   b. B only
   c. Both A and B
   d. Neither A nor B

23. Technician A says that one alignment adjustment will not affect the other alignment adjustments. Technician B says that one alignment adjustment is interrelated with all of the other alignment adjustments and will affect the others. Who is right?
   a. A only
   b. B only
   c. Both A and B
   d. Neither A nor B

24. When working on the steering system, several preliminary checks should be made. Technician A says that the tires should be properly inflated as they may affect the steering system. Technician B says that loose ball joints may affect the steering system. Who is right?
   a. A only
   b. B only
   c. Both A and B
   d. Neither A nor B

25. A squealing sound is heard when the engine is accelerated. Technician A says the problem is a loose power steering belt. Technician B says the problem is loose tie rods. Who is right?
   a. A only
   b. B only
   c. Both A and B
   d. Neither A nor B

Essay

26. Describe how a rack and pinion steering system operates.

27. What is a pitman arm?

28. Define the term steering ratio.

29. List the parts and operation of the standard steering linkage.

30. What is the purpose of the idler arm?

31. What is the purpose of the adjusting sleeve on the tie rods?

32. Define the term caster.

33. Define the term camber.

34. How does a sliding spool valve operate in a power steering system?

CHAPTER 31

Tires and Wheels

INTRODUCTION

The tires on cars today serve several important purposes. They are designed to carry the weight of the vehicle sufficiently, transfer braking and driving torque to the road, and withstand side thrust over varying speeds and conditions. This chapter explains how tires are designed, constructed, sized, and serviced.

OBJECTIVES

After reading this chapter, you will be able to:
- Use tire terminology to define how tires are constructed.
- Identify different characteristics of tires.
- Compare the differences between different types of tires, including ply, radial, and spare tires.
- Identify how tires are sized.
- Analyze the purpose and operation of wheels and rims.
- Analyze several diagnosis and service procedures.

CHAPTER HIGHLIGHTS

31.1 TIRE CONSTRUCTION AND
CHARACTERISTICS
    A. Differences in Tires
    B. Parts of a Tire
    C. Tube and Tubeless Tires
    D. Tire Characteristics
    E. Cords
    F. Radial and Bias Tires
    G. Tread Design
    H. Tire Valves
    I. Puncture Sealing Tires
    J. Compact Spare Tire

31.2 IDENTIFYING TIRES
    A. Metric Tire Sizes
    B. Other Tire Sizing

    C. Tire Placard
    D. UTQG Designation
    E. TPC Specification Number
    F. Load Range

31.3 WHEELS AND RIMS
    A. Dropped Center Wheels
    B. Wheel Sizes
    C. Special Wheels

DIAGNOSIS AND SERVICE

SUMMARY

31.1 TIRE CONSTRUCTION AND CHARACTERISTICS

Differences in Tires

Different tire designs have been used on automobiles over the years to meet many demands. Originally, most vehicles used a tube-type tire. However, tube tires were eventually replaced by tubeless-type tires. From that point on, different tread designs, internal construction, and belts and ply designs have been used on tires.

Parts of a Tire

Although they seem simple, tires have several parts. *Figure 31–1* shows a cutaway view of a tubeless tire. The top or outside of the tire is called the *tread*. Its purpose is to produce the friction for braking and torque for driving. The outside sides of the tire are called the wall. On many tires, the wall is made of a white material and is called the white wall. This is for the external looks of the tire. The undertread is located directly below the tread. The layers of material, called *plies*, are formed over a spacing device and rubberized. The number of piles varies according to the use. For example, most automobile tires have two or four plies. Heavier vehicles such as vans and station wagons may use tires with up to eight plies for strength. The *carcass* is the strong, inner part of the tire that holds the air. The carcass is made of the layers or plies of fabric. It gives the tire its strength.

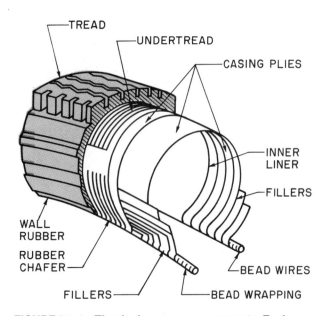

FIGURE 31–1. The tire has many components. Each component plays an important part in tire operation. (Courtesy of Chevrolet Division, General Motors Corporation)

The sidewall and tread material is *vulcanized* into place. Vulcanizing is defined as the process of heating rubber under pressure to mold the rubber into a desired shape. The *bead* wires, casing, and remaining parts are used for strength and durability. The tire is then attached to the wheel rim and an air valve is used to admit the necessary air for pressure.

Tube and Tubeless Tires

The tube-type tire is mounted on the wheel rim with a rubber inner tube placed inside the casing. The inner tube is inflated with air. This causes the tire casing to resist change in shape. The tubeless tire is mounted directly on the rim. The air is retained between the rim and tire casing when inflated.

A tube tire cannot be used without a tube. Some service technicians will, however, use a tube in a tubeless tire in emergency situations. Tubes are sometimes installed by the technician to eliminate the hard-to-find slow leak. Tubes are also used when imbedded dirt or rust prevents a tubeless tire from seating properly on the rim. However, using tubes for these reasons should be avoided if at all possible. Tubes are also useful with wire spoke wheels. The spokes tend to loosen in their sockets with long use, producing small leaks.

Tire Characteristics

Tires have several characteristics that are important in understanding their design. These include:

1. *Tire traction* — *Traction* is defined as a tire's ability to grip the road and move or stop the vehicle.
2. *Ride and handling* — Tires are measured by their ride and handling ability. This is an indication of the degree of comfort a tire delivers to the passenger. It is also a measure of the responsiveness provided to the driver's steering actions.
3. *Rolling resistance* — *Rolling resistance* is a term used to describe the pounds of force required to overcome the resistance of a tire to rotate. As rolling resistance of a tire decreases, fuel mileage typically increases.
4. *Noise* — All tires make a certain amount of noise. Tires can be made "quiet" by scrambling or changing the size, length, and shape of the tread elements. Scrambling prevents the sound frequency buildup that would develop if all of the patterns of the tread were spaced evenly.

Cords

Within the tire, there are layers of plies or belts. These layers have *cords* running through them for strength. *Figure 31–2* shows examples of several cords used inside the plies.

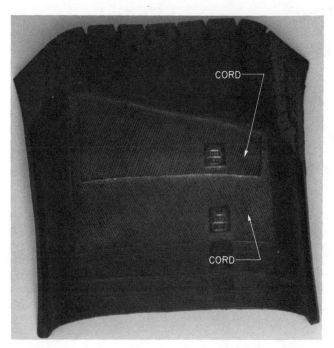

CORD

CORD

FIGURE 31–2. Cords are used inside the plies for strength.

Figure 31–3 illustrates the advantages of different types of cords. Because newer synthetic fibers are constantly being developed, more advantages and better driving characteristics will continually occur in tire design.

Radial and Bias Tires

There are typically three ways that the plies of the tire can be laid down. They can be positioned as a *bias*, belted bias, or belted *radial* tire. *Figure 31–4* shows the differences. The bias ply tire has layers of cord material running at an angle from bead to bead. Each cord used runs opposite to the cord below it. This tire provides strength. The plies tend, however, to work against each other during operation. This action produces heat. In addition, bias ply tires tend to produce a certain amount of rolling resistance. These characteristics increase the wear on the tread and shorten the tire life.

The belted bias tire uses additional belts wrapped around the circumference of the body of the plies. The actual cords within the belts are manufactured at an angle. The design and addition of these belts add strength and stiffness to the tread. This tire typically has less rolling resistance than the bias ply tire. This means that the tire is able to last longer in operation.

In the belted radial or radial ply tire, the cord material in the body runs from bead to bead. It is not at an angle as in the other tires. Additional belts are added for strength and durability. The belted radial tire allows the belts to hold more of the tread on the pavement during cornering and straight driving. In addition, the tire does not *squirm*. Squirm means that as the tire hits the road, it is moving or being pushed together. Less squirm results in better traction, less rolling resistance, less heat buildup, longer life, and better fuel mileage. In addition, in newer radials special materials are used in the cords to increase comfort and driveability.

Types of Tire Cords		
CORD TYPE	**TENSILE STRENGTH, PSI**	**ADVANTAGES**
Rayon	94,000	Soft ride, resilient, inexpensive
Polyester	104,000	Soft ride, more heat resistant, inexpensive
Nylon	122,000	High heat resistance, excellent impact resistance, minimum flex, won't absorb water.
Fiberglass	407,000	Greatest strength, soft ride

FIGURE 31–3. Different types of cords have different advantages. *(Courtesy of Chevrolet Division, General Motors Corporation)*

Bias Ply Tire

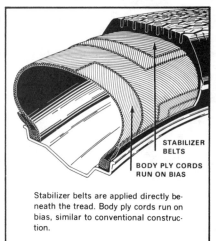

BODY PLY
CORDS RUN
ON BIAS

Body ply cords run on bias from bead to bead. Built with 2 to 4 plies . . . cord angle reversed on each ply. Tread is bonded directly to top ply.

Belted-Bias

STABILIZER
BELTS

BODY PLY CORDS
RUN ON BIAS

Stabilizer belts are applied directly beneath the tread. Body ply cords run on bias, similar to conventional construction.

Radial

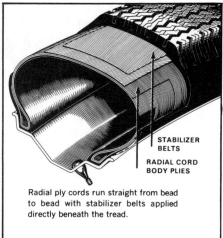

STABILIZER
BELTS

RADIAL CORD
BODY PLIES

Radial ply cords run straight from bead to bead with stabilizer belts applied directly beneath the tread.

FIGURE 31–4. Tires can be bias ply, belted bias, or radial design. The difference is in how the plies are laid down on the tire. *(Courtesy of Firestone Tire and Rubber Company)*

Tread Design

The tread of the tire must be made to work on all types of conditions. For example, cars operate on smooth pavement, gravel, wet pavement, and icy pavement. In addition, steering traction is different than rear wheel drive traction. Today, tire designs are a compromise between these conditions and cost considerations.

Tire treads have been developed to give better traction in wet conditions. The water must be squeezed from the contact area on the road surface. Treads have been developed to move the water away from the tread as efficiently as possible as the tire rolls. Of course, as the tire tread wears, the efficiency is reduced. *Figure 31–5* shows an example of different tread designs.

Technology is gradually changing snow tires as well.

Some manufacturers have developed *hydrophilic* tread compounds. Hydrophilic means attraction to water. A tread composed of part of this compound will literally stick to a wet or icy surface.

Tire Valves

The tire valve is used to admit and exhaust the air into and out of the tire. *Figure 31–6* show a typical tire valve. It has a central core that is spring loaded. This allows air to flow in only one direction, inward. When the small pin is depressed, air flows in the reverse or outward direction. When the valve core become defective, it can be unscrewed for removal and replaced. An airtight cap on the end of the valve produces an extra seal against valve leakage.

DIFFERENT TREAD TYPES

FIGURE 31–5. Tire treads vary according to the needs and conditions of the road.

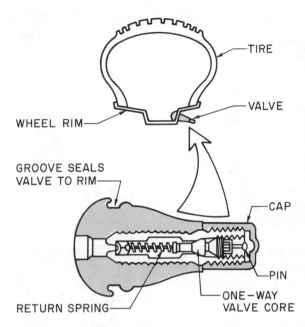

FIGURE 31-6. The tire valve is used to admit or exhaust the air from the tire. A small one-way valve is used to cause air to flow in only one direction.

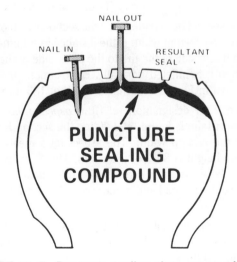

FIGURE 31-7. Puncture sealing tires are used to permanently seal tread punctures up to 3/16 of an inch. *(Courtesy of Oldsmobile Division, General Motors Corporation)*

Puncture Sealing Tires

Puncture sealing tires are made to permanently seal tread punctures up to 3/16 inch in diameter. This is shown in *Figure 31-7*. A resultant sealer is applied to the inside of the tire at the manufacturer. It is a special rubber compound. If a puncturing object penetrates the tire tread into the sealant layer, it picks up a coating of the sealant. When the puncturing object is removed or thrown from the tire by centrifugal force, the sealant adheres to it. The sealant is then pulled into the opening in the tread, causing it to seal.

Compact Spare Tire

Many vehicles today are built smaller than in the past. Smaller cars means that there is less trunk space. Many smaller cars use the compact spare tire to save space and weight. A compact spare tire is shown in *Figure 31-8*. It is called a "temporary use only" spare tire. This tire has a narrow four-inch wide rim. The wheel diameter is usually one inch larger than the road wheels. It should only be used when one of the road tires have failed. Inflation pressure is much higher on the compact spare tire, about 60 psi, and the top rated speed is reduced.

COMPACT SPARE

- Temporary Use Only

- Inflate to 60 PSI

FIGURE 31–8. The compact spare tire is used to save weight and space in smaller vehicles. (Courtesy of Oldsmobile Division, General Motors Corporation)

CAR CLINIC

PROBLEM: CAR PULLS TO LEFT

A vehicle has been pulling to the left recently when the brakes are being applied. The brake fluid seems to be down a little. The vehicle has drum brakes on both the front and rear of the car and has 56,000 miles on it. What could be the problem?

SOLUTION:

The most likely cause of this problem is the front wheel cylinder on the left side leaking. The brake fluid leaking out of the wheel cylinder is getting on the brake shoes and drum. The heat from the friction of the brakes causes the brake fluid to be very sticky. This causes the left front wheel to brake more than the right, causing the vehicle to pull to the left. Remove, service, and replace the front wheel cylinder with a new wheel cylinder kit to eliminate the problem.

31.2 IDENTIFYING TIRES

Metric Tire Sizes

Tires are sized according to the application in which they are to be used and their physical size. The size of the tire must be molded into the side of the tire. Most tires today are sized according to metric standards. In the past, however, there were other ways of identifying tire sizes. *Figure 31–9*

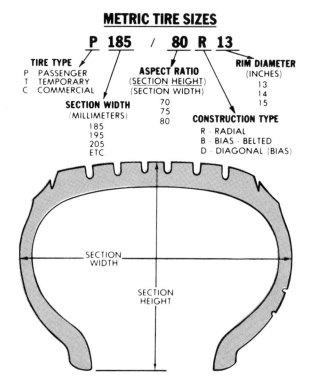

FIGURE 31–9. Tires are sized according to metric sizes. Each designation helps to identify the type and size of tire. (Courtesy of Oldsmobile Division, General Motors Corporation)

shows how metric sizing works. The tire has several designations.

1. The first letter of the size tells if the tire is used for passenger (P), temporary (T), or commercial (C) use.

2. The second designation tells the section width of the tire. The section width is the distance in millimeters from one side of the tire to the other side of the tire when it is inflated normally. For example, the tire shown in Figure 31–9 has a width of 185 mm.

3. The third designation tells the *aspect ratio*. The aspect ratio is found by dividing the section height by the section width. The aspect ratio means the tire's height is 80% of the width. This is called the profile of the tire. Lower profile tires make the car closer to the road and reduce wind drag underneath the car body.

4. The next designation tells the construction type. R means radial tire, B means belted, and D means diagonal (bias).

5. The last designation gives the wheel rim in inches. The most common car rim sizes are 13, 14, and 15 inches in diameter.

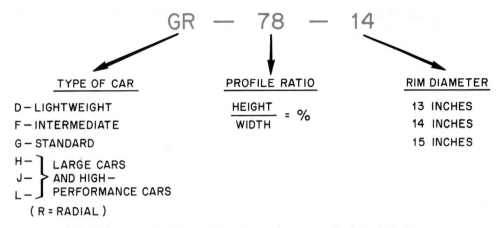

FIGURE 31–10. Several designations may be used to identify tire size.

Other Tire Sizing

Some manufacturers may also use other designations. One common type of designation is called the Numeric identification. For example, a tire sized as 7.75-14 means the approximate cross section of the tire in inches is 7.75. The rim diameter is 14 inches. The Alpha Numeric system is shown in *Figure 31–10*. The first letter or letters indicate the type of car the tire is to be used on. The number 78 is the *profile ratio* (height to width ratio). The third designation is the rim diameter in inches.

Tire Placard

A *tire placard* is permanently located on many vehicles. The placard is normally located on the rear face of the driver's door. Refer to the placard for tire information. It lists the maximum vehicle load, tire size (including spare), and cold inflation pressure (including spare). *Figure 31–11* shows an example of the tire placard.

UTQG Designation

UTQG (Uniform Tire Quality Grading) symbols are required by law to be molded on the sidewall of each new tire sold in the U.S. A typical grading may be 90 CB, 170 BC, or 140 AA. The first number indicates the comparative tread life. A tire marked 140 should wear 40% longer than a tire marked 100. The first letter indicates comparative wet traction. A is best, C is worst. The second letter is a measure of the resistance to heat. Again, A is the best and C is the worst.

Some tire manufacturers use the *DOT* (Department of Transportation) designation. The DOT specification number indicates that the tire has met various tests of quality established by the Department of Transportation. *Figure 31–12* shows an example of the many numbers and ratings molded into the tire sidewall.

TPC Specification Number

On most vehicles originally equipped with radial tires, a TPC (Tire Performance Criteria) Specification Number is molded into the sidewall. This shows that the tire meets rigid size and performance standards developed for that particular automobile. It assures a proper combination of endurance, handling, load capacity, ride, and traction on wet, dry, or snow-covered surfaces.

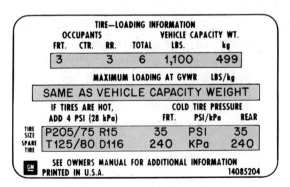

FIGURE 31–11. The tire placard is used to list maximum vehicle load, tire size, and inflation data. *(Courtesy of Oldsmobile Division, General Motors Corporation)*

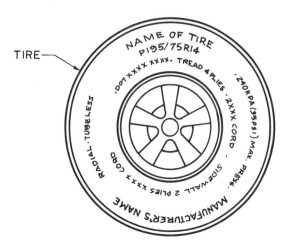

FIGURE 31-12. Several numbers are used to identify the type and style of tire used on the vehicle. These include the tire size, load capacity, name of the tire, manufacturer, number of plies, type of tire (radial or tubeless), and cord information.

Load Range

Tires are measured for the amount of load they can withstand. They are typically identified by letter codes. The codes used are A, B, C, D, E, F, G, H, I, J, L, M, and N. As the size of the tire increases, the *load range* also increases. On some tires, the load range is stated in pounds for a specific tire pressure. For example, the load range molded into the tire may be, "Max Load, 585 kg (1290 lbs) at 240 kpa (35 psi)."

CAR CLINIC ████████

PROBLEM: TIRE WEAR ON OUTSIDE

Tires on a vehicle are wearing on the outside edges. The alignment has been checked and all adjustments are within specifications. The alignment shop says this is normal. Is it normal?

SOLUTION:

Tires should not wear on the outside edges. This type of wear indicates that either the toe-in or the camber is too high. Some vehicles may have to be bent or parts may have to be replaced to adjust the camber. Try a second alignment shop because alignment machines can also be inaccurate if they are not serviced properly.

31.3 WHEELS AND RIMS

Dropped Center Wheels

Tires are mounted on rims that are made of steel, aluminum, or other strong material. The parts of the wheel

are shown in *Figure 31-13*. The mounting holes are used to attach the wheel to the lugs on the axles. The wheel is called a *drop center wheel*. This means that the center of the wheel is made so that it has a smaller diameter. The wheel has a dropped center so that the tire can be easily removed. During removal and installation, the bead of the tire must be pushed into the dropped area. Only then can the other side of the tire be removed over the rim flange.

Wheel Sizes

The most common sizes for wheels are 13, 14, and 15 inch diameters. The exact wheel size used on a vehicle is determined by the automotive manufacturer. Certain cars use different size wheels, which may include 12 or 16 inch wheel rims. The width of the wheel rims is usually 4.5, 5, or 6 inches. The large wheel rim sizes are usually used on heavy-duty and larger vehicles.

Special Wheels

A variety of special wheels are being manufactured today. Steel or aluminum is typically used on these wheels. For example, a very light magnesium metal is used for the wheel and rim of the "mag" wheel. The term *mag* is used today, however, to represent almost any type of wheel that uses special materials and designs.

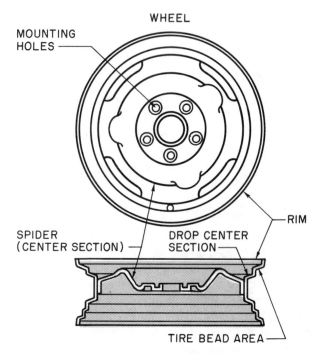

FIGURE 31-13. The wheel is used to support the tire. It has several parts and areas for identification.

NEW ENGINE DESIGNS

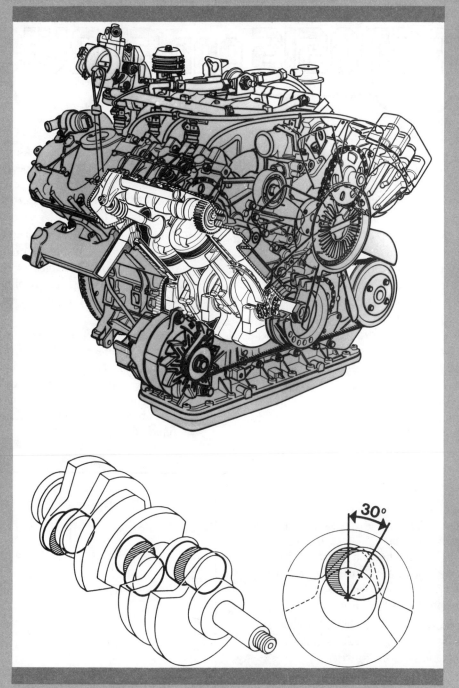

Many new engine designs are being tested to improve efficiency. This V-6 engine uses a 90 degree V rather than the usual 60 degree V configuration. This makes the engine more compact and lowers the center of gravity, which is important for aerodynamics and handling. The crankpins are offset 30 degrees to retain an even firing sequence. (Courtesy of Peugeot Motors of America)

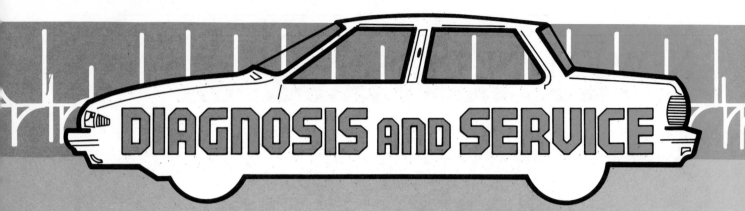

DIAGNOSIS and SERVICE

1. When replacing tires with those not having a TPC Specification Number, use the same size, load range, and construction type (bias, bias belted, or radial) as the original tires on the car. A different type of tire may affect the ride, handling, speedometer/odometer readings, and vehicle ground clearance.
2. Before checking any problem with tires, make sure the front suspension, steering, and brake systems are working and adjusted correctly.
3. Tires must be inflated correctly. The recommended tire pressure is calculated according to the type of tire, weight of the vehicle, and ride. In addition, tire inflation may change as much as 6 psi from hot to cold temperatures.
4. Tire inflation can be measured in pounds per square inch (psi) or kilopascals (kPa). *Figure 31–14* shows the relationship between the two.

5. *Overinflation* of tires increases tire tension and prevents proper deflection of the sidewalls. This results in wear in the center of the tread. The tire also loses its ability to absorb road shocks. See *Figure 31–15*.
6. *Underinflation* of tires distorts the normal contour of the tire body, causing the tire to bulge outward. This wears on the edges of the tread. It also increases internal heat. Heat may weaken the cords and cause the ply to separate. See Figure 31–15.
7. Tires can be checked for excessive wear by observing the tread-wear indicators or *tread bars*. See *Figure 31–16*. They look like narrow strips of smooth rubber across the tread. They will appear on the tire when it is worn down. When these wear bars are visible, the tire is worn out and should be replaced.

INFLATION PRESSURE CONVERSION CHART (KILOPASCALS TO PSI)

kPa	psi	kPa	psi
140	20	215	31
145	21	220	32
155	22	230	33
160	23	235	34
165	24	240	35
170	25	250	36
180	26	275	40
185	27	310	45
190	28	345	50
200	29	380	55
205	30	415	60

Conversion: 6.9 kPa = 1 psi

FIGURE 31–14. Both psi (pounds per square inch) and kPa (kilopascals) can be used to measure the inflation of tires. (Courtesy of Oldsmobile Division, General Motors Corporation)

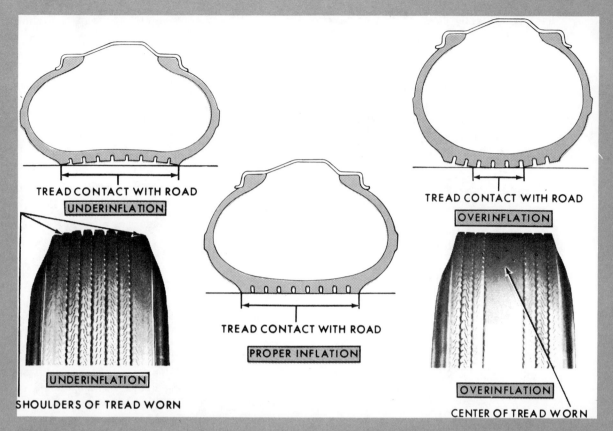

FIGURE 31–15. Both under and overinflation can cause tire damage. Always be sure the tire is inflated to the correct pressure recommended by the manufacturer. *(Courtesy of Chevrolet Division, General Motors Corporation)*

8. Tires can sometimes be rotated from wheel to wheel. There are two acceptable patterns for rotating tires. See *Figure 31–17*. However, always check the owners manual for the correct pattern. *Rotation* is shown for a car with and without a spare tire. (This does not include the "temporary use only" spare tire.)

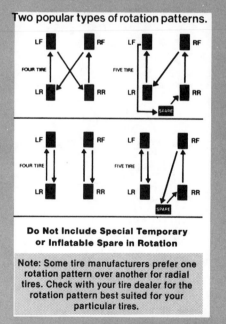

FIGURE 31–17. Tires can be rotated to distribute the wear. Two common rotational patterns are shown. Follow the manufacturer's recommendations when rotating tires. *(Courtesy of Tire Industry Safety Council)*

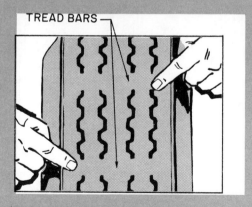

FIGURE 31–16. Always check the wear bars on tires. When the wear bars show on a tire, it is ready to be replaced. *(Courtesy of Tire Industry Safety Council)*

9. It is dangerous to weld a cracked wheel or rim or to heat it for straightening, *Figure 31–18*. Replace the rim rather than trying to repair it.

10. A pull to one side when driving that begins at less than 10 miles per hour may indicate a defective radial tire. Typically a belt has shifted. First make sure that the suspension, steering, and brake systems are adjusted correctly. Now check the tire. Jack up the car and spin the tire. If the tread wobbles from side to side, replace the tire. See *Figure 31–19*.

11. Check tires at least once a month and before long trips. Use an acceptable and accurate pressure gauge. Check the tires when they are cold and remember to check the spare tire as well. *Figure 31–20* shows how tires should look when they are properly inflated.

12. Incorrect front end geometry and operation can cause excessive and uneven wear on the tires. Observe the tire wear patterns to determine the problem as shown in *Figure 31–21*.

13. *Wheel runout* is a measure of the out-of-roundness of the tire. This can be measured by using a dial indicator. Measurements can be taken with the tire on or off of the rim. Both inboard and outboard measurements should be taken on the rim of the wheel. *Figure 31–22* shows where the measurements should be taken. Compare the runout readings to the manufacturer's recommendations and replace the tire if necessary.

14. Tire lug nuts must be tightened in a specific sequence. Using an incorrect tightening sequence may cause the wheel, brake drum, or rotor to bend. See *Figure 31–23*.

15. Tires and wheels are *match-mounted* at the assembly plant on some vehicles. This means the radially stiffest part of the tire or "high spot" is matched to the smallest radius or "low spot" on the wheel. This is done to provide the smoothest possible ride. The high spot is marked with a yellow

FIGURE 31–18. Never try to repair a defective rim that has been bent or broken. Always replace the rim.

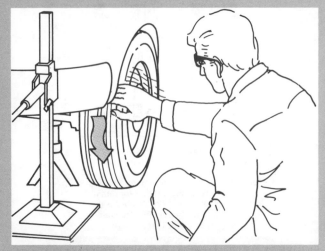

FIGURE 31–19. A defective belt in the tire can be checked by lifting the car and spinning the wheel. The tread will wobble from side to side if the belt is defective.

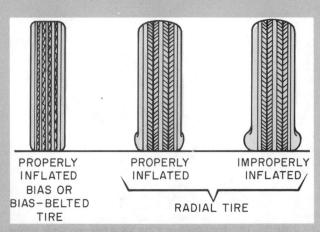

FIGURE 31–20. Different tire types will appear different when properly and improperly inflated.

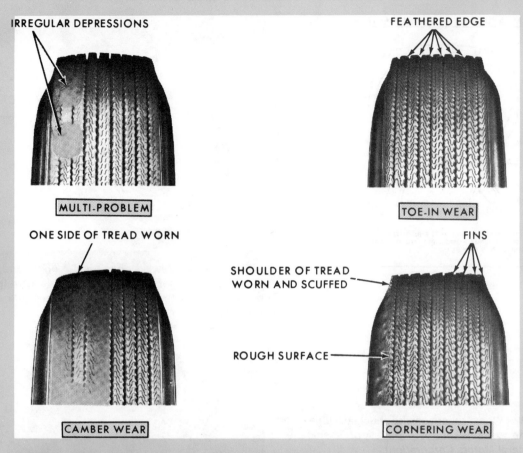

FIGURE 31-21. Tire wear can also be caused by incorrect alignment and front end geometry. *(Courtesy of Chevrolet Division, General Motors Corporation)*

paint mark on the outer sidewall. The low spot of the wheel is at the location of the valve stem. Always mount the wheel so the tire and wheel markings are matched. See **Figure 31-24**.

16. Cold weather reduces tire inflation pressure approximately one pound for every 10 degree drop in temperature. Always check the tire inflation in both summer and winter conditions.

17. If the weight of the vehicle is increased with extra loads or objects, there is an increased risk of tire failure. Always check the owner's manual for the maximum safe load limit the tires can handle.

18. Tires must be balanced correctly in order to run smoothly and evenly at all speeds. There are two types of *balancing*, static and dynamic. *Static* balance is the equal distribution of weight around the axis of rotation. The tire will have no tendency to rotate itself, regardless of position. Wheels that are statically unbalanced cause a hopping or bouncing action. This is called wheel tramp. If wheel tramp is severe, it can cause excessive wear and damage to the tire. *Dynamic balancing* is the equal distribution of weight about the place of rotation. Tires are tested and checked in balancing machines as shown in **Figure 31-25**. When a tire is spinning, it tends to move from side to side if it is out of balance. Wheels that are not dynamically balanced cause the car to vibrate when it is moving.

MEASURING WHEEL RUNOUT

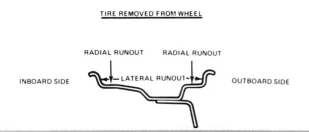

TIRE MOUNTED ON WHEEL

INBOARD SIDE   OUTBOARD SIDE

LATERAL RUNOUT   LATERAL RUNOUT

DIAL INDICATOR   RADIAL RUNOUT*

*IF WHEEL DESIGN MAKES THIS OUTBOARD
MEASUREMENT IMPOSSIBLE, THE INBOARD
SIDE ONLY MAY BE USED.

TIRE REMOVED FROM WHEEL

RADIAL RUNOUT   RADIAL RUNOUT

INBOARD SIDE   LATERAL RUNOUT   OUTBOARD SIDE

FIGURE 31–22. Wheel runout can be checked with a dial indicator at the positions shown. *(Courtesy of Oldsmobile Division, General Motors Corporation)*

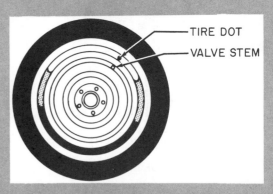

TIRE DOT

VALVE STEM

FIGURE 31–24. Always align the dot on the tire with the valve stem to get the smoothest ride possible after dismounting a tire. *(Courtesy of Oldsmobile Division, General Motors Corporation)*

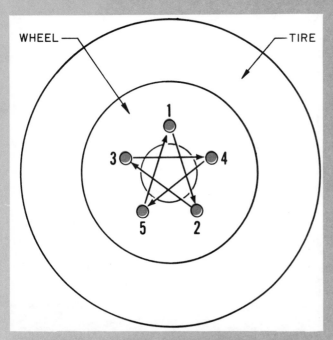

WHEEL   TIRE

FIGURE 31–23. Wheel nuts should be tightened in the correct sequence and to proper torque specifications. *(Courtesy of Oldsmobile Division, General Motors Corporation)*

FIGURE 31–25. Tires are dynamically balanced by spinning them on this tire balancing machine. *(Courtesy of Hunter Engineering Company)*

SUMMARY

This chapter discussed the components and design of tires and wheels. There are many types of tires used on cars today. The differences are made by changing the tread, belts, cords, plies, beads, and materials within the tire. Both tube and tubeless tires are still manufactured. Tires are designed to have several characteristics. These include tire traction, ride and handling, rolling resistance, and noise.

There are three main types of tires. These include the bias ply tire, the belted bias tire, and the belted radial tire. The bias ply tire is designed with layers of cord material running at angles from bead to bead. The belted bias tire has additional belts wrapped around the circumference of the body of the tire. The radial tire has the cords running radially from bead to bead. The radial tire also uses belts made of different materials around the circumference of the tire.

There are many tread designs for tires. Treads are designed mostly to improve the traction on different types of surfaces. This may include wet, dry, gravel, or icy conditions. The air is held inside the tire with the use of a tire valve. It is a one-direction valve that allows air into the tire but not out unless the valve is released.

Tire manufacturers also make tires that are puncture proof and tires that are for temporary use only.

Tires can be identified in a variety of ways. Metric sizes of tires are molded into the sidewall and tell the application, the width in millimeters, the aspect ratio, the type (radial, belted, and so on), and the wheel rim size in inches. Numeric and alpha designations are also used to size tires. In addition, many manufacturers use a tire placard, which is placed on the door of the vehicle, to show maximum vehicle load, tire size, inflation pressure, and so on. Other designations of tires include the UTQG designation, TPC specification number, and load ranges.

The tire is placed on the wheel for support. The design of the wheel is called a "drop center" wheel. This means the center of the wheel has a smaller radius. This makes it easier to remove the tire from the wheel. The most common wheel sizes are 13, 14, and 15 inches.

There is a great deal of information concerning tire diagnosis and service. One of the most important service tips is to keep the tires properly inflated. Under or overinflation will damage the tire very quickly and reduce its mileage. Before diagnosing tires, always make sure the steering, suspension, and brakes are in proper working order. Other service tips include: replace tires when the tread bars are showing, check wheel runout, check for misaligned belts, and balance tires according to the manufacturer's specifications.

TERMS TO KNOW

Can you explain each of the following terms? Review the chapter until you can use each term correctly.

Tread
Plies
Carcass
Vulcanized
Bead
Traction
Rolling resistance
Cords
Bias
Radial
Squirm
Hydrophilic

Aspect ratio
Profile ratio
Tire placard
Department of
  Transportation (DOT)
Load range
Drop center wheel
Tread bars
Static balance
Dynamic balancing
Wheel runout

REVIEW QUESTIONS

Multiple Choice

1. Which part of the tire is used to produce the traction between the tire and road?
   a. Cord
   b. Carcass
   c. Tread
   d. Belts
   e. Bead

2. Which part of the tire is used to hold the air inside the tire housing?
   a. Cord
   b. Carcass
   c. Tread
   d. Belts
   e. Bead

3. As the rolling resistance of a tire increases, the fuel mileage of the vehicle:
   a. Will increase
   b. Will decrease
   c. Will remain the same
   d. Will increase then decrease as speed increases
   e. None of the above

4. _____ are used within the tire plies or belts of a tire to increase the strength of the tire.
   a. Rims
   b. Cords
   c. Tubes
   d. Beads
   e. Valves

5. Which type of tire uses plies that are crisscrossed over each other?
   a. Radial
   b. Radial belted
   c. Bias
   d. Bias radial
   e. All of the above

6. Which type of tire has the lowest rolling resistance?
   a. Bias
   b. Bias ply
   c. Bias belted
   d. Tube bias
   e. Radial belted

7. The _____ tread tire has compounds used to attract water to the tread.
   a. Tubeless
   b. Hydrophilic
   c. Belted
   d. Corded
   e. Tube

8. The compact spare tire should be used:
   a. Whenever possible because it has low rolling resistance
   b. Only on rough surfaces
   c. Only in the city
   d. Only when another tire is damaged and cannot be used
   e. Only on wet surfaces

9. Which of the following is *not* molded into the side of a tire for sizing and identification on metric tire sizes?
   a. Wheel size
   b. Aspect ratio
   c. Application
   d. Tire speed
   e. Construction type (radial, bias, etc.)

10. Most information about a car's tires can be found on the:
    a. Steering wheel
    b. Engine block
    c. Tire placard
    d. Undercarriage of the vehicle
    e. Steering system

11. The maximum load a tire can safely withstand is called the:
    a. Load range
    b. Aspect ratio
    c. Profile
    d. Carcass
    e. Construction type

12. Wheels are made with _____ to aid in removing and installing the tires on the wheels.
    a. Cords
    b. Vulcanized rubber
    c. A dropped center
    d. 16-inch rims
    e. Flexible material

13. Which systems are important to check before diagnosing tire problems?
    a. Steering systems
    b. Brake systems
    c. Suspension systems
    d. All of the above
    e. None of the above

14. Which of the following is very critical for correct tire operation and wear?
    a. Correct tire material
    b. Correct tire inflation
    c. Correct molding on the side of the tire
    d. Correct casing material
    e. None of the above

15. Which of the following can indicate excessive tread wear on a tire?
    a. Tire inflation
    b. Tread bars
    c. Valve condition
    d. Rim condition
    e. Dropped wheel condition

16. Which of the following will affect tire inflation?
    a. Hotter temperatures
    b. Colder temperatures
    c. Amount of pressure inside the tire
    d. All of the above
    e. None of the above

The following questions are similar in format to ASE (Automotive Service Excellence) test questions.

17. The vehicle is pulling to the right side of road. Technician A says that the problem could be a defective radial. Technician B says that the problem could be in the dynamic balancing. Who is right?
    a. A only
    b. B only
    c. Both A and B
    d. Neither A nor B

18. Technician A says that it is not important that the steering system be correctly aligned before diagnosing tire problems. Technician B says that it is very important that the steering system be correctly aligned before diagnosing tire problems. Who is right?
   a. A only
   b. B only
   c. Both A and B
   d. Neither A nor B

19. Technician A says that the "tread bars" are used to show that the tire is in balance. Technician B says that the "tread bars" are used to show out of round on the tire. Who is right?
   a. A only
   b. B only
   c. Both A and B
   d. Neither A nor B

20. Technician A says that the tire pressure need not be changed for colder and/or warmer climates. Technician B says that the tire pressure should be changed for colder and/or warmer climates. Who is right?
   a. A only
   b. B only
   c. Both A and B
   d. Neither A nor B

Essay

21. List the parts of a tire and describe their purpose.

22. Describe at least three characteristics that tires are designed for.

23. Describe the difference between radial and bias tires.

24. Describe how tires are identified.

25. What is the purpose of a tire placard?

CHAPTER 32

Air Conditioning Systems

INTRODUCTION

Many automobiles today have air conditioning installed in the vehicle. Air conditioning is used to remove the warm air inside of the passenger compartment to the outside. This chapter is about air conditioning principles and the parts necessary for air conditioning on cars today.

OBJECTIVES

After reading this chapter, you will be able to:
- Identify the principles of air conditioning.
- Examine the heat and refrigerant flow of an air conditioning unit.
- Describe the purpose and operation of the common parts of an air conditioning system.
- Diagnose and service simple air conditioning problems.

CHAPTER HIGHLIGHTS

32.1 AIR CONDITIONING PRINCIPLES

There are several principles that help to understand air conditioning. These principles deal with heat flow, refrigerant, pressure, vaporization, and condensation.

Purpose of Air Conditioning

During normal summer driving conditions, a great amount of heat enters the passenger compartment. This heat comes from the engine and from the sun or outside air temperature. Air conditioning systems are designed to remove this excess heat so that passengers are comfortable. *Figure 32–1* shows an example of heat flow. Heat is admitted into the passenger compartment from outside. The heat is removed by forcing it to flow through a heat exchanger in the passenger compartment. A fluid then absorbs the heat and transfers it to a second heat exchanger outside of the passenger compartment.

Heat Flow

An air conditioning system is designed to pump heat or BTUs from one point to another. Heat can also be defined as BTUs. BTUs were discussed in the cooling system chapter. All materials or substances have heat in them down to −459 degrees F. At this temperature (absolute zero), there is no more heat. Also, heat always flows from a warmer to a colder object. For example, if one object were at 30 degrees F and another object were at 80 degrees F, heat would flow from the warmer object (80°F) to the colder object (30°F). The greater the temperature difference between the objects, the greater the amount of heat flowing. This is shown in *Figure 32–2*.

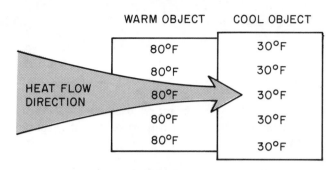

FIGURE 32–2. Heat always flows from a warmer to a colder object. Air conditioning systems are designed to cause air to flow internally from warmer to colder areas.

Heat Absorption

Objects can be in one of several forms. They can either be in a solid, liquid, or gas form. When objects change from one form to another, large amounts of heat can be transferred. For example, an ice cube is a solid form. When an ice cube melts, it absorbs a great amount of heat. In fact, all solids soak up huge amounts of heat without getting warmer when they change from a solid to a liquid.

The same thing happens when a liquid changes to a *vapor*. Large amounts of heat can be absorbed. For example, referring to *Figure 32–3*, this can be shown using a tea kettle with a thermometer set inside. As the burner heats up the water, the temperature of the water starts to rise. It continues to rise until it has reached 212 degrees F. At this point, the temperature will stay at 212 degrees F even when additional heat is applied. The water is changing to a vapor and is soaking or absorbing large quantities of heat. Although this heat does not appear on the thermometer, it is there. It is called *latent heat* or hidden heat.

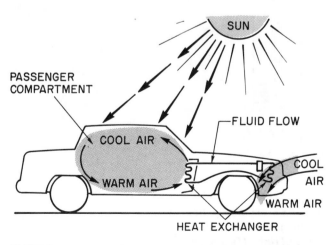

FIGURE 32–1. Heat inside of the passenger compartment is sent through heat exchangers to the outside by use of a fluid.

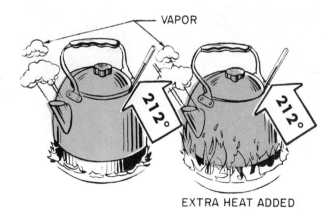

FIGURE 32–3. Heat is absorbed in large quantities when a liquid is changed to a vapor as in the boiling kettle of water. The water in the tea kettle will stay at 212 degrees F even if more heat is applied below the kettle. *(Courtesy of United Delco, Division of General Motors Corporation)*

Condensation

Condensation is the process of changing a vapor back to a liquid. Condensation is usually done by cooling the substance down, below its boiling point. All substances will condense at the same point at which they are boiled. When a vapor is condensed, the heat removed from it is exactly equal to the amount of heat necessary to make it a vapor in the first place.

Pressure and Boiling Points

Pressure also plays an important part in air conditioning. Pressure on a substance such as a liquid changes its boiling point. The greater the pressure on a liquid, the higher the boiling point. If a pressure is placed on a vapor, it will condense at a higher temperature. In addition, as the pressure on a substance is reduced, the boiling point can also be reduced. For example, the boiling point of water is 212 degrees F. The boiling point can be increased by increasing the pressure on the fluid. It can also be decreased by reducing the pressure or placing the fluid in a vacuum.

Pressure on a fluid will also concentrate any heat in the substance. As the pressure on a fluid is increased, the temperature of the fluid tends to increase. For example, refer to **Figure 32–4**. If 37 psi is exerted on a certain volume of vapor, the temperature of the vapor will be 40 degrees F. If the volume is reduced by increasing the pressure, say to 70 psi, the temperature of the vapor will be increased to 70 degrees F, without adding extra heat. This principle is also used in air conditioning systems.

Refrigerant-12 (R-12)

A *refrigerant* is used to transfer heat from inside of the car to outside. It is called R-12. This refrigerant has a boiling point of −21.7 degrees F. This means that it changes from a liquid to a vapor at −21.7 degrees F. If one were to place a flask of R-12 inside a refrigerator, it would boil and draw heat away from everything surrounding it. See **Figure 32–5**. If the refrigerant were then pumped outside (along with the heat it absorbed), the inside of the refrigerator would be cooler. Note that the boiling temperature of R-12

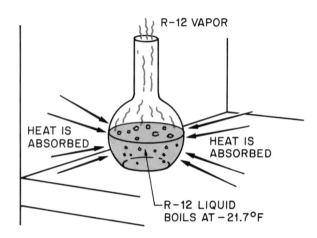

FIGURE 32–5. If R-12 were placed in a flask inside a refrigerator, it would boil at −21.7 degrees F, absorbing all heat surrounding it.

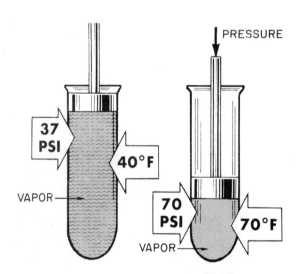

FIGURE 32–4. If a vapor is compressed with pressure from 37 to 70 psi, its temperature will increase from 40 degrees F to 70 degrees F. *(Courtesy of United Delco, Division of General Motors Corporation)*

is too low to operate inside an air conditioning system. The boiling point in an air conditioning system is changed by altering the pressure of R-12. By increasing the pressure, R-12 has a boiling point of from 30–60 degrees F. This is the most appropriate temperature range for freon to operate at.

CAR CLINIC ▮▮▮▮▮▮▮

PROBLEM: ENGINE OVERHEATS

A car has recently been overheating while driving on the highway. The problem started suddenly. One person says the problem is the thermostat, but the thermostat was checked and proved to be OK. What might cause an engine to overheat?

SOLUTION:

One of the more common causes of an engine overheating is a bad radiator pressure cap. The radiator pressure cap is designed to keep 15 pounds of pressure on the cooling system. As the pressure of the cooling system increases, the boiling point also increases. If the pressure on the cooling system has been reduced (bad radiator pressure cap), the boiling point is also reduced. Using a pressure cap and cooling system tester, check the cap. It sounds like the radiator pressure cap is not able to hold pressure on the cooling system.

32.2 REFRIGERATION CYCLE OPERATION

Basic Heat Transfer

Heat exchangers are used to transfer heat in an air conditioning unit. *Figure 32–6* shows an example of how heat

exchangers are used to pump heat from inside the car to outside. When cool refrigerant is sent through the evaporator, which is located inside the passenger compartment, it cools the air passed over the evaporator fins. The refrigerant is boiling and vaporizes, absorbing the hot air from the car. The refrigerant vapor is then pumped outside the passenger compartment to a second heat exchanger. This one is called the condenser. Here, cooler outside air is forced through the fins on the condenser. The air is cooler than the refrigerant. Heat is now transferred from the refrigerant to the passing air, causing it to heat up. The heat in the car has essentially been pumped, via the refrigerant, from the inside to the outside of the car.

Purpose of Compressor

All automotive air conditioning systems have several major parts. These include the compressor, condenser, expansion tube or expansion valve, and evaporator. *Figure 32–7* shows the flow inside of an air conditioning system using all components.

The compressor is the heart of the air conditioning system. It has several purposes. First, the compressor is used to move the refrigerant. Second, the compressor is used to compress the refrigerant to change its boiling point. As the R-12 flows through other components, it undergoes various changes in pressure and temperature. This is required for proper heat transfer to take place.

The compressor is powered by the car engine, through a clutch turned by a belt. It has an intake side (suction) and a discharge side (exhaust). When the R-12 enters the compressor, it is a gas at low temperature and low pressure. When the R-12 leaves the compressor, it is still a gas but it is at a very high pressure. Compressing the gas has also

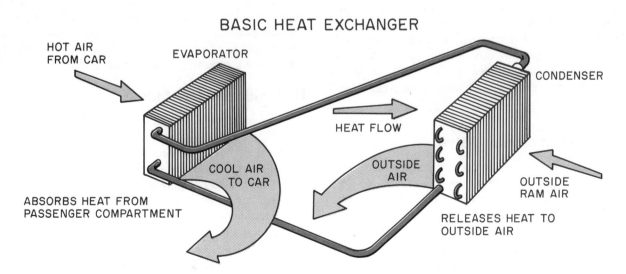

FIGURE 32–6. Heat transfer in an air conditioning system is done by using heat exchangers. The evaporator absorbs heat into a liquid refrigerant. The condenser gives off heat from the refrigerant.

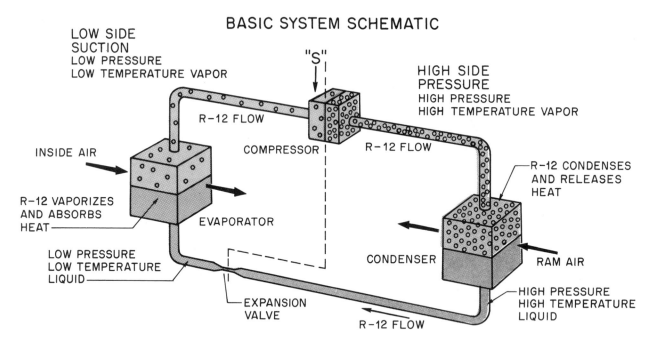

FIGURE 32–7. This schematic shows how the R-12 is changed in temperature and pressure throughout the air conditioning cycle.

increased its temperature well above that of the outside air. This must be done to get the heat to flow in the correct direction, from inside the refrigerant to the outside air.

Purpose of Condenser

The *condenser* is basically a heat exchanger. It consists of a series of tubes and fins. As the hot R-12 gas flows through the tubes, it warms the fins. The fins provide enough surface area to transfer heat effectively. Heat transfer now takes place rapidly between the condenser and the outside air. During this rapid heat loss, R-12 is reduced in temperature below its boiling point. (Remember, the boiling point is well above −21.7 degrees F because the vapor has been pressurized.) R-12 now condenses back into a liquid. As R-12 exits the condenser, it is a liquid at high pressure and high temperature.

Purpose of Expansion Valve

Next the liquid R-12 flows through an *expansion valve*. This valve is a restriction in the R-12 flow. As a hot liquid R-12 enters the valve, the restriction causes the pressure to build up behind it. As the liquid passes through the valve, there is a large pressure drop. This pressure drop changes the boiling point of the R-12. Now the R-12 is at a condition where it is ready to boil or evaporate just before it enters the evaporator.

Purpose of Evaporator

The *evaporator* is another version of a heat exchanger. In the evaporator the heat is transferred from the passenger compartment air to the liquid refrigerant. Since the R-12 entering the evaporator is at a lower temperature, it makes the evaporator fins "cold." Thus warm car air circulating around the cold fins releases heat into the evaporator. As the heat is applied to the refrigerant, the R-12 boils and changes to a vapor again. As it boils (just like the tea kettle earlier), it is able to absorb huge quantities of heat. The result is a rapid heat loss in the inside passenger air. When the R-12 exits the evaporator, it is once again a gas, at low temperature and low pressure. The system then continues to recycle back to the compressor, repeating the cycle.

High Side and Low Side

Every air conditioner has two sides. There is a high side and a low side. The high side refers to that portion of the system where the R-12 is at high pressure. The low side is that portion of the system where the R-12 is at low pressure. A dotted line is used in Figure 32–7 to show the dividing line between the high and low pressures. The high side includes the system between the discharge end of the compressor and the restriction of the expansion valve. It includes the condenser. The low side includes the portion from the restriction or expansion valve, through the evaporator, to the intake end of the compressor.

RADIATOR DESIGN

Air conditioning components require high quality and reliability in manufacturing. These high efficiency radiators can be made as crossflow or down-flow radiators. They are composed of a copper/brass soldered core with soldered-on brass tanks. Product engineering and design personnel use up-to-date methods to mathematically analyze and design radiators for a particular application in a vehicle. (Courtesy of Harrison Radiator Division, General Motors Corporation)

32.3 SYSTEM VARIATIONS

All air conditioning systems use the basic components that were just explained. In addition, several other components are integrated into every system to keep it working properly. These components and their location make up the basic differences between one air conditioning system and another.

Accumulator Orifice Tube System

This system uses an additional *accumulator* and a tube for the expansion valve. The accumulator is located between the output of the evaporator and the input of the compressor. See *Figure 32–8*. The accumulator performs several vital functions in the air conditioning system.

1. It collects excess R-12 liquid, permitting only R-12 gas to enter the compressor.

2. It contains a filtering element with a built-in *desiccant* (moisture-absorbing material) which helps to reduce moisture from the R-12.

3. It contains a filter screen which traps any foreign matter before it can reach the compressor.

In addition, this system uses an orifice tube as the expansion valve. It is used to create the pressure drop needed just before the R-12 enters the evaporator.

Cycling Clutch System

Another type of air conditioning system uses a receiver-filter-drier along with a thermostatically controlled clutch. The receiver-filter-drier is much the same as the accumulator in the preceding system. It is located on the high side of systems. See *Figure 32–9*. In addition to performing filtration and moisture removal, it acts as a reservoir, storing any excess R-12 in the system.

This system also uses an expansion valve rather than an expansion tube to create the required pressure drop. In this system, the expansion valve controls the flow of R-12 entering the evaporator. This metering function is achieved by a sensing bulb. The sensing bulb measures the temperature of the R-12 leaving the evaporator. In turn, based on this temperature, it changes the expansion valve opening to increase or decrease the flow rate.

This system also uses a compressor clutch. The compressor clutch transmits power from the car engine to the compressor. It is turned off and on by a magnetic coil. The magnetic coil is operated by another electrical circuit. The cycling clutch system engages or disengages the compressor according to the demands of the system. The cycle is controlled by a thermostatic switch which senses the temperature at the evaporator. A pressure-sensing switch is used on some models, rather than a temperature-sensing switch.

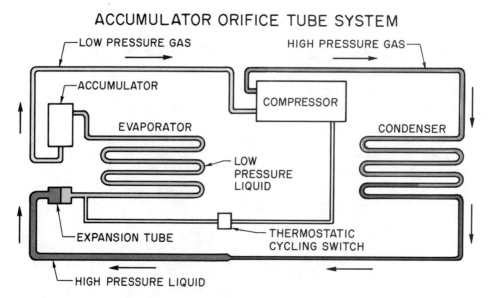

FIGURE 32–8. An accumulator is used on some air conditioning systems to collect excess refrigerant and filter the refrigerant.

BASIC CYCLING CLUTCH SYSTEM

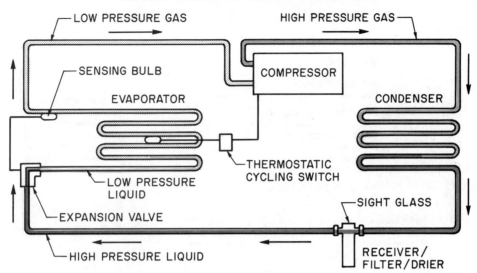

FIGURE 32–9. Certain air conditioning systems use a receiver-filter-drier. This device is placed on the high side of the system and is used to filter, remove moisture, and to act as a reservoir for R-12. A cycling clutch is also used on this system to turn the compressor off and on.

Control Valve System

Certain air conditioning systems use valves to control the operation of the air conditioning unit. *Figure 32–10* shows a typical control valve system. This system uses a non-cycling clutch. The compressor operates continuously. Control devices are used to adjust the temperature, pressure, and flow rate of the R-12 to maintain the required cooling rate. Several control valves are used. At times as the pressure changes inside of the evaporator, it has a tendency to reduce the temperature and freeze up. (The evaporator pressure directly controls the evaporator temperature.) The control valve operates to keep the pressure at a predetermined psi.

In this system, a POA (Pilot Operated Absolute) valve or a STV (Suction Throttling Valve) can be used. In addition, an EPR (Evaporator Pressure Regulator) could be used. The valve used depends upon the vehicle manufacturer. The control valve opens and closes to regulate the flow of R-12. By controlling the pressure, the temperature of the R-12 will also be controlled to slightly above 32 degrees F, as it exits the evaporator.

CAR CLINIC

PROBLEM: STOPPING THE TURBOCHARGER

What is the best way to shut off an engine with a turbocharger? Some say to just shut the engine down after any rpm, while others say to let the engine idle for a short period of time. Who is right?

SOLUTION:

Turbocharged engines should be shut down by first idling the engine for a short period of time. The reason for this procedure is to let the turbocharger in the turbine slow down after being at high speed. During this slow-down time, the engine is still running and is thus sending oil to the turbocharger bearings. If the engine is shut down immediately after high-speed operation, the oil pressure will not be available at the turbocharger bearings, which may cause damage.

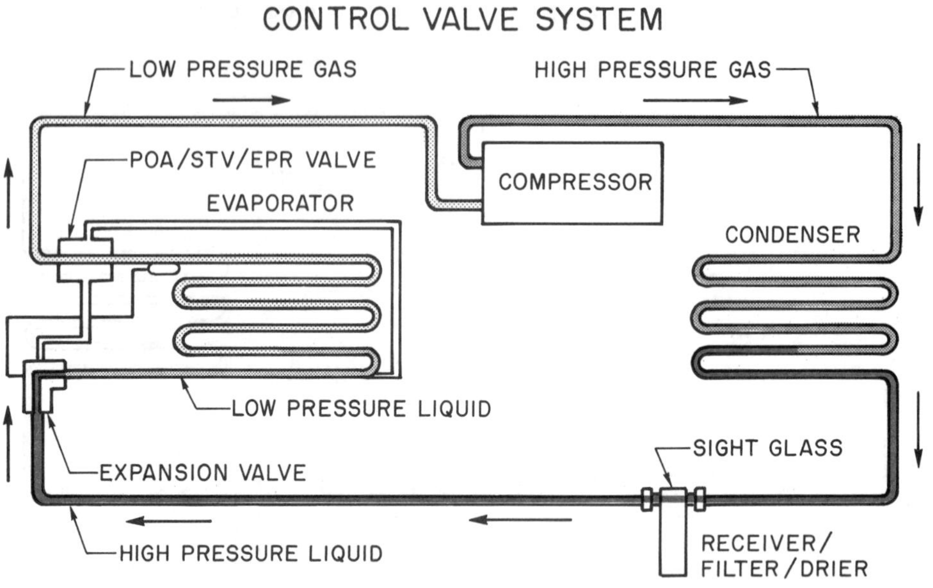

CONTROL VALVE SYSTEM

FIGURE 32–10. Some air conditioning systems use several control valves to operate the system. Both the POA, (Pilot Operated Absolute) valve and the STV (Suction Throttling Valve) are used in this system.

32.4 AIR CONDITIONING PARTS

Once the cycle of operation is understood, each component can be analyzed as to its design and function. This section analyzes the internal operation of several air conditioning components.

Suction Throttling Valve (STV)

Figure 32–11 shows the Suction Throttling Valve (STV). This valve is used to determine the temperature of the evaporator core by controlling the evaporator pressure. In this manner, the valve protects the core against freeze-up. Freeze-up could result in partial or complete loss of cooling capacity.

The valve, which is located in the evaporator outlet line, operates on a spring pressure versus evaporator pressure principle. R-12 vapor flows through the valve inlet. The vapor pressure works against the piston. The piston then pushes against the diaphragm. The spring pressure pushes against the other side of the piston and diaphragm. Evaporator pressure is now working against spring pressure. Any increase in temperature, and thus pressure of R-12 will push the piston to the left, against the spring pressure. This action opens the valve and allows an increase in the amount of vapor flowing to the compressor. In turn, the evaporator pressure will lower and allow the piston to close. The evaporator pressure, and thus the temperature, is controlled to a predetermined setting, with a "throttling" effect.

The temperature lever on the dash may be moved to mix heated air with the maximum cooled air. This will temper the outlet air to a desired temperature. Any movement on the lever also controls a vacuum valve which sends a vacuum to the altitude compensating assembly shown on the STV. Loss of vacuum at this point causes the internal spring pressure to be increased. Now the minimum evaporator pressure will increase. This results in less evaporator cooling capacity. This feature guards against evaporator freeze-up when operating at higher elevations. Two other ports that have *Schrader valves* are also used. One is used to obtain a system pressure, while the other is used to connect to the oil bypass line from the bottom of the evaporator.

Pilot Operated Absolute (POA) Valve

The function of the POA valve is to control the evaporator pressure. See *Figure 32–12*. This is done much the same way as the STV, by restricting the evaporator outlet. Although the end result is the same, this system uses a pilot valve and a bellows to control the pressure. In operation, evaporator pressure is forced into (A) against the piston. The piston is held in place by the piston spring pressure. When evaporator pressure gets too high, the piston slowly lifts and allows evaporator pressure to pass through (B) to the outlet. During this time, evaporator pressure also flows through the piston bleed hole into area (C) and finally to the bellows area

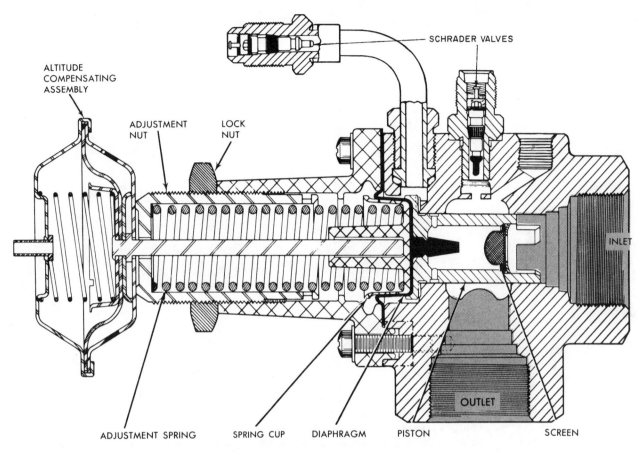

FIGURE 32–11. A Suction Throttling Valve is used to control the temperature and pressure of the R-12. *(Courtesy of United Delco, Division of General Motors Corporation)*

(D). The bellows position controls a needle valve. In operation, the assembly balances out the internal pressures to maintain a predetermined, accurate evaporator pressure. In turn, evaporator temperature will be controlled.

Evaporator Pressure Regulator (EPR) Valve

The main function of the EPR valve is to maintain the evaporator pressure at a sufficient level to avoid the freezing of moisture on the evaporator core. At the same time, it also provides maximum cooling efficiency.

The EPR valve is installed on the suction passage of the compressor. The valve is operated by a gas-filled bellows. As long as the evaporator pressure is above a certain psi, it works against a diaphragm to compress a spring and hold the valve open. When the pressure drops below a certain point, the valve tends to close. This increases evaporator pressure and thus evaporator temperature (preventing evaporator core freeze-up).

Compressors

There are many compressors used in automotive air conditioning systems. The compressor is located in the engine compartment. Its purpose is to draw low-pressure vapor from the evaporator and compress this vapor into a high-temperature, high-pressure vapor. It is also used to circulate the R-12 throughout the system. The compressor is belt-driven by the engine crankshaft.

There are many types of compressors. One type of compressor works on a *swash plate* pump arrangement. Referring to **Figure 32–13**, the swash plate is attached to the center shaft at an angle. As the center shaft turns, the swash plate also turns. This causes six small pistons attached to the swash plate to move back and forth. The pistons are attached to the swash plate using large ball bearings. A suction and pressure are created on the ends of the pistons. A reed valve is used to control the direction of the suction and pressure. An oil pump provides oil pressure to the moving parts.

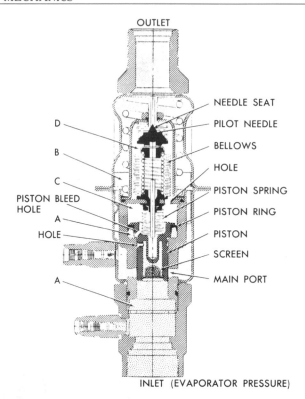

FIGURE 32-12. A Pilot Operated Absolute valve is used to accurately control the evaporator pressure. This is done using a series of bleed holes, bellows, and a needle valve. *(Courtesy of United Delco, Division of General Motors Corporation)*

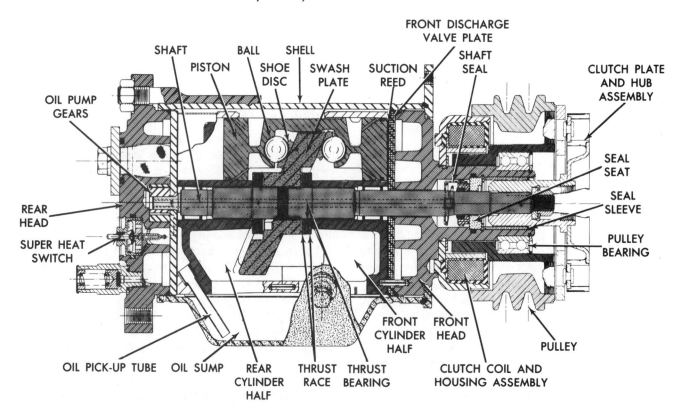

FIGURE 32-13. The compressor in an air conditioning system uses a swash plate pump arrangement to move six pistons back and forth. This produces a suction and pressure to control the flow of R-12. *(Courtesy of United Delco, Division of General Motors Corporation)*

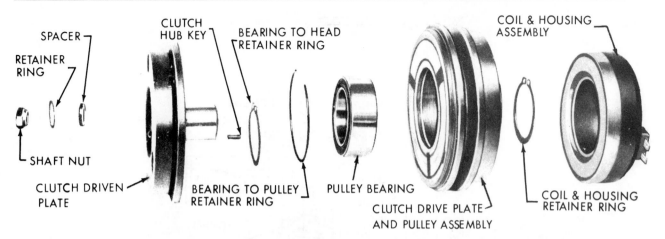

FIGURE 32-14. The main parts of the electromagnetic compressor clutch are shown. When engaged, the drive plate and pulley assembly are forced by magnetism to work against the clutch driven plate. The driven plate drives the compressor main shaft. *(Courtesy of United Delco, Division of General Motors Corporation)*

Compressor Clutch

In all air conditioning systems, the compressor is equipped with an electromagnetic clutch as part of the pulley. It is designed to engage and disengage the pulley to the compressor. Figure 32-13 shows an electromagnetic clutch attached to the compressor. The clutch is engaged by a magnetic field. It is disengaged by springs when the magnetic field is broken. When the controls call for compressor operation, the electrical circuit to the clutch is energized. The pulleys are connected to the compressor shaft. *Figure 32-14* shows the main parts in an electromagnetic compressor clutch.

Receiver-Dehydrator (Drier)

The receiver-dehydrator shown in *Figure 32-15* is a storage tank for liquid refrigerant. R-12 flows from the condenser into the tank. Here a bag of desiccant (moisture-absorbing material) removes moisture from the R-12. The R-12 then flows through a filter screen into the outlet. A sight glass is generally located on the top of the unit. The sight glass shows if there is enough refrigerant in the system. Some sight glasses are located directly in the line from the condenser.

Thermostatic Expansion Valve

The thermostatic expansion valve controls the supply of liquid R-12 to the evaporator. The valve could be considered a variable expansion valve. As shown in *Figure 32-16*, it is controlled by two opposing forces. A spring pressure works against the power element pressure on a diaphragm. The balance between these two forces positions the seat and orifice to a certain size expansion valve. The power element is sensing the temperature of the evaporator outlet.

In operation, when the temperature of the evaporator outlet decreases, this lowers the bulb temperature. When the bulb temperature decreases, the pressure in the power element is reduced. This causes the seat to move closer to the orifice, restricting the flow of R-12 to the evaporator. The operation is reversed when the evaporator temperature increases.

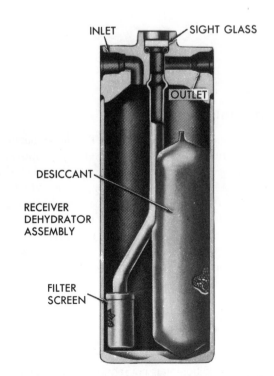

FIGURE 32-15. The receiver-dehydrator assembly is used to filter and remove moisture from the R-12. The desiccant is the moisture-absorbing material. *(Courtesy of United Delco, Division of General Motors Corporation)*

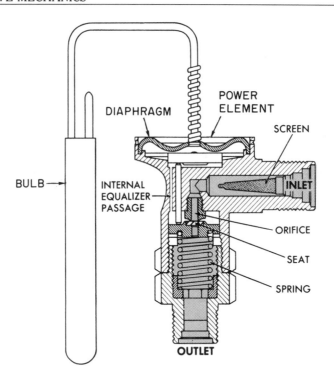

FIGURE 32-16. A thermostatic expansion valve is used to control the size of the expansion valve. The temperature of the evaporator outlet is sensed by the bulb. This causes the power element to move. The pressure from the power element against the spring pressure sets the correct size. *(Courtesy of United Delco, Division of General Motors Corporation)*

Thermostatic Switch

The *thermostatic switch* is used to cycle the electromagnetic clutch off and on. This switch senses evaporator temperature. See *Figure 32–17*. The opening and closing of the internal contacts cycles the compressor. When the temperature of the evaporator approaches the freezing point, the thermostatic switch opens. This action disengages the compressor clutch. The compressor remains off until the evaporator temperature rises to a preset temperature. At this temperature, the switch closes and the compressor resumes operation.

Ambient Switch

The *ambient* switch is used to sense outside air temperature. It is designed to prevent compressor clutch engagement when the air conditioning is not required. The switch is in series with the electromagnetic compressor clutch.

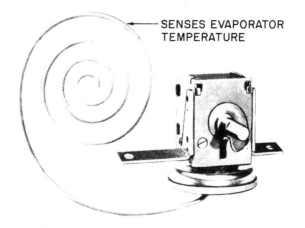

FIGURE 32-17. A thermostatic switch is used to cycle the clutch system on and off. *(Courtesy of United Delco, Division of General Motors Corporation)*

Thermal Limiter and Superheat Switch

The thermal limiter and superheat switch are designed to protect the air conditioning compressor against damage when the refrigerant is partially or totally lost. During this condition, the superheat switch heats up a resistor to melt a fuse in the thermal switch. The compressor ceases to operate and is protected from damage.

Low-pressure Cut-off Switch

The low-pressure cut-off switch is located on the pressure side of the compressor. If low pressure is sensed, the switch opens a set of contact points. The electromagnetic clutch is now inoperative.

Water Control Valve

A water control valve is used in many air conditioning systems. See *Figure 32–18*. Its function is to regulate the flow of engine coolant to the heater core. On most vehicles, the water valve is closed when the air conditioning controls are set at maximum cooling.

Muffler

A muffler, *Figure 32–19*, is used to reduce the compressor noise. It is located on the discharge side of the compressor. A complete air conditioning system, with the location of several common parts highlighted, is shown in *Figure 32–20*.

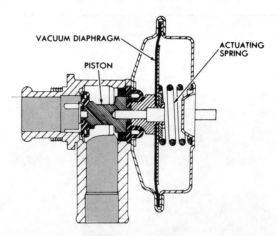

FIGURE 32–18. On some cars a water control valve is used to regulate the flow of engine coolant to the heater core during air conditioning. *(Courtesy of United Delco, Division of General Motors Corporation)*

FIGURE 32–19. A muffler is used on air conditioning systems to reduce the noise of the compressor. *(Courtesy of United Delco, Division of General Motors Corporation)*

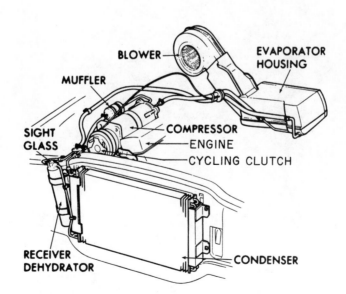

FIGURE 32–20. The parts of the air conditioning system and their locations. *(Courtesy of United Delco, Division of General Motors Corporation)*

AIR CONDITIONING SYSTEM PARTS

These parts are manufactured as complete heating, ventilating, and air conditioning units for specific vehicle applications. Air conditioning assemblies, containing both evaporator and heater cores, provide air conditioning, heating, venting, and defrosting modes. It is also possible to add a side window defogging feature to the assemblies. (Courtesy of Harrison Radiator Division, General Motors Corporation)

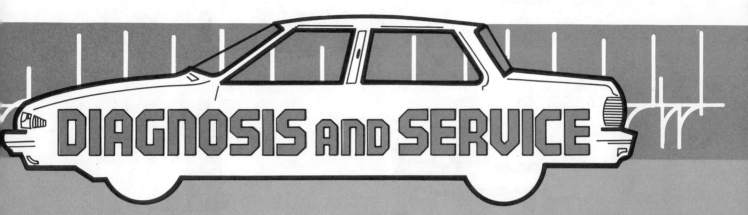

DIAGNOSIS and SERVICE

1. The most common reasons for a compressor failure are inadequate lubrication and contamination. Proper lubrication is critical to correct compressor operation. Contamination usually comes from having moisture in the R-12.

2. The simplest and most effective cure for contamination is system flushing. Flushing will remove moisture, debris, and corrosive materials. Procedures for flushing an air conditioning system are provided in all service manuals.

3. Problems with the compressor clutch can usually be traced to the compressor. A worn or poorly lubricated compressor will put a strain on the clutch. Low voltage can also cause damage to the compressor clutch.

4. Servicing the condenser usually involves cleaning the outside debris trapped in the fins or straightening any bent fins. Any dirt that restricts air flow will reduce the efficiency of the air conditioning unit.

5. Evaporator failures are mostly the result of pinhole leaks which develop in the bottom section from corrosion. Corrosion will result if moisture enters the system and mixes with the R-12. External corrosion can be caused by atmospheric moisture, especially in areas that have high salt content.

6. Whenever an air conditioning system is open to the atmosphere, it will absorb moisture from the air. All moisture must then be evacuated from the system, and the system must be recharged. Follow the procedure listed in the service manual for charging the air conditioning system. ***CAUTION***: *Be careful not to touch or get the R-12 refrigerant in your eyes.*

7. Accumulator/driers frequently need to be replaced. These units become "used up" in normal service. The desiccant becomes saturated with absorbed moisture.

8. Periodic checks and inspections should be performed on any air conditioning system to assure maximum efficiency. These include:
   a. Inspect condenser for plugged and bent fins.
   b. Check evaporator drain tubes for dirt or restrictions.
   c. Check the system at least once a year for proper refrigerant charge.
   d. Every 6,000 miles or less, check the sight glass for low refrigerant indication. ***Note:*** *Not all air conditioning systems have a sight glass.*
   e. Check air discharge temperature and compare it to the manufacturer's specifications.

9. A *manifold gauge set*, ***Figure 32–21***, is used for charging, discharging, evacuating, and diagnosing trouble in air conditioning systems. The left gauge measures the low side. The right gauge is used to

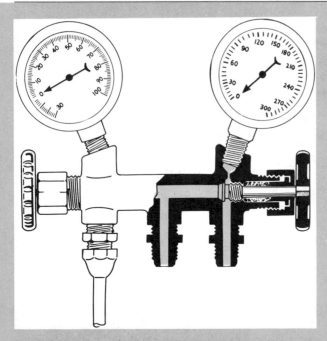

FIGURE 32–21. A manifold gauge set is used to check air conditioning system pressures, to help charge and discharge the system, and to evacuate the system. *(Courtesy of United Delco, Division of General Motors Corporation)*

FIGURE 32–22. A propane leak detector can be used to check for refrigerant leaks. Any refrigerant gas drawn into the sampling tube will cause the torch flame to change colors. *(Courtesy of United Delco, Division of General Motors Corporation)*

check the high side. The center manifold is common to both sides and is used for evacuating or adding refrigerant to the system. To attach the gauge to the air conditioning system, follow the procedures listed in the appropriate service manual.

10. A *propane leak detector* is used to check for refrigerant leaks. See *Figure 32–22*. The tube is placed where there is a suspected leak. Refrigerant gas drawn into the sampling tube attached to the torch causes the torch flame to change color in proportion to the size of the leak.

11. Various electronic leak detectors are available for checking refrigerant leaks. Follow the manufacturer's recommended procedure when checking for leaks. *CAUTION: Be careful not to inhale the gases (Phosgene gas) produced by this process. It is very dangerous to one's health.*

12. The thermostatic switch may fail and cause the compressor to run continuously. This may result in having the evaporator freeze up. Check the switch for correct adjustment according to the specification manual.

13. There may be a suction throttling valve on some older vehicles. The suction throttling valve (STV) may need adjustment. It must keep the evaporator pressure above 29–30 psi. If the pressure falls below the specification, the evaporator core may freeze up.

14. When the evaporator is defective, the trouble shows up as an inadequate supply of cool air. A partially plugged core due to dirt, a cracked case, or a leaking seal is generally the cause.

15. Compressor malfunctions appear in one of four ways: noise, seizure, leakage, or low inlet and discharge pressure.

16. A condenser may malfunction in two ways: it may leak, or it may be restricted.

17. A receiver-dehydrator may fail due to a restriction inside the body of the unit. High pressures on the pressure side indicate a restriction.

18. At temperatures higher than 70 degrees F, the sight glass may indicate whether the refrigerant charge is sufficient. A shortage of liquid refrigerant is indicated after about five minutes of compressor operation by the appearance of slow-moving bubbles (vapor) in the line. A broken column of refrigerant may appear in the glass, indicating an insufficient charge. Continuous bubbles may appear in a properly charged system on a cool day. *Figure 32–23* shows a diagnosis chart for the sight glass check.

19. Expansion valve failures are usually indicated by low suction and discharge pressure, and insufficient cooling.

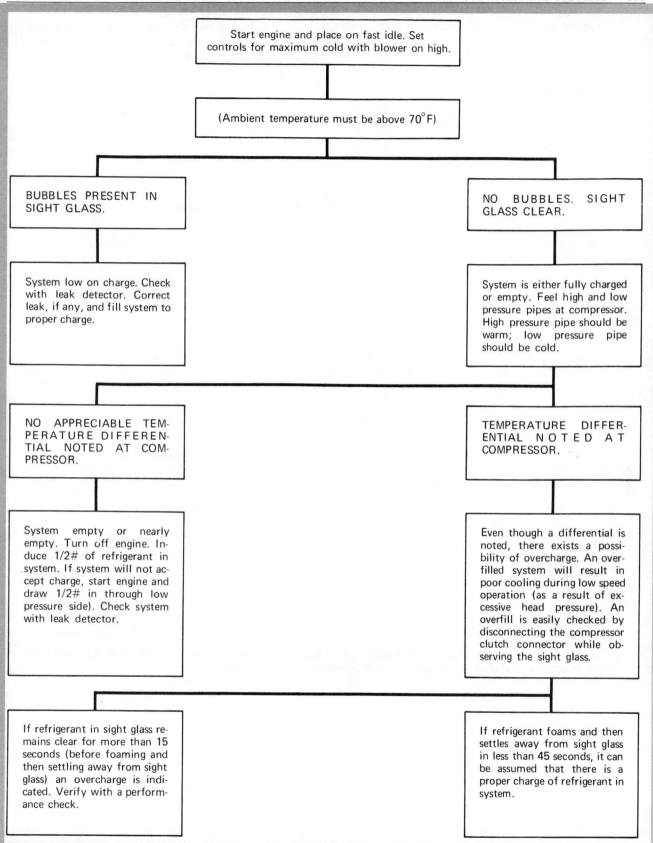

FIGURE 32–23. This diagnosis chart can be used to troubleshoot the air conditioning system by starting with the sight glass. *(Courtesy of United Delco, Division of General Motors Corporation)*

20. *Figure 32-24* shows an example of a diagnosis chart for troubleshooting a compressor that is not engaging. There are many other diagnosis charts available. Refer to the manufacturer's service manuals for more charts.

21. Air conditioning systems are different with each vehicle manufacturer. A great deal of service information is available. Specific service manuals are available for each type of system. Always refer to these manuals for correct service procedures.

22. On certain types of air conditioning compressors, the oil of the compressor must be checked periodically. Follow the service procedure for checking the oil on these compressors.

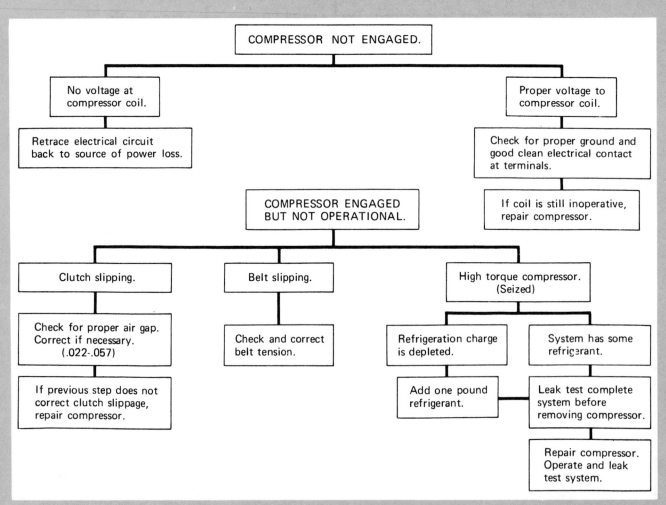

FIGURE 32-24. If the compressor is not operating or engaging, follow this diagnosis chart to determine the problem. *(Courtesy of United Delco, Division of General Motors Corporation)*

SUMMARY

The purpose of this chapter was to discuss and explain the operation of air conditioning systems used in automobiles. When heat gets into the passenger compartment, it is removed by using an air conditioner.

It is necessary to understand several principles of heat flow, pressure, and temperature in order to understand air conditioning. Heat always flows from a hotter to a colder object. Heat can also be absorbed in large quantities when a liquid turns into a gas. This is done by boiling the liquid and evaporating it into a vapor. When the vapor is cooled, it condenses back to a liquid form. Pressure also has a direct effect on the boiling point. As a vapor is compressed under a pressure, its temperature increases.

With these principles in mind, a refrigerant called

R-12 is used. This refrigerant has a boiling point of −21.7 degrees F. Air conditioning systems are designed to change the pressure of the R-12, causing the boiling point to change. When the refrigerant boils, it absorbs large quantities of heat.

Heat transfer in an air conditioning system is done by using heat exchangers. An evaporator and a condenser are normally used. Several other components are also used in an air conditioner. These include a compressor and an expansion valve. Heat is removed from the passenger compartment in the following sequence. As R-12 passes through the evaporator in the passenger compartment, it boils and converts to a vapor. A large amount of heat is absorbed into the R-12. The refrigerant is then sent to the compressor to be compressed. The pressure heats up the R-12. The hot R-12 is sent to the condenser where air flows through the heat exchanger. The air picks up the heat and transfers it outside of the vehicle. The vapor has now been changed to a liquid. The liquid is sent to an expansion valve where the pressure is reduced through the valve. The reduction in pressure causes the boiling point to drop. The liquid is now sent to the evaporator to be boiled again by the heat of the passenger compartment. The refrigerant continues to cycle through the system.

Several systems are used today in air conditioning systems. One system uses an accumulator and an orifice tube. A second system uses a cycling clutch to turn the compressor off and on. A third system uses several controls to operate the system at the maximum efficiency.

Other parts may also be used on an air conditioning system. A Suction Throttling Valve (STV) is used to determine the temperature of the evaporator. This control protects the system from freezing up. The valve uses a spring pressure that works against evaporator pressure. When the valve is operating, the evaporator controls the temperature of the R-12 to maximum efficiency.

A second system uses a POA (Pilot Operated Absolute) valve. It also controls the evaporator outlet. Evaporator pressure works against a piston to open or close the valve. A third system uses an Evaporator Pressure Regulator (EPR) valve. The valve is operated by a gas-filled bellows to control R-12 temperature. Again, this unit is used to prevent core freeze-up.

Many types of compressors are used on air conditioning systems. One popular type uses a swash plate pump to create a suction and pressure on the R-12.

Some air conditioning units also use other controls. The receiver-dehydrator is used to store liquid refrigerant. The thermostatic expansion valve is used to control the size of the expansion valve, based upon the temperature of the evaporator. The thermostatic switch is used to turn the electromagnetic clutch off and on. The ambient switch is used to sense outside air temperature. It controls the time when the compressor cannot be electromagnetically engaged. The thermal limiter and superheat switches are designed to protect the air conditioner compressor when the refrigerant is low. The low-pressure cut-off switch is used to shut off the compressor when there is low refrigerant pressure. The water control valve is used to regulate the engine coolant to the heat core.

Diagnosis and service procedures are very important on the air conditioning system. The two biggest dangers to an air conditioning system are dirt and moisture. Other problems in the air conditioning system include a damaged compressor, controls that are not adjusted correctly, and accumulators and filters being used up. Follow the manufacturer's recommendations and procedures when working with air conditioning equipment.

TERMS TO KNOW

Can you explain each of the following terms? Review the chapter until you can use each term correctly.

Vapor	Accumulator
Latent heat	Desiccant
Condensation	Schrader valve
Refrigerant	Swash plate
Condenser	Thermostatic switch
Expansion valve	Ambient
Evaporator	

REVIEW QUESTIONS

Multiple Choice

1. At what temperature is there a total lack of heat in any substance?
   a. Boiling point
   b. Absolute zero
   c. Freezing point
   d. 88 degrees F
   e. −21.7 degrees F

2. Heat will always flow from a _____ substance.
   a. Colder to a colder
   b. Hotter to a hotter
   c. Hotter to a colder
   d. Colder to a hotter
   e. None of the above

3. At what point can large quantities of heat be absorbed into a liquid or gas?
   a. When a liquid boils to a vapor
   b. When a vapor remains as a vapor
   c. When a liquid freezes
   d. All of the above
   e. None of the above

4. Heat that is hidden is called:
   a. Absolute heat
   b. Warm heat
   c. Specific heat
   d. Latent heat
   e. All of the above

5. When a vapor is changed back to a liquid, it is called:
   a. Evaporation
   b. Boiling point
   c. Condensation
   d. Accumulation
   e. None of the above

6. As the pressure of a liquid increases, the boiling point of the liquid _____.
   a. Decreases
   b. Increases
   c. Remains the same
   d. Drops by only 2 degrees
   e. None of the above

7. What is the boiling point of R-12 that is not under pressure?
   a. −10 degrees F
   b. −21.7 degrees F
   c. −41.7 degrees F
   d. 30 degrees F
   e. 65 degrees F

8. Heat transfer in an air conditioning system is done in the:
   a. Accumulator
   b. Drier
   c. Compressor
   d. Heat exchangers
   e. Thermostatic switch

9. Which of the following processes are happening in the passenger compartment of an air conditioning system?
   a. The liquid R-12 changes to a vapor.
   b. Large amounts of heat are absorbed in the R-12.
   c. The liquid R-12 boils from the passenger compartment heat.
   d. All of the above
   e. None of the above

10. Condensation of the vapor R-12 back to a liquid takes place at the:
    a. Evaporator
    b. Compressor
    c. Condenser
    d. Accumulator
    e. Receiver

11. Which of the following components are on the high side of the air conditioning circuit?
    a. Evaporator
    b. Condenser
    c. Suction side of the compressor
    d. Low-pressure side of the expansion valve
    e. All of the above

12. Which of the following components are on the low side of the air conditioning circuit?
    a. Evaporator
    b. Low-pressure side of the expansion valve
    c. Suction side of the expansion valve
    d. All of the above
    e. None of the above

13. The _____ is used to store liquid refrigerant.
    a. Evaporator
    b. Condenser
    c. Accumulator
    d. Expansion valve
    e. All of the above

14. The _____ is used to reduce the pressure and temperature of the R-12.
    a. Expansion valve
    b. Accumulator
    c. Compressor
    d. Receiver-drier
    e. Pressure regulator

15. The _____ freezes up if it is operated at the wrong pressures and temperatures.
    a. Compressor
    b. Condenser
    c. Evaporator
    d. Expansion valve
    e. None of the above

16. To control evaporator temperature on the air conditioning system, the _____ is controlled.
    a. Pressure
    b. Compressor speed
    c. Engine speed
    d. Condenser speed
    e. Dc volts

17. One common type of compressor uses:
    a. A swash plate
    b. A reed valve
    c. Six pistons
    d. All of the above
    e. None of the above

18. What type of device is used to engage and disengage the compressor?
    a. Friction clutch
    b. Electromagnetic clutch
    c. Centrifugal clutch
    d. Roller clutch
    e. None of the above

19. Which two elements are most dangerous to an air conditioning system?
    a. Dirt and moisture
    b. Carbon and silicon
    c. Dirt and dc voltage signals
    d. Gasoline and oil
    e. All of the above

20. Which of the following is an easy method used to check for refrigerant in the system?
    a. Use a sight glass.
    b. Loosen the system and watch for fluid to flow out.
    c. Apply an excess amount of pressure to get more cooling.
    d. All of the above
    e. None of the above

21. When working on an air conditioning system, a propane torch is used:
    a. To create excess pressure
    b. To check for pressure
    c. To check for R-12 leaks
    d. To solder steel piping
    e. To heat the evaporator during testing

22. A manifold gauge set is used on an air conditioning system to:
    a. Discharge the R-12
    b. Charge the R-12
    c. Evacuate the R-12
    d. All of the above
    e. None of the above

The following questions are similar in format to ASE (Automotive Service Excellence) test questions.

23. Technician A says that the most common problem with air compressors is contamination and lack of lubrication. Technician B says that the most common problem with air compressors is engine vibration. Who is right?
    a. A only
    b. B only
    c. Both A and B
    d. Neither A nor B

24. Technician A says that when the air conditioning system is open to the atmosphere, simply seal the hole and the system can be operated. Technician B says that an opening in the system means moisture will be absorbed into the system. Who is right?
    a. A only
    b. B only
    c. Both A and B
    d. Neither A nor B

25. A Suction Throttling Valve (STV) cannot keep the pressure in the evaporator to the recommended level. Technician A says the system is OK and can still be run. Technician B says that the evaporator will freeze up. Who is right?
    a. A only
    b. B only
    c. Both A and B
    d. Neither A nor B

26. Small air bubbles are noticed in the sight glass on the air conditioning system. Technician A says this is normal. Technician B says that the compressor is overcharging the freon. Who is right?
    a. A only
    b. B only
    c. Both A and B
    d. Neither A nor B

Essay

27. Describe the process of condensation.

28. What happens to the boiling point of a fluid when the pressure increases and/or decreases?

29. Describe the purpose and operation of the expansion valve on an air conditioning system.

30. What is the difference between the high and low sides of an air conditioning system?

31. What is the purpose of an accumulator?

32. Describe the internal operation of how pressure is produced in the air conditioning compressor.

CHAPTER 33

Heating and Ventilation Systems

INTRODUCTION

There are several temperature controls used on the interior of the automobile to maintain a comfortable climate. This chapter discusses the ventilation and heating systems that are used inside the passenger compartment.

OBJECTIVES

After reading this chapter, you will be able to:
■ Identify the purpose of the heating and ventilation systems used in the automobile.
■ Identify common parts used on heating and ventilation systems.
■ Analyze how the passenger ventilation system is designed in the automobile.
■ Compare vacuum and mechanical controls on heating and ventilation systems in the automobile.
■ Identify various diagnosis and service steps used on heating and ventilation systems on automobiles.

CHAPTER HIGHLIGHTS

33.1 HEATING AND VENTILATION SYSTEMS

Purpose of Heating and Ventilation Systems

Heating and ventilation systems are used on the automobile to keep the passenger compartment at a comfortable temperature. Several conditions make the passenger compartment either too hot or too cold. During the winter, cold outdoor temperatures may make the compartment too cold. During the summer, high ambient temperatures may make the compartment too warm. Thus both heating and ventilation are needed to keep the driver comfortable. In addition, as discussed in an earlier chapter, air conditioning may also be used.

Flow-through Ventilation

There are many systems used to heat and vent the passenger compartment. One type is called the flow-through ventilation system. A supply of outside air, which is called *ram air*, flows into the car when it is moving. When the car is not moving, a steady flow of outside air can be produced from the heater fan. *Figure 33–1* shows an example of the flow-through ventilation system. In operation, ram air is forced through an inlet grille. The pressurized air then circulates throughout the passenger and trunk compartment. From there the air is forced outside of the vehicle through an exhaust area.

On certain vehicles, air is admitted by opening or closing two vent knobs under the dashboard. The left knob controls air through the left inlet. The right knob controls air through the right inlet. The air is still considered ram air and is circulated through the passenger compartment.

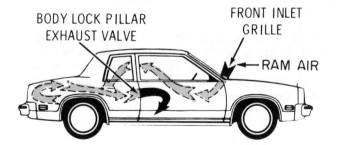

FIGURE 33–1. A flow-through ventilation system is used on some vehicles. Ram air is forced into the inlet grill and sent throughout the passenger and trunk compartment. The air then flows out of the vehicle through exhaust areas. *(Courtesy of Oldsmobile Division, General Motors Corporation)*

Fan Ventilation

Rather than using ram air (especially if the vehicle is stopped), a ventilation fan can be used. The fan is located in the dashboard. It can be accessible from under the dashboard or from inside the engine compartment. *Figure 33–2* shows a typical ventilator assembly. A blower assembly is attached to the motor shaft. The entire unit is placed inside the blower housing. As the *squirrel cage blower* rotates, it produces a strong suction on the intake. A pressure is also created on the output. When the fan motor is energized by using the temperature controls on the dashboard, air is moved through the passenger compartment.

CAR CLINIC ▮▮▮▮▮▮▮▮

PROBLEM: CAR HEATER DOESN'T WORK

A car doesn't heat up in the passenger compartment, especially when it's very cold outside. When the engine is at operating temperature, heat is available. How could the passenger compartment get warmer, especially on short drives?

SOLUTION:

The heater core is on the by-pass circuit around the thermostat when the thermostat is closed. First make sure that the thermostat is operating correctly and is the correct type for the vehicle. Then check to see if the heater core and the hoses to the core are hot before the radiator gets hot. They should be hot before the radiator. Also check to see if there are air bubbles in the coolant. On some vehicles, there may also be a shut-off valve to the heater core. Check the valve for correct operation.

Heater Core

All ventilation and heating systems use a *heater core* to increase the temperature within the passenger compartment. *Figure 33–3* shows an example of a heater core. Hot fluid flowing through the cooling system is tapped off and sent to the heater core. The heater core is much like a small radiator. It is considered a liquid-to-air heat exchanger. As warm or hot water is circulated through the core, air can be heated as it flows through the core fins.

Figure 33–4 shows the flow of coolant from the engine cooling system, through the heater core, and back to the cooling system. Near the top of the engine, just before the thermostat, a coolant line taps off coolant and sends it to the heater core. From the heater core, the coolant is sent back to the suction side of the cooling system. With this system, the thermostat does not have to be open to get heat to the heater core. Different manufacturers may use other circuit connections.

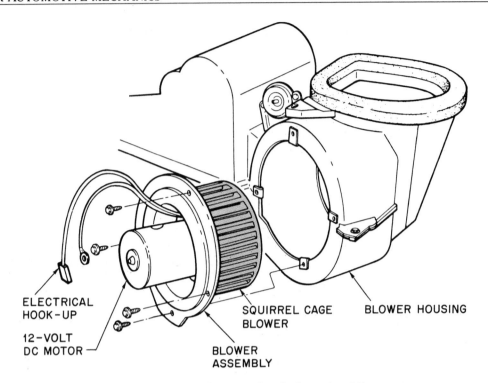

ELECTRICAL
HOOK-UP

12-VOLT
DC MOTOR

SQUIRREL CAGE
BLOWER

BLOWER HOUSING

BLOWER
ASSEMBLY

FIGURE 33-2. A fan is used to help move the air throughout the passenger compartment. It is called a squirrel cage fan. *(Courtesy of Motor Publications, Auto Repair Manual, 1981–1987)*

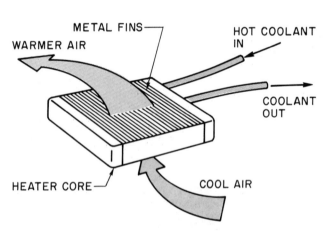

METAL FINS

WARMER AIR

HOT COOLANT IN

COOLANT OUT

HEATER CORE

COOL AIR

FIGURE 33-3. A heater core is used to increase the temperature of the air inside the passenger compartment.

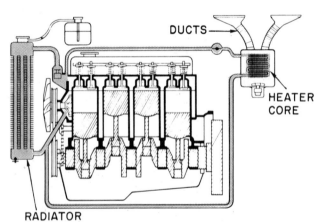

DUCTS

HEATER CORE

RADIATOR

FIGURE 33-4. The coolant for the heater core is sent from the upper portion of the engine, through the heater core, and back to the suction side of the engine coolant system. *(Courtesy of Chevrolet Motor Division, General Motors Corporation)*

Mechanical Duct Controls

One method used to control ventilation and heating systems is by using air ducts with small doors that direct the flow of air. The doors can be controlled either mechanically or by vacuum. Mechanical control of the doors is accomplished by moving the controls on the dashboard. As the control is moved, a cable attached to the control moves the duct door. *Figure 33–5* shows an example of two control cables. One

is used to operate an air duct door for the temperature control. One is used to operate an air duct door for the *defroster*. The cable is made of a strong steel wire that is wrapped within a flexible tube. As the control assembly knob is moved, the steel wire inside the flexible tube moves, which causes the position of the air duct door to change.

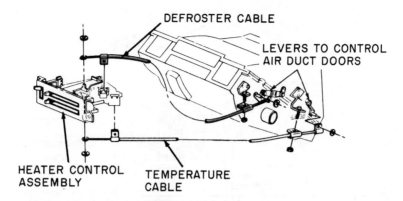

FIGURE 33–5. The ventilation and heating system is controlled by opening and closing air duct doors. These doors can be operated by mechanical cables. Here a temperature and a defroster cable are used to control air flow. As the levers are moved on the control assembly, air duct doors open and close to direct air correctly. *(Courtesy of Oldsmobile Division, General Motors Corporation)*

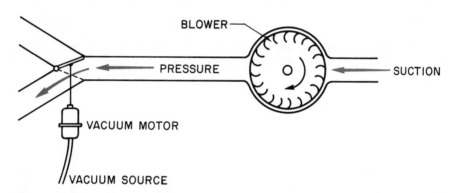

FIGURE 33–6. A vacuum motor can be used to control the operation of air duct doors. When a vacuum is applied or removed from the vacuum motor, a diaphragm moves. This motion is transmitted by a small rod to the air duct door.

Vacuum Duct Controls

A second method used to operate the air duct doors is to use a vacuum-operated motor. *Figure 33–6* shows an example of how the *vacuum motor* operates. The vacuum motor operates much like a solenoid. Instead of using electricity, vacuum is used. When vacuum is applied to the motor, it causes a small diaphragm inside the motor to move. This motion is transmitted by a rod to the air duct door. The air duct door is then moved from one position to another. In some vehicles, there may be several vacuum motors used to operate air duct doors.

Air Outlets in Dashboard

Vehicles today use a variety of ducts and passageways to get the air into the passenger compartment. Each type of vehicle is different. A typical air flow pattern is shown in *Figure 33–7*. Depending upon the temperature controls, air can be directed to the feet, front windshield, center of the passenger compartment, or to the side windows. The temperature controls on the dashboard operate to open and close doors, which direct air to the correct location.

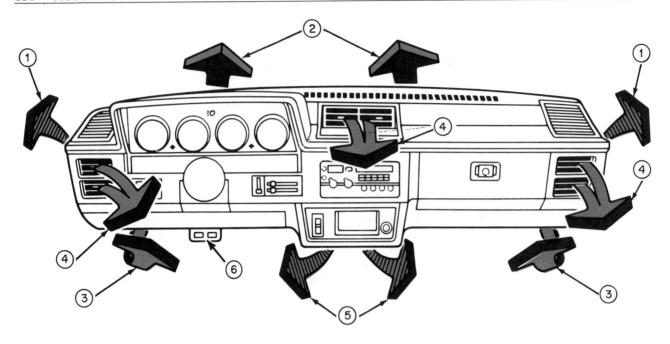

1. SIDE WINDOW DEFOG OUTLETS
2. DEFROST OUTLETS
3. RAM AIR OUTLETS

4. POWER VENT OUTLETS
5. POWER VENT HEAT OUTLETS
6. RAM AIR VENT CONTROLS

FIGURE 33–7. Air can be directed to many locations through the dashboard. Air duct doors, operated both mechanically and by vacuum, open and close to direct air to the correct location. *(Courtesy of Oldsmobile Division, General Motors Corporation)*

Heater Controls

There are numerous types and styles of heater controls. *Figure 33–8* shows an example of a simple dashboard control. Although many controls are much more complex, the functions that can usually be obtained include fan speeds, defrost, heating, venting, and/or air conditioning. The fan control lever is used to control the speed of the fan at different air flow rates. The temperature control lever is used to regulate the temperature of the air sent to the passenger compartment. The *selector control lever* is used to change from heating, to venting, to defrosting.

Mechanical Control Heating Systems

A mechanical heating system is shown in *Figure 33–9*. Air is first pressurized by the blower. The air is then sent either through the heater core, bypassed, or mixed. The temperature door is positioned so that a mix of air can be obtained. In the upward-most position, air flow through the

heater core is blocked. The air is not heated in this mode. When the temperature control knob on the dashboard is adjusted, it moves the temperature door so that more and more air flows through the heater core. The air temperature then increases accordingly. A second air duct door is used to stop all air flow into the passenger compartment. This occurs when the selector control lever is moved to the "off" position on the dashboard. The defroster door is also moved by the selector control lever. In the defrost position, the door is positioned so that air moves to the upper part of the dashboard against the windshield. Otherwise, the air flow is directed to the heat outlets in the center or lower portion of the dashboard.

Vacuum Control Heating Systems

Another method used to control the heating and ventilation system is by using vacuum. *Figure 33–10* shows a vacuum system circuit in the heating mode. There are several additional parts included for correct operation. On

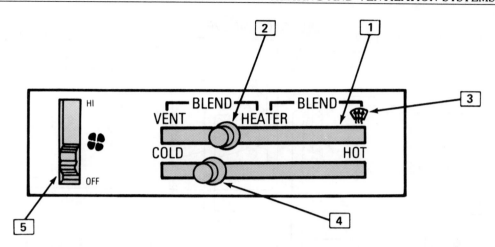

1   THIS POSITION ALLOWS OUTSIDE AIR FLOW TO FLOOR OUTLET. (ADDITIONAL VENTED AIR CAN BE DISTRIBUTED INSIDE CAR BY OPERATING VENT KNOBS.

2   POSITION OF THIS SYSTEM SELECTOR LEVER DETERMINES AIR FLOW FROM FLOOR, INSTRUMENT PANEL OR WINDSHIELD OUTLET—IN "HEATER," FLOW IS ABOUT 80% TO FLOOR AND 20% TO WINDSHIELD OUTLETS (AND SIDE WINDOW DEFOGGERS).

3   THIS POSITION ALLOWS ABOUT 80% AIR FLOW TO WINDSHIELD AND 20% TO FLOOR.

4   TEMPERATURE LEVER POSITION WILL REGULATE OUTLET AIR TEMPERATURE BY BLENDING THE INCOMING OUTSIDE AIR THROUGH/AROUND THE HEATER CORE.

5   THE FAN CONTROL LEVER (OFF - HI) PROVIDES SPEED CONTROL OF THE FAN.

FIGURE 33–8. There are many types of heater controls. There is typically a fan control lever, a temperature control lever, and a selector control lever. *(Courtesy of Oldsmobile Division, General Motors Corporation)*

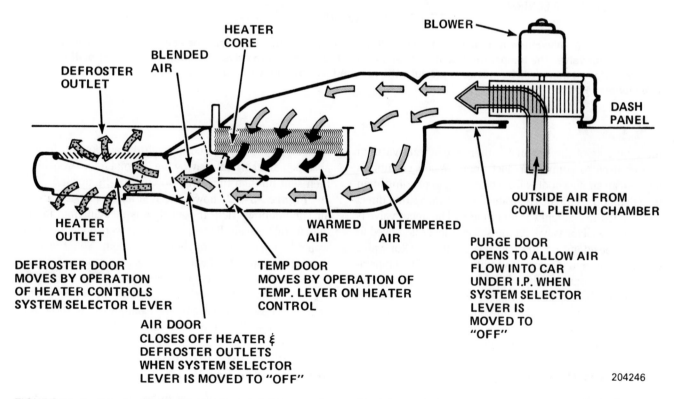

FIGURE 33–9. A typical heating system and air flow are shown. Air from outside is pressurized by the blower. Depending upon the position of the air duct doors, different temperatures can be obtained. *(Courtesy of Oldsmobile Division, General Motors Corporation)*

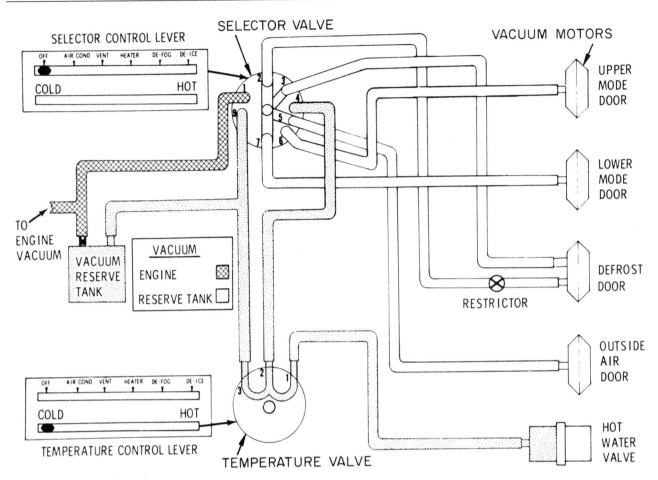

FIGURE 33–10. A vacuum system for controlling air flow is used on some vehicles. The selector control lever controls the selector valve. The temperature control lever controls the temperature valve. Depending upon the position of the valve, different doors will be moved to produce the required air flow and venting. *(Courtesy of United Delco, Division of General Motors Corporation)*

the right side of the schematic, there are several vacuum motors used to control air duct doors. When vacuum is applied, an air duct door will open or close, depending upon the condition required. There are also two valves that direct vacuum to the different vacuum motors. The upper valve is called the selector valve. As the selector lever is moved to different positions on the dashboard, the selector valve directs vacuum to the correct vacuum motor. The temperature valve is controlled by the position of the temperature control lever on the dashboard. A *reserve vacuum tank* also provides extra vacuum when the engine is not running. In addition, there is a hot water valve that opens or closes, allowing engine coolant to circulate through the heater core.

Air Flow with Air Conditioning

If the vehicle has air conditioning, an additional evaporator is placed in the air ducts. See **Figure 33–11**. The evaporator is used to create cool air. As warm passenger compartment air is passed through the evaporator, heat is absorbed and carried away by the internal freon. Cool air is then sent through the remaining part of the ducting.

Figure 33–12 shows a second air flow system using air conditioning. Here the air flow is recirculating, rather than using fresh air. If the fresh air duct were open, fresh air would be circulated throughout the system. Note the location of the evaporator, heater core, vacuum motors, and fan. The exact design of each air flow system is determined by the vehicle manufacturer, the year, and the model of the vehicle. These two designs are examples that show the variation of air flow and supporting controls.

Computerized Automatic Temperature Control

Today certain vehicle manufacturers are using computers to accurately control the air temperature in the passenger compartment. This type of system offers more choices on air

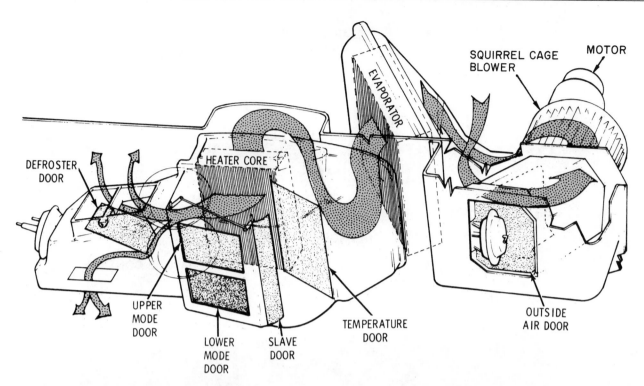

FIGURE 33–11. When an air conditioner is used, an additional evaporator is placed in the air flow. Warm air passing through the evaporator gives up its heat to the internal freon. The cooler air is then sent through the ducting into the passenger compartment. *(Courtesy of United Delco, Division of General Motors Corporation)*

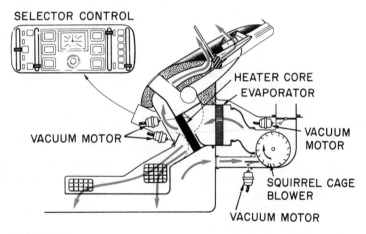

FIGURE 33–12. A typical air flow through a vehicle that has an evaporator used for air conditioning. *(Courtesy of Peugeot Motors of America)*

flow and outlets as compared to previous systems. It operates on a microprocessor with a memory. The system is designed to regulate the temperature of air in the passenger compartment to the desired temperature level. The micro-

processor measures the interior temperature and makes adjustments every seven seconds. *Figure 33–13* shows a *computerized automatic temperature control*, including the major parts and their location.

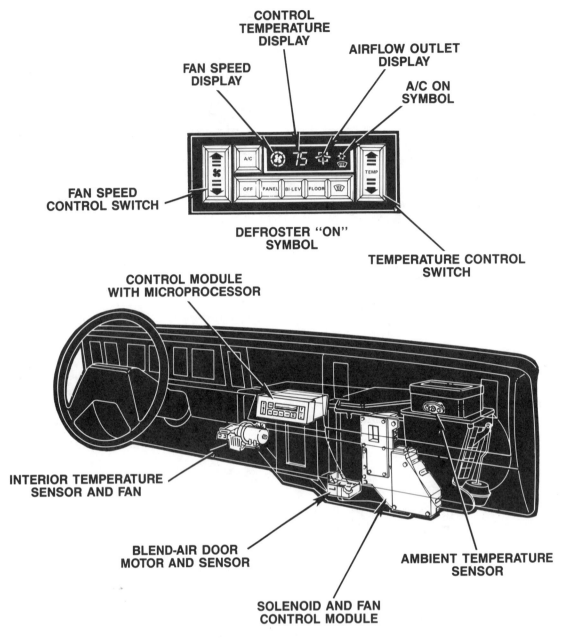

FIGURE 33–13. Certain vehicles are using computers to control the ventilation and heating system. The computer senses the interior air temperature and adjusts the temperature every seven seconds. *(Courtesy of Chrysler Corporation)*

CAR CLINIC

PROBLEM: FRONT END SHIMMY

What are the causes of front end shimmy? A car with 77,000 miles on it seems to have a bad shimmy. What should be checked first?

SOLUTION:

Shimmy can be caused by several problems, including:

1. Not enough caster.
2. Toe-in out of specifications.
3. Loose steering linkage parts.
4. Too much play in the steering gear.
5. Bad shocks.
6. Bad suspension parts.
7. A combination of any of the above.

These items should be checked carefully to determine the cause of shimmy on the front wheels.

ENGINE BALANCE SHAFTS

Smooth engine performance can be produced by using two counter rotating balance shafts. These are designed into the bottom of the engine block. Inter-connected by gears, and driven by a short chain from the crankshaft, the balance shafts turn at two times the engine speed. This moving mass helps to offset the reciprocating mass of the pistons and connecting rods to achieve the desired balancing effect.

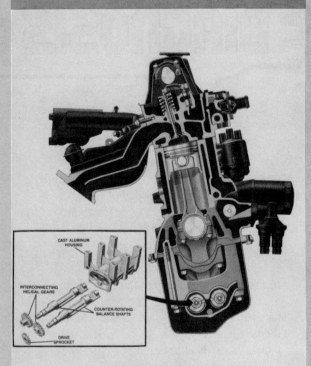

**FOUR-CYLINDER 2.5L ENGINE
WITH BALANCE SHAFTS**

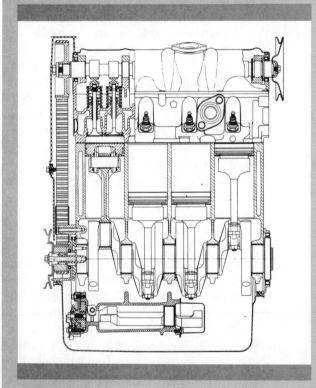

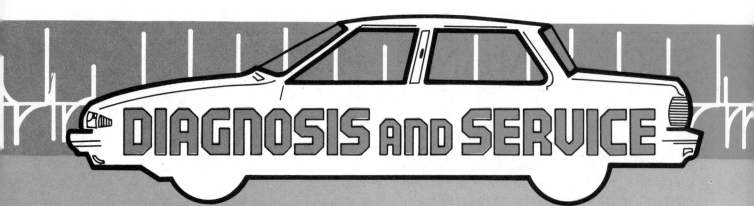

DIAGNOSIS AND SERVICE

1. If the fan blower is inoperative, check the blower fuse. If the fuse is not bad, check for an open circuit between the ignition switch and the blower motor. If the circuitry appears to be correct, check the blower fan switch for damage.

2. Water on the front floor mat may indicate a leaking heater core. Check the liquid to see if antifreeze is mixed in with it. It should be oily because of the antifreeze. If so, check all hose connections to the heater core for leaks. Also, a pressure check on the cooling system may detect a small leak in the heater core. If the heater core is leaking, it must be removed and repaired or replaced. Follow the service manual for the correct removal and replacement procedure.

3. If the heater has a gurgle, whine, or "swish," this may indicate that there is air mixed with the coolant in the cooling system. Check the engine coolant level in the radiator. Also check for obstructions in the heater core or hoses. Repair and/or replace where necessary.

4. On some vehicles, the temperature control cable between the dash adjustment and the temperature door may need adjustment. This can be done by adjusting the cable length shorter or longer. Follow the manufacturer's recommended procedure for adjusting the cable length. There should be a uniform effort from full cold to full hot. In addition, there should be an audible temperature door stop contact upon reaching the end position.

5. *Insufficient heating* in the passenger compartment can be caused by a variety of malfunctions. Refer to **Figure 33-14** to identify the cause and possible correction. Repair and replace as necessary.

6. Heating and ventilation systems that use vacuum hoses for control should be checked by using a vacuum tester. The vacuum tester is used to place a vacuum on the system to observe sealing, leaks, and if the component is operating correctly.

7. An engine that has a damaged cooling system with internal rust and other contaminants will usually develop problems in the heating system. For example, if silicon and calcium deposits form in the radiator, they usually form in the heater core as well. If this is the case, there will be poor heat transfer into the passenger compartment. The heater core will have to be removed and cleaned before it will operate correctly.

8. If there is an inadequate removal of fog or ice from the windshield, there may be several malfunctions. **Figure 33-15** shows an example of some of the common causes and corrections. Service as necessary.

INSUFFICIENT HEATING

Cause	Correction
Slow warming in car.	Incorrect operation of controls. Advise operator of proper operation of heater controls. Explain operation of vents and controls. Low coolant level. Check control cable and blower operation.
Objectionable engine or exhaust fumes in car.	Check for seal between engine compartment and plenum. Check for proper sealing between air inlet duct assembly and cowl. Locate and seal any other air leaks.
Cold drafts on floor.	Check operation and adjustment of vent cables. Advise operator of proper operation of heater system. Advise operator to use blower to force air to rear seat area. Check to be sure front floor mat is under floor mat retainer at cowl.
Insufficient heat to rear seat.	Obstruction on floor, possibly wrinkled or torn insulator material between front seat and floor. Advise operator to use HI blower speed.
Low engine coolant level - drop in heater air temperature at all blower speeds.	Check radiator and cooling system for leaks, correct and fill to proper level. Run engine to clear any air lock.
Failure of engine cooling system to warm up.	Check engine thermostat; replace if required. Check coolant level.
Kinked heater hoses.	Remove kink or replace hose.
Foreign material obstructing water flow through heater core.	Remove foreign material if possible, otherwise, replace core - can usually be heard as squishing noise in core.
Temperature door (valve) improperly adjusted. Air doors do not operate.	Adjust cable. Check installation and/or adjustment of air control or air-defrost cable.

FIGURE 33–14. Insufficient heating can be caused by many damaged or nonoperational parts. *(Courtesy of Oldsmobile Division, General Motors Corporation)*

INADEQUATE REMOVAL OF FOG OR ICE

Cause	Correction
Air door does not open. Defroster door does not open fully.	Check cable operation.
Air door does not open.	Check installation and/or adjustment of air control or air-defrost cable.
Temperature door does not open.	Check and adjust temperature control cable if necessary.
Obstructions in defroster outlets at windshield.	Remove obstruction. Look for and repair loose instrument panel pad cover at defroster outlet.
Damaged defroster outlets.	Reshape outlet flange with pliers. The outlet should have a uniform opening.
Blower motor not connected.	Connect wire. Check ground.
Inoperative blower motor.	Check heater fuse and wiring. Replace motor if necessary. See ETM for blower motor diagnosis.
Inoperative blower motor switch.	Replace switch if necessary.

FIGURE 33–15. Inadequate removal of fog and ice from the windshield can be caused by several malfunctions. *(Courtesy of Oldsmobile Division, General Motors Corporation)*

SUMMARY

Heating and ventilation systems are used on automobiles to keep the passenger compartment comfortable during all weather conditions. There are many designs. Most systems are built to heat, cool, ventilate the passenger compartment, and defrost the windows.

Ventilation inside the passenger compartment is done by several methods. The flow-through ventilation system uses ram air from the forward motion of the car to force air into the passenger area. Blower fans are used when the vehicle is not moving. The blower fan used is typically a squirrel cage design. It is driven by a small dc motor and placed inside a housing under the dashboard.

A heater core is used to get heat into the car. Coolant is tapped off of the cooling system and sent through hoses to the heater core. The heater core is a liquid-to-air heat exchanger. Air passing over the fins picks up the warm air and sends it to the passenger compartment.

Several types of controls are used to direct the air to the proper location. Mechanical duct controls use a steel wire attached to the dashboard controls. As the dashboard controls are moved, a small air duct door is opened or closed to direct air to the proper location. Vacuum duct controls are also used. A vacuum motor (diaphragm inside a housing) is operated by engine vacuum. When vacuum is applied, the diaphragm moves, causing the air duct door to move as well.

Depending upon the vehicle manufacturer, style, and year, air outlets are placed in various positions on the dashboard. When air duct doors are opened or closed, air is directed to the proper locations.

Heater controls are used to create the right temperature of air entering the passenger compartment. Normally an air duct door opens or closes to allow more or less air through the heater core.

Complete systems of heating and ventilation may use a combination of controls. Although each system is different, the basic operational principles remain the same. The selector control lever is used to select the correct mode of operation, defrost, heating, venting, and so on. The temperature control is used to mix the correct amount of heated air with nonheated air. Both mechanical and vacuum controls can be used.

When a heating and ventilation system is used with an air conditioner, an additional evaporator is used in the air flow. As warm passenger compartment air flows through the evaporator, heat is absorbed into the freon and exhausted at the condenser. Cool air (air lacking thermal energy) is then sent into the passenger compartment.

With the increased use of computers on vehicles today, more precise temperature control can be obtained. The computerized automatic temperature control monitors air temperatures inside and outside the vehicle. Based upon these temperatures and the operator's setting, the air temperature is adjusted every seven seconds. An accurate air temperature can thus be easily obtained by using the computer.

There are several diagnosis procedures and service tips that are important for the heating and ventilation system. The most common problems include having a leaky heater core, misadjusted controls, and a faulty cooling system. Check the manufacturer's suggested diagnosis procedure when servicing the heating and ventilation system.

TERMS TO KNOW

Can you explain each of the following terms? Review the chapter until you can use each term correctly.

Ram air	Vacuum motor
Squirrel cage blower	Selector control lever
Heater core	Reserve vacuum tank
Air ducts	Computerized automatic
Defroster	temperature control

REVIEW QUESTIONS

Multiple Choice

1. Many vehicles today are ventilated by using:
   a. The heater core
   b. Flow-through or ram ventilation
   c. The evaporator
   d. Engine coolant
   e. None of the above

2. The fan motor and blower on a typical heating and ventilation system uses a/an:
   a. Squirrel cage blower
   b. Heater core motor
   c. Evaporator to heat the air
   d. Positive displacement cooling pump
   e. Belt-driven pump

3. Heat is put into the air going into the passenger compartment at the:
   a. Evaporator
   b. Vacuum motor
   c. Selector control lever
   d. Heater core
   e. None of the above

4. The actual thermal energy or heat put into the passenger compartment comes from the:
   a. Electrical wires heating up
   b. Evaporator
   c. Coolant from the cooling system
   d. Vacuum motor
   e. Squirrel cage blower

5. To get different temperatures from the temperature control lever on the dashboard:
   a. Cool air is mixed with warm air with through the heater core
   b. The evaporator pressure drops
   c. The electricity on the coils decreases
   d. The selector control lever is put in the "off" position
   e. All of the above

6. The air duct doors can be operated from/by:
   a. The vacuum motors
   b. Mechanical cables
   c. The control levers on the dashboard
   d. All of the above
   e. None of the above

7. The temperature control lever on the dashboard is used to:
   a. Mix air to the correct temperature
   b. Select venting or heating
   c. Change the temperature of the engine coolant
   d. All of the above
   e. None of the above

8. What type of components is/are used on the vacuum control heating system?
   a. Vacuum motors
   b. Selector valve
   c. Vacuum tank
   d. All of the above
   e. None of the above

9. When a car has air conditioning, which component is added into the air flow system?
   a. Heater core
   b. Evaporator
   c. Heater blower
   d. Computer control
   e. Vacuum motor

10. Which type of system monitors air temperature, both inside and outside the vehicle, and makes an adjustment every seven seconds to keep the temperature at the correct setting?
    a. Evaporator system
    b. Computerized automatic temperature control
    c. Heater core automatic
    d. Microprocessor control system
    e. None of the above

11. If there is engine coolant leaking onto the front floor mats, this could be an indication of a:
    a. Leaky vacuum motor
    b. Poor sealing air duct door
    c. Leaky heater core
    d. Leaky control system
    e. All of the above

12. What would need to be done if the heated air was not hot enough and the temperature control lever didn't seem to work smoothly?
    a. Adjust the temperature control lever
    b. Readjust the position of the evaporator
    c. Readjust the position of the heater core
    d. Replace all vacuum motors
    e. Replace the fan motor and blower assembly

The following questions are similar in format to ASE (Automotive Service Excellence) test questions.

13. There is a gurgle or swishing sound coming from under the dashboard near the heating core. Technician A says that there is air in the cooling system. Technician B says that the dash controls are improperly adjusted. Who is right?
    a. A only
    b. B only
    c. Both A and B
    d. Neither A nor B

14. Antifreeze and water are found on the front seat car mat. Technician A says that the heater core may be leaking and should be repaired. Technician B says that the hose connections to the heater core may be leaking. Who is right?
    a. A only
    b. B only
    c. Both A and B
    d. Neither A nor B

15. There is insufficient heating in the passenger compartment. Technician A says that because the cooling system has silicon and calcium built up in it, this may also damage the heater core. Technician B says that the dash controls may need to be adjusted. Who is right?
    a. A only
    b. B only
    c. Both A and B
    d. Neither A nor B

16. The blower fan is inoperative. Technician A says to check the blower fan fuse. Technician B says to check the heater core for signs of being plugged. Who is right?
    a. A only
    b. B only
    c. Both A and B
    d. Neither A nor B

Essay

17. Describe the squirrel cage blower design.
18. What is the purpose of the vacuum motor on heating and ventilation systems?
19. What is the purpose of the reserve vacuum tank?
20. Describe the purpose and operation of the heater core.
21. Define the computerized automatic temperature control system.

CHAPTER 34

Auxiliary and Electrical Systems

INTRODUCTION

This chapter is about various auxiliary and electrical systems that are used on the automobile today. These include systems such as cruise controls, windshield washers, energy absorbers, and horn circuits. These systems are designed in a variety of ways. Different vehicle manufacturers may or may not have similar systems. Many of these systems rely on electrical circuits for their operation and analysis. This chapter is designed to analyze several auxiliary systems as well as to understand some of their basic circuitry.

OBJECTIVES

After reading this chapter, you will be able to:
■ Identify various wiring circuits, schematics, and their symbols.
■ Analyze several electrical circuits, including the headlights, defogger, power seats, and horn.
■ Define the parts and operation of windshield wiper systems.
■ Describe the operation of energy absorbers used on bumpers.
■ Analyze the parts and operation of standard and electronic cruise control systems.
■ Identify common troubleshooting, diagnosis, and service procedures for auxiliary and electrical systems.

CHAPTER HIGHLIGHTS

34.1 READING ELECTRICAL CIRCUITS
    A. Schematics
    B. Electrical Symbols
    C. Reading a Schematic
    D. Fuse Block
    E. Fusible Link

34.2 ELECTRICAL CIRCUITS
    A. Headlights
    B. Defogger
    C. Power Seats
    D. Horn

CHAPTER HIGHLIGHTS (CONTINUED)

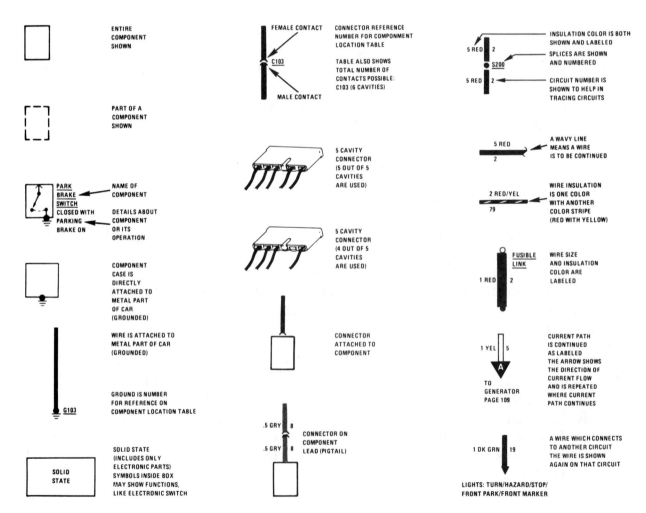

FIGURE 34–1. Various symbols are used to help identify electrical circuits. Becoming familiar with these symbols will help in identifying problems in the systems. (Courtesy of Oldsmobile Division, General Motors Corporation)

34.1 READING ELECTRICAL CIRCUITS

Schematics

Electrical schematics are used to troubleshoot the electrical circuits used on vehicles today. Schematics are used to subdivide the electrical system down into individual parts and circuits. Schematics only show the parts and how electrical current flows. The schematic does not show the actual position or physical appearance of the parts. For example, a four-foot wire is not shown differently than a two-inch wire. All parts are shown as simply as possible, with regard to function only.

Electrical Symbols

The automotive industry uses *electrical symbols* on the schematics to help identify the circuit operation. Not all manufacturers use the same symbols, but they may be similar in nature. *Figure 34–1* shows common symbols that are used by one manufacturer. Many of the symbols are designed to illustrate their meaning by the type of symbol. For example, connector symbols are shaped somewhat like an actual connector. Keep these symbols on hand to be able to read electrical schematics.

Several numbers and identifying characteristics are also shown on schematics. The color of the wire is represented by letters such as PNK (Pink), YEL (Yellow), BLU (Blue), PPL (Purple), ORN (Orange), GRY (Gray), DK GRN (Dark Green), and so on. The size of the wire (both metric and AWG, American Wire Gauge, sizes) and the component location (C 103) are also shown. When the technician is ready to match the schematic parts to the actual hardware, this number is referenced to a *component location table*, along with the schematic. The table tells the technician exactly where to find the component. *Figure 34–2* shows an example of a component location table.

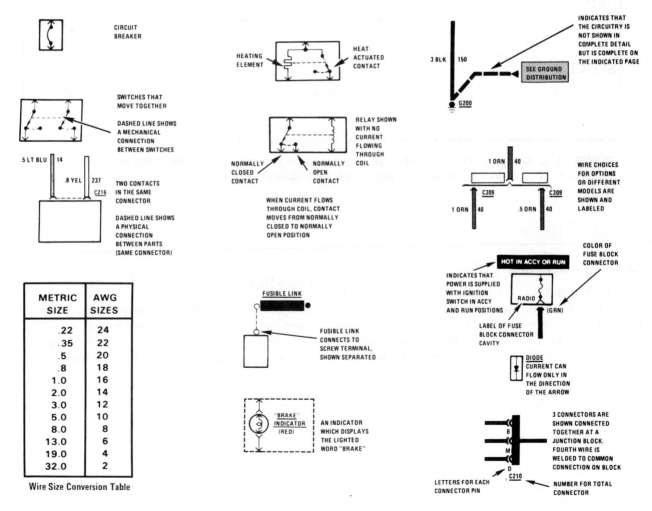

METRIC SIZE	AWG SIZES
.22	24
.35	22
.5	20
.8	18
1.0	16
2.0	14
3.0	12
5.0	10
8.0	8
13.0	6
19.0	4
32.0	2

Wire Size Conversion Table

FIGURE 34–1. (CONTINUED)

COMPONENT LOCATION		Page-Figure

COMPONENTS

Blower Motor	RH rear of engine compartment	201-4-A
Blower Resistors.	On blower housing	201-4-E
Fuse Block	Behind RH side of I/P	201-2-A
Radio Capacitor	Above blower motor.	201-4-E

CONNECTORS

C118 (1 cavity)	Next to blower motor.	201-8-B
C209 (2 cavities)	Behind center of I/P, near control head	201-10-C
C220 (3 cavities)	Behind center of I/P, near grommet	201-10-C
C241 (2 cavities)	Behind I/P, below control head	201-9-A

GROUNDS

G104	On EGR solenoid bracket	201-3-A
G106	RH rear of engine compartment	201-4-A

SPLICES

S106	Engine harness, above water pump	201-18-B

FIGURE 34–2. Component location tables tell the technician the physical location of the actual parts on the schematic. *(Courtesy of Oldsmobile Division, General Motors Corporation)*

Reading a Schematic

Figure 34–3 shows a typical schematic of the heater circuit. This schematic is used here only as an example of how to read it. The schematic is read from top to bottom. With the ignition switch in the "RUN" position, voltage is applied to the fuse block. From the fuse block, electricity flows into the blower switch. The blower switch sets the blower speed by adding resistors in series with the blower motor. In LO two resistors are connected through the YEL wire. In MED one resistor is connected through the BLU wire. In HI full voltage is applied directly to the blower motor through the ORN wire. When the radio is off, electricity flows directly to the blower motor. When the radio is on, electricity flows first through a radio capacitor. The radio capacitor reduces radio noise that is caused by the blower motor.

Fuse Block

Each circuit that is used on the automobile must be fused. Fuses are used so that if the circuit is overloaded, the fuse will melt and open that particular circuit. All fuses are usually placed on a *fuse block*. The fuse block can be located in several areas. On some older cars, the fuse block is located under the dashboard on the left side of the vehicle. On certain newer cars, the fuse block is located in the glove compartment or under the dashboard on the right side. The location depends upon the vehicle manufacturer. There are also several types of fuses that are used. *Figure 34–4* shows an example of several types of fuses used on vehicles today. The amperage (20A) is stamped on the fuse to determine its size.

HEATER

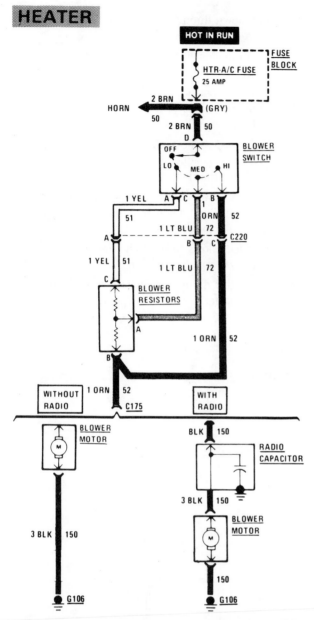

FIGURE 34–3. An example of a simple electrical schematic used to diagnose the heater circuit. Note how the symbols are used to help explain the circuit operation. (Courtesy of Oldsmobile Division, General Motors Corporation)

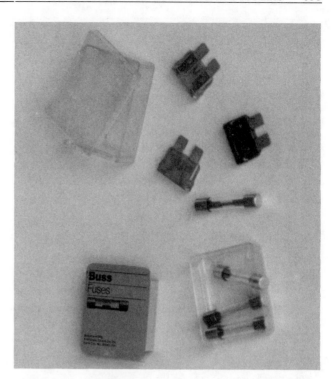

FIGURE 34–4. Different types of fuses can be used in an automobile. The type depends upon the manufacturer, the year, and the exact type of circuit.

CAR CLINIC

PROBLEM: CONVERTER TEMPERATURES

A customer has complained that the catalytic converter is running very hot. The temperature on the converter is about 800–900 degrees F. Is this normal or is something wrong with the engine?

SOLUTION:

Catalytic converters are designed to operate at this temperature and at times even hotter. The heat is needed to help reduce the NO_x emissions. However, if the temperature gets hotter and the catalytic converter glows red, there may be a problem. The first thing to check is a blockage in the catalytic converter. This will cause the temperature to increase significantly.

Fusible Link

On many electrical circuits, there is a *fusible link*. A fusible link is a type of circuit protector in which a special wire melts to open the circuit when the current is excessive. The fusible link acts like an in-line fuse made of wire with a special nonflammable insulation.

34.2 ELECTRICAL CIRCUITS

As the automobile has been developed over the years, more and more electrical circuits have been added to it. An example of some of the electrical circuits used include the clock, defogger, fog lamps, fuel injectors, gauges, headlights, ignition, cigar lighter, dome lights, power windows, horn, power seats, tachometer, tailgate release, and others. All of these circuits can be identified and analyzed by observing their circuit operation on a schematic. This section looks at several electrical circuits that are commonly used on vehicles today.

Headlights

Figure 34–5 shows a headlight circuit. A complete analysis of the headlight circuit can be obtained by tracing the electricity through the schematic. Starting from the top:

1. Electricity is available to the light switch all of the time.

2. The electricity first flows through the circuit breaker.

3. The light switch has three positions. These include OFF, PARK, and HEAD.

4. When the light switch is in the HEAD position, cur-

rent flows through the switch to the headlight *dimmer switch*.

5. The dimmer switch can be in one of two positions: LO or HI.

6. In the LO position, the current flows through the dimmer switch to terminal D. This wire is identified as 1 TAN.

7. The electricity flows through 1 TAN to each of the dual beam headlights. One is for the left side and one is for the right side of the vehicle.

8. There are two electrical circuits inside each dual beam headlight. One circuit is for the high beam and one is for the low beam.

9. Electricity flows through the low beam circuit to ground, completing the circuit.

10. When the dimmer switch is positioned to HI, several other circuits also operate.

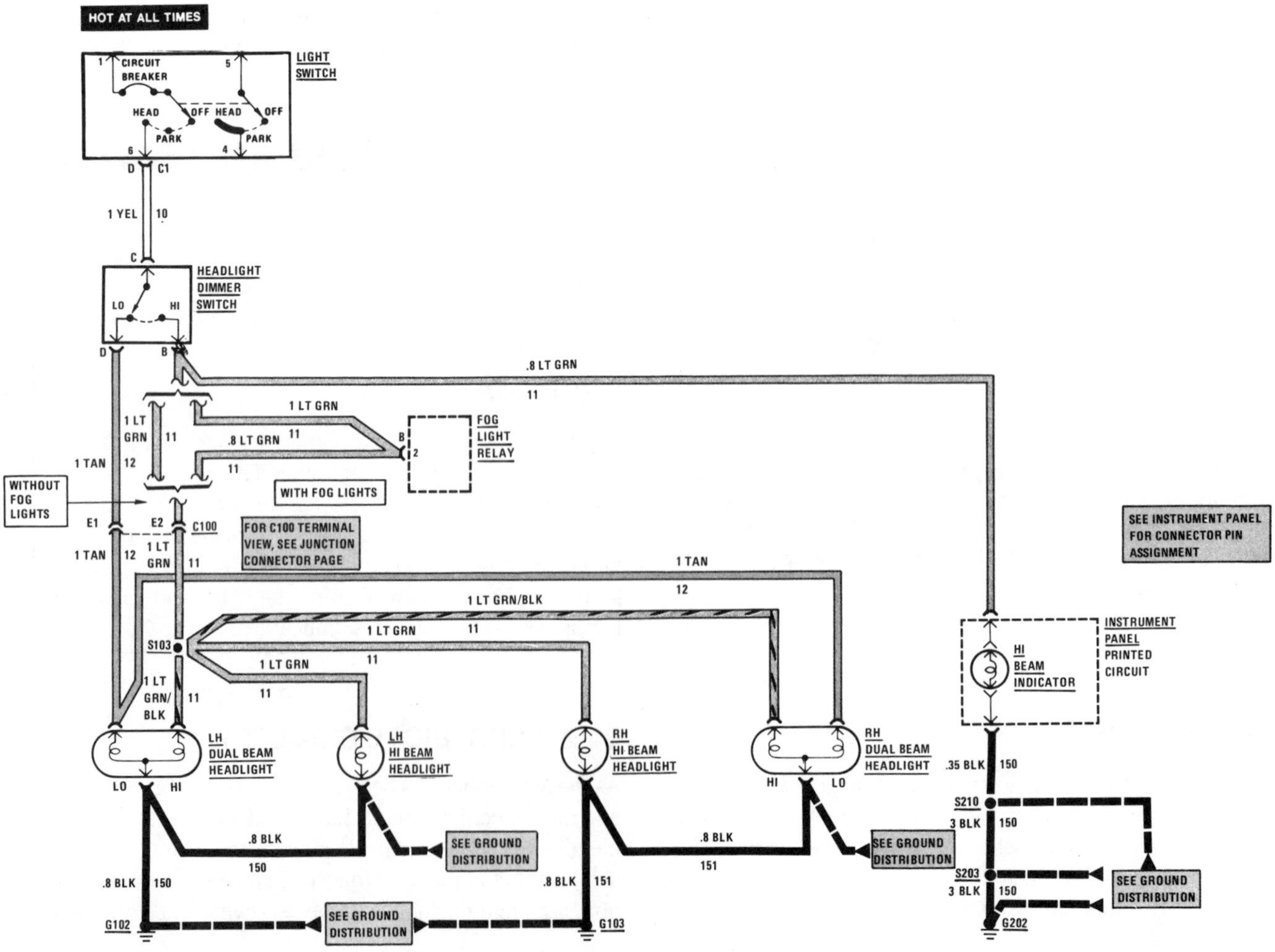

FIGURE 34–5. The headlight circuit for a vehicle. (Courtesy of Oldsmobile Division, General Motors Corporation)

11. One circuit sends electricity to the instrument panel printed circuit to the HI beam indicator light.

12. From point B on the dimmer switch, electricity can flow through a fog light relay or directly to the HI beam circuit in each headlight.

13. After passing through the headlight, the electricity then returns to the battery through ground.

14. Electricity also flows from connector (S103) to a second high beam headlight. Thus when the dim-

mer switch is on HI beam, both front headlights are on.

Defogger

Figure 34–6 shows an example of a defogger electrical circuit. The defogger operates when voltage is applied to the rear window wires. The wires are on the inside surface of the glass. When current flows through them, the wires heat the window to remove fog and ice from the glass.

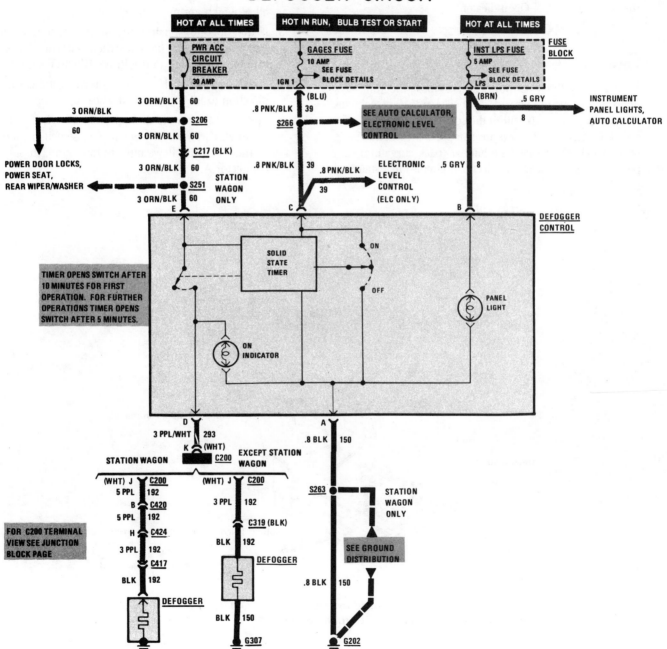

FIGURE 34–6. The defogger is used to remove ice and fog from the rear windows. An electrical wire placed on the inside of the window heats up and removes the ice and fog. *(Courtesy of Oldsmobile Division, General Motors Corporation)*

When the defogger control ON-OFF switch is moved to ON, the defogger control timer is turned on. The defogger control contacts close and voltage is applied to the defogger and the ON indicator. When the instrument panel light circuit provides power, voltage is also applied through the GRY wire to the defogger control panel light.

After the defogger control ON-OFF switch is released, the defogger control timer holds the defogger control contacts closed for 10 minutes for the first operation. The timer holds the contacts closed for 5 minutes for further operation. When the defogger control timer completes its cycle, the contacts open and voltage is removed from the defogger and the defogger control ON indicator.

Power Seats

Many vehicles today use power seats to move the front seats upward, downward, forward, and backward. Many vehicles also use support mats within the seat to fit its shape to the driver. *Figure 34–7* shows an example of the internal parts of a typical power seat. The lumbar (back curvature) supports are also shown. The electrical circuit for a standard power seat system can best be described by referring to *Figure 34–8*. This circuit shows several important parts. The three motors on the bottom are used to move the seat in a specific direction. The motors can run forward or backward, depending upon the direction of current through the windings. The switches for the power seats are in the middle of the schematic. The schematic can be traced from the top to the bottom as follows:

1. Current is available at the circuit breaker at all times.

2. Electricity flows to the upper wire in the LH Power Seat Switch. The current is available at each switch, all of which are normally closed.

3. If the operator pushes the spring-loaded REAR HEIGHT UP switch (far left switch), electricity flows through the switch to the REAR HEIGHT MOTOR.

4. Electricity causes the motor to turn in the correct direction to lift the rear of the seat upward.

5. Electricity then flows through the BLU wire, to the REAR HEIGHT DOWN switch. This switch is open, so the electricity flows directly to ground, through connector A.

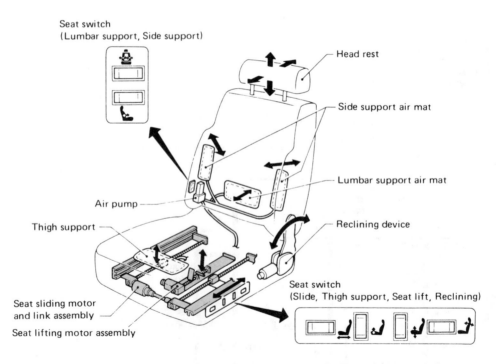

FIGURE 34–7. Power seats are used to adjust the seat and its cushions to best fit the driver. *(Courtesy of Nissan Motor Corporation in USA)*

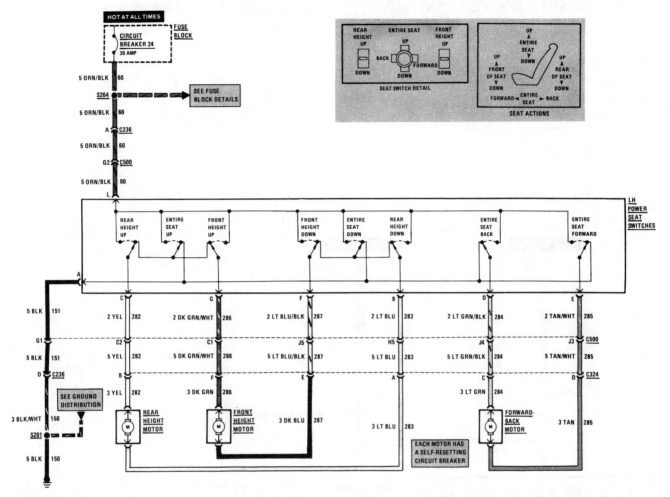

FIGURE 34–8. A typical power seat electrical circuit. *(Courtesy of Oldsmobile Division, General Motors Corporation)*

6. To get the seat to go down, electricity flows in the opposite direction in the motor. The REAR HEIGHT DOWN switch must be closed and the REAR HEIGHT UP switch must be open. This is done when the operator pushes the switch down.

When the other switches are closed or opened, the electricity flows to the correct motor to adjust the seat accordingly.

Horn

The horn is a simple circuit used by the operator of the vehicle to alarm other drivers of danger. The horn circuit is shown in *Figure 34–9*. It consists of several major components. These include two horns, a horn switch, a *horn*

brush/slip ring, and a horn relay. The circuit operates as follows:

1. Electricity is sent from the battery directly to the fuse block and through the fuse.

2. Electricity now flows through the ORN wire to the horn relay.

3. At this point, current wants to flow through the horn brush/slip ring assembly. This assembly is used to keep the horn switch in contact with the horn relay when the steering wheel is turned.

4. When the horn switch is pressed, electrical current flows through the left side of the horn relay.

5. As the horn relay (solenoid) is energized, it closes the horn relay switch, causing the horns to operate.

34.3 WINDSHIELD WIPER SYSTEMS

There are several types of windshield wiper systems. Both rear and front window systems are commonly used. Common components usually include the motor, linkage mechanism, switch, electronic logic circuits, and a washer system.

Windshield Wiper Motor

The motor in a windshield wiper system uses a permanent magnet (PM) type motor. Most front wiper motors have two speeds so that the wiper blade speed can be adjusted. In addition, some vehicles use a variable timer and speed control so the operator can control the wipers more precisely.

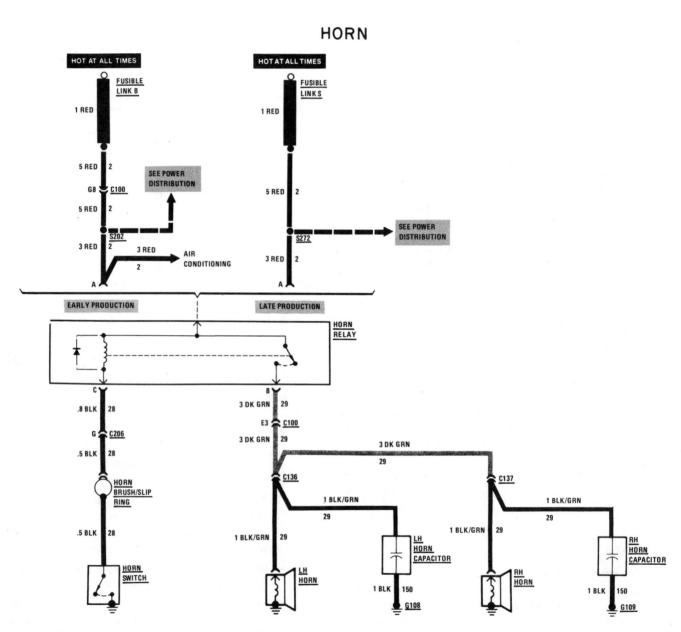

FIGURE 34–9. A horn circuit is used in all automobiles. Two horns, a horn switch, and a horn relay make up the circuit. *(Courtesy of Oldsmobile Division, General Motors Corporation)*

GATHERING TECHNICAL INFORMATION

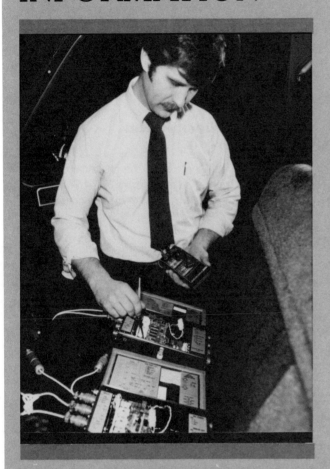

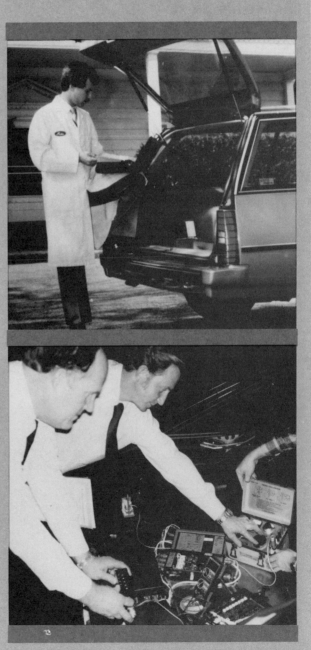

In order to improve the design of vehicles, data must be gathered from actual road experience. Many customers have consented to having their cars fitted with miniaturized electronic data-gathering devices such as strain gages. Strain gages are electromechanical devices that monitor the frequency and magnitude of stresses on the parts to which they are attached. They might be set to record such things as peak loads, speed and acceleration patterns, and frequency of accessory usage. The detectors record signals and store the data in a memory module. After a period of time, the module is removed and evaluated. (Courtesy of General Motors Proving Grounds)

Figure 34–10 shows a typical motor assembly. It includes several components. The motor produces the rotational motion needed. The washer pump is located in the housing for pumping a water spray onto the windshield. An internal mechanism is used to change the rotational motion of the motor to an oscillation motion, which is needed for the wiper blades. The internal mechanism for changing the motion is shown in *Figure 34–11*. As the center motor rotates, the output is an oscillation motion.

Fluid Washer Nozzle

A fluid washer nozzle is used on some cars for the front windshield washer system. The system consists of a fluid container, pump, fluid hoses and pipes, and nozzles or jets. *Figure 34–12* shows a washer system. When the operator turns on the washer, the small pump forces fluid through the hoses and pipes to the nozzle or jet by the windshield. The pump can be located either at the fluid container or wiper motor.

Windshield Wiper Linkage

Several arms and pivot shafts make up the linkage used to transmit the oscillation motion at the motor to the windshield wipers. *Figure 34–13* shows an example of the

linkage. As the wiper motor oscillates, arm A moves from left to right. This moves arm B as well. As arm B moves, it causes the two pivot points to oscillate. The windshield wipers are connected to the two pivot points.

Wiper Arm and Blade

The wiper arms and blades are attached directly to the two pivot points operated by the linkage and motor. The wiper arm transmits the oscillation motion to the wiper blade. The wiper blade wipes the windshield clear of water. *Figure 34–14* shows a wiper arm and blade assembly.

Windshield Wiper System Electrical Circuit

It is not practical to show all of the electrical circuits used on windshield wiper systems. *Figure 34–15* shows a complete circuit with a *pulse wiper system*. The operation of the pulse system is as follows:

1. Voltage is available when the ignition switch is on ACCY (accessory) or RUN.
2. When the wiper/washer switch is in the PULSE posi-

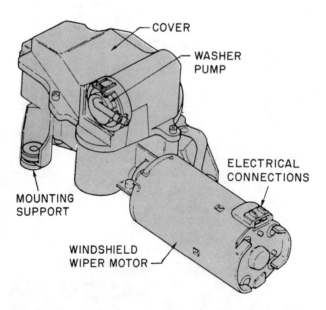

FIGURE 34–10. A motor for a windshield wiper system is shown. The assembly includes the motor, washer, and oscillating mechanism. *(Courtesy of Oldsmobile Division, General Motors Corporation)*

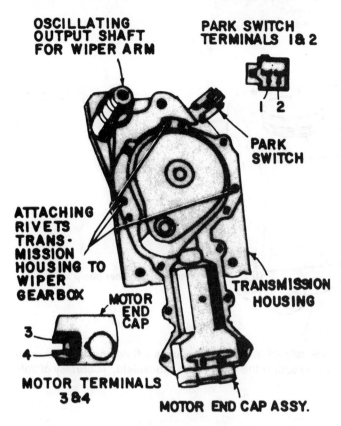

OSCILLATING OUTPUT SHAFT FOR WIPER ARM

PARK SWITCH TERMINALS 1 & 2

I   2

PARK SWITCH

ATTACHING RIVETS TRANS- MISSION HOUSING TO WIPER GEARBOX

MOTOR END CAP

TRANSMISSION HOUSING

3
4

MOTOR TERMINALS 3 & 4

MOTOR END CAP ASSY.

FIGURE 34–11. This oscillation mechanism is used to change rotational motion to oscillation motion on the windshield wiper motor. *(Courtesy of Motor Publications, Auto Repair Manual, 1981–1987)*

tion, voltage is applied to the PNK and GRY wires on the wiper/washer motor module.

3. Voltage is now applied to the solid state control board. Voltage from the control board is sent out and to the coil inside the park relay. The coil pulls the switch to the left.

4. Another voltage from the control board is sent through the park relay switch, through the YEL wire, to run the wiper motor.

5. The park relay switch is held closed by the mechanical arm until the wipers have completed their sweep. The circuit is then opened, and the wipers remain parked until the control board again applies a pulse voltage to the park relay.

6. The length of delay time between sweeps is controlled by the 1.2 megaohm *pulse delay variable resistor* in the wiper/washer switch. The time delay is adjustable from zero to 25 seconds on this circuit.

The LO speed operates as follows:

1. In the LO position, the wiper switch supplies voltage to the DK GRN wire as well as the PNK and GRY wires.

2. The park relay is again energized.

3. Battery voltage is applied continuously to the relay contacts and to the wiper motor. The wiper motor runs continuously at a low speed.

The HI speed operates as follows:

1. Battery voltage is applied directly to the wiper motor through the PPL wire.

2. Voltage is also applied to the DK GRN and the GRY wires to energize the park relay.

3. When turned OFF, the wipers complete the last sweep and park.

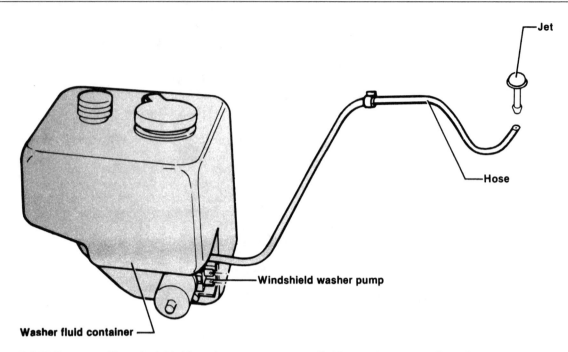

FIGURE 34–12. The windshield washer system uses a fluid container, a small washer pump, and hoses and nozzles (jets) to deliver the wash to the windshield. *(Courtesy of Volkswagen of America, Inc.)*

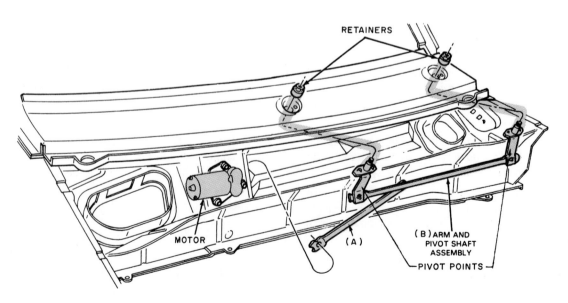

FIGURE 34–13. The oscillation motion from the wiper motor is transferred by a linkage system to the pivot points for the wipers. *(Courtesy of Ford Motor Company)*

The washer operates as follows:

1. When the washer switch is held on for less than one second, voltage is applied to the control board through the PNK and GRY wires.

2. The control board turns on the washer motor for approximately 2 1/2 seconds.

3. The voltage on the GRY wire also operates the park relay.

4. The control board also turns on the wiper motor for about 6 seconds.

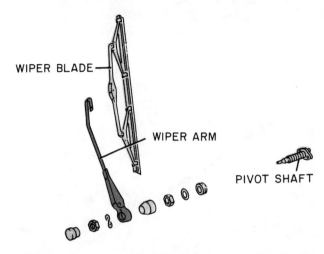

FIGURE 34–14. A wiper arm is used to connect the oscillating motion from the linkage to the wiper blade. (Courtesy of Volkswagen of America, Inc.)

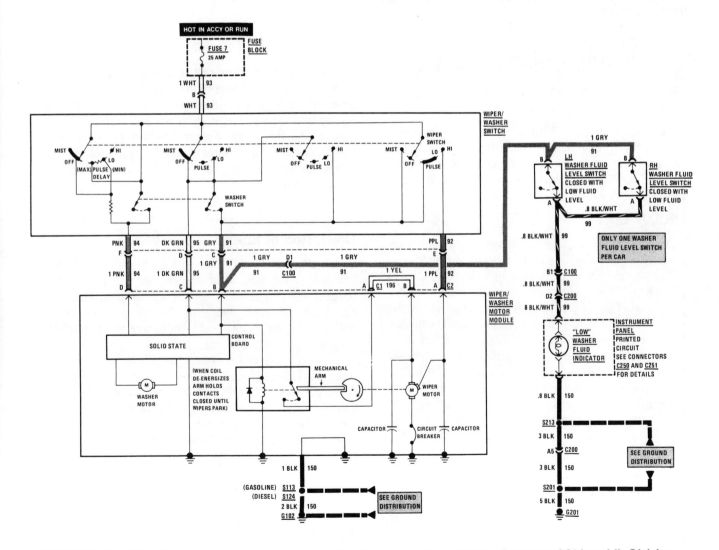

FIGURE 34–15. This circuit shows the complete operation of a pulse wiper system. (Courtesy of Oldsmobile Division, General Motors Corporation)

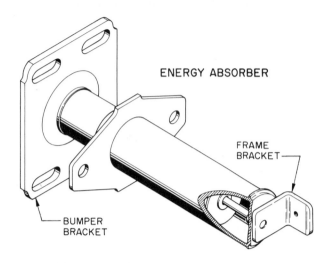

FIGURE 34–16. An energy absorber is used to absorb the energy from a direct impact, caused by a collision upon the bumper. *(Courtesy of United Delco, Division of General Motors Corporation)*

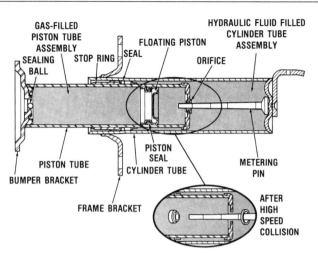

FIGURE 34–17. The parts of an energy absorber. *(Courtesy of United Delco, Division of General Motors Corporation)*

CAR CLINIC

PROBLEM: MULTIPLE ELECTRICAL PROBLEMS

A new car with under 5,000 miles on it has developed unusual electrical problems. For example, certain circuits do not work. These include the automatic door locks, the under-hood light, the automatic antenna, and the radio. What could be the problem?

SOLUTION:

Usually when multiple electrical circuits fail, this indicates the possibility of a burned electrical fusible link. Using the manufacturer's service manual, try to identify which fusible link is common to all of the circuits that are defective. This may take some time. Then check the appropriate fusible link for damage. It should be broken in the center, inside the wire insulation. A fusible link on new cars may be shorted out because the wiring harness rubs against the frame and shorts out. Replace the fusible link, but also identify what caused it to short out; for example, a shorted wiring harness.

34.4 BUMPER ENERGY ABSORBERS

Purpose of Energy Absorbers

Energy absorbers are now being used on the bumper systems of most vehicles. This is primarily a result of safety standards that were incorporated several years ago. These standards specify that protection for safety related items be provided for a predetermined series of barrier impacts. Energy absorbers placed in the front bumper system are used to aid in this system. *Figure 34–16* shows a two-stage energy absorber. The absorber uses a hydraulic principle to absorb the impact energy and restore the bumper to its original position after impact.

Operation of Energy Absorbers

Refer to *Figure 34–17*. There is a gas pressure inside the gas-filled piston tube assembly. Hydraulic fluid fills the inside of the cylinder tube assembly. The gas pressure in the piston tube assembly maintains the unit in an extended position. The stop ring is used to limit travel and to provide extra strength during towing.

During low-speed collisions, hydraulic fluid is forced past the orifice into the piston tube. The metering pin determines the size of the orifice by the position of the bumper and piston tube assembly. The metering and flow of fluid provide the energy absorbing action. The floating piston separates the hydraulic fluid from the gas. After impact, the pressure of the gas behind the floating piston pushes the hydraulic fluid back into the cylinder assembly.

During high-speed collisions, the two-stage orifice shears out of the piston cap. See insert. This allows hydraulic fluid to pass faster into the piston tube. After a high-speed impact, the unit must be replaced.

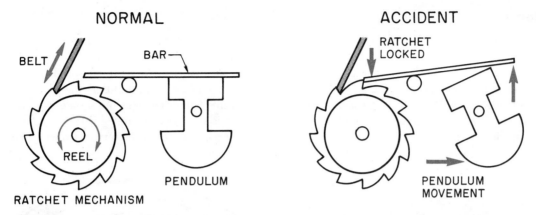

FIGURE 34-18. Some safety belts use a pendulum system to lock up when the vehicle is in a collision.

34.5 SAFETY BELTS

Purpose of Retractors and Reels

In the past, safety belts were very cumbersome and binding. Today, retractors and reels rewind and loosen the belt as needed. The reel allows the safety belt wearer to move around inside the vehicle freely during normal conditions. This freedom makes some people skeptical. They feel the belt may not restrain them in a collision. However, the belts lock solidly when needed.

Operation of Safety Belts

The safety belt shown in **Figure 34-18** is called a car-sensitive belt. It uses a *pendulum* located in the car body and a ratchet mechanism. Under normal conditions, the pendulum and bar are in their resting position. The reel, which holds the belt, is free to rotate. As the occupant leans against the belt, it "gives" or unreels. Under accident conditions, such as a collision, the pendulum tilts toward the force of the impact. This causes the bar to engage the ratchet. The reel and seat belt now lock, restraining the occupant.

34.6 CRUISE CONTROL

Introduction and Purpose

Because of the constant changes and improvements in technology, each cruise control system may be considerably different. There are several types that are used, including the non-resume type, the resume type, and the electronic type.

Cruise control systems are designed to allow the driver to maintain a constant speed without having to apply continual foot pressure to the accelerator pedal. Selected cruise speeds are easily maintained. Speed can be easily changed.

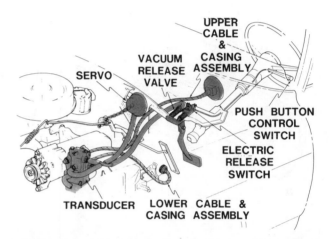

FIGURE 34-19. The parts of a common cruise control system. (Courtesy of General Motors Product Service Training)

Several override systems also allow the vehicle to be accelerated, slowed, or stopped.

When engaged, the cruise control components set the throttle position to the desired speed. The speed will be maintained unless heavy loads and steep hills interfere. The cruise control is disengaged whenever the brake pedal is depressed. **Figure 34-19** shows the common components on a cruise control system.

Cruise Control Switch

The cruise control switch is located on the end of the turn signal or near the center or sides of the steering wheel. There are usually several functions on the switch, including off-on, resume, and engage buttons. The switch is different for resume and non-resume systems. **Figure 34-20** shows a cruise control switch.

Transducer

The *transducer* is a device that controls the speed of the vehicle. When the transducer is engaged, it senses vehicle speed and controls a vacuum source. The vacuum source is used to maintain a certain position on a servo. Locate the transducer on Figure 34–19. Note that speed control is sensed from the lower cable and casing assembly attached to the transmission.

Servo

The *servo* unit is connected to the carburetor or throttle by a rod or linkage, a bead chain, or a *Bowden cable*. The servo unit maintains the desired car speed by receiving a controlled amount of vacuum from the transducer. The variation in vacuum changes the position of the throttle. The servo is shown on Figure 34–19.

An inside view of the servo is shown in *Figure 34–21*.

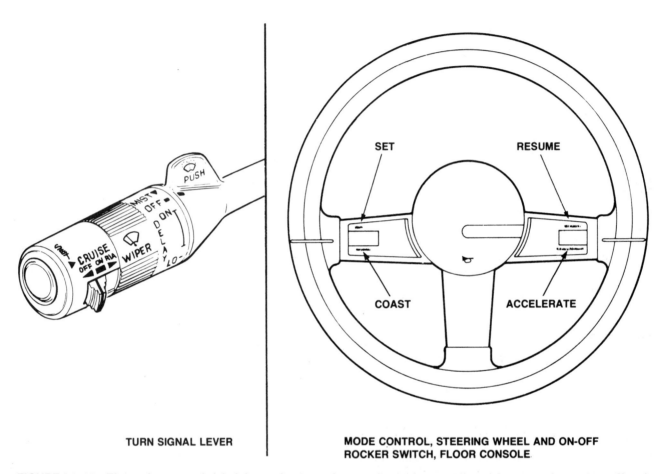

TURN SIGNAL LEVER

MODE CONTROL, STEERING WHEEL AND ON-OFF
ROCKER SWITCH, FLOOR CONSOLE

FIGURE 34–20. The cruise control switch is used to set or increase speed, resume speed, or turn the system off and on. *(Courtesy of General Motors Product Service Training)*

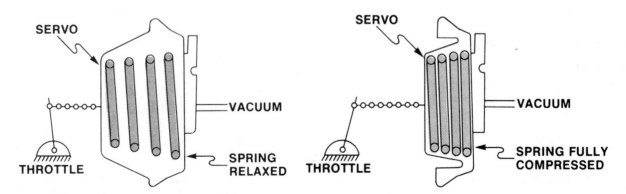

FIGURE 34–21. Inside the servo, vacuum will compress or relax the spring. When the system is on, the throttle is positioned with the spring compressed. *(Courtesy of General Motors Product Service Training)*

When a vacuum is applied, the servo spring is compressed and the throttle is positioned correctly. When the vacuum is released, the servo spring is relaxed and the system is not operating.

Brake-activated Switches

There are two switches that are operated by the position of the brake. When the brake pedal is depressed, the brake release switch disengages the system. A vacuum release valve is also used to disengage the system when the brake pedal is depressed. *Figure 34–22* shows an example of the location of the two switches.

Electrical and Vacuum Circuits

Figure 34–23 shows an electrical and vacuum circuit diagram. The system operates as follows:

1. The object is to energize the solenoid inside the transducer. If the solenoid is energized, the vacuum valve will shift, allowing vacuum to control the servo.

2. When the system is in the off position, current passes through the ignition switch, fuse, and slider in the switch to the resistance wire.

3. The resistance produces a voltage drop before the current gets to the solenoid coil inside the transducer. There is not enough voltage to energize the solenoid coil.

4. When the cruise control is engaged, current flows through the switch to the engage wire. If the vehicle is above 30 mph, the low-speed switch will be closed. Current can then easily pass through the solenoid coil, which, in turn, will control the vacuum.

5. If the brakes are depressed with the cruise control on, electricity will bypass the solenoid, go through the brake light circuit, and be grounded. This will shut off the cruise control by disengaging the solenoid and closing the vacuum to the servo.

6. The object of the vacuum circuit is to get the right amount of vacuum sent to the servo.

7. Vacuum is taken from the manifold and sent through the resume valve. When the cruise control is on, the resume valve is energized and open.

8. Vacuum then flows into the transducer, through the vacuum valve in the transducer, to the servo. Of course, vacuum can only get to the servo if the electrical solenoid is energized.

9. The air control valve inside the transducer acts like a vacuum bleed valve. It is a variable control,

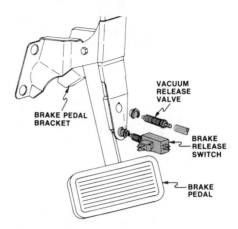

FIGURE 34–22. Two switches, a vacuum release and a brake release switch, disengage the cruise control when the brake pedal is depressed. *(Courtesy of General Motors Product Service Training)*

depending upon the speed of the vehicle. At lower speed settings, the air control valve bleeds off vacuum so less is sent to the servo. At higher speeds, less vacuum is bled off so more vacuum is sent to the servo.

10. When the brake is depressed, disengaging the system, the resume solenoid is closed. When the resume switch is activated, the solenoid is energized to open the vacuum line to the transducer.

Electronic Cruise Control Parts

Cruise control can also be obtained by using electronic components rather than mechanical components. Depending upon the vehicle manufacturer, several additional components may be used. These include:

1. *Electronic control module*, integrated circuitry — used to control the servo unit. The servo unit is again used to control the vacuum which in turn controls the throttle.

2. *Vehicle speed sensor (VSS) buffer amplifier* — used to monitor or sense vehicle speed. The signal created is sent to the electronic control module. A generator speed sensor may also be used in conjunction with the VSS.

3. *Clutch switch* — used on vehicles with manual transmissions to disengage the cruise control when the clutch is depressed.

4. *Accumulator* — used as a vacuum storage tank on vehicles that have low vacuum during heavy load and high road speed.

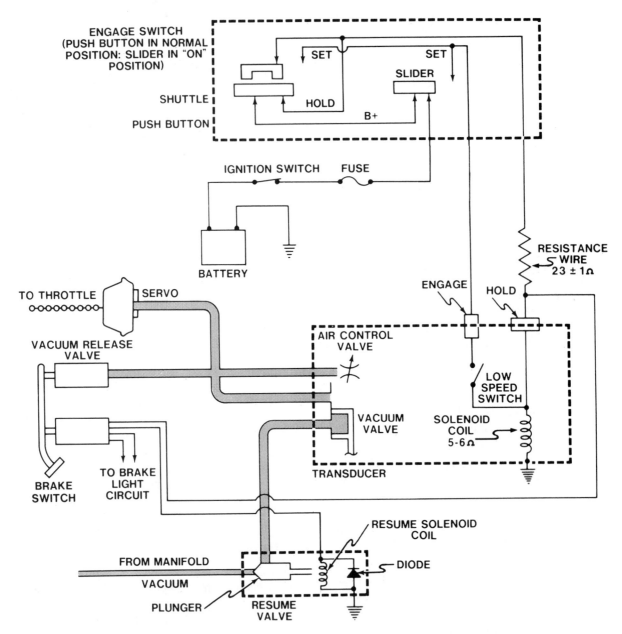

FIGURE 34-23. A cruise control circuit with vacuum and electrical systems. (Courtesy of General Motors Product Service Training)

Electronic Cruise Control Operation

Figure 34-24 shows how electronic cruise control components work together. The throttle position is controlled by the servo unit. The servo unit uses vacuum working against a spring pressure to operate an internal diaphragm. The servo unit vacuum circuit is controlled electronically by the controller. The controller has several inputs that help deter-

mine how it will affect the servo. These inputs include the:

1. *Brake release switch* (clutch release switch)
2. *Speedometer*, buffer amplifier, or generator speed sensor
3. *Turn signal mode switch* (switch used on the turn signal to control the cruise control)

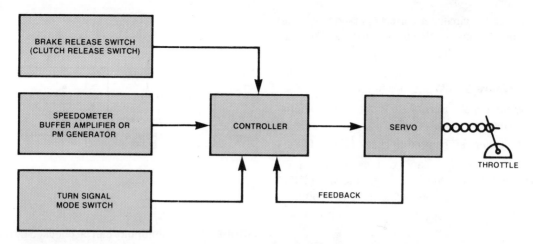

FIGURE 34–24. Electronic cruise control uses an electronic control module (controller) to operate a servo that controls the position of the throttle. *(Courtesy of General Motors Product Service Training)*

SAFETY TIPS

1. *Many electrical wires may have very sharp protrusions on them. When handling wires, be careful not to puncture your hands.*
2. *Always wear safety glasses when working on any electrical circuit.*
3. *Heavy parts are attached to the energy absorbers used on the front of the vehicle. Always be careful to support all parts when working on these absorbers.*
4. *Always use the correct tools when working on the cruise control. This will eliminate the possibility of injury to the hands.*
5. *Remove all metal jewelry (rings, watches, and so on) when working with electrical components.*

1. Use the following troubleshooting procedure when working on *problems in electrical systems.*
   a. Check the problem — Make sure you know what the exact problem is. Don't replace a component that is not faulty.
   b. Refer to the electrical schematic — Read the schematic to make sure you know exactly how the electrical circuit and its components are to operate.
   c. Look for the possible cause — After viewing the schematic, consider possible causes. Check the basic components and those easiest to check, such as the fuse and ground, first.
   d. Test for correct voltages during operation — Using a voltmeter, check for voltages at various parts of the schematic. Based upon this information, again try to determine the cause.
   e. Narrow down the problem to one point in the circuit — After voltage checks have been made, try to narrow down the problem to only one component. Try to isolate the problem.
   f. Find the cause and repair it — Use the component location table to help identify the part.

g. Check the circuit for correct operation — Make sure the cause has been corrected and repaired.

2. When checking electrical circuits, a *jumper wire* can be used to bypass a particular part of the circuit. ***Figure 34–25*** shows a jumper wire.

3. Short finders are available to locate hidden shorts to ground.

4. A fuse tester can be used to check for bad fuses.

5. Testing for voltages can be done as shown in ***Figure 34–26***. One lead is grounded and the other lead is connected to a specific test point.

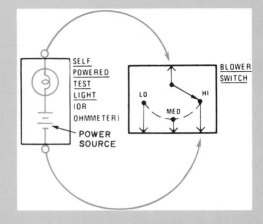

FIGURE 34–27. Continuity can be checked by using a self-powered test light or an ohmmeter. (Courtesy of Oldsmobile Division, General Motors Corporation)

FIGURE 34–25. A jumper wire can be used to bypass a component in an electrical circuit. (Courtesy of Oldsmobile Division, General Motors Corporation)

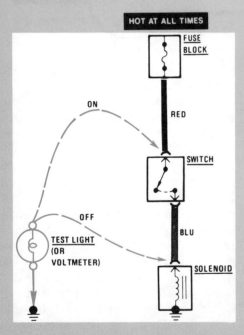

FIGURE 34–26. Voltage can be checked by connecting one lead to ground and the other lead to the test point. (Courtesy of Oldsmobile Division, General Motors Corporation)

6. Test for continuity by using a self-powered test light or ohmmeter. ***Figure 34–27*** shows an example of testing the continuity of a switch.

7. There are many troubleshooting procedures used for windshield wiper systems. The wiper may be inoperative, it may not shut off, it may be sluggish as it moves, or the washer system may not operate. ***Figure 34–28*** shows an example of a troubleshooting procedure for a specific manufacturer. Always refer to the correct manual to use the correct procedure for troubleshooting.

8. Use the following procedure to help *diagnose mechanical problems* in any of the auxiliary or other systems.

a. Operate the system, especially through all conditions such as low, medium, high, off, and on.

b. Observe the malfunction or operational characteristics. Did the problem occur over a period of time or all at once? Ask the operator to give information about its operation. Know the maintenance record of the system.

c. Become completely familiar with how the system should operate. Use operator's manuals, electrical schematics, maintenance manuals, and so on.

d. First check the components that can be easily tested such as fuses, grounds, broken parts, and special noises.

e. Try to isolate the problem to a specific component within the system.

f. Check the suspected component for correct operation.

g. Replace the cause of the problem.

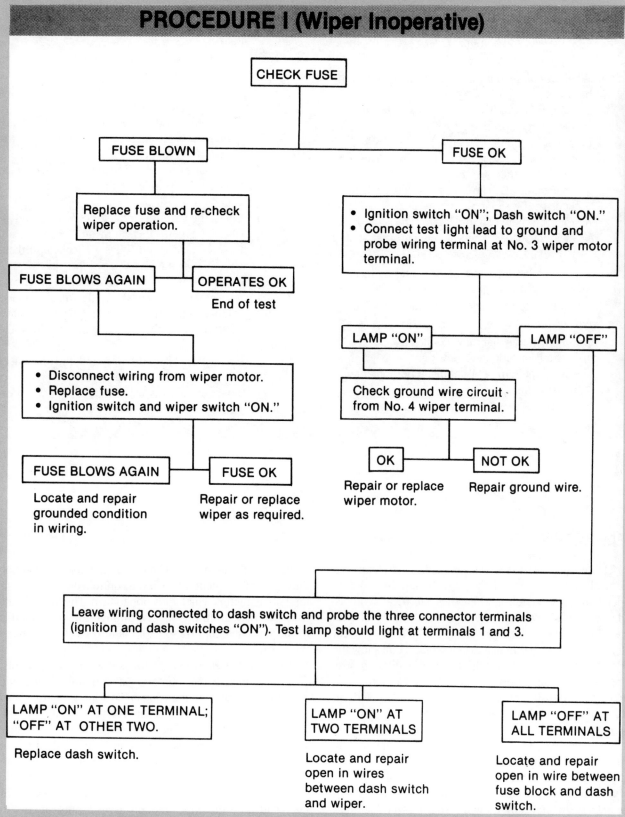

FIGURE 34–28. Troubleshooting procedures are available from the manufacturer for a variety of problems encountered in the windshield wiper, as well as other systems. *(Courtesy of United Delco, Division of General Motors Corporation)*

h. Check the system to make sure the problem has been corrected.

i. If there is still a problem, get a second opinion from another person.

9. When *troubleshooting the energy absorbers* on the bumpers, check for:

a. leakage or oil. If you find a leak or oil, replace the energy absorber.

b. visual distortion or damage to the absorber.

c. a rattling noise as the unit is shaken. This indicates that the metering pin and orifice have been damaged. The orifice has been broken off. Replace if necessary.

d. rotation between the piston tube and the cylinder tube. A good energy absorber should not rotate.

10. The energy absorber can be bench tested by compressing the unit in an arbor press about 3/8 inch. It should restore to its normal position. If it doesn't, replace the unit.

11. Common checks that can be done on a *cruise control* system include the following:

a. Always perform a visual inspection of all components on the cruise control.

b. Check the servo for leaks.

c. Check for blown fuses.

d. Bypass the low-speed switch and check for operation.

e. With the car off and the ignition switch on, engage the cruise control switch. Listen for the solenoid to "thunk." This indicates that the solenoid is operative.

f. Check the brake-activated electric release switch. A small misadjustment can cause the system to be inoperative.

g. Check all electrical connectors for solid contact.

h. Check all electrical components for correct operation.

i. Check the vacuum circuit for correct operation. Check for correct vacuum at the transducer. Observe all vacuum lines for leaks and damaged rubber hoses.

j. Check vacuum hoses that might be pinched.

k. Inspect the mechanical throttle linkage for ease of movement. The bead-chain type should have no slack in it with the engine at the correct idle speed.

l. If no external fault can be found, the transducer may be checked for vacuum leaks or incorrect operation. This is usually done by an authorized transducer repair facility.

SUMMARY

This chapter discussed various auxiliary systems and electrical circuits used on the automobile. In order to understand these systems, schematics, electrical symbols, and various electrical parts needed to be defined. All electrical circuits are analyzed by using a schematic. Current automotive electrical schematics use symbols that show the parts and how electricity flows through the circuit. Wire size, component location, and wire colors are also shown. With the increased number of electrical circuits used on automobiles, a good understanding of schematics is very important.

The headlight circuit can be analyzed by studying the schematic. Electricity is controlled by the light switch. Additional components in the circuit include the dimmer switch, low and high beam in the headlights, and the fuse panel. The defogger operates by using a defogger switch, a timer to hold the defogger on for a certain period of time, and the fuse panel. The power seats operate by using several motors on the bottom of the seats. The motors can operate in either direction. The switches, which are controlled by the operator, tell the motors which direction to turn. The horn circuit uses electricity from the battery, through the horn button, to the horn solenoid. A solenoid is also used to energize the horn.

The windshield wiper system uses several components. The windshield wiper motor is a permanent magnetic motor used to operate the windshield wipers. There is also a mechanism used to change the rotational motion of the motor to oscillation motion for the wipers. A fluid washer nozzle and pump are used to pump cleaning fluid to the windshield. Each vehicle also uses mechanical linkage to connect the oscillation motion of the motor output to the wiper blades. Windshield wiper systems use an electrical circuit to time the wiper motion as in a pulse mode, to operate on LO and HI operation. In addition, the washer system also uses an electrical circuit to operate.

The purpose of using an energy absorber bumper is to guard against vehicle damage in low-speed impacts. A hydraulic absorber is used to cushion the impact. The absorber uses hydraulic fluid and gas inside of a cylinder. During low-speed impacts, hydraulic fluid is forced past small holes and a metering pin inside the cylinder. The metering and flow of the fluid provide the energy absorbing action.

Safety belts are used on all vehicles today. Retractors and reels that rewind and loosen the belt are used for improved passenger comfort. A pendulum located inside the car body and a ratchet mechanism are used to lock the belt during impacts.

Cruise control is used to mechanically or electronically control the position of the throttle during highway operation. These systems help the operator to maintain a constant speed without having to apply foot pressure to the accelerator pedal. A cruise control switch is used to engage or disengage the system. Many types are used.

There are several common parts used on cruise control systems. The servo is connected to the carburetor to control the throttle position. It uses vacuum to move a wire or chain connected to the throttle mechanism. The transducer is used to control the vacuum sent to the servo. A brake-activated switch is used to disengage the system when the brakes are applied. The electrical circuit is used to control valves, which in turn control the vacuum to the transducer.

Electronic cruise control systems use several additional parts. These include the electronic control module, a vehicle speed sensor (VSS), a clutch switch, and an accumulator used to store vacuum.

Diagnosis and service is a very important part of auxiliary and electrical systems. Use a standard troubleshooting procedure to find an electrical problem. For example, always check the problem, don't fix a component that is not faulty, read the schematic, look for possible causes, test for correct voltages, narrow the problem down to one circuit, find the cause, and correct the problem.

Various troubleshooting tools are also used on electrical circuits. These include jumper wires, a short finder, and a fuse tester. A volt ohmmeter can also be used to check for voltages, continuity, and shorts. Always check the easiest components first.

When troubleshooting energy absorber bumpers, check for oil leaking from the absorber, visual distortion, and a rattling noise inside the unit. A rattling noise indicates the orifice and metering pin have been broken due to a high-speed impact.

Common checks that are done on a cruise control system include checking the servo for leaks, checking for blown fuses, checking the brake-activated switch for correct operation, and checking all electrical connectors for solid contact.

TERMS TO KNOW

Can you explain each of the following terms? Review the chapter until you can use each term correctly.

Electrical schematic	Pulse delay variable
Electrical symbol	resistor
Component location	Energy absorber
table	Pendulum
Fuse block	Transducer
Fusible link	Servo
Dimmer switch	Bowden cable
Horn brush/slip ring	Buffer
Pulse wiper system	Jumper wire

REVIEW QUESTIONS

Multiple Choice

1. Electrical schematics are used to:
   a. Show the parts of a circuit
   b. Show the direction of electrical flow
   c. Show the color of wire used in the circuits
   d. All of the above
   e. None of the above

2. To find the exact physical location of a part on the schematic:
   a. Look at the position on the schematic
   b. Look at the component location table
   c. Look at the size of the wire
   d. Look at the length of the wire
   e. None of the above

3. To protect electrical circuits from overloaded conditions, _____ are used.
   a. Fuses
   b. Electrical connectors
   c. Motors
   d. ORN wires
   e. PPL wires

4. Which of the following circuits use a dimmer switch?
   a. Horn
   b. Headlight
   c. Windshield wiper
   d. All of the above
   e. None of the above

5. The defogger system uses an electrical circuit that:
   a. Only operates for a certain period of time
   b. Uses an ON-OFF switch
   c. Uses wires on the inside surface of the window
   d. All of the above
   e. None of the above

6. Electrical seats use motors that:
   a. Have one electromagnet on each motor
   b. Operate in only one direction
   c. Operate in either direction, depending upon the current
   d. Are encased in oil
   e. Operate on 1–2 volts

7. The horn circuit uses:
   a. Seven switches to operate correctly
   b. A solenoid and horn relay switch
   c. Motors that operate in either direction
   d. An electronic control module
   e. All of the above

8. The mechanism in a windshield wiper system for changing rotational motion to oscillation motion is located:
   a. On the windshield wipers
   b. In the washer system
   c. Inside the motor assembly
   d. Under the dashboard
   e. In the trunk

9. Which is *not* a part of a common windshield wiper system?
   a. Fluid washer system
   b. Motor
   c. Wiper blades
   d. Servo
   e. Wiper linkage

10. Which component is used in the electrical circuit to adjust the time delay for pulsing the wipers?
    a. Variable motor
    b. Variable washer nozzle
    c. Linkage
    d. Wiper blades
    e. Variable resistor

11. What is used inside the energy absorber to absorb the shock of impact?
    a. Gas
    b. Springs
    c. Hydraulic fluid
    d. Magnetism
    e. All of the above

12. Which of the following is/are used to keep the energy absorber in place and fully extended during normal conditions?
    a. Gas
    b. Springs
    c. Hydraulic fluid
    d. Magnetism
    e. All of the above

13. Safety belts are locked up during a collision by using a _____.
    a. Magnetic circuit
    b. Spur gear
    c. Pendulum
    d. Clutch
    e. Magnetic clutch

14. Which device on a cruise control system is used to control the amount of vacuum?
    a. Servo
    b. Solenoid coil
    c. Brake-activated switch
    d. Transducer
    e. None of the above

15. Which device on a cruise control system is used to adjust and control the position of the throttle?
    a. Servo
    b. Solenoid
    c. Brake-activated switch
    d. Transducer
    e. Magnetic switch

16. Cruise control circuits operate by using:
    a. Vacuum systems
    b. Electrical systems
    c. Mechanical systems
    d. All of the above
    e. None of the above

17. The servo unit uses vacuum forces working against _____ forces to operate an internal diaphragm.
    a. Gravity
    b. Electrical
    c. Magnetic
    d. Spring
    e. All of the above

18. Which of the following is *not* considered a good troubleshooting practice on electrical circuits?
    a. Refer to an electrical schematic
    b. Test for correct voltages
    c. Narrow down the problem
    d. Check the easiest and basic components first
    e. Replace items and components until the problem is solved

19. Which of the following is considered a way to check the condition of energy absorbers?
    a. Check for leakage or oil
    b. Check for visual distortion and damage
    c. Check for a rattling noise inside the unit
    d. All of the above
    e. None of the above

20. Which of the following is *not* considered a common check on a cruise control unit?
    a. Check the servo for leaks
    b. Check the brake-activated electric release switch
    c. Check the metering pin and orifice
    d. Check all electrical connections for solid contact
    e. Check for solenoid operation

The following questions are similar in format to ASE (Automotive Service Excellence) test questions.

21. An energy absorber on the bumper has a rattling noise when it is shaken. Technician A says that the energy absorber is OK. Technician B says that the absorber simply needs to be readjusted. Who is right?
    a. A only
    b. B only
    c. Both A and B
    d. Neither A nor B

22. Technician A says that a common check on the cruise control is to make sure all vacuum hoses are in good condition. Technician B says that a common check on the cruse control is to make sure that the bead chain on the throttle linkage has no slack in it. Who is right?
    a. A only
    b. B only
    c. Both A and B
    d. Neither A nor B

23. Technician A says that when troubleshooting any electrical circuit, a jumper wire should never be used to bypass a part of the circuit. Technician B says that a voltmeter should be used to test electrical circuits. Who is right?
    a. A only
    b. B only
    c. Both A and B
    d. Neither A nor B

24. When reading an electrical circuit for troubleshooting, Technician A says the flow of electricity normally goes from the left side of the page to the right. Technician B says the flow of electricity normally flows from the top downward through the circuit. Who is right?
    a. A only
    b. B only
    c. Both A and B
    d. Neither A nor B

Essay

25. Describe several electrical symbols used on electrical schematics.

26. What is the fuse block used for in an electrical circuit?

27. Describe the electrical operation of the defogger system used on vehicles.

28. What is the purpose of the component location table used along with certain electrical schematics?

29. Describe the purpose and operation of a fusible link.

30. Describe the operation of the energy absorbing bumpers.

31. Define the purpose and operation of the transducer on a cruise control system.

MONSTER TRUCKS

These monster trucks have specially modified engines to produce high power. The object of the competition is to start at one point, go over the top of several cars, and cross the finish line first. The vehicles are designed to withstand very high torque forces. The names of these two monster trucks are USA 1 and Stomper Bully. (Courtesy of Jim Rennich, Bloomington, MN)

MOTOR MANUALS

DOMESTIC CARS

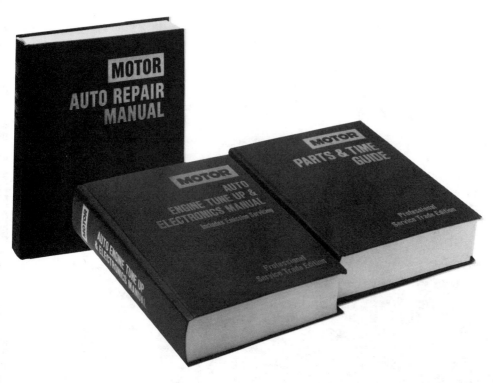

AUTO REPAIR MANUAL

- Step-by-step repair procedures and specifications for mechanical systems
- Multiple-year coverage with shaded areas highlighting new data and manufacturers' changes
- New machine shop section with engine rebuilding specifications

Order Code B

AUTO ENGINE TUNE UP & ELECTRONICS MANUAL

- Electronic and conventional ignition systems
- Computerized engine controls
- Carburetors and fuel injection
- Emission controls, tune up specs
- Electric fuel pumps, turbochargers, diesels
- On-board diagnosis, troubleshooting

Order Code B(1)

PARTS & TIME GUIDE

- Comprehensive indexing with tabs for easy reference
- Part numbers and prices, job operational times on domestic cars
- Labor times for light, medium and heavy trucks
- Engines and transmissions included with each truck chapter

Order Code C

AUTO REPAIR MANUAL — EARLY MODEL AND VINTAGE EDITIONS

	Order Code
• 4th Early Model Edition, 1967–1973	M
• 5th Early Model Edition, 1968–1974	M (1)
• 6th Early Model Edition, 1969–1975	M (2)
• 8th Early Model Edition, 1974–1979	M (5)
• 47th Auto Repair Manual, 1977–1984	M (6)
• Vintage Car Edition, Vol. I, 1935–1953	S
• Vintage Car Edition, Vol. II, 1953–1961	S (1)

GENERAL REPAIR

IMPORTED CARS AND LIGHT TRUCKS

IMPORTED CAR REPAIR MANUAL

- Mechanical repairs and service for Japanese, British, French and Swedish imports, plus Yugo and Hyundai
- Multiple-year coverage with shaded areas highlighting new data and manufacturers' changes
- Fully illustrated, step-by-step procedures, plus specifications

Order Code E

IMPORTED ENGINE TUNE UP & ELECTRONICS MANUAL

- Computerized engine controls, electronic fuel injection systems
- Diagnosis and testing of emission controls
- Electronic and conventional ignition systems
- Specifications and step-by-step service and repair procedures

Order Code E (1)

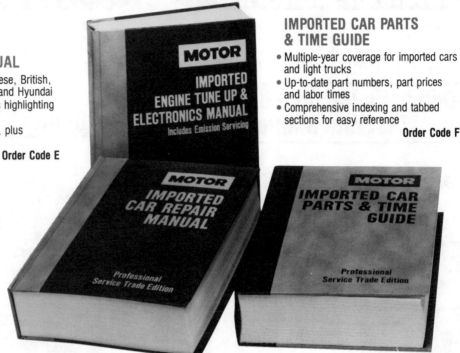

IMPORTED CAR PARTS & TIME GUIDE

- Multiple-year coverage for imported cars and light trucks
- Up-to-date part numbers, part prices and labor times
- Comprehensive indexing and tabbed sections for easy reference

Order Code F

LIGHT, MEDIUM, HEAVY DUTY TRUCKS

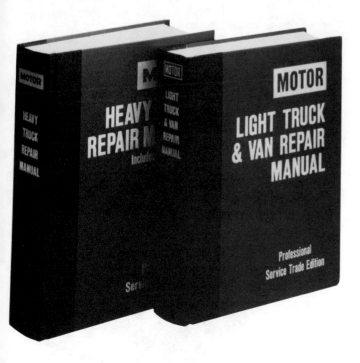

LIGHT TRUCK & VAN REPAIR MANUAL

- Multiple-year coverage for domestic light trucks and vans
- Engines, transmissions, brakes, suspension, steering, air conditioning
- Turbochargers, diesels

Order Code D (1)

HEAVY TRUCK REPAIR MANUAL

- Multiple-year coverage of medium and heavy duty gasoline and diesel trucks
- Tabbed for easy reference
- Cooling & electrical, clutch & transmissions, truck chassis, steering & suspension, drive axle, fuel & exhaust, brakes, diesel engines

Order Code D (3)

Also available:
Heavy Truck Repair Manual — Early Model Edition, 1966–1977   **D (4)**
Truck Engine Tune Up & Electronics Manual, 1977–1984   **D (2)**

MOTOR MANUALS

VACUUM & WIRING DIAGRAM MANUALS

- Car-specific and factory authentic
- Power and ground distribution diagrams
- Location of connectors, splices and grounds
- Essential information for all specialists in electrical work

Domestic Vacuum & Wiring Diagram Manuals

	Order Code
• 11th Edition, 1975–1976 cars	**H**
• 12th Edition, 1977–1978 cars	**H (1)**
• 13th Edition, 1979 cars plus 1980 Citation, Omega, Skylark	**H (2)**
• 14th Edition, 1980–1981 cars, pickups & vans	**H (3)**
• 15th Edition, 1982 cars, pickups & vans	**H (4)**
• General Motors, 1983–1984	**H (7)**
• Ford, Chrysler, AMC, 1983–1984	**H (8)**
• General Motors, 1985	**H (9)**
• Ford, Chrysler, AMC, 1985	**H (10)**
• General Motors, 1986	**H (11)**
• Ford, Chrysler, AMC, 1986	**H (12)**

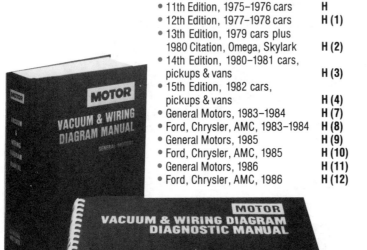

Imported Wiring Diagram Manuals

	Order Code
• 1st Edition, 1977–1982	**H (5)**
• 2nd Edition, 1983–1984	**H (6)**
• 3rd Edition, 1985–1986	**H (13)**

TRANSMISSIONS

- Hydraulic oil circuit diagrams
- Oil pan gasket identification
- Transmission identification code charts
- Computer-controlled torque converter clutches
- Troubleshooting & diagnosis
- Step-by-step service & repair procedures

Domestic Cars & Trucks

Includes light truck and Allison medium duty units

	Order Code
• Automatic Transmission Manual, 9th Edition, 1964–1982	**G**
• Domestic Transmission Manual, 1st Edition	**G (2)**

Imported Cars & Light Trucks

	Order Code
• Imported Transmission Manual, 2nd Edition, 1977–1984	**R**
• Imported Transmission Manual, 3rd Edition, 1982–1987	**R (1)**

SPECIALTY REPAIR

AIR CONDITIONER & HEATER MANUALS

- Factory-installed units, electronic climate controls for cars, pick-ups, vans and trucks
- Leak test, system checks, troubleshooting
- Compressor service, vacuum & wiring diagrams
- Specifications and general service procedures

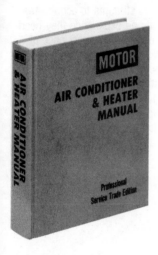

	Order Code
• 5th Edition, domestic cars & trucks, 1974–1980	K
• 6th Edition, domestic cars & trucks, 1981–1983; imports, 1975–1982	K (1)
• 7th Edition, domestic cars & trucks, 1984–1985; imports, 1983–1984	K (2)
• 8th Edition, domestic cars & trucks, 1986; imports, 1985	K (3)
• 9th Edition, domestic cars & trucks, 1987; imports, 1986	K (4)

TECHNICAL SERVICE BULLETINS

- Manufacturers' bulletins "symptom-indexed" to help diagnosis and troubleshooting
- Updates service procedures and specifications

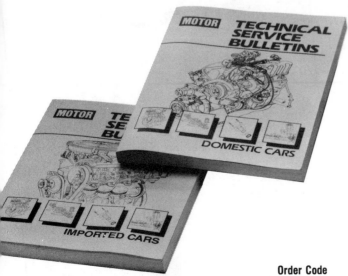

	Order Code
• Domestic Cars, 1980–83	W
• Imported Cars, 1980–83	W (1)
• Domestic Cars, 1984–85	W (2)

EMISSION CONTROL MANUALS

- Complete description of emission controls
- Application charts, vacuum hose routings
- Troubleshooting & diagnosis
- Idle setting procedures, tune up specs

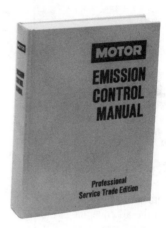

	Order Code
• 3rd Edition, domestic cars & trucks, 1966–1982	J
• 4th Edition, domestic cars & imports, 1982–1983	J (1)
• 5th Edition, domestic cars & imports, 1983–1984	J (2)
• 6th Edition, domestic cars & imports, 1985–1986	J (3)
• 7th Edition, domestic cars & imports, 1986–1987	J (4)

EXCITING *NEW* DELMAR TEXTS . . .

produced in cooperation with

MOTOR

The Delmar learning system

Schwaller's MOTOR

◀

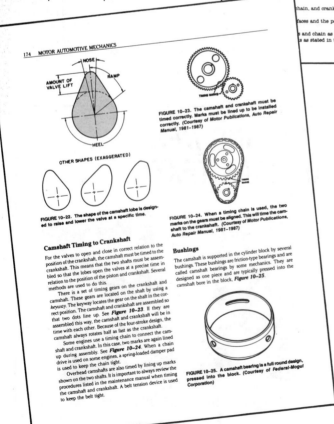

NAME _____
DATE _____
GRADE _____

WORKSHEET
Replacing Timing Gears and Chains
ASE Task #7-8-9

OBJECTIVE: After completing this task, the student will have the hands-on experience to replace timing gears and timing chains.

REFERENCE: Textbook Chapters 9, 10
Motor Auto Repair Manual
Manufacturers' Service Manual(s)

TOOLS NEEDED: Basic hand tools; Vibration damper puller.

SAFETY PRECAUTIONS: Disconnect negative battery cable.

INFORMATION: Timing gears and chains usually wear out before the other internal parts of the engine. Gears can be replaced by removing the front timing cover. This task is done in the same steps on most vehicles; however, use your Service Manual for proper procedures. On most engines, it will be necessary to use a vibration damper puller to remove the damper from the crankshaft.

Vehicle Make _____ Model _____ Year _____

Procedure

The student will:

	NOT ACCOMPLISHED	PARTIALLY ACCOMPLISHED	FULLY ACCOMPLISHED
1. Place fender covers on fenders.	[ ]	[ ]	[ ]
2. Disconnect negative battery cable.	[ ]	[ ]	[ ]
3. Drain cooling system.	[ ]	[ ]	[ ]
4. Remove drive belts and water pump pulleys. NOTE: On some models, it may be necessary to remove the radiator. See your Worksheet on radiator removal.	[ ]	[ ]	[ ]
5. Follow your procedure as outlined in the Service manual for this vehicle.	[ ]	[ ]	[ ]
6. Position the old gears to the proper alignment, as stated in the Manual.	[ ]	[ ]	[ ]
7. Remove the bolts that fasten the camshaft gear to the camshaft.	[ ]	[ ]	[ ]
chain, and crankshaft gear off as a unit.	[ ]	[ ]	[ ]
faces and the parts that will be reassembled.	[ ]	[ ]	[ ]

s and chain as a unit. Align the cam gear and crankshaft
is as stated in the Manual. Torque the bolt(s) th

174 MOTOR AUTOMOTIVE MECHANICS

NOSE
AMOUNT OF VALVE LIFT
RAMP
HEEL

OTHER SHAPES (EXAGGERATED)

FIGURE 10-22. The shape of the camshaft lobe is designed to raise and lower the valve at a specific time.

FIGURE 10-23. The camshaft and crankshaft must be timed correctly. Marks must be lined up to be installed correctly. (Courtesy of Motor Publications, Auto Repair Manual, 1981-1987)

FIGURE 10-24. When a timing chain is used, the two marks on the gears must be aligned. This will time the camshaft to the crankshaft. (Courtesy of Motor Publications, Auto Repair Manual, 1981-1987)

Camshaft Timing to Crankshaft

For the valves to open and close in correct relation to the position of the crankshaft, the camshaft must be timed to the crankshaft. This means that the two shafts must be assembled so that the lobes open the valves at a precise time in relation to the position of the piston and crankshaft. Several methods are used to do this.

There is a set of timing gears on the crankshaft and camshaft. These gears are located on the shaft by using a *keyway*. The keyway locates the gear on the shaft in the correct position. The camshaft and crankshaft are assembled so that two dots line up. See *Figure 10-23*. If they are assembled this way, the camshaft and crankshaft will be in time with each other. Because of the four-stroke design, the camshaft always rotates half as fast as the crankshaft.

Some engines use a timing chain to connect the camshaft and crankshaft. In this case, two marks are again lined up during assembly. See *Figure 10-24*. When a chain drive is used on some engines, a spring-loaded damper pad is used to keep the chain tight.

Overhead camshafts are also timed by lining up marks shown on the two shafts. It is important to always review the procedures listed in the maintenance manual when timing the camshaft and crankshaft. A belt tension device is used to keep the belt tight.

Bushings

The camshaft is supported in the cylinder block by several bushings. These bushings are friction-type bearings and are called camshaft bearings by some mechanics. They are designed as one piece and are typically pressed into the camshaft bore in the block. *Figure 10-25*.

FIGURE 10-25. A camshaft bearing is a full round design, pressed into the block. (Courtesy of Federal-Mogul Corporation)

ACTIVITY II

Directions: Identify the numbered parts shown on the following valve.

PARTS

PARTS

1. _____
2. _____
3. _____
4. _____
5. _____
6. _____
7. _____

ACTIVITY III

Directions: Answer the following questions by circling the letter of the correct response.

Multiple Choice

1. Which of the following are purposes of the cylinder head?
   a. Act as a cap for the combustion
   b. Hold valves
   c. Allow for air flow to cylinder
   d. All of the above
   e. None of the above

42

to support
Automotive Mechanics

CHAPTER NO.	SUBJECT	Auto Repair Manual	Imported Car Repair Manual	Light Truck & Van Repair Manual	Engine Tuneup & Electronics Manual	Imported Engine Tuneup & Electronics Manual	Truck Engine Tuneup & Electronics Manual	Domestic Transmission Manual	Imported Transmission Manual	Air Conditioner & Heater Manual	Emission Control Manual	Vacuum & Wiring Diagram Manual
9		X	X	X								
	Cylinder Block	X	X	X								
	Crankshaft	X	X	X								
	Piston & Rod	X										
	Diagnosis & Service	X	X	X								
10	Cylinder Head	X	X	X								
	Valve Assembly	X	X	X								
	Camshaft	X	X	X								
	Valve Operating Mechanism	X		X								
	Diagnosis & Service	X		X								
11	Lubricating System				X	X	X					
15	Carburetor Systems				X	X	X					
	Types of Carburetors				X	X	X					
	Carburetor Accessories				X	X	X					
	Electronic Carburetor Controls				X	X	X					
16	Injection Systems				X		X					
	Throttle Body Injection				X	X	X					
	Port Fuel Injection				X	X	X					
	High Pressure Diesel Injection				X							
	Diesel Electronic Control				X	X	X					
17	Turbocharging				X	X	X					
20	Ignition Systems				X	X	X					
	Advance Mechanisms				X	X	X					
	Electronic Ignition Systems				X	X	X					
	Engine Control Electronics	X	X	X								
21	Alternators	X	X	X								
22	Starter Motors	X										
	Drive & Clutch											

◀ MOTOR Reference Guide

An excellent link to the extensive **MOTOR** Service Manual Library. This guide heads you and your students in the right direction by cross referencing the topics in each chapter of the text with the appropriate **MOTOR** manual. Complete specifications and detailed shop procedures for specific car models and major foreign manufacturers are included. Updated annually so you'll always have access to the latest information.

Instructor's Guide ▲

This helpful supplement provides answers to text questions, a comprehensive exam, suggestions for classroom instruction, and transparency masters.

◀ Transparency Masters

Over 100 transparency masters support classroom instruction by illustrating the automotive components and systems covered in the text.

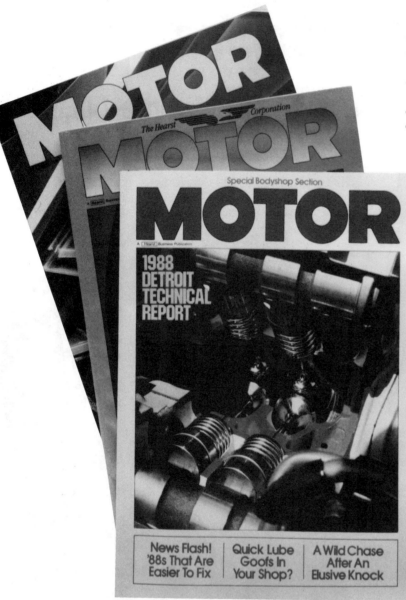
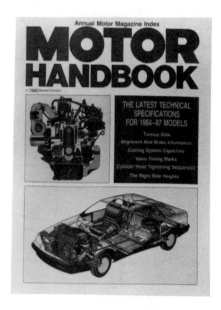

APPENDIX B

A/C   Air Conditioning
ac   Alternating current
AFL   Altitude Fuel Limiter
AIR   Air Injection Reactor
ALCL   Assembly Line Communication Link
API   American Petroleum Institute
ARB   Air Resources Board
ASTM   American Society for Testing of Materials
ATDC   After Top Dead Center
ATF   Automatic Transmission Fluid
AWG   American Wire Gauge
BAT   Battery
Bbl   Barrels (2Bbl carburetor)
BDC   Bottom Dead Center
bhp   Brake horsepower
BLU   Blue wire
BMEP   Brake Mean Effective Pressure
BSFC   Brake Specific Fuel Consumption
BTDC   Before Top Dead Center
Btu   British Thermal Unit
Btu/gal   Btus per gallon
Btu/min   Btus per minute
C   Celsius
C³I   Computer-controlled Coil Ignition System
CAFE   Corporate Average Fuel Economy
cc   Cubic centimeter (cm³)
CCA   Cold Cranking Amperes
CCC   Computer Command Control
CCS   Computer-Controlled Combustion
CID   Cubic Inch Displacement
cm   Centimeter
CO   Carbon monoxide
CO₂   Chemical symbol of carbon dioxide
CV   Constant Velocity
dBA   Decibles
dc   Direct current
DECS   Diesel Electronic Control System

DK GRN   Dark green wire
DOT   Department of Transportation
E   Voltage (electromotive force)
ECCS   Electronic Computer Control System
ECCS   Electronic Constant engine Control System
ECM   Electronic Control Module
EEC   Electronic Engine Control
EFE   Early Fuel Evaporation
EGR   Exhaust Gas Recirculator
emf   Electromotive force
EPA   Environmental Protection Agency
EPR   Evaporative Pressure Regulator valve
ESN   Engine Sequence Number
EST   Electronic Spark Timing
F   Fahrenheit
fhp   Frictional horsepower
ft. lbs.   Foot pounds
g   Gram
g/mi   Grams per mile
Gpm   Gallons per minute
GRY   Gray wire
H   Chemical symbol for hydrogen
H₂O   Chemical symbol for water
H₂SO₄   Chemical symbol for sulfuric acid
HC   Hydrocarbons
HEI   High Energy Ignition system
HO   High Output type engine
Hp   Horsepower
HPAA   Housing Pressure Altitude Advance
HPCA   Housing Pressure Cold Advance
HSC   High Swirl Combustion
I   Amperage (intensity)
IAC   Idle Air Control
ICE   Internal Combustion Engine
ihp   Indicated horsepower
in. lbs.   Inch pounds
k   Kilo

km   Kilometer
kpa   Kilopascals
kw   Kilowatts
L or l   Liter
LPG   Liquid Petroleum Gas
M   Motor octane
M/C   Mixture Control
MA   Mechanical Advantage
MAF   Mass Air Flow
MAP   Manifold Absolute Pressure
MAT   Manifold Air Temperature
MIL   Military
mm   Millimeter
mpg   Miles per gallon
mph   Miles per hour
ms   Milliseconds
MVS   Metering Valve Sensor
N   North pole
N/A   Not Applicable
NAPA   National Auto Parts Association
NC   National Coarse threads
NF   National Fine threads
NO_x Nitrogen Oxide
O_2 Oxygen
OD   Overdrive
OHC   Overhead Camshaft
OHV   Overhead Valves
ORN   Orange wire
OSHA   Occupational Safety and Health Act
P   Power
PbO_2 Chemical symbol for lead peroxide
PCV   Positive Crankcase Ventilation
PFI   Port Fuel Injection
PM   Permanent Magnet
PNK   Pink wire
POA   Pilot-operated Absolute valve
PPL   Purple wire

ppm   Parts per million
PROM   Programmable Read Only Memory
psia   Pressure per square inch absolute
psig   Pressure per square inch gauge
R   Research octane
R   Resistance
R_1, R_2, etc.   Resistor one, resistor two, etc.
rpm   Revolutions per minute
S   South pole
S/V ratio   Surface/Volume ratio
SAE   Society of Automotive Engineers
SEFI   Sequential Electronic Fuel Injection
SFI   Sequential Fuel Injection
SO_4 Chemical symbol for sulfate
STV   Suction Throttling Valve
SUS   Saybolt Universal Seconds
TAMP   Timing, Atomizing, Metering, Pressurizing
TBI   Throttle Body Injection
TDC   Top Dead Center
TEL   Tetraethyl lead
TML   Tetramethyl lead
TPC   Tire Performance Criteria
TPI   Tuned Port Injection
TPS   Throttle Position Sensor
TRS   Transmission Regulated Spark
TSCV   Throttle Solenoid Control Valve
TV   Throttle Valve
U-joint   Universal joint
UNC   Unified National Coarse threads
UNF   Unified National Fine threads
USC   U.S. Customary threads
UTQG   Uniform Tire Quality Grading
VI   Viscosity Index
VIN   Vehicle Identification Number
VSS   Vehicle Speed Sensor
W   Watts
YEL   Yellow wire

GLOSSARY

API American Petroleum Institute (11)

ATF Automatic Transmission Fluid (26)

Abrasive Any material used for grinding; e.g., emery or sand. (13)

Accumulator A device that cushions the motion of a clutch and servo action in an automatic transmission; a component used to store or hold liquid refrigerant in an air conditioning system. (26) (32)

Acidity In lubrication, acidity denotes the presence of acid-type chemicals which are identified by the acid number. Acidity within oil causes corrosion, sludges, and varnish to increase. (11)

Acid rain A form of pollution produced when sulfur and nitrogen are emitted into the air. The mixture of these chemicals with water produces an acid solution which is found in rain. (13)

Actuator A device that causes action or motion on another part. (17)

Additives Chemical compounds added to a lubricant for the purpose of imparting new properties or improving those properties that the lubricant already has. (11)

Adjusting sleeve An internally threaded sleeve located between the tie rod ends. The sleeve is rotated to set toe. (30)

Aeration The process of mixing air with a liquid. Aeration occurs in a shock absorber from rapid fluctuations in movement. (29)

Aerodynamic The ease with which air can flow over the vehicle during higher speed operation. A sound aerodynamic vehicle has very little wind resistance. (8)

Aftermarket Equipment sold to consumers after the vehicle has been manufactured. Aftermarket equipment and parts are sold locally by parts dealers. (17)

Airborne A term used to describe contaminants floating in air moving through the engine. The contaminants are light enough to be suspended in the air stream. (17)

Air cooled Removing heat from the engine by circulating air across the cylinder block and heads. (12)

Air ducts Tubes, channels, or other tubular structures used to carry air to a specific location. (33)

Air-fuel ratio The measure of the amount of air and fuel needed for proper combustion. The most correct air-fuel ratio is 14.7 parts of air to 1 part of fuel. (6)

Airtight container Containers used to hold waste and oily rags, so that spontaneous combustion is eliminated. (2)

Alignment The position of vehicle wheels relative to each other and the car body and frame. Caster, camber, and toe are typically adjusted. (30)

Ambient Surrounding area; e.g., surrounding temperature or circulating air. (13) (32)

Amperage The number of electrons flowing past a given point in one second. (18)

Amperage capacity An indication of the length of time a battery can produce an amperage, or the amount of amperage that a battery can produce before being discharged. (19) (21)

Antifreeze A chemical solution added to the liquid coolant to protect against freezing and to increase the boiling point. (12)

Antifriction bearing A type of bearing that uses balls or rollers between the rotating shaft and the stationary part. (3)

Anti-icers Chemicals added to gasoline to eliminate the freezing of gasoline. (13)

Armature The iron or steel center of an electric motor or solenoid, which is located between two magnets; the movable or rotating part of a generator, which helps to cut the magnetic lines of force; the center moving part or rotational part of the starter. (14) (21) (22)

Asbestos   A fibrous material used for making fire-proof objects. (25)

Aspect ratio   The ratio of the height to width of the tire expressed as a percentage. (31)

Aspirator valve   A device used to draw out fluids by suction. In this case, a pollution device is used to draw fresh air by suction into the exhaust flow to reduce emissions. (24)

Atom   Part of a molecule that has protons, neutrons, and electrons. All things are made of atoms. (18)

Atomization   The breaking down of a liquid into small particles (like a mist) by the use of pressure. (13)

Atomized   A liquid is atomized when it is broken into tiny droplets of the liquid, much like a mist or spray form. (15)

Axial load   A type of load placed on a bearing that is parallel to the axis of the rotating shaft. (3)

BMEP   Brake Mean Effective Pressure. A measure of pressure on top of the piston during the power stroke. (6)

Btu   A unit of thermal or heat energy referred to as a British thermal unit. The amount of heat necessary to raise one pound of water 1 degree F. (13)

Babbitt   A soft metal material on the inside of a bearing insert to allow for embedability (small dirt particles embedding into the metal). (9)

Backing plate   A metal plate that serves as the foundation for the brake shoes and other drum brake hardware. (28)

Back injury   Injury to the back usually caused by lifting heavy objects incorrectly. (2)

Ballast resistor   A resistor in the primary side of the ignition system used to reduce the voltage slightly, approximately 4–5 volts. (20)

Ball joint   A pivot point for turning a front wheel to right or left. Ball joints can be considered either nonloaded or loaded when carrying the car's weight. (29)

Base of 10   The base unit in the metric system is 10. All units are increased or decreased in units of 10. One meter has 10 decimeters, 100 centimeters, and 1,000 millimeters. (4)

Battery cell   That part of a storage battery made from two dissimilar metals and an acid solution. A cell stores chemical energy for use later as electrical energy. (19)

Bead   The edge of a tire's sidewall, usually made of steel wires wrapped in rubber, used to hold the tire to the wheel. (31)

Bearing   A device used to eliminate friction between a rotating shaft and a stationary part. (3)

Bearing cap   A device that retains the needle bearings that ride on the trunnion of a U-joint and is pressed into the yoke. (27)

Bellows   A flexible chamber that can be expanded to draw a fluid in and compressed to pressurize the fluid. (14) (29)

Bias   A diagonal line of direction. In relationship to tires, bias means that belts and plies are laid diagonally or crisscrossing each other. (31)

Bimetal strip   A strip of two different metals connected together. When heated, each metal expands at a different rate. As the bimetal strip is heated, it causes another object to move. (12)

Bimetallic strip   Two pieces of metal, such as brass and steel, attached together. When heat is applied to the strip, each metal expands at a different rate. This causes the metal to bend. A bimetallic strip senses changes in temperature and causes a mechanical movement. (15)

Bleeding   The act of removing air from the brake hydraulic system. (28)

Blow-by   The gases that escape past the rings and into the crankcase area. These gases from the combustion process produce a positive crankcase pressure. (9 (24)

Body and frame   The part of the automobile that supports all other components. The frame supports the engine, drive lines, differential, axles, and so on. The body houses the entire vehicle. (1)

Boiling point   The temperature of a fluid when it changes from a liquid to a gas or vapor. (12) (13)

Bolt hardness   Hardness of a bolt is determined by the number of lines on the head of the bolt. More lines mean a stronger bolt. (3)

Bore   The diameter of the cylinder. (6)

Boss   A cast or forged part of a piston that can be machined for accurate balance. (9)

Bourdon tube   A curved tube that straightens as the pressure inside it is increased. The tube is attached to a needle on a gauge, which senses the movement of the tube and transmits it as a pressure reading. (12)

Bowden cable   A small steel cable inside a flexible tube used to transmit mechanical motion from one point to another. (34)

Brake horsepower   The horsepower available at the rear of the engine. (7)

Brake shoes   The curved metal parts faced with brake lining, which are forced against the brake drum to produce the braking action. (28)

Breaker points   A set of contact points used in the ignition system to open and close the primary circuit. The points act like a switch operated from a distributor cam. (20)

Buffer   Any device used to reduce the shock or motion of opposing forces. (34)

Burner   The combustion chamber where the continuous combustion of the gas turbine occurs. (8)

Bushing   A friction-type bearing usually identified as one piece. (3)

Bushing installers   Tools used to provide correct alignment and applied forces when installing a bushing in a housing. (4)

Bypass tube   A tube directly in front of the thermostat. The coolant bypasses the radiator through this tube when cold. (12)

CAFE   Corporate Average Fuel Economy (23)

Calibrate   To check or measure any instrument. (14)

Caliper   A C-shaped housing that fits over the rotor, holding the pads and containing the hydraulic components that force the pads against the rotors when braking. (28)

Camber   The inward or outward tilt of the wheel at the top. A wheel has a positive camber when the top is tilted out. (30)

Cam ground piston   Piston ground to a cam shape to aid in controlling expansion from heat. (9)

Camshaft   A shaft that is used to open and close the valves. (6)

Capacitor   A device for holding and storing a surge of current. (20)

Capacity specifications   Specifications used to show quantity or amount of liquid in automobile components. (5)

Carbon canister   A canister filled with carbon used to absorb and store fuel vapors that are normally exhausted into the air. (24)

Carbon monoxide   A pollutant in automotive exhaust that is produced when there is insufficient oxygen for combustion. It is a deadly, odorless, tasteless gas. (2) (23)

Carburetor   A device used to mix air and fuel in the correct proportions. (6)

Carcass   The inner part of the tire which holds the air for supporting the vehicle. (31)

Cardan universal joint   A universal joint sometimes known as the four-point or cross-and-bearing-type joint. This joint allows the transmission of power at an angle, but causes rhythmic variations in speed at the output yoke of the joint. (27)

Case-hardened   The outer surface of a metal that has been hardened to reduce the possibility of excessive wear. (9)

Cast   The process of shaping metal by heating to a liquid and pouring the hot metal into a sand mold. (9)

Caster   The backward or forward tilt of the spindle support arm to the top. (30)

Catalyst   A chemical that causes or speeds up a chemical reaction without changing its own composition. (11) (24)

Catalytic converter   A type of emission control device used to change the exhaust emission from the vehicle into harmless chemicals. (24)

Cell density   A measure of how many watts can be discharged per hour, per pound of battery. Cell densities are increasing as batteries are being designed better. (19)

Center link   A steering linkage component connected between the pitman and idler arms. (30)

Centrifugal pump   A pump that draws coolant into its center and then uses centrifugal force to throw the coolant outward and into the cooling system. (12)

Cetane number   The ignition quality of a diesel fuel. The time period or delay between injection and explosion of diesel fuel. (13)

Charging   When the electrical flow of a battery is reversed, the battery is charged. The metals that are alike are converted to lead peroxide and sponge lead, and the electrolyte is now sulfuric acid and water. (19)

Chassis dynamometer   A dynamometer used to measure road horsepower. (7)

Check ball   A ball and spring device in any fluid circuit that allows flow in one direction but stops flow in the other direction. (15)

Clutch   The part of a manual transmission that is used to engage and disengage the transmission from the engine. (25)

Commutator   A device that extracts the electrical energy from the rotating armature. When electrical energy is being extracted from the armature, ac voltage is converted to dc voltage; metal segments that are used on a starter motor to carry electricity to the armature. The commutator also reverses the current flow in the armature at the right time. (21) (22)

Component location table   A table used with an electrical schematic that shows the actual location of the part being investigated. (34)

Compression ratio   A measure of how much the air has been compressed in a cylinder of an engine from TDC to BDC. Compression ratio will usually be from 8–1 to 25–1. (6)

Compression washer   A washer used on an engine to reduce oil leakage when the threads of a bolt are in or near oil. (3)

Compressor   A device on a gas turbine that compresses the air for combustion. (8)

Compensating port   A passage for excess fluid to return to the reservoir when the brakes are released. (28)

Computer-controlled brakes   A system that has a sensor on each wheel feeding electrical impulses into the on-board computer. As the vehicle is stopped, each wheel is stopped or slowed down at the same rate. This condition reduces skidding sideways during rapid braking. (28)

Computerized automatic temperature control   A control system used to monitor and adjust the air temperature inside the passenger compartment. Based upon several inputs, a small microprocessor adjusts air temperature about every 7 seconds. (33)

Condensation   The process of reducing a gas to a more compact state such as a liquid. This is usually done by cooling the substance below its boiling point. For example, the moisture in the air in a fuel tank condenses to water. (14) (32)

Condense   The process of cooling a vapor to below its boiling point. The vapor condenses into a liquid. (13)

Condenser   A capacitor device used to protect the ignition points from corroding; a component in an air conditioning system used to cool a refrigerant below its boiling point. (20) (32)

Conduction   Transfer of heat between two solid objects. (12)

Conductor   Three or less electrons in the valence ring. (18)

Configuration   The figure, shape, or form of an engine. (8)

Conical   Having the shape or form of a cone. (16)

Connecting rod   The connecting link between the piston and crankshaft. (6)

Constant velocity joint (CV joint)   The CV joint consists of balls or tripods, and yokes designed to allow the angular transfer of power without speed variations common to the Cardan universal joint. (27)

Contaminants   Various chemicals in the oil that reduce its effectiveness, including water, fuel, carbon, acids, dust, and dirt particles; also impurities in fuel systems, including dirt, rust, water, and other materials; also chemicals that make the air impure, usually produced from the combustion process. (11) (14) (23)

Continuous combustion   Combustion of air and fuel that continues constantly. (6)

Control arm   The main link between the vehicle frame and the wheels. The control arm acts as a hinge to allow the wheels to go up and down independent of the chassis. (29)

Convection   Transfer of heat by circulation of heated parts of a liquid or gas. (12)

Conventional theory   Flow of electricity from a positive to a negative point. (18)

Conversion factors   Numbers used to convert between USC and the metric system. (4)

Cooling system   The subsystem on an engine used to keep the engine temperature at maximum efficiency. (1)

Cords   The inner materials running through the plies that produce strength in the tire. Common cord materials are fiberglass and steel. (31)

Core   The center of the radiator, made of tubes and fins, used to transfer heat from the coolant to the air. (12)

Core plugs   Plugs inserted into the block that allow the sand core to be removed during casting. Also, at times these plugs will pop out and protect the block if the coolant freezes. (9)

Corrosion   Destruction of a metal in an engine because of chemical or electrochemical reactions with acid. (11)

Corrosive   The eating away of metal within the cooling system. (12)

Counterweight   Weight forged or cast into the crankshaft to reduce vibration. (9)

Crankcase   The area in the engine below the crankshaft. It contains oil and fumes from the combustion process. (9)

Crankcase pressure   The pressure produced in the crankcase from blow-by gases. (24)

Crankshaft   A mechanical device that converts reciprocating motion to rotary motion. (6)

Cross   The cross is the central component of the U-joint connecting the input and output yokes. (27)

Cubic centimeter A unit in the metric system to measure volume. There are 100 cubic centimeters in 1 liter. (4)

Cylinder Internal holes in the cylinder block. (6)

Cylinder block Part of the engine that houses all components. The foundation of the engine. (6)

Cylinder head The top and cover for the cylinder, which houses the valves. (6)

Cylinder sleeve A round cylindrical tube that fits into the cylinder bore. Both wet and dry sleeves are used. (9)

Cylinder taper The shape of the cylinder after it has been worn by the rings. (9)

DOT Department of Transportation (31)

Dashpot A unit using a small cylinder and diaphragm with a small vent hole. It is used to retard, or slow down, movement of some part. (15)

Dealership Privately owned service and sales organization that sells and services vehicles for the automobile manufacturer. (1)

Decibel A unit of sound measurement. Usually 90–100 decibels experienced for a long time can cause hearing damage. (2)

Deep length (well) socket Socket used when a nut is located on a long stud. (4)

Deflection angle The angle at which the oil is deflected inside the torque converter during operation. The greater the angle of deflection, the greater the torque applied to the output shaft. (26)

Defroster Part of the ventilation system on an automobile used to remove ice, frost, or moisture from the front windows. (33)

Desiccant A material that absorbs moisture from a gas or liquid. A desiccant substance is used in an air conditioning system to remove moisture from the refrigerant. (32)

Detergent In a lubrication oil, either an additive or compound having the property of keeping particles suspended so that particles can be filtered. (11)

Diagnosis A term used to describe the process when a mechanic looks for problems in the automobile. When the problem is identified, a solution is selected. (5)

Dial indicator A measuring tool used to adjust small clearances up to 0.001 inch. The clearance is read on a dial. (4)

Diaphragm A partition separating one cavity from another. A fuel pump uses a diaphragm to separate two cavities inside the pump. A diaphragm is usually made of a rubber and fiber material. (14)

Diaphragm spring A type of steel plate spring used in a pressure plate to engage or disengage the friction disc. (25)

Die cast The process of shaping metal by forcing hot metal under pressure into a metal mold. (9)

Diesel engine An intermittent, internal combustion, reciprocating engine that uses the heat of combustion to ignite the fuel. Fuel is supplied by a high-pressure fuel injector instead of a carburetor. (6)

Differential A gear assembly that transmits power from the drive shaft to the wheels. It also allows two opposite wheels to turn at different speeds for cornering and traction. (27)

Differential pressure Difference in pressure on an oil system between the input of a filter and the output of the filter. (11)

Dilution To make thinner or weaker. Oil is diluted by the addition of fuel and water droplets. (11)

Dimmer switch A switch in the headlight circuit used to switch electricity between "LO" and "HI" positions. (34)

Diode Semiconductor in an electrical circuit to allow electrical current to flow in only one direction. (18)

Disc brakes Brakes in which the frictional forces act upon the faces of a disc. (28)

Discharging When an electrical load is put on a battery, the battery is discharged. The internal parts of the battery are chemically changed to the same metals and water. (19)

Dispersed To scatter, spread, or diffuse. (17)

Dispersent A chemical added to motor oil to disperse or keep the particles of dirt from sticking together. (11)

Displacement The volume the piston displaces from BDC to TDC. (6)

Dissipate To become thinner and less concentrated. (23)

Distillation The process at a refinery that separates the hydrocarbons into many products. Usually boiling a liquid to a vapor and condensing it are involved. (13)

Distributor cam Cams on the distributor shaft used to open and close the breaker points. There is one cam for each cylinder of the engine. (20)

Domains Small sections in a metal bar where atoms line up to produce a magnetic field. (18)

Draft The act of drawing or pulling air through a tube. (15)

Drive line Components on the vehicle that transmit the power from the engine to the wheels. (1)

Drive shaft   A metal tube with I-joints or CV joints at each end, used to transmit power from a transmission to the wheels on differentials. (27)

Drop center wheel   A wheel that has its center dropped in with a smaller radius. A drop center wheel is used so that the tire can be easily removed. (31)

Dwell   The length of time in degrees of distributor shaft rotation that the points remain closed. (20)

Dynamic balancing   Equal distribution of weight on each side of a centerline of a wheel. Dynamic means moving or action, and dynamic balancing is done with the wheel moving or spinning. (31)

Dynamometer   A device used to brake or absorb power produced from an engine for testing purposes in a laboratory situation. (7)

EFE   Early Fuel Evaporation (24)

EGR   Exhaust Gas Recirculation (24)

EPA   Environmental Protection Agency (23)

Eccentric   Circles having different center points. (11) (14) (30)

Eddy current   Small circular currents produced inside a metal core in the armature of a starter motor. Eddy currents produce heat and are reduced by using a laminated core. (22)

Efficiency   A ratio of the amount of energy put into an engine as compared to the amount of energy coming out of the engine. Gas engines are about 28% efficient. A measure of the quality of how well a particular machine works. (6) (7)

Electrical schematic   An electrical system layout, showing the parts, the wires, and the electrical flow of the circuit. (34)

Electrical symbol   A symbol used to identify an electrical part in an electrical schematic. (34)

Electricity   The flow of electrons from a negative point to a more positive point. (18)

Electrolyte   A solution of acid and water used as the acid in a battery. Many types of electrolyte are used, but the most common is sulfuric acid and water. (19)

Electromagnetic induction   Producing electricity by passing a wire conductor through a magnetic field, causing the wire to cut the lines of force. (18)

Electromagnetism   Producing magnetism by using electricity flowing through a wire. (18)

Electron   The negative (−) part of the atom. (18)

Electron theory   Electrons will flow from a negative to a positive point. (18)

Emission control   Various devices placed on the vehicle and engine to reduce exhaust pollution. (23)

Emission standards   The Federal Government has established certain emission and pollutant standards on all automobiles. Because of environmental damage in the past, car manufacturers must now meet strict emission standards. (16)

Energy   The ability to do work. (6)

Energy absorber   A type of shock or impact absorber that absorbs impact energy during a collision on the front or rear bumpers. (34)

Engine   The power source that propels the vehicle forward or in reverse. The engine can be of several designs, including the standard gasoline piston engine, diesel engine, and rotary engine. (1)

Engine dynamometer   A dynamometer used to measure brake horsepower. (7)

Ethanol   A hydrocarbon produced from the distillation of corn, wheat, and so on, for use in making gasohol. (13)

Ether   A highly volatile and flammable, colorless liquid used for starting diesel engines in cold weather. (13)

Evaporator   A component in an air conditioning system used to heat a refrigerant above its boiling point. (32)

Expansion valve   A component in an air conditioning system used to create a pressure on one side and reduce the pressure on the other side. (32)

Explosion-proof cabinet   Cabinet used in the automotive shop to store gasoline and other flammable liquids. (2)

External combustion   Combustion of air and fuel externally, or outside of the engine. (6)

Fastener   Objects such as screws, bolts, splines, and so on, which hold together parts of the automobile. (3)

Feeler gauge   Small, thin metal blades or wires, each having a different thickness, used to measure small clearances such as valve clearances. (4)

Field circuit   An electrical circuit in a generator or alternator that causes the north and south poles to be energized. The field circuit can either be stationary, as in a generator, or rotating, as in an alternator. (21)

Fillets   Small, rounded corners machined on the crankshaft for strength. (9)

First aid box A kit made of various first aid bandages, creams, and wraps for treating minor injuries. (2)

Fixed caliper A disc brake caliper design where the caliper is rigidly mounted and creates braking force through opposing pistons. (28)

Fleet service Service given to a fleet of vehicles owned by a particular company. (1)

Floating caliper A moving disc brake caliper with piston(s) on only one side of the rotor. (28)

Fluid coupling A fluid connection between the engine and transmission. The greater the speed the better the fluid coupling between the two. (26)

Fluidity A characteristic of a fluid such as oil, meaning the ease of flow. (11)

Flywheel A heavy circular device placed on the crankshaft. It keeps the crankshaft rotating when there is no power. (6)

Foreman A supervisory position in a dealership, responsible for organizing work schedules and managing the general mechanic. (1)

Forge The process of shaping metal by stamping it into a desired shape. (9)

Four-cycle engine An engine that has intake, compression, power, and exhaust strokes within two revolutions of the crankshaft. (6)

Four-valve head A cylinder head on an engine that has four valves. Two are used for intake and two are used for exhaust. (8)

Franchised dealer A dealership that has a contract with the main car manufacturer to sell and service its automobiles. (1)

Frictional horsepower Horsepower lost due to friction caused by bearings, road resistance, tire rolling resistance, and so on. (7)

Friction The resistance to motion between two bodies in contact with each other. (28)

Friction bearing A type of bearing that uses oil between the rotating shaft and the stationary part. (3)

Friction disc The part of a clutch system that is clamped between the flywheel and the pressure plate. The friction disc is the output of the clutch system. (25)

Fuel consumption The amount of fuel that is consumed or used by the vehicle. Four, six, and eight cylinder engines all have different fuel consumption rates. (8)

Fuel injection Injecting fuel into the engine under a pressure. (6)

Fuel injector Normally the carburetor mixes the air and fuel at a ratio of 14.7 to 1. However, newer vehicles are using fuel injectors. These injectors mix the fuel with the air just before the intake valve. The fuel is injected into the port under a low pressure. Computers control the amount of fuel injected. (14)

Fuel system The subsystem on the engine used to mix the air and fuel correctly. (1)

Fuel vapors When gasoline heats up, it gives off vapors. The vapors take space and, at times, stop the fuel from flowing. Fuel vapors should be sent back to the fuel tank. (14)

Fulcrum The support or point of rest that a lever rests on, also called the pivot point. (25)

Full wave rectification A process of rectifying a voltage from ac to dc by using diodes. The negative half cycle of the voltage is converted to a positive voltage. All cycles are used in full wave rectification. (21)

Fuse block A small plastic block in a vehicle where all the electrical fuses are located. On some vehicles, the turn signal flashers and relays are also connected to the fuse block. (34)

Fusible link A type of electrical circuit protector made from a special wire that melts when the current is excessive. (34)

Galling A displacement of metal, usually caused by a lack of lubrication, too loose fit, or capacity overloads on U-joints. (27)

Gasket A rubber, felt, cork, steel, copper, or asbestos material placed between two parts to eliminate leakage of gases, greases, and other fluids. (3)

Gas turbine engine An engine that uses two turbines, continuous combustion, and operates at high speeds to produce power. (8)

General engine specifications Specifications used to identify a style and type of engine. (5)

General mechanic An automotive career where the person is actively involved in repair and maintenance of the total vehicle. (1)

Governor A device placed on a fuel system to control the amount of fuel being metered into the engine. A governor is only capable of sensing a speed change in the engine. (16)

Harmonics When valve springs are opened and closed rapidly, they may vibrate. Periods of vibration are called harmonics. (10)

Header   A welded steel pipe used as an exhaust manifold. (17)

Heat dam   The narrow groove cut into the top of the piston. It restricts the flow of heat down into the piston. (9)

Heater core   A small radiator-like heat exchanger. Hot coolant from the engine flows through the heater core. Air flow from a fan passes through the fins on the heater core, picking up heat to warm the passenger compartment. (33)

Heat exchanger   A device used to transfer heat from one medium to another. The radiator is a liquid-to-air heat exchanger. (12)

Heat stove   An enclosed area made of thin sheet metal around the exhaust manifold. It preheats air passing over the exhaust manifold before the air enters the air cleaner snorkel. (24)

Heat transfer   The process of moving heat from a warmer object to a colder object. (12)

Heat treated   A process in which a metal is heated to a high temperature, then is quenched in a cool bath of water, oil, and brine (salt water). This process hardens the metal. (9)

Helicoil   A device used to replace a set of damaged threads. (3)

Hemispherical combustion chamber   A type of combustion chamber that is shaped like a half of a circle. This combustion chamber has the valves on either side with the spark plug in the center. (10)

Hi-test   Gasoline that has an octane number near 90–95. (13)

Horn brush slip ring   An electrical contact ring used in the horn circuit. It is located in the steering wheel and is used to maintain electrical contact when the steering wheel is turned. (34)

Horsepower   A measure of work being done per time unit. One hp equals the work done when 33,000 pounds have been lifted one foot in one minute. (7)

Housekeeping   The type of safety in the shop that keeps floors clean, walls and windows clean, lighting proper, containers correct, and tool storage correct. (2)

Hydraulic system   A brake system in which brakes are operated and controlled by using hydraulic fluid under pressure. (28)

Hydrocarbons   A term used to describe the chemical combinations of hydrogen and carbon. A type of automotive pollution produced by incomplete combustion. Hydrocarbons are considered partly burned hydrogen and carbon molecules resulting from combustion. (13) (23)

Hydrometer   An instrument used to measure the specific gravity of battery acid. (19)

Hydrophilic   Attraction to water. Some tires have a hydrophilic tread, which means the tread is attracted to water. (31)

Idler arm   A steering linkage component, fastened to the car frame, which supports the right end of the center link. (30)

Idler gear   A third gear placed in a gear train, usually used to reverse the direction of rotation. (25)

Ignition system   The subsystem on the engine used to ignite the air and fuel mixture efficiently. (1)

I-head   A style of valve arrangement in an engine. I-head refers to the valves being placed directly above the piston in the cylinder head. (8)

Impact socket   Socket used for heavy-duty or high-torque applications. This type of socket is used with an impact wrench. (4)

Impact wrench   An air- or electric-operated power wrench that uses impacts during rotation to loosen or tighten bolts and nuts. (4)

Inclined surface   A slope or slanted surface. An inclined surface is used in valve rotators. (10)

Independent publishers   Publishers that provide service information on automobiles. Examples are Motor Manuals, Mitchell Manuals, Chilton Auto Repair Manual, and so on. (5)

Independent rear suspension   A rear suspension system composed of trailing arms (like control arms) and MacPherson struts or torsion bars. The system allows the rear wheels to move independently of each other. (29)

Independent service   Service provided by independent garages on all types and makes of vehicles. (1)

Indicated horsepower   Theoretical horsepower calculated by the manufacturer of the engine. (7)

Inertia   Objects in motion tend to remain in motion. Objects at rest tend to remain at rest. Inertia is the force keeping these objects at rest or in motion. (22)

Inertia drive   A drive system on a starter motor using inertia to turn a screwsleeve on a spline. (22)

Inhibit   To hold back, restrain, or arrest. (11)

Inhibitor   Any substance that slows down or prevents chemical reactions such as corrosion or oxidation. (11)

In-line   Cylinders in an engine that are in one line or row, such as an in-line four or six cylinder. The cylinders are vertical as well. (8)

Insert guides Valve guides that are small cast cylinders pressed into the cylinder head. (10)

Insulator A material whose atomic structure contains five or more electrons in the valence ring. (18)

Intake manifold vacuum The vacuum produced inside the intake manifold between the valve and the throttle plate. When the throttle plate is closed, there is high intake manifold vacuum. When it is open, there is low intake manifold vacuum. (16)

Integral guides Valve guides that are manufactured and machined as part of the cylinder head. (10)

Integrated circuit A circuit board with many semiconductors forming a complex circuit. (18)

Interference angle When the valve is ground at 45 degrees and the seat is ground at 44 degrees, the two angles will interfere with each other. Interference angles help valves seat faster. (10)

Intermittent combustion Combustion of air and fuel that is starting and stopping. (6)

Internal combustion Combustion of air and fuel inside the engine. (6)

Jack stand A stand used to support the vehicle when working under the car. Always support the car with such stands before working on the underside of the vehicle. (2)

Jounce The action of a car going over bumps in the road. Jounce is similar to bounce. (29)

Jumper wire A wire used when troubleshooting an electrical circuit for bypassing or shorting out a specific component. (34)

Keeper A small, circular, tapered metal piece that keeps the valve retainer attached to the valve. (10)

Keyway A machined slot on a shaft that holds a metal piece called the key. A slot and key used together attach a pulley or hub to a rotating shaft. (3) (10)

Kinetic energy Energy in motion. (28)

Labyrinth seal A type of seal that uses centrifugal forces to eliminate leakage from a rotating shaft. The seal does not touch the rotating part. (3)

Laminated A series of plates of metal placed together and used as a core in a magnetic circuit. (21)

Laminations Thin layers of soft metal used as the core for a magnetic field. (22)

Latent heat Heat that is hidden or not readily observable. (32)

Lean mixture Too much air and not enough fuel for combustion. (6)

Lever A bar supported by a pivot point with a force applied to one end used to move a force exerted on the other. Force times distance moved on one side will always equal the force times distance moved on the other. Typically a lever is used to gain a force with a loss in distance. (25)

L-head A valve arrangement that has the valves located in the block and not in the head. Engines that have L-head designs are commonly called "flat head engines." (8)

Lifter The small component that rides on the camshaft. The camshaft lobe lifts the lifter to aid in opening and closing the valves. (8)

Lines of force Invisible forces around a magnet. (18)

Limited slip differential A differential that uses clutches or cones to lock up the side gears to the differential case. These components eliminate having one wheel spin faster such as on ice. (27)

Line contact The contact made between the cylinder and the torsional rings, usually on one side of the ring. The contact made between the valve and the valve seat. When an interference angle is used, only a small line of contact is produced. (9) (10)

Liquid cooled Removing heat from the engine by circulating liquid coolant throughout the internal parts of the engine. (12)

Liter A unit in the metric system to measure volume. (4)

Load The actual slowing down of the engine output shaft because of a brake applied to the shaft. The load could be that of driving up a hill. The load to the engine is increased in this case. (7)

Load range The amount of load the tire is capable of supporting safely. The value is molded into the tire sidewall and is stated in pounds and/or kilograms. (31)

Lobe The part of the camshaft that raises the lifter. (10)

Lubricating system The subsystem on the engine that is used to keep all moving components lubricated. (1)

MacPherson strut system An independent suspension system consisting of a coil spring, shock absorber, and upper bearing. (29)

Main bearing clearance The clearance between the main bearing journal and the main bearings. (9)

Major and minor thrust The thrust forces applied to the piston on the compression and power strokes. (9)

Manual shift valve   A valve in the automatic transmission which is controlled by the position of the gearshift lever. (26)

Manufacturer's service manual   Service information and technical data supplied by the automotive manufacturer. (5)

Master cylinder   The main unit for displacing brake fluid under pressure in a hydraulic brake system. The master cylinder can be either single or dual design. (28)

Mechanical advantage   A linkage or lever able to gain either distance or force. The rocker arm uses a mechanical advantage to gain distance to open the valve further. (10)

Mechanical efficiency   A measure of the mechanical operation of a machine. (7)

Media   Usually referred to as a certain type of material acting as the environment. In this case, the area where dirt is captured in the fuel filter. (14)

Meter   A device used to measure or control the flow of a liquid such as fuel. (15)

Metered   To control the amount of fuel passing into an injector. Fuel is metered to obtain the correct measured quantity. (16)

Metric system   A system of measurement based on the meter. All other units are derived from the meter. These include centimeter, kilometer, millimeter, liter, cubic centimeter, and so on. (4)

Metric thread   A metric thread is measured by indicating the number of millimeters between each thread. (3)

Micrometer   A measuring tool used to accurately measure length to 0.001 of an inch. (4)

Micron   A distance measurement used to indicate the size of holes in a filter. 1 micron = 0.000039 inch. (14)

Microprocessor   A series of circuits using semiconductors and integrated circuits for computer applications. Microprocessors are also capable of input, storage, and feeding out information to other circuits and systems on the automobile. (18)

Microorganism   A small living organism. (14)

Molecule   The smallest physical unit of a chemical compound. (13)

Monolith   A single body shaped like a pillar or long tubular structure used as a catalyst in a catalytic converter. (24)

Multiple disc clutch   A hydraulic clutch used in the automatic transmission. The clutch uses a series of discs, both friction and smooth metal, to lock up to rotating shafts. (26)

NC   National Coarse or an indicator of the number of threads per inch on a bolt. (3)

NF   National Fine or an indicator of the number of threads per inch on a bolt. (3)

Naturally aspirated   An engine that uses the atmospheric pressure to force the air into the cylinders. (17)

Needle valve   A small valve used in the center of the fuel injector nozzle. The valve is shaped much like a thick needle and controls the opening and closing of ports for fuel injection. (16)

Nitrogen oxide   A type of automotive pollution produced when internal combustion temperatures reach 2200–2500 °F. (13) (23)

OSHA   Occupational Safety and Health Act of 1970. This act provides safety regulations and rules for industry. (2)

Octane number   A number used to identify the resistance to burning of gasoline. (13)

Ohm's law   Voltage equals amperage times resistance ($E = I \times R$). (18)

Oil relief   A small machined area on the side of the lifter that allows oil to circle around the body of the lifter. (10)

Open, shorted, grounded circuits   Conditions in a circuit to render the system inoperative. (18)

Operational specifications   Specifications used to show how the vehicle operates, such as acceleration, tire inflations, and other general information. (5)

Opposed cylinders   An engine that has two rows of pistons that are 180 degrees from each other. (8)

Organic   Pertaining to chemicals that are derived from living things. (11)

Organic material   Another term used for hydrocarbon pollution. (23)

O-ring   A type of static seal used to eliminate leakage between two stationary parts as fluid passes through them. (3)

Oscillation   Fluctuation or variation in motion or in electrical current. When the vehicle hits a bump in the road, the body oscillates up and down. Shocks are used to reduce oscillations. (29)

Overdrive   A gear system used on a transmission to reduce the speed of the input. Normally, the highest gear ratio in a standard transmission is 1–1. Overdrive systems have a higher gear ratio of approximately 0.8–1. (25) (26)

Overhaul and maintenance specifications   Specifications used to service vehicle components such as pistons, crankshafts, rings, bearings, and so on. (5)

Overhead camshaft A camshaft located directly on top of the valves, used on I-head designs. (8)

Overrunning clutch drive A type of drive on a starter motor that uses a series of rollers that lock up to cause the pinion gear to rotate. (22)

Oxidation The process of combining oil molecules with oxygen. (11)

Oxides Chemicals that form when certain pollutants combine with oxygen. (23)

PCV Positive Crankcase Ventilation (24)

PSIG A type of pressure scale read as *p*ounds per *s*quare *i*nch on a *gauge*. (12)

Parallel circuit In this type of circuit, there is more than one path for the current to follow. (18)

Particulates A form of solid air pollution such as microscopic solid or liquid matter that floats in the air. (23)

Parts distribution All serivce shops must have parts available. Parts distribution shops provide a retail business to make available necessary parts for the automobile. (1)

Parts manager The person responsible for making sure the customer's parts are immediately available to the general mechanic. (1)

Passive seal A seal that has no extra springs or tension devices to help make the seal. O-ring seals on valves are called passive valve seals. (10)

Pendulum A swinging device with a weight on one end used to control the movement of a mechanism. Pendulums are used to lock safety belts during impact. (34)

Performance chart A chart that has been produced from a dynamometer. It shows the horsepower, torque, and fuel consumption of an engine at various rpm. (7)

Periphery The external boundary of the torque converter. (26)

Photochemical smog A type of smog produced when hydrocarbons and nitrogen oxides combine with sunlight. (23)

Pinging A sound heard in the automobile engine that is caused by two combustion fronts hitting each other inside the combustion chamber. (13)

Pinion gear The small gear attached to the armature shaft used to crank the flywheel ring gear. The pinion gear is also the smaller of two gears. (22)

Pintle The center pin used to control a fluid passing through a hole; a small pin or pointed shaft used to open or close a passageway. (16) (24)

Piston A cylindrical object that slides in the cylinder. (6)

Piston slap The movement of the piston back and forth in the cylinder in a slapping motion. (9)

Pitch The angle of the valve spring twist. A variable pitch valve spring has unevenly spaced coils. (10)

Pitman arm A steering linkage component that connects the steering gear to the linkage at the left end of the center link. (30)

Pivot point The center point on the rocker arm. (10)

Planetary carrier The part of a planetary gear system that connects the axis of the planet gears together. (26)

Planetary gear systems A gear assembly that includes a sun gear, planet gears, and a ring gear. By locking up one gear, various gear ratios and speeds can be produced. (26)

Pliers A tool used to grip or cut various objects when working on the automobile. (4)

Plies Layers of material that wrap around a tire. (31)

Pole shoes Soft iron pieces that wire is wrapped around inside the starter motor. (22)

Pollution Addition of harmful products to the environment. Types of pollution include air, water, noise, chemical, thermal, and nuclear. (23)

Poppet-type valve Mechanical equipment uses many types of valves. One valve is called the poppet valve. Poppet valves are those that operate from a camshaft and open and close a port. Automobile intake and exhaust valves are poppet valves. (10)

Ported vacuum Vacuum taken from the carburetor slightly above the throttle plate. (20)

Positive displacement pump A type of pressure pump classification that pumps an exact amount of fluid for each revolution. (24)

Power A measure of work being done. (6)

Precombustion chamber A second combustion chamber placed directly off the main combustion chamber. The precombustion chamber is used to ignite a rich mixture of air and fuel. This mixture then ignites a lean mixture in the main combustion chamber. (10)

Prefix A term used to indicate how many units the meter is increased or decreased. One thousand meters is equal to one kilometer. Kilo is the prefix. (4)

Preignition The process of a glowing spark or deposit igniting the air-fuel mixture before the spark plug. (13)

Prematurely Occurring too soon. (14)

Pressure The exertion of force upon a body in contact with it. Pressure is developed within the cooling system and is measured in pounds per square inch on a gauge. (12)

Pressure gauge   A gauge used to read various pressures such as fuel pump, transmission oil, and fuel injection pressures. (4)

Pressure plate   The part in a clutch system used to squeeze or clamp the clutch disc between it and the flywheel. (25)

Pressure regulator valve   A valve used in an automatic transmission to regulate the pressure of the oil inside the valve body. (26)

Primary battery   A type of battery that cannot be recharged after use. (19)

Primary circuit   A circuit in the ignition system that uses 12 volts to operate. It includes the ignition switch, ballast resistor or resistive wire, primary coil wires, condenser, and contact points. (20)

Primary shoe   The forward shoe on a two-shoe drum brake system, often having shorter linings than the other. (28)

Propane   One of four gases found in natural gas. Methane, ethane, propane, and butane are in natural gas. Propane and butane have the highest amount of energy and can be made into a liquid by being put into a pressurized container. (13)

Property class   A number stamped on the end of a metric bolt to indicate the hardness of the bolt. (3)

Proportioning valve   A valve in the brake hydraulic system that reduces pressure to the rear wheels to achieve better brake balance. (28)

Proton   The positive (+) part of the atom. (18)

Puller   A tool attached to a shaft and gear, used to remove the gear from the shaft by applying certain pressures. (4)

Pulse delay variable resistor   A resistor in the wiper system used to time delay the wiper motion from zero to twenty-five seconds time delay. (34)

Pulse width   A term used to identify the length of time an injector will inject fuel. Large pulse width means more fuel being metered into the engine. (16)

Pulse wiper system   A wiper system using electronic circuits that cause the wipers to pulse or turn on one time, then off for a certain number of seconds. (34)

Purge   To separate or clean by carrying off gasoline fumes. The carbon canister has a purge line to remove impurities. (24)

Pushrod   Connector between the lifter and the rocker arm. (8)

Quenching   The cooling of the gases by pressing the gas volume out into a thin area. Quenching occurs inside the wedge-type combustion chamber in the quench area. (10)

RPM   Revolutions per minute on any rotating shaft. (7)

Rack and pinion steering   A steering system consisting of a flat gear (rack) and a mating gear (pinion). The pinion meshes with teeth on the rack causing the rack to move left or right. This motion moves tie rods and the spindle arm to steer the front wheels. (30)

Radial   Something that radiates from a center point. Radial tires have cord materials running in a direction from the center point of the tire, usually from bead to bead. (31)

Radiation   Transfer of heat by converting heat energy to radiant energy. (12)

Ram air   Air that is forced into the engine or passenger compartment by the force of the vehicle moving forward. (33)

Reciprocating   An up and down or back and forth motion. (6)

Rectify   To change one type of voltage to another. Usually ac voltage is rectified to dc voltage. (21)

Refrigerant   A liquid capable of vaporizing at low temperatures, such as ammonia or freon. (32)

Regenerator   A device placed on a gas turbine engine to take the heat of exhaust and put it into the intake of the engine. (8)

Regional offices and distributorships   Offices owned and operated by the automobile company. They are considered to be the link between the automobile manufacturer and the dealerships. (1)

Regular gasoline   Gasoline that has an octane number near 85–90. (13)

Relay   An electromagnetic device by which the opening or closing of one circuit operates another device. A relay in a voltage regulator uses a set of points that are opened and closed by magnetic forces. The opening or closing of the points controls another circuit, commonly the field circuit. (21)

Reluctor   In an electronic ignition system, a metal wheel with a series of tips used to produce the signal for the transistor. (20)

Reserve vacuum tank   A small vacuum storage tank used on vacuum-operated ventilation and heating controls. (33)

Residual   Remaining or left over pressure. (28)

Resistance   The part in an electrical circuit that holds back the electrons, also called the load. (18)

Resonator   A device used in an exhaust system to reduce noise, usually used in conjunction with a muffler. (17)

Rich mixture   Too much fuel and not enough air for combustion. (6)

Road horsepower Horsepower available at the drive wheels of the vehicle. (7)

Rocker arm An arm that has a pivot point in the center. One side is lifted by the camshaft movement and the other side moves down, opening the valves. (8)

Rolling resistance A term used to describe the amount of resistance a tire has to rolling on the road. Tires that have a lower rolling resistance usually get better gas mileage. Typically radial tires have lower rolling resistance. (31)

Rotary A circular motion. (6)

Rotary engine An engine that uses a rotor rather than pistons to produce power. It is an intermittent, internal combustion engine and is rotary, not reciprocating. (6)

Rotor The rotating component in a generator or alternator. (21)

Runner A cast tube on an intake or exhaust manifold used to carry air in or out of the engine. (17)

Running gear Component on the automobile that is used to control the vehicle. This includes braking systems, wheels, and tires. (1)

SAE Society of Automotive Engineers (11)

Safety glasses Glasses to be worn at all times when in the automotive shop. They should be designed with safety glass and side protectors, and they should be comfortable. (2)

Saybolt Universal Viscosimeter A meter used to measure the time in seconds required for 60 cubic centimeters of a fluid to flow through a hole on the meter at a given temperature under specified conditions. (11)

Schrader valve A spring-loaded valve or directional valve used to admit pressure into a sealed system, used on both air conditioning units and tires. (32)

Screw-pitch gauge A gauge used to measure the number of threads per inch on a bolt. (3)

Scrub radius The distance between the centerline of the ball joints and the centerline of the tire at the point when the tire contacts the road surface. (30)

Scuffing Scraping and heavy wear from the piston on the cylinder walls. (9)

Seal A device used on rotating shafts to keep oil or other fluid on one side of the seal, thus eliminating leakage. (3)

Seating When two metals must seal gases and liquids, they must be worked together to make a good seal. This process of getting two metal surfaces to seal is called seating. (10)

Secondary circuit A circuit in the ignition system that uses 20,000 or more volts to operate. It includes the secondary coil windings, the rotor, distributor cap, coil and spark plug wires, and spark plugs. (20)

Secondary shoe The rear shoe on a two-shoe drum brake system, often having a longer lining than the primary shoe. (28)

Selector control valve A lever located on the dashboard used to select one of several heating and ventilation modes. (33)

Self-energizing A drum brake arrangement where the braking action pulls the shoe lining tighter against the drum. (28)

Semiconductor A material having exactly four electrons in the valence ring. (18)

Series circuit A circuit in which there is only one path for the current to follow. (18)

Service bulletin Technical service information provided by the manufacturer, used as updates for the service manuals. (5)

Service manager The person responsible for the entire service operation of the dealership. (1)

Service manual A manual provided by the manufacturer or other publisher that describes service procedures, troubleshooting and diagnosis, and specifications. (5)

Service procedures A set of listed steps used to disassemble, assemble, or repair an automotive component. (5)

Service representative A person who works in the area of providing service to the dealership from the car manufacturer. (1)

Servo A hydraulically operated component that operates or controls the operation of the transmission band on the automatic transmission; a device used on a cruise control system to maintain the speed of the vehicle. (26) (34)

Servo brake A drum brake arrangement where the action of one shoe reinforces the action of the other shoe. (28)

Shackle The small arm between the frame and one end of the leaf spring. It is used to allow the spring to shorten and lengthen during normal driving conditions. (29)

Shank The diameter of the bolt, usually measured in fractions of an inch or in millimeters. (3)

Shift valve A valve used in an automatic transmission that controls the oil flow to the clutches and transmission bands. (26)

Shims Metal or plastic spacers of various thicknesses used to adjust caster and/or camber. (30)

Shock absorber A device used on a suspension system to dampen the oscillations or jounce of the springs when the car goes over bumps in the road. (29)

Shroud An object that covers the area between the fan and the radiator. (12)

Shrouding When a valve is placed close to the side of the combustion chamber, the air and fuel may be restricted by the side of the chamber. This restriction is referred to as shrouding. (10)

Shunt More than one path for current to flow, such as a parallel part of a circuit. (22)

Siamese ports Intake or exhaust ports inside the cylinder head where two cylinders are feeding through the one port. (10)

Slant An in-line cylinder arrangement that has been placed at a slant. This arrangement makes the engine have a lower profile for aerodynamic design. (8)

Slave cylinder A type of hydraulic cylinder used as a means to actuate the clutch mechanism. As the clutch pedal is pushed down, the hydraulic pressure produced in the slave cylinder is used to move the clutch mechanism. (25)

Slip joint A splined shaft that can slide in a mating shaft to allow changes in drive shaft length. (27)

Slipper skirt A piston that has a cutaway skirt so that the piston can come closer to the counter-weights. This makes the overall size of the engine smaller. (9)

Slip rings A type of commutator used on an alternator made of two copper rings that are split in half. (21)

Slotted frame Slotted holes on the frame used to reposition parts for camber/caster adjustment. (30)

Sludges Material formed as a result of oil in the presence of various acids. (11)

Smoking rules Only smoke in designated "smoking" areas. Dangerous explosive fuels in the shop may be ignited if this rule is not followed. (2)

Snap rings Small rings, either external or internal, that are used to prevent gears and pulleys from sliding off the shaft. (3)

Snorkel tube A long, narrow tube attached to the air cleaner, used to direct air into the air filter. (24)

Socket points The number of points inside the socket head. Six, 8, and 12 points are most common. In applications where only a small amount of rotation of the ratchet is possible, use a 12-point socket. (4)

Solenoid A coil of wire wound around a movable core. When voltage is applied to the coil, the magnetic field causes the metal core to move. A solenoid converts electrical energy into mechanical energy. A device that converts electrical energy to mechanical energy using magnetism. Used in a starter system and other electrical circuits. (15) (22)

Specialty shops Service shops that specialize in certain components of the automobile. Some include carburetor shops, body shops, transmission shops, muffler shops, and so on. (1)

Specifications Any technical data, numbers, clearances, and measurements used to diagnose and adjust automobile components. They are also called specs. (5)

Specific gravity The weight of a solution as related to water. Water has a specific gravity of 1.000. Sulfuric acid, being heavier than water, has a specific gravity of 1.835. (19)

Splines External or internal teeth cut into a shaft, used to keep a pulley or hub secured on a rotating shaft. (3)

Spool valve A cylindrically shaped rod with different size diameters. Usually the cylindrical valve is placed inside a bore inside the transmission valve body. As the valve is moved in and out, different hydraulic circuits are operated. (26)

Sprag A pointed steel piece inside an overrunning clutch mechanism that allows rotation in one direction but locks up rotation in the opposite direction. (22)

Squirm To wiggle or twist about a body. When applied to tires, squirm is the wiggle or movement of the tread against the road surface. Squirm increases tire wear. (31)

Squirrel cage blower A type of air pressure fan shaped like a squirrel cage, used to move air throughout a system. The squirrel cage fan is run by a motor and placed inside a housing to improve its efficiency of operation. (33)

Stabilizer bar A reinforcement component on a suspension system that prevents the body from diving or leaning on turns. (29)

Stance The manner of standing or being placed. A vehicle's stance refers to the level or evenness of its position. (29)

Standard bolt and nut torque specifications A chart showing the standard torque for common sizes of bolts. (3)

Starting and charging system The subsystem on the engine used to start the engine and charge the battery. (1)

Static balance Equal distribution of weight around a center point. Static means stationary and static balancing is done with the wheels stationary. (31)

Stator The stationary part in an alternator, which cuts the magnetic lines of force; the part of a torque converter that is stationary, used to direct the flow of fluid back to the rotary pump at the correct angle. (21) (26)

Steering axis inclination The inward tilt of the spindle support arm ball joints at the top. (30)

Steering knuckle A part of the front suspension that connects the wheel to the suspension system and the tie rod ends for steering. The wheel spindle is also attached to the steering knuckle. (29)

Stellite A very hard metal made from cobalt, chromium, and tungsten, used for insert-type valve seats. (10)

Stirling engine An external combustion, continuous combustion engine, having four cylinders that operate in a particular sequence. (8)

Stoichiometric ratio A 14.7 to 1 air-fuel ratio. This is the best ratio to operate an internal combustion engine. (16)

Stratified To layer or have in layers. (8)

Stratified charge engine An engine that has an additional small combustion chamber. The air and fuel in this chamber is very rich. The air and fuel in the regular chamber is leaner. The small chamber ignites the larger chamber mixture, reducing emissions. (8)

Stroke The distance from TDC to BDC of piston travel. (6)

Strut rod A rod on the suspension system located ahead of or behind a lower control arm to retain the arm in its intended position. (29)

Sulfur A chemical compound in diesel fuel that produces pollution. When mixed with oxygen and water, sulfur produces a strong acid. (13)

Sump A pit or well where a fluid is collected. Oil is collected in the oil sump. (11)

Supercharger A device placed on a vehicle to increase the amount of air, and therefore the amount of fuel that is sent into the engine. (8)

Supercharging The process of forcing air into an engine cylinder with an air pump. The forced air can come from a blower or turbocharger. (17)

Supporting career areas The automotive industry supports careers in the following areas: claims adjusting, vocational teaching, auto body repairing, frame and alignment repair, and specialty shops. (1)

Surging A sudden rushing of water from the water pump. (12)

Suspendability The ability of a fluid to suspend heavier dirt particles within the oil, rather than falling to the bottom. (11)

Suspension systems Components that support the total vehicle, including springs, shock absorbers, torsion bars, axles, and connecting linkages. (1)

Swash plate An angular plate attached to the bottom of the four pistons on a Stirling engine. As the pistons move downward, the swash plate is turned. A mechanical system that is used for pumping. An angled plate is attached to a center shaft and pistons are attached to the plate along the axis of the shaft. As the shaft rotates, the pistons move in and out of a cylinder, producing a suction and pressure. (8) (32)

Sway bar A bar on the suspension system that connects the two sides together. It is designed so that during cornering, forces on one wheel are shared by the other. (29)

Synchronizer An assembly in a manual transmission used to make both gears rotate at the same speed before meshing. (25)

Synthetic A product made by combining various chemical elements (a man-made product). (11)

TDC Position of the piston. TDC means the piston is at top dead center. (6)

TRS Transmission Regulated Spark (24)

Tandem Meaning one object behind the other. (28)

Tang A projecting piece of metal placed on the end of the torque converter on automatic transmissions. The tangs are used to rotate the oil pump. (26)

Tappets Another term for valve lifters. (10)

Tensile strength The amount of pressure per square inch the bolt can withstand just before breaking when being pulled apart. (3)

Tetraethyl lead A chemical added to gasoline to increase the octane and aid in lubrication of the valves. (13)

Thermal efficiency A measure of how effectively an engine converts heat energy in fuel into mechanical energy at the rear of the engine. (7)

Thermostatic switch A heat-sensitive switch used to turn on and off an air conditioning compressor. (32)

Threaded fasteners A type of fastener such as bolts, studs, setscrews, cap screws, machine screws, and self-tapping screws that has thread on it. All are considered threaded fasteners. (3)

Threads per inch A number used to identify bolts, showing the number of threads per inch on the bolt. (3)

Three phase Voltages produced from a generator or alternator can either be single phase or three phase. Three-phase voltages are electrically 120 degrees apart from each other. (21)

Throttle body The part of a fuel system where fuel is injected into the air stream. The throttle body is located in the same place as the carburetor on a standard fuel system. (16)

Throttle plate The plate or circular disk that controls the amount of air going into an engine. It is usually controlled by the position of the operator's foot. (16)

Throttle valve A valve used in the automatic transmission that changes oil flow based upon the throttle position of the engine. (26)

Thrust Another name for axial load. (3)

Thrust bearing An antifriction bearing designed to absorb any thrust along the axis of the rotating shaft. (3)

Thrust plate The plate used to bolt the camshaft to the block, which absorbs camshaft thrust. (10)

Timing The process of identifying when air, fuel, and ignition occur in relation to the crankshaft rotation. (6)

Timing diagrams A graphical method used to identify the time in which all of the events of the four-stroke cycle engine operate. (6)

Tire placard A permanently located sticker on the vehicle that gives tire information such as load, pressure, and so on. (31)

Toe (in, out) The inward or outward pointing of the front wheels as measured in inches or millimeters. (30)

Torque A twisting force applied to a shaft or bolt. (3) (7)

Torque converter The coupling between the engine and transmission on an automatic transmission. It is also used to multiply torque at lower speeds. (26)

Torque specifications Specifications used to tell the service mechanic the exact torque that should be applied to bolts and nuts. (5)

Torque wrench A wrench used to measure the amount of torque or twisting force applied to a bolt or nut. (4)

Torsional rings Rings that have a slight twist when placed within the cylinder wall. These are made by adding a chamfer or counterbore on the ring. (9)

Torsional vibration A vibration produced in a spinning shaft, caused by torque applied to the shaft. (25)

Torsion bar A steel shaft that serves the same purpose as a coil spring. It is located between the lower control arm and the frame, and it twists when the vehicle moves up and down. (29)

Traction A tire's ability to hold or grip the road surface. (31)

Transaxle A type of transmission used on front wheel drive vehicles where the engine is crosswise. The transmission is designed so the differential is built in the transmission and the output goes directly to the front wheels. (25)

Transducer A device that senses pressure in an exhaust manifold, used to control another circuit. A device that transmits energy from one system to another. A transducer is used in cruise control systems to control vacuum based upon vehicle speed. (24) (34)

Transformer A set of coils that use magnetism to change one voltage to another. The ignition transformer changes 12 volts to 20,000 volts. (20)

Transistor A semiconductor used in circuits to turn off or on a second circuit, also used for amplification of signals. (18)

Transmission band A type of hydraulic clutch that uses a metal band fitted around a clutch housing. As the band is tightened, the housing rotation is stopped. (26)

Tread The outer surface of a tire used to produce friction with the road for starting and stopping. (31)

Tread bars Narrow strips of rubber molded into the tread. When the tread bars show, the tire is worn enough to be replaced. (31)

Tripod The tripod is the central part of certain CV joints. It has three arms or trunnions with needle bearings and rollers running in grooves or races in the assembly. (27)

Troubleshooting Another term used for diagnosis. (5)

Trunnion The arm or arms of the four-point U-joint, which serves as the inner bearing surface or race. (27)

Tuned ports Intake ports used on fuel injection engines, designed to produce equal and minimum restriction to the air flow. (16)

Tune-up specifications Specifications primarily used during a tune-up on an automobile. (5)

Turbine A component in a gas turbine engine that changes the energy in the gases into rotary motion for power. A vaned type of wheel being turned by a fluid such as exhaust gases passing over it. The rotary part or vaned wheel inside a torque converter, used to turn the transmission. (8) (17) (26)

Turbocharged An engine that uses the exhaust gases to turn a turbine. The turning turbine forces in extra fresh air for more performance. (17)

Turbulence A term used to describe combustion chambers. It means rapid movement and mixing of air and fuel inside the combustion chamber. (10)

Turning radius The amount (in degrees) that one front wheel turns more sharply than the other front wheel during a turn. (30)

Type A fire A fire resulting from the burning of wood, paper, textiles, and clothing. (2)

Type B fire A fire resulting from the burning of gasoline, greases, oils, and other flammable liquids. (2)

Type C fire A fire resulting from the burning of electrical equipment, motors, and switches. (2)

USC measurements Standard English measurements, including feet, inches, miles, pounds, ounces, and so on. (4)

Vacuum An enclosed space in which the pressure is below zero psig. (12) (15)

Vacuum diaphragm A device that has a spring on one side of a flexible material and vacuum on the other side sealed in a housing. The position of the flexible material controls the movement of another system. (15)

Vacuum gauge A gauge designed to read various vacuum readings on an engine, the most common being intake manifold vacuum, and pollution control equipment. (4)

Vacuum modulator A diaphragm used on automatic transmissions that controls the throttle valve based upon engine intake manifold vacuum. (26)

Vacuum motor A small diaphragm inside a housing operated by vacuum working against a spring pressure. When vacuum is applied, the diaphragm moves. This movement is then used to open or close small air doors or other apparatus. (33)

Valence ring The outer orbit of electrons in an atom. (18)

Valve A device used to open and close a port to let intake and exhaust gases in and out of the engine. (6)

Valve body The part of an automatic transmission used to direct the oil flow to different parts of the transmission. The valve body also houses most of the valve for control. (26)

Valve bounce When a valve is forced to close because of spring pressure, the valve may bounce when it closes. This action can damage the seats or break the valve in two. (10)

Valve clearance The clearance or space between the valve and the rocker arm. As the parts heat up, the clearance is reduced because of expansion. This keeps the valves from remaining open when the engine is hot. (8)

Valve face The part of a poppet valve that actually touches the seat for sealing in the cylinder head. (10)

Valve float If a valve spring is not strong enough to close the valve, the valve may float or stay open slightly longer than designed. This condition will limit the maximum rpm an engine can develop. (10)

Valve guide The part in the cylinder head that holds the stem of the valve. (10)

Valve train clearance The clearance between the lifters, rocker arms, and valves. This clearance is necessary because as the parts heat up, they will expand. The valve train clearance allows for this expansion. (10)

Vapor A substance in a gaseous state. Liquids become a vapor when they are brought above their boiling point. (32)

Vaporize The process of passing from a liquid to a gas. Fuel is vaporized when it is heated. (13) (15)

Vapor lock Vapor buildup that restricts the flow of gasoline through the fuel system. Vapor lock occurs from heating the fuel, causing it to turn to a vapor. (13) (14)

Varnish A deposit in an engine lubrication system resulting from oxidation of the motor oil. Varnish is similar to, but softer than, lacquer. (11)

V-configuration   A style of engine that has two rows of cylinders that are approximately 90 degrees apart and in a V shape. (8)

Venturi   A restriction in a tube where air or liquid is flowing. A venturi always causes a vacuum to be created at the point of greatest restrictions. (15)

Vernier caliper   A measuring tool used to accurately measure length to 0.001 inch. (4)

Vernier scale   A scale for measuring in which two lines are adjusted to line up vertically with each other. (4)

Viscosity   A fluid property that causes resistance to flow. The higher the viscosity, the greater the resistance to flow. The lower the viscosity, the easier for the fluid to flow. (11)

Viscosity index   A common term used to measure a fluid's change of viscosity with a change in temperature. The higher the viscosity index, the smaller the relative change in viscosity with temperature. (11)

Voltage   The push or pressure used to move electrons along a wire. (18)

Voltage drop   Voltage lost at each resistor, usually defined as I × R drop. (18)

Volumetric efficiency   A measure of how well air flows in and out of an engine. (7)

Vulcanized   A process of heating rubber under pressure to mold it into a special shape. (31)

Watt's law   Power equals voltage times amperage $(P = E \times I)$. (18)

Wedge-shaped combustion chamber   A type of combustion chamber that is shaped similar to a wedge or V. This chamber is designed to increase the movement of air and fuel to aid in mixing. (10)

Wheel cylinder   A unit for converting hydraulic fluid pressure to mechanical force for brake applications. (28)

Wheel runout   A measure of the out-of-roundness of a wheel or tire. (31)

Wheel spindle   The short shaft on the front wheel upon which the wheel bearings ride and to which the wheel is attached. (29)

Work   Work is defined as the result of a force applied to a mass, moved a certain distance. Work = Force × Distance. (7)

Yoke   The Y-shaped metal device that is attached to the drive shaft. (27)

Zener diode   A type of diode that requires a certain amount of voltage before it will conduct electricity. This voltage is used to control transistors in voltage regulators. (21)

Zerk   A lubrication fitting through which grease is applied to a steering joint with a grease gun. (30)

INDEX